Lecture Notes in Artificial Intelligence 4798

Edited by J. G. Carbonell and J. Siekmann

Subseries of Lecture Notes in Computer Science

Zili Zhang Jörg Siekmann (Eds.)

Knowledge Science, Engineering and Management

Second International Conference, KSEM 2007
Melbourne, Australia, November 28-30, 2007
Proceedings

Volume Editors

Zili Zhang
Deakin University
School of Engineering and Information Technology
Geelong, VIC 3217, Australia
E-mail: zzhang@deakin.edu.au

Jörg Siekmann
German Research Center of Artificial Intelligence (DFKI), Germany
E-mail: Siekmann@dfki.de

Library of Congress Control Number: Applied for

CR Subject Classification (1998): I.2.6, I.2, H.2.8, H.3-5, F.2.2, K.3

LNCS Sublibrary: SL 7 – Artificial Intelligence

ISSN 0302-9743
ISBN-10 3-540-76718-5 Springer Berlin Heidelberg New York
ISBN-13 978-3-540-76718-3 Springer Berlin Heidelberg New York

Springer is a part of Springer Science+Business Media

springer.com

Printed in Germany

Typesetting: Camera-ready by author, data conversion by Scientific Publishing Services, Chennai, India
Printed on acid-free paper SPIN: 12190126 06/3180 5 4 3 2 1 0

Preface

The second international conference on Knowledge Science, Engineering and Management (KSEM2007) was held in picturesque Melbourne, Australia, during November 28–30 2007, and hosted by Deakin University.

The aim of this interdisciplinary conference is to provide a forum for researchers in the broad areas of knowledge science, knowledge engineering and knowledge management to exchange ideas and to report state-of-the-art research results. Recent years have seen the growing importance of the synergism of knowledge science, engineering and management to provide stronger support for complex problem solving and decision making. KSEM aims at bridging the three areas and promoting their synergism.

KSEM2007 attracted 124 submissions from 21 countries/regions around the world. All submitted papers were reviewed by at least two PC members or external reviewers. The review process was very selective. From the 124 submissions, 42 (33.8%) were accepted as regular papers, and another 28 (22.6%) as short papers. Authors of accepted papers came from 16 countries/regions. This volume of the proceedings contains the abstracts of five invited talks, two of them with extended versions, and all the regular and short papers. The regular papers were categorized into three broad sections, that is, knowledge science, knowledge engineering, and knowledge management.

The technical program featured five invited talks, one panel discussion, and all the accepted papers. The five distinguished invited speakers were John Debenham, Andreas Dengel, Lakhmi Jain, WB Lee, and Ling Zhang.

The success of KSEM2007 was assured by team effort from the sponsors, organizers, reviewers, and participants. We would like to thank the three Area Chairs, Zhi Jin, David Bell, and Eric Tsui, for coordinating and monitoring the whole paper review process. We would like to acknowledge the contribution of the individual Program Committee members and thank the external reviewers. Thanks to Publicity Chair Dongmo Zhang, and Organizing Co-chairs Shang Gao and Shui Yu for their great efforts. Thanks also go to Robert Ruge for maintaining the conference management system. Special thanks to Ruqian Lu, Kate Smith-Miles, and Chengqi Zhang for their valuable advice and suggestions. Our sincere gratitude goes to the participants and all authors of the submitted papers.

We are grateful to our sponsors: Air Force Office of Scientific Research, Asian Office of Aerospace Research and Development (AFOSR/AOARD), (AFOSR/AOARD support is not intended to express or imply endorsement by the U.S. Federal Government.); The School of Engineering and Information Technology and the Faculty of Science and Technology at Deakin University; The Hong Kong Polytechnic University; German Research Center for Artificial Intelligence (DFKI); and Zhuhai Overseas Professional Placement Office Melbourne.

We wish to express our gratitude to the Springer team directed by Alfred Hofmann for their help and cooperation. Thanks to Springer for their special contribution – The Student Best Paper Award at KSEM2007.

November 2007

Zili Zhang
Jörg Siekmann

Organization

KSEM2007 was hosted and organized by the School of Engineering and Information Technology, Deakin University, Australia. The conference was held at Novotel Melbourne on Collins, Melbourne, November 28–30, 2007.

Conference Committee

Conference Co-chairs	Kate Smith-Miles (Deakin University, Australia) Ruqian Lu (Chinese Academy of Sciences, China)
Program Co-chairs	Zili Zhang (Deakin University, Australia/Southwest University, China) Jörg Siekmann (German Research Centre of Artificial Intelligence, Germany)
Area Chairs	Knowledge Science: Zhi Jin (Chinese Academy of Sciences, China) Knowledge Engineering: David Bell (Queen's University Belfast, UK) Knowledge Management: Eric Y.H. Tsui (The Hong Kong Polytechnic University, China)
Organizing Co-chairs	Shang Gao (Deakin University, Australia) Shui Yu (Deakin University, Australia)
Publicity Chair	Dongmo Zhang (University of Western Sydney, Australia)

Program Committee

Klaus-Dieter Althoff (University of Hildesheim, Germany)
Nathalie Aussenac-Gilles (IRIT, CNRS / Paul Sabatier University, France)
Philippe Besnard (IRIT-CNRS, France)
Cungen Cao (Chinese Academy of Sciences, China)
Laurence Cholvy (ONERA Toulouse, France)
Daniel Crabtree (Victoria University of Wellington, New Zealand)
Jim Delgrande (Simon Fraser University, Canada)
Xiaotie Deng (City University of Hong Kong, China)
Kevin C. Desouza (University of Washington, USA)
Rose Dieng-Kuntz (INRIA Sophia Antipolis, France)
Patrick Doherty (Linkoping University, Sweden)
Xiaoyong Du (Renmin University of China, China)
Jim Duggan (National University of Ireland, Ireland)
Martin Dzbor (Open University, UK)
Leif Edvinsson (Lund University, Sweden)

Thomas Eiter (Vienna University of Technology (TU Wien), Austria)
Scott E. Fahlman (Carnegie Mellon University, USA)
Xiaoying Gao (Victoria University of Wellington, New Zealand)
Hector Geffner (ICREA and Universitat Pompeu Fabra, Spain)
Lluis Godo (IIIA, Spanish Council for Scientific Research, CSIC, Spain)
Nicola Guarino (ISTC-CNR, Italy)
Gongde Guo (Fujian Normal University, China)
Suliman Hawamdeh (The University of Oklahoma, USA)
Minghua He (Aston University, UK)
Andreas Herzig (IRIT, CNRS / Université Paul Sabatier, France)
Knut Hinkelmann (University of Applied Sciences Northwestern Switzerland)
Jun Hong (Queen's University Belfast, UK)
Zhisheng Huang (Vrije University Amsterdam, The Netherlands)
Anthony Hunter (University College London, UK)
Toru Ishida (Kyoto University, Japan)
David Israel (SRI International, USA)
Gabriele Kern-Isberner (Universität Dortmund, Germany)
John Kidd (Aston Business School, UK)
Patrick Lambe (Straits Knowledge, Singapore)
Jérôme Lang (CNRS / Universitè Paul Sabatier, France)
Li Li (Swinburne University of Technology, Australia)
Stephen Shaoyi Liao (City University of Hong Kong, China)
Wei Liu (The University of Western Australia, Australia)
Weiru Liu (Queen's University Belfast, UK)
James Lu (Emory University, USA)
Dickson Lukose (MIMOS Bhd., Malaysia)
Xudong Luo (The University of Birmingham, UK)
Michael Madden (National University of Ireland, Ireland)
Simone Marinai (University of Florence, Italy)
Pierre Marquis (Université d'Artois, France)
John-Jules Meyer (Utrecht University, The Netherlands)
Yuan Miao (Victoria University, Australia)
Vibhu Mittal (Google, Inc., USA)
Kenneth Murray (SRI International, USA)
Yoshitero Nakamori (Japan Advanced Institute of Science and Technology, Japan)
Patricia Ordonez de Pablos (The University of Oviedo, Spain)
Ewa Orlowska (Institute of Telecommunications, Poland)
Maurice Pagnucco (University of New South Wales, Australia)
Deepak Ramachandran (University of Illinois at Urbana-Champaign, USA)
Ulrich Reimer (University of Applied Sciences St. Gallen, Switzerland)
Ulrich Remus (University of Canterbury, New Zealand)
Torsten Schaub (Universität Potsdam, Germany)
Choon Ling Sia (City University of Hong Kong, China)
Andrew Skabar (La Trobe University, Australia)
Heiner Stuckenschmidt (Universität Mannheim, Germany)

Kaile Su (Peking University, China)
Jigui Sun (Jilin University, China)
Mirek Truszczynski (University of Kentucky, USA)
Abel Usoro (University of Paisley, UK)
Leon van der Torre (University of Luxembourg, Luxembourg)
Huaiqing Wang (City University of Hong Kong, China)
Hui Wang (University of Ulster, UK)
Ju Wang (Guangxi Normal University, China)
Kewen Wang (Griffith University, Australia)
Zhongtuo Wang (Dalian University of Technology, China)
Qingxiang Wu (Queen's University Belfast, UK)
Dongming Xu (University of Queensland, Australia)
Yue Xu (Queensland University of Technology, Australia)
Mingsheng Ying (Tsinghua University, China)
Jia-Huai You (University of Alberta, Canada)
Qingtian Zeng (Shangdong University of Science and Technology, China)
Chunxia Zhang (Beijing Institute of Technology, China)
Mengjie Zhang (Victoria University of Wellington, New Zealand)
Mingyi Zhang (Guizhou Academy of Sciences, China)
Shichao Zhang (Guangxi Normal University, China)
Yan Zhang (University of Western Sydney, Australia)
Aoying Zhou (Fudan University, China)
Xiaofang Zhou (University of Queensland, Australia)
Zhi-Hua Zhou (Nanjing University, China)
Hong Zhu (Fudan University, China)
Zhaohui Zhu (Nanjing University of Aeronautics and Astronautics, China)
Sandra Zilles (University of Alberta, Canada)
Meiyun Zuo (Renmin University of China, China)

External Reviewers

Yukika Awazu
Kerstin Bach
Hung Bui
Yixiang Chen
Chris Connolly
Sylvie Coste-Marquis
Jan-Oliver Deutsch
Laurent Garcia
Alexandre Hanft
Aimin Hao
Laura Hollink
He Hu
Linpeng Huang
Philip Hutto
Grzegorz Majewski
Régis Newo
Gaston Tagni
Yisong Wang
Chinthake Wijesooriya
Xiao-Bing Xue
Kang Ye
Yao Zhu

Table of Contents

Invited Talks

Regular Papers

Knowledge Science

Knowledge Engineering

Knowledge Management

Short Papers

Building Relationships and Negotiating Agreements in a Network of Agents

John Debenham

Faculty of Information Technology
University of Technology, Sydney
Australia
debenham@it.uts.edu.au

Abstract. Fully automated trading, such as e-procurement, is virtually unheard of today. Trading involves the maintenance of effective business relationships, and is the complete process of: need identification, product brokering, supplier brokering, offer-exchange, contract negotiation, and contract execution. Three core technologies are needed to fully automate the trading process. First, real-time data mining technology to tap information flows and to deliver timely information at the right granularity. Second, intelligent agents that are designed to operate in tandem with the real-time information flows from the data mining systems. Third, virtual institutions in which informed trading agents can trade securely both with each other and with human agents in a natural way. This discussion focusses on the second technology the design of "information driven" trading agents. The "information-based agency" is being developed principally by John Debenham [UTS, Australia] and Carles Sierra [IIIA, Spain]. Published work to date has considered argumentation frameworks between a set of (often just two) agents, and has not taken into account the relationships that the various agents have with other agents. Drawing ideas from the business research literature we are constructing a model of multi-agent interaction that accommodates the sorts of inter-agent agreements found in the business world. The working hypothesis is that the information-based approach will prove to be a suitable conceptual framework through which networks of agents, interacting under the constraints of various styles of business relationship, may be implemented.

Z. Zhang and J. Siekmann (Eds.): KSEM 2007, LNAI 4798, p. 1, 2007.

Knowledge Technologies for the Social Semantic Desktop

Andreas R. Dengel

Knowledge Management Dept, DFKI &
University of Kaiserslautern
Germany
andreas.dengel@dfki.de

Abstract. The web of information has turned into a web of people and is slowly evolving into a web of meaning. Desktop computing of the future will build on WWW-standards like RDF and RDFS allowing to evolutionary gather knowledge through lightweight interfaces by observing the user and learn from his/her actions to work with information within collaborative processes. Thus, the Social Semantic Desktop will allow both, personal information management and secure information sharing within trusted communities. In this keynote paper, I will present some of our ideas to implement such a forthcoming office workspace. I will start with a typical example of how people manage their own information today, depicting deficits and explaining how semantic services could help out. I will give examples of semantic services and show how they can be used for collaboration and community building.

1 Introduction

Tony is a leading expert in civil engineering and architecture. He is known as a somehow extraverted but highly creative engineer. He is currently working on various challenging projects in Barcelona, Spain, to which all his intuition and energy is dedicated. This is not only because he is Spaniard and he likes this city very much but also because he has noticed some detractors and antagonists who try hard to creep in his efforts.

But Tony is well prepared to meet these objections. He has established a personal archive representing his unique experience of how to organize information necessary to do his job in the best possible way. In his calendar, Tony organizes all his important meetings, events as well as the agendas of the various projects. His file system captures not only all relevant draft plans, sketches, and concepts as well as address data about relevant craftsmen and factories, but also knowledge about static issues and alternative materials. Tony has collected all correspondence with his client, local partners, potential suppliers, and others involved in his work. In order to have appropriate and fast access to relevant information items, he specifically uses different systems meeting his mindset to archive all of his documents.

Z. Zhang and J. Siekmann (Eds.): KSEM 2007, LNAI 4798, pp. 2–9, 2007.

Because of Tony's genius, the projects in Barcelona are running very well. His concepts are full of architectural surprises, improvisations, and special features. Most important, his clients become increasingly in favor of his ideas. However, because of the communication intensity and several redesigns, the deliverables become very complex and it becomes more and more difficult to explain the details to others. Thus, Tony is afraid because he uses so many different systems to organize his documents, tasks, processes, and contacts, and the individual work packages take more time then he expected, too much time. Tony is killed in an accident before the project is finished. Unfortunately, none of the other project members involved is expert enough to understand the taxonomies and systematics of Tony's repositories or has the time to penetrate the archive. Thus, the projects are close to fail and the client threatens to terminate the contract.

This is a typical example of how projects are accomplished for many centuries. The named Tony is really Antoni Gaudi y Cornet who lived from 1852 to 1926 and who created examples of architecture which were not known before, such as Casa Battlo, Casa Vicens and the dome Sagrada Familia in Barcelona [10]. He knocked over by a tram before his major project Sagrada Familia could be finished. In addition, Antoni Gaudi's drawings, drafts and models burned during the Spanish Civil War and thus were lost for the ensuing ages.

Considering the above-described scenario, it becomes apparent that many things have not changed until today. Although a computer may have eased Tony's communication, drafting and workflow activities, there is still a lack in how to concisely record information and how to provide intuitive insights for other individuals taking into account their perspective needs regarding their role, task or interest. Indexes and taxonomies may help but they are not an invention of computer science. They already existed in the library of Alexandria more than 2300 years ago.

In the following, I present a new technological approach for designing an office workspace of the future as a vivid interaction environment assisting Tony in expressing complex issues and relationships of his ideas in mind.

2 Personal Information Management

Where Tony used a pen and paper to capture all his visions and creative power, we are nowadays using the computer. It is the primary means of the information society to collect information of any kind. Nearly everything we find in the Web is input by human beings through a computer: editors fill news channels, sales people present electronic product catalogues, and researchers publish their papers via the Internet. Unfortunately, since information in the Web is mostly represented in natural language, the available documents are only fully understandable by human beings. In order to bridge this gap, many researchers follow the mission to transform the Web of links into a Web of meaning or a Semantic Web. It is built on ontologies in which documents are described by means of a standardized vocabulary providing machine understandable semantics. Annotating documents with ontological concepts may simplify automatic processing and provide a software agent with background knowledge in the form of the ontology itself to infer facts not explicitly part of the annotated data. However, the question is how we can generate an appropriate vocabulary as a means to build an ontology to be shared with others. Let us start with the individual workspace.

Like Tony, I do my very best to organize my information on my desktop for a later reuse, and as many users I use taxonomies in order to describe hierarchically ordered categories, some of which are shown in Fig. 1. Without explanation, the tacit relationships I had in mind when I organized my information items are not transparent to anybody, especially if there is no shared context. The example covers different directories I am using: the fragment of the file system on the left hand side shows a hierarchy of folders which in the end contain documents categorized according to different topics, such as "Hybrid Classification" or "Semantic Desktop". The Outlook folders in the centre of Fig. 1 present folders denoting my contacts, e.g. "Miles" or "Zhang" with whom I interchange emails while the bookmarks on the right hand side represent a collection of important links to web sites, e.g. tourist information or conference announcements I am interested in.

Since the content of all respective information objects are unknown, the names themselves are not sufficient to provide insights about implicit relationships to another person but myself. Especially, it is not transparent that the folder "Semantic Desktop" contains a keynote paper that I have submitted to the KSEM07 held in Melbourne, Australia, which was published using Springer LNAI formatting guidelines. For preparing the paper, the PC chair Dr. Zhang and I had some discussions via email and since I participated in the conference, I visited the web page of the conference hotel in Melbourne.

Although all of these information objects obviously do have relationships with one another, they are filed in different and independent directories; the relationships are invisible, only existing in my own neural memory. The example reveals the weakness of existing repositories to make conceptual relationships explicit in order to foster association and memory. It also underlines how hard it is for non-insiders to comprehend my way of thinking. Thus, in the current state, a future data archaeologist will be as clueless trying to reconstruct those relationships existing between the entries in different directories as Tony's successors were trying to realize his concept of Sagrada Familia after his death.

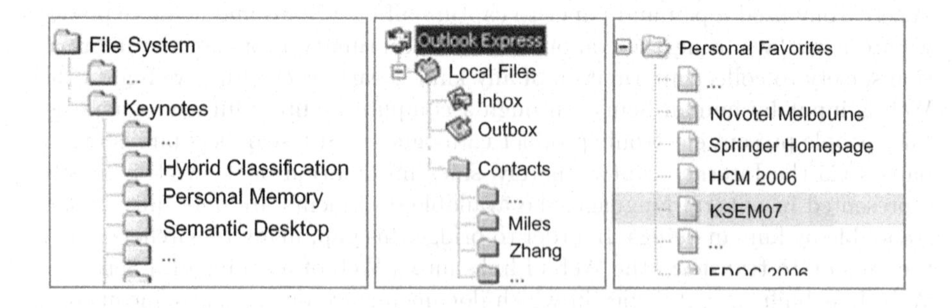

Fig. 1. A typical example of personal information management

3 Towards the Semantic Desktop

In contrast to the above example, ontologies offer the possibility to explicitly express concepts and their relationships according to the user's preferences (for instance, there might be concepts like *person* or *organization* with a relationship *works-for*) thus defining a common vocabulary and representing the user's individual mental models at the same time. Hence, ontologies function like subjective eye glasses for the user's perception of the information universe. The glasses only may offer a cutout but the corresponding information items are semantically linked to ontological concepts.

However, the success of semantic technologies strongly depends on convincing as many users as possible to participate in distributed but social ontology generation. For that reason we focus on high potential domains of knowledge workers, like Tony, and develop intuitive interfaces, stimulating creative work and allowing to share the resulting "Personal Semantic Web" with others. It is obvious that bonding ontologies into the information processes is the biggest challenge. We therefore have to consider different aspects, like

- Making use of individual ontologies, e.g. people generate folders and name them as categorical concepts
- Allowing for the integration of existing public ontologies and combining them with the native structures
- Providing intuitive means to establish and maintain existing ontologies for any domain

One option to evolutionary generate ontological concepts is to observe the user and learn from his/her actions to work with information within processes. To help knowledge workers like Tony, our approach is to bring the Semantic Web to the individual desktop in order to develop a Semantic Desktop [1,3].

A Semantic Desktop is a means to manage all personal information across application borders based on Semantic Web standards such as RDF (Resource Description Framework) and RDFS (RDF Schema) allowing a formal and unique description of any resource and its relationships. Meaning is encoded in sets of triples being like elementary sentences composed of subject, verb, and object. Subject and object are each identified by a URI (Universal Resource Identifier) while the verb describes properties of the subject. The URIs ensure that concepts are not just words, like in Fig. 2, but are tied to a unique definition that everyone could find in the Web. Networks of triples act as an extended personal memory assisting users to file, relate, share, and access all digital information like documents, multimedia, and messages. For that purpose, a user may employ a system of categories, i.e. person, event, or organization, to build his own Personal Information Model (PIMO). For that purpose, he may access ontologies shared within a company or a social group, or create instances, classes, and properties (on the fly), and annotate the resources respectively [5,8]. This process is as simple to use as tagging and is assisted by user observation when interacting with documents. Fig. 2 shows an graphical representation of a PIMO-cutout

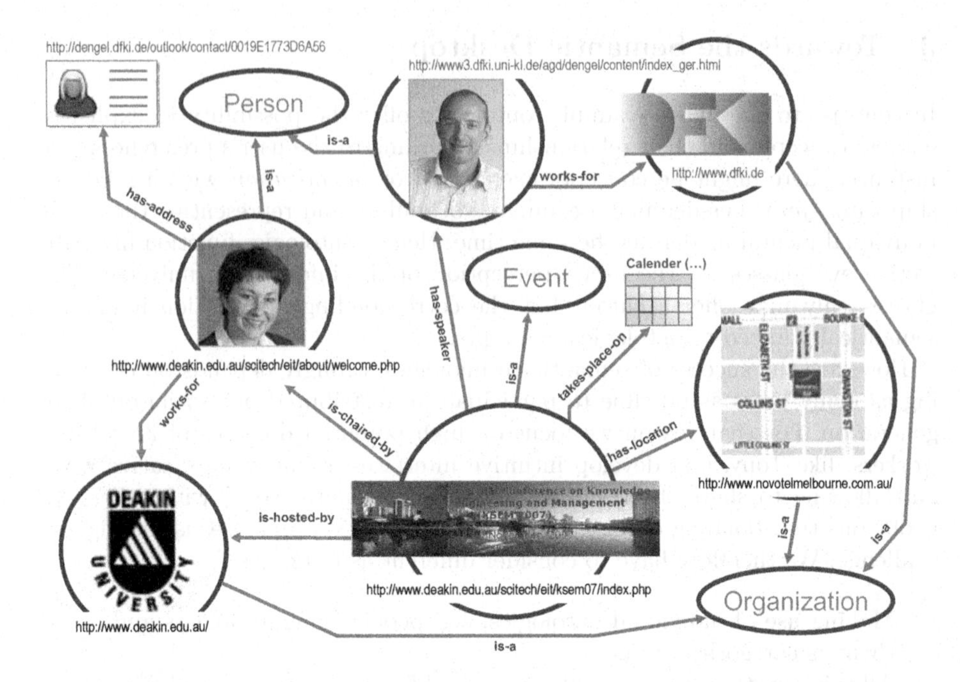

Fig. 2. Exemplary PIMO-cutout representing my talk at KSEM07

modeling some aspects of my relationship with KSEM07 as well as shared resources connected to it.

As shown in the example, triples of RDF form webs of information about related things. Documents are linked across applications and browsing through the personal information space is now possible. Consequently, addresses, documents, photos, appointments, or emails represented by their URIs that have been spread in the local data jungle can be linked conveniently, weaving a personal semantic web at a user's desktop. Data structures are not changed; existing applications are extended but not replaced. In order to avoid multiple appearances of the same entity, smushing techniques are used for identifying the synonymous resources and for aggregating the appropriate data. In conclusion, the PIMO reflects experience and typical user behavior. It may be processed by a computer in order to provide proactive and adaptive information support, and it allows personalized semantic search.

4 Semantic Services

The Semantic Desktop is built on a middleware platform that allows combining information from native applications like the file-system, Firefox, Thunderbird or MS-Outlook [6]. As a consequence proactive services may by used, such as an assisted mail classification in which the content of an email is related to known entities in the PIMO, or "save-as" actions may be backed by proposed related categories, i.e. folders, to which the content fits best. A user may complement the filing by adding relations and text annotations.

Furthermore, activities of knowledge workers are observed and collected in order to understand contextual behavior, i.e. browsing and clicking, reading and writing, task related actions, etc. which lead to individualized context-aware services [7]. In particular, reading an email or browsing a web site creates an "information-push" from the PIMO to an assisting sidebar, the so-called Miniquire, in which all current tasks and context-relevant information, categories, persons and projects are listed for the user. In addition a user may link the considered resource to one of the system-proposed concepts. Furthermore, changing an application, i.e. moving the mouse from the web browser to MS-Outlook in order to write a new email, leads to a context switch to which the sidebar information is adapted [3]. Fig. 3 shows an example for such a scenario.

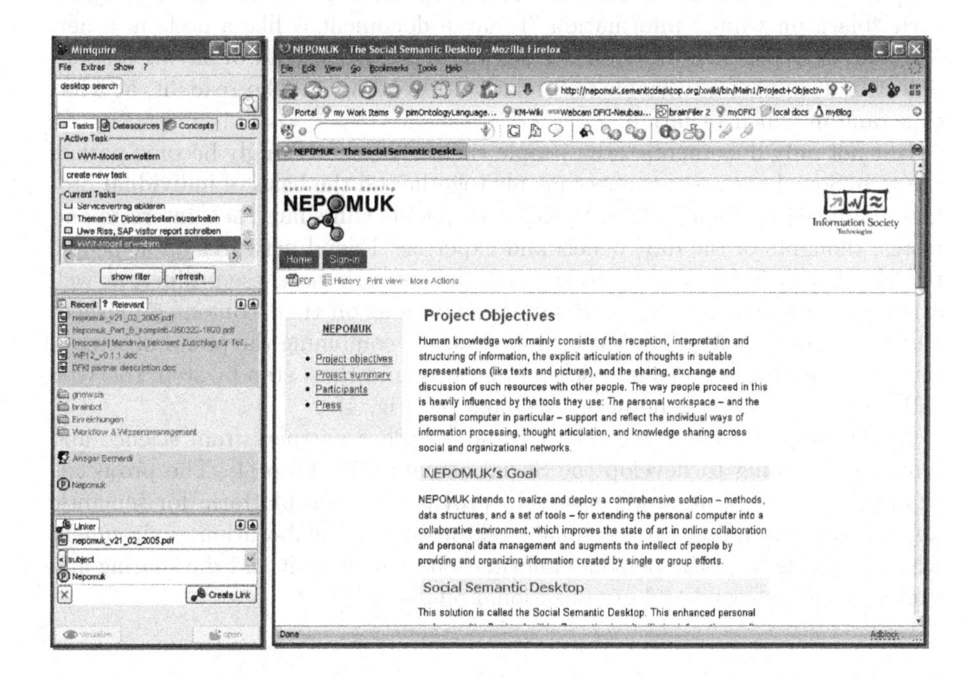

Fig. 3. Context-aware user support by the Miniquire sidebar

5 The Influence of Web 2.0

The Semantic Desktop specifies a driving paradigm for desktop computing using standards of the Semantic Web but integrating native office applications and data. It allows that individual trains of thoughts lead to multi-dimensional perspective organization of contents and thus to a dematerialization of the classical archive. We are convinced that based on the needs and expectations of users today the software industry will evolve to a future way of semantic desktop computing supporting the processing of ontologies by the operating system. End users will benefit as they integrate and also communicate better than today's desktop applications (based on ontologies and Semantic Web standards). In addition, it is not just technology that may change our way of how to deal with information. The traditional Web has recently undergone an orthogonal shift into a Web of People/Web 2.0 where the focus is set on folksonomies, collective intelligence, and the wisdom of trusted communities which influences office work as well. Thus, only if we combine information, technology, and new social standards, we can understand the dimensions of the workspace of the future [9].

The Web and all its connected devices are considered to be one global platform of reusable services and data where one can build on the work of others using open standards. The Web became part of our thinking and part of our workspace, and the documents we generate at our workspace become part of the Web. When Tony collected and organized his documents he had to deal with physical objects with "black on white" information. Today a document is like a node in a net, a system of hyper-links to books, texts, pictures, etc. Instead of having a static object, a document is variable and relative depending on who reads it, at what time, and in which situation.

But not only documents, even people themselves increasingly become part of the web. Social networks connect people together on the basis of individual profiles expressed by their PIMO. In this way, people will share their interests and tastes, thoughts of the day, beliefs and expertise. Friend networks allow people to link with their friends and to traverse the network via these profiles, as well as to give comments, votes, and recommendations on their content published. Therefore, the Semantic Desktop will be "social" combining the power of semantic web technologies and the wisdom of groups, and thus, step by step, the Web will evolve into a new generation: Web 3.0 (cf. Fig. 4).

In the project NEPOMUK, DFKI and 15 other partners from science and industry are going to develop the Social Semantic Desktop [4]. The project is running for three years and intends to provide an open platform for semantic applications extending the personal desktop into a collaboration environment which supports both the personal information management and the sharing and exchange across social and organizational relations [2].

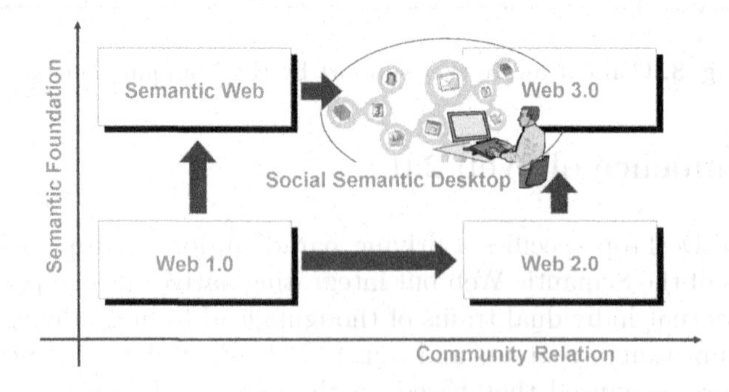

Fig. 4. Towards Web 3.0

In my keynote I will further deepen our approach of how to develop a Social Semantic Desktop. I will specifically propose a set of semantic services as key features for future office work in order to enhance individual productivity as well as organizational intelligence. They will provide the basis to foster collaboration, to share experience along supply chains, to proactively support distributed engineering workflows, or simply to socialize knowledge within communities.

References

1. Dengel, A.R.: Six Thousand Words about Multi-Perspective Personal Document Management. In: EDOCW 2006. Proc. of the 10th IEEE on International Enterprise Distributed Object Computing Conference Workshops, p. 62. IEEE Computer Society Press, Washington, DC, USA (2006)
2. Groza, T., Handschuh, S., Moeller, K., Grimnes, G., Sauermann, L., Minack, E., Mesnage, C., Jazayeri, M., Reif, G., Gudjonsdottir, R.: The NEPOMUK project - On the way to the Social Semantic Desktop. In: 3rd Int. Conf. on Semantic Technologies (I-SEMANTICS) (2007)
3. Holz, H., Maus, H., Bernardi, A., Rostanin, O.: From Lightweight, Proactive Information Delivery to Business Process-Oriented Knowledge Management. Journal of Universal Knowledge Management 0(2), 101–127 (2005)
4. NEPOMUK Consortium, http://nepomuk.semanticdesktop.org/
5. Sauermann, L., Bernardi, A., Dengel, A.: Overview and outlook on the semantic desktop. In: Decker, S., et al. (eds.) Proc. of Semantic Desktop Workshop at the ISWC, Galway, Ireland (November 2005)
6. Sauermann, L., Grimnes, G.A., Kiesel, M., Fluit, C., Maus, H., Heim, D., Nadeem, D., Horak, B., Dengel, A.: Semantic Desktop 2.0: The Gnowsis Experience. In: Cruz, I., Decker, S., Allemang, D., Preist, C., Schwabe, D., Mika, P., Uschold, M., Aroyo, L. (eds.) ISWC 2006. LNCS, vol. 4273, pp. 887–900. Springer, Heidelberg (2006)
7. Schwarz, S.: A Context Model for Personal Knowledge Management Applications. In: Roth-Berghofer, T.R., Schulz, S., Leake, D.B. (eds.) MRC 2005. LNCS (LNAI), vol. 3946, pp. 18–33. Springer, Heidelberg (2006)
8. Sintek, M., van Elst, L., Scerri, S., Handschuh, S.: Distributed knowledge representation on the Social Semantic Desktop: Named graphs, views and roles in NRL. In: ESWC. Proc. of the 4th European Semantic Web Conference (2007)
9. Wahlster, W., Dengel, A.: Web 3.0: Convergence of Web 2.0 and the Semantic Web. Technology Radar II, 1–23 (2006)
10. Wikipedia. Article on Antoni Gaudi, http://en.wikipedia.org/wiki/Antoni_Gaudi

Knowledge-Based Intelligent Engineering Systems in Defence and Security

Lakhmi Jain

School of Electrical and Information Engineering
University of South Australia
Adelaide, South Australia SA 5095, Australia
Lakhmi.jain@unisa.edu.au

Abstract. The Knowledge-Based Intelligent Engineering Systems Centre (KES) is focused on modelling, analysis and design in the areas of Intelligent Information Systems, Physiological Sciences Systems, Electronic commerce and Service Engineering. The Centre aims to provide applied research support to the Information, Defence and Health Industries. The overall goal will be to synergise contributions from researchers in the diverse disciplines such as Engineering, Information Technology, Science, Health, Commerce and Security Engineering. The research projects undertaken in the Centre include adaptive mobile robots, aircraft landing support, learning paradigms, teaming in multi-agent systems, target detection, image registration and detection in intelligent environment, unmanned air vehicles (UAVs) and simulation, intelligent decision support systems, neuro-fuzzy systems, medical diagnostic systems in intelligent environment and so on. This talk will focus on Knowledge-Based Intelligent Engineering Systems in Defence and Security applications.

Z. Zhang and J. Siekmann (Eds.): KSEM 2007, LNAI 4798, p. 10, 2007.

Auditing and Mapping the Knowledge Assets of Business Processes – An Empirical Study

W.B. Lee*, Vivien Shek, and Benny Cheung

Knowledge Management Research Centre,
Department of Industrial and Systems Engineering,
The Hong Kong Polytechnic University, Hong Kong
wb.lee@inet.polyu.edu.hk

Abstract. A systematic, contextual and action-oriented methodology was developed and has been adopted in a large business unit in a public utility company in Hong Kong to map out the company's explicit and tacit knowledge assets based on input from both structured questionnaires and interactive workshops conducted in an open and participative manner. Outputs from the knowledge audit include a critical knowledge inventory list, an assessment of the level of diffusion and codification of the knowledge sources, and a knowledge map for each of the business processes. The inquiry process makes the data collection transparent and also facilitates knowledge sharing, interaction, mutual understanding and consensus among workshop participants.

Keywords: Knowledge audit, Knowledge assets, Knowledge mapping, Intellectual capital.

1 Background of the Study

The integration of knowledge management (KM) into business processes has been identified by KM experts in an international Delphi-study on the future of knowledge management as one of the most pressing and practical research issues [1]. As the effective management of an organization's knowledge assets is recognized to be a critical success factor in business performance, this points to the importance of the knowledge audit as the first step in understanding into how knowledge is handled in mission critical business processes in an organization. In spite of the fact that knowledge assets are known to determine the success or failure of a business, these are seldom found in a company's book. A knowledge audit provides an evidence based assessment of the knowledge assets within an organization as well as an assessment of the KM effort on which the company should focus. A knowledge audit involves a complete analysis and investigation of the company in terms of what knowledge exists in the company, where it is, who owns it and how it is created. However, there is no standard framework among KM practitioners as to how a knowledge audit should be constituted and conducted.

* Corresponding author.

Z. Zhang and J. Siekmann (Eds.): KSEM 2007, LNAI 4798, pp. 11–16, 2007.

Boisot [2] was among the first to come up with a theoretical framework to define alternative types of knowledge as positioned in a three dimensional epistemological space which he called Information-Space (I-Space) so as to relate the functional relationship between codification, abstraction and diffusion of various types of knowledge. These concepts are found to be most useful in understanding the movement of knowledge in the I-Space and its implications in the social and cultural dimensions. However, very few practical cases have been reported in the literature where such a framework has been applied in a practical business context.

Apart from the analysis of knowledge types and of the movement of knowledge, most knowledge management fails not only because of the failure to identify the mission critical knowledge resources (people, document, and skill sets etc.) that need to be retained and developed, but also because the ways in which this information is collected, analyzed and evaluated can be highly subjective and often political. Although different approaches to a knowledge audit have been described in the literature, most of them are more related to a knowledge management audit. Similar to the evaluation of intellectual capital, there is no universally accepted approach as how organizational knowledge assets should be analyzed. The most commonly used audit tools are based on interviews, focus groups and survey questionnaires. It has been found that most of the knowledge audit methods reported are mainly of the rating type, limited in their scope of study and lacking in interaction with the employees of the company being audited. These factors reduce the efficiency and validity of the knowledge audit.

2 The Audit Methodology

In view of the importance of the knowledge audit and of the deficiencies of the current audit methods, a systematic, contextual and action-oriented methodology called STOCKS (Strategic Tool to Capture Critical Knowledge and Skills) has been developed by the authors to map out the knowledge asset of organizations based on input from both structured questionnaires and interactive workshops conducted in an open and participative manner. It is composed of seven phases which are: selection and prioritization of mission critical business processes, work flow study, collection of data through form filling, and STOCKS workshop, building knowledge inventory, analysis as well as in-depth interview and data validation followed by recommendations of knowledge management strategies (Figure 1). Data and information are collected through the filling in of specially designed forms. The objective of the STOCKS workshop is for the knowledge workers of the organization to consolidate and validate the data collected from the filled STOCKS forms. A STOCKS schema is prepared before running the workshop. As shown in Figure 2, the schema contains the fields which include a selected business process, its process flow (i.e. tasks), industrial technology, documents and tacit knowledge. The naming and grouping of the knowledge items, the choice of words, their relationship and hierarchy, all have to be agreed among workshop participants in the development of the controlled vocabulary, thesaurus and taxonomy.

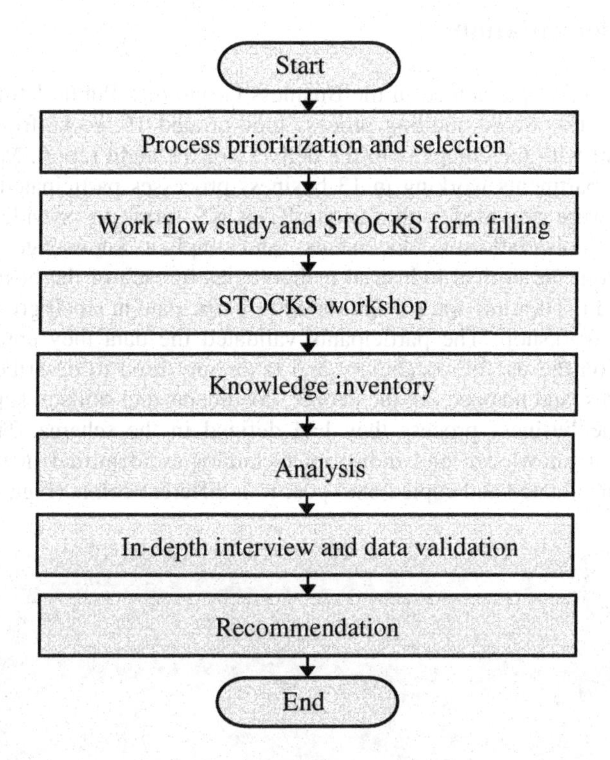

Fig. 1. The STOCKS audit process

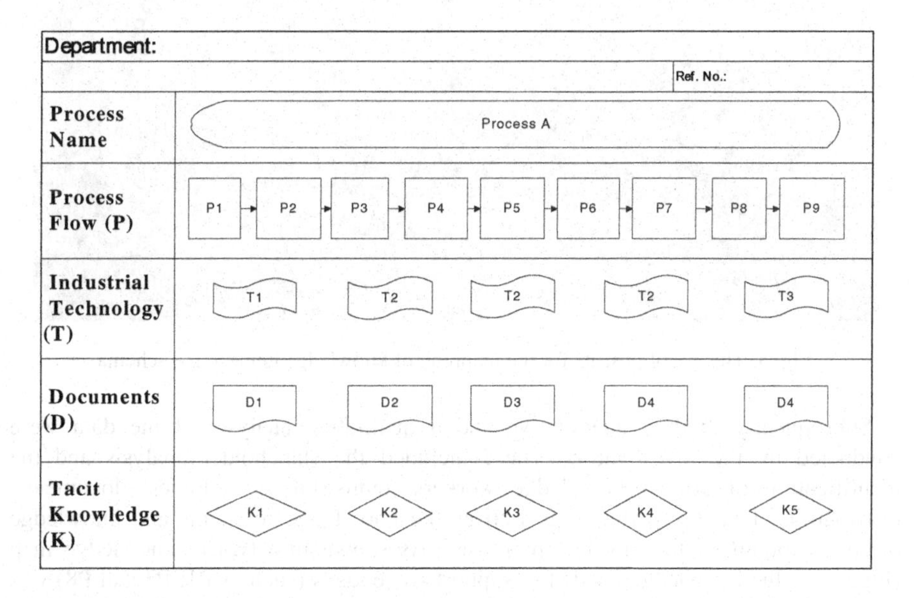

Fig. 2. The schema for the mapping of various knowledge assets onto the business process

3 The Implementation

STOCKS was trial implemented in the Business Group of a Public Utility Company in Hong Kong. The whole auditing process took around 16 weeks from setting the scope alignment with the company to the delivery of the audit report. More than 110 staff from 5 departments working in 13 business processes participated in the audit project. They were provided with various STOCKS forms to provide information about the IT tools/platforms, documents, and implicit knowledge, as well as information about the critical industrial technologies in each of the business process selected for audit. The staff were then invited to participate in small groups in a half-day STOCKS workshop. The participants validated the data they entered into the forms, agreed on the use of vocabulary and taxonomy used to describe the various knowledge items, and mapped out the people, documents and skills to support each of the tasks in the business process they had defined in the schema. The names of documents, tacit knowledge and industrial technologies identified in the STOCKS forms were consolidated and copied onto post-it in different colors (Figure 3).

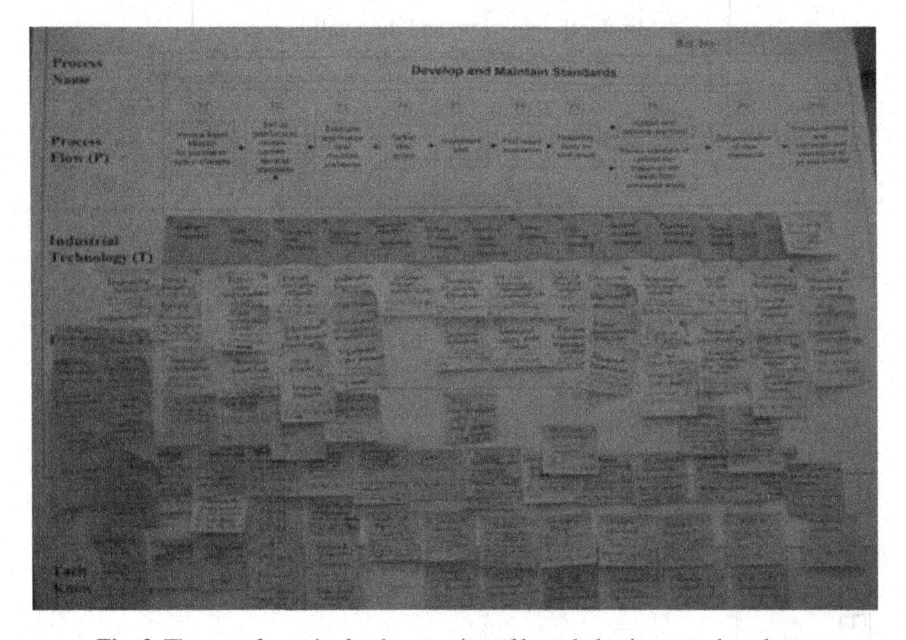

Fig. 3. The use of post-its for the mapping of knowledge items to the schema

Subsequently, both a quantitative and a qualitative analysis of the data were conducted by the investigators. These included the stakeholder analysis and the identification of critical knowledge workers, industrial technologies, documents, distribution of tacit knowledge as well as the knowledge fountains and knowledge discovery points of the business processes. As shown in a typical knowledge map (Figure 4), the knowledge needed to support some tasks (such as P1, P7 and P8) was found to have not been codified and the tacit knowledge assets would be lost if the staff involved left the company.

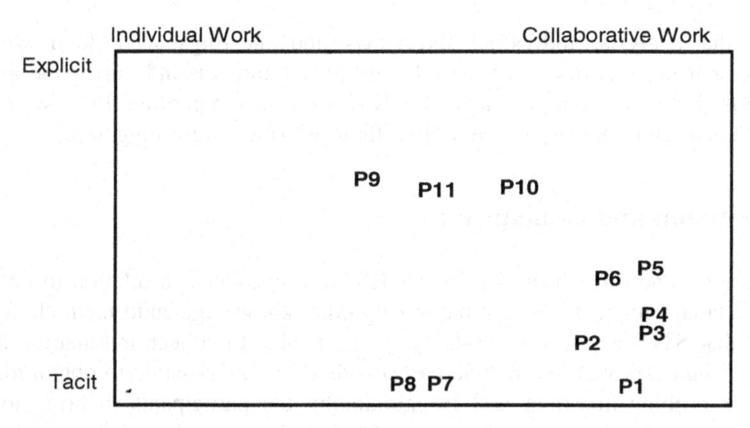

Fig. 4. Distribution of knowledge in tasks P in the two dimensional I-Space

In the STOCKS forms, relevant information about a codified or implicit knowledge item such as its source, user, format, location and medium of communication is identified and is then consolidated onto a list, a knowledge inventory. Based on this information, the corresponding suppliers and customers of knowledge in a business process are plotted on a knowledge map (Figure 5) to show the relationships and connectivity. Very often, the map may also reveal the existence of an isolated network, any duplication of knowledge sources and also the intensity of knowledge exchange between internal and external parties of the organization.

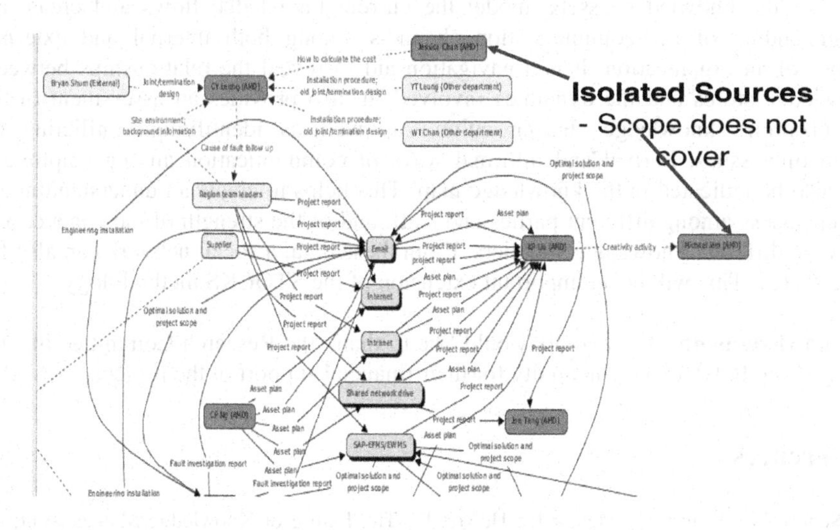

Fig. 5. A typical knowledge map revealing isolated tacit knowledge sources with high risk of being lost

After the analysis, individual interviews and meetings were held with the respective participants in order to clarify any uncertainties found during the analysis stage. Based on the concrete data, the KM strategies regarding Process, People, Content and Technology aspects as well as their priorities, were suggested.

4 Discussion and Conclusion

The outcomes and effectiveness of STOCKS were evaluated in relation to both KM and intellectual capital. When compared with other knowledge audit methodologies, it is found that STOCKS is more systematic, and is able to collect and analyze a large amount of data. Through the interactive workshop, it was possible to obtain ideas by the use of probing questions and by stimulating the participants to bring to mind matters they were not consciously aware of before (i.e. they did not know what they knew). More important, STOCKS creates an interactive and non-threatening environment in which staff can share and prioritize the knowledge which they perceive as important. The workshop also enhances communication among all the different staff in the organization. With most intellectual capital tools, there is often a pre-set taxonomy of the capital items to be audited and rated. The workflow of the business process and the specific knowledge needs are often not the subject of the audit study. STOCKS helps to identify the critical organizational knowledge that needs to be captured, and retained for the healthy operation and sustainability of the business. STOCKS is a process-oriented tool which links up information management, business process management and strategic management in an organization.

The visual display of the information that is captured and the display of the relationships in the form of a knowledge map help to uncover the primary sources and users of the knowledge assets, model the current knowledge flows and create an understanding of the communication channels among both internal and external parties of an organization. It is a navigation aid to reveal the relationships between knowledge stores and the dynamics involved. It also provides an assessment of an organization's knowledge sharing culture as well as identifying conflicting or competing issues. Formal and informal ways of communication among employees can also be indicated in the knowledge map. This helps to create an understanding of the closeness among different parties and staff, and of the strength of the connections between different knowledge suppliers. From these data, a social network can also be constructed. This will be an important extension of the STOCKS methodology.

Acknowledgments. The authors would like to thank the Research Committee of The Hong Kong Polytechnic University for their financial support of the project.

References

1. Scholl, W., König, C., Meyer, B., Heisig, P.: The Future of Knowledge Management: an International Delphi Study. Journal of Knowledge Management 8(2), 19–35 (2004)
2. Boisot, M.H.: Knowledge Assets: Securing Competitive Advantage in the Information Economy. Oxford University Press, Oxford (1998)

Quotient Space Based Multi-granular Analysis

Ling Zhang[1] and Bo Zhang[2]

[1] Artificial Intelligence Institute
Anhui University, Hefei
Anhui, China
zling@ahu.edu.cn

[2] Department of Computer Science and Technology
Tsinghua University
Beijing, China
dcszb@tsinghua.edu.cn

Abstract. We presented a quotient space model that can represent a problem at different granularities; each model has three components: the universe X, property f and structure T. So a multi-granular analysis can be implemented based on the model. The basic properties among different quotient spaces such as the falsity preserving, the truth preserving properties are discussed. There are three quotient-space model construction approaches, i.e., the construction based on universe, based on property and based on structure. Four examples are given to show how a quotient space model can be constructed from a real problem and how benefit we can get from the multi-granular analysis. First, by adding statistical inference method to heuristic search, a statistical heuristic search approach is presented. Due to the hierarchical and multi-granular problem solving strategy, the computational complexity of the new search algorithm is reduced greatly. Second, in the collision-free paths planning in robotics, the topological model is constructed from geometrical one. By using the truth preserving property between these two models, the paths planning can be implemented in the coarser and simpler topological space so that the computational cost is saved. Third, we discuss the quotient space approximation and the multi-resolution signal analysis. And the second-generation wavelet analysis can be obtained from quotient-space based function approximation. It shows the equivalence relation between the quotient space model based analysis and wavelet transform. Fourth, in the automatic assembly sequence planning of mechanical product, we mainly show how a quotient structure can be constructed from the original one. By using the simpler quotient structure, the assembly sequence planning can be simplified greatly. In conclusion, the quotient-space model enables us to implement a multi-granular analysis. And we can get great benefit from the analysis.

Z. Zhang and J. Siekmann (Eds.): KSEM 2007, LNAI 4798, p. 17, 2007.

An Ontology-Based Reasoning Framework for Reaction Mechanisms Simulation

Y.C. Alicia Tang[1], Sharifuddin Mohd. Zain[2], Noorsaadah Abdul Rahman[2], and Rukaini Abdullah[3]

[1] Department of Computer Science, University of Tenaga Nasional
Jalan Kajang-Puchong, 43009 Selangor, Malaysia
[2] Department of Chemistry, University of Malaya, 50603 Kuala Lumpur, Malaysia
[3] Department of Artificial Intelligence, University of Malaya
50603 Kuala Lumpur, Malaysia
aliciat@uniten.edu.my, {smzain,noorsaadah,rukaini}@um.edu.my

Abstract. Many chemistry students have difficulty in understanding an organic chemistry subject called reaction mechanisms. Mastering the subject would require the application of chemical intuition and chemical commonsense adequately. This work discusses a novel framework using Qualitative Reasoning (QR) to provide means for learning reaction mechanisms through simulation. The framework consists of a number of functional components. These include substrate recognizer, qualitative model constructor, prediction engine, molecule update routine, explanation generator, and a knowledge base containing essential chemical facts and chemical theories. Chemical processes are represented as qualitative models using Qualitative Process Theory (QPT) ontology. The construction of these models is automated based on a set of QR algorithms. We have tested the framework on the S_N1 and the S_N2 reaction mechanisms. Representative cases of reaction simulation and causal explanation are also included to demonstrate how these models can serve as a cognitive tool fostering the acquisition of conceptual understanding via qualitative simulation.

Keywords: qualitative reasoning, reaction mechanisms, QPT, ontology.

1 Introduction

Qualitative Reasoning (QR) is a branch of Artificial Intelligence (AI) research that attempts to model behavior of dynamic physical systems without having to include a bunch of formulas and quantitative data in the system. Qualitative representation captures the intuitive and causal aspects of many human mental models. The research spans all aspects of the theory and applications of qualitative reasoning about physical systems. Qualitative Process Theory (QPT)[1] is one of the prominent QR ontology that is widely used to represent the behavior of dynamical systems. CyclePad [2] that teaches analysis and design of thermal cycles is the first smart educational software that employed QPT. QALSIC

Z. Zhang and J. Siekmann (Eds.): KSEM 2007, LNAI 4798, pp. 18–29, 2007.

is among the earliest applications of QPT in inorganic chemistry for qualitative analysis of a limited set of chemical reactions [3]. The chemical processes described in this paper were modeled using QPT ontology.

In the study of science subjects such as chemistry, it is believed that students should deeply understand the qualitative principles that govern the subject, including the physical processes and the causal relationships before they are immersed in complex problem solving. A reaction mechanism describes the sequence of steps that occur during the conversion of reactants to product. Examples of reaction mechanisms are S_N1, S_N2, electrophilic addition, and eliminations. Most of the time, the organic chemists could work out the mechanisms by only using commonsense developed from their chemical intuition and knowledge. A large number of chemistry students had difficulty in understanding reaction mechanisms. They learn the subject by memorizing the steps involved in each reaction. As a result, most students are unable to solve new problems. This finding initiated the work described in this paper. Even though there are many applications of AI techniques in organic chemistry, none has used QPT as the knowledge capture tool. This paper discusses the first use of the QR approach coupled with the QPT ontology to develop a framework that is able to simulate processes such as "make-bond" and "break-bond" in order to reproduce the chemical behaviors of organic reaction mechanism. The paper also introduced OntoRM, which is a set of ontology specifically for use with reaction mechanisms simulation. The framework will later on be transformed into a learning tool called Qualitative Reasoning in Organic Mechanisms (QRIOM).

We have reported in [4] about the modeling decisions and problems faced when trying to cast the expert knowledge into qualitative models. In [5], we justified the problem as a suitable domain by comparing inorganic chemical reactions and organic reaction mechanisms. In the work, we have also grouped all reacting species as either a nucleophile (charged/neutral) or an electrophile (charged/neutral), upon which chemical processes are selected. In [6], we provided guidelines for chemical properties abstraction and a description of how the QPT is used for modeling. In [7], we further ascertained that there are two main reusable processes, namely "make-bond" and "break-bond", for the entire reaction mechanisms, specifically on S_N1 and S_N2. An algorithm for automating the "make-bond" and the "break-bond" processes in QPT terms has also been discussed in the paper, where the initiation of the entire process is from a simple substrate.

In Sect. 2, we provide the methodology of our work. These include data sets, algorithms and functional components used by the framework. Section 3 discusses the simulation algorithm together with a process reasoning scenario and an explanation generation example. Section 4 concludes the work.

2 Methodology

The following data, modeling constructs, algorithms, and components are needed in order to simulate and reproduce the behavior of reaction mechanisms.

2.1 QPT as the Knowledge Capture Tool

Among the well-known QR ontology are component-centered [8], constraint-based [9] and process-centered [1]. QPT is a process-centered ontology. The theory provides the necessary means for representing qualitative knowledge, and the notion of processes needed in expressing chemical reaction steps (E.g. "protonation" and "dissociation" processes). In QPT, a description of the model is given by a set of *individual views* and *processes.* The *individual views* (E.g. a nucleophile) describe objects and their general characteristics while the *processes* (E.g. "make-bond") support changes in system behavior. A process is described by five slots: *Individuals*, *Preconditions*, *Quantity-conditions*, *Relations* and *Influences.* The *Quantity-conditions* slot contains inequalities involving *quantities.* A *quantity* is used to represent an object's characteristic, which is crucial in determining the status of a process (active/inactive). The statements in the Relations slot defined functional dependencies among *quantities.* Other important design constructs are the *qualitative proportionalities* and the *quantity spaces* (see Sect. 2.2). Note: words typed in italics are QPT modeling constructs. Further discussion about the ontology is beyond the scope of this paper. Readers may refer to Forbus [1] for a complete description of the ontology.

2.2 Qualitative Proportionality, Direct Influence, and Quantity Space

This section discusses the main modeling constructs used in the framework. *Qualitative proportionalities* are responsible for propagating the effects of processes execution. For example: lone-pair-electron (O) P^{+}_{-} no-of-bond (O) means "an increase in covalent bond on the 'O' atom will cause a decrease in the number of lone-pair electrons on it". In QPT, the dynamic aspects are expressed by the notion of direct influence, represented in the slot called *Influences* as either I+ or I-. For example, I+(no-of-bond(O),Am[bond-activity]) indicates that in a chemical process, the direct effect on 'O' is the extra covalent bond it would gain. This effect will be propagated to other statements in the *Relations* slot. Examples of *quantities* are number of covalent bond, lone-pair electrons and nucleophilic reactivity. A quantity consists of two parts: amount (magnitude) and derivative (sign). *Quantity space* is a collection of numbers which form a partial order. Processes start and stop when orderings between the *quantities* change. Table 1 gives the three main *quantities* used in the problem. The main task of qualitative simulation based on this formalism is to keep track of the changing states in each quantity for each individual view used in a reaction, in order to explain why and how a particular process happens or ends.

2.3 Data Sets

Reaction formulas tested by the framework include the production of alkyl halide from tertiary alcohol and the production of alcohol from tertiary alkyl halide.

Table 1. Quantities and quantity spaces used in the framework for modeling the behaviors of nucleophiles and electrophiles

Quantity	Quantity Space	Remarks
Charges	[negative, neutral, positive]	At any time the charge on any atom can either be negative, neutral or positive.
no-of-bond	[one, two, three, four]	We consider only the important atoms for nucleophilic substitution reaction. For example, the 'four' goes to carbon; the 'one' is for hydrogen when they are in the most stable state.
lone-pair-electrons	[zero, one, two, three, four]	The maximum value 'four' is for halide ions. The minimum 'zero' goes to hydrogen ion.

Examples of the reaction formulas used in modeling and simulation are (1) and (2).

$$(CH_3)_3COH + HCl \rightarrow (CH_3)_3CCl + H_2O. \tag{1}$$

$$(CH_3)_3CBr + H_2O \rightarrow (CH_3)_3COH + HBr. \tag{2}$$

Both the equations can be explained by the S_N1 mechanism. In general, S_N1 is a two-stage mechanism. In the first stage, the alcohol oxygen (the 'O' from the 'OH' group) is protonated. Meaning, the 'O' captures the electrophile ('H^+'). This is to make the 'O^+H2' as a good leaving group, in order to break the bond between the 'C' and the 'O^+H2'. Once broken, a carbocation will be produced. In the second stage, the incoming nucleophile ('X^-') can bond to the carbocation to form a neutral and stable final product. In any chemical reaction, some bonds are broken and new bonds are made. Often, these changes are too complicated to happen in one simple stage. Usually, a reaction may involve a series of small changes one after the other. Thus, (1) can be subdivided into a series of small step, as shown in Fig. 1. The main ideas of the reactions in Fig. 1 will be modeled as QPT processes. These processes are automatically generated by the Qualitative Model Constructor (module 2, Fig. 4). In which, the chemical properties of each organic reaction are represented as chemical theories using the qualitative proportionality construct. To avoid being too technical in terms of chemistry contents, we provided in this paper only the results (qualitative models in QPT syntax), rather than the entire modeling activity. A qualitative simulation scenario for the first reaction step in (1) is demonstrated in Sect. 3.

2.4 Knowledge Validation

The chemical knowledge used by QRIOM has two-tier architecture (Fig. 2). We are exploring the development of OntoRM, working on top of the QPT. It is to

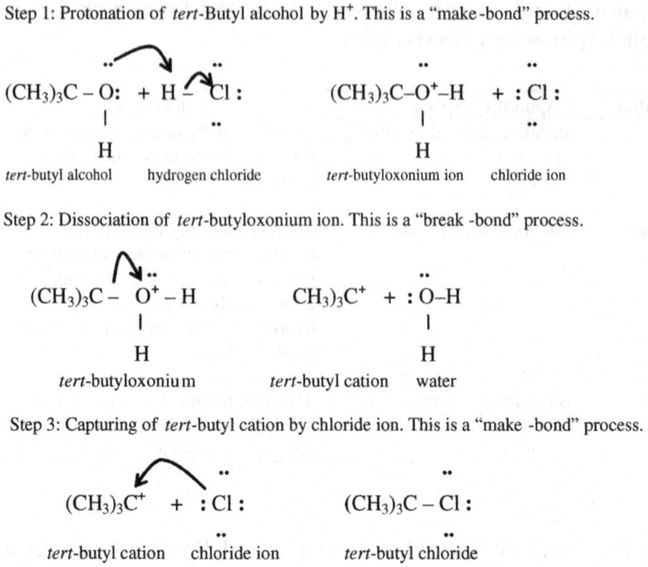

Step 1: Protonation of *tert*-Butyl alcohol by H^+. This is a "make-bond" process.

$$(CH_3)_3C-\ddot{O}(H): + H-\ddot{Cl}: \quad \longrightarrow \quad (CH_3)_3C-\ddot{O}^+(H)-H + :\ddot{Cl}:$$

tert-butyl alcohol hydrogen chloride *tert*-butyloxonium ion chloride ion

Step 2: Dissociation of *tert*-butyloxonium ion. This is a "break-bond" process.

$$(CH_3)_3C-\ddot{O}^+(H)-H \quad \longrightarrow \quad (CH_3)_3C^+ + :\ddot{O}(H)-H$$

tert-butyloxonium *tert*-butyl cation water

Step 3: Capturing of *tert*-butyl cation by chloride ion. This is a "make-bond" process.

$$(CH_3)_3C^+ + :\ddot{Cl}: \quad \longrightarrow \quad (CH_3)_3C-\ddot{Cl}:$$

tert-butyl cation chloride ion *tert*-butyl chloride

Fig. 1. The production of alkyl halide can be explained by a series of three reaction steps

facilitate knowledge validation during simulation such as to constrain the use of the chemical knowledge base. As an example, the OntoRM can be used to check if a primary alcohol can undergo a S_N1 reaction. Representative examples of OntoRM are given in Fig. 3. Further discussion about OntoRM is not the focus of this paper.

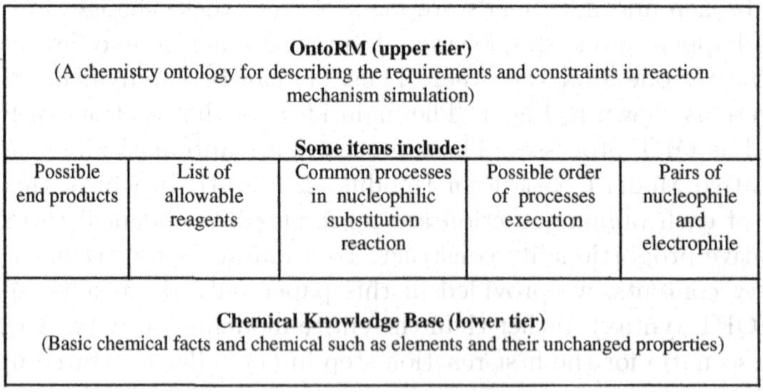

OntoRM (upper tier)
(A chemistry ontology for describing the requirements and constraints in reaction mechanism simulation)

Some items include:

Possible end products	List of allowable reagents	Common processes in nucleophilic substitution reaction	Possible order of processes execution	Pairs of nucleophile and electrophile

Chemical Knowledge Base (lower tier)
(Basic chemical facts and chemical such as elements and their unchanged properties)

Fig. 2. Knowledge-base used by QRIOM

```
ReactionMechanism
Sn1 [
        hasAlias =>> STRING;
        hasReactants =>> FuncUnit;
        hasReactantNames =>> STRING;
        hasProduct =>> PROD_STRING;
        hasProductNames =>> STRING;
        hasDegreeSubstituent =>> NUMBER;
        hasReactivity =>> BOOLEAN;
        hasRateDetermineStep =>> WHAT_STEP_STRING;
        hasProcessOrder =>> PROCESS_ORDER_STR;
        hasViewsPairConstraint =>> SPECIES_TYPE;
        hasSpecialCause = >> SOLVENT_TYPE;   ]

ElectrophileView[
        hasName =>> STRING;
        hasNeutral =>> Electrophile;
        hasCharge =>> Electrophile;
        hasBond =>> NUMBER;
        hasRsDegree =>> NUMBER;
        hasCarbocationStability=>>FUZZY_VALUE;
        hasLonePair =>> NUMBER;
        hasReactivity =>>BOOLEAN;
        hasChargeOperator =>> PLUS_MINUS;
        hasBondOperator =>> ADD_REMOVE;      ]
```

Fig. 3. Definitions of S_N1 mechanism and electrophile in OntoRM

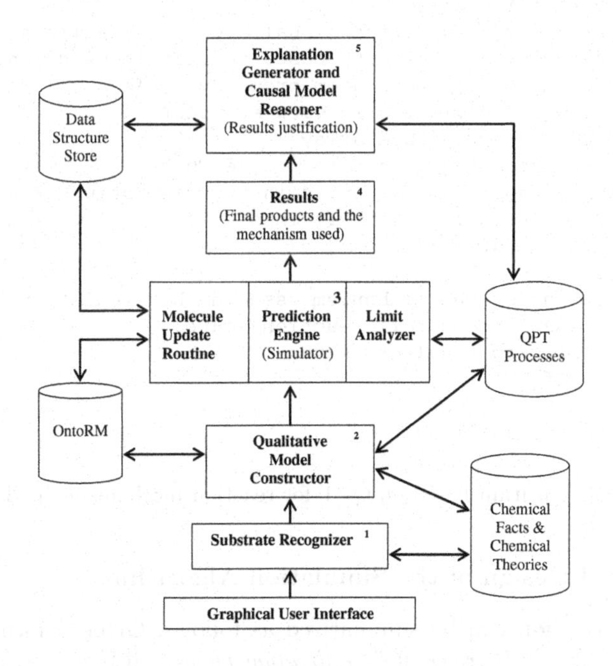

Fig. 4. Functional components of the framework

2.5 Functional Components

The QR framework is presented in Fig. 4. The framework consists of a set of reusable components such as the substrate recognizer, qualitative models, OntoRM, prediction engine, and causal explanation generator. The roles of the prediction engine and the explanation generator are discussed in Sect. 3.

The modules in Fig. 4 serve as embedded intelligence to the simulator. When used, it is expected to generate the following outputs: Final products; intermediates produced at each step; the sequence of use of the chemical processes; the name of the mechanism used; the structural change of the substrate; and the parameters change of each nucleophile and electrophile.

3 Qualitative Simulation

The QR algorithm for reaction mechanism simulation is outlined in Fig. 5.

```
QUALITATIVE SIMULATION ALGORITHM
Simulation(substrate, reagent, OUTPUT)
1. Recognize substrate
2. Construct individual views
3. Determine candidate processes
4. Construct QPT process
5. Perform processes reasoning
      Store process's quantity from the direct influence slot
      Perform limit analysis
      Check qualitative proportionalities in Relation-slot
      Store propagated effects based on quantities dependency
      Update atom table and atom property table
      Store new individuals in view structure
      Update view structure array
6. If process_stopping_condition = true Then
      Check if any reactive units in the view structure
      If reactive units <> EMPTY Then
           Go to step 3
      Else
           Suggest the mechanism used in the simulation
           Show the overall reaction route
           Display final products
        End_If
   End_If
7. Generate explanations
```

Fig. 5. QR algorithm based on QPT for reaction mechanism simulation

3.1 Top Level Design of the Simulation Algorithm

The overall simulation can be summarized as follows: Given a formula in the form "A (substrate) + B (reagent)", *individual views* will be constructed based on their chemical properties. These *views* will be stored in Instance Structure

(IS). Next, it is the checking of what processes can be used. When a process is in active state, reasoning will begin (details are given in Sect. 3.2). Briefly, the reasoning engine will keep track of the values of the affected *quantities*, starting from the first process until the entire reaction ends. A process will stop when the statements in its *quantity-condition* slot are invalid. If there are still reactive units (E.g. charged species or species that have not completed their valences), the reasoning process will be repeated. The entire reaction will end when there is no more *views* in the IS. When a reaction ends, outputs are displayed, together with all steps/processes involved in producing the outputs. If a user needs an explanation for the results or has a question regarding the behavior of a *quantity*, then the explanation module will be run.

3.2 Process Reasoning

A qualitative model for the "protonation" process is shown in Fig. 6. All complexities in constructing a qualitative model are hidden from the users, since the QPT model construction process is automated.

Process "Protonation" (e.g. $((CH_3)_3C\text{-}OH)$ is protonated by H^+)

Individuals
;there is an electrophile (charged)
1. H *; hydrogen ion*
; there is a nucleophile (neutral) that has lone pairs electron
2. O *; alcohol oxygen*

Preconditions
3. A_m [no-of-bond(O)] = TWO
4. is_reactive(R_3C-OH)
5. leaving_group(OH, poor)

Quantity-Conditions
6. A_m[non-bonded-electron-pair(O)] >= ONE
7. charges(H, positive)
8. electrophile(H, charged)
9. nucleophile(O, neutral)
10. charges(O, neutral)

Relations
11. D_s[charges(H)]= -1
12. D_s[charges(O)]= 1
13. lone-pair-electron(O) P± no-of-bond(O)
14. charges(O) P∓ lone-pair-electron(O)
15. lone-pair-electron(H) P no-of-bond(H)
16. charges(H) P± no-of-bond(H)

Influences
17. I_+ (no-of-bond(O), A_m[bond-activity])

Fig. 6. A chemical process modeled using QPT ontology

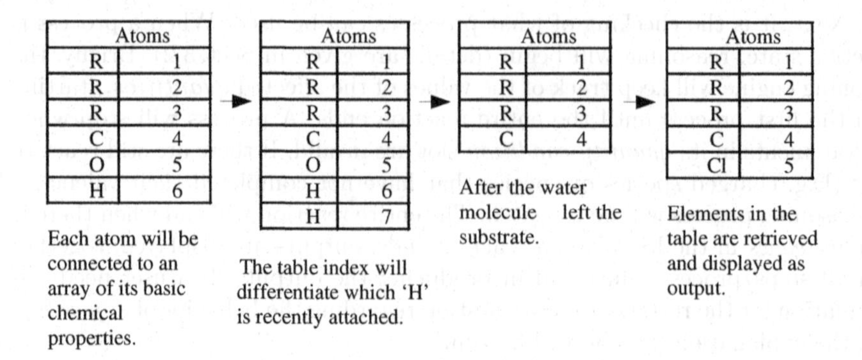

Fig. 7. A snapshot of the contents of an atom array during simulation. $R = CH_3$.

"Protonation" is the first reaction step of the S_N1 mechanism for predicting the final product for $(CH_3)_3COH + HCl$. Prediction begins with the *Influences* slot where the number of covalent bond on the 'O' will increase (Line 17). Such effect will propagate to other dependent *quantities*. For example, the number of lone-pair electrons will decrease when more covalent bonds are made on the 'O' via the inverse *qualitative proportionality* (Line 13). When the lone-pair electron of 'O' decreases, the charges on 'O' will also increase (Line 14). This will make the 'O' become positively charged and having an extra covalent bond (hence it is unstable). When 'O' is protonated, the 'H' is no longer positively charged (Line 16), thus violating the statement in the *quantity-conditions* slot. All values assigned

Table 2. Atom property array keeps track of the step-by-step changing of values in various quantities

	Charges	No. of covalent bond	Lone -pair electron
C	Neutral	4	0
O	Neutral	2	2
H	Neutral	1	0

(+)

	Charges	No. of covalent bond	Lone -pair electron
H	Positive	0	0

(=)

	Charges	No. of covalent bond	Lone -pair electron
C	Neutral	4	0
O	Positi ve	3	1
H	Neutral	1	0
H	Neutral	1	0

to each individual view are taken from the *quantity spaces* by the limit analyzer, in that it keeps track of the current values of each *quantity* and their direction of change. The running result of the limit analyzer, acted on the atom array will be used to produce the molecule structure for the final product. An atom array is a table that stores the elements of a substrate during reasoning in order to produce the structure of the final product. Figure 7 depicts the contents of an atom array during processes simulation for reaction formula in (1). When performing limit analysis, the atom property array will also be updated. This is when the processes move from one to another until the entire reaction ended. A snapshot of the atom property array during process reasoning is given in Table 2. The above is achieved through constant updating of the *qualitative proportionality* statements (Lines 13-16) using values in the *quantity spaces.* In QRIOM, qualitative models can be inspected by learners at any stage of the learning process. This helps to sharpen a learner's logical and critical thinking in the way that the learner has to think hard for why the statements in each slot are relevant or negligible.

3.3 Explanation Generation

The ability to generate causal explanation has been one of the promises of the QR approach. Causality is normally used to impose order. For example, when given 'P causes Q', we believe that if we want to obtain Q we would bring about P. As such, when we observe Q we will think that P might be the reason for it. We will demonstrate how the modeling constructs of QPT can provide this nature of explanation. Figure 8 shows a partial causal diagram derived from Fig. 6.

Two qualitative proportionalities (abstracted from the left branch of Fig. 8) are presented (3) and (4) to manifest the explanation generation ability of the approach.

$$lone\,pair\,electron(O)\ \ P^{+}_{-}\ \ no\,of\,bond(O). \tag{3}$$

$$charges(O)\ \ P^{-}_{+}\ \ lone\,pair\,electron(O). \tag{4}$$

Based on the above parameters dependency statements, a set of hypothetical Q&A and answers can be devised, as follows:

- Question 1: How would you explain a decrease in the lone-pair-electron on the 'O' atom?
- Answer 1: The immediate cause of the process is the number of covalent bond on 'O' will increase. This quantity will influence the lone-pair-electron on 'O', and the influence is strictly decreasing through the inverse proportionality relationship. Thus, a decrease in the lone-pair-electron on 'O' is observed.
- Question 2: How would the above qualitative proportionalities explain the 'O' atom become positively charged?
- Answer 2: The number of lone-pair-electron will decrease when more covalent bonds are made on 'O' via the inverse proportionality defined in (3). In (4), when the lone-pair-electron on 'O' decreases, the charge on it will increase.

The inspection of cause effect chain can help a learner to sharpen his or her reasoning ability, in that the learner is able to pick up the underlying concept better than merely memorizing the reaction steps and formulas.

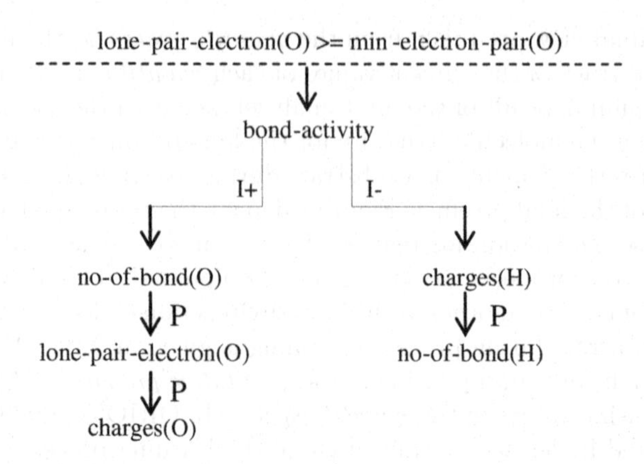

Fig. 8. The inequality above the dotted line is the entry condition to the process. Effects propagation is modeled using the 'Is' and the 'Ps' of the QPT design constructs.

4 Conclusion

We proposed a QR framework using the QPT ontology to systematically gather, reuse, and explain the knowledge about reaction mechanisms. The ontology provides the means to describe processes in conceptual terms, and embody notions of causality which is important to explain the behavior of chemical systems. The framework can provide a learning environment that assists learners in understanding the 'How', 'Why', 'Why-not', and 'What' aspects of the general principles of organic reactions. Our approach enables prediction (for the outputs) to be made, as well as causal explanation generation about theories of chemical phenomenon. The system can be expected to provide explanation to the following questions: (1) What are the chemical processes used? (2) What was their sequence of use? (3) What happened to each functional group (E.g. a particular nucleophile) in a reaction? With this promise, students can better understand the underlying chemical concepts. Their critical thinking can be improved especially from inspecting the cause-effect chain that explains an aspect of system behavior using only the ontological primitives. We envisage that the tool will improve a student's intuitive learning, in that it can lead to deeper and systematic understanding of chemical processes.

References

1. Forbus, K.: Qualitative Process Theory. Artificial Intelligence 24, 85–168 (1984)
2. Forbus, K., Whalley, P., Everett, J., Ureel, L., Brokowski, M., Baher, J., Kuehne, S.: CyclePad: An Articulate Virtual Laboratory for Engineering Thermodynamics, Artificial Intelligence Journal 114, 297–347 (1999)
3. Syed Mustapha, S.M.F.D., Pang, J.S., Zain, S.M.: QALSIC: Towards Building an Articulate Educational Software using Qualitative Process Theory Approach in Inorganic Chemistry for High School Level. International Journal of Artificial Intelligence in Education 15(3), 229–257 (2005)

4. Alicia Tang, Y.C., Zain, S.M., Abdul Rahman, N., Abdullah, R.: Modelling of Organic Reaction Mechanism Using Qualitative Process Theory: A Modeller's Experience. In: ATC. Conference on Intelligent System and Robotics, Advanced Technology Congress, Malaysia (2005)
5. Alicia Tang, Y.C., Syed Mustapha, S.M.F.D.: Representing SN1 Reaction Mechanism using the Qualitative Process Theory. In: Bailey-Kellogg, C., Kuipers, B. (eds.) International Workshop on Qualitative Reasoning 2006, Hanover, USA, pp. 137–147 (2006)
6. Alicia Tang, Y.C., Zain, S.M., Abdul Rahman, N., Abdullah, R.: Knowledge Representation and Simulation of Nucleophilic Substitution Reactions using Qualitative Reasoning Approach. In: IEEE TENCON, Hong Kong (2006) ISBN: 1-4244-0549-1
7. Alicia Tang, Y.C., Syed Mustapha, S.M.F.D., Abdullah, R., Zain, S.M., Abdul Rahman, N.: Towards Automating QPT Model Construction for Reaction Mechanisms Simulation. In: Price, C., Snooke, N. (eds.) International Workshop on Qualitative Reasoning 2007, Aberystwyth, United Kingdom (2007) (poster)
8. de Kleer, J., Brown, J.S.: Qualitative Physics Based on Confluences. Artificial Intelligence Journal 24, 7–83 (1984)
9. Kuipers, B.J.: Qualitative Simulation. Artificial Intelligence Journal 29, 289–338 (1986)

Identifying Dependency Between Secure Messages for Protocol Analysis

Qingfeng Chen[1], Shichao Zhang[2], and Yi-Ping Phoebe Chen[1]

[1] Faculty of Science and Technology
Deakin University, VIC 3125, Australia
{qifengch, phoebe}@deakin.edu.au
[2] Faculty of Information Technology
University of Technology Sydney, PO Box 123, Broadway NSW 2007, Australia
zhangsc@it.uts.edu.au

Abstract. *Collusion attack* has been recognized as a key issue in e-commerce systems and increasingly attracted people's attention for quite some time in the literatures of information security. Regardless of the wide application of security protocol, this attack has been largely ignored in the protocol analysis. There is a lack of efficient and intuitive approaches to identify this attack since it is usually hidden and uneasy to find. Thus, this article addresses this critical issue using a compact and intuitive Bayesian network (BN)-based scheme. It assists in not only discovering the secure messages that may lead to the attack but also providing the degree of dependency to measure the occurrence of collusion attack. The experimental results demonstrate that our approaches are useful to detect the collusion attack in secure messages and enhance the protocol analysis.

1 Introduction

Security has become a high profile problem in e-commerce systems. Usually, it is achieved by security protocols, secure and unobstructed communication channel and trustworthy principals [1]. A secure e-commerce system relies on their valid and flawless combination. However, it is not easy to guarantee their correctness due to the increasingly complicated security protocols and hostile environment [3].

A number of security protocols have been reported with subtle flaws. The application of formal analysis for security protocols starts with the analysis of key distribution protocols for communication between two principals [2]. Despite the seeming simplicity of this problem, it is in fact not easy to handle because, in a hostile environment, the intruders may intercept, alter, or delete messages. Thus, the previous studies of protocol analysis focus on this topic. Ideally, they assume the principals are trustworthy and the communication is secure.

The types of applications to which a security protocol can be put however become more varied and complex, such as financial transactions [2]. Network must not only defend against intruders who may impersonate honest principal or attempt to learn secrets (*external threats*), but they also must be robust against a group of dishonest

Z. Zhang and J. Siekmann (Eds.): KSEM 2007, LNAI 4798, pp. 30–38, 2007.

principals who may collude together to uncover secrets (*internal threat*). Although the latter is viewed as a danger, most companies feel reasonably safe that the internal threat can be controlled through the corporate policies and internal access control. Thus, they concentrate on the unknown outside user who may obtain unauthorized access to corporation's sensitive assets. Although it is difficult for an individual to break the protection over secrets, it is feasible to deploy an attack in collusion with a certain number of dishonest principals [4, 10].

The security association instead of keys has become a potential way to detect threats. There have been considerable efforts to ensure the digital data is secure in the context of collusion. A general fingerprinting solution to detect any unauthorized copy is presented in [5]. A novel collusion-resilience mechanism using pre-warping was proposed to trace illegal unwatermarked copy [6]. Unfortunately, they may be too expensive and difficult to use.

To identify collusion attacks, we need a numeric estimation of the occurrence of the attack, which can correctly capture the threat. The main idea is a collusion attack is usually arisen from the attacks on encryption keys and delivered messages. Suppose a messages m_1 is shared by principal P_1 and P_2, and a message m_2 is shared by P_3 and P_4. If m_1 and m_2 are revealed to a hostile intruder, the attack may occur. Thus, the dependencies can be used to evaluate the occurrence of the attack. Bayesian networks that have been widely used to represent the probabilistic relationships among variables and do probabilistic inference with them are eligible to fulfill this role [7]. The transaction databases of principals provide valuable data to perform Bayesian inference.

This article uses Bayesian networks to learn the dependency model (*probabilistic dependence*) between secure messages. It measures the collusion attack by observing the decrease of probability in case of removing the corresponding arcs of the model. This assists in discovering the collusion threat and enhancing the protocol analysis.

The rest of this paper is organized as follow. Section 2 presents some basic concepts. A framework based on Bayesian networks is proposed in Section 3. Experiments are conducted in section 4. Section 5 gives a summary to this paper.

2 Preliminaries

This article aims to develop methods for detecting the collusion attack held by a group of dishonest principals. Although it is not easy for an intruder to completely break the underlying cryptography algorithms, they may utilize any flaw or weakness of security protocols to uncover confidential information combining keys and messages from several dishonest principals. To help understand the problem, we begin with introducing some basic concepts.

In this article, k represents an encryption key, $E(m, k)$ represents m is encrypted by k and $S(m, k^{-})$ represents m is signed by private key k^{-}, and $(m_1, \ldots, m_n)$ represent a combination of messages. For an electronic transaction including n principals, a principal P_i shares a set of secure messages m_{ij}, $1 \le i \ne j \le k$, with the principal P_j, in a probability p_{ij}. The probability p_{ij} denotes the degree to which a message is shared between principals. For example, suppose the message m_1 and m_2 are shared by P_1, P_2 and P_3. Thus, the probability that (m_1, m_2) is revealed by P_1 and P_2 is equal to 33%.

Suppose all the principals who know *X*'s public key are called *X*'s neighbors. Figure 1 presents principal *A*'s and *B*'s neighbors. Let *N*(*X*) and *U*(*X*) be the set of neighbors of *X* and the set of usable pairwise key of *X*, respectively. Thus, we have *N*(*A*) = {*C*, *D*} and *N*(*B*) = {*E*, *F*}. *G* is not a neighbor of *A* because he/she shares a symmetric key *k* rather than a pairwise key with *A*. If *A* and *B* share each other's secrets, then *A* and *B* can communicate with each other's neighbours. In the same way, *A* can communicate with *E* and *F* by pretending to be *B*, and *B* can communicate with *C* and *D* by pretending to be *A*. $U(A) = \{K_{CA}, K_{DA}\}$ and $U(B) = \{K_{EB}, K_{FB}\}$ prior to collusion, but $U(A) = U(B) = \{K_{CA}, K_{DA}, K_{EB}, K_{EB}\}$ via collusion. Thus, a certain number of dishonest principals can obtain unexpected secrets via collusion.

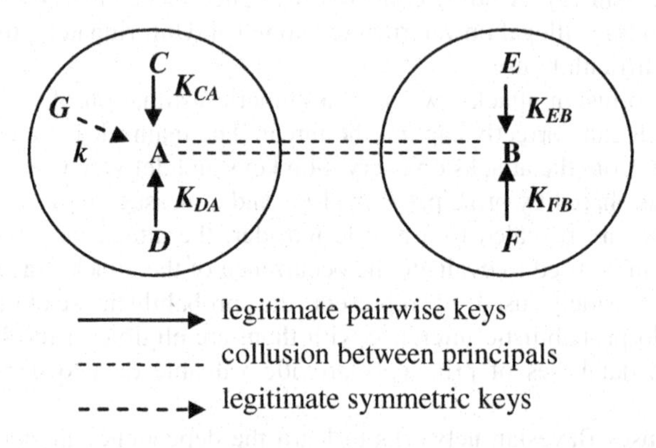

Fig. 1. An example of principal *A*'s and *B*'s neighbours

It is observed that there should be dependencies between the messages shared by principals. Although it is not easy to exactly confirm the principals who can conduct the attack, it is feasible to find the probable messages causing the attack and evaluate its occurrence probability. Bayesian network is appropriate to perform this role. With the prior knowledge (the probability of divulgence of secure messages) from the transaction databases, we can work out the likelihood of collusion attack.

3 Identifying Collusion Attack Using Bayesian Network

3.1 Structure

A Bayesian network comprises a set of nodes and a set of directed links. Each node represents a variable from the domain and each link connects pairs of nodes, representing the direct dependencies between variables. Usually, the strength of the relationship between variables is quantified by conditional probability distribution associated with each node. Only nodes that take discrete values are considered in this article. Thus, a given node (secure message) represents proposition, taking the binary values *known* (*known* by the principal) and *unknown* (*unknown* by the principal). To represent the described dependencies, we construct a Bayesian network using a directed acyclic graph (DAG). This includes three primary steps.

1. Identifying the variables of interest by answering the question what are the nodes to represent and what values they can take.
2. Generating the network structure by deciding the parent nodes and child nodes.
3. Qualifying the relations between connected nodes.

Suppose a transaction T consists of a set of principals $P = \{P_1, \ldots, P_n\}$ and a collection of secure messages $M = \{m_1, \ldots, m_k\}$. Each principal P_i has a set of secure messages $M(P_i) \subseteq M$. Each message (variable) in the transaction is viewed as a potential node in the network and has the binary values *known* and *unknown*. There will be one edge from m_i to m_j if m_i is a direct cause of m_j. For example, k and $E(m_3, k)$ may be the direct cause of m_3. Thus, there may be an edge from k to m_3 and another edge from $E(m_3, k)$ to m_3. Figure 2 shows a DAG regarding k, $E(m_3, k)$ and m_3. A node is a parent of a child, if there is an edge from the former to the latter.

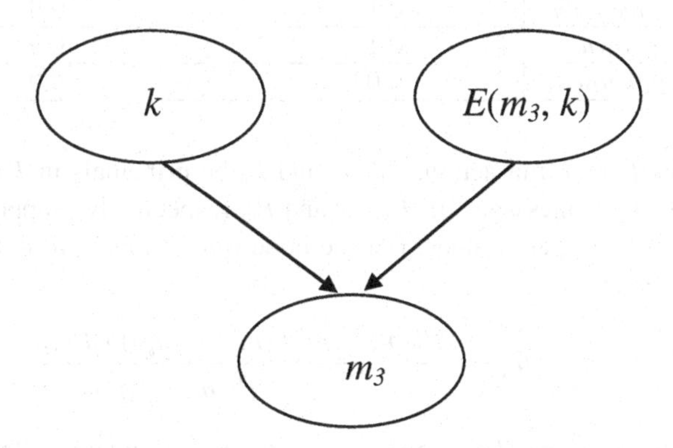

Fig. 2. An example of DAG including k, $E(m_3, k)$ and m_3

In addition, most messages are transited by cipher text to defend them against malicious attacks. If an intruder knows a key and the encrypted message using the key, he/she shall know the plain text of the message. In reality, the network can become more complex in case of collusion attack because more dependencies are considered and the principals can communicate with neighbors of the other side. Finally, we can construct the whole DAG. This actually answers the second question. The remaining work is how to qualify the dependencies between linked nodes.

3.2 Ascertaining the Probability of Variables

To measure the collusion attack in an intuitive way, it is necessary to work out the probabilities for each variable in the obtained DAG. The probability is actually a conditional probability, which relies on all its parent nodes in DAG that connect to this node. We need to look at all possible combinations of values of those parent nodes (instantiation of the parent set) and compute the conditional probabilities of variables in DAG.

After specifying the topology of the BN, the next step is to qualify the dependencies between linked nodes. As we are only considering discrete variables at this stage, it has the form of a conditional probability table (CPT). For example, suppose the parents of node (m_1, m_2) are m_1 and m_2 and it has the possible joint values {*<known, known>*, *<known, unknown>*, *<unknown, known>*, *<unknown, unknown>*}. Table 1 specifies in order the probability of divulgence (*D*) of (m_1, m_2) for each of these cases to be <0.2, 0.1, 0.1, 0.02>. Thus, the probability of without divulgence (*WD*) of (m_1, m_2) is defined as one minus the above probabilities in each case, namely <0.8, 0.9, 0.9, 0.98>.

Table 1. A CPT of the node (m_1, m_2)

m_1	m_2	$P((m_1, m_2)=D \mid m_1, m_2)$	$P((m_1, m_2)=WD \mid m_1, m_2)$
known	*known*	0.2	0.8
Known	*unknown*	0.1	0.9
unknown	*known*	0.1	0.9
unknown	*unknown*	0.02	0.98

Definition 1. Let *T* be a transaction, P_1, … and P_n be principals in *T* and $M(P_1)$, … and $M(P_n)$ be the set of messages of P_1, … and P_n, respectively. Suppose m_1, … and m_k are the direct cause (parent nodes) of the node (m_1, …, m_k), $m_i \in M(P_1) \cup \ldots \cup M(P_n)$. Thus, we have

$$P((m_1,\cdots,m_k) \mid m_1,\cdots,m_k) = \frac{P(m_1,\cdots,m_k \mid (m_1,\cdots,m_k)) * P((m_1,\cdots,m_k))}{P(m_1,\cdots,m_k)} \quad (1)$$

There are three probabilities that need to be calculated. The probability of $P(m_1,\cdots,m_k)$ can be obtained by multiplying the probability of the variables, namely $\prod_{i=1}^{k} P(m_i)$ since they are independent. In the same way, $P(m_i)$ can be derived by computing the conditional probability on its parent nodes. The computation is repeated until a root node of the alternative network is reached.

The formula (1) presents the computation of conditional probabilities. However, most of transaction data is encrypted during transmission. Unlike the general messages, an encrypted message needs an encryption key when coding and requires a decryption key when decoding. In particular, the conditional probabilities of the nodes of pairwise keys are only relevant to the principal's neighbors. Thus, we have

$$P(\{m\}_k \mid m,k) = \frac{P(\{m\}_k) * P(m,k \mid \{m\}_k)}{P(m,k)} \quad (3)$$

In the formula (3), $P(m,k)$ can be obtained by multiplying $P(m)$ by $P(k)$; $P(\{m\}_k)$ represents the probability of that the cipher text $\{m\}_k$ is broken; and $P(m,k \mid \{m\}_k)$ represents the conditional probability of that the principals who know $E(m, k)$ also know the key k and the message m. Decryption is a reverse procedure in contrast to encryption. Thus, we have

$$P(m \mid \{m\}_k, k^-) = \frac{P(\{m\}_k, k^- \mid m) * P(m)}{P(\{m\}_k, k^-)} \tag{4}$$

In the formula (4), $P(\{m\}_k, k^-)$ cannot be derived by simply multiplying $P(\{m\}_k)$ by $P(k)$ because $\{m\}_k$ and k^- are not independent. It can be obtained by computing the probability of that the principals who know both $\{m\}_k$ and k^-. $P(\{m\}_k, k^- \mid m)$ represents the conditional probability of that the intruder who knows m also knows $\{m\}_k$ and k^-. There are two options for the intruder to know $\{m\}_k$ and k^-. One is the intruder knows $\{m\}_k$ by obtaining k and m together, and the others is the intruder knows $\{m\}_k$ but has no knowledge about m and k at all.

Consider a BN including n nodes, Y_1 to Y_n, taken in that order. A particular value in the joint probability distribution is represented by $P(Y_1 = y_1, Y_2 = y_2, \ldots, Y_n = y_n)$, or more compactly, $P(y_1, y_2, \ldots, y_n)$. In addition, the value of particular node is conditional only on the values of its parent nodes according to the structure of a BN. Based on the chain rule of probability theory, we thus have

$$P(x_1, x_2, \cdots, x_n) = \prod_i P(x_i \mid Parents(x_i)) \tag{5}$$

The above describes the computation of conditional probabilities of variables in the network, in which a secure message transmitted in a transaction T may be linked by a number of connected nodes in DAG. The derived network can be used to reason about the domain. It can be conditioned upon any subset of their variables and support any direction of reasoning when we observe the value of some variables. In other words, the measure of collusion attacks can be transferred to the dependencies between connected nodes.

4 Experiments

4.1 Data Derivation

The experiment uses a simulated data set of electronic transaction[1]. Each principal has a set of secure messages. A principal X may use one public key to communicate with all other principals, or use different public keys to communicate with different principals. Only the principal who knows the public key can communicate with X.

It is important to determine the possibility of that the messages are obtained. Table 2 presents an instance of the data, in which k_i indicates symmetric keys. The variables take two values namely *known* and *unknown*, which are replaced by 1 and 0, respectively in the real data set. The format of the data this system accepts is simple text file, in which the first row contains names of the variables and the subsequent

[1] http://www.deakin.edu.au/~qifengch/ksem/data2.txt

rows contain the data one case in each row. The names of the variables in the first row and the fields in data rows should be separated by tabulators. There are 32 cases in the data file, each of which had 19 variables.

Table 2. Messages of principals in a transaction *T*

m_1	m_2	m_3	m_4	k_1	k_2
known	*known*	*unknown*	*unknown*	*known*	*unknown*
known	*unknown*	*known*	*unknown*	*known*	*known*
known	*known*	*known*	*unknown*	*known*	*known*

4.2 Analysis

We use B-Course [8] that is a web-based data analysis tool for Bayesian modeling, to build a dependency model and discover interesting relations out of the data set in this article. The data set can be uploaded online using the D-trail of B-Course.

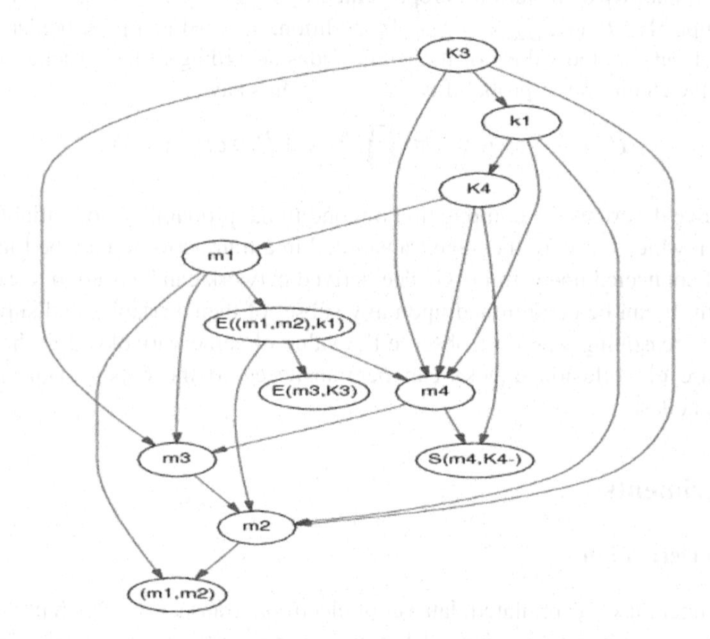

Fig. 3. A dependency model of the data set

B-Course allows users to control which variables are included in the analysis. k_2, k_3, K_1, K_2, K_5, K_6, K_7 and K_8 are excluded since they appear to be irrelevant to this transaction. Thus, only 11 variables are considered in the dependency model. We continue to monitor the status of the search, and finally stop the search when the continuing search does not seem to result in better result. The most probable model for the data set is showed in the Figure 3, in which the arc (*dependency*) is measured by observing how much the probability of the model is changed by removing the arc.

If the removed arc makes the model less probable, it can be viewed as a strong dependency; otherwise a weak dependency. Below you can see a list of selected statements describing how removing an arc affects the probability of the model. The details can be seen in [9].

Table 3 shows the dependencies and the ratios of the probability. The dependencies are classified by *strong* (solid line), *weak* (sparse dash line) and *very weak* (dense dash line), respectively. For example, removing any of strong arcs would result in a model with probability less than **one millionth** of that of the original model and removing any of the weak arcs from the chosen model would decrease the probability of the model to less than **one thousandth** of the probability of the original model.

Table 3. Strength of dependencies

ID	Dependency	Ratio of the probability
1	$m_1 \rightarrow (m_1, m_2)$	1 : one millionth
2	$K_3 \dashrightarrow m_3$	1 : 5965
3	$m_2 \dashrightarrow (m_1, m_2)$	1 : 425
4	$E((m_1, m_2), k_1) \dashrightarrow m_2$	1 : 306
5	$m_4 \dashrightarrow S(m_4, K_4^-)$	1 : 231
6	$K_4 \dashrightarrow m_4$	1 : 117
7	$m_1 \dashrightarrow E((m_1, m_2), k_1)$	1 : 12
8	$K_4 \dashrightarrow S(m_4, K_4^-)$	1 : 5.03

Looking at the dependency 1 and dependency 3, the former is a strong arc in comparison with the latter. Although the dependency 3 is a very weak arc, it still shows the dependency between m_2 and (m_1, m_2). The very weak dependency 3 may be caused due to the tampered or missing m_2. Thus, the collusion attack on the combination of m_1 and m_2 may happen if m_1 and m_2 are shared by multiple principals.

Looking at the dependency 4 and the dependency 7, the former shows a dependency from cipher text to plain text, whereas the latter is in the opposite way. In conjunction with the above dependency 1 and dependency 3, they can lead to a chained dependency containing cipher text and plain text. This provides varied ways for the intruder to make collusion attacks from any place in the chain. For example, the intruder can know $E((m_1, m_2), k_1)$ by either obtaining m_1 and m_2 or obtaining $E((m_1, m_2), k_1)$ directly.

Looking at the dependency 2, this is a weak dependency. It represents that m_3 is dependent on K_3. Thus, if K_3 is known by the intruder, the intruder has an opportunity to know m_3. Nevertheless, the intruder needs to obtain the cipher text of m_3 as well.

Looking at the dependency 5, dependency 6 and dependency 8, they are actually relevant to the encryption and decryption of m_4. In the same manner, they can create a chained dependency starting from m_4, via $S(m_4, K_4^-)$ and K_4, and back to m_4. Any variable in the chain can be used by the intruder to conduct a collusion attack. For example, the intruder can know m_4 by either obtaining $S(m_4, K_4^-)$ and K_4 or obtaining m_4 directly.

5 Conclusion

Collision attack has been viewed as a potential threat to e-commerce systems. The previous efforts to avoid this attack using key predistribution scheme have showed their limitations in many aspects. In addition, there is a lack of intuitive way to measure the occurrence of collusion attack. This article aims to establish a dependency model by using Bayesian network and provide a numerical estimation of the dependencies. We use B-course to analyze a simulated transaction data set and obtain an intuitive dependency model, from which the potential collusion attacks can be identified. The experimental results demonstrate that it is able to assist in identifying the collusion attack and enhancing the protocol analysis.

References

1. Ettinger, J.E.: Information security. Chapman & Hall, Sydney (1993)
2. Meadows, C.: Formal methods for cryptographic protocol analysis: emerging issues and trends. IEEE Journal on Selected Areas in Communications 21(1), 44–54 (2003)
3. Abadi, M.: Secret by typing in security protocols. Journal of the ACM 46(5), 749–786 (1999)
4. Du, W., Deng, J., Han, Y., Varshney, P., Katz, J., Khalili, A.: A pairwise key predistribution scheme for wireless sensor networks. ACM Transactions on Information and System Security 8(2), 228–258 (2005)
5. Boneh, D., Shaw, J.: Collusion-secure fingerprinting for digital data. IEEE Transactions on Information Theory 44(5), 1897–1905 (1998)
6. Celik, M.U., Sharma, G., Tekalp, A.M.: Collusion-resilient fingerprinting using random pre-warping. In: Proceeding of IEEE International Conference of Image Processing, pp. 509–512 (2003)
7. Richard, E.N: Learning Bayesian networks. Prentice Hall, Englewood Cliffs (2004)
8. http://b-course.cs.helsinki.fi/obc/
9. http://www.deakin.edu.au/ qifengch/ksem/dependence.doc
10. Chen, Q., Chen, Y., Zhang, S., Zhang, C.Q.: Detecting Collusion Attacks in Security Protocols. In: Zhou, X., Li, J., Shen, H.T., Kitsuregawa, M., Zhang, Y. (eds.) APWeb 2006. LNCS, vol. 3841, pp. 297–306. Springer, Heidelberg (2006)

A Diagrammatic Reasoning System for $\mathcal{ALC}$

Frithjof Dau and Peter Eklund

Faculty of Informatics
University of Wollongong
Wollongong, NSW, 2522 Australia
{dau,peklund}@uow.edu.au

Abstract. Description logics (DLs) are a well-known family of knowledge representation (KR) languages. The notation of DLs has the style of a variable-free first order predicate logic. In this paper a diagrammatic representation of the DL $\mathcal{ALC}$– based on Peirce's existential graphs – is presented and a set of transformation rules on these graphs provided. As the transformation rules modify the diagrammatic representation of $\mathcal{ALC}$ this produces a diagrammatic calculus. Some examples present in the paper illustrate the use and properties of this calculus.

1 Introduction

Description logics (DLs) [1] are a well-known and understood family of knowledge representation (KR) languages tailed to express knowledge about concepts and concept hierarchies. The basic building blocks of DLs are atomic concepts, atomic roles and individuals that can be composed by language constructs such as intersection, union, value or number restrictions (and more) to build more complex concepts and roles. For example, if MAN, FEMALE, MALE, RICH, HAPPY are concepts and if HASCHILD is a role, we can define

$$\text{MAN} \sqcap \exists \text{HASCHILD}.\text{FEMALE} \sqcap \exists \text{HASCHILD}.\text{MALE} \sqcap \forall \text{HASCHILD}.(\text{RICH} \sqcup \text{HAPPY})$$

as the concept of men who have both male and female children where all children are rich or happy. Let us call this concept HAPPYMAN.

The formal notation of DLs has the style of a variable-free first order predicate logic (FOL) and DLs correspond to decidable fragments of FOL. Like FOL, DLs have a well-defined, formal syntax and Tarski-style semantics, and they exhibit sound and complete inference features. The variable-free notation of DLs makes them easier to comprehend than the common FOL formulas that include variables. Nevertheless without training the symbolic notation of FOL can be hard to learn and difficult to comprehend.

A significant alternative to symbolic logic notation has been the development of a diagrammatic representation of DLs. It is well accepted that diagrams are, in many cases, easier to comprehend than symbolic notations [2,3,4], and in particular it has been argued that they are useful for knowledge representation systems [5,6]. This has been acknowledged by DL researchers and is a common view among the broader knowledge representation community [7].

Z. Zhang and J. Siekmann (Eds.): KSEM 2007, LNAI 4798, pp. 39–51, 2007.

A first attempt to experiment with diagrammatic KR can be found in [5], where a graph-based representation for the textual DL CLASSIC is elaborated. In [8], a specific DL is mapped to the diagrammatic system of conceptual graphs [9]. In [10], a UML-based representation for a DL is provided. In these treatments the focus is on a graphical *representation* of DL, however *reasoning* is a distinguishing feature of DLs. Correspondences between graphical representation of the DL and the DL reasoning system are therefore important inclusions in any graphical representation. However, to date they remain largely unelaborated.

On the other hand there are diagrammatic reasoning systems that have the expressiveness of fragments of FOL or even full FOL. Examples are the various classes of spider- and constraint diagrams [11,12], which are based on Euler circles and Venn-Peirce-diagrams, or the system of Sowa's conceptual graphs [9], which are based on Peirce's existential graphs. Contemporary elaborations of these systems include a well-defined syntax, extensional semantics and/or translations to formulas of FOL, and – most importantly for the goal of this paper – sound and complete calculi which can be best understood as manipulations of the diagrams.

This paper presents a diagrammatic representation of the DL $\mathcal{ALC}$ in the style of Peirce's existential graphs (EGs) [13,4]. An diagrammatic calculus for $\mathcal{ALC}$, based on Peirce's transformation rules, is provided. The DL $\mathcal{ALC}$ is the smallest propositionally closed DLwhich renders it a good starting point for developing DLs as diagrammatic reasoning systems. More expressive DLs are targeted for further research. Note also that it is well known that $\mathcal{ALC}$ is a syntactical variant of the multi-modal logic **K**, thus the results of this paper can be reused in modal logics. The reasons for choosing EGs are given in [14].

Reasoning with DLs is usually carried out by means of tableau algorithms. The calculus of this paper differs significantly from this approach in two respects. First, the rules of the calculus are *deep-inference* rules, as they modify deep nested sub-formulas, whereas tableau algorithms (similar to other common calculi) only modify formulas at their top-level. Secondly, the rules can be best understood to modify the diagrammatic Peirce-style representations of $\mathcal{ALC}$, i.e., the calculus is a genuine *diagrammatic* calculus.

The paper is structured as follows. First, an introduction existential graphs is provided in Section 2. In Section 3 the syntax and semantics of the DL $\mathcal{ALC}$ as we use it in this paper is introduced. In Section 4, the diagrammatic calculus for $\mathcal{ALC}$ is presented. Due to space limitations, the proof of its soundness and completeness is omitted: this result can be found in [14]. In Section 5, some examples and meta-rules for the calculus are provided. Finally, Section 6 provides a summary of this research and its significance.

2 Existential and Relation Graphs

Existential graphs (EGs) are a diagrammatic logic invented by C.S. Peirce (1839-1914) at the turn of the 20th century. We briefly introduce a fragment of EGs called BETA, corresponding to first order logic.

The graphs of Beta are composed of predicate names of arbitrary arity, heavily drawn lines, called LINES OF IDENTITY, are used to denote both the existence of objects and the identity between objects. Closed curves called CUTS are used to negate the enclosed subgraph. The area where the graph is written or drawn is what Peirce called the SHEET OF ASSERTION. Consider the following EGs.

cat —1— on —2— mat   (cat —1— on —2— mat)   cat —(1 on 2 — mat)

The first graph contains two lines of identity, hence it denotes two (not necessarily different) objects. They are attached to the unary predicates 'cat' and 'mat', respectively, and both are attached to the binary predicate 'on'. The meaning of the graph is therefore 'there is a cat and a mat such that the cat is on the mat', or in short: 'a cat is on a mat'. In the second graph, the cut encloses our first example completely. Hence its meaning is 'it is not true that there is a cat on a mat'. In the third graph, the left line of identity begins on the sheet of assertion. Hence, the existence of the object is asserted and not denied. For this reason this graph is read 'there is a cat which is not on any mat'.

Lines of identity may be connected to networks that are called LIGATURES. In the two left-most graphs in the figures below ligatures are used. The meaning of these graphs is 'there exists a male, human african', and 'it is not true that there is a pet cat such that it is not true that it is not lonely and owned by somebody', i.e., 'every pet cat is owned by someone and is not lonely'.

EGs are evaluated to true or false. Nonetheless, they can be easily extended to RELATION GRAPHS (RGS) [15,16] which are evaluated to relations instead. This is done by adding a syntactical device that corresponds to free variables. The diagrammatic rendering of free variables can be done via numbered question markers. The rightmost graph below is a relation graph with two free variables. This graph describes the binary relation *is_stepmother_of*.

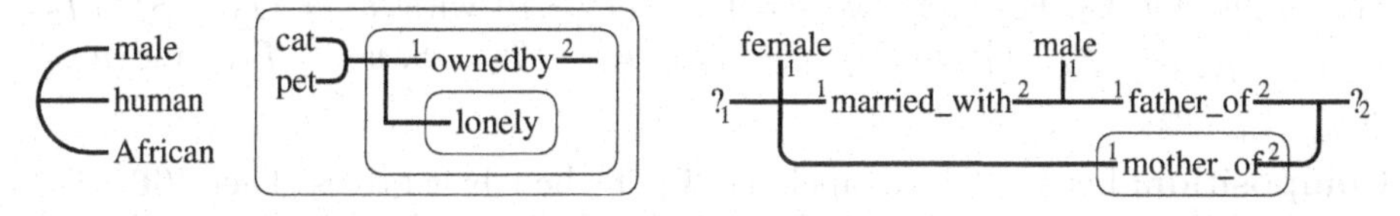

3 The Description Logic $\mathcal{ALC}$

The vocabulary $(\mathcal{A}, \mathcal{R})$ of a DL consists of a set $\mathcal{A}$ of (ATOMIC) CONCEPTS, denoting sets of individuals, and a set $\mathcal{R}$ (ATOMIC) ROLES, denoting binary relationships between individuals. Moreover, we consider vocabularies that include the universal concept $\top$. From these atomic items more complex concepts and roles are built with constructs such as intersection, union, value and number restrictions, etc. For example, if C, C_1, C_2 are concepts, then so are $C_1 \sqcap C_2$, $\neg C$, $\forall R.C$, $\exists R.C$, or $\leq nR.C$ (these constructors are called conjunction, negation, value restriction, existential restriction, and qualified number restriction).

In this paper we focus on the description logic $\mathcal{ALC}$. For our purpose we consider $\mathcal{ALC}$ to be composed of conjunction, negation and existential restriction.

In contrast to the usual approach, in our treatment the concepts of $\mathcal{ALC}$ are introduced as labeled trees. This is more convenient for defining the rules of the calculus, and the labeled trees are conveniently close to Peirce's notion of graphs.

An INTERPRETATION is a pair $(\Delta^{\mathcal{I}}, \mathcal{I})$, consisting of an nonempty DOMAIN $\Delta^{\mathcal{I}}$ and INTERPRETATION FUNCTION $\mathcal{I}$ which assigns to every $A \in \mathcal{A}$ a set $A^{\mathcal{I}} \subseteq \Delta^{\mathcal{I}}$ and to role $R \in \mathcal{R}$ a relation $R^{\mathcal{I}} \subseteq \Delta^{\mathcal{I}} \times \Delta^{\mathcal{I}}$. We require $\top^{\mathcal{I}} = \Delta^{\mathcal{I}}$.

Trees can be formalized either as special graphs or as special posets. We adopt the second approach, i.e., a tree is a poset $(T, \geq)$, where $s \geq t$ can be understood as 's is an ancestor of t'. A LABELED TREE is a structure $\mathbf{T} := (T, \leq, \nu)$, where $(T, \leq)$ is a tree and $\nu : T \to L$ is a mapping from the set of nodes to some set L of labels. The greatest element of T is the ROOT of the tree. As is usual, each node v gives rise to a SUBTREE $\mathbf{T}_v$ ($\mathbf{T}_v = (T_v, \geq |_{T_v \times T_v}, \nu|_{T_v})$ with $T_v := \{w \in T \mid v \geq w\}$). We write $\mathbf{T}' \subseteq \mathbf{T}$, if $\mathbf{T}'$ is a subtree of $\mathbf{T}$. Isomorphic labeled trees are implicitly identified.

Next we introduce some operations to inductively construct labeled trees. We assume to have a set L of labels.

Chain: Let $l_1, \ldots, l_n \in L$. With $l_1 l_2 \ldots l_n$ we denote the labeled tree $\mathbf{T} := (T, \geq, \nu)$ with $T := \{v_1, \ldots, v_n\}$, $v_1 > v_2 > \ldots > v_n$ and $\nu(v_1) = l_1, \ldots, \nu(v_n) = l_n$. That is, $l_1 l_2 \ldots l_n$ denotes a CHAIN, where the nodes are labeled with $l_1, l_2, \ldots, l_n$, respectively. We extend this notation by allowing the last element to be a labeled tree: If $l_1 l_2 \ldots l_n \in L$ and if $\mathbf{T}'$ is a labeled tree, then $l_1 l_2 \ldots l_n \mathbf{T}'$ denotes the labeled tree $\mathbf{T} := (T, \geq, \nu)$ with $T := T' \cup \{v_1, \ldots, v_n\}$, $v_1 > v_2 > \ldots > v_n$ and $v_i > v$ for each $i = 1, \ldots, n$ and $v \in T'$, and $\nu := \nu' \cup \{(v_1, l_1), \ldots, (v_n, l_n)\}$. That is, $\mathbf{T}$ is obtained by placing the chain $l_1 l_2 \ldots l_n$ above $\mathbf{T}'$.

Substitution: Let $\mathbf{T}_1, \mathbf{T}_2$ be labeled trees and $\mathbf{S} := (S, \geq_s, \nu_s)$ a subtree of $\mathbf{T}_1$. Then $\mathbf{T} := \mathbf{T}_1[\mathbf{T}_2 / \mathbf{S}]$ denotes the labeled tree obtained from $\mathbf{T}_1$ when $\mathbf{S}$ is substituted by $\mathbf{T}_2$. Formally, we set $\mathbf{T} := (T, \geq, \nu)$ with $T := (T_1 - S) \cup T_2$, $\geq := \geq_1 |_{T_1 - S} \cup \geq_2 \cup \{(w_1, w_2) \mid w_1 > v, w_1 \in T_1 - S, w_2 \in T_2\}$, and $\nu := \nu_1 |_{(T_1 - S)} \cup \nu_2$.

Composition: Let $l \in L$ be a label and $\mathbf{T}_1, \mathbf{T}_2$ be labeled trees. Then $l(\mathbf{T}_1, \mathbf{T}_2)$ denotes the labeled tree $\mathbf{T} := (T, \geq, \nu)$, where we have $T := T_1 \cup T_2 \cup \{v\}$ for a fresh node v, $\geq := \geq_1 \cup \geq_2 \cup (\{v\} \times (T_1 \cup T_2))$, and $\nu := \nu_1 \cup \nu_2 \cup \{(v, l)\}$. That is, $\mathbf{T}$ is the tree having a root labeled with l and having $\mathbf{T}_1$ and $\mathbf{T}_2$ as subtrees.

Using these operations, we can now define the tree-style syntax for $\mathcal{ALC}$.

Definition 1 ($\mathcal{ALC}$-Trees). *Let a vocabulary $(\mathcal{A}, \mathcal{R})$ be given with $\top \in \mathcal{A}$. Let '$\sqcap$' and '$\neg$' be two further signs. Let $(\Delta^{\mathcal{I}}, \mathcal{I})$ be a interpretation for the vocabulary $(\mathcal{A}, \mathcal{R})$. We inductively define the elements of $\mathcal{ALC}^{Tree}$ as labeled trees $\mathbf{T} := (T, \geq, \nu)$, as well as the interpretation $\mathcal{I}(\mathbf{T})$ of $\mathbf{T}$ in $(\Delta^{\mathcal{I}}, \mathcal{I})$.*

Atomic Trees: *For each $A \in \mathcal{A}$, the labeled tree A (i.e. the tree with one node labeled with A), as well as $\top$ are in $\mathcal{ALC}^{Tree}$. We set $\mathcal{I}(A) = A^{\mathcal{I}}$ and $\mathcal{I}(\top) = \Delta^{\mathcal{I}}$.*

Negation: *Let $\mathbf{T} \in \mathcal{ALC}^{Tree}$. Then the tree $\mathbf{T}' := \neg\mathbf{T}$ is in $\mathcal{ALC}^{Tree}$. We set $\mathcal{I}(\mathbf{T}') = \Delta^{\mathcal{I}} - \mathcal{I}(\mathbf{T})$.*

Conjunction: *Let* $\mathbf{T}_1, \mathbf{T}_2 \in \mathcal{ALC}^{Tree}$*. Then the tree* $\mathbf{T} := \sqcap(\mathbf{T}_1, \mathbf{T}_2)$ *is in* $\mathcal{ALC}^{Tree}$*. We set* $\mathcal{I}(\mathbf{T}) = \mathcal{I}(\mathbf{T}_1) \cap \mathcal{I}(\mathbf{T}_2)$*.*

Exists Restriction: *Let* $\mathbf{T} \in \mathcal{ALC}^{Tree}$*, let* R *be a role name. Then* $\mathbf{T}' := R\mathbf{T}$ *is in* $\mathcal{ALC}^{Tree}$*. We set* $\mathcal{I}(\mathbf{T}') = \{x \in \Delta^{\mathcal{I}} \mid \exists y \in \Delta^{\mathcal{I}} : xRy \wedge y \in \mathcal{I}(\mathbf{T})\}$*.*

The labeled trees of $\mathcal{ALC}^{Tree}$ *are called* $\mathcal{ALC}$-TREES*. Let* $\mathbf{T} := (T, \geq, \nu) \in \mathcal{ALC}^{Tree}$*. An element* $v \in T$ *respectively the corresponding subtree* $\mathbf{T}_v$ *is said to be* EVENLY ENCLOSED*, iff* $|\{w \in T \mid w > v \text{ and } \nu(w) = \neg\}|$ *is even. The notation of* ODDLY ENCLOSED *is defined accordingly.*

Of course, $\mathcal{ALC}$-trees correspond to the formulas of $\mathcal{ALC}$, as they are defined in the usual linear fashion. For this reason, we will sometimes mix the notation of $\mathcal{ALC}$-formulas and $\mathcal{ALC}$-trees. Particularly, we sometimes write $\mathbf{T}_1 \sqcap \mathbf{T}_2$ instead of $\sqcap(\mathbf{T}_1, \mathbf{T}_2)$. Moreover, the conjunction of trees can be extended to an arbitrary number of conjuncts, i.e.: If $\mathbf{T}_1, \ldots, \mathbf{T}_n$ are $\mathcal{ALC}$-trees, we are free to write $\mathbf{T}_1 \sqcap \ldots \sqcap \mathbf{T}_n$. We agree that for $n = 0$, we set $\mathbf{T}_1 \sqcap \ldots \sqcap \mathbf{T}_n := \top$.

Next, a diagrammatic representation of $\mathcal{ALC}$-trees in the style of Peirce's RGs is provided. As $\mathcal{ALC}$-concepts correspond to FOL-formulas with exactly one free variable, we we assign to each $\mathcal{ALC}$-tree $\mathbf{T}$ a corresponding RG $\Psi(\mathbf{T})$ with exactly one (now unnumbered) query marker. Let A be an atomic concept, R be a role name, let $\mathbf{T}$, $\mathbf{T}_1$, $\mathbf{T}_2$ be $\mathcal{ALC}$-trees where we already have defined $\Psi(\mathbf{T}) = ?\!\!-\!\!-\!G$, $\Psi(\mathbf{T}_1) = ?\!\!-\!\!G_1$, and $\Psi(\mathbf{T}_2) = ?\!\!-\!\!G_2$, respectively. Now Ψ is defined inductively as follows:

$$\Psi(\top) := ?\!\!-\!\!- \qquad \Psi(A) := ?\!\!-\!\!-A \qquad \Psi(R\mathbf{T}) := ?\!\!-\!\!-R\!\!-\!\!-G$$

$$\Psi(\mathbf{T}_1 \sqcap \mathbf{T}_2) := ?\!\!-\!\!\left[\begin{array}{l} G_1 \\ G_2 \end{array}\right. \qquad \Psi(\neg\mathbf{T}) := ?\!\!-\!\!\boxed{\!-G}$$

Considering our HAPPYMAN-example given in the introduction, the corresponding $\mathcal{ALC}$-tree, and a corresponding RG, is provided below. The rules of the forthcoming calculus can be best understood to be carried out on Peirce RGs.

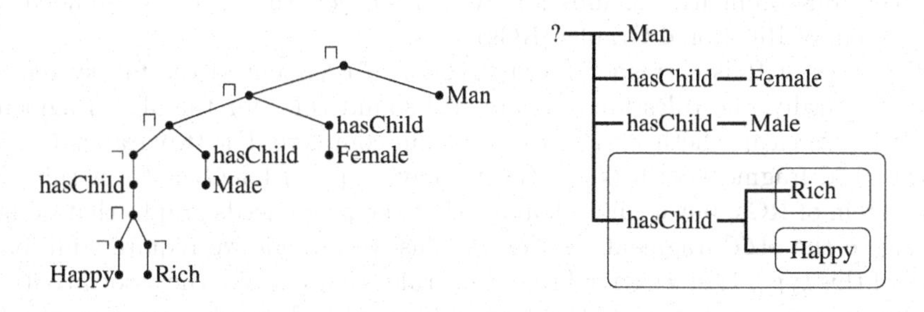

It might be argued that having only constructors for conjunction, negation and existential restriction is a downside of this system. Although we argued in the introduction that diagrams are easier to comprehend than the symbolic notation for DLs, reading the diagrams also requires effort. Some reading heuristics help in the understanding RGs. For example, subgraphs of the form

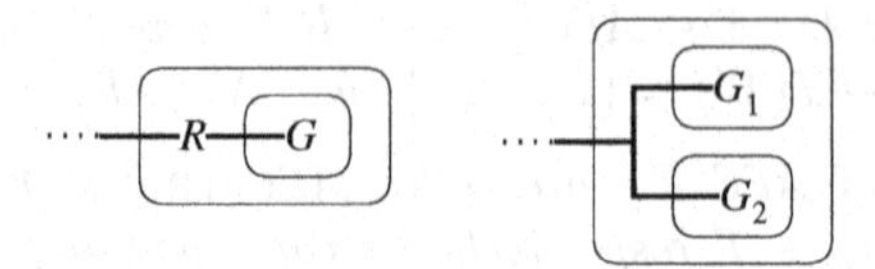

are literally read $\ldots\neg\exists R.\neg C$ and $\ldots\neg(\neg C_1 \sqcap \neg C_2)$, respectively, but it is more convenient to read them as $\forall\exists R.C$ resp., $C_1 \sqcup C_2$. Shin [4] coined the term *multiple reading* for this approach. In her book, she elaborates this for EGs by proving a translation from EGs to FOL that assigns to each EG a *set* of (equivalent) FOL-formulas (unfortunately, her translations are slightly flawed: see [17] for a discussion and correction of her reading algorithm). Shin argues thoroughly in [4] that this multiple reading is a unique feature of the Peirce's graphs, a feature that distinguishes them from the usual symbolic notation in mathematical logic. This feature is not a drawback, but an advantage of Peirce's system.

Finally, we define semantic entailment between $\mathcal{ALC}$-trees.

Definition 2 (Semantics). *Let* $\mathfrak{T} \subseteq \mathcal{ALC}^{Tree}$ *and let* $\mathbf{T} \in \mathcal{ALC}^{Tree}$. *We set*

$$\mathfrak{T} \models \mathbf{T} \;:\Longleftrightarrow\; \bigcap_{i \in I} \mathcal{I}(\mathbf{T}_i) \subseteq \mathcal{I}(\mathbf{T}) \text{ for each interpretation } (\Delta^{\mathcal{I}}, \mathcal{I})$$

For $I = \emptyset$, *we set* $\bigcap_{i \in I} \mathcal{I}(\mathbf{T}_i) := \Delta^{\mathcal{I}}$ *for the respective model, and write* $\models \mathbf{T}$. *For* $|I| = 1$, *we write* $\mathbf{T}' \models \mathbf{T}$, *omitting the curly set brackets, or we write* $\mathbf{T}' \sqsubseteq \mathbf{T}$ *(adopting the common DL-notation), and we say that that* $\mathbf{T}'$ SUBSUMES $\mathbf{T}$ *resp. that* $\mathbf{T}$ IS SUBSUMED BY $\mathbf{T}'$.

4 The Calculus for $\mathcal{ALC}^{Tree}$

Peirce provided a set of five rules for the system of existential graphs, termed *erasure*, *insertion*, *iteration*, *deiteration*, *double cut*. They form a sound and complete diagrammatic calculus for EGs. Moreover, they can be extended for the system of Relational Graphs (RGs).

The class of RGs corresponding to $\mathcal{ALC}$ is a fragment of the full system of RGs. Naturally, the rules for RGs are still sound rules for the $\mathcal{ALC}$-fragment, but it is less clear whether these rules remain complete. For two graphs G_1, G_2 of the $\mathcal{ALC}$-fragment with $G_1 \models G_2$, we have a proof for $G_1 \vdash G_2$ within the full system of RGs, but it might happen that the proof needs graphs that do not belong to the $\mathcal{ALC}$-fragment. In the calculus we provide we require additional rules of this type. Besides trivial rules, like rules that capture the associativity of conjunction, we need special rules for handling roles. The rules *iteration of roles into even* and *deiteration of roles from odd* are the most important examples.

Next in the presentation the Peirce style rules for $\mathcal{ALC}^{Tree}$ are provided. These rules transform a given $\mathcal{ALC}$-tree into a new $\mathcal{ALC}$-tree. In order to make the calculus more understandable, we provide, within the rule definitions, some examples and diagrams illustrating them. For each rule name we provide an abbreviation that will be used in the proofs.

Definition 3 (Calculus). *The calculus for $\mathcal{ALC}$-Trees over a given vocabulary $(\mathcal{A}, \mathcal{R})$ consists of the following rules:*

Addition and Removal of $\top$ ***($\top$-add. and $\top$-rem.):*** *Let* $\mathbf{T}$ *be an* $\mathcal{ALC}$*-tree, let* $\mathbf{S} \subseteq \mathbf{T}$ *be a subtree. For* $\mathbf{T}' := \mathbf{T}[\,\mathbf{S} \sqcap \top \,/\, \mathbf{S}\,]$ *we set* $\mathbf{T} \dashv\vdash \mathbf{T}'$ *(*$\mathbf{T} \dashv\vdash \mathbf{T}'$ *abbreviates* $\mathbf{T} \vdash \mathbf{T}'$ *and* $\mathbf{T}' \vdash \mathbf{T}$*). We say that* $\mathbf{T}'$ *is derived from* $\mathbf{T}$ *by* ADDING A $\top$-NODE, *and* $\mathbf{T}$ *is derived from* $\mathbf{T}'$ *by* REMOVING A $\top$-NODE. *For the Peirce graphs, this rule corresponds to adding or removing a branch to/from a heavily drawn line. A simple example is given below. These rules are "technical helper" rules that will be often combine with other rules that add or remove subtrees.*

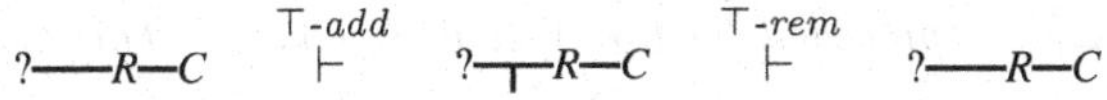

Addition and Removal of Roles (R-add. and R-rem.): *Let* $\mathbf{T}$ *be an* $\mathcal{ALC}$*-tree with a subtree* $\mathbf{S} \subseteq \mathbf{T}$. *Let* R *be a role name. Then for* $\mathbf{T}[\,\neg R \neg \top \,/\, \top\,]$ *we set* $\mathbf{T} \dashv\vdash \mathbf{T}'$. *We say that* $\mathbf{T}'$ *is derived from* $\mathbf{T}$ *by* ADDING THE ROLE R, *and* $\mathbf{T}$ *is derived from* $\mathbf{T}'$ *by* REMOVING THE ROLE R. *An example for this rule is given below. Due to the symmetry of the rules, the inverse direction is a proof as well.*

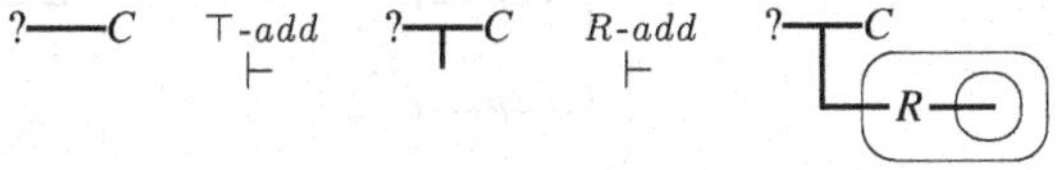

Associativity of Conjunction (conj.): *Let* $\mathbf{T}$ *be an* $\mathcal{ALC}$*-tree with a subtree* $\mathbf{S}_1 \sqcap (\mathbf{S}_2 \sqcap \mathbf{S}_3)$. *For* $\mathbf{T}' := \mathbf{T}[\,(\mathbf{S}_1 \sqcap \mathbf{S}_2) \sqcap \mathbf{S}_3 \,/\, \mathbf{S}_1 \sqcap (\mathbf{S}_2 \sqcap \mathbf{S}_3)\,]$ *we set* $\mathbf{T} \dashv\vdash \mathbf{T}'$. *We say that* $\mathbf{T}'$ *is derived from* $\mathbf{T}$ *resp.* $\mathbf{T}$ *is derived from* $\mathbf{T}'$ *by* USING THE ASSOCIATIVITY OF CONJUNCTION.

Addition and Removal of a Double Negation (dn): *Let* $\mathbf{T} := (T, \geq, \nu)$ *be an* $\mathcal{ALC}$*-tree, let* $\mathbf{S} \subseteq \mathbf{T}$ *be a subtree. Then for* $\mathbf{T}' := \mathbf{T}[\,\neg\neg\mathbf{S} \,/\, \mathbf{S}\,]$ *we set* $\mathbf{T} \dashv\vdash \mathbf{T}'$. *We say that* $\mathbf{T}'$ *is derived from* $\mathbf{T}$ *by* ADDING A DOUBLE NEGATION *and* $\mathbf{T}$ *is derived from* $\mathbf{T}'$ *by* REMOVING A DOUBLE NEGATION.

Consider the four graphs below. The second and the third graph can be derived from the first by adding a double negation. Inferences in the opposite direction can also be carried out. The fourth graph is a result of adding a double negation in the general *theory of RGs, but in the system of* $\mathcal{ALC}$*-trees, this is even not a diagram of an* $\mathcal{ALC}$*-tree, as the cuts cross more than one heavily drawn line.*

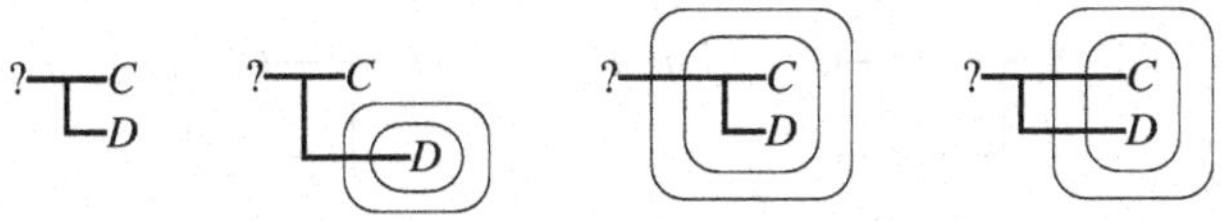

Erasure from even, Insertion into odd (era. and ins.): *Let* $\mathbf{T} :=$ *be an* $\mathcal{ALC}$*-tree with a positively enclosed subtree* $\mathbf{S} \subseteq \mathbf{T}$. *Then for* $\mathbf{T}' := \mathbf{T}[\,\top \,/\, \mathbf{S}\,]$ *we set* $\mathbf{T} \vdash \mathbf{T}'$. *We say that* $\mathbf{T}'$ *is derived from* $\mathbf{T}$ *by* ERASING $\mathbf{S}$ FROM EVEN.

Vice versa, let $\mathbf{T} =$ *be an* $\mathcal{ALC}$*-tree with an negatively enclosed subtree* $\top \subseteq \mathbf{T}$*. Let* $\mathbf{S} \in \mathcal{ALC}^{Tree}$*. Then for* $\mathbf{T}' := \mathbf{T}[\,\mathbf{S}\,/\,\top\,]$ *we set* $\mathbf{T} \vdash \mathbf{T}'$*. We say that* $\mathbf{T}'$ *is derived from* $\mathbf{T}$ *by* INSERTING $\mathbf{S}$ INTO ODD.

This is another set of rules that often hold together with the addition and removal of $\top$*. Examples will be given later.*

Iteration and Deiteration (it. and deit.): *Let* $\mathbf{T} := (T, \geq, \nu)$ *be an* $\mathcal{ALC}$*-tree with with a subtree* $\mathbf{S} := (S, \geq_S, \nu_S) \subseteq \mathbf{T}$*. Let* s *be the greatest element of* $\mathbf{S}$*, let* t *be the parent node of* s *in* $\mathbf{T}$*. Let* $\nu(t) = \sqcap$*, let* $v \in T$ *be a node with* $v < t$*,* $v \notin S$*,* $\nu(v) = \top$*, such that for each node* w *with* $t > w > v$ *we have* $\nu(w) = \neg$ *or* $\nu(w) = \sqcap$*. Then for* $\mathbf{T}' := \mathbf{T}[\,\mathbf{S}\,/\,\top\,]$ *we set* $\mathbf{T} \dashv\vdash \mathbf{T}'$*. We say that* $\mathbf{T}'$ *is derived from* $\mathbf{T}$ *by* ITERATING $\mathbf{S}$ *and* $\mathbf{T}$ *is derived from* $\mathbf{T}'$ *by* DEITERATING $\mathbf{S}$.

Iteration and Deiteration often combine with the addition and removal of $\top$*, and they are the most complex rules. Consider the following six Peirce graphs. The second and the third graph can be derived from the first graph by iterating the subgraph* ?—(R_1—C_1) *(preceded by the* $\top$*-addition rule). The next three graphs are* not *results from the iteration rule. In the fourth graph, the condition that* $\nu(w) = \neg$ *or* $\nu(w) = \sqcap$ *holds for each node* w *with* $t > w > v$ *is violated. The fifth graph violates* $v < t$*. Finally, the sixth graph violates* $v \notin S$*.*

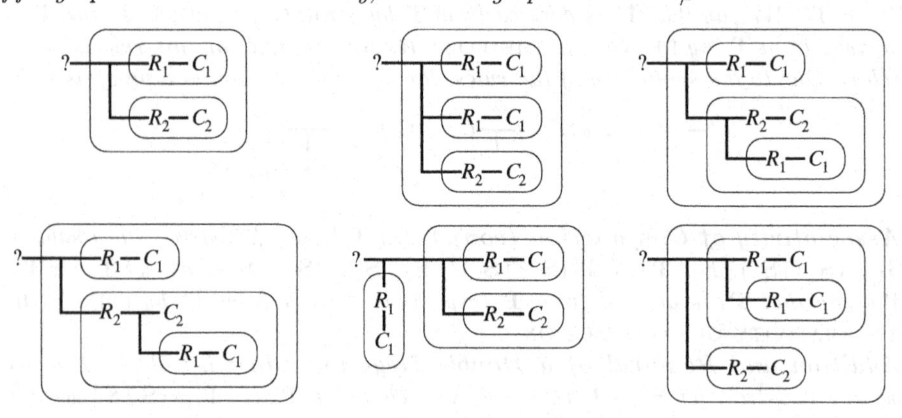

Iteration of Roles into even, Deiteration of Roles from odd (R-it. and R-deit.): *Let* $\mathbf{T}$ *be an* $\mathcal{ALC}$*-tree. Let* $\mathbf{S}_a, \mathbf{S}_b, \mathbf{S}_1, \mathbf{S}_2$ *be* $\mathcal{ALC}$*-trees with* $\mathbf{S}_a := R\mathbf{S}_1 \sqcap \neg R\mathbf{S}_2$ *and* $\mathbf{S}_b := R(\mathbf{S}_1 \sqcap \neg\mathbf{S}_2)$*. If* $\mathbf{S}_a \subseteq \mathbf{T}$ *is positively enclosed, for* $\mathbf{T}' := \mathbf{T}[\,\mathbf{S}_b\,/\,\mathbf{S}_a\,]$ *we set* $\mathbf{T} \vdash \mathbf{T}'$*, and we say that* $\mathbf{T}'$ *is derived from* $\mathbf{T}$ *by* DEITERATING THE ROLE R FROM ODD. *Vice versa, if* $\mathbf{S}_b \subseteq \mathbf{T}$ *is negatively enclosed, for* $\mathbf{T}' := \mathbf{T}[\,\mathbf{S}_a\,/\,\mathbf{S}_b\,]$ *we set* $\mathbf{T} \vdash \mathbf{T}'$*, and we say that* $\mathbf{T}'$ *is derived from* $\mathbf{T}$ *by* ITERATING THE ROLE R INTO EVEN. *Below a simple example is provided.*

?—R—C_1 / (R—C_2) $\quad \overset{R\text{-}deit}{\vdash} \quad$?—R—C_1 / (C_2)

Definition 4 (Proof). *Let* $\mathbf{T}_a, \mathbf{T}_b$ *be two* $\mathcal{ALC}$*-Trees. A* PROOF FOR $\mathbf{T}_a \vdash \mathbf{T}_b$ *is a finite sequence* $(\mathbf{T}_1, \mathbf{T}_2, \ldots, \mathbf{T}_n)$ *with* $\mathbf{T}_a = \mathbf{T}_1$*,* $\mathbf{T}_b = \mathbf{T}_n$*, where each*

$\mathbf{T}_{i+1}$ is obtained from $\mathbf{T}_i$ by applying one of the rules of the calculus. Let $\mathfrak{T}$ be a set of $\mathcal{ALC}$-Trees and let $\mathbf{T}$ be an $\mathcal{ALC}$-Tree. We set $\mathfrak{T} \vdash \mathbf{T}$ if and only if there are $\mathbf{T}_1, \ldots, \mathbf{T}_n \in \mathfrak{T}$ with $\mathbf{T}_1 \sqcap \ldots \sqcap \mathbf{T}_n \vdash \mathbf{T}$.

As proved in [14], the calculus is sound and complete.

Theorem 1 (Soundness and Completeness). *Let $\mathfrak{T}$ be a set of $\mathcal{ALC}$-Trees and let $\mathbf{T}$ be an $\mathcal{ALC}$-Tree. Then we have $\mathfrak{T} \models \mathbf{T} \iff \mathfrak{T} \vdash \mathbf{T}$.*

5 Metarules and Examples

In this section we firstly present two helpful metarules and then some examples to illustrate the Peirce-style calculus for $\mathcal{ALC}$.

Each rule of the calculus is basically the substitution of a subtree of a given $\mathcal{ALC}$-tree by another subtree. Each rule can be applied to arbitrarily deeply nested subtrees. Moreover, if we have a rule that can be applied to positively enclosed subtrees, then we always have a rule in the converse direction that can be applied to negatively enclosed subtrees (and visa versa). Due to these structural properties of rules, we immediately obtain the following helpful lemma (adapted from [9]).

Lemma 1 (Cut-and-Paste). *Let $\mathbf{S}_a, \mathbf{S}_b$ be two $\mathcal{ALC}$-trees with $\mathbf{S}_a \vdash \mathbf{S}_b$. Let $\mathbf{T}$ be an $\mathcal{ALC}$-tree. Then if $\mathbf{S}_a \subseteq \mathbf{T}$ is a positively enclosed subtree of $\mathbf{T}$, we have $\mathbf{T} \vdash \mathbf{T}[\,\mathbf{S}_b \,/\, \mathbf{S}_a\,]$. Visa versa, if $\mathbf{S}_b \subseteq \mathbf{T}$ is negatively enclosed, we have $\mathbf{T} \vdash \mathbf{T}[\,\mathbf{S}_a \,/\, \mathbf{S}_b\,]$.*

An immediate consequence of the lemma is: If we have two $\mathcal{ALC}$-trees $\mathbf{T}_a, \mathbf{T}_b$ and if $(\mathbf{T}_1, \ldots, \mathbf{T}_n)$ is a proof for $\mathbf{T}_a \vdash \mathbf{T}_b$, then $(\neg\mathbf{T}_n, \ldots, \neg\mathbf{T}_1)$ is a proof for $\neg\mathbf{T}_b \vdash \neg\mathbf{T}_a$. This will be used later.

For $\mathcal{ALC}$, the full deduction theorem holds.

Theorem 2 (Deduction Theorem). *Let $\mathfrak{T}$ be a set of $\mathcal{ALC}$-trees, let $\mathbf{T}_1, \mathbf{T}_2$ be two $\mathcal{ALC}$-trees. Then we have $\mathfrak{T} \cup \{\mathbf{T}_1\} \vdash \mathbf{T}_2 \iff \mathfrak{T} \vdash \neg(\mathbf{T}_1 \sqcap \neg\mathbf{T}_2)$.*

Proof: See [14].

In the following, some examples that correspond to distributing quantifiers in $\mathcal{ALC}$-concepts are provided. We start by moving existential quantifiers inwardly, when followed by a conjunction. In the symbolic notation of $\mathcal{ALC}$, we have the following subsumption relation: $\exists R.(C_1 \sqcap C_2) \sqsubseteq \exists R.C_1 \sqcap \exists R.C_2$. A proof for this relation by means of $\mathcal{ALC}$-trees is given below. In this and the next proof we assume $\Psi(C_1) = ?\!\!-\!\!\!-G_1$ and $\Psi(C_2) = ?\!\!-\!\!\!-G_2$.

?—R—[G1, G2] ⊢ (it) ?—[R—[G1, G2], R—[G1, G2]] ⊢ (2× era) ?—[R—[G1], R—[G2]] ⊢ (2×T-rem) ?—[R—G1, R—G2]

Next, we consider the equivalence of $\forall R.(C_1 \sqcap C_2)$ and $\forall R.C_1 \sqcap \forall R.C_2$. Below, a formal proof with Peirce's graphs is provided. The first and the (identical) last graph correspond to the concept $\forall R.(C_1 \sqcap C_2)$, and the fourth graph corresponds to the concept $\forall R.C_1 \sqcap \forall R.C_2$. So the first three steps prove the subsumption relation $\forall R.(C_1 \sqcap C_2) \sqsubseteq \forall R.C_1 \sqcap \forall R.C_2$, and the last six steps prove the subsumption relation $\forall R.C_1 \sqcap \forall R.C_2 \sqsubseteq \forall R.(C_1 \sqcap C_2)$.

As we have already said, the deiteration-rule and the erasure rule are usually followed by the $\top$-removal rule, and conversely, the iteration rule and the insertion rule are usually preceded by the $\top$-addition rule. In the proof, these two steps are combined without explicitly mentioning the $\top$-removal/addition rule.

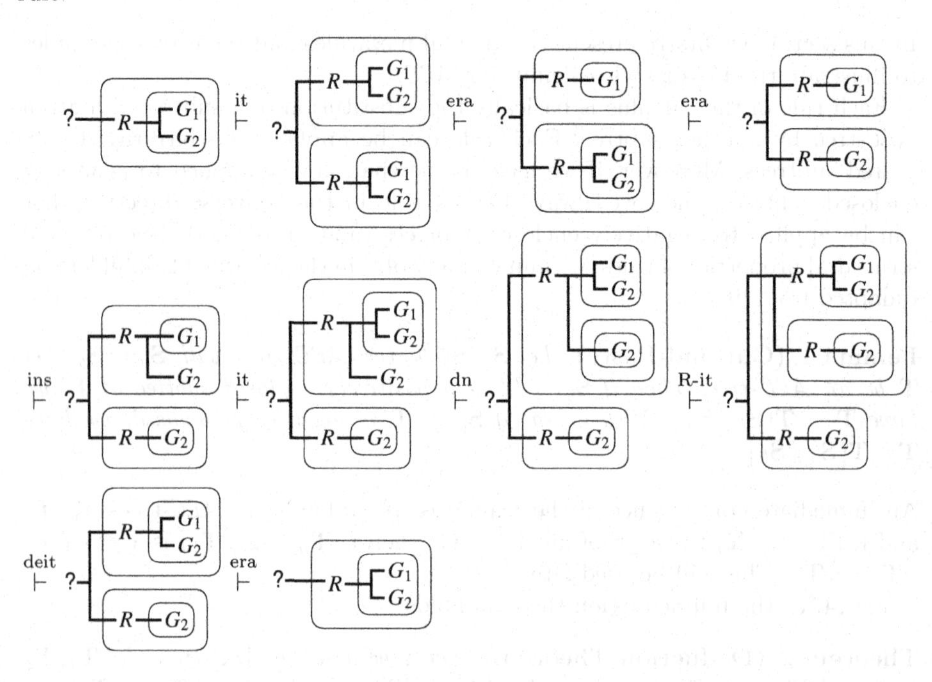

A similar example is the equivalence of $\exists R.(C_1 \sqcup C_2)$ and $\exists R.C_1 \sqcup \forall R.C_2$. Compared to the previous example, we exchanged the quantifiers $\forall$ and $\exists$ and the junctors $\sqcap$ and $\sqcup$. The symmetry of these examples is (thanks to Lem. 1), reflected by the calculus. The proof for $\exists R.(C_1 \sqcup C_2) \sqsubseteq \exists R.C_1 \sqcup \forall R.C_2$ is given below. In order to render this proof more understandable, we now set $\Psi(C_1) = ?\!\!-\!\!-H_1$ and $\Psi(C_2) = ?\!\!-\!\!-H_2$. The interesting part of the proof is the middle step, i.e., $(*)$. According to the remark after Lem. 1, we can carry out the last 6 steps in the previous proof in the inverse direction, if each graph in this proof is additionally enclosed by a cut (then the proof consists of the rules insertion, deiteration, R-deiteration, double negation, deiteration and erasure). When we replace in this proof each subgraph $?\!\!-\!\!-G_1$ by $?\!\!-\!\!\!\!\left(H_1\right)$ and each subgraph $?\!\!-\!\!-G_2$ by $?\!\!-\!\!\!\!\left(H_2\right)$, we obtain the proof for $(*)$. The steps before and after $(*)$ are simple helper steps in order to add or remove some double negations.

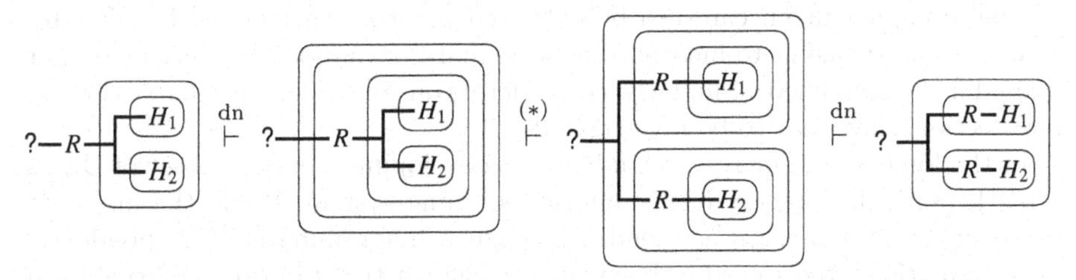

We see that the the proof for $\exists R.(C_1 \sqcup C_2) \sqsubseteq \exists R.C_1 \sqcup \forall R.C_2$ is essentially the inverse direction of the proof for $\forall R.C_1 \sqcap \forall R.C_2 \sqsubseteq \forall R.(C_1 \sqcap C_2)$. The proof for $\exists R.C_1 \sqcup \forall R.C_2 \sqsubseteq \exists R.(C_1 \sqcup C_2)$ can be obtained similarly from the proof for $\forall R.(C_1 \sqcap C_2) \sqsubseteq \forall R.C_1 \sqcap \forall R.C_2$. This shows some benefit of the symmetry of Peirce's rules.

The final example, the *mad cow ontology*, is a popular example for $\mathcal{ALC}$-reasoning. Consider the following $\mathcal{ALC}$-definitions:

$$Cow \equiv Animal \sqcap Vegetarian \qquad Sheep \equiv Animal \sqcap hasWool$$
$$Vegetarian \equiv \forall eats.\neg Animal \qquad MadCow \equiv Cow \sqcap \exists eats.Sheep$$

A question to answer is whether this ontology is consistent. Such a question can be reduced to rewriting the ontology to a single concept $MadCow \equiv Animal \sqcap \forall eats.\neg Animal \sqcap \exists eats.(Animal \sqcap hasWool)$ and investigate whether this concept is satisfiable, i.e., whether there exists as least one interpretation where this concept is interpreted by a non-empty set. We will show that this is not the case by proving with our calculus that the concept entails the absurd concept. The proof is given below. Again, each application of the erasure-rule is followed by an application of the $\top$-removal rule (although this is not explicitly mentioned in the proof).

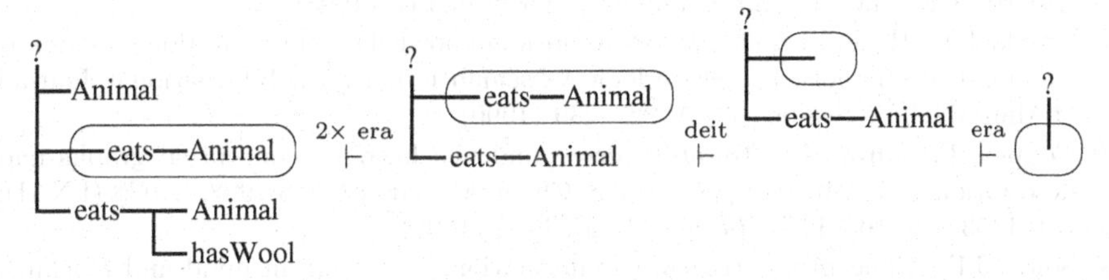

We started with the Peirce graph for the given concept and derived the absurd concept, thus the ontology is not satisfiable. The madcow ontology is therefore inconsistent and this is proved using the calculus developed,

6 Conclusion and Further Research

This paper provides steps toward a diagrammatic representation of DLs, including importantly diagrammatic inference mechanisms. To the best of our knowledge this is the first attempt to providing *diagrammatic* reasoning facilities for DLs. The results presented in this paper show promise in investigating RGs further as diagrammatic versions of corresponding DLs.

The approach taken can also be extended to other variants of DL. For instance, a major task is to incorporate individuals, or number restrictions (either unqualified or qualified). Similarly, constructors on roles, like inverse roles or role intersection, have also to be investigated.

In the long term, our research advocates developing a major subset of DL as a mathematically precise diagrammatic reasoning system. While the intention is to render DLs more user-friendly through a diagrammatic correspondence, diagrammatic systems need to be evaluated against the traditional textual form of DL in order to measure readability improvement. Cognition and usability experiments with such a evaluation in mind are planned as future work.

References

1. Baader, F., Calvanese, D., McGuinness, D.L., Nardi, D., Patel-Schneider, P.F. (eds.): The Description Logic Handbook: Theory, Implementation, and Applications. Description Logic Handbook. Cambridge University Press, Cambridge (2003)
2. Larkin, J.H., Simon, H.A.: Why a diagram is (sometimes) worth ten thousand words. Cognitive Science 11(1), 65–100 (1987)
3. Shimojima, A.: On the Efficacy of Representation. PhD thesis, The Department of Philosophy, Indiana University (1996), Available at http://www.jaist.ac.jp/~ashimoji/e-papers.html
4. Shin, S.J.: The Iconic Logic of Peirce's Graphs. Bradford Book, Massachusetts (2002)
5. Gaines, B.R.: An interactive visual language for term subsumption languages. In: IJCAI, pp. 817–823 (1991)
6. Kremer, R.: Visual languages for konwledge representation. In: KAW 1998. Proc. of 11th Workshop on Knowledge Acquisition, Modeling and Management, Banff, Alberta, Canada, Morgan Kaufmann, San Francisco (1998)
7. Nosek, J.T., Roth, I.: A comparison of formal knowledge representation schemes as communication tools: Predicate logic vs semantic network. International Journal of Man-Machine Studies 33(2), 227–239 (1990)
8. Coupey, P., Faron, C.: Towards correspondence between conceptual graphs and description logics. In: Mugnier, M.-L., Chein, M. (eds.) ICCS 1998. LNCS (LNAI), vol. 1453, pp. 165–178. Springer, Heidelberg (1998)
9. Sowa, J.F.: Conceptual structures: information processing in mind and machine. Addison-Wesley, Reading, Mass (1984)
10. Brockmans, S., Volz, R., Eberhart, A., Löffler, P.: Visual modeling of owl dl ontologies using uml. In: McIlraith, S.A., Plexousakis, D., van Harmelen, F. (eds.) ISWC 2004. LNCS, vol. 3298, pp. 198–213. Springer, Heidelberg (2004)
11. Stapleton, G.: Reasoning with Constraint Diagrams. PhD thesis, Visual Modelling Group, Department of Mathematical Sciences, University of Brighton (2004), Available at http://www.cmis.brighton.ac.uk/Research/vmg/GStapletonthesis.html
12. Stapleton, G., Howse, J., Taylor, J.: Spider diagrams. LMS Journal of Computation and Mathematics 8, 145–194 (2005)
13. Zeman, J.J.: The Graphical Logic of C. S. Peirce. PhD thesis, University of Chicago (1964), Available at http://www.clas.ufl.edu/users/jzeman/

14. Dau, F., Eklund, P.: Towards a diagrammatic reasoning system for description logics. Journal of Visual Languages and Computing (submitted, 2006), `www.kvocentral.org`
15. Burch, R.W.: A Peircean Reduction Thesis: The Foundation of Topological Logic. Texas Tech. University Press, Texas, Lubbock (1991)
16. Pollandt, S.: Relation graphs: A structure for representing relations in contextual logic of relations. In: Priss, U., Corbett, D.R., Angelova, G. (eds.) ICCS 2002. LNCS (LNAI), vol. 2393, pp. 24–48. Springer, Heidelberg (2002)
17. Dau, F.: Fixing shin's reading algorithm for peirce's existential graphs. In: Barker-Plummer, D., Cox, R., Swoboda, N. (eds.) Diagrams 2006. LNCS (LNAI), vol. 4045, pp. 88–92. Springer, Heidelberg (2006)

Fuzzy Constraint Logic Programming with Answer Set Semantics

Jie Wang and Chunnian Liu

Key Laboratory of Multimedia and Intelligent Software,
Beijing University of Technology, Beijing 100022, P.R. China
{wj,ai}@bjut.edu.cn

Abstract. In this paper, we present a new framework of fuzzy constraint logic programming language including negation as failure which is a combination of fuzzy logic and fuzzy relation. We give the answer set semantics which is based on the method of stable model. Although much work have been done for answer set programming, no work addressed the answer set semantics for fuzzy constraint logic programming with negation as failure. Also, we give an example to illustrate the idea. Compared to Zadeh's compositional rule of inference for approximate reasoning used in [9], we find an effective and efficient computational procedure for fuzzy constraint logic programming by using answer set semantics.

Keywords: Fuzzy Constraint, Answer Set Semantics, Logic Program.

1 Introduction

Logic programming [10,1] is a powerful tool for knowledge representation and knowledge reasoning in artificial intelligence. As an important formalism for reasoning, logic programming started in the early 1970's [7] based on earlier work in automated theorem proving, and began to flourish especially with the spreading of PROLOG. The answer set programming (ASP) paradigm [4] has gained a lot of popularity in the last years, due to its truly declarative non-monotonic semantics, which has been proven useful in a number of interesting applications. Although ASP provides a powerful solution for knowledge representation and nonmonotonic reasoning, it also has some drawbacks regarding the configure ability of the semantics w.r.t. the type of application under consideration, as witnessed by the large number of extensions, both syntactically and semantically, e.g., most ASP semantics demand that a solution to a program satisfies all the rules. Further, the literals available in the program, i.e. the building blocks of rules, can only be true or false, and classical consistency is mandatory. Sometimes however, it is impossible to find a solution that fully satisfies all rules of the program. In this case, one might still wish to look for a solution satisfying the program at least to a reasonably high degree. At other times, it may not even be required to obtain a solution that satisfies a program fully. That is, one might be more interested in a solution satisfying the program to a satisfactory high

Z. Zhang and J. Siekmann (Eds.): KSEM 2007, LNAI 4798, pp. 52–60, 2007.

degree, especially if this solution comes at a lower cost. In [9] the author present a framework to deal with this problem above by integrating fuzzy logic programming and fuzzy mathematical programming [8]. While fuzzy relation combines fuzzy predicate and fuzzy constraint, the compositional rule of inference, when viewed in the right way, bridges the gap between a fuzzy logic clause and a fuzzy optimization formulation. This framework can also be seen as a fuzzification of the constraint logic programming paradigm based on approximate reasoning.

However, the calculus in [9] has the problem that the results do not always agree to intuition, we will see examples later in Section 4. In order to further improve the power of fuzzy knowledge representation, this paper presents a fuzzy constraint logic programming language. Its syntax part is similar that of [9] (both fuzzy predicate and fuzzy relation are combined under the basic concept of fuzzy constraint). However, the semantics of fuzzy constraint logic programming described in this paper is very different from [9]. Actually, ours is an extension of answer set semantics for logic programs proposed in [3,4]. The answer set semantics defines when a set S of ground literals is an answer set of a given program and give a query a rational answer.

The main contributions of this paper can be summarized as follows:

Firstly, we present an alternative of fuzzy constraint logic programming languages with negation as failure, and give its answer set semantics. Although much work have been done for answer set programming, no work addressed the answer set semantics for fuzzy constraint logic programming with negation as failure.

Secondly, compared to Zadeh's compositional rule of inference for approximate reasoning used in [9], we find an effective and efficient computational procedure by using answer set semantics for fuzzy constraint logic programming.

2 Fuzzy Constraint Satisfaction

In this section, we recall the necessary concepts and notations related to fuzzy constraint satisfaction problems [14,2,12,11,17], fuzzy logic and the cut set technique [6,13] in fuzzy set theory.

Firstly, an n-ary fuzzy relation R is a fuzzy subset of $U_1 \times \cdots \times U_n$, where U_i is the domain of variable x_i, defined by the membership function:

$$\mu_R : U_1 \times U_2 \times \cdots U_n \longrightarrow [0, 1].$$

It simply extends the domain of a fuzzy set to the Cartesian product of the domain universes.

Example 1. *The relational concept "very Similar" between two sisters can be defined as follows:*

$$\mu_{very-similar}(Mary, Rose) = 0.8,$$
$$\mu_{very-similar}(Mary, Alice) = 0.2.$$

The fuzzy predicate "X likes Y" can be defined as

$$\mu_{like}(John, Mary) = 0.8,$$
$$\mu_{like}(Mary, Jone) = 0.2. \cdots etc.$$

The fuzzy constraint "the sum of X and Y is almost zero"(i.e..$x+y \doteq 0$) can be defined as

$$\mu_{x+y\doteq 0}(3,-2.9)=0.7,$$
$$\mu_{x+y\doteq 0}(9.1,-7)=0.1.$$

Crisp relations are special cases of fuzzy relations, for example:

$$\mu_{\underline{R}}:U_1\times\cdots\times U_n\longrightarrow\{0,1\}$$

i.e., $\mu_{\underline{R}}(X)=1$ and $\mu_{\underline{R}}(X)=0$ are interpreted as $X=(x_1,\cdots,x_n)\in R$ and $(x_1,\cdots,x_n)\bar{\in} R$, respectively.

Second, we recall the concept of fuzzy logic. Predicates in classical logic can be viewed in similar way: a n-ary predicate, $p(X_1,\cdots,X_n)$ on domain $U_1\times\cdots\times U_n$, defined as a truth value mapping from the Cartesian product of the domain universes to a truth value set T, i.e.,

$$t:U_1\times\cdots\times U_n\longrightarrow T$$

where T={0,1} can be induced naturally from $\mu_{\underline{R}}(X)$ and vice versa:

$$\underline{t}(\mathbf{X})=\begin{cases}1, & if \quad \mu_{\underline{R}}(\mathbf{X})=1;\\ 0, & if \quad \mu_{\underline{R}}(\mathbf{X})=0.\end{cases} \tag{1}$$

i.e., a predicate $p(X_1,\cdots,X_n)$ is true if and only if $(X_1,\cdots,X_n)$ is in the relation R, otherwise false. Thus, in fuzzy logic, T is extend to [0,1] and treat $t(X)$ as degree of truth of the predicate.

Third, we recall fuzzy constraint problems.

Definition 1. *A fuzzy constraint satisfaction problem (FCSP) is a 3-tuple(X,D,C), where sets:*

1. *$X=\{x_i|i=1,\cdots,n\}$is a finite set of variables.*
2. *$D=\{d_i|i=1,\cdots,n\}$is a the set of domains. Each domain d_i is a finite set containing the possible values for the corresponding variables x_i in X.*
3. *$C_f=\{R_i^f \mid \mu_{R_i^f}:(\prod_{x_j\in var(R_i^f)} d_j \longrightarrow [0,1], i=1,\cdots,m\}$ is a set of fuzzy constraints. Here $var(R_i^f)$ denotes the variable set in constraint R_i^f:*
   $$var(R_i^f)=\{x'_{1_{R_i^f}},\cdots,x'_{k_{R_i^f}}\}\subseteq X$$

Definition 2. *A label of a variable x is an assignment of a value to the variable, denoted as v_x. A compound label $v_{X'}$ of all variables in set $X'=\{x'_1,\cdots,x'_m\}\subseteq X$ is a simultaneous assignment of values to all variables in set X':*

$$v_{X'}=(v_{x'_1},\cdots,v_{x'_m}) \tag{2}$$

We interpret C_f is a set of fuzzy constraint as degree of satisfaction of a label $(a_1,\cdots,a_n)$ to a constraint R_i^f.

By using the cut-set technique in fuzzy mathematics [6], a fuzzy constraint can induce a crisp constraint.

Definition 3. *Given the cut level $\sigma \in [0,1]$, the induced crisp constraint R_c of a fuzzy constraint R_f is defined as:*

$$\mu_{R_C}(v_{var(R_f)}) = \begin{cases} 1 & if \quad \mu_{R_f}(var_{(R_f)}) \geq \sigma, \\ 0 & otherwise. \end{cases} \tag{3}$$

Intuitively, a cut level for a fuzzy constraint is a kind of threshold: if the satisfaction degree to which a compound label satisfies a fuzzy constraint is not less than the threshold, the label is regarded as satisfactory with respect to the constraint; otherwise, it is regarded as unsatisfactory. A solution to a FCSP (X, D, C^f) is a compound label of all variables in X such that $\mu_{R_C}(v_{var_{(R_f)}}) = 1$.

3 Fuzzy Constraint Logic Programs

In this section, we extend the Horn clauses to fuzzy constraint Horn clauses. The main ideas are: (i) at the predicate level, a crisp relation is extended to fuzzy relation to encompass both fuzzy predicate and fuzzy constraint (in particular, equality and inequality); (ii) at the clause level, the *and* operation is generalized to the *confluence* operation as in the fuzzy decision making context.

Definition 4. *A (well-formed) fuzzy constraint formula is defined inductively as follows:*

1. *If p is an n-ary fuzzy predicate defined on variable $x_1, \cdots, x_n$, then $p(x_1, \cdots, x_n)$ is a fuzzy constraint formula (called an atomic formula, or simply an atom), $\neg p(x_1, \cdots, x_n)$ is the (classic) negation of atom: a fuzzy relation defined by membership function $\mu_{\neg p} = 1 - \mu_p$.*
2. *If f and g are functions defined over variables $(x_1, \cdots, x_m)$ and $(y_1, \cdots, y_n)$ respectively, then $f(x_1, \cdots, x_m) \Diamond g(y_1, \cdots, y_n)$ is a fuzzy constraint formulas (called an atomic formula, or simply an atom), where $\Diamond \in \{\prec, \preceq, \simeq, \succ, \succeq\}$ (the set of fuzzy relational operators).*
3. *If F is a fuzzy constraint formula and x is a variable occurring in F, then $(\forall x \in F)$ and $(\exists x \in F)$ are also fuzzy constraint formulae.*

Definition 5. *A fuzzy constraint program clause is a fuzzy constraint formula of the form*

$$\forall x_1, \cdots, x_s(A \longleftarrow (B_1, B_2, \cdots, B_n; \neg C_1, \cdots, \neg C_m))$$

where A is an atom B_i, C_i are literals (atom or negation of atom); $x_1, \cdots, x_s$ are all the variables occurring in A, B_i and C_i; $\neg$ is called negation as failure. For notational convenience, we can drop the symbol $\forall$ and simplify it as

$$A \longleftarrow B_1, \cdots, B_n, \neg C_1, \cdots, \neg C_m.$$

In the above, the left hand of $\longleftarrow$ is called head and the right one is called the body of the fuzzy constraint program clause.

When the body is empty, the form of a fuzzy constraint formula

$$A \longleftarrow$$

is called a fuzzy constraint unit clause.

When the head is empty

$$\longleftarrow B_1, \cdots, B_n, \neg C_1, \cdots, \neg C_m$$

is called a fuzzy constraint goal clause (each B_i is called the subgoal of the goal clause).

Definition 6. *A fuzzy constraint logic program is a finite set of fuzzy constraint program clauses.*

In order to define the semantics of fuzzy constraint logic programming, we will use the following definitions.

Definition 7. *If f and g are functions defined over variables $(x_1, \cdots, x_m)$ and $(y_1, \cdots, y_n)$ respectively, then for arbitrary atom of the form*

$$A = f(x_1, \cdots, x_m) \Diamond g(y_1, \cdots, y_n), \text{ where } \Diamond \in \{\prec, \preceq, \simeq, \succ, \succeq)\},$$

we define

$$\mu(A) = \begin{cases} 1 \ \ if & T(A) = true, \\ 0 \ \ if & T(A) = false. \end{cases} \tag{4}$$

Definition 8. *If a literal B has a membership degree α, then the membership degree of $\neg B$ is $1 - \alpha$ and for a given cut level σ, we have*

$$\mu_\sigma(B) = \begin{cases} 1 \ \ if & \alpha \geq \sigma, \\ 0 \ \ if & \alpha < \sigma. \end{cases} \tag{5}$$

$$\mu_\sigma(\neg B) = \begin{cases} 1 \ \ if & (1-\alpha) \geq \sigma, \\ 0 \ \ if & (1-\alpha) < \sigma. \end{cases} \tag{6}$$

4 Answer Set Semantics

Logic programming with the stable model semantics [3,4] have emerged as a viable model for solving constraint satisfaction problems. In this section, we present the stable model semantics for the fuzzy constraint logic programming (FCLP) framework that we have described above. We improve upon the power of knowledge representation by extending the semantics with new types of atom, e.g., atom with fuzzy relational operator. The extended semantics is also based on subset-minimal models as general logic programming.

The semantics of fuzzy constraint logic programs treats a rule with variables as shorthand for the set of its ground instances. It is more sufficient than defining answer set for fuzzy constraint programs without variables. This can be done in three steps.

1. Let Π be the fuzzy constraint logic program without variables that, in addition, does not contain $\neg$, and let X be the sets of ground literals in the language of Π. Each ground literal B has a initial satisfaction degree μ_B based on the membership function of the corresponding fuzzy relation, and a fuzzy constraint clause $A \leftarrow Body$ may update μ_B.

   Given a cut level σ, let M be the subset of ground literal X whose satisfaction degree $\mu \geqslant \sigma$, we can compute satisfaction degrees and construct M recursively as follows:

   M:= those ground literals whose initial satisfaction degree $\mu \geqslant \sigma$;
   repeat for each fuzzy constraint clause $B \leftarrow B_1, \cdots, B_m$,
   if $A \leftarrow B_1, \cdots, B_m \in M$
   then $\mu^{`}(B) := (\mu(B_1) + \cdots + (B_m))/m$;
   $M := M\{B\}$;
   $\mu(B) := max(\mu(B), \mu^{`}(B))$,
   until no change in M and μ

   (a) For any fuzzy constraint rule from Π
   $$A \longleftarrow B_1, \cdots, B_n$$
   if $B_1, \cdots, B_n \in M$, then $M = M \cup A$, $\mu(A) = (\mu(B_1) + \cdots + \mu(B_n))/n$; otherwise $N = N \cup A$.

   Especially, For any fuzzy constraint rule
   $$A \longleftarrow (B_1, \cdots, B_n)$$
   $$A \longleftarrow (C_1, \cdots, C_m)$$
   If $B_1, \cdots, B_n \in M$, $C_1, \cdots, C_m \in M$,
   then $M = M \cup A$, and
   $\mu(A) = \{max((\mu(B_1) + \cdots + \mu(B_n))/n, (\mu(C_1) + \cdots + \mu(C_m))/m)$.
2. Now let Π be the fuzzy constraint logic program without variables that, in addition, does not contain $\neg$ and let X be the sets of ground literals in the language of Π. M be the set of ground literal we have get by 1. The answer set of Π is the minimal subset S of X such that:

   (b) For any fuzzy constraint rule from Π
   $$A \longleftarrow (B_1, \cdots, B_n)$$
   if $B_1, \cdots, B_n \in S$ and $B_1, \cdots, B_n \in M$,
   then $A \in S$ and $\mu(A) = (\mu(B_1) + \cdots + \mu(B_n))/n$.

   We denote the answer sets of a program Π that does not contain negation-as-failure by $\alpha(\Pi)$.
3. Now let Π be any fuzzy constraint logic program without variables. By X we again denote the set of ground literal in the language of Π. For any set $S \in X$, let Π^S to be fuzzy constraint program obtained from Π by deleting:

   (c) each rule that has a formula $\neg C$ in its body with $C \in M$ and
   (d) all formulas of the form $\neg C$ in the bodies of the remaining rules.

Clearly, Π^S does not contain $\neg$, so that its answer set is already defined. If this answer set coincides with S, then we say that S is an answer set of Π. In other words, the answer sets of Π are characterized by the equation

$$S = \alpha(\Pi^S)$$

The answer sets of Π are possible sets of beliefs that a rational agent may hold on the basis of the information expressed by the rule Π. If S is the set of ground literals that the agent believes to be true, then any rule that has a subgoal $\neg C$ with $C \in S$ will be of not use to him, and he will view any subgoal $\neg C$ with $C \in S$ as trivial. Thus he will be able to replace the set of rules Π by the simplified set of rules Π^S. If the answer set of Π^S coincides with A, then the choice of S as the set of beliefs is rational.

Now we illustrate our idea using the same example in [9]. Suppose a hypothetical match-making system, the following fuzzy constraint Horn clauses state the requirement of a good male candidate M, for a female applicant F.

Example 2. *An excerpt of hypothetical match-making system. Rule is as follows:*

$good - candidate(M, F) \longleftarrow \wedge(\neg smokes(M), tall(M).$
$income(M) \succ income(F), kind(M)).$
$kind(X) \longleftarrow likes(XPet).$
$kind(X) \longleftarrow likes\_social\_work(X).$

And facts are:

smokes(Tom)[1.0], smokes(John)[0.0], smokes(David)[0.0], tall(Tom)[0.9], tall(John)[0.8], tall(David)[0.7], income(Tom)[4000], income(John)[3000], income(David)[2400], income(Jane)[2000], likes(Tom,cat)[0.9], likes(Tom,dog)[0.8], likes(John,dog)[0.7], likes(David,dog)[0.7], likes_social_work(Tom)[0.7], likes_social_work(John)[0.1], likes_social_work(David)[0.5].

In the above, M, F, X, Y and Pet are variables, smokes(X) is a crisp predicate and tall(X), kind(x), likes(X,Y), likes_social_work(X) are fuzzy predicates. Income(X) is a predefined function which returns the income of person X in dollars, enumerated in square brackets in this example.

We now discuss the above example by using our idea. Given $\sigma = 0.5$, from the rule and the facts, we can get by the definition in Sections 2, 3 and the first step of stable model semantics, and then we can get the stable model of the example step by step. The answer that the program is supposed to return for a ground query is yes, no or unknown depending on whether the stable model contains the query, the negation of the query or neither. Suppose the system is given the query "$good\_candidate(M, Jane)$", i.e., who is the better candidate for Jane, the system will check if $good - candidate(X, Jane)$, $X \in \{Tom, John, David\}$ belong to the stable model.

From the example above, we can get answers

$$good\_candidate(David, Jane)[0.85]$$
$$good\_candidate(John, Jane)[0.875]$$

Compared to [9] for the example, our result is different. The reason is that the author uses the Zadeh's compositional rule of inference for approximate reasoning base on fuzzy relation. So, in the processing of approximate reasoning, the other result $good\_candidate(David, Jane)$ is discarded. compared to Zadeh's compositional rule of inference for approximate reasoning, we find an effective and efficient computational procedure by using answer set semantics for fuzzy constraint logic programming.

5 Conclusion

We have presented fuzzy constraint logic programs which is a combination of fuzzy logic programming and fuzzy mathematics programming. Furthermore, the syntax include negation as failure. We give the stable model semantics for such fuzzy constraint logic programming which is a generalization of answer set semantics of extended logic programming. Compared to Zadeh's compositional rule of inference for approximate reasoning [9], we find an effective and efficient computational procedure by using answer set semantics for fuzzy constraint logic programming.

Acknowledgments. This work is supported by the National Natural Science Foundation of China under Grand No. 60496322, Beijing municipal foundation for talent under Grand No. 20061D0501500206, open project of Beijing Multimedia and Intelligent Software Key laboratory in Beijing University of Technology, start-up fund and youth fund from Beijing University of Technology.

References

1. Apt, K.R.: Logic programming, Handbook of Theoretical Computer Science, Ch. 10, vol. B, pp. 493–574. MIT Press, Cambridge (1990)
2. Dubois, D., Fargier, H., Prade, H.: Propagation and satisfaction of flexible constraints. In: Yager, R., Zadeh, L. (eds.) Fuzzy sets, Neural Networds and Soft Computing, pp. 166–187. Van Nostrand Reinhold, New York (1994)
3. Gelfond, M., Lifschitz, V.: The stable model semantics for logic programming. In: Logic programming: Proc. of the Fifthe int. Conf. and Symp., pp. 1070–1080 (1988)
4. Gelfond, M., Lifschitz, V.: Classical negation in logic programs and disjunctive databases. New generation computing, pp. 579–597 (1990)
5. Joxan, J., Michael, J.M.: Constraint Logic Programming: A Survey. Journal of Logic Programming (1994)
6. Klir, G.J., Yuan, B.: Fuzzy Sets and Fuzzy Logic: Theory and applications. Prentice-Hall, Englewood Cliffs, NJ (1995)
7. Kowalski, R.A.: Predicated logic as a programming language. In: Information processing 1974, pp. 569–574 (1974)
8. Lai, U.J., Hwang, C.L.: Fuzzy Mathematical Programming: Methods and Applications. In: Lecture Notes in Economics and Mathematical Systems, Springer, Heidelberg (1992)
9. Lim, H.: Fuzzy Constraint Logic Programming, Fuzzy Systems. In: International Joint Conference of the Fourth IEEE International Conference on Fuzzy Systems, vol. 4, pp. 20–24 (1995)
10. Lloyd, J.: Foundations of logic programming. Springer, Heidelberg (1984)
11. Luo, X., Jennings, N.R., Shadbolt, N., Leung, H.F., Lee, J.H.M.: A fuzzy constraint based model for bilateral, multi-issue negotiation in semi-competitive environments. Artificial Intelligence 148(1-2), 53–102 (2003)
12. Luo, X., Lee, J.H.M., Leung, H.F., Jennings, N.R.: Prioritised Fuzzy Constraint Satisfaction Problems: Axioms, Instantiation and Validation. Fuzzy Sets and Systems 136(2), 155–188 (2003)

13. Luo, X., Zhang, C., Jennings, N.R.: A hybrid model for sharing information between fuzzy, uncertain and default reasoning models in multi-agent systems. International Journal of Uncertainty, Fuzziness and Knowledge-Based Systems 10(4), 401–450 (2002)
14. Machworth, A.K.: Consistency in networks of relations. Artificial Intelligence 8(1), 99–118 (1977)
15. Mukaidono, M., Kikuchi, H.: Foundations of fuzzy logic programming. In: Wang, P.Z, Loe, K.F. (eds.) Between mind and computer, pp. 225–244. World Scientific, Singapore (1993)
16. Tsang, E.P.K.: Foundations of constraint Satisfaction. Academic Press, New York (1993)
17. Wang, J., Liu, C.: Agent Oriented Probabilistic Logic Programming with Fuzzy Constraints. In: Shi, Z.-Z., Sadananda, R. (eds.) PRIMA 2006. LNCS (LNAI), vol. 4088, pp. 664–672. Springer, Heidelberg (2006)
18. Zadeh, L.A.: Fuzzy Sets. Inform. and Control 8, 338–353 (1965)

Prime Implicates for Approximate Reasoning

David Rajaratnam and Maurice Pagnucco

National ICT Australia* and University of New South Wales
Sydney, Australia
{daver,morri}@cse.unsw.edu.au

Abstract. Techniques for improving the computational efficiency of inference have held a long fascination in computer science. Two popular methods include *approximate logics* and *knowledge compilation*. In this paper we apply the idea of *approximate compilation* to develop a notion of prime implicates for the family of classically sound, but incomplete, approximate logics S-3. These logics allow for differing levels of approximation by varying membership of a set of propositional atoms. We present a method for computing the prime S-3-implicates of a clausal knowledge base and empirical results on the behaviour of prime S-3-implicates over randomly generated 3-SAT problems. A very important property of S-3-implicates and our algorithm for computing them is that decreasing the level of approximation can be achieved in an incremental manner without re-computing from scratch (Theorem 7).

1 Introduction

Approximate logics and knowledge compilation have been developed as a response to the problem of the computational intractability of classical logic. In the case of approximate logics, Cadoli and Schaerf have developed a dual family of logics to enable successive levels of approximation [1]. The family S-3 is classically sound but incomplete while S-1 provides for complete but unsound reasoning.

An orthogonal direction of research aimed at tackling the intractability of classical logic is that of knowledge compilation (see [2] for a recent survey). Here the focus is on transforming the knowledge base with a view to making certain operations more efficient. One common technique is that of prime implicates. The prime implicates of a propositional formula represent a logically equivalent clausal characterisation of the formula that allows for polynomial time entailment checking. Prime implicates have been studied extensively in the knowledge compilation literature. Developing prime S-3-implicates can therefore be understood in the broader context of *approximate knowledge compilation* techniques.

The contributions of this paper are twofold. Firstly, we introduce the notion of prime S-3-implicates and show that they are exactly characterised by iteratively

* NICTA is funded by the Australian Government's Department of Communications, Information Technology, and the Arts and the Australian Research Council through *Backing Australia's Ability* and the ICT Research Centre of Excellence programs.

Z. Zhang and J. Siekmann (Eds.): KSEM 2007, LNAI 4798, pp. 61–72, 2007.

resolving between clauses, where the biform variables are restricted to a given set of atoms S (the *relevance* set). As a result, well-known resolution based algorithms for computing prime implicates, such as Tison's method, can now be understood in terms of computing a succession of prime S-3-implicates. This enables the adaptation of these algorithms for prime S-3-implicate computation.

Finally, we present empirical results examining the behaviour of prime S-3-implicates over randomly generated 3-SAT problems. Of particular interest here is the prior research examining the behaviour of classical prime implicates for 3-SAT instances, and the identification of peaks in the number of prime implicates of varying lengths [3]. We show that the peak in the number of prime S-3-implicates varies uniformly for differing sizes of relevance set and discuss its implications for approximate knowledge compilation.

2 Background

We consider an underlying propositional language $\mathcal{L}$ over a set of propositional atoms (variables) $\mathcal{P}$, with the constant $\Box$ to signify the empty clause.

The approximate logics under consideration will be defined for a propositional language restricted to formulas in negation normal form (NNF). An NNF language is a fragment of a full propositional language restricted to the logical operators $\neg$, $\vee$, and $\wedge$, with the further restriction that the negation operator $\neg$ can only appear next to a propositional atom. We will also be particularly interested in formulas that are in conjunctive normal form (CNF), which is a restricted form of NNF. A CNF consists of a conjunction of clauses, where a *clause* consists of a disjunction of literals.

The notion of a *conjugate pair* is used to refer to positive and negative literals with the same propositional atom (the notation $\bar{l}$ is used to refer to the conjugate of a literal l). This notation is extended to arbitrary NNF formulas in the obvious way. Finally, we denote with $letters(\gamma)$ the set of propositional atoms occurring within the formula γ.

2.1 Prime Implicates

An *implicate* of a formula is a non-tautological clause that is implied by that formula. For example, given formula Γ and clause γ, then γ is an implicate of Γ if and only if $\Gamma \models \gamma$. A clause is a *prime implicate* of a formula if it is an implicate of that formula and there is no other implicate that entails it. Testing entailment between two non-tautological clauses can be achieved through subsumption checking. A clause γ is said to *subsume* a clause δ if the literals of γ are a subset of the literals of δ. For example, $p \vee q$ subsumes $p \vee q \vee r$. The set of prime implicates represents a canonical compiled form of the original formula.

Prime implicates have been well studied in the *knowledge compilation* literature and there are efficient algorithms that can perform subsumption checking in polynomial time with respect to the size of the knowledge base and query [4]. Unfortunately, it is well known that there can potentially be an exponential number of prime implicates with respect to the original knowledge base.

2.2 The S-3 Approximate Logics

The following results are due to Cadoli and Schaerf [1,5]. We consider a semantics with two truth values 0 and 1, intuitively *true* and *false* respectively. A *truth assignment* is a function mapping literals into the set $\{0,1\}$, and an S-3 interpretation is then defined in terms of such a mapping.

Definition 1 (S-3-Interpretation). *An S-3-interpretation of $\mathcal{L}$ is a truth assignment mapping every atom l in $S \subseteq \mathcal{P}$ and its negation $\neg l$ into opposite truth values. Moreover, it does not map both l in $\mathcal{P} \backslash S$ and its negation $\neg l$ into 0.*

Truth values are assigned to arbitrary NNF formulas using two rules:

$\vee - rule$ $I \models \alpha \vee \beta$ iff $I \models \alpha$ or $I \models \beta$.
$\wedge - rule$ $I \models \alpha \wedge \beta$ iff $I \models \alpha$ and $I \models \beta$.

A formula α is *S-3-satisfiable* if there is a S-3-interpretation I such that I maps α into the truth value 1. *S-3-entailment* is defined as follows: K *S-3-entails* α (written $K \models_S^3 \alpha$) iff every S-3-interpretation satisfying K also satisfies α. Note, $\models_S^3$ is sound but not complete with respect to $\models$ (classical entailment).

Example 1. Consider the knowledge base $K = \{\neg d \vee c, \neg c \vee m, \neg m \vee v\}$ and $S = \{c\}$. Now $K \models_S^3 \neg d \vee m$, because $c \in S$ therefore c and $\neg c$ must take on opposite truth values. However $K \not\models_S^3 \neg d \vee v$, even though $K \models \neg d \vee v$, because m and $\neg m$ can be simultaneously *true* under a S-3-interpretation where $m \notin S$.

Theorem 1 (Monotonicity). *For any S, S' such that $S \subseteq S' \subseteq \mathcal{P}$ and NNF formulas K and α, if $K \models_S^3 \alpha$ then $K \models_{S'}^3 \alpha$ (hence $K \models \alpha$.)*

Theorem 2 (Uniform Complexity). *Given NNF formula K, CNF formula α, and $S \subseteq \mathcal{P}$, then there exists an algorithm for deciding if $K \models_S^3 \alpha$ which runs in $\mathcal{O}(|K|.|\alpha|.2^{|S|})$ time.*

Theorem 1 establishes the soundness of S-3-entailment with respect to classical entailment, and shows how the logic becomes more complete as the set S grows. Theorem 2 establishes the attractive computational attributes of the logic, showing that the degree of approximation determines the tractability of its computation. Importantly, the attractive complexity result is dependent on the query being in CNF (established in [6]).

Theorem 3 (Reducing S-3-Entailment to S-3-Unsatisfiability). *Given NNF formulas K and α, such that letters(α) $\subseteq S \subseteq \mathcal{P}$, then $K \models_S^3 \alpha$ iff $K \cup \{\overline{\alpha}\}$ is not S-3-satisfiable.*

Theorem 3 shows that the relationship between S-3-entailment and S-3-unsatisfiability is similar to the classical case, albeit with some restrictions.

Now, the case when the set S is empty is of particular interest as S-3 reduces to Levesque's logic for reasoning about explicit beliefs [7]. In this case, we can omit the prefix S and simply refer to *3-satisfiability* and *3-entailment* (written as $\models^3$) or a *3-interpretation*.

Theorem 4 (From S-3-Entailment to 3-Entailment). *Let $S \subseteq \mathcal{P}$ be the set $\{p_1, \ldots, p_n\}$ and K and α be formulas in NNF. $K \models^3_S \alpha$ iff $K \models^3 \alpha \vee (p_1 \wedge \neg p_1) \vee \ldots \vee (p_n \wedge \neg p_n)$. Equivalently $K \models^3_S \alpha$ iff $K \models^3 \alpha \vee (l_1 \vee \ldots \vee l_n)$ holds for any combination $\{l_1, \ldots, l_n\}$, where each l_i $(1 \leq i \leq n)$ is either p_i or $\neg p_i$.*

Theorem 4 shows how a S-3-entailment relationship has a corresponding 3-entailment relationship. It is particularly important for our purposes as it forms the basis for proving a number of the properties presented later in this paper.

3 Prime S-3-Implicates

The previous section summarised the properties of Cadoli and Schaerf's family of S-3 logics. In this section we introduce our formalisation of the notion of prime S-3-implicates which serves as a bridge to S-3 entailment and then classical (propositional) entailment. Some of the properties of prime S-3-implicates mirror the properties of classical prime implicates, such as reducing entailment checking to subsumption checking. However, other properties relate prime S-3-implicates for differing relevance sets and therefore have no classical counterparts. In particular it is shown how the prime S-3-implicates for a set S effects subsequent S-3-entailment checking for a different relevance set.

Definition 2 (Prime S-3-Implicates). *Let Γ be a formula in NNF, γ be a clause, and $S \subseteq \mathcal{P}$, then γ is a* S-3-implicate *of Γ iff $\Gamma \models^3_S \gamma$. Further, γ is a* prime S-3-implicate *of Γ iff γ is an S-3-implicate of Γ and there is no other S-3-implicate ϕ of Γ such that $\phi \models^3_S \gamma$. We define $\Pi^3_S(\Gamma)$ as the set of prime S-3-implicates of Γ; $\Pi^3_S(\Gamma) = \{\gamma | \gamma$ is a prime S-3-implicate of $\Gamma\}$.*

This definition follows closely its classical counterpart. In fact, due to the monotonicity property (Theorem 1), it can be seen as a generalisation of the classical prime implicate property, (albeit restricted to a NNF language). The classical implicates then are simply the special case where S contains all propositional atoms in the language. The following results motivate the notion of S-3 implicates.

Theorem 5 (Reduction to Subsumption Checking). *Given NNF formula Γ, CNF formula $\gamma_1 \wedge \ldots \wedge \gamma_n$ where each γ_i, $0 \leq i \leq n$ is a clause, and $S \subseteq \mathcal{P}$, then $\Gamma \models^3_S \gamma_1 \wedge \ldots \wedge \gamma_n$ iff for each γ_i $(1 \leq i \leq n)$, either γ_i contains a conjugate pair of literals or γ_i is subsumed by some clause in $\Pi^3_S(\Gamma)$.*

Theorem 6 (Prime S-3-Implicate Equivalence). *Given NNF formulas Γ and γ, and $S \subseteq \mathcal{P}$, then $\Gamma \models^3_S \gamma$ iff $\Pi^3_S(\Gamma) \models^3_S \gamma$.*

Theorems 5 and 6 mirror the equivalent prime implicate properties. Proving these properties follows in the same manner as their classical counterparts. Theorem 5 shows how the prime S-3-implicates enable subsumption based S-3-entailment checking. Theorem 6 shows the logical equivalence of the prime S-3-implicates to its original formulation.

Further properties highlight the relationship between different S-3 logics and their prime S-3-implicates. These follow largely from Theorem 4, which provides a way to reduce arbitrary S-3-entailment relations to 3-entailment, to be subsequently lifted to a combined S-3-entailment relation.

Theorem 7 (Partial Compilation). *For any S, S' such $S \subseteq \mathcal{P}$ and $S' \subseteq \mathcal{P}$, and NNF formula Γ, then $\Pi^3_{S' \cup S}(\Gamma) = \Pi^3_{S'}(\Pi^3_S(\Gamma)) = \Pi^3_S(\Pi^3_{S'}(\Gamma))$.*

Theorem 7 shows that computing the prime S-3-implicates of a set of prime S-3-implicates (for differing S) can be precisely characterised in terms of another set of prime S-3-implicates. In effect prime S-3-implicates can be viewed as a *partial compilation* to some larger superset. The significance of this property lies in its application to the iterative compilation of knowledge bases. Furthermore, it is this property that distinguishes prime S-3-implicates from those approximate compilation techniques, such as Horn approximations, which cannot necessarily be compiled further to achieve a better approximation.

Theorem 8 (S-3-Entailment for a Partial Compilation). *For any S, S' such $S \subseteq \mathcal{P}$ and $S' \subseteq \mathcal{P}$, and NNF formulas Γ and γ, then $\Gamma \models^3_{S' \cup S} \gamma$ iff $\Pi^3_S(\Gamma) \models^3_{S' \setminus (S \setminus letters(\gamma))} \gamma$.*

In essence Theorem 8 establishes an equivalence between a S-3-entailment query of a knowledge base and a (possibly) larger S-3-entailment query of a partially compiled knowledge base. It shows that a partially compiled knowledge base has the potential to reduce the size of the relevance set needed to answer a given query. It is important to note that the partial compilation does not always reduce the size of the relevance set. This is a result of the interplay between the relevance set and the actual query.

Theorem 8 shows the potential for a partially compiled knowledge base to reduce the size of the relevance set needed for subsequent queries. Therefore, when limited to CNF queries, it can also serve to reduce the complexity of reasoning for these queries. To appreciate this we can consider the case where $S' = \mathcal{P}$; $S' \cup S$ reduces to $\mathcal{P}$ and we observe that $\models^3_{\mathcal{P}}$ is simply classical entailment. Therefore Theorem 8 specialises to $\Gamma \models \gamma$ iff $\Pi^3_S(\Gamma) \models^3_{\mathcal{P} \setminus (S \setminus letters(\gamma))} \gamma$. Now, when γ is in CNF, the complexity of checking S-3-entailment is bounded by the size of the relevance set (Theorem 2) therefore we can test classical entailment in $\mathcal{O}(|\Pi^3_S(\Gamma)|.|\gamma|.2^{|\mathcal{P} \setminus (S \setminus letters(\gamma))|})$. Provided that the size of the query is significantly smaller than the size of the knowledge base then the complexity approaches $\mathcal{O}(|\Pi^3_S(\Gamma)|.2^{|\mathcal{P} \setminus S|})$. That is, the cost of reasoning can be directly offset against the cost of generating $\Pi^3_S(\Gamma)$ and its resulting size. Of course the size of $\Pi^3_S(\Gamma)$, and its relationship to the size of Γ, is a critical factor; a factor that is considered later in this paper.

4 Computing Prime S-3-Implicates

Having defined prime S-3-implicates for NNF formulas we now limit ourselves to the CNF case and present a resolution-based method for generating the prime

S-3-implicates of a set of clauses. Note, the fact that resolution can be used for computing prime S-3-implicates is not entirely surprising, as resolution-based methods have previously been used for testing S-3-satisfiability [5]. The formalisation developed here mirrors that of classical resolution as presented in [8].

Definition 3 (Resolvent). *For any pair of clauses $A = k_1 \vee \ldots \vee k_n \vee p$ and $B = l_1 \vee \ldots l_m \vee \neg p$, where k_i $(1 \leq i \leq n)$ and l_i $(1 \leq i \leq m)$ are literals and p is a propositional atom, the* resolvent *of A and B with respect to p, written $res(A, B; p)$, is defined as $k_1 \vee \ldots . \vee k_n \vee l_1 \vee \ldots \vee l_m$.*

Lemma 1 (S-Resolution Principle). *For any pair of clauses $A = k_1 \vee \ldots \vee k_n \vee p$ and $B = l_1 \vee \ldots l_m \vee \neg p$, and set S of propositional atoms such that $p \in S$ then $\{A, B\} \models^3_S res(A, B; p)$.*

Definition 4 (S-Resolvent Closure). *For any set A of clauses and set S of propositional atoms define $R_S(A) = A \cup \{C | C = res(C_1, C_2; p)$ where $C_1, C_2 \in A$, $C_1 \neq C_2$, $p \in S$, $C_1 = k_1 \vee \ldots \vee k_n \vee p$, and $C_2 = l_1 \vee \ldots l_m \vee \neg p\}$.*

$R^n_S(A)$ is defined inductively by: $R^0_S(A) = A$, $R^{i+1}_S(A) = R_S(R^i_S(A))$.

We define the resolvent closure *of A under S as $R^*_S(A) = \cup_{n \in N} R^n_S(A)$.*

Theorem 9 (Adequacy of S-Resolution). *For any set of clauses A and set S of propositional atoms, A is S-3-unsatisfiable iff $\Box \in R^*_S(A)$.*

The adequacy of the S-resolution property shows that if a knowledge base is S-3-unsatisfiable then iteratively applying S-resolution will eventually derive the empty clause (i.e., a contradiction). Further if a contradiction is derived after applying S-resolution to the knowledge base then it is indeed S-3-unsatisfiable. If we consider the special case where $S = \mathcal{P}$ then the principle reduces to the familiar classical resolution property.

Definition 5 (Residue of Subsumption). *For set A of clauses, the* residue of subsumption *of A, written $RS(A)$, is defined as $RS(A) = \{C \in A | C$ is not subsumed by any other clause of A and C does not contain a conjugate pair$\}$.*

Theorem 10 (Adequacy). *For any set A of clauses and set S of propositional atoms $\Pi^3_S(A) = RS(R^*_S(A))$.*

This theorem shows that S-resolution is adequate to capture the computation of all the prime S-3-implicates of a set of clauses. This mirrors the classical case where resolution is adequate for generating all the prime implicates of a set of clauses. It is therefore straightforward to see that resolution-based mechanisms for generating prime implicates can be adapted to generating prime S-3-implicates by simply limiting resolution to biform variables occurring in S.

Tison's Method [9] is one of the simplest algorithms for generating prime implicates from a set of clauses. It therefore serves as a useful method for showing the adaptation to prime S-3-implicate computation. The difference between Algorithm 1 and the classical Tison's Method is minor; occuring on line 4 of our algorithm with the restriction that the biform variable be in the set S. In the classical case the set S is generated internally to the algorithm by combining all the atoms from the input clauses, and therefore contains all propositional atoms.

Algorithm 1. Adapted Tison's Method for Computing Prime S-3-implicates

function S3-TISON(*clauses*, *S*, *PS3I*)

Input: *clauses*, set of input clauses
S, set of propositional atoms
Output: *PS3I*, set of prime S-3-implicates

1: $PS3I \leftarrow clauses$
2: **for all** $p \in S$ **do**
3: **for all** pairs of clauses $\gamma_1, \gamma_2 \in PS3I$ **do**
4: **if** $l_1 \in \gamma_1$ and $l_2 \in \gamma_2$ and $l_1 = \overline{l_2}$ and $letters(l_1) = p$ **then**
5: $PS3I \leftarrow PS3I \cup \{res(\gamma_1, \gamma_2; p)\}$
6: **end if**
7: **end for**
8: REMOVE-SUBSUMED(PS3I)
9: **end for**

Here REMOVE-SUBSUMED(PS3I) is a procedure that removes subsumed clauses from the set PS3I, and has been omitted for brevity.

Theorem 11 (Correctness of S3-TISON). *For any set of clauses A and set S of propositional atoms, the S3-TISON algorithm computes the set $\Pi^3_S(A)$.*

We can understand this algorithm by observing that for a set of propositional atoms $S = \{p_n, \ldots, p_1\}$, $\Pi^3_S(A) = \Pi^3_{\{p_n\}}(\Pi^3_{\{p_{n-1}\}}(\ldots \Pi^3_{\{p_2\}}(\Pi^3_{\{p_1\}}(A))\ldots))$. The outer loop of the algorithm (lines 2 - 9) can be seen as providing the enumeration of S, while the inner loop (lines 3 - 7) successively computes the prime S'-3-implicates of A for ever increasing subsets S' of S.

Alternative resolution-based techniques, in particular the Incremental Prime Implicate Algorithm (IPIA) algorithm [10], can also be adapted for computing prime S-3-implicates. The IPIA algorithm allows for efficient incremental recompilation of prime implicates. Due to space considerations we do not present this adapted IPIA algorithm here, and instead limit ourselves to providing only the intuition behind the adaptation.

The adaptation of IPIA for computing prime S-3-implicates is based on the observation that for any set of atoms S, set of clauses A, and clause γ, then $\Pi^3_S(A \cup \{\gamma\}) = \Pi^3_{\{letters(\gamma) \cap S\}}(\Pi^3_S(A) \cup \{\gamma\})$. Intuitively it means that to add a new clause to an already compiled knowledge base we need only consider recompilation with respect to the atoms from S that appear in the new clause. A total compilation of a knowledge base can be achieved through an iterative recompilation process. The clauses in the knowledge base are treated sequentially and at each step the compiled set is recompiled with a new clause being added.

It is worth noting an important difference between the operations of the Tison and IPIA algorithms. In Tison's method it is the enumeration of propositional *atoms* that drives the algorithm. On the other hand, for the IPIA algorithm it is the enumeration of *clauses* in the knowledge base that dominates. The reason for this is that clauses in a knowledge base tend to be relatively small compared to the total size of the knowledge base and therefore each recompilation step will be limited to only a small number atoms. As a result, the ordering of clauses is more important than any particular enumeration of atoms. The significance of this observation will become evident when discussing the empirical results.

5 Empirical Results

The previous sections provided the formal development of prime S-3-implicates as a method for approximate compilation and showed a close relationship to existing resolution-based techniques. In this section experimental results are considered. There are two basic motivations for these experiments. Firstly, to give an indication of the practicality of using prime S-3-implicates in approximate knowledge compilation. Secondly, to provide insight into the operating behaviour of traditional algorithms for computing classical prime implicates.

The data set used consisted of 3-SAT instances, randomly generated using a fixed clause-length model (also known as Random K-SAT [11]) with 10 variables, and varying the number of clauses. For each data point, determined by the clause to variable ratio, 1000 3-SAT instances were generated. For each 3-SAT instance the prime S-3-implicates were computed for every subset S of the 10 variables.

In general there can be a number of factors that would determine the choice of relevance set; such as the relevance to the expected query, the cost of computing the partially compiled knowledge base, and the size of the partially compiled knowledge base. In this paper we only consider the size of the compiled knowledge base. In particular we are interested in understanding what could be the most compact approximate compilations that one could hope to achieve. For our randomly generated data set, Figure 1 represents this best outcome, Figure 2 the average, and Figure 3 the worst case outcome. To not clutter the presentation not all set sizes are displayed. Importantly, the set sizes 0 and 10 represent the uncompiled knowledge base and the classically compiled prime implicates respectively, and are therefore unchanged throughout the 3 figures.

Figure 1 presents the mean of the minimum number of prime S-3-implicates for relevance sets of particular sizes for each 3-SAT instance. To clarify the method, for every set size (other than 0 and 10) there will be multiple compiled instances to consider. In this figure we take the instance with the minimum value. So, for example with the sets of size 9, for each 3-SAT instance we consider only those prime S-3-implicates generated from a relevance set of size 9 and take the single instance with the minimum number of prime S-3-implicates. We choose the minimum here to obtain a sense of the best possible result. Finally, for each data point we take the mean of these minimum values.

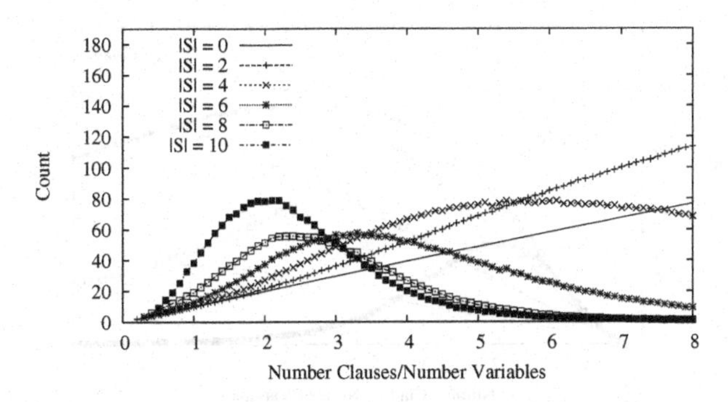

Fig. 1. Mean of the minimum aggregate number of prime S-3-implicates for problem size of 10 variables

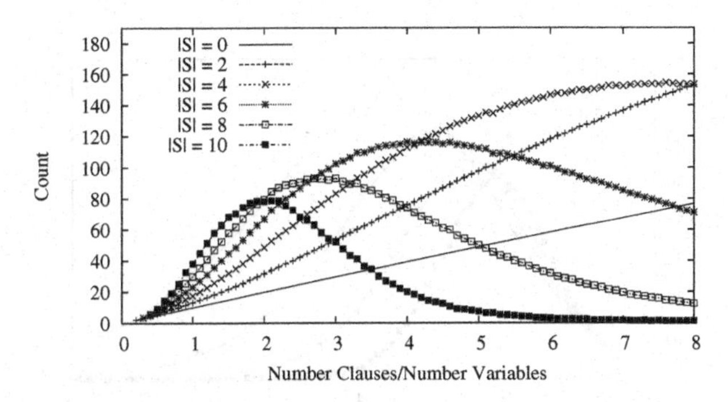

Fig. 2. Mean of the aggregate number of prime S-3-implicates for problem size of 10 variables

Figure 2 presents the mean number of prime S-3-implicates for relevance sets of particular sizes for each 3-SAT instance. Figure 3 examines the mean of the maximum aggregate number of prime S-3-implicates (the worst case).

It is clear from the figures that S-3 logics of differing sizes of relevance sets will have the number of their prime S-3-implicates peak at differing critical ratios. These peaks will occur increasingly to the right (i.e., increasing critical ratios) of the classical prime implicate peak, as the size of the relevance set decreases. For example, the peak for the set size 8 is to the right of the peak for set size 9 and the peak for set size 7 is to the right of that for 8. This change in peaks of the aggregate data for differing set sizes is also reflected in the peaks for individual clause lengths. Figure 4 shows the varying peaks for differing relevance set sizes for prime S-3-implicate clauses of length 3.

Explaining this behaviour can best be achieved with reference to Schrag and Crawford's explanation of prime implicate peaks [3]. They point out that the number of prime implicates of size 0 corresponds to the transition between a satisfiable and unsatisfiable set of formulas, since an unsatisfiable formula has a single prime implicate of size 0. Hence the number of prime implicates of size 0 changes dramatically from 0 to 1 at the 50%-satisfiability point (near the ratio

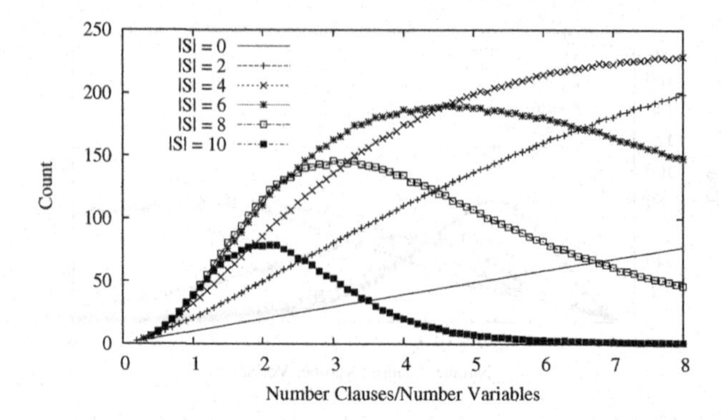

Fig. 3. Mean of the maximum aggregate number of prime S-3-implicates for problem size of 10 variables

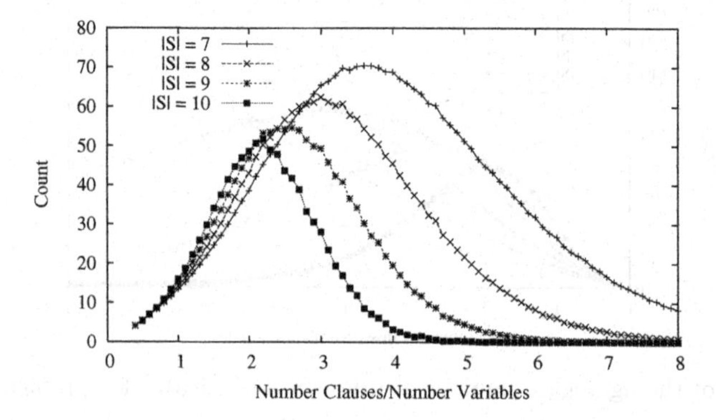

Fig. 4. Mean of the prime S-3-implicates for clause length of size 3 and varying relevance sets. The set size of 10 corresponds to the the classical prime implicate case.

4.25). Peaks in the number of prime implicates of clause sizes other than 0 occur to the left of this point and move further left as the clause size increases. For example the peak in clauses of size 3 occurs at a smaller critical ratio than the peak in clauses of size 2. The aggregate peak in prime implicates is then the composite of these various peaks.

Now we consider the satisfiability of S-3 logics. A S-3-satisfiable formula may be classically unsatisfiable, as the logic is not complete. As the size of the relevance set decreases the number of formulas that are S-3-satisfiable increases. The transition point between the prime S-3-implicate clauses of length 0 occurs at a critical point that is to the right of the classical case and this moves further to the right as the size of the relevance set decreases. The peaks in the prime S-3-implicates for the other clause lengths follows a similar pattern. Figure 4 shows this behaviour for clauses of length 3 . Hence the aggregate peaks also follow this pattern.

Having explained the behaviour it is interesting to analyse its practical implications. Firstly, regarding the potential use of prime S-3-implicates as an approximate compilation technique, the results are mixed. The minimum aggregate values show that genuine reductions in the size of the compiled knowledge base can be achieved. However this will occur around the peak in the size of the classical prime implicates. As the critical ratio increases the classical prime implicates will reduce much more quickly than their various approximations. So at a higher critical ratio having a poorer approximation (i.e., a smaller relevance set) can result in a significantly larger compiled knowledge base.

For resolution-based algorithms our results also offer some insight. Optimisations of Tison's method primarily focus on improved methods for subsumption checking and improved selection of resolution clauses to reduce the number of subsumption checks that need to be made. Essentially they have largely focused on improvements in the inner loop of the algorithm (lines 3 - 7 of Algorithm 1). However our results show that the choice of enumeration of variables plays an important role in reducing the number of intermediate clauses. An optimal choice will result in intermediate clauses that correspond to minimal sets of prime S-3-implicates for increasing set sizes. Unfortunately, it also shows that at higher critical ratios an increase in intermediate clauses will be unavoidable for any enumeration of variables, highlighting a fundamental weakness of Tison's method.

The IPIA algorithm on the other hand is less vulnerable to this weakness because the prime implicates are constructed by enumerating the clauses in the knowledge base, and variable enumeration takes place over a relatively small number of variables. Therefore one might expect that as each new clause is integrated into the compiled knowledge base the size of this knowledge base will correspond closely to the classical prime implicate curves of Figures 1, 2, and 3. So, while under some circumstances Tison's method may outperform the IPIA, under most circumstances the IPIA will exhibit more stable behaviour.

6 Conclusion and Future Work

In this paper we developed the notion of prime S-3-implicates for the S-3 family of logics, examining three key areas: establishing the theoretical foundations, developing computational methods, and analysing empirical results.

The foundational properties established that prime S-3-implicates are a generalisation of the classical concept of prime implicates. We further showed that they can be treated as both a form of approximate and partial compilation.

In developing computational methods it was shown that prime S-3-implicates can be characterised through a generalised form of classical resolution. This allowed for the adaptation of classical techniques as a computation method, and also enabled greater understanding of the behaviour of these classical algorithms.

Experimental results highlighted the potential benefits, and difficulties, of using prime S-3-implicates for approximate knowledge compilation. They further highlighted the need for greater consideration of variable enumeration in the classical prime implicate algorithms.

Koriche [12] introduces *prime S-implicates* which is a notion that differs to the current work. Koriche's concept is developed for a logic with modal operators to capture the notion of sound but incomplete reasoning, and complete but unsound reasoning. Further, it is intended as a tool for specifying the behaviour of an anytime reasoner; as opposed to being a logic for use by such a reasoner. As such, it is developed differently to prime S-3-implicates introduced here.

This work opens up a number of interesting avenues for future research. Complexity results would need to be proven, expected to be exponential in the size of the relevance set. Further experimentation would be needed to extend practical understanding; examining benchmark 3-SAT instances of varying problem sizes. Recently, Finger and Wassermann have broadened the research on approximate logics: extending S-3 to a full propositional language, introducing fine-grained control of inference, and providing a tableau based proof system [13]. Approximate compilation techniques should similarly be developed for this extension.

Finally, future work is needed to establish a map of approximate compilation methods that mirrors the work being undertaken in knowledge compilation for classical logic [2]. While significant to *approximate* compilation techniques, it could also provide broader insights for knowledge compilation in general.

References

1. Cadoli, M., Schaerf, M.: Approximate entailment. In: Ardizzone, E., Sorbello, F., Gaglio, S. (eds.) Trends in Artificial Intelligence. LNCS, vol. 549, pp. 68–77. Springer, Heidelberg (1991)
2. Darwiche, A., Marquis, P.: A knowledge compilation road map. Journal of Artificial Intelligence Research 17, 229–264 (2002)
3. Schrag, R., Crawford, J.M.: Implicates and prime implicates in random 3-SAT. Artificial Intelligence 81(1-2), 199–222 (1996)
4. de Kleer, J.: An improved incremental algorithm for generating prime implicates. In: Rosenbloom, P., Szolovits, P. (eds.) Proc. of the Tenth National Conf. on Artificial Intelligence, pp. 780–785. AAAI Press, Stanford (1992)
5. Cadoli, M.: Tractable Reasoning in Aritificial Intelligence. LNCS, vol. 941. Springer, Heidelberg (1995)
6. Cadoli, M., Schaerf, M.: On the complexity of entailment in propositional multivalued logics. Annals of Math. and Artificial Intelligence 18(1), 29–50 (1996)
7. Levesque, H.J.: A knowledge-level account of abduction. In: Proceedings of the 11th International Joint Conference on Artificial Intelligence, pp. 1061–1067 (1989)
8. Singh, A.: Logics For Computer Science. Prentice Hall of India, New Delhi (2003)
9. Tison, P.: Generalization of consensus theory and application to the minimalization of boolean functions. IEEE Trans on Elec Computers EC-16(4), 446–456 (1967)
10. Kean, A., Tsiknis, G.K.: An incremental method for generating prime implicants/impicates. Journal of Symbolic Computation 9(2), 185–206 (1990)
11. Mitchell, D.G., Selman, B., Levesque, H.J.: Hard and easy distributions for SAT problems. In: Rosenbloom, P., Szolovits, P. (eds.) Proc. of the Tenth National Conf. on Artificial Intelligence, pp. 459–465. AAAI Press, Menlo Park, California (1992)
12. Koriche, F.: A logic for anytime deduction and anytime compilation. In: Dix, J., Fariñas del Cerro, L., Furbach, U. (eds.) JELIA 1998. LNCS (LNAI), vol. 1489, pp. 324–341. Springer, Heidelberg (1998)
13. Finger, M., Wassermann, R.: Approximate and limited reasoning: Semantics, proof theory, expressivity and control. J. Log. Comput. 14(2), 179–204 (2004)

Distributed Constraint Satisfaction for Urban Traffic Signal Control

Kazunori Mizuno[1] and Seiichi Nishihara[2]

[1] Department of Computer Science, Takushoku University
Hachioji, Tokyo 193-0985, Japan
mizuno@cs.takushoku-u.ac.jp
[2] Department of Computer Science, University of Tsukuba
Tsukuba, Ibaraki 305-8573, Japan
nishihara@cs.tsukuba.ac.jp

Abstract. Urban traffic problems including traffic accidents and traffic congestion or jams have been very serious for us. Urban traffic flow simulation has been important for making new control strategies that can reduce traffic jams. In this paper, we propose a method that can dynamically control traffic signals by equivalently representing as the constraint satisfaction problem, CSP. To solve local congestion in each intersection, we define the whole system as multi-agent systems where the represented CSP is extended to distributed CSP, DCSP, in each of which variable is distributed to each intersection agent. Each intersection agent determines some signal parameters by solving the DCSP. The proposed method is implemented on our separately developed agent-oriented urban traffic simulator and applied to some roadnetworks, whose experimental simulations demonstrated that our method can effectively reduce traffic jams even in the roadnetworks where traffic jams are liable to occur.

1 Introduction

Recent remarkable progress of our traffic environments including the increase in cars and improvement of roadnetworks have given us the good results industrially and economically. In another aspect, however, urban traffic problems, including traffic jams and accidents, and environmental problems, including noise and air pollution, have been serious for us. In particular, for traffic jams, which are inveterately happening in urban areas, it can be expected to reduce traffic jams by taking measures to meet the situation.

On the other hand, due to improvement of the computer performances and technologies, there have been many studies on urban traffic flow simulations within a PC [1,3,4,5,6,8,9,13,14,16], enabling us to try to verify the effective measures for reducing traffic jams in advance.

To reduce traffic jams, controlling traffic signals can be one of the most effective measures [2,11,20,22]. The present control methods in Japan, based on selection out of some patterns that are made by the experts beforehand, may be difficult to control traffic signals dynamically and flexibly. It also seems to need much costs for the increase in the number of intersections because of the centralized control method.

Z. Zhang and J. Siekmann (Eds.): KSEM 2007, LNAI 4798, pp. 73–84, 2007.

In this paper, we propose the method that can dynamically and distributively control traffic signals according to the current local traffic situation. We represent determining the parameters for controlling as the constraint satisfaction problem, CSP. Assuming that traffic jams can reduce by solving local congestion at each intersection, we also define the whole system as the multi-agent systems [10, 15], where the represented CSP is extended to the distributed CSP, DCSP, in which variables and constraints are distributed among intersection agents. Each intersection agent dynamically determines the parameters of its own traffic signals by solving the CSP while receiving congestion information from road agents connected to itself. The proposed method is implemented in our separately developed agent-oriented urban traffic simulator, where we experimentally demonstrate that our method can reduce traffic jams.

In section 2, after describing traffic signals and their parameters for controlling, we refer to the present control method in Japan. In section 3, the proposed control method is given: we first introduce our developed traffic simulator. Then, we represents controlling signals as the CSP and give the procedure for constraint satisfaction processes. In section 4, we show experimental results where our method is applied to some roadnetworks. In section 5, we briefly review related works for controlling traffic signals and the intelligent transportation system, ITS, in Japan.

2 Distributed Constraint Satisfaction and Urban Traffic Signal

2.1 Distributed Constraint Satisfaction

A constraint satisfaction problem, CSP, involves finding values for problem variables which are subject to constraints specifying the acceptable combinations of values. Such combinatorial search problems are ubiquitous in artificial intelligence and pattern analysis, including scheduling and planning problems. A CSP is defined as (V, D, C), where V is a set of variables, D is a set of values to be assinged to variables, and C is a set of constraints required among variables. A distributed CSP, DCSP, is the CSP in which variables and constraints are distributed among multiple agents [7]. Various application problems in multi-agent systems, e.g., the distributed resource allocation problem, the distributed scheduling problem, the distributed interpretation task, and the multiagent truth maintenance task can be represented as DCSPs.

2.2 Traffic Signal and Parameters for Controlling

A traffic signal signs one of three colors, *green*, *yellow*, and *red*, and also works cyclically at all times. To control traffic signal, it is necessary to fix three parameters, *cyclelength*, *split*, and *offset*, after revelation of signals is determined.

- *cyclelength*: time length where the color of the signal goes around once,
- *split*: time distribution of green time length,
- *offset*: time lag of the green time period between signals at the adjacent intersections.

2.3 The Present Control Method

In Japan, two methods, the pattern selection method and the MODERATO method, are mainly used to control traffic signals. The former selects one of several patterns fixed by the experts which investigate traffic situations in advance. The latter fixes $cyclelength$ and $split$ by calculating a pattern that minimizes a delay time in which cars pass through intersections. The $offset$ parameter selects a pattern same as the former.

These methods have some drawbacks to dynamically control signals. It is difficult to flexibly cope the traffic conditions that are not assumed since parameter patterns have to be fixed in advance. Constructing the patterns also needs an enormous cost and advanced knowledge for traffic. Since traffic flow tends to be changeable over time, the patterns should be adjusted and reconstructed periodically.

3 Constraint-Based Traffic Signal Control

3.1 Basic Ideas

Solving local congestion at each intersection can make traffic jams reduce in the entire roadnetwork. We therefore propose the method for controlling signals at each intersection. The basic ideas of the proposed method are as follows.

i) we define the problem for fixing the signal parameters described in 2.2 as the CSP. The CSP is extended to DCSP, in each of which local variable is distributed to each intersection agent.
ii) The proposed method is implemented in the our separately developed traffic simulator, each of which intersection agent receives congestion information from road agents connected to the intersection.
iii) Since each intersection agent has multiple local variables, the search algorithm to solve the DCSP is based on the algorithm described in [21].

3.2 Our Agent-Oriented Traffic Flow Simulator

We have developed a agent-oriented traffic flow simulator [18]. In this system, four elements, i.e., vehicles, intersections, traffic signals, and roads, of which traffic flow is composed, are modelled as agent sets to be cooperated, that is, interaction among these agents causes several traffic phenomena.

The system is defined as a triple, $Sys =< Car, Intersection, Road >$, where $Car (= \{c_1, \ldots, c_N\})$ is the set of vehicle agents that directly make up traffic flow. $Intersection$ and $Road$ are sets of intersection and road agents, respectively, in each of which element provides various types of information in deciding actions in vehicle agents. A roadnetwork is given as a graph structure, where $Intersection$ and $Road$ correspond to sets of vertices and edges respectively. Fig. 2 gives data sets that intersection and road agents should hold. $Intersection$, which has information, e.g., existence of signal agents and pairlink (See the below definition and Fig. 3), etc., can get congestion information from connected road agents. According to the degree of congestion, signal parameters are fixed, being sent to signal agents. Signal agents sign colors of signals.

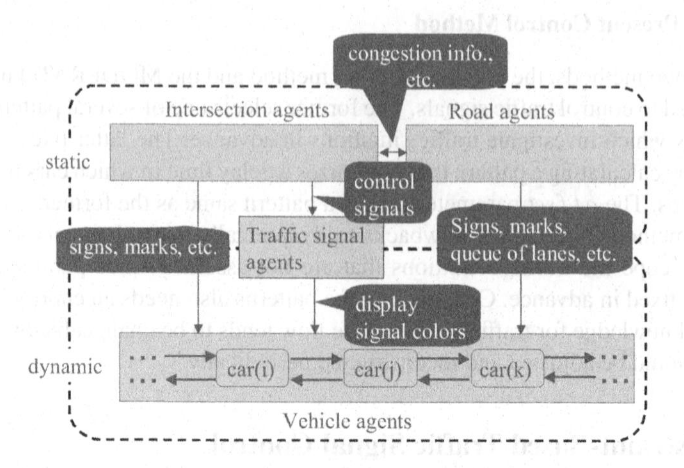

Fig. 1. The model of our traffic flow simulator

Intersection_agent{
/* common data */
P; (coordinates)
Sig; (No. of signal agents belonged to)
S_j; (signal agent ID, $j = 1, \cdots, Sig$)
Cl; (No. of road agents connected to)
Cid_k; (connected road agent ID, $k = 1, \cdots, Cl$)
p_link_l; (list of pairlink, $l = 1, 2$)
$Wmax, Wmin$; (congestion degree in connected roads)
$sline$; (existence of stop line)
etc...
/* Traffic signal agent */
Signal_agent{
cid (intersection ID belonged to);
color; (current color indication, $red, yellow, green$)
etc...
}
Road_agent{
N_j; (connecting intersection ID, $j = 0, 1$);
Tp; (kind of road)
Wd; (breadth)
Ct; (kind of the center line)
Ln; (No. of lanes)
Sn; (No. of road signs)
$cqueue_k$; (queue of each lane, $k = 1, \cdots, Ln$)
J (amount of congestion)
etc...
}

Fig. 2. An example of data sets for intersection and road agents

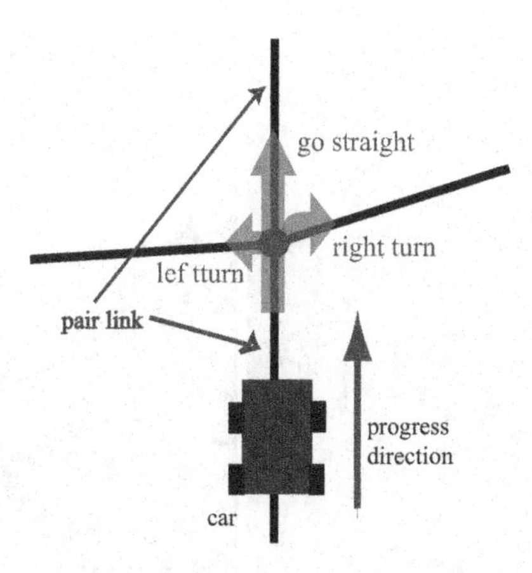

Fig. 3. An example of a pairlink

Definition 1. ***pairlink***
A pair of roads, or links, out of four roads which are connected to the same intersection, where passing through is allowed by the green signal. In the intersection without signals, a pair of roads both of which are as straight as possible.

Road provides several information, e.g., road marks and signs, bus stops, and lane queues, etc., that should be recognized by vehicle agents.

Car, or vehicle agent sets, can update their position, direction and speed by deciding its own next action by using two types of knowledge, or basic rules and individual rules [18].

We have developed our traffic flow simulator as a 3D animation system to visualize more realistic traffic phenomena, where we integrate our simulator with our separately developed virtual city generation system [19]. Fig. 4 shows execution examples of our simulator, where the example (b), in particular, gives a result where traffic flow appears in a virtual city.

3.3 Representation

Fixing signal control parameters is represented as a triple $P = (V, D, C)$.

$$
\begin{aligned}
P &= (V, D, C), \\
V &= \{cycle_{ik}, gstart_{ik}, gtime_{ik}\}, (i = 1, \ldots, n,\ k = 1, 2) \\
D &= \{D_{cycle}, D_{gstart}, D_{gtime}\}, \\
C &= \{C_{in}, C_{adj}, C_{road}\},
\end{aligned}
$$

where V is a set of variables, in which elements, $cycle_{ik}$, $gstart_{ik}$, and $gtime_{ik}$, correspond to time length of one cycle, start time period of the green sign in the cycle, and time length of the green sign, respectively, in the k-th signal belonging to the i-th intersection agent, as shown in Fig. 5. D is a set of values to be assigned to variables,

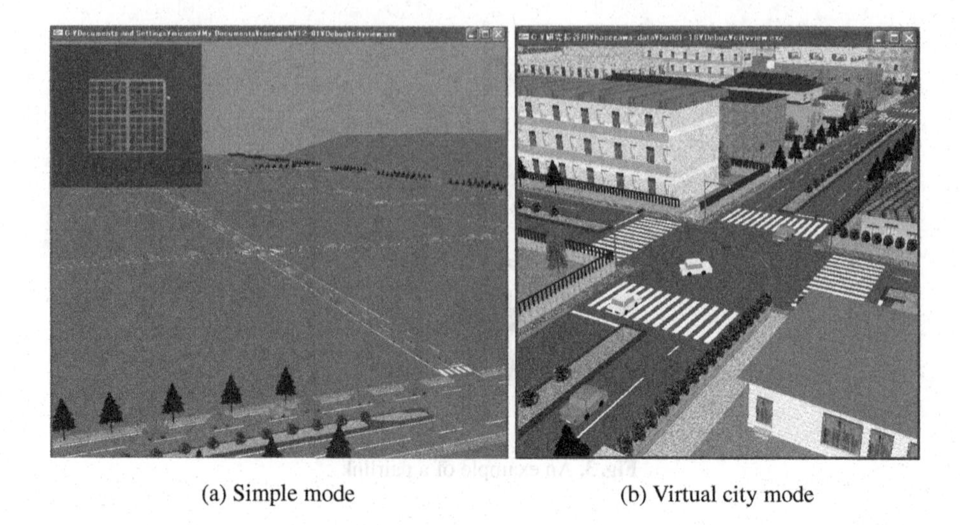

(a) Simple mode (b) Virtual city mode

Fig. 4. Example screens on our developed urban traffic flow simulator

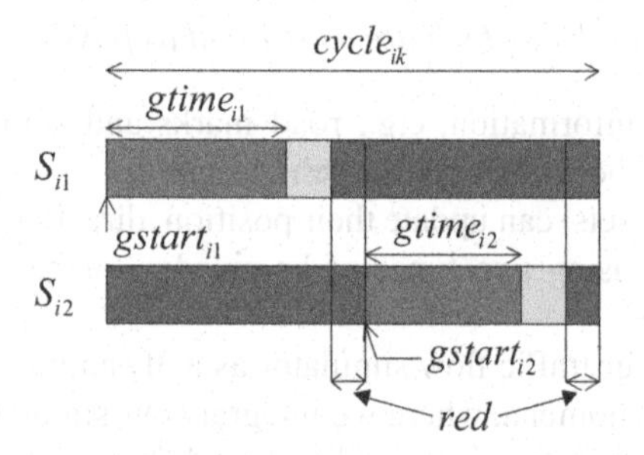

Fig. 5. Variables which the intersection agent *i* has

where $D_{cycle} = \{40,\ldots,140\}$ and $D_{gstart} = D_{gtime} = \{0,\ldots,140\}$. Therefore, a signal has cycle time length from 40 to 140 and green time length from 0 to 140. C is a set of constraints composed of subsets, C_{in}, C_{adj}, and C_{road}.

(1) C_{in}: constraints required inside one intersection agent
C_{in} is the subset of constraints to be satisfied between signals, S_{i1} and S_{i2}, belonging to the same intersection *i*. The constraints are very hard, which must be satisfied not to make traffic interrupt. Constraints in C_{in} are as follows.

- Cycle time length of S_{i1} and S_{i2} must be same, that is,
  $cycle_{i1} = cycle_{i2}$,
- S_{i1} and S_{i2} must not be the green sign in the same time, that is,
  $gstart_{i2} = gstart_{i1} + gtime_{i1} + yellow + red \times 2$,
  $gstart_{i1} = gstart_{i2} + gtime_{i2} + yellow + red \times 2$,

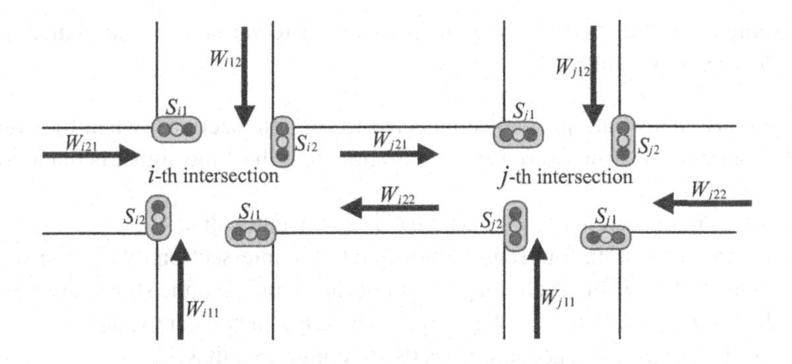

Fig. 6. An example of adjacent signals and the degree of congestion

- Both green time length of S_{i1} and S_{i2} should be length from 20% to 80% of cycle time length, that is,
  $cycle_{ik} \times 0.2 < gtime_{ik} < cycle_{ik} \times 0.8$.

(2) C_{adj}: constraints on signals between the adjacent intersection agents
C_{adj} is the subset of constraints on signals, S_{ik} and S_{jk}, between adjacent intersections, i and j. The constraints are soft, which are desirable to be satisfied to make traffic smooth.

Definition 2. ***adjacent signals***
When two intersections, i and j, with signal agents, S_{ik} and S_{jk} are adjacent in the sense of graph structures and length of the link (i, j) is less than a constant value, S_{i1} and S_{j1} at the same direction are adjacent signals.

Constraints in C_{adj} are as follows.

- Cycle time length of adjacent signals may be same, that is,
  $cycle_{ik} = cycle_{jk}$,
- Start time period of *green* of S_{ik} may be shifted from that of S_{jk}, that is,
  $gstart_{ik} = gstart_{jk} + Dist_{ij}/LV_{ij}$,
  where $Dist_{ij}$ and LV_{ij} denote the distance and speed limit at the road between i and j.

(3) C_{road}: constraints decided by congestion information from road agents
The constraints can play the role in resolving local congestion, in which values are re-assigned to variables according to congestion information from road agents. The degree of congestion for each road is defined by $W = CL/RL$, where CL denotes length of car rows and RL length of the road. The maximum and minimum degrees, $Wmax_{ikl}$ and $Wmin_{ikl}$, of congestion are also calculated for each lane, l, of roads approaching to intersections, where $Wmax_{ikl}$ is the degree of congestion when the color of signal is just changed to *green* from *red* and $Wmin_{ikl}$ is one when the color is just changed to *red* from *yellow*. We define the priority value, P_value_{ik} of signal, S_{ik} for priorly assigning values to variables in CSPs, calculated as $P\_value_{ik} = Wmin_{ikl} + Wmax_{ikl}$.

According to W and P_value, assignment of values to variables is reassigned subject to the following constraints.

- If there are many cars in roads connected to the intersection i when the signal is just changed to *yellow* from *green*, cycle time length of the signal is made longer, that is,
  $Wmin_{ikl} > 0.4 \Rightarrow cycle_{ik} + = C$, where C is a constant value,
- if there are few cars in four roads connected to the intersection i when the signal is just changed to *yellow*, cycle time length of the signal is made shorter, that is,
  for all l, $Wmin_{ikl} < 0.4 \Rightarrow cycle_{ik} - = C$, where C is a constant value,
- the offset value is taken account if roads are congested, that is,
  $Wmax_{ikl} > 0.4 \Rightarrow gstart_{ik} = gstart_{jk} + Dist_{ij}/LV_{ij}$,
- green time length of each signal is decided accoring to its road congestion, that is,

$$gtime_{ik} = \frac{Wmax_{ik1} + Wmax_{ik2}}{\sum_{k=1}^{2}\sum_{l=1}^{2} Wmax_{ikl}}.$$

3.4 Algorithm

Fig. 7 gives the procedure for solving CSP defined in section 3.3. The procedure is basically executed for each group of adjacent signals. Parameters fixed by the procedure are applied to each signal after the current cycle is terminated. In each intersection agent, checking constraints in C_{road} is first executed while congestion information is received from connected road agents. Then, checking constraints in C_{in} and C_{adj} are executed according to the priority value, P_value_{ik} for each signal, S_{ik}.

procedure constraint_check()
  for (all intersections with signals)
    check constraints in C_{road};
  end for
  while (all constraints are not satisfied)
    check constraints in C_{in} according to the priority value;
    check constraints in C_{adj} according to the priority value;
  end while
  apply the arranged values to signals;
end.

Fig. 7. The constraint satisfaction process

4 Experiments

We implement the proposed control method in our developed traffic simulator and conduct some experiments. In experiments, we use two roadnetworks as shown in Fig. 8, where the roadnetwork A contains 25 nodes, or intersections, and 40 edges, or roads and the roadnetwork B contains 121 nodes and 202 edges. The distance between intersections is 100 meters in both roadnetworks. Table 1 summarizes each experimental

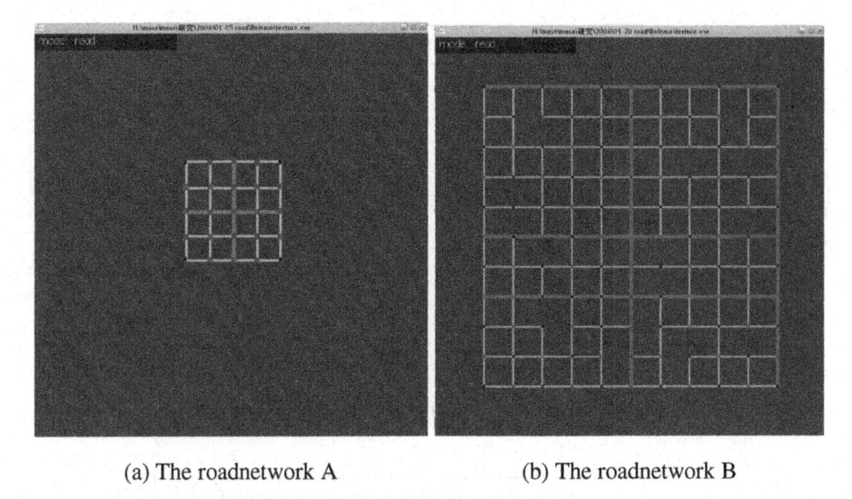

(a) The roadnetwork A (b) The roadnetwork B

Fig. 8. Roadnetworks used in experiments

Table 1. Conditions of experiments

roadnetwork	No. of pairs of signals	No. of cars	execution time
roadnetwork A	21	100, 200	60*min*
roadnetwork B	60, 113	1000, 1300	30*min*

condition. As for evaluation, we measure length of traffic jams in the entire roadnetwork for each 30 second for the execution time in Table 1. We define length of traffic jams as total length of cars with a speed of 0 kilometers an hour, which approach to intersections.

Fig. 9 gives the experimental result in the roadnetwork A. Figs. 10 and 11 give the experimental results in the roadnetwork B. All results show that traffic jams could be reduced due to applying our method. Comparing Fig. 10 which has 60 pairs of signals with Fig. 11 which has approximately doubled pairs of signals, 113, an increase in the number of signals tends to increase traffic jams when no control methods are applied. In contrast, by applying our method as a control strategy, length of traffic jams seems not to increase, although it can be supposed that an increase in the number of signals causes traffic jams.

5 Related Work

As for traffic signal control methods adopting intelligent approaches, there has been some studies. In [2], as well as our study, controlling signals is defined as multi-agent systems, each of which intersection agent determines *cyclelength* and *split* by making use of some rules. It may be difficult for this method to fix *offset* because the intersection agent is not communicated to the other one. In [20], genetic algorithms, GA,

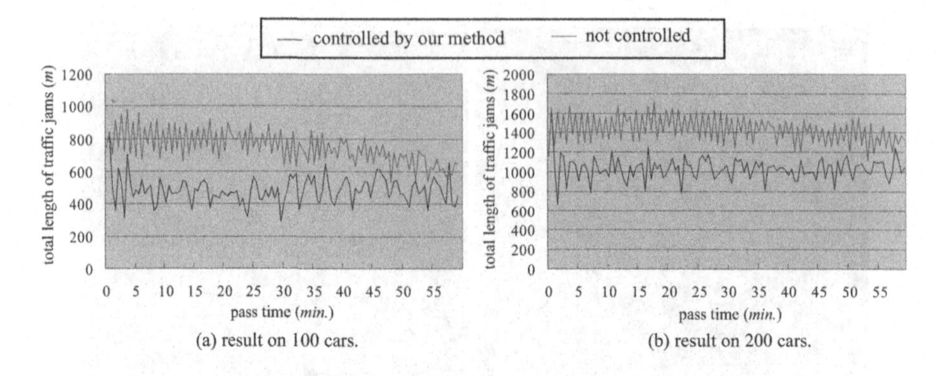

Fig. 9. Experimental result on the roadnetwork A

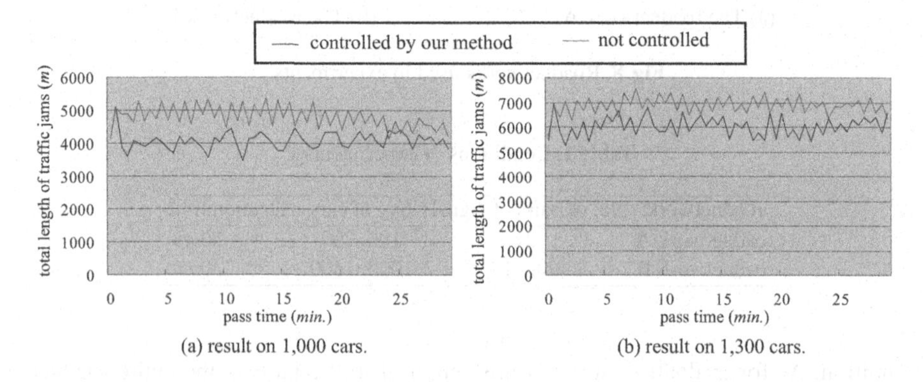

Fig. 10. Experimental result on the roadnetwork B, where there are 60 pairs of signals, i.e. 60 intersections have signals

are used to fix the parameters. This method puts states, or colors of signals, of each time period of all signals into a chromosome code and tries to minimize the number of cars which is stopping. Therefore, increasing in the number of signals causes inefficient search performance because of the enormous size of the chromosome.

On the other hand, in Japan, due to development and introduction of intelligent transportation systems, ITS [12], advanced traffic management by getting more precise traffic data has been becoming feasible. It may also be easy to control traffic by universal traffic management systems, UTMS [17]. UTMS, which is composed of some subsystems, is a kind of ITS systems and can communicate with individual cars through sensors or radio beacons put at roads, not only providing traffic information to drivers but also aggressively managing travel and distribution information.

Usual methods to control traffic signals are conducted based on data obtained for a few minutes ago. Therefore, these methods can be inconvenience because there is difference between control measures to be conducted and current traffic conditions, i.e., delays of control happen to a greater or lesser extent. UTMS can be available to solve the problem of delays. The characteristics of controlling signals using UTMS are

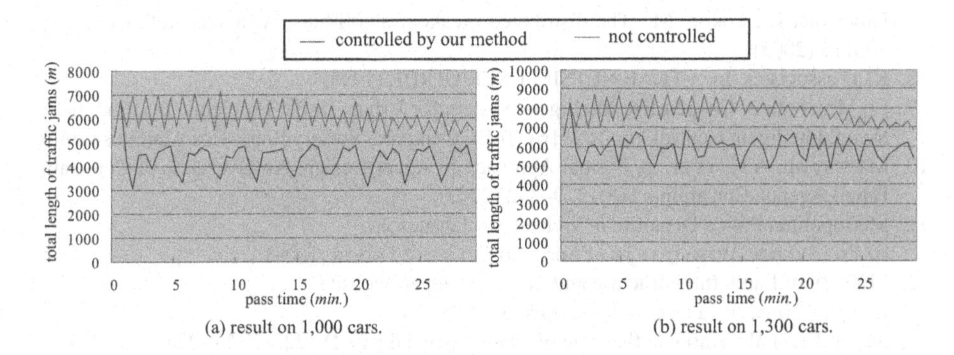

(a) result on 1,000 cars.

(b) result on 1,300 cars.

Fig. 11. Experimental result on the roadnetwork B, where there are 113 pairs of signals, i.e. 113 intersections have signals

'prediction of traffic demands', 'real-time control', and 'distributed-type decision of control'. The control methods using UTMS are discussed and trially applied to some areas in Japan [11].

6 Conclusion

We have proposed the method that dynamically control urban signals according to the congestion situation in roadnetworks. Our method represents adjusting control parameters as the CSP. The CSP is extended to distributed CSP, in which intersection agents have some local variables and attempt to resolve local congestion, resulting in reducing entire jams. We implemented our method to the agent-oriented traffic flow simulator, in which experimental simulation demonstrated that our method can reduce traffic jams. Our future works should consist in applying our mehotd to more realistic roadnetworks and extending our method by including deciding revelation of signals. It should be important to make our method cooperate UTMS or ITS systems.

References

1. Barrett, C.L., et. al.: TRANSIMS (TRansportation ANalysis SIMulation) volume 0 – Overview. Tech. Rep., LA-UR-99-1658 (1999)
2. Chen, R.S., Chen, D.K., Lin, S.Y.: ACTAM: Cooperative Multi-Agent System Architecture for Urban Traffic Signal Control. IEICE Trans. Inf. & Syst. E88-D(1), 119–126 (2005)
3. Dia, H.: An agent-based approach to modelling driver route choice behaviour under the influence of real-time information. Transportation Research Part C 10, 331–349 (2002)
4. Ehlert, P.A.M., Rothkrantz, L.J.M.: Microscopic traffic simulation with reactive driving agents. In: 2001 IEEE Intelligent Transportation Systems Conference Proceedings, pp. 861–866. IEEE Computer Society Press, Los Alamitos (2001)
5. Fernandes, J.M., Oliveira, E.: TraMas: Traffic Control through Behaviour-based Multi-Agent System. In: Proc. PAAM 1999, pp. 457–458 (1999)
6. Helbing, D., et al.: Micro- and Macro-Simulation of Freeway Traffic. Mathematical and Computer Modelling 35, 517–547 (2002)

7. Hirayama, K., Yokoo, M.: The distributed breakout algorithms. Artficial Intelligence 161, 89–115 (2005)
8. KLD associates, Inc.: TRAF NETSIM USER GUIDE (1995)
9. Li, M., et al.: A Cooperative Intelligent System for Urban Traffic Problems. In: Proc. IEEE Symp. Intelligent Control, pp. 162–167. IEEE Computer Society Press, Los Alamitos (1996)
10. Mackworth, A.K., et al.: A Formal Approach to Agent Design: An Overview of Constraint Based Agents. Constraints 8(3), 229–242 (2003)
11. Metropolitan Police Department Website (in Japanese), `http://www.npa.go.jp/koutsuu/kisei/index.html`
12. Ministry of Land, Infrastructure and Transport Japan Website, `http://www.mlit.go.jp/road/ITS/`
13. Nagel, K., et al.: Traffic at the edge of chaos. Artficial Life IV, 222–235 (1994)
14. Nagel, K., et. al.: TRANSIMS for Urban planning. Tech. Rep., LA-UR-98-4389 (1999)
15. Nareyek, A.: Constraint-Based Agents. LNCS (LNAI), vol. 2062. Springer, Heidelberg (2001)
16. Raney, B., et al.: Large Scale Multi-Agent Transportation Simulations. In: 42nd ERSA (European Regional Science Association) Congress (2002)
17. Universal Traffic Manegement Society of Japan Website, `http://www.utms.or.jp/english/index.html`
18. Yamada, M., Mizuno, K., Fukui, Y., Nishihara, S.: An Agent-Based Approach to Urban Traffic Flow Simulation. In: IWAIT 2006. Proc. International Workshop on Advanced Image Technology 2006, pp. 502–507 (2006)
19. Yamada, K., Samamaki, M., Mizuno, K., Fukui, Y., Nishihara, S.: Automatic Generation of Building Shapes for City Views. In: IWAIT 2006. Proc. International Workshop on Advanced Image Technology 2006, pp. 484–489 (2006)
20. Yamamoto, N., Morishita, S.: Simulation of Traffic Flow by Cellular Automata and its Control (in Japanese). Journal of The Japan Society of Mechanical Engineers (C) 65(637), 3553–3558 (1999)
21. Yokoo, M., Hirayama, K.: Distribute Constraint Satisfaction Algorithm for Complex Local Problems (in Japanese). Journal of Japan Society for Artificial Intelligence 12(2), 348–354 (2000)
22. Wiering, M., et al.: Intelligent Traffic Light Control. Tech. Rep., UU-CS-2004-029 (2004)

Convergence Analysis on Approximate Reinforcement Learning

Jinsong Leng[1], Lakhmi Jain[1], and Colin Fyfe[2]

[1] School of Electrical and Information Engineering,
Knowledge Based Intelligent Engineering Systems Centre,
University of South Australia, Mawson Lakes SA 5095, Australia
Jinsong.Leng@postgrads.unisa.edu.au,
Lakhmi.Jain@unisa.edu.au

[2] Applied Computational Intelligence Research Unit,
The University of Paisley, Scotland
Colin.Fyfe@paisley.ac.uk

Abstract. Temporal difference (TD) learning is a form of approximate reinforcement learning using an incremental learning updates. For large, stochastic and dynamic systems, however, it is still on open question for lacking the methodology to analyse the convergence and sensitivity of TD algorithms. Meanwhile, analysis on convergence and sensitivity of parameters are very expensive, such analysis metrics are obtained only by running an experiment with different parameter values. In this paper, we utilise the TD(λ) learning control algorithm with a linear function approximation technique known as tile coding in order to help soccer agent learn the optimal control processes. The aim of this paper is to propose a methodology for analysing the performance for adaptively selecting a set of optimal parameter values in TD(λ) learning algorithm.

Keywords: Approximate reinforcement learning, Agent, Convergence.

1 Introduction

An agent can be defined as a hardware and/or software-based computer system displaying the properties of autonomy, social adaptness, reactivity, and proactivity [11]. Most of agent based systems can be modeled as Markov Decision Processes (MDPs) [3,6]. Learning what action to take via interacting with the environment without external instruction is a basic ability for an agent.

TD learning [8] is a kind of unsupervised learning method for pursuing a goal through interacting with the environment. TD depends on the ideas of asynchronous dynamic programming (DP) [4]. The performance of TD algorithms is sensitive not only to the generation of the linear approximation function, but to the parameter values as well. The success and efficiency of TD learning algorithms depend on the way for representing the state space and the values of the parameters. The convergence of reinforcement learning (RL) techniques is an open problem.

Z. Zhang and J. Siekmann (Eds.): KSEM 2007, LNAI 4798, pp. 85–91, 2007.

In this paper, we adopt SoccerBots [1] as the learning environment. Sarsa(λ) learning algorithm with a linear function approximation technique known as tile coding [2] is used to help soccer agents learn the competitive or cooperative skills. The aim of this paper is to derive a methodology for adaptive selecting the optimal parameter values. The contribution of this paper is to propose a methodology for finding the optimal parameter values, which can result in significantly enhancing the performance of TD learning algorithms.

The rest of the paper is organised as follows: Section 2 introduces the TD techniques and state space representation, and then we introduce the methodology to derive the optimal parameter values. Section 3 presents the major properties of simulation and the experimental results. Finally, we discuss future work and conclude the paper.

2 Approximate Learning Techniques and Convergence Analysis

2.1 Approximate Learning Techniques

DP is a model-based approach, which means the state space and state transition probabilities are known in advance. DP uses stochastic optimisation techniques to compute the long term value function, and then deriving the policy to map states into actions. For most RL problems, however, the model of the environment is unknown in priori. Monte Carlo method is a model-free approach updating the state value using sample episodes, which enables an agent to learn in unknown environments. The value function is updated at the end of each episode.

TD(λ) computes the accumulated value (the mean or discounted rewards) using sample backup, which is based on dynamic programming and Monte Carlo techniques. Normally, TD(λ) represents that TD is combined with another mechanism known as eligibility traces. The one-step update at a given time t and state s is shown in equation (1).

$$V_{t+1}(s) = V_t(s) + \alpha\left[r_{t+1}(s) + \gamma V_{t+1}(s') - V_t(s)\right] \tag{1}$$

where the parameter γ is the discount rate, $0 \leq \gamma \leq 1$. The rewards in the future are geometrically discounted by the parameter γ.

Utilising the eligibility traces is another important mechanism for enhancing learning performance [8,10]. The parameter λ indicates the traces decay factor is considered for every sample episode. For the backward view TD prediction, the accumulating traces can be defined as:

$$e_t(s,a) = \begin{cases} \gamma\lambda e_{t-1}(s,a), & \text{if } s \neq s_t \\ \gamma\lambda e_{t-1}(s,a) + 1, & \text{if } s = s_t \end{cases} \tag{2}$$

whereas replacing traces use $e_t(s,a) = 1$ for the second update. In most cases, replacing traces can cause better performance significantly than accumulating traces [9].

Then the incremental state value equation (1) can be updated as follow:

$$\Delta V_t(s_t) = \alpha\big[r_{t+1} + \gamma V_t(s_{t+1}) - V_t(s_t)\big] e_t(s) \tag{3}$$

The value function is enumerated by accumulating temporal difference errors until the value function converges.

The tile coding method updates the state values using a linear approximation function, in which includes a number of tilings that they are offset over the state space. Each tile splits the state space into cells and only can have one cell to be activated. The value function is a summation of all tilings.

2.2 Convergence Analysis

The performance metrics are obtained by simulating an experiment a number of replications, which can be used for analysing the speed of convergence or the eventual convergence to optimality.

The convergence of TD(λ) has been proved by Dayan and Sejnowski [5], indicating that TD(λ) converges with probability one, under the standard stochastic convergence constraints on the learning rate and some conditions. The convergence may drastically depend on the parameter values in the learning algorithms, such as the learning rate α, the discount rate γ, the decay-rate λ, and the greedy action selection policy ϵ [7].

The numerical data about the performance can only be obtained via rerunning the simulator with different parameter values. Such performance data can be treated as performance metrics for analysing convergence and sensitivity. These datasets include the eventual state value the accumulated rewards during the learning period.

Ideally, the dataset can be generated using all possible parameter values and then be used as performance metrics to detect the optimal parameter values. However, the performance metrics are very expensive to be obtained. To obtain a data needs to run an experiment with a number of replications. The performance metrics are interpreted to analyse the quality of convergence.

3 Case Study

We utilise SoccerBots [1] as the testbed. We simplify the simulation environment and define a learning problem called ball interception skill. The state space is represented using the distance from the ball to the player, the angle from ball to the player, the velocity of the ball and the angle of the ball. We use the linear approximation technique known as tile coding to compute the state values. In this paper, we use Sarsa (λ) algorithm in conjunction with tile coding to learn the interception ball skill.

Sarsa (λ) algorithm has four parameters, e.g. ϵ, α, γ, and λ. For simplicity, we set ϵ to 0.1 and λ to 0.9 for all the experiments. The dataset of rewards is generated by changing the values of α, γ, as shown in Table 1.

By running the simulation system 2500 times for every set of parameters, we obtain the dataset by tuning the parameters α and γ, as illustrated in Table 1.

Table 1. α, γ and rewards

α, γ	0.001	0.005	0.01	0.015	0.02
0.99	-30.58	-31.84	-29.47	-33.25	-35.64
0.97	-39.5	-27.36	-29.06	-29.24	-30.49
0.95	-26.14	-23.54	-24.46	-25.01	-26.15
0.93	-24.48	-21.62	-23.78	-25.17	-24.3
0.91	-37.37	-22.63	-33.14	-24.48	-28.89

We draw the three dimension diagram and find the approximate optimal parameter values using Meshc function. The maximum value area is indicated by the contour on the parameters α and γ plane, shown in Fig. 1.

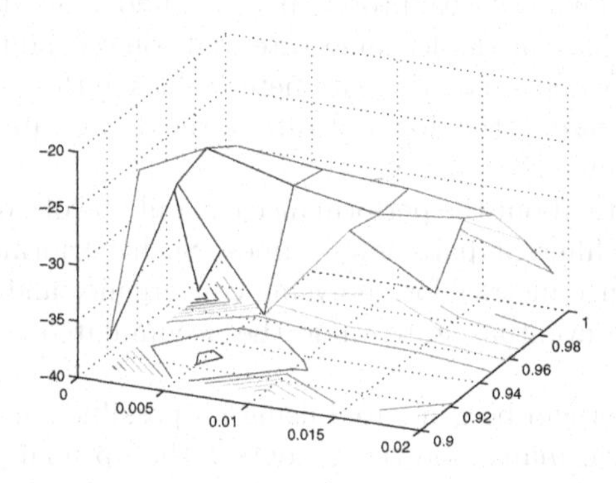

Fig. 1. Sarsa(λ): the relationship among α, γ, and rewards

The diagram illustrates that the approximate optimal values of α and γ are 0.005 and 0.93 respectively.

Next, we reconstruct the data in Table 1 and obtain a new three columns table (α, γ, R). The new correlation coefficients metric between α, γ and reward R is calculated as follow:

$$\begin{pmatrix} 1.0000 & -0.0000 & 0.0838 \\ -0.0000 & 1.0000 & -0.4084 \\ 0.0858 & -0.3979 & 1.0000 \end{pmatrix} \quad (4)$$

where $r_{\alpha\gamma}$ = -0.0000, $r_{\alpha R}$ = 0.0838, and $r_{\gamma R}$ = -0.4084.

We set the parameter baseline as: $\alpha = 0.005$, $\gamma = 0.093$, $\lambda = 0.9$, and $\epsilon = 0.1$. Because there is non-linear relationship between α and γ, we can refine the optimal parameter values separately by changing one parameter values.

Firstly, we generate the dataset by adjusting the values of α, as shown in Table 2.

Table 2. α and rewards

0.001	0.003	0.005	0.007	0.01	0.013	0.015	0.02
-24.48	-25.98	-21.62	-20.53	-23.78	-22.09	-25.17	-24.3

Fig. 2 (a) illustrates the polynomial relationship between α and rewards.

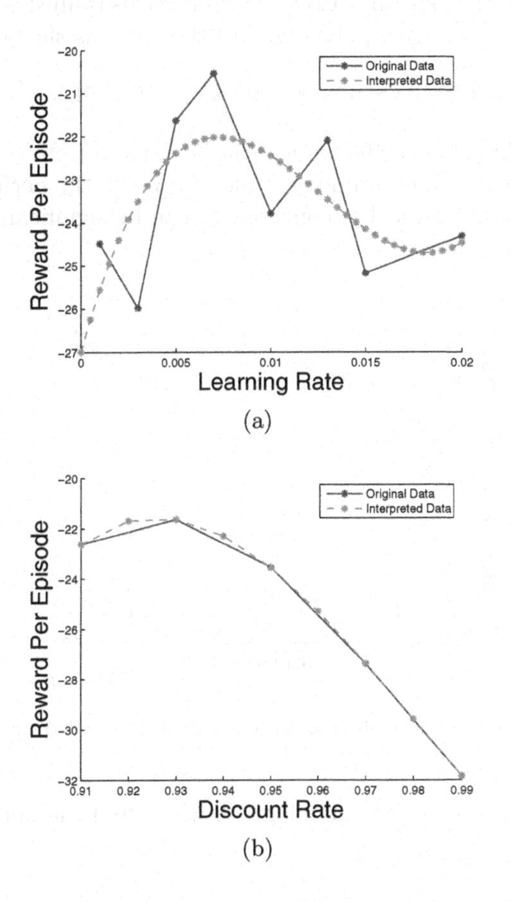

Fig. 2. The polynomial relationship between (a) α and rewards. (b) γ and rewards.

We can use the third-degree polynomial function to model the dataset, as shown in (5):

$$R(\alpha) = 10^6 \times (4.0376 \times \alpha^3 - 0.1540 \times \alpha^2 + 0.0016 \times \alpha^1) \tag{5}$$

The optimal value of α is 0.007, which can be derived from (5).

Secondly, we can obtain the dataset by adjusting the values of γ, as shown in Table 3.

Table 3. γ and rewards

0.91	0.93	0.95	0.97	0.99
-22.63	-21.63	-23.54	-27.36	-31.84

The polynomial relationship between γ and rewards is illustrated in Fig. 2 (b). We can fit the third-degree polynomial to the data, as shown in (6):

$$R(\gamma) = 10^4 \times (2.3438 \times \gamma^3 - 6.9095 \times \gamma^2 + 6.7671 \times \gamma^1 - 2.2047) \qquad (6)$$

The optimal value of γ is 0.93, which can be derived from (6).

By ultilising the optimal parameter values obtained, the performance such as the speed and eventual reward of convergence can be significantly improved, as indicated in Fig. 3.

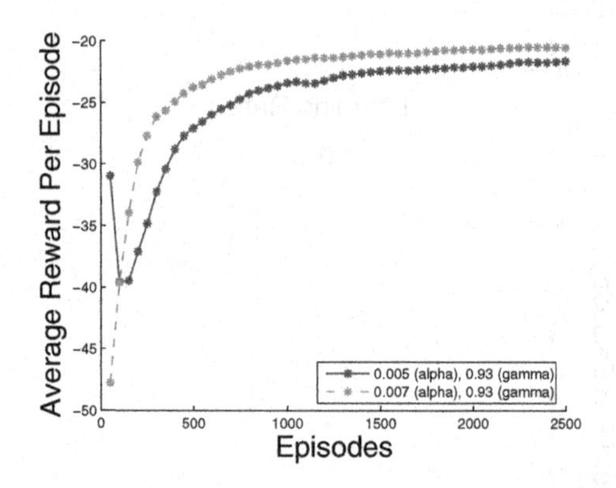

Fig. 3. Sarsa(λ): the convergence with different α and γ

Similarly, we can use the methodology above to find the optimal values of λ and ϵ.

4 Conclusion and Future Work

The performance metrics are not only expensive to be obtained, but also difficult to be interpreted. This paper provides a methodology to analyse and distinguish the optimal parameter values from those performance metrics. The experimental results have demonstrated that the converging speed and eventual reward to optimality can be improved by using the optimal parameter values found. The ultimate goal is to develop a methodology for adaptively choosing the parameter values, so as to enhance the performance and efficiency of the learning algorithms.

References

1. Teambots (2000), http://www.cs.cmu.edu~trb/Teambots/Domains/SoccerBots
2. Albus, J.S.: A Theory of Cerebellar Function. Mathematical Biosciences 10, 25–61 (1971)
3. Bellman, R.: A Markovian Decision Process. Journal of Mathematics and Mechanics 6 (1957)
4. Bellman, R.: Dynamic Programming. Princeton University Press, Princeton, NJ (1957)
5. Dayan, P., Sejnowski, T.J.: TD(λ) Converges with Probability 1. Machine Learning 14(1), 295–301 (1994)
6. Howard, R.A.: Dynamic Programming and Markov Processes. MIT Press, Cambridge (1960)
7. Leng, J., Jain, L., Fyfe, C.: Simulation and Reinforcement Learning with Soccer Agents. In: Journal of Multiagent and Grid systems, vol. 4(4), IOS Press, The Netherlands (to be published in 2008)
8. Sutton, R.S.: Learning to Predict by the Method of Temporal Differences. Machine Learning 3, 9–44 (1988)
9. Sutton, R.S., Barto, A.G.: Reinforcement Learning: An Introduction. MIT Press, Cambridge (1998)
10. Watkins, C.J.C.H.: Learning from Delayed Rewards. PhD thesis, Cambridge University, Cambridge, England (1989)
11. Wooldridge, M., Jennings, N.: Intelligent Agents: Theory and Practice. Knowledge Engineering Review 10(2), 115–152 (1995)

Combinative Reasoning with RCC5 and Cardinal Direction Relations*

Juan Chen, Dayou Liu**, Changhai Zhang, and Qi Xie

College of Computer Science and Technology, Jilin University, Changchun 130012, China
Key Laboratory of Symbolic Computation and Knowledge Engineering of Ministry of Education, Jilin University, Changchun 130012, China
dyliu@jlu.edu.cn

Abstract. It is inadequate considering only one aspect of spatial information in practical problems, where several aspects are usually involved together. Reasoning with multi-aspect spatial information has become the focus of qualitative spatial reasoning. Most previous works of combing topological and directional information center on the combination with MBR based direction model or single-tile directions. The directional description is too approximate to do precise reasoning. Different from above, cardinal direction relations and RCC5 are introduced to represent directional and topological information. We investigate the mutual dependencies between basic relations of two formalisms, discuss the heterogeneous composition and give the detail composing rules. Then point out that only checking the consistency of topological and directional constraints before and after entailing by the constraints of each other respectively will result mistakes. Based on this, an improved constraint propagation algorithm is presented to enforce path consistency. And the computation complexities of checking the consistency of the hybrid constraints over various subsets of RCC5 and cardinal direction relations are analyzed at the end.

1 Introduction

Spatial representation and reasoning are important in many application areas such as geographic information systems, robotics, and computer vision etc. While pure quantitative approach can be beneficial in many cases, it is an inappropriate representation of human cognition and spatial reasoning. This is partially because humans usually use qualitative information, and the numeric information is usually unnecessary and unavailable at human level. The qualitative approach to spatial

* Supported by NSFC Major Research Program 60496321, National Natural Science Foundation of China under Grant Nos. 60373098, 60573073, 60603030, the National High-Tech Research and Development Plan of China under Grant No. 2006AA10Z245, the Major Program of Science and Technology Development Plan of Jilin Province under Grant No. 20020303, the Science and Technology Development Plan of Jilin Province under Grant No. 20030523, European Commission under Grant No. TH/Asia Link/010 (111084).
** Corresponding author.

Z. Zhang and J. Siekmann (Eds.): KSEM 2007, LNAI 4798, pp. 92–102, 2007.

reasoning is known as Qualitative Spatial Reasoning (QSR for short)[1]. QSR usually use algebraic or logistic method to describe the topology, distance, direction, shape and size information of spatial objects; furthermore it also study composite reasoning, constraint satisfaction problem (CSP) and qualitative simulation etc. reasoning problems. Most previous work focused on single aspect of spatial objects and developed a set of formal models, but real problems often involve several aspects of space. It is inadequate considering only one aspect of spatial information in practical problems, moreover different aspects of space are often inter-dependent, so reasoning with multi-aspect spatial information has become the focus of QSR.

To combine topological and directional information, Sharma [2] approximated the region as its minimal bounding box (MBR) and represented the topological and directional relations as interval relations formed by the projections of the MBRs along two axes, identified the composition tables between topological and directional relations. Sistla and Yu [3] defined a mixed model only including the rudimental topological and directional relations: {*left_of*, *right_of*, *behind*, *in_front_of*, *above*, *below*, *inside*, *outside*, *overlaps*}. Recently Li [4] [5] combined RCC5 and RCC8 with DIR9 direction calculus respectively, discussed the CSP and analyzed the reasoning complexity. It worth to mention that all the direction descriptions in these works are constrained in the direction calculi based on the MBRs or single-tile directions which is too approximate to do precise reasoning. We based our work on a different direction model from above, the cardinal direction relation (CDR) [6] [7]. Since CDR is border insensitive, RCC5 is selected to describe the topological information.

In this paper we analyze the dependency between the basic relations of two formalisms, discuss the heterogeneous composition of RCC5 and CDR, and present an improved path-consistent algorithm. The rest of paper is organized as follow: Section 2 and 3 introduce the notion and terminologies of CDR and RCC5 respectively. Section 4 is the analysis of the dependency between the two formalisms. Section 5 elaborates the detail composing rules of CDR and RCC5. Section 6 discusses the reasoning problems. Without any special notation, all regions mentioned below are simple regions.

2 Cardinal Direction Relations

Direction is an important aspect of space and has been investigated a lot, such as the Frank's cone-based and projection-based methods [8] and Freska's double-cross method[9] for describing the direction between points; Balbiani's rectangle algebra [10]and Li's DIR9 model [4] [5] for describing the direction between rectangles. Another approach representing the direction between extended object is CDR [6] [7]. Unlike above approaches, this approach doesn't approximate a region as a point or a MBR, which makes it more expressive.

A cardinal direction relation is a binary relation involving a target object and a reference object, and a symbol that is a non-empty subset of nine atomic relations whose semantics are motivated by the nine tiles divided by the MBR of the reference object b: northwest--$NW(b)$, north--$N(b)$, northeast--$NE(b)$, west--$W(b)$, east--$E(b)$, southwest--$SW(b)$, south--$S(b)$, southeast--$SE(b)$ and same--$O(b)$, i.e., the MBR of b,. And only 218 basic direction relations, denoted by $\mathcal{D}$, can be realized out of 2^9=512 possible combinations for the connected regions, which is a set of JEPD (Jointly

Exhaustive and Pairwise Disjoint) relations. A JEPD relation means that for any two definite regions, only one relation can be satisfied in the relation set. Given region *a*, *b*, let *dir*(*a*, *b*) represent the direction of *a* with respect to *b* as shown in Fig. 1 *dir*(*a*, *b*)= *NW*:*N*:*W*.

Definition 1. A basic cardinal direction relation is an expression R_1:...:R_k where

- $1 \leq k \leq 9$
- $R_i \neq R_j$ ($1 \leq i, j \leq k$)
- $R_1, \ldots, R_k \in \{NW, N, NE, W, E, SW, S, SE, O\}$;
- There exists simple regions $a_1, \ldots, a_k$ such that $a_1 \in R_1(b), \ldots, a_k \in R_k(b)$ and $a = a_1 \cup \ldots \cup a_k$ is still a simple region for any simple reference region *b*.

A basic cardinal direction R_1:...:R_k is called single-tile or atomic if k=1, otherwise it is called multi-tile.

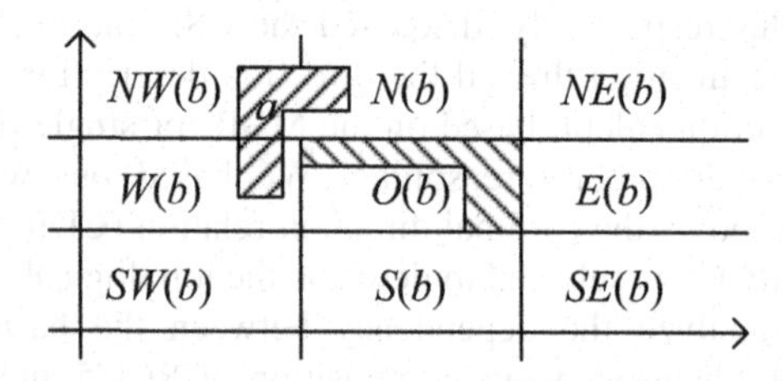

Fig. 1. Pictorial example of *dir*(*a*, *b*)=*NW*:*N*:*W*

3 The Region Connection Calculus RCC5

Topology is a fundamental aspect of space. A well known framework is the so-called RCC (Region Connection Calculus) [11], and in particular RCC8 and RCC5 are the two best known approaches. Since the direction model selected here doesn't discriminate the boundary of spatial entities. The boundary insensitive RCC5 is selected to describe the topological information.

For two regions *a* and *b*, a^o and b^o denote the interior of *a* and *b* respectively. Then have

- *a* is equal to *b*, denoted by *a EQ b* iff $a=b$
- *a* is part of *b*, denoted by *a P b* iff $a \subseteq b$
- *a* is proper part of *b*, denoted by *a PP b* iff $a \subset b$
- *a* overlaps *b*, denoted by $a^o \cap b^o \neq \emptyset$
- *a* is discrete from *b*, denoted by *a DR b* iff $a^o \cap b^o = \emptyset$
- *a* partially overlaps *b*, denoted by *a PO b* iff $a \subsetneq b$, $b \subsetneq a$ and $a^o \cap b^o \neq \emptyset$

The subalgebra generated by above relations is known as RCC5 algebra containing five basic relations: *a EQ b* (*a* is equal to *b*), *a DR b* (*a* is discrete from *b*), *a PO b* (*a* partially overlaps *b*), *a PP b* (*a* is a proper part of *b*) and *a PPI b* (*b* is a proper part of *a*), the inverse of *PP*; which is the subalgebra of RCC8. An RCC5 relation is written as a set of basic relations, e.g., {*PO*, *PP*} denotes the union of the two basic relations *PO* and *PP*. The universal relation, denoted by $\mathcal{T}$, is the union of all basic relations.

4 Mutual Dependencies Between Basic Relations of RCC5 and CDR

The topological and direction relations of two regions are not independent but influenced reciprocally. For example, two regions *a* and *b*, if *a PP b* is satisfied then the direction *dir(a, b)=O*; while if *dir(a, b)=NW:N*, *a DR b* must be satisfied. According to the definitions of RCC5 and CDR, it is clear that the main interaction of two formalisms focuses on the direction tile *O*, which is the MBR of reference region. We define the concept of including and intersecting of CDR to discriminate basic cardinal direction relations. Table1 gives the possible basic directional (topological respectively) relations between region *a*, *b* when a certain topological (directional respectively) relation is satisfied.

Definition 2. Let $R_1=R_{11}:\ldots:R_{1n}$ and $R_2=R_{21}:\ldots:R_{2m}$ be two basic cardinal direction relations; R_2 includes R_1, denoted by $R_1 \subseteq R_2$ iff $\{R_{11},\ldots, R_{1n}\} \subseteq \{R_{21},\ldots, R_{2m}\}$holds; R_2 properly includes R_1,denoted by $R_1 \subset R_2$ iff$\{R_{11},\ldots, R_{1n}\} \subset \{R_{21},\ldots, R_{2m}\}$holds.

Example 1. The basic cardinal direction relation *NW:W:N* includes and properly includes relation *NW:W*.

Definition 3. Let $R_1=R_{11}:\ldots:R_{1n}$ and $R_2=R_{21}:\ldots:R_{2m}$ be two basic cardinal direction relations; R_3 is the intersection of R_1 and R_2 denoted by $R_1 \cap R_2=R_3$ iff $\{R_{11},\ldots, R_{1n}\} \cap \{R_{21},\ldots, R_{2m}\}=\{R_{31},\ldots, R_{3k}\}$ and $R_3=R_{31}:\ldots:R_{3k}$ holds.

Example 2. The intersection of basic cardinal direction relations *NW:N:O* and *W:N:O* is *N:O*.

Table 1. The interactive table between basic relations of RCC5(*t*)and CDR(*d*)

t	*Direl(t)*	*d*	*Toprel(d)*
a DR b	$\models dir(a, b) = \mathcal{D}$	$dir(a, b)=O$	$\models \mathcal{T}$
a PO b	$\models O \subseteq dir(a, b)$	$O \subset dir(a, b)$	$\models$ *DR, PO, PPI*
a PP b	$\models dir(a, b)= O$	$O \subsetneq dir(a, b)$	$\models$ *DR*
a PPI b	$\models O \subseteq dir(a, b)$		
a EQ b	$\models dir(a, b)=O$		

5 Heterogeneous Compositions

Originate from Allen [12] first using composition table to process time interval constraints, composing has become the key technique in QSR. Given a set *Rels* of basic relation symbols, composition is a function: $Rels \times Rels \rightarrow 2^{Rels}$. Let R_1 and R_2 be the relations of *Rels*, the composition of R_1 and R_2, denoted by $R_1 \circ R_2$ is another relation from 2^{Rels} which contains all relations T such that there exist objects a, b and c such that $a\ R_1\ b$, $b\ R_2\ c$ and $a\ T\ c$ hold. If the two relations come from two different domains, we call it heterogeneous composition such as composing RCC5 with CDR. The results can be not only topological relations but also the cardinal direction relations. When the result is topological, composing is easy to do; while the result is directional, the directional relations entailing from topological relations are too many to operate. So it is necessary to give simplified composing rules. To facilitate illustration, introduce following notions:

- $\delta(R_1,.., R_k)$ represents all rational basic cardinal directions formed by atomic relations $R_1,..,R_k$, such as $\delta(W, NW, SW)=\{W, NW, SW, W{:}NW, W{:}SW, W{:}NW{:}SW\}$, $NW{:}SW$ is not in the set for it cannot hold when the regions are simple.
- $\sigma(S_1,..,S_m)$ represents all rational basic cardinal directions formed by the cross joining the relation sets, such as $\sigma(\{O\},\{W{:}NW, W\}, \{SW,W\})=\{O{:}W{:}NW{:}SW, O{:}W{:}NW, O{:}W{:}SW, O{:}W\}$.
- a, b, c, x and y represent the simple regions.
- $inf_x(a)$and $inf_y(a)$ ($sup_x(a)$and $sup_y(a)$) represent the maximum lower bound (minim upper bound)of the projection of region a on x and y axis.

5.1 CDR$\circ$RCC5

Proposition 1. If $dir(a, b)=R$, $b\ DR\ c$, then $R \circ DR=\mathcal{D}$ holds.

Proof: Obvious. ∎

Proposition 2. If $dir(a, b)=R$, $b\ EQ\ c$, then $R \circ EQ = R$ holds.

Proof: Obvious. ∎

Proposition 3. If $dir(a, b)=R$, $b\ PP\ c$, then $R \circ PP=\sigma(Op(R), Np(R), Sp(R), Wp(R), Ep(R), NWp(R), NEp(R), SWp(R), SEp(R))$ holds. Where the lowercase p in $Np(R)$ represents R is composing with PP, similar with other symbols.

Proof: Since $b\ PP\ c$, it must have following order constraints: $sup_x(b) \le sup_x(c) \wedge sup_y(b) \le sup_y(c) \wedge inf_x(b) \ge inf_x(c) \wedge inf_y(b) \ge inf_y(c)$

- If $O \subseteq R$, which means $\exists x \subseteq a$ such that $inf_x(b) \le inf_x(x) \wedge sup_x(x) \le sup_x(b) \wedge inf_y(b) \le inf_y(x) \wedge sup_y(x) \le sup_y(b)$ then according to above order constraints $inf_x(c) \le inf_x(x) \wedge sup_x(x) \le sup_x(c) \wedge inf_y(c) \le inf_y(x) \wedge sup_y(x) \le sup_y(c)$ must be satisfied, which means $dir(x,c)=O \subseteq dir(a,c)$.
- If $N \subseteq R$, which means $\exists\ x \subseteq a$ such that $sup_x(x) \le sup_x(b) \wedge sup_y(b) \le inf_y(x) \wedge inf_x(b) \le inf_x(x)$ hold; then basing on above order constraints $sup_x(x) \le sup_x(c) \wedge inf_x(x) \ge inf_x(c) \wedge inf_y(c) < inf_y(x)$ must be satisfied, which means $dir(x, c)=\delta(N, O) \subseteq$

$dir(a, c)$. Analogous results: $\delta(S,O) \subseteq dir(a,c)$, $\delta(W,O) \subseteq dir(a,c)$, $\delta(E,O) \subseteq dir(a, c)$ can get; when *S*, *W* or *E* is included in *R*.

- If $NW \subseteq R$, which means $\exists\ x \subseteq a$ such that $sup_x(x) \leq inf_x(b) \wedge sup_y(b) \leq inf_y(x)$ hold; then according to the order constraints of *b PP c*, $sup_x(x) < sup_x(c) \wedge inf_y(c) < inf_y(x)$ must be satisfied, which means $dir(x,c) = \delta(NW, W, N, O) \subseteq dir(a, c)$. Analogous results: $\delta(SW, W, S, O) \subseteq dir(a,c)$, $\delta(NE, E, N, O) \subseteq dir(a,c)$, $\delta(SE, S, E, O) \subseteq dir(a,c)$ can get; When *SW*, *NE* or *SE* is included in *R*.

To sum up, let $Np(R) = \varnothing$, when $N \subsetneq R$; otherwise $N \subseteq R$, $Np(R) = \delta(N, O)$. Similarly, we can define $NEp(R)$, $Wp(R)$, $Op(R)$, $Ep(R)$, $SWp(R)$, $Sp(R)$, $SEp(R)$. Then according to the definition of cardinal direction relations, $R \circ PP = \sigma(Op(R), Np(R), Sp(R), Wp(R), Ep(R), NWp(R), NEp(R), SWp(R), SEp(R))$ holds. ∎

Example 3. Let $R = NW{:}O{:}W$, then $O(R) = \{O\}$, $W(R) = \delta(W,O)$, $NW(R) = \delta(NW, W, N, O)$, $SW(R) = N(R) = S(R) = E(R) = NE(R) = SE(R) = \varnothing$, so $R \circ PP = \sigma(\{O\}, \delta(W,O), \delta(NW, W, N, O)) = \{O, O{:}W, O{:}N, O{:}NW{:}W, O{:}NW{:}N, O{:}W{:}N, O{:}NW{:}W{:}N\}$.

Proposition 4. If $dir(a, b) = R$, *b PPI c*, then $R \circ PPI = \sigma(Oi(R), Ni(R), Si(R), Wi(R), Ei(R), NWi(R), NEi(R), SWi(R), SEi(R))$ holds. Where the lowercase *i* in $Ni(R)$ represents *R* is composing with *PPI*, similar with other symbols.

Proof: Since *b PPI c*, it must have following order constraints $sup_x(c) \leq sup_x(b) \wedge sup_y(c) \leq sup_y(b) \wedge inf_x(c) \geq inf_x(b) \wedge inf_y(c) \geq inf_y(b)$.

- If $O \subseteq R$, which means $\exists x \subseteq a$ such that $inf_x(b) \leq inf_x(x) \wedge sup_x(x) \leq sup_x(b) \wedge inf_y(b) \leq inf_y(x) \wedge sup_y(x) \leq sup_y(b)$ holds, it can not give any constraints to the direction of *a* w.r.t *c*, so $dir(a, c) = \mathcal{D}$.
- If $NW \subseteq R$, which means $\exists x \subseteq a$ such that $sup_x(x) \leq inf_x(b) \wedge sup_y(b) \leq inf_y(x)$ holds; then $sup_x(x) \leq inf_x(c) \wedge sup_y(c) \leq inf_y(x)$ is satisfied, which means $dir(x, c) = NW \subseteq dir(a, c)$. Analogous results: $SW \subseteq dir(a, c)$, $NE \subseteq dir(a, c)$ and $SE \subseteq dir(a, c)$ can get when *SW*, *NE* or *SE* is included in *R*.
- If $N \subseteq R$, which means $\exists x \subseteq a$ such that $sup_x(x) \leq sup_x(b) \wedge sup_y(b) \leq inf_y(x) \wedge inf_x(b) \leq inf_x(x)$ holds; then $sup_y(c) \leq inf_y(x)$ is satisfied, which means $dir(x, c) = \delta(N, NW, NE) \subseteq dir(a, c)$. Analogous results: $\delta(S, SW, SE) \subseteq dir(a,c)$, $\delta(NW, N, NE) \subseteq dir(a, c)$ and $\delta(NE, E, SE) \subseteq dir(a, c)$ can get when *S*, *W* or *E* is included in *R*.

To sum up, let $Ni(R) = \varnothing$, when $N \subsetneq R$; otherwise $Ni \subseteq R$, $Ni(R) = \delta(NW, N, NE)$. Similarly, we can define $NEi(R)$, $Wi(R)$, $Oi(R)$, $Ei(R)$, $SWi(R)$, $Si(R)$, $SEi(R)$. Then according to the definition of cardinal direction relations, $R \circ PPI = \sigma(Oi(R), Ni(R), Si(R), Wi(R), Ei(R), NWi(R), NEi(R), SWi(R), SEi(R))$holds. ∎

Example 4. Let $R = NW{:}W{:}SW$, then $Wi(R) = \delta(W, NW, SW)$, $NWi(R) = \{NW\}$, $SWi(R) = \{SW\}$, $Oi(R) = Ni(R) = Si(R) = Ei(R) = NEi(R) = SEi(R) = \varnothing$; so $R \circ PPI = \sigma(\delta(W, NW, SW), \{NW\}, \{SW\}) = \{W{:}NW{:}SW\}$.

Proposition 5. If $dir(a, b)=R$, b PO c, then $R \circ PO = \sigma(Oo(R), No(R), So(R), Wo(R), Eo(R), NWo(R), NEo(R), SWo(R), SEo(R))$ holds. Where the lowercase o in $No(R)$ represents R is composing with PO, similar with other symbols.

Proof: Since b PO c, it must exist region $x \subset b \wedge x \subset c$ such that $inf_x(b) \le inf_x(x) \wedge inf_x(c) \le inf_x(x) \wedge inf_y(b) \le inf_y(x) \wedge inf_y(c) \le inf_y(x) \wedge sup_x(x) \le sup_x(b) \wedge sup_x(x) \le sup_x(c) \wedge sup_y(x) \le sup_y(b) \wedge sup_y(x) \le sup_y(c)$ holds.

- If $NW \subseteq R$, which means $\exists y \subseteq a$ such that $sup_x(y) \le inf_x(b) \wedge sup_y(b) \le inf_y(y)$ holds, then $sup_x(y) < sup_x(c) \wedge inf_y(c) < inf_y(y)$ i.e., $dir(y, c) = \delta(NW, N, W, O) \subseteq dir(a, c)$. Analogous results: $\delta(SW, S, W, O) \subseteq dir(a, c)$, $\delta(SE, S, E, O) \subseteq dir(a, c)$, $\delta(NE, N, E, O) \subseteq dir(a, c)$ hold, when SW, SE or NE is included in R.
- If $N \subseteq R$, which means $\exists y \subseteq a$ such that $sup_x(y) \le sup_x(b) \wedge sup_y(b) \le inf_y(y) \wedge inf_x(b) \le inf_x(y)$ holds, then $inf_y(c) < inf_y(y)$ i.e., $dir(y, c) = \delta(NW, N, NE, W, O, E) \subseteq dir(a, c)$. Analogous results $\delta(SW, S, SE, W, O, E) \subseteq dir(a, c)$, $\delta(NW, W, SW, N, O, S) \subseteq dir(a, c)$, $\delta(NE, E, SE, N, O, S) \subseteq dir(a, c)$ hold when S, W or E is included in R.
- If $O \subseteq R$, which means $\exists y \subseteq a$ such that $inf_x(b) \le inf_x(y) \wedge sup_x(y) \le sup_x(b) \wedge inf_y(b) \le inf_y(y) \wedge sup_y(y) \le sup_y(b)$ holds. It hasn't any constraints to the direction of a w.r.t c, so $dir(a, c) = \mathcal{D}$.

To sum up, let $No(R) = \varnothing$, when $N \subsetneq R$; otherwise $N \subseteq R$, $No(R) = \delta(NW, N, NE, W, O, E)$. Similarly, we can define $NEo(R)$, $Wo(R)$, $Oo(R)$, $Eo(R)$, $SWo(R)$, $So(R)$, $SEo(R)$. Then according to the definition of cardinal direction relations, $R \circ PO = \sigma(Oo(R), No(R), So(R), Wo(R), Eo(R), NWo(R), NEo(R), SWo(R), SEo(R))$. holds. ∎

Example 5. Let $R=NW{:}N$, then $NWo(R)=\delta(NW, N, W, O)$, $No(R)= \delta(NW, N, NE, W, O, E)$, $Oo(R)=SWo(R)=So(R)=Eo(R)=NEo(R) =SEo(R)=\varnothing$; so $R \circ PO=\sigma(\delta(NW, N, W, O), \delta(NW, N, NE, W, O, E))$.

5.2 RCC5 ∘ CDR

Proposition 6. If a DR b, $dir(b, c)=R$, then $DR \circ R = \mathcal{D}$ holds.

Proof: Obvious.

Proposition 7. If a EQ b, $dir(b, c)=R$, then $DR \circ R = R$ holds.

Proof: Obvious.

Proposition 8. If a PP b, $dir(b, c)=R$, then $PP \circ R = \{R' | R' \subseteq R, R' \ne \varnothing, R' \in \mathcal{D}\}$ holds.

Proof: a PP b means $a \subset b$. If R is single-tile, then $b \in R(c)$ the direction tile divided by reference object c, so $a \in R(b)$ i.e., $dir(a, c)=R$. Otherwise R is multi-tile, where k is the number of tiles included in R, there must exist a partition of b: $b_1, b_2, \ldots, b_k$ satisfies each atomic relation in R. Because a is the proper subset of b, there must exist some partitions: $\{b_{i1},\ldots,b_{in}\} \subseteq \{b_1,\ldots,b_k\}$ such that $(b_{i1}\cap a)\cup\ldots\cup(b_{in}\cap a)=a$, which means a satisfy the corresponding atomic relations that $b_{i1},\ldots,b_n$ defined. To sum up, when R is single-tile $PP \circ R = R$, otherwise R is multi-tile; $PP \circ R$ includes each CDR which is non-empty combinations of the atomic relations included in R. ∎

Example 6. Let *a PP b*, *dir(b, c)=NW:W:SW*, then *dir(a, c)={NW, W, SW, NW:W, W:SW, NW:W:SW}*. *a*, *b* and *c* are simple regions, so *NW:SW* is not rational.

Proposition 9. If *a PPI b*, *dir(b, c)=R*, then $PPI \circ R = \{R' \mid R' \supseteq R, R' \in \mathcal{D}\}$ holds.

Proof: Direct deduce from proposition 8. ∎

Proposition 10. If *a PO b*, *dir(b, c)=R*, then $PO \circ R = \{R' \mid R' \cap R \neq \emptyset, R' \in \mathcal{D}\}$ holds.

Proof: Since *a PO b*, there must exist a subset x, $x \subset c$ and $x \subset b$. With the proof of Proposition 8, $dir(x, c)=\{R' \mid R' \subseteq R\}$ must hold, then the intersection of *dir(a, c)* and *dir(b, c)* must nonempty. ∎

Example 7. Let *a PO b* and *dir(b, c)=NW:W:SW* then $dir(a, c)=\{R' \mid NW \subseteq R' \vee W \subseteq R' \vee SW \subseteq R', R' \in \mathcal{D}\}$.

6 Combinative Reasoning with RCC5 and CDR

Constraint satisfaction problem is a researching focus in spatial and temporal field. The most common CSP in QSR is the problem *RSAT*, i.e. checking the consistency of a set of given spatial constraints with the form of *x R y*, where *x*, *y* denote spatial objects and *R* is the relation between them. If the consistency of this set of constraints can be decided in polynomial time, it is called tractable. As the key technique in solving CSP, path-consistent algorithm has received lots attention in QSR. Renz [11] and Narrate [13] gave the tractable subsets $\hat{\mathcal{H}}_5$, $\mathcal{D}_{rec}$ in RCC5 and CDR respectively, whose consistency can be decided by path-consistent algorithm.

Example 8. Consider composite constraints between region *a*, *b* and *c*: $\theta=\{a\ PPI\ b, b\ PPIc, a\ PPI\ c\}$and directional constraints $\Sigma=\{dir(a, b)=W{:}O{:}E{:}SW{:}SE, dir(b, c)=W{:}O{:}E{:}SW{:}S{:}SE, dir(a, c)=W{:}O{:}E{:}SW{:}SE\}$.

θ and Σ represent topological and directional constraints over the same set of variables, the problem is to decide the path consistency of the composite constraints, i.e. $RSAT(\theta\cup\Sigma)$. Only checking path consistency of θ and Σ before and after entailing by the constraints of each other will result mistakes. As shown in Example 8, both θ and Σ are independently consistent before and after the entailments, but their union is not consistent: according to *a PPI b* and *dir(b, c)=W:O:E:SW:S:SE*, it can deduce that $dir(a, c) \in \{R' \mid W{:}O{:}E{:}SW{:}S{:}SE \subseteq R', R' \in \mathcal{D}\}$, while *W:O:E:SW:SE* does not belong to this set. Based on this, an improved algorithm of path-consistency is presented as follows

Algorithm: BIP-CON
Input: A set θ of RCC5 constraints, and a set Σ of cardinal direction constraints over the variables $x_1,\ldots, x_n$, besides x_i *EQ* x_i and $dir(x_i, x_i)=O, i=1,\ldots,n$.
Output: *false*, if $\theta\cup\Sigma$ is not path-consistent; *true*, otherwise $\theta\cup\Sigma$ is path-consistent
$Q\leftarrow\{(i,j)|i<j\}$;(i indicates variable x_i of θ, analogous for j, C_{ij} is the union of topological constraints t_{ij} and directional constraints d_{ij} between i, j.
while $Q\neq\varnothing$ *do* select and delete an arc (i, j) from Q;
    for $k\neq i, k\neq j$ ($k\in\{1,..,n\}$) *do*
        if BIRE (i, j, k) *then*
            if $C_{ik}=\varnothing$ *then* return false;
                else add(i, k) to Q
        if BIRE (k, i, j) *then*
            if $C_{kj}=\varnothing$ *then* return false;
                else add(k, j) to Q
Function: BIRE (i, k, j)
Input: three region variables i, k and j.
Output: *true*, if C_{ij} is revised; *false*, otherwise.
oldt:=t_{ij}; oldc:=d_{ij}
$t_{ij} := (t_{ij}\cap Toprel(d_{ij}))\cap((t_{ik}\cap Toprel(d_{ik}))\circ(t_{kj}\cap Toprel(d_{kj})))$
$d_{ij}:= d_{ij}\cap Direl(t_{ij})\cap(d_{ik}\circ t_{kj})\cap(t_{ik}\circ d_{kj})\cap(d_{ik}\circ d_{kj})$
$t_{ij}:=t_{ij}\cap Toprel(d_{ij})$ $d_{ij}:=d_{ij}\cap Direl(t_{ij})$
if (t_{ij}= oldt) *and* (d_{ij} = oldc) *then* return *false*;
$t_{ij}:= Converse(t_{ji})\cap t_{ij}$; $d_{ij}:=Converse(d_{ji})\cap d_{ij}$
Return *true*. ∎

Algorithm BIP-CON is formally analogous to classic path-consistent algorithm. If only consider one type of constraints (topological or directional constraints), BIP-CON also can decide the path-consistency. The main novelty of our algorithm lies in the function BIRE, the directional relation d_{ij} is not only revised by the CDR entailments of t_{ij} but also the heterogeneous compositions $t_{ik}\circ d_{kj}$ and $d_{ik}\circ t_{kj}$. It can easily get that the time and space complexity of BIP-CON is $O(n^3)$ and $O(n^2)$ respectively, where n is the number of variable involved in θ and Σ.

Theorem 1. Given a set θ of arbitrary constraints over RCC5 and a set Σ of constraints of $\mathcal{D}_{rec}$ on the variables involving in θ, if for any topological constraints $t_{ij}\in RCC5\backslash\hat{\mathcal{H}}_5$, the directional constraint d_{ij} doesn't contain basic cardinal direction relation O, then $RSAT(\theta\cup\Sigma)$ can be decided in $O(n^3)$ time.

Proof: According to Table 1, any topological relation entailed by cardinal direction relation in $\mathcal{D}_{rec}$ is contained in $\hat{\mathcal{H}}_5$. There are only four relations in $RCC5\backslash\hat{\mathcal{H}}_5$, they are $\{PP, PPI\}$, $\{DR, PP, PPI\}$, $\{PP, PPI, EQ\}$ and$\{DR, PP, PPI,EQ\}$. If given $dir(a, b)\supset O$ then the above four relations can be entailed to$\{PPI\}$, $\{DR, PPI\}$, $\{PPI\}$, $\{DR, PPI\}$ respectively, which are included in $\hat{\mathcal{H}}_5$. Else given $dir(a, b)\not\supseteq O$, the above four relations can be entailed to $\varnothing$,$\{DR\}$, $\varnothing$, $\{DR\}$ respectively, included in $\hat{\mathcal{H}}_5$ too. Besides, their union is also included in $\hat{\mathcal{H}}_5$. Since $\mathcal{D}_{rec}$ and $\hat{\mathcal{H}}_5$ are the two tractable subsets whose consistency can be decided by path-consistent algorithm, $RSAT(\theta \cup \Sigma)$ can be decided in $O(n^3)$.

Theorem 2. The complexity of $RSAT(\mathcal{D} \cup \hat{\mathcal{H}}_5)$ is $O(n^4)$.

Proof: Checking the consistency of topological constraints based on $\hat{\mathcal{H}}_5$ is in $O(n^3)$ time, and checking the consistency of directional constraints based on $\mathcal{D}$ has been shown to be $O(n^4)$[14]. From Table 1, all possible basic cardinal direction relations can only entail $\mathcal{T}$, {*DR, PO, PPI*} or {*DR*} which belong to the maximal tractable subset $\hat{\mathcal{H}}_5$. So $RSAT(\mathcal{D} \cup \hat{\mathcal{H}}_5)$ is polynomial. First run algorithm BIP-CON to enforce the path consistency, and then employ algorithm REG-BCON[14] to check the consistency for cardinal direction constraints. Obviously the complexity is $O(n^4)$.

7 Conclusions

This paper investigates the problem of combining topological and directional information for QSR, where RCC5 represents the topology and CDR describes the directional part. The mutual dependencies between RCC5 and CDR have been given by the interactive table. Then elaborate the heterogeneous composition of CDR with RCC5, RCC5 with CDR, and present an improved constraint propagation algorithm, besides analyze the complexity of the combinative reasoning. Since the proof of above approach is based on set theory, it can get similar results in three-dimensional space. For the future work, integrative reasoning with more spatial aspects or temporal information is a working direction. Due to the existing indeterminacy everywhere of real world, modeling the integrative uncertain reasoning is another direction.

References

1. Cohn, A.G., Hazarika, S.M.: Qualitative spatial representation and reasoning: An overview. Fundamenta Informaticae. 43, 2–32 (2001)
2. Sharma, J.: Integrated spatial reasoning in geographic information systems: combining topology and direction. Ph.D. Thesis, Department of Spatial Information Science and Engineering, University of Maine, Orono, ME (1996)
3. Sistla, A.P., Yu, C.: Reasoning about qualitative spatial relationships. Journal of Automated Reasoning. 25, 291–328 (2000)
4. Li, S.: Combining Topological and Directional Information: First Results. In: Lang, J., Lin, F., Wang, J. (eds.) KSEM 2006. LNCS (LNAI), vol. 4092, pp. 252–264. Springer, Heidelberg (2006)
5. Li, S.: Combining Topological and Directional Information for Spatial Reasoning. In: Veloso, M.M. (ed.) 20th International Joint Conference on Artificial Intelligence, Hyderabad, India, pp. 435–440 (2007)
6. Skiadopoulos, S., Koubarakis, M.: Composing cardinal direction relations. Artificial Intelligence 152, 143–171 (2004)
7. Skiadopoulos, S., Koubarakis, M.: On the consistency of cardinal direction constraints. Artificial Intelligence 163, 91–135 (2005)

8. Frank, A.U.: Qualitative spatial reasoning about cardinal directions. In: Mark, D., White, D. (eds.) Proc. of the 7th Austrian Conf. on Artificial Intelligence, pp. 157–167. Morgan Kaufmann, Baltimore (1991)
9. Freksa, C.: Using orientation information for qualitative spatial reasoning. In: Frank, A.U., Formentini, U., Campari, I. (eds.) Theories and Methods of Spatio-Temporal Reasoning in Geographic Space. LNCS, vol. 639, pp. 162–178. Springer, Heidelberg (1992)
10. Balbiani, P., Condotta, J.-F., Cerro, L.F.d.: A Model for Reasoning about Bidmensional Temporal Relations. In: Cohn, A.G., Schubert, L.K., Shapiro, S.C. (eds.) Proc. of Principles of Knowledge Representation and Reasoning (KR) Principles of Knowledge Representation and Reasoning, pp. 124–130. Morgan Kaufmann, Trento (1998)
11. Renz, J.: Qualitative Spatial Reasoning with Topological Information. LNCS (LNAI), vol. 2293. Springer, Heidelberg (2002)
12. Allen, J.: Maintaining Knowledge about Temporal Intervals. Communications of the ACM. 26, 832–843 (1983)
13. Navarrete, I., Sciavicco, G.: Spatial Reasoning with Rectangular Cardinal Direction Relations. In: ECAI 2006, Workshop on Spatial and Temporal Reasoning, Riva del Garda, Italy, pp. 1–9 (2006)
14. Navarrete, I., Morales, A., Sciavicco, G.: Consistency Checking of Basic Cardinal Constraints over Connected Regions. In: Veloso, M.M. (ed.) 20th International Joint Conference on Artificial Intelligence, Hyderabad, India, pp. 495–500 (2007)

A Merging-Based Approach to Handling Inconsistency in Locally Prioritized Software Requirements

Kedian Mu[1], Weiru Liu[2], Zhi Jin[3], Ruqian Lu[3], Anbu Yue[2], and David Bell[2]

[1] School of Mathematical Sciences
Peking University, Beijing 100871, P.R. China
[2] School of Electronics, Electrical Engineering and Computer Science
Queen's University Belfast, BT7 1NN, UK
[3] Academy of Mathematics and System Sciences
Chinese Academy of Sciences, Beijing 100080, P.R. China

Abstract. It has been widely recognized that the relative priority of requirements can help developers to resolve inconsistencies and make some necessary trade-off decisions. However, for most distributed development such as Viewpoints-based approaches, different stakeholders may assign different levels of priority to the same shared requirements statement from their own perspectives. The disagreement in the local priorities assigned to the same shared requirements statement often puts developers into a dilemma during inconsistency handling process. As a solution to this problem, we present a merging-based approach to handling inconsistency in the Viewpoints framework in this paper. In the Viewpoints framework, each viewpoint is a requirements collection with local prioritization. Informally, we transform such a requirements collection with local prioritization into a stratified knowledge base. Moreover, the relationship between viewpoints is considered as integrity constraints. By merging these stratified knowledge bases, we then construct a merged knowledge base with a global prioritization, which may be viewed as an overall belief in these viewpoints. Finally, proposals for inconsistency handling are derived from the merged result. The global prioritization as well as the local prioritization may be used to argue these proposals and to help developers make a reasonable trade-off decision on handling inconsistency.

Keywords: Inconsistency, Knowledge bases merging, Requirements engineering, Viewpoints, Local prioritization.

1 Introduction

For any complex software system, the development of requirements typically involves many different stakeholders with different concerns. Then the software requirements specifications are increasingly developed in a distributed fashion. The Viewpoints framework [1] has been developed to represent and analyze the

Z. Zhang and J. Siekmann (Eds.): KSEM 2007, LNAI 4798, pp. 103–114, 2007.

different perspectives and their relationships during the requirements stage. A viewpoint is a description of the system-to-be from a particular stakeholder, or a group of stakeholders. It reflects concerns of a particular stakeholder. The requirements specification of the system-to-be comprises a structured collection of loosely coupled, locally managed, distributable viewpoints, with explicit relationships between them to represent their overlaps [2]. These viewpoints may overlap, complement, or contradict each other. Then it makes inconsistency management more necessary during the requirements stage [2,3].

It has been recognized that the relative priority of requirements can help project managers resolve conflicts and make some necessary trade-off decisions [4,5]. However, different stakeholders may assign different levels of priority to the same shared requirements statement in the distributed development such as Viewpoints-based approaches. Actually, for a shared requirements statement, each priority given by a particular stakeholder is a measure of its relative importance only from the perspective of the stakeholder. Moreover, only these local priorities are available in many cases. The disagreement in these local priorities assigned to the same shared requirements statement often puts developers into a dilemma. To make a reasonable trade-off decision on resolving inconsistency, developers need to know global prioritization as well as local prioritization.

As a solution to this problem, we provide a merging-based approach to handling inconsistency in the Viewpoints framework in this paper. Merging is viewed as an usual way to globalization from a set of local information. Informally speaking, we construct a merged requirements specification with global prioritization by merging locally prioritized requirements collections of viewpoints based on the merging operators presented in [6]. The relationship between viewpoints is considered as integrity constraints during the merging process. Then we derive proposals for inconsistency handling from the merged result. Moreover, the global prioritization as well as the local priorities can be used to argue the proposals and help developers make trade-off decisions.

The rest of this paper is organized as follows. Section 2 gives an introduction to the Viewpoints framework and merging operators presented in [6], respectively. Section 3 provides a merging-based approach to handling inconsistency in the Viewpoints framework. Section 4 gives some comparison and discussion about the merging-based approach. Finally, we conclude this paper in Section 5.

2 Preliminaries

2.1 Logical Representation of Viewpoints

Although heterogeneity of representation allows different viewpoints to use different representations to describe their requirements [2], first order logic is appealing for formal representation of requirements statements since most tools and notations for representing requirements could be translated into formulas of first order logic [7]. Moreover, in a logic-based framework for representing requirements, reasoning about requirements is always based on some facts that describe a certain scenario [7]. It implies that checking the consistency of requirements collections only considers ground formulas[1] rather than unground formulas. Furthermore, if we restrict the first order language to propositional case, it may render consistency checking decidable. This gives some computational advantages. For these reasons, we assume a classical first order language without function symbols and quantifiers. This classical first order logic is the most convenient to illustrate our approach, as will be shown in the rest of the paper.

Let $\mathcal{L}_{\Phi_0}$ be the language composed from a set of classical atoms Φ_0 and logical connectives $\{\vee, \wedge, \neg, \rightarrow\}$ and let $\vdash$ be the classical consequence relation[2]. Let $\alpha \in \mathcal{L}_{\Phi_0}$ be a classical formula and $\Delta \subseteq \mathcal{L}_{\Phi_0}$ a set of formulas in $\mathcal{L}_{\Phi_0}$. In this paper, we call Δ *a set of requirements statements (or a requirements collection)* while each formula $\alpha \in \Delta$ represents a requirements statement.

An usual approach to prioritizing a requirements collection is to group requirements statements into several priority categories, such as the most frequent three-level scale of "*High*", "*Medium*", "*Low*" [8]. Let m, a natural number, be the scale of the priority and L be $\{l_0^m, \cdots, l_{m-1}^m\}$, a totally ordered finite set of m symbolic values of the priorities, i.e. $l_i^m < l_j^m$ iff $i < j$. Furthermore, each symbolic value in L could associate with a linguistic value. For example, for a three-level priority set, we have a totally ordered set L as $L = \{l_0^3, l_1^3, l_2^3\}$ where l_0^3 : *Low*, l_1^3 : *Medium*, l_2^3 : *High*. For example, if we assign l_2^3 to a requirements statement α, it means that α is one of the most important requirements statements. Prioritization over Δ is in essence to establish a prioritization function $P : \Delta \rightarrow L$ by balancing the business value of each requirements statement against its cost and technique risk. Actually, for every Δ, prioritization provides a priority-based partition of Δ, $< \Delta^{m-1}, \cdots, \Delta^1, \Delta^0 >$, where $\Delta^k = \{\alpha | \alpha \in \Delta, P(\alpha) = l_k^m\}$, for $k = m-1, \cdots, 0$. We then use $< \Delta^{m-1}, \cdots, \Delta^1, \Delta^0 >$ or (Δ, P) to denote a prioritized requirements collection in this paper. Note that different viewpoints may use different scales of the priority in the Viewpoints framework.

In the Viewpoints framework, let $V = \{v_1, \cdots, v_n\}(n \geq 2)$ be the set of viewpoints and L_i the scale of the priority used by viewpoint v_i, $i \in [1, \cdots, n]$. Then the requirements specification could be represented by a $n+1$ tuple $< (\Delta_1, P_1), \cdots, (\Delta_n, P_n), R >$, where Δ_i and P_i $(1 \leq i \leq n)$ are the set of requirements statements and the prioritization mapping of viewpoint v_i, respectively, and R is a set of relationships for consistency checking between these viewpoints, such as the relationships to represent their overlaps.

Because we use the classical logic as the uniform representation of viewpoints, an individual relationship between v_i and v_j could be also explicitly represented by special formulas associated with some formulas in Δ_i and Δ_j. These may be added to the requirements set $\Delta_i \cup \Delta_j$ if necessary. For example, for the relationship of total overlaps defined in [9], we may use a set of formulas in the form of $\phi(a) \leftrightarrow \psi(a)$ to denote the notation ψ and ϕ overlap totally, where

[1] There is no variable symbol appearing in the ground formula. For example, $user(John)$ is a ground atom, and $user(x)$ is not a ground atom.

[2] We view each ground atomic predicate formula as a propositional atom.

a is a constant. For the sake of simplicity, we use $\Delta_{(v_{i_1},\cdots,v_{i_k})}$ to represent the relationships among viewpoints $v_{i_1},\cdots,v_{i_k}$. Moreover, we assume that we should check the consistency of $\Delta_{(v_{i_1},\cdots,v_{i_k})} \cup (\bigcup_{j=1}^{k} \Delta_{i_j})$ if $\Delta_{(v_{i_1},\cdots,v_{i_k})} \in R$.

It is not surprising that some stakeholders are more important than others. Let L_V be a r-level priority set used in prioritizing viewpoints. Then prioritizing viewpoints is to establish a prioritization mapping $P_V : V \mapsto L_V$. In the rest of this paper, we use (V, P_V) to denote a set of prioritized viewpoints.

A logical contradiction is any situation in which some fact α and its negation $\neg\alpha$ can be simultaneously derived from the same set of formulas Δ. Some works about inconsistency handling in requirements engineering [7,10] refer to *the logical contradiction* as *the inconsistency.* In this paper, we only consider this type of inconsistency in requirements engineering.

Now we give an example to illustrate this representation.

Example 1. Consider the following scenario in eliciting demands about an user interface of a game system. Just for convenience, we assume that the three-level priority set is adopted to prioritize viewpoints as well as requirements of each viewpoint. *Alice* is a delegate of players of an earlier game system. She gives three requirements as follows:

(a1) *The style (sty) of user interface should be more fashionable(FAS) than that of the earlier system.*;
(a2) *The user interface should provide flexible (FLE) choice of settings(set) to players*;
(a3) *The elements(ele) of user interface should be familiar(FAM) to all the players.*

Then she assigns the level of *high* and the level of *medium* to (a1-2) and (a3), respectively.

Bob is a delegate of potential players of the system-to-be. He gives three demands as follows:

(b1) *The style of user interface should be very fashionable*;
(b2) *The user interface should provide flexible choice of settings to players*;
(b3) *The elements of user interface should be unexpected (UNE) to all players.*

He assigns the level of *high* to (b1) and (b2). (b3) is viewed as a requirements statement with the level of *medium*.

John is a consultant in the user interface of game systems. He gives the same demands as Alice. The main difference between *Alice* and *John* is that he assigns the level of *low* to the third. In addition, *Bob* is one of the most important stakeholders. *Alice* and *John* are two important stakeholders. Their priorities are *High*, *Medium*, and *Medium*, respectively. Obviously, the three stakeholders should reach agreement on the user interface.

Let v_A be the viewpoint of *Alice*, then $P_V(v_A) = l_1^3$ and

$$\Delta_A = \{\mathrm{FAS(sty)}, \mathrm{FLE(set)}, \mathrm{FAM(ele)}\}.$$
$$P_A(\mathrm{FAS(sty)}) = l_2^3,\ P_A(\mathrm{FLE(set)}) = l_2^3,\ P_A(\mathrm{FAM(ele)}) = l_1^3.$$
$$(\Delta_A, P_A) =< \{\mathrm{FAS(sty)},\mathrm{FLE(set)}\}, \{\mathrm{FAM(ele)}\}, \emptyset > .$$

Let v_B be the viewpoint of *Bob*, then $P_V(v_B) = l_2^3$ and

$$\Delta_B = \{\text{FAS(sty)},\ \text{FLE(set)}, \text{UNE(ele)}\}.$$
$$P_B(\text{FAS(sty)}) = l_2^3,\ P_B(\text{FLE(set)}) = l_2^3,\ P_B(\text{UNE(ele)}) = l_1^3.$$
$$(\Delta_B, P_B) =< \{\text{FAS(sty)},\ \text{FLE(set)}\}, \{\text{UNE(ele)}\}, \emptyset > .$$

Let v_J be the viewpoint of *John*, then $P_V(v_J) = l_1^3$ and

$$\Delta_J = \{\text{FAS(sty)},\ \text{FLE(set)}, \text{FAM(ele)}\}.$$
$$P_J(\text{FAS(sty)}) = l_2^3,\ P_J(\text{FLE(set)}) = l_2^3,\ P_J(\text{FAM(ele)}) = l_0^3.$$
$$(\Delta_J, P_J) =< \{\text{FAS(sty)},\ \text{FLE(set)}\}, \emptyset, \{\text{FAM(ele)}\} > .$$

Obviously, $R = \{\Delta_{(v_A,v_B,v_J)}\}$ and $\Delta_{(v_A,v_B,v_J)} = \{\text{FAM(ele)} \leftrightarrow \neg\text{UNE(ele)}\}$. Then the partial requirements specification comprising viewpoints $\{v_A, v_B, v_J\}$ is $< (\Delta_A, P_A), (\Delta_B, P_B), (\Delta_J, P_J), R >$. Moreover, we can conclude that

$$\Delta_A \cup \Delta_B \cup \Delta_J \cup \Delta_{(v_A,v_B,v_J)} \vdash \text{UNE(ele)} \wedge \neg\text{UNE(ele)}.$$

We will come back to this example in section 3.

2.2 Knowledge Bases Merging

Merging is an usual approach to fusing a set of heterogeneous information. The gist of knowledge base merging is to derive an overall belief set from a collection of knowledge bases. A flat knowledge base K is a set of formulas in $\mathcal{L}_{\Phi_0}$. An interpretation is a total function from Φ_0 to $\{0, 1\}$, denoted by a bit vector whenever a strict total order on Φ_0 is specified. Ω is the set of all possible interpretations. An interpretation ω is a model of a formula φ, denoted $\omega \models \varphi$, iff $\omega(\varphi) = 1$. Then K is consistent iff there exists at least one model of K.

A stratified knowledge base is a finite set K of formulas in $\mathcal{L}_{\Phi_0}$ with a total pre-order relation $\preceq$ on K. Intuitively, if $\varphi \preceq \psi$ then φ is regarded as more preferred than ψ. From the pre-order relation $\preceq$ on K, K can be stratified as $K = (S_1, \cdots, S_n)$, where S_i contains all the minimal propositions of set $\bigcup_{j=i}^{n} S_j$ w.r.t $\preceq$. Each S_i is called a stratum of K and is non-empty. We denote $\bigcup K = \bigcup_{j=1}^{n} S_j$. A prioritized knowledge profile E is a multi-set of stratified knowledge bases, i.e. $E = \{K_1, \cdots, K_n\}$.

Many model-based as well as syntax-based merging operators have been presented to merge either flat or stratified knowledge bases. Informally, syntax-based operators aim to pick some formula in the union of the original bases. It may result in lossing of some implicit beliefs during merging. In contrast, model-based merging operators aim to select some interpretations that are the closest to the original bases. They retain all the original knowledge and may also introduce additional formulas. Most merging operators just generate a flat base as the result. At present, only the merging operators presented in [6] can be used to construct a stratified merged knowledge base. In this paper, we adopt the syntax-based operators presented in [6] to merge inconsistent requirements collections.

Given a stratified knowledge base K, its models are defined as minimal interpretations with regard to a total pre-order relation $\preceq_X$ on interpretations that is induced from K by an ordering strategy X. The three widely used ordering strategies are the *best out* ordering [11], the *maxsat* ordering [12] and the *leximin* ordering [11]. In this paper, we use the *maxsat* ordering, though it is not obligatory.

Definition 1 (Maxsat Ordering [12]). *Given* $K=(S_1,\cdots,S_n)$. *Let* $r_{MO}(\omega)=\min\{i : \omega \models S_i\}$, *for* $\omega \in \Omega$. *By convention,* $\min\{\emptyset\} = +\infty$. *Then the maxsat ordering* $\preceq_{maxsat}$ *on* Ω *is defined as:* $\omega \preceq_{maxsat} \omega'$ *iff* $r_{MO}(\omega) \leq r_{MO}(\omega')$.

Given a stratified knowledge base K, from the pre-order relation $\preceq_{maxsat}$ induced from K on Ω, the interpretations in Ω can also be stratified as $\Omega_{K,maxsat} = (\Omega_1,\cdots,\Omega_m)$.

Yue et al. [6] argued that if the knowledge bases are designed independently, then only the relative preference between interpretations induced from a knowledge base by some ordering strategy is meaningful in a merging process.

Definition 2 (Relative Preference Relation [6]). *Let* $\{\Omega_{K_1,X_1},\cdots,\Omega_{K_n,X_n}\}$ *be a multi-set. A binary relative preference relation* $R \subseteq \Omega \times \Omega$ *is defined as:* $R(\omega,\omega')$ *iff* $|\{\Omega_{K_i,X_i} \text{ s.t. } \omega \prec_i \omega'\}| > |\{\Omega_{K_i,X_i} \text{ s.t. } \omega' \prec_i \omega\}|$, *where* $\prec_i$ *is the strict partial order relation induced from* Ω_{K_i,X_i}.

$R(\omega,\omega')$ means that more knowledge bases prefer ω than ω'.

Definition 3 (Undominated Set [6]). *Let* R *be a relative preference relation over* Ω *and let* Q *be a subset of* Ω. Q *is called an undominated set of* Ω, *if* $\forall\omega \in Q$, $\forall\omega' \in \Omega \setminus Q$, $R(\omega',\omega)$ *does not hold.* Q *is a minimal undominated set of* Ω *if for any undominated set* P *of* Ω, $P \subset Q$ *does not hold.*

We denote the set of minimal undominated sets of Ω w.r.t R as U^R_Ω. Then we can stratify the interpretations as follows:

Definition 4 (Stratification of Ω Obtained from R [6]). *Let* R *be a relative preference relation. A stratification of interpretations* $\Omega = (\Omega_1,\cdots,\Omega_n)$ *can be obtained from* R *such that* $\Omega_i = \cup Q$, *where* $Q \in U^R_{\Omega-\cup_{j=1}^{i-1}\Omega_j}$.

Definition 5 (Maxsat-Dominated Construction [6]). *Let* $\Omega=(\Omega_1,\cdots,\Omega_n)$ *be a stratification of interpretation and* S *be a set of propositions. A stratified knowledge base* $K_S^{maxsat,\Omega} = (S_1,\cdots,S_m)$ *is a maxsat-dominated construction from* S *w.r.t* Ω *if* $\bigcup_{i=1}^{m} S_i \subseteq S$ *and* $\Omega_{K_S^{maxsat,\Omega},maxsat} = \Omega$.

Yue et al. [6] has also shown how to construct a maxsat-dominated construction as a stratified merged result from the original bases based on the stratification of Ω obtained from R.

Proposition 1. *Let* $\Omega = (\Omega_1,\cdots,\Omega_n)$ *be a stratification of interpretation and* S *be a set of propositions. If there exists a stratified knowledge base* K *s.t.*

$\Omega_{K,maxsat} = \Omega$ *and* $\bigcup K \subseteq S$, *then* $K_S^{maxsat,\Omega} = (S_1, \cdots, S_n)$ *is a maxsat-dominated construction from* S *w.r.t* Ω, *where* $S_i = \{\varphi \in S | \forall \omega \in \Omega_i, \omega \models \varphi\} - \bigcup_{j=1}^{i-1} S_j$ *and* $S_i \neq \emptyset$.

Actually, if there is an integrity constraint μ during the merging process, then we only need to use Ω^μ instead of Ω in the definitions above, where Ω^μ is the set of all the models of μ.

3 A Merging-Based Approach to Handling Inconsistent Requirements with Local Prioritization

We start this section with consideration of *Example* 1. Intuitively, developers should persuade someone to abandon some requirements statements so as to retain more important requirements from a global perspective. Consider the local priorities of the two requirements involved in the inconsistency:

$$P_A(\text{FAM(ele)}) = l_1^3,\ P_J(\text{FAM(ele)}) = l_0^3,\ P_B(\text{UNE(ele)}) = l_1^3.$$

As a shared requirements statement, $FAM(ele)$ has two different priorities given by *Alice* and *John*, respectively. To determine whether $UNE(ele)$ is more important than $FAM(ele)$ from a global perspective, it is necessary to derive a merged requirements collection with global prioritization based on the requirements collections with local prioritization.

3.1 Merging an Ordered Knowledge Profile

Merging provides a promising way to extract an overall view from distributed viewpoints. Intuitively, each of viewpoints involved in inconsistencies may be viewed as a stratified knowledge base. The knowledge profile consisting of these knowledge bases should be ordered since some viewpoints are more important than others.

An ordered knowledge profile is a finite set E of knowledge bases with a total pre-order relation $\leq_E$ on E. Intuitively, if $K_i \leq_E K_j$ then K_i is regarded as more important than K_j. From the pre-order relation $\leq_E$ on E, E can be stratified as $E = (T_1, \cdots, T_m)$, where T_i contains all the minimal knowledge bases of set $\bigcup_{j=i}^{m} T_j$ with regard to $\leq_E$. Generally, the pre-order relation on E should be considered during the merging process. Actually, as mentioned in [6], only the *relative preference relation* over interpretations is meaningful in the merging process. Consequently, we will integrate the pre-order relationship over a profile into the relative preference relation defined in *Definition 1.*

Definition 6 (Level Vector Function). *Let* $E = (T_1, \cdots, T_m)$ *be an ordered knowledge profile. Level vector function* s *is a mapping from* E *to* $\{0,1\}^m$ *such that* $\forall K \in E$, *if* $K \in E_i$ $(1 \leq i \leq m)$, *then* $s(K) = (a_1, \cdots, a_m)$, *where* $a_i = 1$ *and* $a_j = 0$ *for all* $j \in [1, m]$, $j \neq i$.

Given a total ordering relation $\leq_s$ on N^m as follows: $\forall(a_1,\cdots,a_m),(b_1,\cdots,b_m)\in\{0,1\}^m$, $(a_1,\cdots,a_m)\leq_s(b_1,\cdots,b_m)$ iff $a_i=b_i$ for all i, or $\exists i$ s.t $a_i>b_i$ and $a_j=b_j$ for all $j<i$. Further, $(a_1,\cdots,a_m)<_s(b_1,\cdots,b_m)$ iff $(a_1,\cdots,a_m)\leq_s(b_1,\cdots,b_m)$ and $(b_1,\cdots,b_m)\not\leq_s(a_1,\cdots,a_m)$. Obviously,$K_i\leq_E K_j$ iff $s(K_i)\leq_s s(K_j)$. It means $s(K)$ gives a numerical measure of the relative importance of K w.r.t $\leq_E$.

Then we give an alternative definition of relative preference relation over interpretations as follows:

Definition 7 (Relative Preference Relation). *Let $E=\{K_1,\cdots,K_n\}$ be an ordered knowledge profile and $\{\Omega_{K_1,X_1},\cdots,\Omega_{K_n,X_n}\}$ be a multi-set. A binary relative preference relation $R_s\subseteq\Omega\times\Omega$ is defined as*

$$R_s(\omega,\omega')\ \textit{iff}\ \sum_{\Omega_{K_i,X_i}\ s.t.\ \omega\prec_i\omega'} s(K_i)<_s\sum_{\Omega_{K_j,X_j}\ s.t.\ \omega'\prec_j\omega} s(K_j),$$

where $\prec_i$ is the strict partial order relation induced from Ω_{K_i,X_i}.

Essentially, by introducing level vector function s, R_s considers $\leq_E$ as well as $\prec_i$ for each i. In the rest of this paper, we adopt R_s instead of R to construct a stratified merged knowledge base from an ordered knowledge profile.

Example 2. Consider an ordered knowledge profile $E=(K_1,K_2)$, where $K_1=(\{p\},\{\neg p\})$ and $K_2=(\{\neg p\},\{p\})$. The set of interpretations is $\Omega=\{\omega_1=1,\omega_2=0\}$. Then $r_{MO,K_1}(\omega_1)=1$, $r_{MO,K_1}(\omega_2)=2$; $r_{MO,K_2}(\omega_1)=2$, $r_{MO,K_2}(\omega_2)=1$. So, $\omega_1\prec_{K_1,maxsat}\omega_2$ and $\omega_2\prec_{K_2,maxsat}\omega_1$. If we do not consider $\leq_E$, neither $R(\omega_1,\omega_2)$ nor $R(\omega_2,\omega_1)$ holds. Then $\Omega=(\{\omega_1,\omega_2\})$ signifies that there is no meaningful merged result. In contrast, if we consider $\leq_E$ on E, then $s(K_1)=(1,0)$ and $s(K_2)=(0,1)$. So, $R_s(\omega_1,\omega_2)$ holds. The stratification of interpretations is $\Omega=(\{\omega_1\},\{\omega_2\})$. We get a maxsat-dominated construction $K=(\{p\},\{\neg p\})$. It is an intuitive result of merging.

3.2 Handling Inconsistent Requirements Collections with Local Prioritization

The gist of this paper is to provide a merging-based approach to handling inconsistent viewpoints, as shown in figure 1. Informally speaking, we first transform each requirements collection with a local prioritization involved in inconsistencies to a stratified knowledge base (SKB). The relationship between corresponding viewpoints is viewed as an integrity constraint during the merging process. Then we construct a stratified merged knowledge base based on the merging operators presented in [6]. The merged result can be considered as a overall view of these viewpoints. Moreover, the ordering relation over the merged knowledge base could be viewed as a global prioritization on the merged requirements collection. Finally, we derive proposal candidates for handling inconsistency from the stratified merged knowledge base. The global prioritization as well as the local prioritization may be used to argue these proposals and help developers make some trade-off decisions. If a proposal is acceptable to all the viewpoints involved in inconsistencies, the viewpoints will be modified according to the proposal.

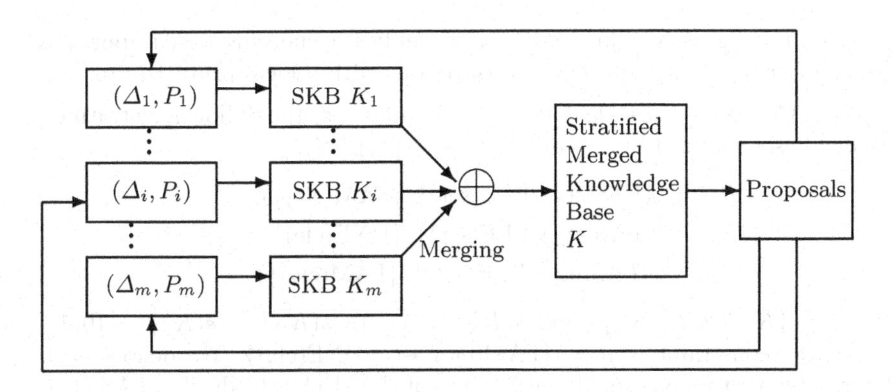

Fig. 1. A Merging-base Approach to Handling Inconsistency

From Viewpoints To Stratified Knowledge Bases: Let (Δ_i, P_i) be a requirements collection of viewpoint v_i $(1 \leq i \leq n)$. Then a stratified knowledge base induced by (Δ_i, P_i), denoted K_i, is defined as follows:

- $K_i = \Delta_i$; A total pre-order relationship $\preceq_i$ on K_i is presented as:

$$\forall \alpha,\ \beta \in K_i,\ \alpha \preceq_i \beta \text{ iff } P_i(\alpha) \geq P_i(\beta).$$

- K_i is stratified as $K_i = (S_{i_1}, \cdots, S_{i_m})$, where $S_{i_1}, \cdots, S_{i_m}$ is given by deleting all $\emptyset$ from $\Delta_i^{m-1}, \cdots, \Delta_i^0$.

Constructing A Stratified Merged Knowledge Base: Suppose that $v_{i_1}, \cdots,$ v_{i_k} are the viewpoints involved in inconsistency. Let $E = \{K_{i_1}, \cdots, K_{i_k}\}$ be a knowledge profile, where K_{i_l} is the stratified knowledge base induced by (Δ_{i_l}, P_{i_l}) for all $1 \leq i_l \leq i_k$. Let Ω be the set of interpretations. Then we

- define an ordering relation $\leq_E$ on E s.t. $K_{i_l} \leq_E K_{i_j}$ iff $P_V(v_{i_l}) \geq P_V(v_{i_j})$;
- compute the level vector function s based on stratification of E w.r.t $\leq_E$.
- consider $\Delta_{(v_{i_1}, \cdots, v_{i_k})}$ as an integrity constraint μ and compute $\Omega^\mu = \{\omega \in \Omega, \omega \models \mu\}$;
- find $\Omega^u_{K_{i_l}, maxsat}$ for all i_l.
- based on $\{\Omega^u_{K_{i_1}, maxsat}, \cdots, \Omega^u_{K_{i_k}, maxsat}\}$, construct stratification of interpretations $\Omega^u = (\Omega^u_1, \cdots, \Omega^u_m)$ by using relative preference relation R_s over Ω^u.
- get a maxsat-dominated construction K based on *Proposition* 1.

Deriving A Proposal Candidate To Handling Inconsistency: The preference relation on the maxsat-dominated construction K describes the relative importance of requirements from a global perspective. Then it naturally derives proposals that the requirements with lower global priorities should be abandoned so as to resolve the inconsistencies. However, these proposals are just recommendations. Stakeholders will make further trade-off decisions based on the global prioritization as well as the local prioritization.

We give an example to illustrate how to apply the merging-based approach to handling inconsistent requirements collections with a local prioritization.

Example 3. Consider *Example* 1 again. We may get the following stratified knowledge bases induced by v_A, v_B, and v_J respectively:

$$K_A = (\{\text{FAS(sty)}, \text{FLE(set)}\}, \{\text{FAM(ele)}\});$$
$$K_B = (\{\text{FAS(sty)}, \text{FLE(set)}\}, \{\text{UNE(ele)}\});$$
$$K_J = (\{\text{FAS(sty)}, \text{FLE(set)}\}, \{\text{FAM(ele)}\}).$$

Then $E = (\{K_B\}, \{K_A, K_J\})$ and $s(K_B) = (1, 0)$, $s(K_A) = s(K_J) = (0, 1)$. The integrity constraint is $\mu = \{\text{FAM(ele)} \leftrightarrow \neg\text{UNE(ele)}\}$. We denote each model by a bit vector consists of truth values of (FAM(ele), UNE(ele), FAS(sty), FLE(set)). Then $\Omega^\mu = \{\omega_1 = 0100, \omega_2 = 0101, \omega_3 = 0110, \omega_4 = 0111, \omega_5 = 1000, \omega_6 = 1001, \omega_7 = 1010, \omega_8 = 1011\}$. r_{MO} is given in table 1.

Table 1. Ranks of interpretations given by the maxsat ordering strategy

ω	K_A	K_B	K_J
0100	$+\infty$	2	$+\infty$
0101	$+\infty$	2	$+\infty$
0110	$+\infty$	2	$+\infty$
0111	1	1	1
1000	2	$+\infty$	2
1001	2	$+\infty$	2
1010	2	$+\infty$	2
1011	1	1	1

Then we can get

$$\Omega^\mu_{K_A,maxsat} = (\{\omega_8, \omega_4\}, \{\omega_5, \omega_6, \omega_7\}, \{\omega_1, \omega_2, \omega_3\});$$
$$\Omega^\mu_{K_B,maxsat} = (\{\omega_8, \omega_4\}, \{\omega_1, \omega_2, \omega_3\}, \{\omega_5, \omega_6, \omega_7\});$$
$$\Omega^\mu_{K_J,maxsat} = (\{\omega_8, \omega_4\}, \{\omega_5, \omega_6, \omega_7\}, \{\omega_1, \omega_2, \omega_3\}).$$

Furthermore, we get a stratification of Ω^μ based on the relative preference relation R_s on Ω^μ as $\Omega^\mu = (\{\omega_8, \omega_4\}, \{\omega_1, \omega_2, \omega_3\}, \{\omega_5, \omega_6, \omega_7\})$. According to Proposition 1, we get a maxsat-dominated stratified construction

$$K = (\{\text{FAS(sty)}, \text{FLE(set)}\}, \{\text{UNE(ele)}\}, \{\text{FAM(ele)}\}).$$

This merged result implies that FAM(ele) is *less important than* UNE(ele) from a global perspective.

The proposal for handling inconsistency derived from this result of merging, denoted π, is that *the developer had better persuade Alice and John to abandon their shared demand about elements of user interface.* If the proposal π is acceptable to the three stakeholders, then

$$\Delta^\pi_A = \Delta_A - \{\text{FAM(ele)}\},\ \Delta^\pi_B = \Delta_B,\ \Delta^\pi_J = \Delta_J - \{\text{FAM(ele)}\},$$

where Δ_i^π is the modification of Δ_i by performing π. The inconsistency disappears in $\Delta_A^\pi \cup \Delta_B^\pi \cup \Delta_J^\pi \cup \Delta_{(v_A,v_B,v_J)}$.

4 Discussion and Comparison

The disagreement in the local prioritization over shared requirements often leads inconsistency handling to a dilemma. It may be viewed as a promising way to identify appropriate proposals for handling inconsistency from a global perspective. But this does not mean that the global prioritization is more crucial than the local prioritization. We argue that both the global prioritization and the local prioritization play important roles in resolving inconsistencies. For example, the proposal π in *Example* 3 is considered appropriate to handling the inconsistency *from a global perspective.* Obviously, *John* maybe accept the proposal, since for the demand to be abandoned, $P_J\{\text{FAM(ele)}\} = Low$. But we can't assure that *Alice* agrees to abandon the shared demand since $P_A\{\text{FAM(ele)}\} = Medium$. In summary, the local prioritization has an impact on the acceptance of merged result to viewpoints. How to identify appropriate common proposals for inconsistency handling based on the local prioritization as well as the merged preference is still one of issues in our future work.

On the other hand, we adopt the syntax-based merging operators presented in [6] during merging process. The syntax-based merging operator aims to pick out some formulas from original knowledge bases. Then the merged result can be explained clearly. But it is possible that we can not get a stratified merged requirements collection in some case. However, introducing model-based merging operators also leads to a problem of how to explain additional formulas in the merged result in terms of viewpoints demands. It seems to be a dilemma.

5 Conclusions

Identifying appropriate actions or proposals for handling inconsistency is still a big problem in requirements engineering. The relative priority of requirements is considered as a useful clue to resolving conflicts and making trade-off decisions. However, in distributed development of requirements specifications such as the Viewpoints framework, the disagreement in local priorities of shared requirements statements often leads inconsistency handling to a dilemma.

The main contribution of this paper is to provide a merging-based approach to handling inconsistency in locally prioritized software requirements. Given an inconsistency, each viewpoint involved in the inconsistency is transformed into a stratified knowledge base, whilst the relationship between these viewpoints is considered as an integrity constraint. Based on the merging operators presented in [6], we construct a stratified merged knowledge base as an overall view of these inconsistent viewpoints. The ordering relationship over this stratified merged knowledge base could be considered as a global prioritization over the requirements specification. Generally, the requirements with lower merged preference may be considered as requirements to be abandoned. Then we may derive some proposals for handling the inconsistency from the merged result. The global prioritization as well as the local prioritization may be used to argue these proposals and help developers identifying acceptable common proposals.

Acknowledgements

This work was partly supported by the National Natural Science Fund for Distinguished Young Scholars of China under Grant No. 60625204, the Key Project of National Natural Science Foundation of China under Grant No. 60496324, the National Key Research and Development Program of China under Grant No. 2002CB312004, the National 863 High-tech Project of China under Grant No. 2006AA01Z155, the Knowledge Innovation Program of the Chinese Academy of Sciences, and the NSFC and the British Royal Society China-UK Joint Project.

References

1. Finkelsetin, A., Kramer, J., Nuseibeh, B., Finkelstein, L., Goedicke, M.: Viewpoints: A framework for integrating multiple perspectives in system development. International Journal of Software Engineering and Knowledge Engineering 2, 31–58 (1992)
2. Nuseibeh, B., Kramer, J., Finkelstein, A.: Viewpoints: meaningful relationships are difficult? In: Proceedings of the 25th International Conference on Software Engineering, pp. 676–681. IEEE CS Press, Los Alamitos (2003)
3. Finkelstein, A., Gabbay, D., Hunter, A., Kramer, J., Nuseibeh, B.: Inconsistency handling in multiperspective specifications. IEEE Transactions on Software Engineering 20, 569–578 (1994)
4. Wiegers, K.: First things first:prioritizing requirements. Software Development 7, 48–53 (1999)
5. Davis, A.: Just Enough Requirements Management: Where Software Development Meets Marking. Dorset House, New York (2005)
6. Yue, A., Liu, W., Hunter, A.: Approaches to constructiing a stratified merged knowledge base. In: Mellouli, K. (ed.) ECSQARU 2007. LNCS, vol. 4724, pp. 54–65. Springer, Heidelberg (2007)
7. Hunter, A., Nuseibeh, B.: Managing inconsistent specification. ACM Transactions on Software Engineering and Methodology 7, 335–367 (1998)
8. Wiegers, K.: Software Requirements, 2nd edn. Microsoft Press, Redmond (2003)
9. Spanoudakis, G., Finkelstein, A., Till, D.: Overlaps in requirements engineering. Automated Software Engineering 6, 171–198 (1999)
10. Gervasi, V., Zowghi, D.: Reasoning about inconsistencies in natural language requirements. ACM Transactions on Software Engineering and Methodologies 14, 277–330 (2005)
11. Benferhat, S., Cayrol, C., Dobois, D., Lang, J., Prade, H.: Inconsistency management and prioritized syntax-based entailment. In: Proceedings of IJCAI 1993, pp. 640–647. Morgan Kaufmann, San Francisco (1993)
12. Brewka, G.: A rank-based description language for qualitative preferences. In: Proc. of ECAI 2004, pp. 303–307. IOS Press, Amsterdam (2004)

A Dynamic Description Logic for Representation and Reasoning About Actions

Liang Chang[1,2], Fen Lin[1,2], and Zhongzhi Shi[1]

[1] The Key Laboratory of Intelligent Information Processing,
Institute of Computing Technology,
Chinese Academy of Sciences,
PO Box 2704-28, Beijing, 100080, China
[2] Graduate University of Chinese Academy of Sciences
{changl,linf,shizz}@ics.ict.ac.cn

Abstract. We present a dynamic description logic for representation and reasoning about actions, with an approach that embrace actions into the description logic $\mathcal{ALCO}$@. With this logic, description logic concepts can be used for describing the state of the world, and the preconditions and effects of atomic actions; Complex actions can be modeled with the help of standard action operators, such as the test, sequence, choice, and iteration operators; And both atomic actions and complex actions can be used as modal operators to construct formulas. We develop a terminable and correct algorithm for checking the satisfiability of formulas. Based on the algorithm, many reasoning tasks on actions are effectively carried out, including the realizability, executability, projection and planning problems.

1 Introduction

Description logics are a well-known family of formalisms for representing knowledge about static application domains. They are playing an important role in the Semantic Web, acting as the basis of the W3C-recommended Web ontology language OWL [1,9].

The study of integrating description logics and action formalisms is driven by two factors. One is the demand to represent and reason about semantic web services [11], for which an obvious concern is to combine in some way the static descriptions of the information provided by ontologies with the dynamic descriptions of the computations provided by web services [4]. The other factor is the fact that there is an expressive gap between existing action formalisms: they are either based on first- or higher-order logics and do not admit decidable reasoning, like the Situation Calculus [12] and the Fluent Calculus [14], or are decidable but only propositional, like those based on propositional dynamic logics [6,7] or based on propositional temporal logics [3].

One approach to integrate description logics with action formalisms was proposed by Shi [13]. This approach was characterized by constructing dynamic description logics, in such a way that actions are embraced into description

Z. Zhang and J. Siekmann (Eds.): KSEM 2007, LNAI 4798, pp. 115–127, 2007.

logics and treated as citizens. Based on the description logic $\mathcal{ALC}$, a dynamic description logic was constructed in that paper. But for that logic an efficient decision algorithm that capable for the open world is still an open problem.

In the present paper, following the approach of [13], we construct a dynamic description logic $\mathcal{D}_{\mathcal{ALCO}@}$ for representation and reasoning about actions. This logic holds the following features.

- Atomic actions are represented over ontologies expressed in the description logic $\mathcal{ALCO}@$. I.e., the state of the world and the preconditions and effects of atomic actions are described by formulas of $\mathcal{ALCO}@$.
- Based on formulas and atomic actions, complex actions can be constructed with the help of four standard action operators: the test, sequence, choice, and iteration operators.
- Both atomic actions and complex actions can be used as modal operators to construct formulas of the form $< \pi > \varphi$ and $[\pi]\varphi$, which respectively state that "the action π can be executed with the formula φ be true after the execution" and "whenever π is executed the formula φ must be true after the execution".
- A terminable, sound and complete satisfiability-checking algorithm that capable for the open world is provided. Based on the algorithm, many reasoning tasks on actions can be effectively carried out, such as the realizability, executability, projection and planning problems.

The syntax and semantics of this logic is introduced in the next section. A satisfiability-checking algorithm for the logic is presented in Section 3 and its termination and correctness is also demonstrated. Four important reasoning tasks on actions are studied in Sections 5. We conclude the paper with a discussion of related work and some suggestions concerning future research.

2 Syntax and Semantics

Primitive symbols of $\mathcal{D}_{\mathcal{ALCO}@}$ are a set N_R of *role names*, a set N_C of *concept names*, a set N_I of *individual names* and a set N_A of *atomic action names*. Starting with them, concepts, formulas, and actions can be built with the help of a set of operators.

Concepts of $\mathcal{D}_{\mathcal{ALCO}@}$ are generated with the following syntax rule:

$$C, C' \longrightarrow C_i \mid \{p\} \mid @_pC \mid \neg C \mid C \sqcup C' \mid \forall R.C \quad (1)$$

where $C_i \in N_C$, $p \in N_I$, $R \in N_R$. Concepts of the form $\{p\}$, $@_pC$, $\neg C$, $C \sqcup C'$ and $\forall R.C$ are respectively named as *nominal*, *at*, *negation*, *disjunction* and *value restriction* concepts.

Concepts of the form $C \sqcap C'$, $\exists R.C$, $\top$ and $\bot$ are introduced respectively as abbreviations of $\neg(\neg C \sqcup \neg C')$, $\neg(\forall R.\neg C)$, $C \sqcup \neg C$ and $\neg\top$, and are respectively named as *conjunction*, *exists restriction*, *top* and *bot* concepts.

A *concept definition* is of the form $D \equiv C$ for a concept name D and a concept C. A *TBox* $\mathcal{T}$ is a finite set of concept definitions with unique left-hand sides. $\mathcal{T}$ is acyclic if and only if there are no cyclic dependencies between these definitions.

Given an acyclic TBox $\mathcal{T}$, the set N_C of concept names is divided into two disjointed sets N_{CD} and N_{CP}, where N_{CD} is the set of *defined concept names*, i.e., concept names occurring on the left-hand side of concept definitions that contained in $\mathcal{T}$, and N_{CP} is the set of *primitive concept names*, i.e., concept names that are not defined in $\mathcal{T}$.

Formulas of $\mathcal{D}_{\mathcal{ALCO}@}$ are generated with the following syntax rule:

$$\varphi, \varphi' \longrightarrow C(p) \mid R(p,q) \mid <\pi> \varphi \mid \neg\varphi \mid \varphi \vee \varphi' \tag{2}$$

where $p, q \in N_I$, $R \in N_R$, C is a concept, and π is an action. Formulas of the form $C(p)$, $R(p,q)$, $<\pi> \varphi$, $\neg\varphi$ and $\varphi \vee \varphi'$ are respectively named as *concept assertion*, *role assertion*, *diamond assertion*, *negation formula* and *disjunction formula*.

Formulas of the form $[\pi]\varphi$, $\varphi \wedge \varphi'$, $\varphi \rightarrow \varphi'$, *true* and *false* are introduced respectively as abbreviations of $\neg <\pi> \neg\varphi$, $\neg(\neg\varphi \vee \neg\varphi')$, $\neg\varphi \vee \varphi'$, $\varphi \vee \neg\varphi$ and $\neg$*true*, and are respectively named as *box assertion*, *conjunction formula*, *implication formula*, *tautology* and *contradiction*.

An *atomic-action definition* is of the form $A(v_1, ..., v_n) \equiv (P, E)$, where,

- A is an atomic action name; $v_1, ..., v_n$ is a finite sequence of all the individual names occurring in P and E;
- P is a finite set of formulas for describing the preconditions;
- E is a finite set of effects, with each effect be of form $P(v_k)$, $\neg P(v_k)$, $R(v_k, v_j)$ or $\neg R(v_k, v_j)$, where $v_j, v_k \in N_I$, $P \in N_{CP}$ and $R \in N_R$; and
- let P=$\{\varphi_1, \ldots, \varphi_n\}$ and E=$\{\phi_1, \ldots, \phi_m\}$, then P and E subject to the constraint that $\varphi_1 \wedge \ldots \wedge \varphi_n \rightarrow \neg\phi_k$ for each k with $1 \leq k \leq m$.

Given a finite set of atomic-action definitions $\mathcal{A}_C$, an atomic action name A is called *defined* w.r.t. $\mathcal{A}_C$ if A occurring in the left-hand side of an atomic-action definition. We call $\mathcal{A}_C$ an *ActionBox* if no atomic action name is defined more then once.

Let $A(v_1, ..., v_n) \equiv (P, E)$ be an atomic-action definition and $p_1, ..., p_n$ a sequence of individual names, then $A(p_1, ..., p_n)$ is called an *atomic action* that defined by $(P_{\{p_1/v_1,...,p_n/v_n\}}, E_{\{p_1/v_1,...,p_n/v_n\}})$. Where $P_{\{p_1/v_1, \ldots, p_n/v_n\}}$ is generated by replacing each occurrence of v_k in P with p_k, respectively for each $1 \leq k \leq n$; and $E_{\{p_1/v_1,...,p_n/v_n\}}$ is similarly generated. We also call that $A(p_1, ..., p_n)$ *is defined according to the atomic-action definition* $A(v_1, ..., v_n) \equiv (P, E)$. For simplicity, sometimes we also use $(P_{\{p_1/v_1,...,p_n/v_n\}}, E_{\{p_1/v_1,...,p_n/v_n\}})$ to denote the atomic action $A(p_1, ..., p_n)$.

Actions of $\mathcal{D}_{\mathcal{ALCO}@}$ are generated with the following syntax rule:

$$\pi, \pi' \longrightarrow A(p_1, ..., p_n) \mid \varphi? \mid \pi \cup \pi' \mid \pi; \pi' \mid \pi^* \tag{3}$$

where φ is a formula. Actions of the form $\varphi?$, $\pi \cup \pi'$, $\pi; \pi'$ and π^* are respectively named as *test-*, *choice-*, *sequence-* and *iteration-* actions.

Especially, we introduce *if* φ *then* π *else* π' as abbreviation of $(\varphi?; \pi) \cup ((\neg\varphi)?; \pi')$, and *while* φ *do* π as abbreviation of $(\varphi?; \pi)^*$; $(\neg\varphi)?$.

A semantic *model* for $\mathcal{D}_{\mathcal{ALCO}@}$ is a pair M=(W, I), where W is a set of states, I associates with each state $w \in W$ an interpretation $I(w) = (\Delta^I, C_0^{I(w)}, \ldots, R_0^{I(w)}, \ldots, p_0^I, \ldots)$, with $C_i^{I(w)} \subseteq \Delta^I$ for each $C_i \in N_c$, $R_i^{I(w)} \subseteq \Delta^I \times \Delta^I$ for each $R_i \in N_R$, and $p_i^I \in \Delta^I$ for each $p_i \in N_I$. Based on the interpretation of all the states, each action π is interpreted indirectly by I as a binary relation $\pi^I \subseteq W \times W$.

It should be noted that the interpretation p_i^I of each individual name p_i will not vary according to different states.

Given a model M=(W, I) and a state $w \in W$, the value $C^{I(w)}$ of a concept C, the truth-relation $(M, w) \models \varphi$ (or simply $w \models \varphi$ if M is understood) for a formula φ, and the relation π^I for an action π are defined inductively as follows:

(1) $\{p\}^{I(w)} = \{p^I\}$;
(2) If $p^I \in C^{I(w)}$ then $(@_pC)^{I(w)} = \Delta^I$, else $(@_pC)^{I(w)} = \emptyset$;
(3) $(\neg C)^{I(w)} = \Delta^I - C^{I(w)}$;
(4) $(C \sqcup D)^{I(w)} = C^{I(w)} \cup D^{I(w)}$;
(5) $(\forall R.C)^{I(w)} = \{x \mid \forall y.((x, y) \in R^{I(w)}$ implies $y \in C^{I(w)})\}$;
(6) $(M, w) \models C(p)$ *iff* $p^I \in C^{I(w)}$;
(7) $(M, w) \models R(p, q)$ *iff* $(p^I, q^I) \in R^{I(w)}$;
(8) $(M, w) \models <\pi> \varphi$ *iff* $\exists w' \in W.((w, w') \in \pi^I$ and $(M, w') \models \varphi)$;
(9) $(M, w) \models \neg\varphi$ *iff* $(M, w) \models \varphi$ not holds;
(10) $(M, w) \models \varphi \vee \psi$ *iff* $(M, w) \models \varphi$ or $(M, w) \models \psi$;
(11) Let S be a formula set, then, $(M, w) \models S$ *iff* $(M, w) \models \varphi_i$ for all $\varphi_i \in S$;
(12) Let $A(p_1, ..., p_n)$ be an atomic action defined by preconditions P and effects E, with E=$\{\phi_1, \ldots, \phi_m\}$, then $A(p_1, ..., p_n)^I = (P, E)^I = \{(w_1, w_2) \in W \times W \mid (M, w_1) \models P$, $C^{I(w_2)} = C^{I(w_1)} \cup C^+ - C^-$ for each primitive concept name $C \in N_{CP}$, and $R^{I(w_2)} = R^{I(w_1)} \cup R^+ - R^-$ for each role name $R \in N_R\}$, where,

- C^+= $\{ p_k^I \mid C(p_k) \in E\}$,
- C^-=$\{ p_k^I \mid \neg C(p_k) \in E\}$,
- R^+= $\{ (p_j^I, p_k^I) \mid R(p_j, p_k) \in E \}$,
- R^-= $\{ (p_j^I, p_k^I) \mid \neg R(p_j, p_k) \in E \}$;

(13) $(\varphi?)^I = \{(w_1, w_1) \in W \times W \mid (M, w_1) \models \varphi\}$;
(14) $(\pi \cup \pi')^I = \pi^I \cup \pi'^I$;
(15) $(\pi; \pi')^I = \pi^I \circ \pi'^I$, where $\circ$ is the composite operation on binary relations;
(16) $(\pi^*)^I$ = reflexive transitive closure of π^I.

The interpretation of atomic actions adopts the possible models approach [15] and follows the style introduced in [2].

A model M=(W, I) *satisfies a concept definition* $D \equiv C$, in symbols $M \models D \equiv C$, if and only if for every state $w \in W$ it is $D^{I(w)} = C^{I(w)}$.

M *is a model of a TBox* $\mathcal{T}$, in symbols $M \models \mathcal{T}$, if and only if M satisfies all the concept definitions contained in $\mathcal{T}$.

Let φ be a formula such that each atomic actions occurring in φ is defined according to atomic-action definitions contained in an ActionBox $\mathcal{A}_C$. Then, φ is *satisfiable* w.r.t. $\mathcal{T}$ and $\mathcal{A}_C$, if and only if there exists a model $M = (W, I)$ of $\mathcal{T}$ and a state $w \in W$ such that $(M, w) \models \varphi$.

φ is *valid* w.r.t. $\mathcal{T}$ and $\mathcal{A}_C$, if and only if for every model $M = (W, I)$ of $\mathcal{T}$ and any state $w \in W$: we have $(M, w) \models \varphi$. It is obvious that φ is valid w.r.t. $\mathcal{T}$ and $\mathcal{A}_C$ iff $\neg\varphi$ is unsatisfiable w.r.t. $\mathcal{T}$ and $\mathcal{A}_C$.

3 Algorithm for Satisfiability Checking

The satisfiability-checking algorithm presented here is in fact based on an elaborated combination of (i) a tableau for the description logic $\mathcal{ALCO}@$, (ii) a prefixed tableau for propositional dynamic logic [8], and (iii) the embodiment of the possible models approach for interpreting actions.

A *prefix* $\sigma.\varepsilon$ is composed of a sequence-action σ and a set of effects ε, and are formed with the following syntax rule:

$$\sigma.\varepsilon \quad \longrightarrow \quad (\emptyset, \emptyset).\emptyset \quad | \quad \sigma; (P, E).(\varepsilon - \{\neg\varphi | \varphi \in E\}) \cup E \tag{4}$$

where $(\emptyset, \emptyset)$ and (P, E) are atomic actions, $\sigma; (P, E)$ is a sequence-action, $(\varepsilon - \{\neg\varphi | \varphi \in E\}) \cup E$ is a set of effects. We also use $\sigma_0.\varepsilon_0$ to denote the initial prefix $(\emptyset, \emptyset).\emptyset$.

A *prefixed formula* is a pair $\sigma.\varepsilon : \varphi$, where $\sigma.\varepsilon$ is a prefix, φ is a formula.

The intuition of the definition of prefixes is that: let φ be the formula for checking satisfiability, and let φ be satisfied by a state w_0 of a model $M = (W, I)$, i.e., $(M, w_0) \models \varphi$. Then, we can construct a function $\imath$ to map each prefix into a state in W, such that: (i) The initial prefix $(\emptyset, \emptyset).\emptyset$ is mapped to the state w_0, i.e., $\imath(\sigma_0.\varepsilon_0) = w_0$; (ii) For each prefix $\sigma.\varepsilon$, the sequence-action σ is used to record the track of atomic actions executed from the state w_0 to the state $\imath(\sigma.\varepsilon)$. (iii) For each prefix $\sigma.\varepsilon$, the set ε is used to record the essential differences between the state w_0 and the state $\imath(\sigma.\varepsilon)$.

A *branch* $\mathcal{B}$ is a union of three sets: a set $\mathcal{B}_{PF}$ of prefixed formulas, a set $\mathcal{B}_I$ of inequalities on individual names and a set $\mathcal{B}_E$ of eventualities. Each *inequality* is of form $p \neq q$ for two individual names p, q. Each *eventuality* is of form $X \doteq < \pi^* > \varphi$ with X a propositional variable and $< \pi^* > \varphi$ a formula.

Tableau expansion rules on $\mathcal{D}_{\mathcal{ALCO}@}$ concepts and formulas are presented in Figure 1 and Figure 2. They are similar to tableau rules for $\mathcal{ALCO}@$, except that prefixes are embedded here. It should be noted that the $\neg\forall$- and $\forall$-rule can only be triggered by formulas prefixed with $\sigma_0.\varepsilon_0$.

Figure 3 presents tableau expansion rules on sequence-, test- and choice-actions. Inspired by De Giacomo's approach [8], we introduce some propositional variables to detect the presence of π loops which do not fulfill $< \pi^* > \varphi$. More precisely, in the case that a prefixed iterated eventuality $\sigma.\varepsilon :< \pi^* > \varphi$ appear, the $*_{<>}$-rule will be applied to introduce a new variable X and add a record $X \doteq < \pi^* > \varphi$ into $\mathcal{B}$. Then the X-rule will be used for further expansion.

The possible models approach for interpreting actions is captured as relationships between prefixes, and embodied in rules of Figure 4. The $atom_{<>}$-rule will generate new prefixes, on which some constraints will be added by the $\neg atom_{<>}$-rule. If formulas of the form $\forall R.C$ or $\neg\forall R.C$ are generated and are not prefixed by $\sigma_0.\varepsilon_0$, then the B_2-rule will be applied to map them into formulas prefixed by the initial prefix. Similarly, if primitive formulas are generated and are not prefixed by $\sigma_0.\varepsilon_0$, then the B_1-rule will be applied to map them backward.

$\neg\neg_C$-rule:	If $\sigma.\varepsilon : (\neg(\neg C))(x) \in \mathcal{B}$, and $\sigma.\varepsilon : C(x) \notin \mathcal{B}$, then set $\mathcal{B} := \mathcal{B} \cup \{\sigma.\varepsilon : C(x)\}$.
$\{\}$-rule:	If $\sigma.\varepsilon : \{q\}(x) \in \mathcal{B}$, and $\{x \neq q, q \neq x\} \cap \mathcal{B} = \emptyset$, then set $\mathcal{B} := \mathcal{B}[q/x]$, where $\mathcal{B}[q/x]$ is obtained by replacing each occurrence of x in $\mathcal{B}$ with q.
$\neg\{\}$-rule:	If $\sigma.\varepsilon : (\neg\{q\})(x) \in \mathcal{B}$, and $\{x \neq q, q \neq x\} \cap \mathcal{B} = \emptyset$, then set $\mathcal{B} := \mathcal{B} \cup \{x \neq q\}$.
$\neg@$-rule:	If $\sigma.\varepsilon : (\neg @_q C)(x) \in \mathcal{B}$, and $\sigma.\varepsilon : (\neg C)(q) \notin \mathcal{B}$, then set $\mathcal{B} := \mathcal{B} \cup \{\sigma.\varepsilon : (\neg C)(q)\}$.
$@$-rule:	If $\sigma.\varepsilon : (@_q C)(x) \in \mathcal{B}$, and $\sigma.\varepsilon : C(q) \notin \mathcal{B}$, then set $\mathcal{B} := \mathcal{B} \cup \{\sigma.\varepsilon : C(q)\}$.
$\neg\sqcup$-rule:	If $\sigma.\varepsilon : (\neg(C_1 \sqcup C_2))(x) \in \mathcal{B}$, and $\{\sigma.\varepsilon : (\neg C_1)(x), \sigma.\varepsilon : (\neg C_2)(x)\} \nsubseteq \mathcal{B}$, then set $\mathcal{B} := \mathcal{B} \cup \{\sigma.\varepsilon : (\neg C_1)(x), \sigma.\varepsilon : (\neg C_2)(x)\}$.
$\sqcup$-rule:	If $\sigma.\varepsilon : (C_1 \sqcup C_2)(x) \in \mathcal{B}$, and $\{\sigma.\varepsilon : C_1(x), \sigma.\varepsilon : C_2(x)\} \cap S = \emptyset$, then set $\mathcal{B} := \mathcal{B} \cup \{\sigma.\varepsilon : C(x)\}$ for some $C \in \{C_1, C_2\}$.
$\neg\forall$-rule:	If 1. $\sigma_0.\varepsilon_0 : \neg(\forall R.C)(x) \in \mathcal{B}$, and 2. there is no y such that $\sigma_0.\varepsilon_0 : R(x, y) \in \mathcal{B}$ and $\sigma_0.\varepsilon_0 : (\neg C)(y) \in \mathcal{B}$, then introduce a new individual name y that has never occurred in $\mathcal{B}$, and set $\mathcal{B} := \mathcal{B} \cup \{\sigma_0.\varepsilon_0 : R(x, y),\ \sigma_0.\varepsilon_0 : (\neg C)(y)\}$.
$\forall$-rule:	If 1. $\sigma_0.\varepsilon_0 : (\forall R.C)(x) \in \mathcal{B}$, and 2. there is a y with $\sigma_0.\varepsilon_0 : R(x, y) \in \mathcal{B}$ and $\sigma_0.\varepsilon_0 : C(y) \notin \mathcal{B}$, then set $\mathcal{B} := \mathcal{B} \cup \{\sigma_0.\varepsilon_0 : C(y)\}$.

Fig. 1. The tableau expansion rules on $\mathcal{D}_{\mathcal{ALCO}@}$ concepts

$\neg_f$-rule:	If $\sigma.\varepsilon : \neg(C(x)) \in \mathcal{B}$, and $\sigma.\varepsilon : (\neg C)(x) \notin \mathcal{B}$, then set $\mathcal{B} := \mathcal{B} \cup \{\sigma.\varepsilon : (\neg C)(x)\}$.
$\neg\neg_f$-rule:	If $\sigma.\varepsilon : \neg(\neg\varphi) \in \mathcal{B}$, and $\sigma.\varepsilon : \varphi \notin \mathcal{B}$, then set $\mathcal{B} := \mathcal{B} \cup \{\sigma.\varepsilon : \varphi\}$.
$\neg\vee$-rule:	If $\sigma.\varepsilon : \neg(\varphi \vee \psi) \in \mathcal{B}$, and $\{\sigma.\varepsilon : \neg\varphi, \sigma.\varepsilon : \neg\psi\} \nsubseteq \mathcal{B}$, then set $\mathcal{B} := \mathcal{B} \cup \{\sigma.\varepsilon : \neg\varphi, \sigma.\varepsilon : \neg\psi\}$.
$\vee$-rule:	If $\sigma.\varepsilon : \varphi \vee \psi \in \mathcal{B}$, and $\{\sigma.\varepsilon : \varphi, \sigma.\varepsilon : \psi\} \cap \mathcal{B} = \emptyset$, then set $\mathcal{B} := \mathcal{B} \cup \{\sigma.\varepsilon : \phi\}$ for some $\phi \in \{\varphi, \psi\}$.

Fig. 2. The tableau expansion rules on $\mathcal{D}_{\mathcal{ALCO}@}$ formulas

$\neg;_{<>}$-rule: If $\sigma.\varepsilon : \neg < \pi_1; \pi_2 > \varphi \in \mathcal{B}$, and $\sigma.\varepsilon : \neg < \pi_1 >< \pi_2 > \varphi \notin \mathcal{B}$,
then set $\mathcal{B} := \mathcal{B} \cup \{\sigma.\varepsilon : \neg < \pi_1 >< \pi_2 > \varphi\}$.

$;_{<>}$-rule: If $\sigma.\varepsilon :< \pi_1; \pi_2 > \varphi \in \mathcal{B}$, and $\sigma.\varepsilon :< \pi_1 >< \pi_2 > \varphi \notin \mathcal{B}$,
then set $\mathcal{B} := \mathcal{B} \cup \{\sigma.\varepsilon :< \pi_1 >< \pi_2 > \varphi\}$.

$\neg ?_{<>}$-rule: If $\sigma.\varepsilon : \neg < \phi? > \varphi \in \mathcal{B}$, and $\{\sigma.\varepsilon : \neg\phi,\ \sigma.\varepsilon : \neg\varphi\} \cap \mathcal{B} = \emptyset$,
then set $\mathcal{B} := \mathcal{B} \cup \{\sigma.\varepsilon : \psi\}$ for some $\psi \in \{\neg\phi,\ \neg\varphi\}$.

$?_{<>}$-rule: If $\sigma.\varepsilon :< \phi? > \varphi \in \mathcal{B}$, and $\{\sigma.\varepsilon : \phi,\ \sigma.\varepsilon : \varphi\} \nsubseteq \mathcal{B}$,
then set $\mathcal{B} := \mathcal{B} \cup \{\sigma.\varepsilon : \phi,\ \sigma.\varepsilon : \varphi\}$.

$\neg\cup_{<>}$-rule: If $\sigma.\varepsilon : \neg < \pi_1 \cup \pi_2 > \varphi \in \mathcal{B}$, and $\{\sigma.\varepsilon : \neg < \pi_1 > \varphi,\ \sigma.\varepsilon : \neg < \pi_2 > \varphi\} \nsubseteq \mathcal{B}$,
then set $\mathcal{B} := \mathcal{B} \cup \{\sigma.\varepsilon : \neg < \pi_1 > \varphi,\ \sigma.\varepsilon : \neg < \pi_2 > \varphi\}$.

$\cup_{<>}$-rule: If $\sigma.\varepsilon :< \pi_1 \cup \pi_2 > \varphi \in \mathcal{B}$, and $\{\sigma.\varepsilon :< \pi_1 > \varphi,\ \sigma.\varepsilon :< \pi_2 > \varphi\} \cap \mathcal{B} = \emptyset$,
then set $\mathcal{B} := \mathcal{B} \cup \{\sigma.\varepsilon : \psi\}$ for some $\psi \in \{< \pi_1 > \varphi,\ < \pi_2 > \varphi\}$.

$\neg *_{<>}$-rule: If $\sigma.\varepsilon : \neg < \pi^* > \varphi \in \mathcal{B}$, and $\{\sigma.\varepsilon : \neg\varphi,\ \sigma.\varepsilon : \neg < \pi >< \pi^* > \varphi\} \nsubseteq \mathcal{B}$,
then set $\mathcal{B} := \mathcal{B} \cup \{\sigma.\varepsilon : \neg\varphi,\ \sigma.\varepsilon : \neg < \pi >< \pi^* > \varphi\}$.

$*_{<>}$-rule: If $\sigma.\varepsilon :< \pi^* > \varphi \in \mathcal{B}$, and
there is no variable X such that $X \doteq < \pi^* > \varphi \in \mathcal{B}$ and $\sigma.\varepsilon : X \in \mathcal{B}$,
then introduce a new variable X, set $\mathcal{B} := \mathcal{B} \cup \{X \doteq < \pi^* > \varphi,\ \sigma.\varepsilon : X\}$.

X-rule: If $\sigma.\varepsilon : X \in \mathcal{B}$ with $X \doteq < \pi^* > \varphi \in \mathcal{B}$, and $\{\sigma.\varepsilon : \varphi,\ \sigma.\varepsilon :< \pi > X\} \cap \mathcal{B} = \emptyset$,
then set $\mathcal{B} := \mathcal{B} \cup \{\sigma.\varepsilon : \varphi\}$, or set $\mathcal{B} := \mathcal{B} \cup \{\sigma.\varepsilon : \neg\varphi,\ \sigma.\varepsilon :< \pi > X\}$.

Fig. 3. The tableau expansion rules on $\mathcal{D}_{\mathcal{ALCO}@}$ actions (*Part I*)

atom$_{<>}$-rule: If $\sigma.\varepsilon :< (P, E) > \varphi \in \mathcal{B}$, and $\{\sigma.\varepsilon : \phi \mid \phi \in P\} \nsubseteq \mathcal{B}$ or there is no prefix
$\sigma_i.\varepsilon_i$ such that both $\varepsilon_i = (\varepsilon - \{\neg\varphi | \varphi \in E\}) \cup E$ and $\sigma_i.\varepsilon_i : \varphi \in \mathcal{B}$,
then, introduce a prefix $\sigma'.\varepsilon' := \sigma; (P, E).(\varepsilon - \{\neg\varphi | \varphi \in E\}) \cup E$, and
set $\mathcal{B} := \mathcal{B} \cup \{\sigma.\varepsilon : \phi \mid \phi \in P\} \cup \{\sigma'.\varepsilon' : \varphi\} \cup \{\sigma'.\varepsilon' : \psi \mid \psi \in \varepsilon'\}$.

$\neg$*atom*$_{<>}$-rule: If $\sigma.\varepsilon : \neg < (P, E) > \varphi \in \mathcal{B}$ with $\{\sigma : \neg\phi \mid \phi \in P\} \cap \mathcal{B} = \emptyset$, and there is
a prefix $\sigma_i.\varepsilon_i$ such that $\varepsilon_i = (\varepsilon - \{\neg\varphi | \varphi \in E\}) \cup E$ and $\sigma_i.\varepsilon_i : \neg\varphi \notin \mathcal{B}$,
then set $\mathcal{B} := \mathcal{B} \cup \{\sigma_i.\varepsilon_i : \neg\varphi\}$,
or set $\mathcal{B} := \mathcal{B} \cup \{\sigma.\varepsilon : \neg\phi\}$ for some $\phi \in P$.

B_1-rule: If $\sigma.\varepsilon : \varphi \in \mathcal{B}$, φ is of form $R(x, y)$, $\neg R(x, y)$, $C(x)$ or $(\neg C)(x)$, where $C \in N_{CP}$,
and both $\varphi \notin \varepsilon$ and $\sigma_0.\varepsilon_0 : \varphi \notin \mathcal{B}$,
then set $\mathcal{B} := \mathcal{B} \cup \{\sigma_0.\varepsilon_0 : \varphi\}$.

B_2-rule: If $\sigma.\varepsilon : D(x) \in \mathcal{B}$, D is of form $\forall R.C$ or $\neg\forall R.C$, $\sigma.\varepsilon$ is not $\sigma_0.\varepsilon_0$,
and $\sigma_0.\varepsilon_0 : D^{Regress(\sigma.\varepsilon)}(x) \notin \mathcal{B}$,
then set $\mathcal{B} := \mathcal{B} \cup \{\sigma_0.\varepsilon_0 : D^{Regress(\sigma.\varepsilon)}(x)\}$.

Fig. 4. The tableau expansion rules on $\mathcal{D}_{\mathcal{ALCO}@}$ actions (*Part II*)

Given a concept D and a prefix $\sigma.\varepsilon$, the concept $D^{Regress(\sigma.\varepsilon)}$ present in B_2-rule of Figure 4 is constructed as follows:

1. Let σ be of form $\sigma_0; (P_1, E_1); \ldots; (P_n, E_n)$. Construct $n+1$ sets ρ_0 ,..., ρ_n, such that:
   - $\rho_n := \emptyset$;
   - $\rho_i := (\rho_{i+1} - E_{i+1}) \cup \{\neg\varphi | \varphi \in E_{i+1}\}$ for each i with $n > i \geq 0$.
2. Set $\varepsilon' := \varepsilon - \rho_0$. Let $Obj(\varepsilon')$ be all the individual names occurring in ε'. Then, construct $D^{Regress(\sigma.\varepsilon)}$ inductively as follows:
   - For primitive concept name $C_i \in N_{CP}$, $C_i^{Regress(\sigma.\varepsilon)} := C_i \sqcup \bigsqcup_{C_i(p)\in\varepsilon'} \{p\} \sqcap \bigsqcap_{\neg C_i(p)\in\varepsilon'} \neg\{p\}$;
   - $\{p\}^{Regress(\sigma.\varepsilon)} := \{p\}$;
   - $(@_pC)^{Regress(\sigma.\varepsilon)} := @_pC^{Regress(\sigma.\varepsilon)}$;
   - $(\neg C)^{Regress(\sigma.\varepsilon)} := \neg C^{Regress(\sigma.\varepsilon)}$;
   - $(C \sqcup D)^{Regress(\sigma.\varepsilon)} := C^{Regress(\sigma.\varepsilon)} \sqcup D^{Regress(\sigma.\varepsilon)}$;
   - $(\forall R.C)^{Regress(\sigma.\varepsilon)} := (\bigsqcup_{p\in Obj(\varepsilon')} \{p\} \sqcup \forall R.C^{Regress(\sigma.\varepsilon)})$ $\sqcap \forall R.(\bigsqcup_{q\in Obj(\varepsilon')} \{q\} \sqcup C^{Regress(\sigma.\varepsilon)})$ $\sqcap \bigsqcap_{p,q\in Obj(\varepsilon'), R(p,q)\notin\varepsilon', \neg R(p,q)\notin\varepsilon'} (\neg\{p\} \sqcup \forall R.(\neg\{q\} \sqcup C^{Regress(\sigma.\varepsilon)}))$ $\sqcap \bigsqcap_{R(p,q)\in\varepsilon'} (\neg\{p\} \sqcup @_qC^{Regress(\sigma.\varepsilon)})$;

Inspired by Liu's ABox updating algorithm [10], the algorithm here is technically designed to guarantee the following property:

Lemma 1. *For any model $M = (W, I)$ and any states $w, w' \in W$, if $(w, w') \in \sigma^I$, then $(D^{Regress(\sigma.\varepsilon)})^{I(w)} = D^{I(w')}$.*

Due to space constraints, detail proofs must be omitted here.

An eventuality $X \doteq < \pi^* > \varphi$ is *fulfilled* in a branch $\mathcal{B}$ iff there exists prefixes $\sigma.\varepsilon$ and $\sigma'.\varepsilon'$ such that $\varepsilon = \varepsilon'$, $\sigma.\varepsilon : X \in S$, and $\sigma'.\varepsilon' : \varphi \in S$.

A branch $\mathcal{B}$ is *completed* iff no tableau expansion rules can be applied on it.

A branch $\mathcal{B}$ is *ignorable* iff (i) it is completed, and (ii) it contains an eventuality X which is not fulfilled.

A branch $\mathcal{B}$ is *contradictory* iff one of the following conditions holds:

- there exists a formula φ and prefixes $\sigma.\varepsilon$ and $\sigma'.\varepsilon'$, such that $\sigma.\varepsilon : \varphi \in S$, $\sigma'.\varepsilon' : \neg\varphi \in S$ and $\varepsilon = \varepsilon'$;
- there exists a concept C, an individual name p, and prefixes $\sigma.\varepsilon$ and $\sigma'.\varepsilon'$, such that $\sigma.\varepsilon : C(p) \in S$, $\sigma'.\varepsilon' : (\neg C)(p) \in S$ and $\varepsilon = \varepsilon'$;
- there exists two individual names p, q and a prefix $\sigma.\varepsilon$, such that $\sigma.\varepsilon : \{q\}(p) \in \mathcal{B}$ and $\{p \neq q, q \neq p\} \cap \mathcal{B} \neq \emptyset$;
- there exists an inequality $p \neq p \in \mathcal{B}$ for some individual name p.

We are now ready to finish the description of *the satisfiability-checking algorithm.*

Given an acyclic TBox $\mathcal{T}$ and an ActionBox $\mathcal{A}_C$, the satisfiability of a formula φ w.r.t. $\mathcal{T}$ and $\mathcal{A}_C$ is decided with the following steps:

1. Transform the formula φ into an normal form $nf(\varphi)$ with the following steps:
   (a) Replace each occurrence of atomic actions with their definitions; and
   (b) Replace each occurrence of defined concept names with their definitions.
2. Construct a branch $\mathcal{B} := \{\sigma_0.\varepsilon_0 : nf(\varphi)\}$.
3. If the tableau expansion rules can be applied to $\mathcal{B}$ in such a way that they yield a completed branch which is neither contradictory nor ignorable, then the algorithm returns "φ is satisfiable", and "φ is unsatisfiable" otherwise.

Lemma 2. *Given an acyclic TBox $\mathcal{T}$, an ActionBox $\mathcal{A}_C$ and a formula φ, the satisfiability-checking algorithm terminates.*

A basic evidence for this lemma is that the number of all the prefixes introduced by the $atom_{<>}$-rule is bounded by $2^{|CL(\varphi)|+|Eff(\varphi)|}$. Where $CL(\varphi)$ is the set of all subformulas of φ. $Eff(\varphi)$ is the union of the effects E_i for all the atomic actions (P_i, E_i) occurring in φ.

For demonstrating the soundness and completeness, we introduce a *mapping* function to act as a bridge between prefixes and states:

A *mapping* $\imath$ with respect to a branch $\mathcal{B}$ and a model $M = (W, I)$ is a function from prefixes occurring in $\mathcal{B}$ to states in W, such that for all prefixes $\sigma.\varepsilon$ and $\sigma;(P,E).(\varepsilon - \{\neg\varphi|\varphi \in E\}) \cup E$ present in $\mathcal{B}$ it is:

$$(\,\imath(\sigma.\varepsilon),\ \ \imath(\sigma;(P,E).(\varepsilon - \{\neg\varphi|\varphi \in E\}) \cup E)\,) \in (P,E)^I.$$

Then, such a mapping holds the following property:

Lemma 3. *For a mapping $\imath$ with respect to a branch $\mathcal{B}$ and a model $M = (W, I)$, we have:*

- $C^{I(\imath(\sigma.\varepsilon))} = C^{I(\imath(\sigma_0.\varepsilon_0))} \cup \{p^I \mid C(p) \in \varepsilon\} - \{p^I \mid \neg C(p) \in \varepsilon\}$ *for each primitive concept name $C \in N_{CP}$;*
- $R^{I(\imath(\sigma.\varepsilon))} = R^{I(\imath(\sigma_0.\varepsilon_0))} \cup \{(p^I, q^I) \mid R(p,q) \in \varepsilon\} - \{(p^I, q^I) | \neg R(p,q) \in \varepsilon\}$ *for each role name $R \in N_R$.*

Based on the mapping function, we introduce the satisfiability of branches:

A branch $\mathcal{B}$ is *satisfiable* iff there is a model $M = (W, I)$ and a mapping $\imath$ such that: (i) for every prefixed formula $\sigma.\varepsilon : \varphi \in \mathcal{B}$ it is $(M, \imath(\sigma.\varepsilon)) \models \varphi$, and (ii) for every inequality $p \neq q \in \mathcal{B}$ it is $p^I \neq q^I$.

Then, we can demonstrate that our algorithm holds the following properties.

Lemma 4. *For the application of any expansion rule, a branch $\mathcal{B}$ is satisfiable before the application if and only if there is a satisfiable branch after the application.*

Lemma 5. *A branch $\mathcal{B}$ is satisfiable if it is completed and neither contradictory nor ignorable.*

Lemma 6. *A branch $\mathcal{B}$ is unsatisfiable if it is contradictory or ignorable.*

As an immediate consequence of Lemmas 2, 4, 5, and 6, the satisfiability-checking algorithm always terminates, and answers with "φ is satisfiable" iff φ is satisfiable w.r.t $\mathcal{T}$ and $\mathcal{A_C}$.

Theorem 1. *It is decidable whether or not a $\mathcal{D}_{\mathcal{ALCO}@}$ formula is satisfiable w.r.t an acyclic TBox and an ActionBox.*

4 Reasoning About Actions

In this section we investigate the realizability, executability, projection and planning problems on actions.

In the following we let $\mathcal{T}$ and $\mathcal{A_C}$ be an acyclic TBox and an ActionBox respectively. For actions and formulas that we will refer to, let all the atomic actions occurring in them be defined according to $\mathcal{A_C}$.

The *realizability problem* is to check whether an action makes sense. Formally, an action π is *realizable* w.r.t. $\mathcal{T}$ and $\mathcal{A_C}$ iff there exists a model $M = (W, I)$ of $\mathcal{T}$ and two states w, w' in W such that $(w, w') \in \pi^I$.

Realizability of actions can be checked according to the following result:

Theorem 2. *An action π is realizable w.r.t. $\mathcal{T}$ and $\mathcal{A_C}$ iff the formula $< \pi >$ true is satisfiable w.r.t. $\mathcal{T}$ and $\mathcal{A_C}$.*

The *executability problem* is to decide whether an action is executable on states described by certain formula. Formally, an action π is *executable* on states described by a formula φ iff, for any model $M = (W, I)$ of $\mathcal{T}$ and any state $w \in W$ with $(M, w) \models \varphi$: a model $M' = (W', I')$ of $\mathcal{T}$ can always be constructed by introducing a finite number of states w_1', ..., w_n' into M, with constraints $W' = W \cup \{w_1', ..., w_n'\}$ and $I'(w_i) = I(w_i)$ for each $w_i \in W$, such that $(M', w) \models \varphi$ and there exists a state $w' \in W'$ with $(w, w') \in \pi^{I'}$.

Executability of actions can be checked according to the following theorem:

Theorem 3. *Let (P_1, E_1), ..., (P_n, E_n) be all the atomic actions occurring in π and φ. Then, π is executable on states described by φ iff the following formula is valid w.r.t. $\mathcal{T}$ and $\mathcal{A_C}$:*

$$[((P_1, E_1) \cup ... \cup (P_n, E_n))^*]\Pi \rightarrow (\varphi \rightarrow < \pi > true)$$

where Π is the formula $(\phi_1 \rightarrow < (P_1, E_1) > true) \wedge ... \wedge (\phi_n \rightarrow < (P_n, E_n) > true)$, with ϕ_i be the conjunction of all formulas in P_i, for each $1 \leq i \leq n$.

The *projection problem* is to decide whether a formula really holds after executing an action on certain states. Formally, a formula ψ is a *consequence of applying* an action π in states described by a formula φ iff, for any model $M = (W, I)$ of $\mathcal{T}$ and any states $w, w' \in W$ with $(M, w) \models \varphi$ and $(w, w') \in \pi^I$: we have $(M, w') \models \psi$.

The projection problem can be reasoned according to the following theorem:

Theorem 4. *ψ is a consequence of applying π in states described by φ iff the formula $\varphi \rightarrow [\pi]\psi$ is valid w.r.t. $\mathcal{T}$ and $\mathcal{A}_{\mathcal{C}}$.*

Given an initial state and a goal statement, the *planning problem* is to find an action sequence that will lead to states in which the goal will be true. Formally, let Σ be a set of actions, let φ and ψ be two formulas that respectively describe the initial state and the goal condition; Then, the planning problem is to find an action sequence that can be executed on any states satisfying φ and will achieve the goal ψ at the resulted states. Such an action sequence is also called a *plan*.

For the planning problem we have the following results:

Theorem 5. *Given a set of actions $\Sigma = \{\pi_1, ..., \pi_m\}$ and two formulas φ and ψ, let (P_1, E_1), ..., (P_n, E_n) be all atomic actions occurring in Σ, φ and ψ. Then, there exists a plan to achieve the goal ψ starting from initial states described by φ iff the following formula is valid w.r.t. $\mathcal{T}$ and $\mathcal{A}_{\mathcal{C}}$:*

$$[((P_1, E_1) \cup ... \cup (P_n, E_n))^*]\Pi \rightarrow (\varphi \rightarrow < (\pi_1 \cup ... \cup \pi_m)^* > \psi)$$

where Π is the formula $(\phi_1 \rightarrow < (P_1, E_1) > true) \wedge ... \wedge (\phi_n \rightarrow < (P_n, E_n) > true)$, with ϕ_i be the conjunction of all formulas in P_i, for each $1 \leq i \leq n$.

Furthermore, a working plan can be generated from the proof if the formula is proved to be valid.

Finally, given an action sequence π_1, ..., π_n, we can also check whether it is a *plan* by deciding whether the following formula is valid w.r.t. $\mathcal{T}$ and $\mathcal{A}_{\mathcal{C}}$:

$$(\varphi \rightarrow [\pi]\psi) \wedge ([((P_1, E_1) \cup ... \cup (P_n, E_n))^*]\Pi \rightarrow (\varphi \rightarrow < \pi > true))$$

where π is the sequence-action $\pi_1; ...; \pi_n$.

5 Conclusion

The dynamic description logic $\mathcal{D}_{\mathcal{ALCO}@}$ provides a powerful language for representing and reasoning about actions based on the description logic $\mathcal{ALCO}@$. On the one hand, actions in the logic are represented over ontologies expressed by the description logic $\mathcal{ALCO}@$, and can be constructed with the help of standard action operators. On the other hand, actions can be used as modal operators to construct formulas; Based on the algorithm for checking the satisfiability of formulas, many reasoning tasks on actions can be effectively carried out.

Another approach to integrate description logics with action formalisms was proposed by Baader [2], and was characterized by constructing action formalisms based on description logics. The projection and executability problems were studied in that paper and reduced to standard reasoning problems on description logics. A major limitation of that formalisms is that actions are restricted to be either atomic actions or finite sequences of atomic actions. Compared with Baader's DL-based action formalism, our logic provides a more powerful language for representing and reasoning about actions.

A dynamic description logic named $\mathcal{PDLC}$ was firstly studied by Wolter [16]. In that logic, actions were used as modal operators to construct both concepts and formulas, so that concepts with dynamic meaning could be described. But actions were only treated as abstract modal operators, and could not be further specified and reasoned. Furthermore, efficient decision algorithm for that logic is still an open problem.

Following Shi's approach [13], a dynamic description logic was constructed in our previous work [5], in which actions could be used as modal operators to construct both concepts and formulas. However, the iteration of actions were not allowed, so that actions such as "*while* φ *do* π" could not be described and reasoned. Furthermore, compared with [5], another contribution of the present paper is that four reasoning tasks on actions were effectively carried out.

Acknowledgments. This work was partially supported by the National Science Foundation of China (No. 90604017), 863 National High-Tech Program (No. 2007AA01Z132), and National Basic Research Priorities Programme (No. 2007CB311004).

References

1. Baader, F., Horrocks, I., Sattler, U.: Description Logics as Ontology Languages for the Semantic Web. In: Hutter, D., Stephan, W. (eds.) Mechanizing Mathematical Reasoning. LNCS (LNAI), vol. 2605, pp. 228–248. Springer, Heidelberg (2005)
2. Baader, F., Lutz, C., Milicic, M., Sattler, U., Wolter, F.: Integrating Description Logics and Action Formalisms: First Results. In: Veloso, M., Kambhampati, S. (eds.) Proceedings of the 12th Nat. Conf. on Artif. Intell., pp. 572–577. AAAI Press, Stanford (2005)
3. Calvanese, D., De Giacomo, G., Vardi, M.: Reasoning about Actions and Planning in LTL Action Theories. In: Fensel, D., Giunchiglia, F., McGuinness, D., Williams, M. (eds.) 8th Int. Conf. on Principles and Knowledge Representation and Reasoning, pp. 593–602. Morgan Kaufmann, San Francisco (2002)
4. Calvanese, D., De Giacomo, G., Lenzerini, M., Rosati, R.: Actions and Programs over Description Logic Ontologies. In: Calvanese, D., Franconi, E., Haarslev, V., Lembo, D., Motik, B., Turhan, A., Tessaris, S. (eds.) 2007 Int. Workshop on Description Logics. CEUR-WS, pp. 29–40 (2007)
5. Chang, L., Lin, F., Shi, Z.: A Dynamic Description Logic for Semantic Web Service. In: 3rd Int. Conf. on Semantics, Knowledge and Grid, IEEE Press, Los Alamitos (2007)
6. Foo, N., Zhang, D.: Dealing with the Ramification Problem in the Extended Propositional Dynamic Logic. In: Wolter, F., Wansing, H., Rijke, M., Zakharyaschev, M. (eds.) Advances in Modal Logic, vol. 3, pp. 173–191. World Scientific, Singapore (2002)
7. De Giacomo, G., Lenzerini, M.: PDL-based Framework for Reasoning about Actions. In: Gori, M., Soda, G. (eds.) Topics in Artificial Intelligence. LNCS, vol. 992, pp. 103–114. Springer, Heidelberg (1995)
8. De Giacomo, G., Massacci, F.: Tableaux and Algorithms for Propositional Dynamic Logic with Converse. In: McRobbie, M.A., Slaney, J.K. (eds.) CADE 1996. LNCS, vol. 1104, pp. 613–627. Springer, Heidelberg (1996)

9. Horrocks, I., Patel-Schneider, P.F., Harmelen, F.V.: From SHIQ and RDF to OWL: the Making of a Web Ontology Language. J. Web Sem. 1(1), 7–26 (2003)
10. Liu, H., Lutz, C., Milicic, M., Wolter, F.: Updating Description Logic ABoxes. In: Doherty, P., Mylopoulos, J., Welty, C. (eds.) 10th Int. Conf. on Principles of Knowledge Representation and Reasoning, pp. 46–56. AAAI Press, Stanford (2006)
11. McIlraith, S., Son, T., Zeng, H.: Semantic Web Services. IEEE Intelligent Systems. 16(2), 46–53 (2001)
12. Reiter, R.: Knowledge in Action: Logical Foundations for Specifying and Implementing Dynamical Systems. MIT Press, Cambridge (2001)
13. Shi, Z., Dong, M., Jiang, Y., Zhang, H.: A Logic Foundation for the Semantic Web. Science in China, Series F 48(2), 161–178 (2005)
14. Thielscher, M.: Introduction to the Fluent Calculus. Electron. Trans. Artif. Intell. 2(3-4), 179–192 (1998)
15. Winslett, M.: Reasoning about Action Using a Possible Models Approach. In: 7th Nat. Conf. on Artif. Intell., pp. 89–93. AAAI Press, Stanford (1988)
16. Wolter, F., Zakharyaschev, M.: Dynamic Description Logic. In: Zakharyaschev, M., Segerberg, K., Rijke, M., Wansing, H. (eds.) Advances in Modal Logic, vol. 2, pp. 431–446. CSLI Publications, Stanford, CA (1998)

An Argumentative Reasoning Service for Deliberative Agents*

Alejandro J. García, Nicolás D. Rotstein, Mariano Tucat, and Guillermo R. Simari

Consejo Nacional de Investigaciones Científicas y Técnicas (CONICET)
Department of Computer Science and Engineering
Universidad Nacional del Sur, Bahía Blanca, Argentina
{ajg,ndr,mt,grs}@cs.uns.edu.ar

Abstract. In this paper we propose a model that allows agents to deliberate using defeasible argumentation, to share knowledge with other agents, and to represent individual knowledge privately. We describe the design and implementation of a Defeasible Logic Programming Server that handles queries from several remote client agents. Queries will be answered using public knowledge stored in the Server and individual knowledge that client agents can send as part of a query, providing a particular context for it. The Server will answer these contextual queries using a defeasible argumentative analysis. Different types of contextual queries are presented and analyzed.

1 Introduction

Deliberative agents that take part of a Multi-agent System (MAS) usually reason by using two sources of knowledge: public knowledge they share with other agents, and individual or private knowledge that arise in part from their own perception of the environment. In this paper we propose a model that allows agents to deliberate using defeasible argumentation, to share knowledge with other agents, and to represent individual knowledge privately. We focus on the design and implementation of a client-server approach based on Defeasible Logic Programming that provides a knowledge representation formalism and a Defeasible Argumentation reasoning service. Thus, agents can reason with the mentioned formalism, using both private and shared knowledge, by means of this external service.

In our approach, a Defeasible Logic Programming Server (DeLP-server) will answer queries received from client agents that can be distributed in remote hosts. Public knowledge can be stored in the DeLP-server represented as a Defeasible Logic Program. To answer queries, the DeLP-server will use this public knowledge together with individual knowledge that clients might send, creating a particular *context* for the query. These *contextual queries* will be answered

* Partially supported by Universidad Nacional del Sur, CONICET (PIP 5050), and Agencia Nacional de Promoción Científica y Tecnológica.

Z. Zhang and J. Siekmann (Eds.): KSEM 2007, LNAI 4798, pp. 128–139, 2007.

using a defeasible argumentative analysis. Several DeLP-servers can be used simultaneously, each of them providing a different shared knowledge base. Thus, several agents can consult the same DeLP-server, and the same agent can consult several DeLP-servers. Our approach do not impose any restriction over the type, architecture, or implementation language of the client agents.

In our model, both public knowledge stored in the server and contextual knowledge sent by the agents are used for answering queries, however, no permanent changes are made to the stored program. The temporal scope of the contextual information sent in a query is limited and it will disappear with the finalization of the process performed by the DeLP-server to answer that query. Since agents are not restricted to consult a unique server, they may perform the same contextual query to different servers, and they may share different knowledge with other agents through different servers. Thus, several configurations of agents and servers can be established (statically or dynamically). For example, special-purpose DeLP-servers can be used, each of them representing particular shared knowledge of a specific domain. Thus, like in other approaches, shared knowledge will not be restricted to be in a unique repository and therefore, it can be structured in many ways.

2 DeLP Basis

In our approach, both the individual knowledge of an agent and the public knowledge loaded in a DeLP-Server are represented using a defeasible logic program. A brief description of Defeasible Logic Programming (DeLP) is included in this section –for a detailed presentation see [1]. DeLP is a formalism that combines logic programming and defeasible argumentation [2,3]. In DeLP, knowledge is represented using facts, strict rules or defeasible rules. *Facts* are ground literals representing atomic information or the negation of atomic information using the strong negation "$\sim$". *Strict Rules* are denoted $L_0 \leftarrow L_1, \ldots, L_n$ and represent firm information, whereas *Defeasible Rules* are denoted $L_0 \prec L_1, \ldots, L_n$ and represent defeasible knowledge, *i.e.*, tentative information. In both cases, the *head* L_0 is a literal and the *body* $\{L_i\}_{i>0}$ is a set of literals. In this paper we will consider a restricted form of program that do not have strict rules.

Definition 1. *A restricted defeasible logic program (DeLP-program for short) $\mathcal{P}$ is a set of facts and defeasible rules. When required, $\mathcal{P}$ is denoted (Π,Δ) distinguishing the subset Π of facts and the subset Δ of defeasible rules.*

Strong negation is allowed in the head of program rules, and hence may be used to represent contradictory knowledge. From a program (Π,Δ) contradictory literals could be derived, however, the set Π (which is used to represent non-defeasible information) must possess certain internal coherence. Therefore, Π has to be non-contradictory, *i.e.*, no pair of contradictory literals can be derived from Π. Given a literal L the complement with respect to strong negation will be denoted $\overline{L}$ (*i.e.*, $\overline{a}{=}{\sim}a$ and $\overline{\sim a}{=}a$). Adding facts to a DeLP-program can produce a contradictory set Π, producing a non-valid program. However, defeasible rules can be added without any restriction.

Observation 1. *Let (Π,Δ) be a DeLP-program, Δ_1 a set of defeasible rules and Π_1 a set of facts. The pair (Π,($\Delta \cup \Delta_1$)) is a valid DeLP-program, but ($\Pi \cup \Pi_1$, Δ) may not because $\Pi \cup \Pi_1$ can be a contradictory set. Nevertheless, if $\Pi \cup \Pi_1$ is a non-contradictory set then ($\Pi \cup \Pi_1$, $\Delta \cup \Delta_1$) is a valid DeLP-program.*

Definition 2 (DeLP-query). *A* DeLP-query *is a ground literal that DeLP will try to warrant. A query with at least one variable will be called* schematic query *and will represent the set of DeLP-queries that unify with the schematic one.*

To deal with contradictory and dynamic information, in DeLP, *arguments* for conflicting pieces of information are built and then compared to decide which one prevails. The prevailing argument provides a *warrant* for the information that it supports. In DeLP a query L is *warranted* from a program (Π,Δ) if a *non-defeated* argument $\mathcal{A}$ supporting L exists. An *argument* $\mathcal{A}$ for a literal L [1], denoted $\langle \mathcal{A}, L \rangle$, is a minimal set of defeasible rules $\mathcal{A} \subseteq \Delta$, such that $\mathcal{A} \cup \Pi$ is non-contradictory and there is a derivation for L from $\mathcal{A} \cup \Pi$.

Example 1. Consider the DeLP-program $\mathcal{P}_1 = (\Pi_1, \Delta_1)$, where $\Pi_1 = \{q, s, t\}$ and $\Delta_1 = \{(r \prec q)(\sim r \prec q, s)(r \prec s)(\sim r \prec t)(\sim a \prec q)(a \prec s)\}$. From $\mathcal{P}_1$ the following arguments can be built:
$\langle \mathcal{R}_1, \sim r \rangle = \langle \{\sim r \prec t\}, \sim r \rangle$ $\langle \mathcal{R}_2, r \rangle = \langle \{r \prec q\}, r \rangle$ $\langle \mathcal{R}_3, r \rangle = \langle \{r \prec s\}, r \rangle$
$\langle \mathcal{R}_4, \sim r \rangle = \langle \{\sim r \prec q, s\}, \sim r \rangle$ $\langle \mathcal{A}_1, \sim a \rangle = \langle \{\sim a \prec q\}, \sim a \rangle$ $\langle \mathcal{A}_2, a \rangle = \langle \{a \prec s\}, a \rangle$

To establish if $\langle \mathcal{A}, L \rangle$ is a non-defeated argument, *defeaters* for $\langle \mathcal{A}, L \rangle$ are considered, *i.e.*, counter-arguments that by some criterion are preferred to $\langle \mathcal{A}, L \rangle$. It is important to note that in DeLP the argument comparison criterion is modular and thus, the most appropriate criterion for the domain that is being represented can be selected. In the examples in this paper we will use *generalized specificity* [4], a criterion that favors two aspects of an argument: it prefers (1) a *more precise* argument (*i.e.*, with greater information content) or (2) a *more concise* argument (*i.e.*, with less use of rules). Using this criterion in Example 1, $\langle \mathcal{R}_4, \sim r \rangle$ is preferred to $\langle \mathcal{R}_2, r \rangle$ (more precise).

A defeater $\mathcal{D}$ for an argument $\mathcal{A}$ can be *proper* ($\mathcal{D}$ is preferred to $\mathcal{A}$) or *blocking* (same strength). A defeater can attack the conclusion of another argument or an inner point of it. Since defeaters are arguments, there may exist defeaters for them, and defeaters for these defeaters, and so on. Thus, a sequence of arguments called *argumentation line* [1] can arise. Clearly, for a particular argument there might be more than one defeater. Therefore, many argumentation lines could arise from one argument, leading to a tree structure called *dialectical tree* [1]. In a dialectical tree (see Fig. 1), every node (except the root) is a defeater of its parent, and leaves are non-defeated arguments. A dialectical tree provides a structure for considering all the possible acceptable argumentation lines that can be generated. In a dialectical tree every node can be marked as *defeated* or *undefeated*: leaves are marked as undefeated nodes, and inner nodes are marked defeated when there is at least a child marked undefeated, or are marked undefeated when all its children are marked defeated. Figure 1 shows three different marked dialectical trees, where white triangles represent undefeated nodes, black triangles defeated ones, and arrows the defeat relation.

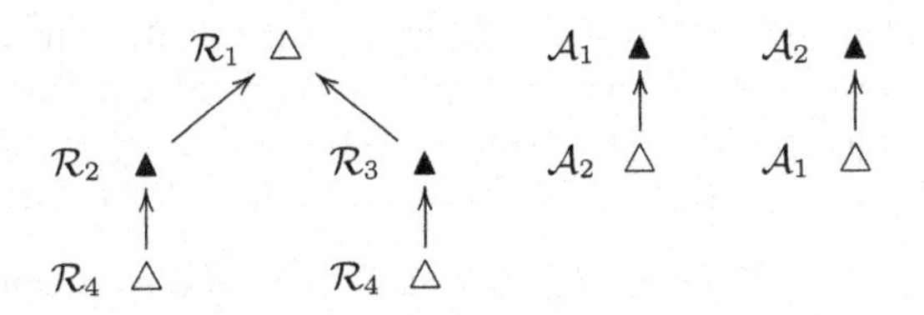

Fig. 1. Dialectical trees of Ex. 1

Definition 3 (Warranting a DeLP-query). *A DeLP-query Q is warranted from a DeLP-program $\mathcal{P}$ if there exists an argument $\mathcal{A}$ supporting Q such that $\mathcal{A}$ is the root of a dialectical tree and the root is marked as undefeated.*

Definition 4 (Answer for a DeLP-query). *The answer for a query Q from a DeLP-program $\mathcal{P}$ is either:* YES, *if Q is warranted from $\mathcal{P}$;* NO, *if the complement of Q is warranted from $\mathcal{P}$;* UNDECIDED, *if neither Q nor its complement are warranted from $\mathcal{P}$; or* UNKNOWN, *if Q is not in the language of the program $\mathcal{P}$.*

Consider again Ex. 1. From $\mathcal{P}_1$ the answer for $\sim r$ is YES whereas for r is NO (Fig. 1 shows the dialectical tree that provides a warrant for $\sim r$). The answer for a is UNDECIDED and the answer for $\sim a$ is also UNDECIDED (observe that $\mathcal{A}_1$ and $\mathcal{A}_2$ block each other). Finally, the answer for z is UNKNOWN.

3 Contextual Queries

As stated above, several DeLP-Servers can be used in a MAS, and each of them provides a defeasible argumentation reasoning service for other agents. Public knowledge (represented as a DeLP-program) can be stored in each DeLP-Server, and will be used for answering DeLP-queries. For example, if the query $\sim r$ is sent to a DeLP-Server where the program (Π_1, Δ_1) of Ex. 1 is stored, then the answer YES will be returned.

Besides public knowledge, agents may have their own private knowledge. Therefore, the proposed model not only allows agents to perform queries to the program stored in a DeLP-Server, but also permits the inclusion in the query of private pieces of information related to the agent's particular context. This type of query will be called *contextual query* and will be introduced next.

Definition 5 (Contextual query). *Given a DeLP-program $\mathcal{P}$=(Π,Δ), a contextual query for $\mathcal{P}$ is a pair $[\Phi,Q]$ where Q is a DeLP-query, and Φ is a non-contradictory set of literals such that $\Pi \cup \Phi$ is a non-contradictory set.*

Definition 6 (Warrant for a contextual query). *A contextual query $[\Phi,Q]$ is warranted from the DeLP-program (Π,Δ) if Q is warranted from $(\Pi \cup \Phi,\Delta)$.*

Note that Def. 5 requires $\Pi \cup \Phi$ to be non-contradictory due to the problem addressed in Obs. 1. The effect of performing a contextual query $[\Phi, Q]$ to a DeLP-program (Π,Δ) (Def. 6) will be to add all the literals in Φ to (Π,Δ)

building (temporarily) a new program ($\Pi \cup \Phi$,Δ), which will be used for returning the appropriate answer. Note that although some new information not present in (Π,Δ) may be considered for warranting a contextual query [Φ,Q], this process will not affect the content of the stored program (Π,Δ) for future queries.

Observation 2. *A DeLP-query is a particular case of a contextual query* [Φ,Q] *where* $\Phi = \emptyset$. *Given* [$\emptyset$,Q] *and a program* (Π,Δ), *it is easy to show that* [$\emptyset$,Q] *is warranted from* (Π,Δ), *iff* Q *is warranted from* (Π,Δ).

Since contextual queries are performed following a client/server model, the sender of a query (client) may not know the content of the program in a Server. Therefore, it may not be possible for a client to know in advance whether $\Pi \cup \Phi$ is non-contradictory. Hence, we will extend the notion of answer (Def. 4) to return INVALID and the subset of Φ that produces the contradictory set.

Definition 7 (Answer for a contextual query). *An answer for a contextual query* [Φ,Q] *from a DeLP-program* (Π,Δ) *is a pair* (Ans,S), *where* $Ans \in$ {YES, NO, UNDECIDED, UNKNOWN, INVALID} *and* S *is a set of literals. If* $\Pi \cup \Phi$ *is contradictory, then* Ans = INVALID *and* $S = \{L \in \Pi \cup \Phi \mid \overline{L} \in \Pi \cup \Phi\}$. *Otherwise, the answer will be* (YES,{}) *if* Q *is warranted from* ($\Pi \cup \Phi$,Δ); (NO,{}) *if* $\overline{Q}$ *is warranted from* ($\Pi \cup \Phi$,Δ); (UNDECIDED,{}) *if neither* Q *nor* $\overline{Q}$ *are warranted from* ($\Pi \cup \Phi$,Δ); *and* (UNKNOWN,{}), *if* Q *is not in the language of* ($\Pi \cup \Phi$,Δ).

Example 2. Consider the DeLP-program $\mathcal{P}_2 = (\Pi_2, \Delta_2)$, where:
$\Pi_2 = \{r\}$ and $\Delta_2 = \{(m \multimap c), (\sim m \multimap r), (c \multimap s), (\sim c \multimap s, p)\}$.
Here, the letters m, r, c, s and p are the abbreviations of *use_metro*, *rush_hour*, *closed_roads*, *snow* and *snowplow* respectively. From $\mathcal{P}_2$, the answer for the DeLP-query c is UNDECIDED, whereas the answer for the contextual query $[\{s\}, c]$ is (YES,{}) (s activates an argument for c). Note that the answer for $[\{s, p\}, c]$ is (NO,{}) because now p activates a proper defeater. Observe that from $\mathcal{P}_2$, the answer for $\sim m$ is YES, however, the answer for the contextual query $[\{s\}, \sim m]$ is (UNDECIDED,{}) and the answer for $[\{s, p\}, \sim m]$ is (YES,{}). Finally, the answer for $[\{\sim m, p, \sim r\}, r]$ is (INVALID,{r}). Since these contextual queries do not actually make a change on program $\mathcal{P}_2$, after performing them, the DeLP-query $\sim m$ will still remain warranted from $\mathcal{P}_2$ and c will not be warranted from $\mathcal{P}_2$.

Proposition 1. *Given a DeLP-program* (Π,Δ) *and a contextual query* [Φ,Q], *if* $Q \in \Pi \cup \Phi$ *then the answer for* [Φ,Q] *is* (YES,{}). *If* $\overline{Q} \in \Pi \cup \Phi$ *then the answer for* [Φ,Q] *is* (NO,{}). Proof: By the proposition 5.1 given in [1], if $Q \in \Pi \cup \Phi$ then Q is warranted from ($\Pi \cup \Phi$,Δ). Hence, by Definition 7, the answer for [Φ,Q] is (YES,{}). By the same cited proposition, if $\overline{Q} \in \Pi \cup \Phi$ then $\overline{Q}$ is warranted from ($\Pi \cup \Phi$,Δ) and hence, the answer for [Φ,Q] is (NO,{}).

3.1 Generalized Contextual Queries

In a contextual query [Φ,Q], imposing that $\Pi \cup \Phi$ must be a non-contradictory set may be in some cases too restrictive. Although the agent will be informed in case of an invalid query (Definition 7), the agent should have to reformulate the query in order to obtain an answer. Therefore, it would be useful to have a more general type of query that, without imposing the mentioned restriction, resolves the problem of a contradictory set before the query is processed.

In this section we propose three generalized types of contextual queries, each one with a different policy for handling the mentioned problem. In the first one (called non-prioritized contextual query), given a query $[\Phi,Q]$ and a program (Π,Δ), if $\Pi\cup\Phi$ is contradictory, then the DeLP-Server will assign less priority to the knowledge of the agent (Φ) than to the one in the Server (Π). Hence, those literals from Φ that are problematic will be ignored when answering that query. In contrast, in the third one (called prioritized contextual query), when $\Pi\cup\Phi$ is contradictory, more priority will be assigned to the literals in Φ than those in Π. Therefore, the literals in Π that produce the contradiction will be temporarily ignored for answering that query. The second generalized type (called restrictive contextual query) has the effect of temporarily ignore literals from Π while processing the query. In this case, no policy is needed since no literals are added to Π and a contradiction can never occur. The symbols "$*$", "$+$", "$-$" will be used to distinguish each type of query.

Definition 8 (Non-prioritized contextual query)
Given a DeLP-program $\mathcal{P}$, a non-prioritized contextual query for $\mathcal{P}$ is denoted $[\Phi^*,Q]$, where Q is a DeLP-query and Φ^* is a non-contradictory set of literals.

Definition 9 (Warrant for a non-prioritized contextual query)
A query $[\Phi^*,Q]$ is warranted from (Π,Δ) if Q is warranted from $((\Pi \otimes \Phi^*),\Delta)$. Let $C(\Pi) = \{\overline{L} \text{ if } L \in \Pi\}$, then $(\Pi \otimes \Phi^*) = (\Phi^* \setminus C(\Pi)) \cup \Pi$.

In contrast with Def. 5, in a non-prioritized contextual query $[\Phi^*,Q]$ the set $\Pi\cup\Phi^*$ can be contradictory. However, the set Φ^* is required to be non-contradictory as a minimal coherence principle. This type of query is useful for agents that need to assign a preference to the information stored in the Server. The operator $\otimes$ resembles a non-prioritized merge operator that assigns a preference to the elements of Π when merging Π and Φ. Thus, the effect of performing a non-prioritized contextual query $[\Phi^*,Q]$ to a program (Π,Δ) will be to temporarily consider in (Π,Δ) only those literals from Φ^* that are not problematic. Thus, if $\Pi\cup\Phi^*$ is contradictory, those literals from Φ^* that produce contradictions will be ignored by the server. The subset of ignored literals will be returned with the answer to inform that they have not been used for the dialectical analysis. If $[\Phi^*,Q]$ returns $(\cdot,\{\})$, all the literals of Φ were considered.

Definition 10 (Answer for a non-prioritized contextual query)
The answer for $[\Phi^*,Q]$ from (Π,Δ) is a pair (Ans,S), where $Ans \in$ {YES, NO, UNDECIDED, UNKNOWN, INVALID} is computed as shown in Def. 7, and the set $S \subseteq \Phi^*$ will contain those literals that have been ignored by the DeLP-Server while processing Q, (i.e., $L \in S$ if $L \in \Phi^*$ and $\overline{L} \in \Pi$). If Φ^* is contradictory, then $Ans =$ INVALID and $S = \{L \in \Phi^* \mid \overline{L} \in \Phi^*\}$.

Example 3. Consider the DeLP-program $\mathcal{P}_3 = (\Pi_3, \Delta_3)$, where $\Pi_3 = \{r, p, {\sim}f\}$ and $\Delta_3 = \{({\sim}m \prec r), (m \prec c, r), (t \prec {\sim}c)\}$. The letters m, r, c, p, f and t are the abbreviations of *use_metro*, *rush_hour*, *closed_roads*, *park_closed*, *fog* and *use_taxi* respectively. From $\mathcal{P}_3$ the answer for the query $\sim m$ is YES whereas for the non-prioritized contextual query $[\{c\}^*, \sim m]$ the answer is (NO,{}). Observe that the query $[\{\sim r\}^*, \sim m]$ returns the answer (YES,$\{\sim r\}$), since $r \in \Pi$ is preferred. The DeLP-query t has the answer NO, whereas $[\{\sim c, r, \sim p, f\}^*, t]$ returns the answer (YES,$\{\sim p, f\}$). Note that $[\{\sim c, \sim t\}^*, t]$ returns the answer (NO,{}) because $\{\sim c, \sim t\} \cup \Pi_3$ is non-contradictory and hence, the literal $\sim t$ is included for answering the query. Although there exists a defeasible derivation for t with $\sim c$ and $(t \prec \sim c)$, if $\sim t$ is present as a fact, in DeLP there is no argument for t.

Proposition 2. *Given a DeLP-program* (Π,Δ) *and a non-prioritized contextual query* $[\Phi^*,Q]$ *it holds that:*

1. *if* $Q \in \Pi$ *then the answer for* $[\Phi^*,Q]$ *will always be* (YES,S).
2. *if* $Q \in \Phi^*$ *and* $\overline{Q} \notin \Pi$ *then the answer for* $[\Phi^*,Q]$ *will always be* (YES,S).
3. *if* $\overline{Q} \in \Pi$ *then the answer for* $[\Phi^*,Q]$ *will always be* (NO,S).
4. *if* $\overline{Q} \in \Phi^*$ *and* $Q \notin \Pi$ *will always be for* $[\Phi^*,Q]$ *will be always* (NO,S).
   (Proof is omitted due to space restrictions)

Observation 3. *A contextual query* $[\Phi,Q]$ *is a particular case of a non-prioritized contextual query* $[\Phi^*,Q]$ *where* $(\Pi \cup \Phi^*)$ *is a non-contradictory set.*

The next type of contextual query will allow an agent to temporarily ignore some literals from the program stored in a DeLP-Server with the purpose of warranting Q. Since no literals will be added to the program in the server, then no restriction over the literals included in the query is needed.

Definition 11 (Restrictive contextual query)
Given a DeLP-program $\mathcal{P}=(\Pi,\Delta)$, *a restrictive contextual query for* $\mathcal{P}$ *is denoted* $[\Phi^-,Q]$ *where* Q *is a DeLP-query and* Φ^- *is an arbitrary set of literals.*

Definition 12 (Warrant for a restrictive contextual query). *The query* $[\Phi^-,Q]$ *is warranted from* (Π,Δ) *if* Q *is warranted from* $((\Pi \setminus \Phi^-),\Delta)$.

Definition 13 (Answer for a restrictive contextual query)
The answer $[\Phi^-,Q]$ *from* (Π,Δ) *is a pair* (Ans,S), *where* $Ans \in$ {YES, NO, UNDECIDED, UNKNOWN} *is computed as shown in Def. 7, and* $S = \Phi^- \cap \Pi$, *i.e.,* S *will contain those literals that have been effectively ignored by the DeLP-Server.*

Example 4. Consider $\mathcal{P}_3$ from Ex. 3. The answer for the restrictive contextual query $[\{r\}^-, \sim m]$ will be (UNDECIDED,{r}). Observe that a query $[\Phi^-,Q]$ might include in Φ^- literals that are not in Π. This superfluous literals will have not effect to the query, and the agent will know them because they will not be returned in the answer. For instance, the answer for $[\{r, z\}^-, \sim m]$ is (UNDECIDED,{r}) .

Observation 4. *Given a restrictive contextual query* $[\Phi^-,Q]$ *to a DeLP-program* (Π,Δ), *in the particular case that* $(\Pi \cap \Phi^-)$ *is empty, then the restrictive contextual query behaves like a contextual query.*

As stated above, the type of contextual query we propose next will assign a preference to the information sent by the agent. Thus, if the Server has a program

(Π,Δ) and the query includes a set Φ such that $(\Pi \cup \Phi)$ is a contradictory set, then, some literals from Π will be ignored for answering the query.

Definition 14 (Prioritized contextual query)
Given a DeLP-program $\mathcal{P}$, a prioritized contextual query for $\mathcal{P}$ is denoted $[\Phi^+,Q]$ where Q is a DeLP-query and Φ^+ is a non-contradictory set of literals.

Definition 15 (Warrant for a prioritized contextual query)
The query $[\Phi^+,Q]$ is warranted from (Π,Δ) if Q is warranted from $((\Pi \oplus \Phi^+),\Delta)$. Let $C(\Phi^+) = \{\overline{L} \text{ if } L \in \Phi^+\}$, then $(\Pi \oplus \Phi^+) = (\Pi \setminus C(\Phi^+)) \cup \Phi^+$.

The operator $\oplus$ resembles a prioritized merge operator. It removes (temporarily) from Π the complement of the literals that belongs to Φ^+ and then (temporarily) adds Φ^+. Both operations (the subtraction and then the addition) are made in the same "transaction". As remarked above, all of these changes to the set Π are local to the warrant process of the contextual query, and they will not affect other queries nor make permanent changes to Π.

Definition 16 (Answer for a prioritized contextual query)
The answer for $[\Phi^+,Q]$ from (Π,Δ) is a pair (Ans,S). The first element $Ans \in$ {YES, NO, UNDECIDED, UNKNOWN, INVALID} is computed as shown in Def. 7, and the set $S \subseteq \Pi$ will contain those literals that have been ignored by the DeLP-Server while processing Q (i.e., $L \in S$ if $L \in \Pi$ and $\overline{L} \in \Phi^+$). If Φ^+ is contradictory, then $Ans=$ INVALID and $S=\{L \in \Phi^+ \mid \overline{L} \in \Phi^+\}$.

Example 5. Consider again $\mathcal{P}_3$ of Ex. 3. From (Π_3, Δ_3) the answer for the query $[\{c\}^+, \sim m]$ is (NO,$\{\}$). In contrast with the Ex. 3, the answer for $[\{\sim r, c\}^+, \sim m]$ is (UNDECIDED,$\{r\}$), because $\sim m$ is warranted from $(\{p, \sim f, \sim r\}, \Delta_3)$ since $\sim r$ is preferred over $r \in \Pi$. The answer for $[\{\sim c, r, \sim p, f\}^+, t]$ is (YES,$\{p, \sim f\}$).

Proposition 3. *Given a DeLP-program (Π,Δ) and a prioritized contextual query $[\Phi^+,Q]$ it holds that:*
1. *if $Q \in \Phi^+$ then the answer for $[\Phi^+,Q]$ will always be* (YES,S).
2. *if $Q \in \Pi$ and $\overline{Q} \notin \Phi^+$ then the answer for $[\Phi^+,Q]$ will always be* (YES,S).
3. *if $\overline{Q} \in \Phi^+$ then the answer for $[\Phi^+,Q]$ will always be* (NO,S).
4. *if $\overline{Q} \in \Pi$ and $Q \notin \Phi^+$ then the answer for $[\Phi^+,Q]$ will always be* (NO,S).

The proof is analogous to Proposition 2 and is not included due to space reasons.

Observation 5. *The effect of considering a negated literal $\sim L$ to answer a query differs from the effect of removing its complement L (see Ex. 6).*

Example 6. Consider $\Pi_6 = \{x\}$ and $\Delta_6 = \{(a \prec x), (b \prec \sim x)\}$. From (Π_6, Δ_6) the answer for the DeLP-query a is YES, and the answer for b is UNDECIDED. Observe that the answer for $[\{\text{x}\}^-,a]$ is (UNDECIDED,$\{\}$) –since there are no arguments for nor against a– and the answer for $[\{\text{x}\}^-,b]$ is also (UNDECIDED,$\{\}$). Nevertheless, the answer for $[\{\sim\text{x}\}^+,a]$ is (UNDECIDED,$\{\text{x}\}$), and the answer for $[\{\sim\text{x}\}^+,b]$ is (YES,$\{\text{x}\}$). Observe finally that the answer for $[\{\sim\text{x}\}^*,a]$ is (YES,$\{\sim\text{x}\}$), and the answer for $[\{\sim\text{x}\}^*,b]$ is (UNDECIDED,$\{\sim\text{x}\}$).

Observation 6. *A contextual query $[\Phi,Q]$ is a particular case of a prioritized contextual query $[\Phi^+,Q]$ where $(\Pi \cup \Phi^+)$ is a non-contradictory set.*

3.2 Extended and Combined Contextual Queries

Extending contextual queries to consider not only facts but also defeasible rules is straightforward, because (as stated in Observation 1) if (Π,Δ) is a DeLP-program then $(\Pi,(\Delta \cup \Delta_1))$ is also a DeLP-program for any set of defeasible rules Δ_1. The next example shows why this kind of queries are useful.

Example 7. Let $\Pi_7 = \{f, h\}$ and $\Delta_7 = \{(a \prec g), (\sim d \prec h, g)\}$, from (Π_7, Δ_7) the answer for the DeLP-query a is UNDECIDED, and the answer for $\sim d$ is UNDECIDED. Consider an agent that has the DeLP-program $(\{g\}, \{d \prec a\})$ as part of its private knowledge. From (Π_7, Δ_7) the answer for $[\{g\}, a]$ will be (YES,{}). Then, since the agent has the defeasible rule $(d \prec a)$ it could assume that if a is warranted then d will be warranted. However, from $(\Pi_7 \cup \{g\}, \Delta_7 \cup \{d \prec a\})$ the answer for d is NO, because there is a proper defeater that attacks d.

Definition 17 (Extended contextual query). *Given a DeLP-program (Π,Δ), an extended contextual query for (Π,Δ) is a pair $[\mathcal{P},Q]$ where Q is a DeLP-query and $\mathcal{P}=(\Phi,\Delta_1)$ is a DeLP-program such that $\Pi \cup \Phi$ is a non-contradictory set.*

Definition 18 (Warrant for an extended contextual query). *The query $[(\Phi,\Delta_1),Q]$ is warranted from (Π,Δ) if Q is warranted from $(\Pi \cup \Phi, \Delta \cup \Delta_1)$.*

The definition of answer for $[(\Phi,\Delta_1),Q]$ will be the same as Def. 7 where Δ is replaced by $\Delta \cup \Delta_1$. Thus, a contextual query $[\Phi,Q]$ is a particular case of an extended contextual query $[(\Phi,\{\}),Q]$, where the set of defeasible rules is empty.

Extending non-prioritized, restrictive and prioritized contextual queries to include defeasible rules is straightforward. The corresponding definitions will not be included because they are analogous to the previous ones.

Although several types of contextual queries have been defined, there are some modifications to the program in the server that cannot be performed (see Ex. 8). Therefore, we propose a new type of query that is a generalization of the previous ones: the *combined contextual query.* This type of query will allow an application of successive changes to a DeLP-program.

Example 8. Consider the DeLP-program (Π_8, Δ_8), where $\Pi_8 = \{a, b\}$ and $\Delta_8 = \{(z \prec y), (y \prec \sim b), (\sim y \prec a), (\sim z \prec \sim a)\}$ Here, the answer for $[\{\sim b\}^+, z]$ is (UNDECIDED,$\{b\}$) because the argument $\{(z \prec y), (y \prec \sim b)\}$ has $\{\sim y \prec a\}$ as a defeater. Suppose that an agent wants to add $\sim b$ but to simultaneously remove a in order to "deactivate" the mentioned defeater. If it uses first $[\{\sim b\}^+, z]$ and then $[\{a\}^-, z]$, the literal $\sim b$ will not be part of the program when the second query is processed. Inverting the order of the queries does not solve the problem because a will be present for the second query. On the other hand, if it submits $[\{\sim b, \sim a\}^+, z]$, the query will remove a but the literal $\sim a$ will be added (see Obs. 5). Therefore, the answer for $[\{\sim b, \sim a\}^+, z]$ will be (UNDECIDED,$\{b, a\}$) because the argument $\{\sim z \prec \sim a\}$ defeats $\{(z \prec y), (y \prec \sim b)\}$.

Definition 19 (Combined contextual query)
Given a DeLP-program $\mathcal{P}=(\Pi,\Delta)$, a combined contextual query for $\mathcal{P}$ is a pair $[(Seq, \Delta_1),Q]$ where Q is a DeLP-query, Δ_1 is a set of defeasible rules and Seq is a sequence of non-contradictory sets of literals, marked with $*$, $+$ or $-$.

Thus, a combined contextual query allows to apply several types of changes to a program. This resolves issues like the one mentioned in Ex. 8. For example, $[([\{a\}^-,\{\sim b\}^+],\{\}),\, z]$ will modify (Π_8, Δ_8) by removing a, and then adding $\sim b$ and removing b, before computing the DeLP-query z over the modified program.

The effect of performing a combined contextual query $[(Seq,\ \Delta_1),Q]$ to a DeLP-program (Π,Δ) will be to consider the changes defined in the sequence Seq, apply them in the given order, and then compute the answer for the DeLP-query Q from the modified program. The program will be modified according to Definitions 9, 12 and 14.

Example 9. Consider the program in Example 8, if we perform the combined query $[([\{a,\sim b\}^-,\{x\}^*,\{\sim b\}^+],\{a \mathrel{-\!\!\prec} x\}),z]$, then a and $\sim b$ will be removed from Π_8 ($\sim b \notin \Pi_8$, so its exclusion has no effect); next, x is added to Π_8 (since $\sim x \notin \Pi_8$); finally, $\sim b$ is added to Π_8 and, collaterally, b is dropped from Π_8. Therefore, the program over which query z will be performed is (Π_9, Δ_9), where $\Pi_9 = \{x, \sim b\}$ and $\Delta_9 = \{(z \mathrel{-\!\!\prec} y),\ (y \mathrel{-\!\!\prec} \sim b),\ (\sim y \mathrel{-\!\!\prec} a), (\sim z \mathrel{-\!\!\prec} \sim a),\ (a \mathrel{-\!\!\prec} x)\}$. Here, there is one argument for z: $\{(z \mathrel{-\!\!\prec} y),\ (y \mathrel{-\!\!\prec} \sim b)\}$, which is defeated by an argument for $\sim y$: $\{(\sim y \mathrel{-\!\!\prec} a),\ (a \mathrel{-\!\!\prec} x)\}$. Hence, the answer for $[([\{a,\sim b\}^-,\{x\}^*,\{\sim b\}^+],\{a \mathrel{-\!\!\prec} x\}),z]$ is (UNDECIDED,$\{a,b\}$).

4 Usage and Implementation

Several DeLP-Servers can be used simultaneously, each of them providing a different shared knowledge base. Hence, agents may perform the same contextual query to different servers, and they may share different knowledge with other agents through different servers. Thus, several configurations of agents and servers can be established (statically or dynamically). One possible configuration is to have a single DeLP-Server with all the public knowledge of the system loaded in it. This has the advantage of having the shared knowledge concentrated in one point, but it has the disadvantages of any centralized configuration. In a MAS with many client agents, several copies of the same DeLP-Server (providing the same knowledge) can be created, and each agent could select a different server for sending its queries. Having several copies of the same DeLP-Server is also useful in the case of time-consuming queries, because one agent may distribute or parallelize its own queries.

Since public knowledge will not be restricted to be in a unique repository, then it can be structured in many ways. For example, several servers with different shared knowledge can be created. Thus, special-purpose DeLP-Servers can be implemented, each of them representing knowledge of a specific domain. The MAS may then have several groups of agents with specific tasks, and each group can share specific knowledge through each special-purpose DeLP-Server.

Observe that, by using extended contextual queries, an agent may send a complete DeLP-program (Π,Δ) to a server in order to obtain answers for its queries. Thus, in a MAS it will be possible to have some DeLP-Servers with no public knowledge stored, which will receive pairs $[(\Pi,\Delta),\ Q]$ and will return the answer for Q from (Π,Δ). This particular kind of servers will act only as reasoning services without using public knowledge. As mentioned above, having more than one copy of this particular server without pre-loaded knowledge will be useful for one agent to distribute or parallelize several queries.

It is important to note that, in our model, the public program stored in a DeLP-server is loaded upon server creation and it remains unaffected thereafter. That is, a client agent cannot make permanent changes to it. Our design choice differs from other approaches that use a public shared space for storing knowledge (as a blackboard or a tuple space) where every agent has permission to change it. If client agents in our model were allowed to make permanent changes to the public knowledge stored in a DeLP-server, then the model would collapse to a shared memory repository, and several issues would have to be addressed. For example, a permission model would have to be defined because one agent can change information that is needed by other agents, a policy for updating information that changes dynamically could be needed, etc.

The DeLP-Server was implemented as a stand-alone program that runs in a host and interacts with client agents using the TCP protocol. It supports the connection of several agents, which can be either in the same host, or in different ones. Furthermore, an agent may connect to several DeLP-Servers, and send queries to all of them. The DeLP-Server is implemented in SWI-PROLOG [5] and client agents can be written in any language that supports TCP connections.

5 Conclusions and Related Work

In this paper we propose a model that allows agents to deliberate using defeasible argumentation, to share knowledge with other agents, and to represent individual knowledge in a private manner. We have considered the design and implementation of a Defeasible Logic Programming Server that handles queries from several client agents distributed in remote hosts. Queries are answered using public knowledge stored in the Server and individual knowledge that client agents can send as part of a query, providing a particular context for it. Different types of contextual queries were presented and analyzed.

Our approach was in part inspired by Jinni/Bin-Prolog [6], where several servers can be created, and knowledge can be shared through them. Nevertheless, both approaches have several differences. Their knowledge representation language and their reasoning formalism is PROLOG, whereas we use DeLP. In contrast with us, they share knowledge through blackboards where agents can make permanent changes. Finally, we provide several types of contextual queries, whereas they use PROLOG queries and Linda operations.

In [7] a description of a light-weight Java-based argumentation engine is given that can be used to implement a non-monotonic reasoning component over the Internet or agent-based applications. In contrast with us, they provide an application programming interface (API) that exposes key methods to allow an agent or Internet application developer to access and manipulate the knowledge base to construct rules, specify and execute queries and analyze results. Therefore, their engine allows the agent or Internet application developer to change the program or knowledge base. This capability, in a scenario where several agents try to access to the same knowledge base requires that several issues have to be considered (*e.g.*, synchronization and access/update permissions). Another difference with our approach is that their engine is implemented in Java and presented as a self-contained component that can be integrated into applications. In contrast, a DeLP-Server is an independent stand-alone program that runs in a host and interact with agents using the TCP protocol. In [7], it is not clear what are the possible answers that an agent will obtain from the argumentation engine.

In [8], a query answering system employing a backward-chaining approach is introduced. This query answering system, Deimos, is a suite of tools that supports Defeasible Logic [9,10]. Although related, Defeasible Logic differs from Defeasible Logic Programming (see [1] for a comparison of both approaches). Similar to us, the system is prepared to receive and compute queries using rules representing defeasible knowledge. In contrast with us, the query answering system is accessible through a command line interface and a CGI interface. Thus, it must be used by a human being, whereas, in our approach, a software agent or any kind of application is allowed to interact with the Server, including a human being through a proper program interface. Another difference with our approach is that in Deimos there is no notion of argument. In order to decide between two contradictory conclusions, Deimos compares only one pair of rules, whereas in DeLP the two arguments that support those conclusions are compared. Comparing only a pair of rules may be problematic as we show in [1].

References

1. García, A.J., Simari, G.R.: Defeasible logic programming: An argumentative approach. Theory and Practice of Logic Programming 4(1), 95–138 (2004)
2. Chesñevar, C.I., Maguitman, A.G., Loui, R.P.: Logical Models of Argument. ACM Computing Surveys 32(4), 337–383 (2000)
3. Prakken, H., Vreeswijk, G.: Logical systems for defeasible argumentation. In: Handbook of Philosophical Logic, 2nd edn., Kluwer Academic Pub., Dordrecht (2000)
4. Stolzenburg, F., García, A.J., Chesñevar, C.I., Simari, G.R.: Computing generalized specificity. Journal of Aplied Non-Classical Logics 13(1), 87–113 (2003)
5. (SWI-Prolog), `http://www.swi-prolog.org`
6. Tarau, P.: (Jinni 2002: A High Performance Java and .NET based Prolog for Object and Agent Oriented Internet Programming) (2002), `http://www.cs.unt.edu/~tarau/`
7. Bryant, D., Krause, P.J., Vreeswijk, G.: Argue tuProlog: A Lightweight Argumentation Engine for Agent Applications. In: COMMA 2006. Proc. of 1st Int. Conference on Computational Models of Argument, pp. 27–32. IOS Press, Amsterdam (2006)
8. Maher, M., Rock, A., Antoniou, G., Billington, D., Miller, T.: Efficient defeasible reasoning systems. Int. Journal on Artificial Intell. Tools 10(4), 483–501 (2001)
9. Nute, D.: Defeasible logic. In: Handbook of Logic in Artificial Intelligence and Logic Programming, vol. 3, pp. 355–395. Oxford University Press, Oxford (1994)
10. Governatori, G., Maher, M., Antoniou, G., Billington, D.: Argumentation semantics for defeasible logic. Journal of Logic and Computation 14, 675–702 (2004)

On Defense Strength of Blocking Defeaters in Admissible Sets

Diego C. Martínez, Alejandro J. García, and Guillermo R. Simari

Comisión Nacional de Investigaciones Científicas y Técnicas - CONICET
Artificial Intelligence Research and Development Laboratory (LIDIA)
Department of Computer Science and Engineering, Universidad Nacional del Sur
Av. Alem 1253 - (8000) Bahía Blanca - Bs. As. - Argentina
{dcm,ajg,grs}@cs.uns.edu.ar

Abstract. Extended argumentation framework is a formalism where defeat relations are determined by establishing a preference between arguments involved in symmetric conflicts. This process possibly leads to blocking situations, where conflicting arguments are found to be incomparable or equivalent in strength. In this work we introduce new argumentation semantics for extended frameworks, by taking into account the strength of an argument defense. Each of these new admissibility notions relies on the presence of blocking defeaters as argument defenders.

1 Introduction

Argumentation has become an important subject of research in Artificial Intelligence and it is also of interest in several disciplines, such as Logic, Philosophy and Communication Theory. This wide range of attention is due to the constant presence of argumentation in many activities, most of them related to social interactions between humans, as in civil debates, legal reasoning or every day dialogues. Basically, an argument is a piece of reasoning that supports a claim from certain evidence. The tenability of this claim must be confirmed by analyzing other arguments for and against such a claim. In formal systems of defeasible argumentation, a claim will be accepted if there exists an argument that supports it, and this argument is acceptable according to an analysis between it and its counterarguments. After this dialectical analysis is performed over the set of arguments in the system, some of them will be *acceptable*, *justified* or *warranted* arguments, while others will be not. The study of the acceptability of arguments is the search for rationally based positions of acceptance in a given scenario of arguments and their relationships. It is one of the main concerns in Argumentation Theory.

Abstract argumentation systems [1,2,3,4] are formalisms for argumentation where some components remain unspecified, being the structure of an argument the main abstraction. In this kind of system, the emphasis is put on the semantic notion of finding the set of accepted arguments. Most of them are based on the single abstract concept of *attack* represented as an abstract relation, and extensions are defined as sets of possibly accepted arguments. For two arguments

Z. Zhang and J. Siekmann (Eds.): KSEM 2007, LNAI 4798, pp. 140–152, 2007.

$\mathcal{A}$ and $\mathcal{B}$, if $(\mathcal{A}, \mathcal{B})$ is in the attack relation, then the acceptance of $\mathcal{B}$ is conditioned by the acceptance of $\mathcal{A}$, but not the other way around. It is said that argument $\mathcal{A}$ attacks $\mathcal{B}$, and it implies a priority between conflicting arguments. It is widely understood that this priority is related to the argument strengths. Several frameworks do include an argument order [3,5,6], although this order is used at another level, as the classic attack relation is kept.

In [7,8] an extended abstract argumentation framework is introduced, where two kinds of defeat relations are present. These relations are obtained by applying a preference criterion between conflictive arguments. The conflict relation is kept in its most basic, abstract form: two arguments are in conflict simply if both arguments cannot be accepted simultaneously. The preference criterion subsumes any evaluation on arguments and it is used to determine the direction of the attack. This argument comparison, however, is not always succesful and therefore attacks, as known in classic frameworks, are no longer valid. For example, consider the following argument, called NS_1:

> NS_1: *Buy new ski tables, because they are specially shaped to improve turns.*

The claim of NS_1 is that new skis must be bought, for a reason about skiing performance. Now suppose a new argument, called OS is introduced

> OS: *Buy used old ski tables, because they are cheap and your skills are good enough to face any turn.*

This argument exposes two reasons for buying old skis. Clearly, OS is in conflict with NS_1 as both arguments cannot be accepted altogether: we buy new skis, or we buy old ones. Given this dichotomy, both argument must be compared to each other. The claim of argument OS is supported by more reasons than NS_1, and one of them is also about skiing performance. Therefore, it is valid to state that OS is preferable to NS_1, and then the former became a *proper defeater* of the latter. This is the strongest form of attack. Now consider two additional arguments, called NS_2 and NS_3:

> NS_2: *Buy new ski tables, as they are not expensive, and old used ski tables are not helping to improve your skills.*
> NS_3: *Buy new ski tables, because the colors and graphic styles are cool.*

Both arguments are supporting the same claim ("*Buy new ski tables*"), and therefore they are in conflict with OS. Again, a comparison is needed. The argument NS_2 states that new skis are not expensive and they can improve skills. As it refers to price and performance, the information exposed in NS_2 is as strong as the one exposed in OS. In fact, both arguments can be considered equally strong when supporting its claims. Therefore, NS_2 and OS are blocking each other, and then both arguments are said to be *blocking defeaters*. On the other hand, NS_3 is based on style taste to support its claim. No reference to the money invested or to personal skills is made. Although they are in conflict, NS_3 and OS cannot be compared to each other. Arguments NS_3 and OS are also *blocking defeaters*, but of different nature, as they are incomparable.

Extended argumentation frameworks (EAF) are suitable to model situations like above. In [8] a simple semantics based on suppression of arguments is presented, and in [9] the well-formed structure of argumentation lines is introduced. These are two characterizations of unique sets of accepted arguments, according to specific dialectical interpretations. However, under the classic acceptability notions [1] in abstract frameworks it is possible to have several alternative sets of acceptance (extensions) for a given argument scenario. This is a common outcome of attack cycles ($\mathcal{A}_1$ is attacked by an argument $\mathcal{A}_2$, that is attacked by $\mathcal{A}_3$,..., attacked by $\mathcal{A}_n$, which in turn attacks $\mathcal{A}_1$). In extended abstract frameworks this is also present: every argument in the ski example is defeated by another argument. In this case, when searching for accepted arguments it is not easy to find a starting point, as everyone has a defeater. Therefore, several extensions arises. Using extended frameworks, however, it is more clear how defeat occurs regarding to argument strength.

In this work we explore new semantic considerations based on the quality of a defense. In the previous example, arguments NS_2 and NS_3 are defeaters of OS, which is a defeater of NS_1. Therefore, NS_1 is said to be defended by NS_2 and NS_3. The ideal defense is achieved by stronger arguments, but in this case a defender is considered as strong as OS, and the other is actually incomparable. This scenario leads to several argument extensions. However, in extended abstract frameworks the strength of defenses is a pathway to compare extensions, in a search for a position of acceptance following rational criteria. The fact that arguments play the role of defenders with different strengths, and how it can be used to define new extensions, is the motivation of this work.

This paper is organized as follows. In the next section, the extended argumentation frameworks are formally introduced. In Section 3 a simple notion of admissibility regarding blocked defeaters is presented. In Section 4 and Section 5 new admissibility sets are characterized taking the strength of defense into account. Finally, the conclusions and future work are presented in Section 6.

2 Extended Argumentation Framework

In our extended argumentation framework three relations are considered: *conflict*, *subargument* and *preference* between arguments. The definition follows:

Definition 1. *An extended abstract argumentation framework (EAF) is a quartet* Φ=$\langle AR, \sqsubseteq, \mathbf{C}, \mathbf{R}\rangle$, *where* AR *is a finite set of arguments, and* $\sqsubseteq$, $\mathbf{C}$ *and* $\mathbf{R}$ *are binary relations over* AR *denoting respectively subarguments, conflicts and preferences between arguments.*

Arguments are abstract entities, as in [1], that will be denoted using calligraphic uppercase letters, possibly with indexes. In this work, the subargument relation is not relevant for the topic addressed. Basically, it is used to model the fact that arguments may include inner pieces of reasoning that can be considered arguments by itself, and it is of special interest in dialectical studies [9]. Hence, unless explicitly specified, in the rest of the paper $\sqsubseteq = \emptyset$. The conflict relation

C states the incompatibility of acceptance between arguments. Given a set of arguments S, an argument $\mathcal{A} \in S$ is said to be in conflict in S if there is an argument $\mathcal{B} \in S$ such that $\{\mathcal{A}, \mathcal{B}\} \in \mathbf{C}$. The relation **R** is introduced in the framework and it will be used to evaluate arguments, modeling a preference criterion based on a measure of strength.

Definition 2. *Given a set of arguments AR, an* argument comparison criterion *$\mathbf{R}$ is a binary relation on AR. If $\mathcal{A}\mathbf{R}\mathcal{B}$ but not $\mathcal{B}\mathbf{R}\mathcal{A}$ then $\mathcal{A}$ is strictly preferred to $\mathcal{B}$, denoted $\mathcal{A} \succ \mathcal{B}$. If $\mathcal{A}\mathbf{R}\mathcal{B}$ and $\mathcal{B}\mathbf{R}\mathcal{A}$ then $\mathcal{A}$ and $\mathcal{B}$ are indifferent arguments with equal relative preference, denoted $\mathcal{A} \equiv \mathcal{B}$. If neither $\mathcal{A}\mathbf{R}\mathcal{B}$ or $\mathcal{B}\mathbf{R}\mathcal{A}$ then $\mathcal{A}$ and $\mathcal{B}$ are incomparable arguments, denoted $\mathcal{A} \bowtie \mathcal{B}$.*

For two arguments $\mathcal{A}$ and $\mathcal{B}$ in AR, such that the pair $\{\mathcal{A}, \mathcal{B}\}$ belongs to **C** the relation **R** is considered. In order to elucidate conflicts, the participant arguments must be compared. Depending on the preference order, two notions of argument defeat are derived.

Definition 3. *Let $\Phi = \langle AR, \sqsubseteq, \mathbf{C}, \mathbf{R}\rangle$ be an EAF and let $\mathcal{A}$ and $\mathcal{B}$ be two arguments such that $(\mathcal{A}, \mathcal{B}) \in \mathbf{C}$. If $\mathcal{A} \succ \mathcal{B}$ then it is said that $\mathcal{A}$ is a proper defeater of $\mathcal{B}$. If $\mathcal{A} \equiv \mathcal{B}$ or $\mathcal{A} \bowtie \mathcal{B}$, it is said that $\mathcal{A}$ is a blocking defeater of $\mathcal{B}$, and viceversa. An argument $\mathcal{B}$ is said to be a defeater of an argument $\mathcal{A}$ if $\mathcal{B}$ is a blocking or a proper defeater of $\mathcal{A}$.*

Example 1. Let $\Phi_1 = \langle AR, \sqsubseteq, \mathbf{C}, \mathbf{R}\rangle$ be an EAF where $AR = \{\mathcal{A}, \mathcal{B}, \mathcal{C}, \mathcal{D}, \mathcal{E}\}$, $\sqsubseteq = \emptyset$, $\mathbf{C} = \{\{\mathcal{A}, \mathcal{B}\}, \{\mathcal{B}, \mathcal{C}\}, \{\mathcal{C}, \mathcal{D}\}\}, \{\mathcal{C}, \mathcal{E}\}\}$ and $\mathcal{A} \succ \mathcal{B}, \mathcal{B} \succ \mathcal{C}$, $\mathcal{E} \bowtie \mathcal{C}$, $\mathcal{C} \equiv \mathcal{D}$.

Extended abstract frameworks can also be depicted as graphs, with different types of arcs, called *EAF-graphs.* We represent arguments as black triangles. An arrow ($\rightarrow$) is used to denote proper defeaters. A double-pointed straight arrow ($\longleftrightarrow$) connects blocking defeaters considered equivalent in strength, and a double-pointed zig-zag arrow ($\leftrightsquigarrow$) connects incomparable blocking defeaters. In Figure 1, the framework Φ_1 is shown. Argument $\mathcal{A}$ is a proper defeater of $\mathcal{B}$. Argument $\mathcal{B}$ is a proper defeater of $\mathcal{C}$, and $\mathcal{E}$ is an incomparable blocking defeater of $\mathcal{C}$ and viceversa. Argument $\mathcal{D}$ and $\mathcal{C}$ are blocking defeaters being equivalent in strength.

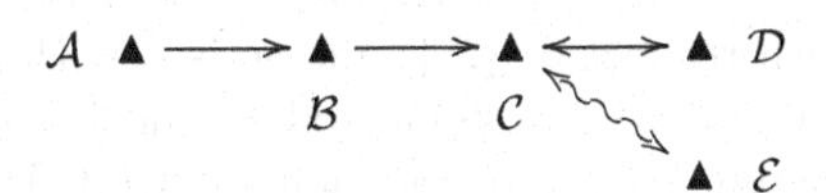

Fig. 1. EAF-graph of framework Φ_1

Example 2. The EAF of the introductory example about buying ski tables is $\Phi_{ski} = \langle AR, \sqsubseteq, \mathbf{C}, \mathbf{R}\rangle$ where
$AR = \{NS_1, OS, NS_2, NS_3\}$, $\sqsubseteq = \emptyset$, $\mathbf{C} = \{\{NS_1, OS\}, \{OS, NS_2\}, \{OS, NS_3\}\}$ and $OS \succ NS_1, OS \equiv NS_2$, $OS \bowtie NS_3$. The framework is shown in Figure 2.

The central element of attention in this work is blocking defeat. This kind of defeat arises when no strict preference is established between conflictive arguments.

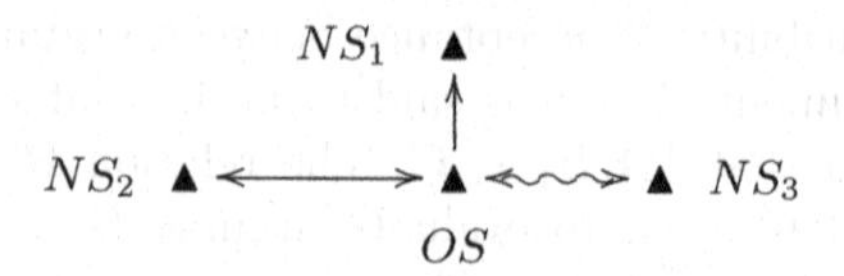

Fig. 2. EAF-graph of framework Φ_{ski}

Even then, blocked arguments are able to defend other arguments, although this defense is achieved with different strenght, as in Example 2. When this happens, a good position to be adopted is the identification of good and bad defenses. In the next section, we present new considerations about acceptability [1] regarding blocking defeaters.

3 Admissibility

As said before, argumentation semantics is about argument classification through several rational positions of acceptance. A central notion in most argument extensions is *acceptability.* A very simple definition of acceptability in extended abstract frameworks is as follows.

Definition 4. *Let $\Phi = \langle AR, \sqsubseteq, \mathbf{C}, \mathbf{R} \rangle$ be an EAF. An argument $\mathcal{A} \in AR$ is acceptable with respect to a set of arguments $S \subseteq AR$ if and only if every defeater $\mathcal{B}$ of $\mathcal{A}$ has a defeater in S.*

Fig. 3. Simple extended abstract framework

Defeaters mentioned in Definition 4 may be either proper or blocking ones. In the Figure 3, argument $\mathcal{A}$ is acceptable with respect to $\{\mathcal{C}\}$. Argument $\mathcal{C}$ is acceptable with respect to the empty set. Argument $\mathcal{D}$ is acceptable with respect to $\{\mathcal{F}\}$, $\mathcal{F}$ is acceptable with respect to $\{\mathcal{F}\}$, and so is $\mathcal{E}$ with respect to $\{\mathcal{E}\}$. Note that an argument that is not strictly preferred to another conflicting argument is considered strong enough to stand and defend itself. In Figure 3, $\mathcal{F}$ is defending itself against $\mathcal{E}$. Because of this intrinsic property of blocked arguments, a stronger notion of acceptability can be defined, by excluding self-defended arguments.

Definition 5. *Let $\Phi = \langle AR, \sqsubseteq, \mathbf{C}, \mathbf{R} \rangle$ be an EAF. An argument $\mathcal{A} \in AR$ is x-acceptable with respect to a set of arguments $S \subseteq AR$ if it is acceptable with respect to $S - \{\mathcal{A}\}$.*

Clearly, the focus is put in those arguments that need their own defense. In the framework of Figure 3, argument $\mathcal{D}$ is x-acceptable with respect to $\{\mathcal{F}\}$.

Argument $\mathcal{F}$ is not x-acceptable with respect to $\{\mathcal{F}\}$, as this argument lacks of other defense than the one provided by itself. In the framework AF_{ski}, argument NS_2 is x-acceptable with respect to $\{NS_2, NS_3\}$.

Following the usual steps in argumentation semantics, the notion of acceptability leads to the notion of admissibility. This requires the definition of conflict-free set of arguments. A set of arguments $S \subseteq AR$ is said to be *conflict-free* if for all $\mathcal{A}, \mathcal{B} \in S$ it is not the case that $\{\mathcal{A}, \mathcal{B}\} \in \mathbf{C}$.

Definition 6. *Let $\Phi = \langle AR, \sqsubseteq, \mathbf{C}, \mathbf{R}\rangle$ be an EAF. A set of arguments $S \subseteq AR$ is said to be admissible if it is conflict-free and every argument in S is acceptable with respect to S.*

An admissible set is able to defend any argument included in that set. As stated before, some arguments can only be defended by themselves, although they can be defenders of other arguments. Therefore, using the refined version of acceptability of Definition 5, the corresponding admissibility notion is as follows.

Definition 7. *Let $\Phi = \langle AR, \sqsubseteq, \mathbf{C}, \mathbf{R}\rangle$ be an EAF. A set of arguments $S \subseteq AR$ is said to be x-admissible if it is conflict-free and every argument in S is x-acceptable with respect to S.*

An x-admissible extension S may include a blocked argument $\mathcal{A}$ only if the intrinsic self-defense of $\mathcal{A}$ is superfluous. Every x-admissible extension is admissible by definition. In the framework of Figure 3, the set $\{\mathcal{D}, \mathcal{F}\}$ is admissible, but not x-admissible. In the framework Φ_{ski}, the sets $\{NS_1, NS_2\}$ and $\{NS_1, NS_3\}$ are admissible but not x-admissible. The set $\{NS_1, NS_2, NS_3\}$ is admissible and x-admissible, as arguments NS_2 and NS_3 are defenders of each other.

When blocking defeat is present, x-admissiblity is a stronger notion than classic admissibility. If a set S is admissible but not x-admissible, then at least one argument $\mathcal{A}$ in S has a defender that can only be defended by itself. This may be considered as a sign of weakness of S. Consider the framework of Figure 4. The admissible sets are $\emptyset$, $\{\mathcal{A}\}$, $\{\mathcal{B}\}$, $\{\mathcal{C}\}$ and $\{\mathcal{A}, \mathcal{C}\}$. According to x-admissibilty semantics, only the empty set and $\{\mathcal{A}, \mathcal{C}\}$ are valid extensions.

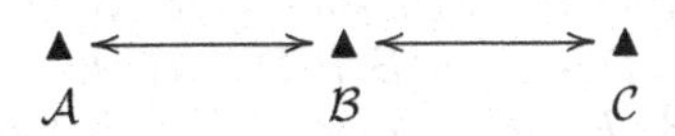

Fig. 4. $\{\mathcal{A}, \mathcal{C}\}$ is x-admissible

An x-admissible set may be a good position of acceptance when trying to avoid arguments that cannot be defended but by themselves. Sometimes, however, these kind of arguments is mandatory for a defense, as in Figure 2 for argument NS_1. The first step is to distinguish an acceptance set due to defense based on proper-defeaters. In the following section a new semantic notion characterizes a set of accepted arguments only when they benefit from a strictly preferred defense. In Section 5, an argument extension based on the strongest argument defense is introduced.

4 Acceptance Through Strongest Defense

When conflicts and preferences are considered as separate elements in the frameworks, a deeper evaluation of extensions can be considered. Lets analyze the situation depicted in Figure 5(a) where $\mathcal{D} \bowtie \mathcal{B}$ and $\mathcal{E} \bowtie \mathcal{C}$.

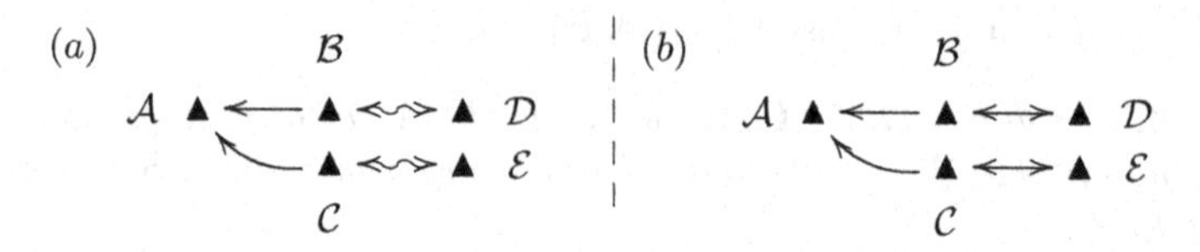

Fig. 5. Argument $\mathcal{A}$ is defended by $\mathcal{D}$ and $\mathcal{E}$

According to Definition 4, argument $\mathcal{A}$ is acceptable with respect to $\{\mathcal{D}, \mathcal{E}\}$. The set $\{\mathcal{A}, \mathcal{D}, \mathcal{E}\}$ is admissible and it is also a complete extension (in the sense of [1]). However, $\{\mathcal{A}, \mathcal{D}, \mathcal{E}\}$ it is not x-admissible, as $\mathcal{D}$ and $\mathcal{E}$ are blocked arguments. In fact, no set including these arguments is x-admissible. Note that the defense of $\mathcal{A}$ is a blockade-based one, that is, it relies on blocking defeat. Now consider the same framework where $\mathcal{D} \equiv \mathcal{B}$ and $\mathcal{E} \equiv \mathcal{C}$, as shown in Figure 5($b$). There is still a blockade of defeaters of $\mathcal{A}$, but it is due to an equivalence of strength between arguments. Again, $\mathcal{A}$ is defended by $\{\mathcal{D}, \mathcal{E}\}$, but the position is more cohesive. The strenght of arguments can be measured, although only to conclude that it is the same. This is a better defense than blockade by incomparability as before. The quality of a defense determines the quality of an admissible set. For an argument $\mathcal{A}$, being acceptable with respect to a set S does not necessarily means that is strongly defended by S. Following the previous example, it is clear that the best defense of $\mathcal{A}$ occurs when $\mathcal{D} \succ \mathcal{B}$ and $\mathcal{E} \succ \mathcal{C}$. In this scenario, x-admissible sets are not a suitable semantics if a credulous position is desired, as the only x-admissible set is the empty set. Thus, the ability to distinguish quality among admissible sets is interesting.

Definition 8. *Let $\Phi = \langle AR, \sqsubseteq, \mathbf{C}, \mathbf{R} \rangle$ be an EAF. Let $\mathcal{A} \in AR$ be an argument with defeater $\mathcal{B} \in AR$ and let $\mathcal{D} \in AR$ such that $\{\mathcal{B}, \mathcal{D}\} \in \mathbf{C}$. Then argument $\mathcal{A}$ is said to be* confirmed *by $\mathcal{D}$ against $\mathcal{B}$ if $\mathcal{D} \succ \mathcal{B}$. It is said that $\mathcal{A}$* sustained *by $\mathcal{D}$ against $\mathcal{B}$ if $\mathcal{D} \equiv \mathcal{B}$. Finally, $\mathcal{A}$ is said to be* held up *by $\mathcal{D}$ against $\mathcal{B}$ if $\mathcal{D} \bowtie \mathcal{B}$. In all cases $\mathcal{D}$ is said to be a defender of $\mathcal{A}$.*

Example 3. Consider the EAF of Figure 6. As several blocking situations are present, many cases of defense can be found. For example, argument $\mathcal{A}$ is confirmed by $\mathcal{D}$ against $\mathcal{B}$. It is held up by $\mathcal{F}$ against $\mathcal{B}$, and sustained by $\mathcal{E}$ against $\mathcal{C}$. Argument $\mathcal{C}$, in turn, is confirmed by $\mathcal{B}$ against $\mathcal{A}$. Argument $\mathcal{G}$ is sustained by itself against $\mathcal{H}$.

In the framework of Figure 6, argument $\mathcal{A}$ is acceptable with respect to $S_1 = \{\mathcal{E}, \mathcal{F}\}$. It is also acceptable with respect to $S_2 = \{\mathcal{E}, \mathcal{D}\}$. However, $\mathcal{A}$ is not confirmed by any argument in S_1, as it is done in S_2. Therefore, $\{\mathcal{A}\} \cup S_2$ may

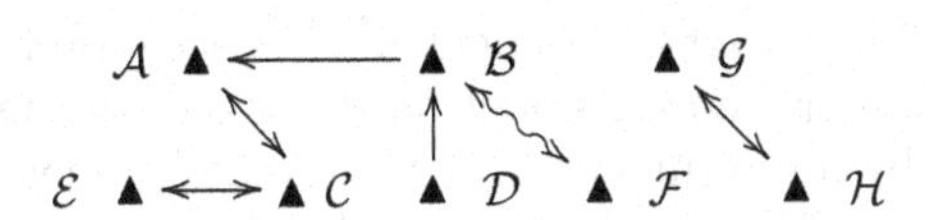

Fig. 6. Argument defense

be considered a stronger admissible set than $\{\mathcal{A}\} \cup S_1$. This is reinforced by the fact that $\mathcal{D}$ does not need defense.

Definition 9. *Let $\Phi = \langle AR, \sqsubseteq, \mathbf{C}, \mathbf{R} \rangle$ be an EAF. An argument $\mathcal{A}$ is strong-acceptable with respect to a set of arguments $S \subseteq AR$ if $\mathcal{A}$ is confirmed by an argument in S against every defeater of $\mathcal{A}$.*

In the extended framework of Figure 6, argument $\mathcal{F}$ is strong-acceptable with respect to $\{\mathcal{D}\}$. Argument $\mathcal{A}$ is not strong-acceptable with respect to any set, because it cannot be confirmed against $\mathcal{C}$. Argument $\mathcal{D}$ is strong-acceptable with respect to the empty set, as it has no defeaters.

Definition 10. *Let $\Phi = \langle AR, \sqsubseteq, \mathbf{C}, \mathbf{R} \rangle$ be an EAF. A set of arguments $S \subseteq AR$ is strong-admissible if every argument in S is strong-acceptable w.r.t. S.*

In the *EAF* of Figure 6, the only strong-addmissible sets are $\{\}$, $\{\mathcal{D}\}$, and $\{\mathcal{D}, \mathcal{F}\}$. Note that argument $\mathcal{A}$ is excluded because its only defense against $\mathcal{C}$ is a blockade-based one.

Example 4. Consider the EAF of Figure 7. The sets $\{\mathcal{B}, \mathcal{D}\}$ and $\{\mathcal{A}, \mathcal{C}\}$ are admissible, but only $\{\mathcal{B}, \mathcal{D}\}$ is strong-admissible. This is because $\mathcal{C}$ is not strong-acceptable with respect to $\{\mathcal{A}\}$. The set $\{\mathcal{B}, \mathcal{D}\}$ is also x-admissible, as $\mathcal{B}$ is acceptable with respect to $\{\mathcal{D}\}$ and $\mathcal{D}$ is acceptable with respect to $\{\mathcal{B}\}$.

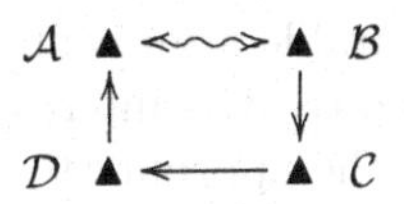

Fig. 7. The set $\{\mathcal{B}, \mathcal{D}\}$ is strong-admissible

Strong-admissibility is a more skeptical criterion for argument acceptance than classical admissible sets. Clearly, an admissible set may be not strong admissible, as in previous examples. Also an x-admissible set is not necessarily strong-admissible: in the framework of Figure 4, the set $\{\mathcal{A}, \mathcal{C}\}$ is x-admissible, but no proper defeat is present and therefore it is not strong-admissible.

Proposition 1. *If a set S is strong-admissible, then it is admissible and x - admissible.*

Proof. Let S be a strong-admissible set. If an argument $\mathcal{A}$ is strong-acceptable with respect to a set S, then it is acceptable with respect to S, as a collective defense is achieved. Thus, every argument in S is acceptable w.r.t S, and then S is admissible. Even more, no argument needs its own defense: every argument $\mathcal{X} \in S$ is confirmed by an argument $\mathcal{Y} \in S$. As no argument can be confirmed by itself, then $\mathcal{X} \neq \mathcal{Y}$ and then $\mathcal{X}$ is acceptable with respect to $S - \{\mathcal{X}\}$. Thus, S is x-admissible. □

A strong admissible set is perhaps the most reliable defense that an argument may acquire, as it is based only on strictly preferred arguments. It may be considered that an argument in a strong-admissible set is safely justified. However, it is not always possible to count with this kind of defense. In the framework Φ_{ski} the only strong-admissible set is the empty set. In the framework of Figure 6 argument $\mathcal{A}$ is somehow defended, but it is rejected according to strong-admissibility. Sometimes it is desirable to adopt a more credulous position of acceptance. Therefore, it is interesting to evaluate the quality of a defense and its alternatives when incomparable or equivalent-in-force defenders are involved. This is addressed in the following section.

5 Weighing Up Defenses

An argument $\mathcal{A}$ may be collectively defended by several sets of arguments. In Φ_{ski}, argument NS is defended by $\{NS_1\}$ and it is also defended by $\{NS_2\}$. As a consequence, the union of these sets constitutes a defense for $\mathcal{A}$. The final set of accepted arguments depends on the position adopted by the rational agent, for which the knowledge is modeled by the framework . Sometimes, maximality is required, a position in which an agent accepts all it can defend. Another interesting position is to accept a small set of arguments and its strongest defenders. This corresponds to an agent that, given a special situation where arguments are defended in different manners, decides to accept a minimal set where, if not needed, weakest defenders are discarded. In particular, in extended abstract frameworks, defense can occur with different levels of strength, and when a lot of equivalent-in-force or incomparable arguments are involved in the defeat scenario (leading to several extensions) it is interesting to target the construction of valid extensions according to the appropriate defense.

This evaluation of defenses can be achieved by considering a rational, implicit ordering of argument preferences. This is a common sense property: given two arguments for which a preference must be made, the best scenario is where one of them is effectively preferred to the other. If this is not the case, then at least is desirable to acknowledge the equivalence in strength, so it is clear that both arguments are somehow related. The worst case is to realize that both arguments are not related enough even to evaluate a difference in force. This is formalized in the next definition.

Definition 11. *Let $\Phi = \langle AR, \sqsubseteq, \mathbf{C}, \mathbf{R} \rangle$ be an EAF. Let $\mathcal{A}$ and $\mathcal{B}$ be two arguments in AR The function $pref : AR \times AR \to \{0, 1, 2\}$ is defined as follows*

$$pref(\mathcal{A}, \mathcal{B}) = \begin{cases} 0 & \text{if } \mathcal{A} \bowtie \mathcal{B} \\ 1 & \text{if } \mathcal{A} \equiv \mathcal{B} \\ 2 & \text{if } \mathcal{A} \succ \mathcal{B} \end{cases}$$

The simple graduation of preference conclusions stated in Definition 11 allows the comparison of individual defenses. For an argument $\mathcal{A}$, the strength of its defenders is evaluated as stated in Definition 12.

Definition 12. *Let $\Phi = \langle AR, \sqsubseteq, \mathbf{C}, \mathbf{R} \rangle$ be an EAF. Let $\mathcal{A} \in AR$ be an argument with defeater $\mathcal{B}$, which is defeated, in turn, by arguments $\mathcal{C}$ and $\mathcal{D}$. Then*

1. *$\mathcal{C}$ and $\mathcal{D}$ are equivalent in strength defenders of $\mathcal{A}$ if $pref(\mathcal{C}, \mathcal{B}) = pref(\mathcal{C}, \mathcal{D})$.*
2. *$\mathcal{C}$ is a stronger defender than $\mathcal{D}$ if $pref(\mathcal{C}, \mathcal{B}) > pref(\mathcal{C}, \mathcal{D})$. It is also said that $\mathcal{D}$ is a weaker defender than $\mathcal{C}$*

In the framework Φ_{ski}, argument NS_2 is a stronger defender of NS than NS_3. In the framework of Figure 6, for argument $\mathcal{A}$, the argument $\mathcal{F}$ is a weaker defender than $\mathcal{D}$ and also than $\mathcal{E}$. The evaluation of a collective defense follows from Definition 12.

Definition 13. *Let $\Phi = \langle AR, \sqsubseteq, \mathbf{C}, \mathbf{R} \rangle$ be an EAF. Let $\mathcal{A} \in AR$ be an argument acceptable with respect to $S_1 \subseteq AR$. A set of arguments $S_2 \subseteq AR$ is said to be a stronger collective defense of $\mathcal{A}$ if $\mathcal{A}$ is acceptable with respect to S_2, and*

1. *There does not exists two argument $\mathcal{X} \in S_1$ and $\mathcal{Y} \in S_2$ such that $\mathcal{X}$ constitutes a stronger defense than $\mathcal{Y}$*
2. *For at least one defender $\mathcal{X} \in S_1$ of $\mathcal{A}$, there exists an argument $\mathcal{Y} \in S_2$ which constitutes a stronger defense of $\mathcal{A}$.*

A set of arguments S_2 is a stronger collective defense of $\mathcal{A}$ than the set S_1 if the force of defense achieved by elements in S_2 is equal or stronger than those in S_1. Informally, every argument $\mathcal{X}$ in S_1 has a competitor $\mathcal{Y}$ in S_2 that is a stronger or equivalent in strength defender. Note that it is possible that $\mathcal{X} = \mathcal{Y}$.

In the framework Φ_{ski}, the set $\{NS_2\}$ is a stronger defense than $\{NS_3\}$. The set $\{NS_2, NS_3\}$ is not considered a stronger collective defense than $\{NS_2\}$ (maximality is not relevant under this notion). On the other hand, $\{NS_2, NS_3\}$ is considered a stronger collective defense than $\{NS_3\}$, because of the inclusion of a stronger defender.

The strength of a defense is a pathway to evaluate admissible sets. In Dung's classic abstract framework, admissible sets may be compared, for instance, by set inclusion, and then maximal extensions are of interest. Even then, two admissible sets with the same number of arguments may be considered indistinguishable alternatives for acceptance, as in the Nixon Diamond example [1]. In extended abstract frameworks, defeat may occur in different ways, according to preference criterion $\mathbf{R}$, and this can be used to evaluate the inner composition of an admissible set.

Example 5. Consider the EAF of Figure 8. The admissible sets are $\emptyset$ (trivial), every singleton set, $\{\mathcal{A}, \mathcal{D}\}$ and $\{\mathcal{A}, \mathcal{C}\}$. Argument $\mathcal{A}$ is defended by sets $\{\mathcal{D}\}$ and $\{\mathcal{C}\}$, but the first one is a stronger collective defense than the second one. Then $\{\mathcal{A}, \mathcal{D}\}$ is an admissible set with stronger inner defenses than $\{\mathcal{A}, \mathcal{C}\}$.

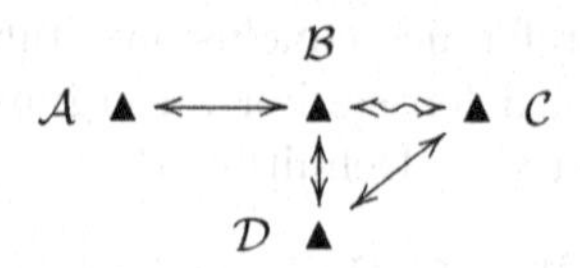

Fig. 8. $\{\mathcal{D}\}$ is a stronger defense of $\mathcal{A}$ than $\{\mathcal{C}\}$

Proposition 2. *Let $\langle AR, \sqsubseteq, \mathbf{C}, \mathbf{R}\rangle$ be an EAF. If an argument $\mathcal{A} \in AR$ is strong-acceptable with respect to a set of arguments $S \subseteq AR$, then no other set $S' \subseteq AR$ constitutes a stronger collective defense than S.*

Proof. Trivial, as every individual defense cannot be strengthened by another argument because it is achieved by proper defeaters.

Definition 14. *An admissible set of arguments S is said to be top-admissible if, for any argument $\mathcal{A} \in S$, no other admissible set S' provides a stronger defense of $\mathcal{A}$ than S.*

In Figure 6, the set $\{\mathcal{A}, \mathcal{D}, \mathcal{E}\}$ is top-admissible, while the set $\{\mathcal{A}, \mathcal{D}, \mathcal{F}\}$ is not. In the framework AF_{ski}, the admissible set $\{NS, NS_1\}$ is top-admissible. The set $\{NS, NS_2\}$ is not, as NS can be benefited from a better defense. In the EAF of Figure 8, the set $\{\mathcal{A}, \mathcal{D}\}$ is top-admissible. Top-admissibility is also a measure of strength, other than cardinality of admissible sets. Although every maximal (with respect to set inclusion) admissible set is top-admissible, the converse is not true. In Figure 6, the set $\{\mathcal{A}, \mathcal{D}, \mathcal{F}\}$ is top-admissible, but not maximal.

Definition 14 is a measure of quality. What top-admissibility semantics requires is that best admissible sets are selected, according to the strength of every defense. Note that every strong-admissible set is top-admissible, but a top-admissible set may not be strong-admissible as the former is allowing blockade-based defenses.

6 Conclusions and Related Work

In this work, new restricted notions of admissibility were introduced by taking into account the strength of an argument defense in *extended abstract frameworks* [8,9]. Each of the new admissibility notions presented relies on the presence of blocking defeaters in the argument scenario. A set S is x-admissible if no blocked argument in S needs its own defense. A set is strong-admissible if every argument in that set is defended by arguments strictly preferred to its defeaters, discarding any blockade-based defense. A set is top-admissible if no inner defense can be individually strengthened. As said before, the final set of accepted arguments depends on the position adopted by the rational agent. The aim of this work was to provide new views for selecting an admissible set of arguments in extended abstract frameworks.

Acceptability notions are defined in [1], and many interesting works are based on these concepts. Other extensions based on acceptability of arguments were defined, although using classic frameworks, and then usually maximal or minimal sets are relevant. In [10] minimality is explored as a property of defenses by distinguishing two types of arguments, one of them mandatory as a defense, while the other used only when is needed. Thus, the defense is also weighed out according to relevance of defenders. Despite the framework used, weaker defenders in EAF may be considered as a form of restricted arguments in [10].

Preferences between arguments are also present in other works, notably in [4], where the notion of *strict-defense* influenced *strong-acceptability* in EAF. Also in that work an argument may defend itself, although in a different way: it is preferable to its attackers. In EAF self-defended arguments are related to symmetric blocking and each argument's context. Preference is used in [6], where value-based argumentation frameworks are presented. In that work, defeat is derived by comparing arguments. Similar to [4], an attack is overruled (*i.e.* it does not succeed) if the attacked argument is preferred to its attacker. Otherwise, a defeat relation arises, even when both argument may promote equal or unrelated values. In EAF this leads to blocking defeat, the basis for distinguishing grades on defense, as it is done in the present work. In [11] principles for the evaluation of extensions are presented; skepticism among them. According to the *weak skepticism relation*, a strong-admissible set is more skeptical than admissible sets. Also top-admissibility is more skeptical than classical admissibility.

References

1. Dung, P.M.: On the Acceptability of Arguments and its Fundamental Role in Nomonotonic Reasoning and Logic Programming. In: Proc. of the 13th. IJCAI 1993, pp. 852–857 (1993)
2. Vreeswijk, G.A.W.: Abstract argumentation systems. Artificial Intelligence 90(1-2), 225–279 (1997)
3. Amgoud, L., Cayrol, C.: On the acceptability of arguments in preference-based argumentation. In: UAI 1998. 14th Conference on Uncertainty in Artificial Intelligence, pp. 1–7. Morgan Kaufmann, San Francisco (1998)
4. Amgoud, L., Cayrol, C.: A reasoning model based on the production of acceptable arguments. Annals of Mathematics and Artificial Intelligence 34(1-3), 197–215 (2002)
5. Amgoud, L., Perrussel, L.: Arguments and Contextual Preferences. In: CD 2000. Computational Dialectics-Ecai workshop, Berlin (2000)
6. Bench-Capon, T.: Value-based argumentation frameworks. In: Proc. of Nonmonotonic Reasoning, pp. 444–453 (2002)
7. Martínez, D., García, A., Simari, G.: Progressive defeat paths in abstract argumentation frameworks. In: Proceedings of the 19th Conf. of the Canadian Society for Computational Studies of Intelligence 2006, pp. 242–253 (2006)
8. Martínez, D., García, A., Simari, G.: On acceptability in abstract argumentation frameworks with an extended defeat relation. In: COMMA 2006. Proc. of I. Intl. Conf. on Computational Models of Arguments, pp. 273–278 (2006)

9. Martínez, D., García, A., Simari, G.: Modelling well-structured argumentation lines. In: Proc. of XX IJCAI-2007, pp. 465–470 (2007)
10. Cayrol, C., Doutre, S., Lagasquie-Schiex, M.C., Mengin, J.: "minimal defence": a refinement of the preferred semantics for argumentation frameworks. In: NMR, pp. 408–415 (2002)
11. Baroni, P., Giacomin, M.: Evaluation and comparison criteria for extension-based argumentation semantics. In: COMMA 2006. Proc. of I. International Conf. on Computational Models of Arguments, pp. 157–168 (2006)

Ontology-Based Inference for Causal Explanation

Ph. Besnard[1], M.-O. Cordier[2], and Y. Moinard[2]

[1] IRIT, CNRS, Université Paul Sabatier
118 route de Narbonne, 31062 Toulouse cedex, France
besnard@irit.fr
[2] IRISA, INRIA, Université Rennes I
Campus de Beaulieu, 35042 Rennes cedex, France
{cordier,moinard}@irisa.fr

Abstract. We define an inference system to capture explanations based on causal statements, using an ontology in the form of an *IS-A* hierarchy. We first introduce a simple logical language which makes it possible to express that a fact causes another fact and that a fact explains another fact. We present a set of formal inference patterns from causal statements to explanation statements. These patterns exhibit ontological premises that are argued to be essential in deducing explanation statements. We provide an inference system that captures the patterns discussed.

1 Introduction

We are aiming at a logical formalization of explanations from causal statements. For example, it is usually admitted that fire is an explanation for smoke, on the grounds that fire causes smoke. In other words, fire causes smoke is a premise from which it can be inferred that fire is an explanation for smoke. In this particular example, concluding from cause to explanation is immediate but such is not always the case, far from it. In general, the reasoning steps leading from cause to explanation are not so trivial:

Example. *We consider two causal statements:*

(i) Any ship that is about to sink causes her crew to launch some red rocket(s)

(ii) In France, July the 14th *causes colourful rockets to be launched (fireworks)*

So, if the place is a coastal city in France, on July the 14th*, then red rockets being launched could be explained either by some ship(s) sinking or by the French national day.*

In this example, it is needed to acknowledge the fact that a red rocket is a kind of (colourful) rocket in order to get the second explanation, which makes sense.

Example (con'd). *Suppose that we now add the following statement:*

(i) Seeing a red rocket being launched triggers a rescue process

Now, a possible explanation for the triggering of the rescue process, as happens in practice, is that we are on July the 14th *in a coastal city in France.*

Z. Zhang and J. Siekmann (Eds.): KSEM 2007, LNAI 4798, pp. 153–164, 2007.

In this paper, we define a dedicated inference system to capture explanations based on causal statements and stress that the rôle of ontology-based information is essential. In the second section, we introduce the logical language that we propose to use. In the third section, we define the set of patterns dedicated to inferring explanations from causal statements and ontological information. In the fourth section, we give the formal inference system allowing us to derive explanations from a given set of formulas. We conclude by discussing the rôle of ontology and by giving some perspectives.

2 Language

We distinguish various types of statements in our formal system:

C: A theory expressing causal statements. E.g., $On(alarm)$ *causes* $Heard(bell)$.
O: An ontology listing entities, in the form of an IS-A hierarchy which defines classes and sub-classes. E.g., *loud_bell is a bell* and *soft_bell is a bell.*
W: A classical first order theory expressing truths (i.e., incompatible facts, co-occurring facts, ...). E.g., $Heard(soft\_bell) \to \neg Heard(loud\_bell)$.

We assume a logical language whose alphabet consists of constants $a, b, c, \ldots$ and unary predicates such as $P, \ldots$ Intuitively, constants denote objects or entities, including classes and sub-classes to be found in the IS-A hierarchy. The predicates (unary for simplicity, they could be n-ary in theory) are used to express facts or events on these objects or entities as $Heard(soft\_bell)$ or $Own(blue\_car)$. The causal statements express causal relations between facts or events expressed by these predicates as in $On(alarm)$ *causes* $Heard(bell)$.

Moreover, we assume that these unary predicates "inherit upwards" through the IS-A hierarchy in the following sense: If b IS-A c then $P(b)$ entails $P(c)$. Consider owning as an example, together with small car and car. Of course, a small car is a car and "I own a small car" entails ""I own a car". This property happens to be fundamental when designing inference patterns (next section) as it allows us to apply inheritance properties between entities to facts and events on these entities.. It also means that we restrict ourselves to those predicates that "inherit upwards", which precludes for instance the predicate *Dislike* as $Dislike(small\_car)$ does not imply $Dislike(car)$. In the following, we say that $Heard(soft\_bell)$ is a specialization of $Heard(bell)$ and that $Heard(bell)$ is a generalization of $Heard(soft\_bell)$.

The formal system we introduce below is meant to infer (this inference will be noted $\vdash_C$), from such premises $C \cup O \cup W$, formulas denoting explanations.

In the sequel, $\alpha, \beta, \ldots$ denote the so-called sentential atoms (i.e., ground atomic formulas) and $\Phi, \Psi, \ldots$ denote sets thereof.

Atoms

1. *Sentential atoms*: $\alpha, \beta, \ldots$ (Ground atomic formulas)
2. *Causal atoms*: α *causes* β.
3. *Ontological atoms*: $b \to_{IS-A} c$.
4. *Explanation atoms*: α *explains* β *because_possible* Φ.

An ontological atom reads: b is a c.

An explanation atom reads: α *is an explanation for* β *because* Φ *is possible.*

Notation: α *explains* β *bec_poss* Φ *abbreviates* α *explains* β *because_possible* Φ.

Formulas

1. *Sentential formulas:* Boolean combinations of sentential atoms.
2. *Causal formulas:* Boolean combinations of causal atoms and sentential atoms.

The premises of the inference $\vdash_C$, namely $C \cup O \cup W$, consist of sentential formulas as well as causal formulas and ontological *atoms* (no ontological formula). Notice that explanation atoms cannot occur in the premises.

The properties of causal and ontological formulas we consider are as follows.

1. **Properties of the causal operator**
   (a) *Entailing the conditional*: If α *causes* β, then $\alpha \to \beta$.
2. **Properties of the ontological operator**
   (a) *Upward inheritance*: If $b \to_{IS-A} c$, then $\alpha[b] \to \alpha[c]$[1].
   (b) *Transitivity*: If $a \to_{IS-A} b$ and $b \to_{IS-A} c$, then $a \to_{IS-A} c$.
   (c) *Reflexivity*: $c \to_{IS-A} c$.

Reflexivity is unconventional a property for an *IS-A* hierarchy. It is included here because it helps keeping the number of inference schemes low (see later).

W is supposed to include (whether explicitly or via inference) all conditionals induced by the ontology O. For example, if *loud_bell* $\to_{IS-A}$ *bell* is in O then *Heard(loud_bell)* $\to$ *Heard(bell)* is in W. Similarly, W is supposed to include all conditionals induced by the causal statements in C. For example, if *On(alarm) causes Heard(bell)* is in C, then *On(alarm)* $\to$ *Heard(bell)* is in W.

3 Patterns for Inferring Explanations

A set of patterns is proposed to infer explanations from premises $C \cup O \cup W$.

3.1 The Base Case

A basic idea is that what causes an effect can always be suggested as an explanation when the effect happens to be the case:

$$\text{If} \begin{pmatrix} \alpha \textit{ causes } \beta \\ \text{and} \\ W \not\models \neg\alpha \end{pmatrix} \text{then} \quad \alpha \textit{ explains } \beta \textit{ because\_possible } \{\alpha\}$$

[1] $\alpha[b]$ denotes the atomic formula α and its (only) argument b.

Example. *Consider a causal model such that $W \not\vdash \neg On(alarm)$ and O is empty whereas*

$$C = \{On(alarm)\ causes\ Heard(bell)\}$$

Then, the atom

$$On(alarm)\ explains\ Heard(bell)\ because\_possible\ \{On(alarm)\}$$

is inferred. That is, $On(alarm)$ is an explanation for $Heard(bell)$.

By the way, "is an explanation" must be understood as provisional. Inferring that $On(alarm)$ is an explanation for $Heard(bell)$ is a tentative conclusion: Should $On(alarm)$ be ruled out, e.g., $\neg On(alarm) \in W$, then $On(alarm)$ is not an explanation for $Heard(bell)$.

Formally, with $Form = On(alarm)\ explains\ Heard(bell)\ bec\_poss\ \{On(alarm)\}$: $C \cup O \cup W \vdash_C Form$; $C \cup O \cup W \cup \{\neg On(alarm)\} \not\vdash_C Form$

3.2 Wandering the IS-A Hierarchy: Going Upward

What causes an effect can be suggested as an explanation for any consistent ontological generalization of the effect:

$$\text{If} \begin{pmatrix} \alpha\ causes\ \beta[b] \\ \text{and} \\ b \rightarrow_{IS-A} c \\ \text{and} \\ W \not\models \neg\alpha \end{pmatrix} \text{then} \quad \alpha\ explains\ \beta[c]\ because\_possible\ \{\alpha\}$$

Example. $C = \{On(alarm)\ causes\ Heard(bell)\}$ *and* $O = \{bell \rightarrow_{IS-A} noise\}$. *$W$ contains no statement apart from those induced by C and O, that is:*

$$W = \{On(alarm) \rightarrow Heard(bell), Heard(bell) \rightarrow Heard(noise)\}$$

Inasmuch as noise could be bell, $On(alarm)$ then counts as an explanation for $Heard(noise)$.

$$C \cup O \cup W \vdash_C On(alarm)\ explains\ Heard(noise)\ bec\_poss\ \{On(alarm)\}$$

Again, it would take $On(alarm)$ to be ruled out for the inference to be prevented.

Example. $C = \{On(alarm)\ causes\ Heard(bell)\}$

$$O = \{bell \rightarrow_{IS-A} noise, hooter \rightarrow_{IS-A} noise\}$$

W states that a hooter is heard (and that $Heard(bell)$ is not $Heard(hooter)$) and additionally expresses the conditionals induced by C and O, that is:

$$W = \left\{ \begin{array}{c} Heard(hooter) \\ \neg(Heard(bell) \leftrightarrow Heard(hooter)) \\ On(alarm) \rightarrow Heard(bell) \\ Heard(hooter) \rightarrow Heard(noise) \\ Heard(bell) \rightarrow Heard(noise) \end{array} \right\}$$

Even taking into account the fact that bell is an instance of noise, it cannot be inferred that $On(alarm)$ is an explanation for $Heard(noise)$. The inference fails because it would need noise to be of the bell kind (which is false, cf hooter). Technically, the inference fails because $W \vdash \neg On(alarm)$.

The next example illustrates why resorting to ontological information is essential when attempting to infer explanations: the patterns in sections 3.2-3.3 extend the base case for explanations to *ontology-based* consequences, not any consequences.

Example. *Rain makes me growl. Trivially, I growl only if I am alive. However, rain cannot be taken as an explanation for the fact that I am alive.*

$$C = \{Rain\ causes\ I\_growl\}$$
$$W = \{I\_growl \rightarrow I\_am\_alive\}$$

$C \cup O \cup W \nvdash_C$ *Rain explains I_am_alive bec_poss* $\{Rain\}$

3.3 Wandering the IS-A Hierarchy: Going Downward

What causes an effect can presumably be suggested as an explanation when the effect takes place in one of its specialized forms:

$$\text{If} \begin{pmatrix} \alpha\ causes\ \beta[c] \\ \text{and} \\ b \rightarrow_{IS-A} c \\ \text{and} \\ W \not\models \neg(\alpha \wedge \beta[b]) \end{pmatrix} \text{then}\ \ \alpha\ \ explains\ \ \beta[b]\ \ because\_possible\ \ \{\alpha, \beta[b]\}$$

Example. *Consider a causal model with C and O as follows:*

$$C = \{On(alarm)\ causes\ Heard(bell)\}\ \ and\ \ O = \left\{ \begin{array}{l} loud\_bell \rightarrow_{IS-A} bell \\ soft\_bell \rightarrow_{IS-A} bell \end{array} \right\}$$

O means that loud_bell is more precise than bell. Since $On(alarm)$ is an explanation for $Heard(bell)$, it also is an explanation for $Heard(loud\_bell)$ and similarly

Heard(soft_bell). This holds inasmuch as there is no statement to the contrary: The latter inference would not be drawn if for instance $\neg Heard(soft\_bell)$ *or* $\neg(Heard(soft\_bell) \to On(alarm))$ *were in* W. *Formally, with* $Form(loud) =$ $On(alarm)$ *explains* $Heard(loud\_bell)$ *bec_poss* $\{On(alarm), Heard(loud\_bell)\}$ *and* $Form(soft) =$
$On(alarm)$ *explains* $Heard(soft\_bell)$ *bec_poss* $\{On(alarm), Heard(soft\_bell)\}$*:*
$C \cup O \cup W \vdash_C Form(loud)$, $C \cup O \cup W \vdash_C Form(soft)$, *and*
$C \cup O \cup W \cup \{\neg(Heard(soft\_bell) \to On(alarm))\} \not\vdash_C Form(soft)$

3.4 Transitivity of Explanations

We make no assumption as to whether the causal operator is transitive (from α *causes* β and β *causes* γ does α *causes* γ follow?). However, we do regard inference of explanations as transitive which, in the simplest case, means that if α explains β and β explains γ then α explains γ.

The general pattern for transitivity of explanations takes two causal statements, α *causes* β and β' *causes* γ where β and β' are ontologically related, as premises in order to infer that α is an explanation for γ.

In the first form of transitivity, β' is inherited from β by going upward in the *IS-A* hierarchy.

$$\text{If} \quad \begin{pmatrix} \alpha \textit{ causes } \beta[b], \\ \beta[c] \textit{ causes } \gamma, \\ b \to_{IS-A} c, \\ \text{and} \\ W \not\models \neg\alpha \end{pmatrix} \quad \text{then} \quad \alpha \textit{ explains } \gamma \textit{ because\_possible } \{\alpha\}$$

Example. *Sunshine makes me happy. Being happy is why I sing. Therefore, sunshine is a plausible explanation for the case that I am singing.*

$$C = \left\{ \begin{array}{c} Sunshine \textit{ causes } I\_am\_happy \\ I\_am\_happy \textit{ causes } I\_am\_singing \end{array} \right\}$$

$$W = \left\{ \begin{array}{c} Sunshine \to I\_am\_happy \\ I\_am\_happy \to I\_am\_singing \end{array} \right\}$$

So, we get:

$$C \cup O \cup W \vdash_C Sunshine \textit{ explains } I\_am\_singing \textit{ bec\_poss } \{Sunshine\}.$$

The above example exhibits transitivity of explanations for the simplest case that $\beta = \beta'$ in the pattern α *causes* β and β' *causes* γ entail α *causes* γ (trivially, if $\beta = \beta'$ then β and β' are ontologically related). Technically, $\beta = \beta'$ is obtained by applying reflexivity in the ontology. This is one illustration that using reflexivity in the ontology relieves us from the burden of tailoring definitions to capture formal degenerate cases.

Example. *Let* $O = \{bell \rightarrow_{IS-A} noise\}$ *and*

$$C = \left\{ \begin{array}{c} On(alarm)\ causes\ Heard(bell) \\ Heard(noise)\ causes\ Disturbance \end{array} \right\}$$

W states the facts induced by C and O, that is:

$$W = \left\{ \begin{array}{c} On(alarm) \rightarrow Heard(bell) \\ Heard(noise) \rightarrow Disturbance \\ Heard(bell) \rightarrow Heard(noise) \end{array} \right\}$$

So, we get:

$$C \cup O \cup W \vdash_C On(alarm)\ explains\ Disturbance\ bec\_poss\ \{On(alarm)\}.$$

In the second form of transitivity, β' is inherited from β by going downward in the IS-A hierarchy.

$$\text{If} \left(\begin{array}{l} \alpha\ causes\ \beta[c], \\ \beta[b]\ causes\ \gamma, \\ b \rightarrow_{IS-A} c, \\ \text{and} \\ W \not\models \neg(\alpha \wedge \beta[b]) \end{array} \right) \quad \text{then} \quad \alpha\ explains\ \gamma\ because\_possible\ \{\alpha, \beta[b]\}$$

Example. $O = \{loud\_bell \rightarrow_{IS-A} bell\}$

$$C = \left\{ \begin{array}{c} On(alarm)\ causes\ Heard(bell) \\ Heard(loud\_bell)\ causes\ Deafening \end{array} \right\}$$

$$W = \left\{ \begin{array}{c} Heard(loud\_bell) \rightarrow Heard(bell) \\ On(alarm) \rightarrow Heard(bell) \\ Heard(loud\_bell) \rightarrow Deafening \end{array} \right\}$$

$On(alarm)$ does not cause $Heard(loud\_bell)$ (neither does it cause $Deafening$), but it is an explanation for $Heard(loud\_bell)$ by virtue of the upward scheme. Due to the base case, $Heard(loud\_bell)$ is in turn an explanation for $Deafening$. In fact, $On(alarm)$ is an explanation for $Deafening$ by virtue of transitivity.

Considering a causal operator which is transitive would give the same explanations but is obviously more restrictive as we may not want to endorse an account of causality which is transitive. Moreover, transitivity for explanations not only seems right in itself but it also means that our model of explanations can be plugged with any causal system whether transitive or not.

3.5 Explanation Provisos and Their Simplifications

Explanation atoms are written α *explains* β *because_possible* Φ as the definition is intended to make the atom true just in case it is successfully checked that the proviso is possible: An explanation atom is not to be interpreted as a kind of conditional statement. Indeed, we do not write *"if_possible"*. The argument in *"because_possible"* gathers those conditions that must be possible together if α is to explain β (there can be others: α can also be an explanation of β wrt other arguments in *"because_possible"*).

Using $\bigwedge\Phi$ to denote the conjunction of the formulas in the set Φ, the following scheme amounts to simplifying the proviso attached to an explanation atom.

$$\text{If} \begin{pmatrix} W \models \bigwedge\Phi \rightarrow \bigvee_{i=1}^{n} \bigwedge\Phi_i \quad \text{and} \\ \text{for all } i \in \{1, \cdots, n\} \alpha \textit{ explains } \beta \textit{ because\_possible } (\Phi_i \cup \Phi) \end{pmatrix}$$
$$\text{then} \quad \alpha \textit{ explains } \beta \textit{ because\_possible } \Phi$$

4 A Formal System for Inferring Explanations

The above ideas are embedded in a short proof system extending classical logic:

1. ***Causal formulas***
   (a) $(\alpha \textit{ causes } \beta) \rightarrow (\alpha \rightarrow \beta)$

2. ***Ontological atoms***
   (a) If $b \rightarrow_{IS-A} c$ then $\alpha[b] \rightarrow \alpha[c]$
   (b) If $a \rightarrow_{IS-A} b$ and $b \rightarrow_{IS-A} c$ then $a \rightarrow_{IS-A} c$
   (c) $c \rightarrow_{IS-A} c$

3. ***Explanation atoms***

   (a) $$\text{If} \begin{pmatrix} \alpha \textit{ causes } \beta[b], \\ a \rightarrow_{IS-A} b, \qquad a \rightarrow_{IS-A} c \\ W \not\models \neg(\alpha \wedge \beta[a]) \end{pmatrix}$$
   $$\text{then} \quad \alpha \textit{ explains } \beta[c] \textit{ because\_possible } \{\alpha, \beta[a]\}$$

   (b) $$\text{If} \begin{pmatrix} \alpha \textit{ explains } \beta \textit{ because\_possible } \Phi \\ \beta \textit{ explains } \gamma \textit{ because\_possible } \Psi \end{pmatrix}$$
   $$\text{then} \quad \alpha \textit{ explains } \gamma \textit{ because\_possible } (\Phi \cup \Psi)$$

   (c) $$\text{If} \begin{pmatrix} W \models \bigwedge\Phi \rightarrow \bigvee_{i=1}^{n} \bigwedge\Phi_i \qquad \text{and} \\ \text{for all } i \in \{1, \cdots, n\} \ \alpha \textit{ explains } \beta \textit{ because\_possible } (\Phi_i \cup \Phi) \end{pmatrix}$$
   $$\text{then} \quad \alpha \textit{ explains } \beta \textit{ because\_possible } \Phi$$

These schemes allow us to obtain the inference patterns described in the previous section. E.g., the base case for explanation is obtained by combining (2c) with (3a) (yielding another illustration of reflexivity in the ontology relieving us from the burden of introducing further formal material) prior to simplifying by means of (3c). Analogously, the upward case is obtained by applying (2c) upon (3a) before using (3c).

A more substantial application is:

$$C = \left\{ \begin{array}{c} \alpha \textit{ causes } \beta[b] \\ \beta[c] \textit{ causes } \gamma \end{array} \right\}$$

$$O = \{b \rightarrow_{IS-A} c\}$$

$$W = \left\{ \begin{array}{c} \alpha \rightarrow \beta[b] \\ \beta[b] \rightarrow \beta[c] \\ \beta[c] \rightarrow \gamma \end{array} \right\}$$

The first form of transitivity in Section 3.4 requires that we infer:

$$\alpha \textit{ explains } \gamma \textit{ because\_possible } \{\alpha\}$$

Let us proceed step by step:

$$\begin{array}{rl} \alpha \textit{ explains } \beta[c] \textit{ because\_possible } \{\alpha\} & \text{by (3a) as upward case} \\ \beta[c] \textit{ explains } \gamma \textit{ because\_possible } \{\beta[c]\} & \text{by (3a) as base case} \\ \alpha \textit{ explains } \gamma \textit{ because\_possible } \{\alpha, \beta[c]\} & \text{by (3b)} \\ \alpha \textit{ explains } \gamma \textit{ because\_possible } \{\alpha\} & \text{by (3c) simplifying the proviso} \end{array}$$

5 A Generic Diagram

Below an abstract diagram is depicted that summarizes many patterns of inferred explanations from various cases of causal statements and $\rightarrow_{IS-A}$ links.

In this example, for each pair of symbols $(\sigma 1, \sigma 2)$, there is only one "explanation path". E.g., we get α_1 *explains* γ_4 *because_possible* $\{\alpha_1, \gamma_1, \gamma_4\}$, through α_3 and γ_1. Indeed, from α_1 *explains* α_3 *because_possible* $\{\alpha_1\}$ as well as α_3 *explains* γ_1 *because_possible* $\{\alpha_3, \gamma_1\}$ and γ_1 *explains* γ_4 *because_possible* $\{\gamma_1, \gamma_4\}$, we obtain α_1 *explains* γ_4 *because_possible* $\{\alpha_1, \alpha_3, \gamma_1, \gamma_4\}$ (using transitivity twice). Lastly, we simplify the condition set by virtue of $W \vdash \alpha_1 \rightarrow \alpha_3$.

In other examples, various "explanation paths" exist. It suffices that the inference pattern 3a can be applied with more than one "a", or that transitivity (3b) can be applied with more than one β. We have implemented a program in DLV [8] (an implementation of the Answer Set Programming formalism [1]) that takes only a few seconds to give all the results $s1$ *explains* $s2$ *bec_poss* Φ, for all examples of this kind, including the case of different explanation paths (less than one second for the diagram depicted below).

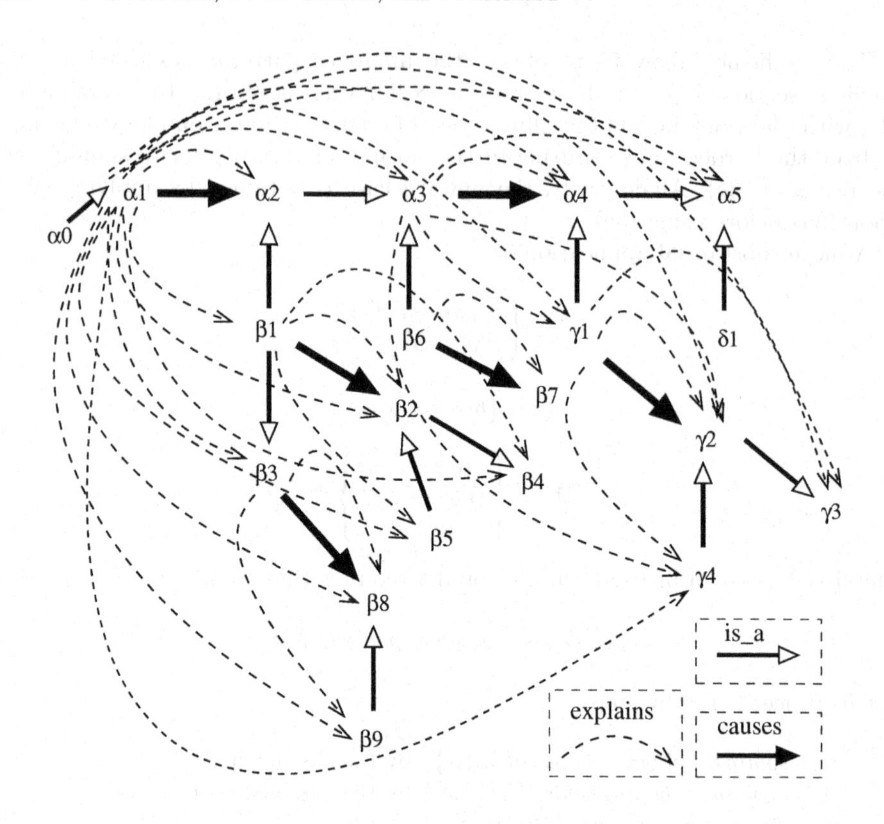

6 Back to Ontology

The explanation inferences that we obtain follow the patterns presented before. The inference pattern (3a) is important, and, in it, the direction of the $\rightarrow_{IS-A}$ links $a \rightarrow_{IS-A} b$ and $a \rightarrow_{IS-A} c$ is important: Unexpected conclusions would ensue if other directions, e.g., $b \rightarrow_{IS-A} a$ and $c \rightarrow_{IS-A} a$ were allowed.

We have considered ontological information in the most common form, as a $\rightarrow_{IS-A}$ hierarchy. A closer look at the process of inferring explanations reveals that $\rightarrow_{IS-A}$ links serve as a means to get another kind of ontological information by taking advantage of the property of the unary predicates to "inherit upwards". The basis for an explanation are causal statements, and these apply to facts or events [9]. Indeed, the ontological information which is eventually used in the process of inferring explanations is about facts or events, through the "inherit upwards" property. We eventually resort to an ontology over events.

Example. *Getting cold usually causes Mary to become active. I see Mary jogging. So, Mary getting cold might be taken as an explanation for her jogging. This holds on condition that the weather is possibly cold, otherwise the inference fails to ensue: In the presence of the information that the weather is warm, Mary getting cold is inconsistent.*

$$C = \{Mary\_was\_getting\_cold \text{ causes } Mary\_is\_moving\_up\}$$

$$O = \{Mary\_is\_jogging \rightarrow_{IS-A} Mary\_is\_moving\_up\}$$

$$W = \left\{ \begin{array}{c} Mary\_was\_getting\_cold \rightarrow Mary\_is\_moving\_up \\ Mary\_is\_jogging \rightarrow Mary\_is\_moving\_up \end{array} \right\}$$

If allowing such an ontology over events as in O, we could use an extended version of our proof system to infer the atom

$Mary\_was\_getting\_cold$
explains $Mary\_is\_jogging$ *bec_poss* $\{Mary\_was\_getting\_cold\}$.

That is to say, $Mary\_was\_getting\_cold$ *would be inferred as an explanation for* $Mary\_is\_jogging$*. Also, the inference would break down if for example both* $Warm\_weather$ *and* $Mary\_was\_getting\_cold \rightarrow \neg Warm\_weather$ *were in* W.

7 Conclusion

We have provided a logical framework allowing predictive and abductive reasoning from *causal* information. Indeed, our formalism allows to express causal information in a direct way. Then, we deduce so-called *explanation atoms* which capture what might explain what, in view of some given information. We have resorted to *ontological* information. Not only is it generally useful, it is key in generating sensible explanations from causal statements.

In our approach, the user provides a list of *ontological atoms* $a \rightarrow_{IS-A} b$ intended to mean that object a "is_a" b. The basic terms, denoted α, β are then concrete atoms built with unary predicates, such as $P(a)$. The user also provides causal information in an intuitive form, as causal atoms α *causes* β (which can occur in more complex formulas).

This makes formalization fairly short and natural. The ontology provided is used in various patterns of inference for explanations. In our approach, such information is rather easy to express, or to obtain in practice due to existing ontologies and ontological languages. If we were in a purely propositional setting, the user should write $Own\_small\_car \rightarrow_{IS-A} Own\_car$, $Own\_big\_car \rightarrow_{IS-A} Own\_car$, and, when necessary $Heard\_small\_car \rightarrow_{IS-A} Heard\_car$, $Heard\_big\_car \rightarrow_{IS-A} Heard\_car$, and so on. This would be cumbersome, and error prone. This contrasts with our setting, which, moreover, is "essentially propositional" in that, for what concerns the causal atoms, it is as if $Own(small\_car)$ were a propositional symbol Own_small_car, while, for what concerns the ontology, we really use the fact that $Heard$ and Own are predicates.

As always with knowledge representation, some care must be exercized as to the vocabulary used. E.g., minimization formalisms such as circumscription or predicate completion (see logic programming and its "offspring" answer set programming) require to distinguish between "positive notions" (to be minimized) from negative notions: Writing Fly vs. *not* Fly yields a different behavior than writing *not* $Unfly$ and $Unfly$. Here, we have a similar situation, since the inference patterns would not work properly if we were to use predicates that do not "inherit upwards" with respect to the ontology provided: We should not expect to infer $Dislike\_small\_car \rightarrow_{IS-A} Dislike\_car$, since *Dislike* obviously fails to "inherit upwards" when the ontology contains $small\_car \rightarrow_{IS-A} car$.

It does seem to us, that what we propose here is a good compromise between simplicity, as well as clarity, when it comes to describing a situation, and efficiency and pertinence of the results provided by the formalism.

Our work differs from other approaches in the literature in that it strictly separates causality, ontology and explanations. The main advantages are that information is more properly expressed and that our approach is compatible with various accounts of these notions, most notably causality. In particular, we need no special instances of α *causes* α to hold nor γ (equivalent with β) to be an effect of α whenever α *causes* β holds (contrast with [3,4,5,6,7,10] although in the context of actions such confusion is less harmful). In our approach, these are strictly limited to being where they belong, i.e., explanations. Space restriction prevent us from giving more details on the differences with other approaches.

Acknowledgements

It is a pleasure for us to thank the referees for their constructive remarks.

References

1. Baral, C.: Knowledge representation, reasoning and declarative problem solving. Cambridge University Press, Cambridge (2003)
2. Besnard, Ph., Cordier, M.-O.: Inferring Causal Explanations. In: Hunter, A., Parsons, S. (eds.) ECSQARU 1999. LNCS (LNAI), vol. 1638, pp. 55–67. Springer, Heidelberg (1999)
3. Bell, J.: Causation as Production. In: Brewka, G., Coradeschi, S., Perini, A., Traverso, P. (eds.) ECAI 2006, pp. 327–331. IOS Press, Riva del Garda, Italy (2006)
4. Bochman, A.: A Logic for Causal Reasoning. In: Gottlob, G., Walsh, T. (eds.) IJCAI 2003, pp. 141–146. Morgan Kaufmann, Acapulco, Mexico (2003)
5. Giunchiglia, E., Lee, J., Lifschitz, V., McCain, N., Turner, H.: Nonmonotonic Causal Theories. Artificial Intelligence 153(1-2), 49–104 (2004)
6. Halpern, J., Pearl, J.: Causes and Explanations: A Structural-Model Approach. Part I: Causes. In: Breese, J.S., Koller, D. (eds.) UAI 2001, pp. 194–202. Morgan Kaufmann, Seattle, Wa. (2001)
7. Halpern, J., Pearl, J.: Causes and Explanations: A Structural-Model Approach - Part II: Explanations. In: Nebel, B. (ed.) IJCAI 2001, pp. 27–34. Morgan Kaufmann, Seattle, Wa. (2001)
8. Leone, N., Pfeifer, G., Faber, W., Eiter, T., Gottlob, G., Perri, S., Scarcello, F.: The DLV System for Knowledge Representation and Reasoning. ACM Trans. on Computational Logic (TOCL) 7(3), 499–562 (2006)
9. Mellor, D.H.: The Facts of Causation. Routledge, London (1995)
10. Shafer, G.: Causal Logic. In: Prade, H. (ed.) ECAI 1998, pp. 711–720. Wiley, Brighton, UK (1998)

Predicting Partners' Behaviors in Negotiation by Using Regression Analysis

Fenghui Ren and Minjie Zhang

School of Computer Science and Software Engineering
University of Wollongong, Australia
{fr510,minjie}@uow.edu.au

Abstract. Prediction partners' behaviors in negotiation has been an active research direction in recent years. By employing the estimation results, agents can modify their own ways in order to achieve an agreement much quicker or to look after much higher benefits for themselves. Some of estimation strategies have been proposed by researchers to predict agents' behaviors, and most of them are based on machine learning mechanisms. However, when the application domains become open and dynamic, and agent relationships are complicated, it is difficult to train data which can be used to predict all potential behaviors of all agents in a multi-agent system. Furthermore because the estimation results may have errors, a single result maybe not accurate and practical enough in most situations. In order to address these issues mentioned above, we propose a power regression analysis mechanism to predict partners' behaviors in this paper. The proposed approach is based only on the history of the offers during the current negotiation and does not require any training process in advance. This approach can not only estimate a particular behavior, but also an interval of behaviors according to an accuracy requirement. The experimental results illustrate that by employing the proposed approach, agents can gain more accurate estimation results on partners' behaviors by comparing with other two estimation functions.

1 Introduction

Negotiation is a means for agents to communicate and compromise to reach mutually beneficial agreements [1] [2]. However, in most situations, agents do not have complete information about their partners' negotiation strategies, and may have difficulties to make a decision on future negotiation, such as how to select suitable partners for further negotiation [3] [4] or how to generate a suitable offer in next negotiation cycle [5]. Therefore estimation approaches which can predict uncertain situations and possible changes in future are required for helping agents to generate good and efficient negotiation strategies. Research on partners' behaviors estimation has been a very active direction in recent years. Several estimation strategies are proposed [6] [7] by researchers. However, as these estimation strategies are used in real applications, some of limitations are emerged.

Z. Zhang and J. Siekmann (Eds.): KSEM 2007, LNAI 4798, pp. 165–176, 2007.

Machine learning is one of the popular mechanisms adopted by researchers in agents' behaviors estimation. In general, this kind of approaches have two steps in order to estimate the agents' behaviors properly. In the first step, the proposed estimation function is required to be well trained by training data. Therefore, in a way, the performance of the estimation function is almost decided by the training result. In this step, data are employed as many as possible by designers to train a system. The training data could be both synthetic or collected from the real world. Usually, the synthetic data are helpful in training a function to enhance its problem solving skill for some particular issues, while the real world collected data can help the function to improve its ability in complex problem solving. After the system being trained, the second step is to employ the estimation function to predict partners' behaviors in future. However, no matter what and how many data are employed by designers to train the proposed function, it is unsuspicious to say that the training data will be never comprehensive enough to cover all situations in reality. Therefore, even though an estimation function is well trained, it is also very possible that some estimation results do not make sense at all for some kind of agents whose behaviors' records are not included in the training data. At present stage, as the negotiation environment becomes more open and dynamic, agents with different kinds of purposes, preferences and negotiation strategies can enter and leave the negotiation dynamically. So the machine learning based agents' behaviors estimation functions may not work well in some more flexible application domains for the reasons of (1) lacking of sufficient data to train the system, and (2) requesting plenty of resources in each training process.

In order to address those issues mentioned above, in this paper we propose a power regression approach to analyze and estimate partners' behaviors in negotiation. According to our knowledge, this is the first time that the regression analysis approach is employed to estimate partners' behaviors in negotiation. By comparing with machine learning mechanisms, the proposed approach only uses the historical offers in the current negotiation to estimate partners' behaviors in future negotiation and does not require any additional training process. So the proposed approach is very suitable to work under an open and dynamic negotiation environment, and to make a credible judgements on partners' behaviors timely. Also, because the proposed approach does not make any strict assumption on agents' purposes, preferences and negotiation strategies, it can be employed widely in negotiation by different types of agents. Furthermore, the proposed approach not only represents the estimation results in the form of interval, but also gives the probability that each individual situation may happen in future. By employing the proposed representation format, agents can have an overview on partners' possible behaviors and their favoritism easily, and then modify their own strategies based on these distributions. Therefore, the proposed approach provides more flexible choices to agents when they make decision in negotiation.

The rest paper is organized as follows: Section 2 introduces both the background and assumption of our proposed approach; Section 3 introduces the way that how the proposed regression analysis works; Section 4 introduces the ways to predict partners' behaviors under different accuracy requirements based on the regression analysis results; Section 5 illustrates the performance of the proposed power regression function through experiments and advantages of the proposed function by comparing with other two estimation approaches; and Section 7 concludes this paper and outlines future works.

2 Background

In this section, we introduce the background about the proposed power regression function for partners' behaviors estimation in negotiation. The regression analysis is a combination of mathematics and probability theory which can estimate the strength of a modeled relationship between one or more dependent variables and independent variables. In order to simplify the complexity for the proposed regression analysis approach, we make a simple assumption about the negotiation as follows:

Assumption: The benefits which an agent gains from its partner in each negotiation cycle should be a serial in monotone ascending order.

The reason behind this assumption is that partners will not break their promises and the new offer will be no less than the existing one. According to the assumption above, we propose a power regression function to predict the partners' behaviors. The proposed power regression function is given as follows:

$$u = a \times t^b \tag{1}$$

where u is the expected utility gained from a partner, t $(0 \leq t \leq \tau)$ is the negotiation cycle and both a and b $(a \geq 0,\ b \geq 0)$ are parameters which are independent on t. Based on our assumption and the proposed power regression function, it is noticed that four types of partners' behaviors [8] are distinguished based on the range of parameter b.

- Boulware: when $b > 1$, the rate of change in the slope is decreasing, corresponding to smaller concession in the early cycles but large concession in later cycles.
- Linear: when $b = 1$, the rate of change in the slope is zero, corresponding to making constant concession throughout the negotiation.
- Conceder: when $0 < b < 1$, the rate of change in the slope is increasing, corresponding to large concession in early cycle but smaller concession in later cycles.
- When $b = 0$, the rate of change of the slope and the slope itself are always zero, corresponding to not making any concession throughout the entire negotiation.

In the following Section, we will introduce the proposed power regression function to analysis and estimate partners' possible behaviors in a more efficient and accurate way.

3 Regression Analysis on Partners' Behaviors

In this section, we introduce the proposed power regression analysis approach on the negotiation between two players. We do the equivalence transformation on Equation 1 as follows:

$$\ln(u) = \ln(a \times t^b) = \ln(a) + b \times \ln(t) \tag{2}$$

Then let $u^* = \ln(u)$, $a^* = \ln(a)$ and $t^* = \ln(t)$, Equation 2 can be rewritten as $u^* = a^* + b \times t^*$. The new function indicates a linear relationship between the variables t^* and u^*. Both parameters a^* and b are independent on t^*. Since the above equation only estimates partners' possible behaviors, the difference (ε) between partners' real behaviors ($\hat{u^*}$) and the expected behaviors (u^*) should obey the Gaussian distribution which is $\varepsilon \sim N(0, \sigma^2)$, where $\varepsilon = \hat{u^*} - a^* + b \times t^*$.

Let pairs $(t_0, \hat{u_0}), \ldots, (t_n, \hat{u_n})$ are the instances in the current negotiation, where $t_i (t_i < t_{i+1})$ indicates the negotiation cycle and $\hat{u_i} (\hat{u_i} \leq \hat{u_{i+1}})$ indicates the real utility value the agent gained from its partners. We transform all pairs $(t_i, \hat{u_i}) (i \in [0, n])$ to $(t_i^*, \hat{u_i^*})$ by using $t_i^* = \ln(t_i)$ and $\hat{u_i^*} = \ln(u_i)$.

Because for each $\hat{u_i^*} = a^* + b \times t_i^* + \varepsilon_i$, $\varepsilon_i \sim N(0, \sigma^2)$, $\hat{u_i^*}$ is distinctive, then the joint probability density function for $\hat{u_i^*}$ is:

$$\begin{aligned} L &= \prod_{i=1}^{n} \frac{1}{\sigma\sqrt{2\pi}} \exp[-\frac{1}{2\sigma^2}(\hat{u_i^*} - a^* - bt_i^*)^2] \\ &= (\frac{1}{\sigma\sqrt{2\pi}})^n \exp[-\frac{1}{2\sigma^2}\sum_{i=1}^{n}(\hat{u_i^*} - a^* - bt_i^*)^2] \end{aligned} \tag{3}$$

In order to make L to achieve its maximum, obviously $\sum_{i=1}^{n}(\hat{u_i^*} - a^* - bt_i^*)^2$ should achieve its minimum value. Let $Q(a^*, b) = \sum_{i=1}^{n}(\hat{u_i^*} - a^* - bt_i^*)^2$, we calculate the first-order partial derivative for $Q(a^*, b)$ on both a^* and b, and let the results equal to zero, which are shown as follows:

$$\begin{cases} \frac{\partial Q}{\partial a^*} = -2\sum_{i=1}^{n}(\hat{u_i^*} - a^* - bt_i^*) = 0, \\ \frac{\partial Q}{\partial b} = -2\sum_{i=1}^{n}(\hat{u_i^*} - a^* - bt_i^*)t_i^* = 0. \end{cases} \tag{4}$$

Then it equals:

$$\begin{cases} na^* + (\sum_{i=1}^{n} t_i^*)b = \sum_{i=1}^{n} \hat{u_i^*}, \\ (\sum_{i=1}^{n} t_i^*)a^* + (\sum_{i=1}^{n} t_i^{*2})b = \sum_{i=1}^{n} t_i^* \hat{u_i^*}. \end{cases} \tag{5}$$

Because Equation 5's coefficient matrix is:

$$\begin{vmatrix} n & \sum_{i=1}^{n} t_i^* \\ \sum_{i=1}^{n} t_i^* & \sum_{i=1}^{n} t_i^{*2} \end{vmatrix} = n\sum_{i=1}^{n} t_i^{*2} - (\sum_{i=1}^{n} t_i^*)^2 = n\sum_{i=1}^{n}(t_i^* - \bar{t})^2 \neq 0 \tag{6}$$

So parameters a^* and b have their unique solutions, which is

$$\begin{cases} b = \frac{n\sum_{i=1}^{n} t_i^* \hat{u_i^*} - (\sum_{i=1}^{n} t_i^*)(\sum_{i=1}^{n} u_i^*)}{n\sum_{i=1}^{n} t_i^{*2} - (\sum_{i=1}^{n} t_i^*)^2}, \\ a^* = \frac{1}{n}\sum_{i=1}^{n} u_i^* - \frac{b}{n}\sum_{i=1}^{n} t_i^*. \end{cases} \quad (7)$$

In order to simplify the solution, let

$$\begin{cases} S_{xx} = \sum_{i=1}^{n} t_i^{*2} - \frac{1}{n}(\sum_{i=1}^{n} t_i^*)^2, \\ S_{yy} = \sum_{i=1}^{n} \hat{u_i^{*2}} - \frac{1}{n}(\sum_{i=1}^{n} \hat{u_i^*})^2, \\ S_{xy} = \sum_{i=1}^{n} t_i^* \hat{u_i^*} - \frac{1}{n}(\sum_{i=1}^{n} t_i^*)(\sum_{i=1}^{n} \hat{u_i^*}). \end{cases} \quad (8)$$

Then a^*, b are represented as:

$$\begin{cases} b = \frac{S_{xy}}{S_{xx}}, \\ a^* = \frac{1}{n}\sum_{i=1}^{n} \hat{u_i^*} - (\frac{1}{n}\sum_{i=1}^{n} t_i^*)b. \end{cases} \quad (9)$$

and finally $a = e^{a^*}$.

4 Partners' Behaviors Prediction

In last section, we proposed a power regression function to predict partners' behaviors, and also specified how to decide parameters a and b. However, it has to be mentioned that the proposed power regression function can only provide an estimation on partners' possible behaviors, which might not exactly accord with the partners' real behaviors. In reality, the real behaviors should similar to the estimation behaviors, and the more similar to the estimated behaviors, the higher probability it may happen. So we can deem that the differences (ε) between the estimation behaviors and the real behaviors obey the Gaussian distribution $N(\varepsilon, \sigma^2)$. Thus, if the deviation σ^2 can be calculated, we can make a precise decision on the range of partners' behaviors. It is known that there is more than 99% probability that partners' expected behaviors locate in the interval $[u - 3\sigma, u + 3\sigma]$. In this section, we introduce the proposed way to calculate the deviation σ and to estimate the interval for partners' behaviors.

In order to calculate the deviation σ, we firstly calculate the distance between the estimation results (u_i) and the real results on partners' behaviors ($\hat{u_i}$) as $d_i = \hat{u_i} - u_i$. It is known that all d_i ($i \in [1, n]$) obey the Gaussian distribution $N(0, \sigma^2)$, where $\sigma = \sqrt{\frac{\sum_{i=1}^{n}(d_i - \overline{d})^2}{n}}$ and $\overline{d} = \frac{1}{n}\sum_{i=1}^{n} d_i$. By employing the Chebyshev's inequality, we can calculate (1) the interval of partners' behaviors according to any accuracy requirements; and (2) the probability that any particular behavior may happen on potential partners in future.

5 Experiments

In this section, we demonstrate several experiments to test our proposed regression analysis approach. We display an overview table and figure about all estimated results in each negotiation cycle, and a particular curve to show the situation in the last negotiation cycle. Also, we compare the proposed power regression approach with the Tit-For-Tat approach [8] and random approach. The experimental results illustrate the outstanding performance of our proposed approach. In order to simplify the implementation process, all agents in our experiment employ the NDF [9] negotiation strategy. The partners' behaviors cover all possible situations in reality, which are conceder, linear and boulware.

5.1 Scenario 1

In the first experiment, a buyer want to purchase a mouse pad from a seller. The acceptable price for the buyer is in $[\$0, \$1.4]$. The deadline for buyer to finish this purchasing process is 11 cycles. In this experiment, the buyer adopts conceder behavior in the negotiation, and the seller employs the proposed approach to estimate the buyer's possible price in the next negotiation cycle. The estimated results in each negotiation cycle are displayed in Table 1. Each row in Table 1 illustrates the estimation result in each negotiation cycle in the form of estimated power regression function, estimation results (μ), and deviation (σ). For example, it can be seen in the $7th$ negotiation cycle, according to instances, the proposed approach estimates a price of \$1.26 from the buyer in next cycle. Then according to the historical record in the $8th$ cycle, the real price given by the buyer in this cycle is \$1.27 which is very close to the estimation price. In Figure 1, we illustrate the situation on the tenth negotiation cycle. It can be seen that the accuracy of the estimation is very high because all estimated prices are in the interval of $[\mu - 2\sigma, \mu + 2\sigma]$ except the $4th$ and $5th$ negotiation cycle, and all estimated results are almost exactly same as the real prices. For example, the estimation prices for $2th$, $3th$ and $8th$ are \$1.06, \$1.12 and \$1.26, respectively. While the real prices given by buyer in these cycles are \$1.07,\$1.13, and 1.26, which almost have no difference with the real ones. Lastly, we also did a comparison between the proposed approach and other two estimation approaches (Tit-For-Tat and random approach). In Figure 2, the comparison results are illustrated. It can be seen that even though the Tit-For-Tat approach can follow the trend of buyer's price changing, the errors (10% of acceptable price span in average) are also very big. The random approach even cannot catch the main trend. The experimental results convince us that the proposed approach outperforms both Tit-For-Tat and random approaches very much when a buyer adopts conceder negotiation behavior.

5.2 Scenario 2

In the second experiment, a buyer wants to buy a keyboard from a seller. The desired price for the buyer is in the interval of $[\$0, \$14]$. We let the buyer to employ the linear negotiation strategy, and still set the deadline to 11 cycles. The estimated results are displayed in Table 2. For example, in the $6th$ cycle, the estimated price is \$7.57 and the real price is \$7.4. In Figure 3, the situation

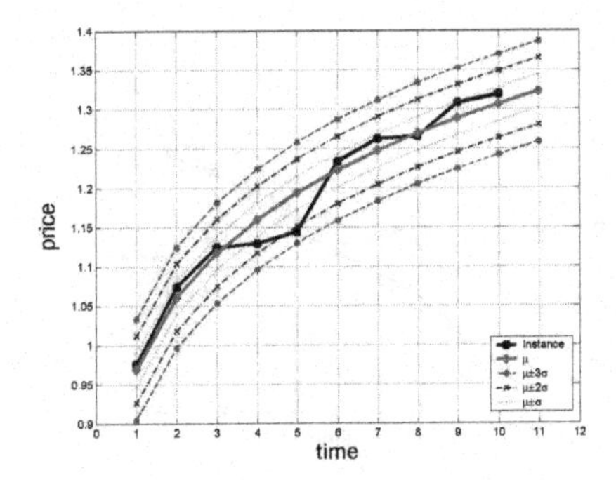

Fig. 1. Prediction results for scenario 1

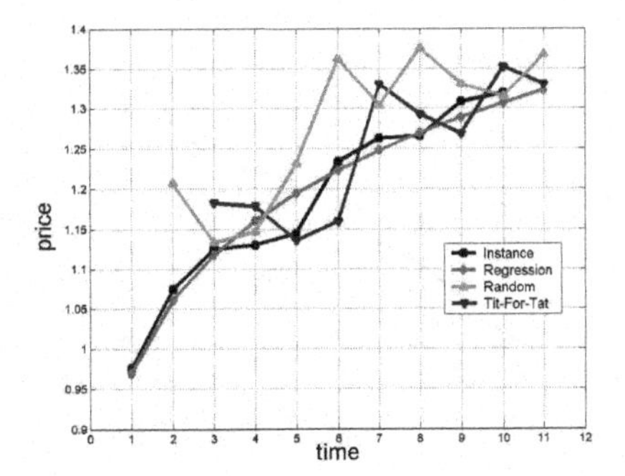

Fig. 2. Results comparison for scenario 1

Table 1. Prediction results for scenario 1

Cycle	Instance	Power regression function	Estimated (μ, σ)
2	(0.98, 1.07)	price=$0.98t^{0.14}$	(1.14, 0.00)
3	(0.98, 1.07, 1.12)	price=$0.98t^{0.13}$	(1.17, 0.00)
4	(0.98, 1.07, 1.12, 1.13)	price=$0.98t^{0.11}$	(1.18, 0.01)
5	(0.98, 1.07, 1.12, 1.13, 1.14)	price=$0.99t^{0.10}$	(1.18, 0.02)
6	(0.98, 1.07, 1.12, 1.13, 1.14, 1.23)	price=$0.98t^{0.11}$	(1.22, 0.02)
7	(0.98, 1.07, 1.12, 1.13, 1.14, 1.23, 1.26)	price=$0.98t^{0.12}$	(1.26, 0.02)
8	(0.98, 1.07, 1.12, 1.13, 1.14, 1.23, 1.26, 1.27)	price=$0.97t^{0.12}$	(1.28, 0.02)
9	(0.98, 1.07, 1.12, 1.13, 1.14, 1.23, 1.26, 1.27, 1.30)	price=$0.97t^{0.13}$	(1.30, 0.02)
10	(0.98, 1.07, 1.12, 1.13, 1.14 1.23, 1.26, 1.27, 1.30, 1.32)	price=$0.97t^{0.13}$	(1.32, 0.02)

on the 10*th* cycle is displayed. It can be seen that the estimated power regression function fits the real prices very well. In Figure 4, the comparison results among Tit-For-Tat approach, random approach and our proposed approach is illustrated. It can be seen that errors between the estimated prices and the real prices is the smallest value by employing the proposed approach, while it is the biggest value by employing the random approach. Even the Tit-For-Tat approach can fellow partner's trend, but distances between the estimated prices and the real prices are also very big. The second experimental results demonstrate that when partners perform as the linear behaviors, the proposed approach also outperforms other two approaches used for agents' behavior estimation.

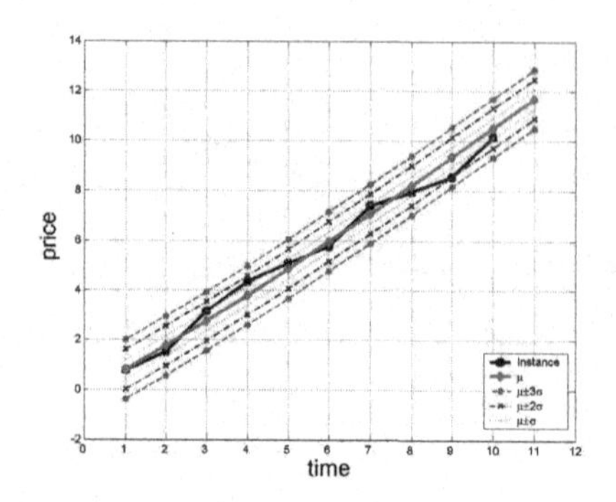

Fig. 3. Prediction results for scenario 2

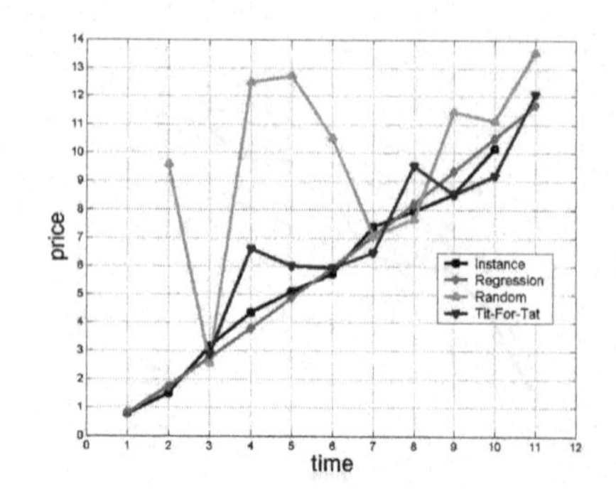

Fig. 4. Results comparison for scenario 2

Table 2. Prediction results for scenario 2

Cycle	Instance	Power regression function	Estimated (μ, σ)
2	(0.79, 1.50)	price=$0.79t^{0.94}$	(2.20, 0.00)
3	(0.79, 1.50, 3.15)	price=$0.74t^{1.23}$	(4.09, 0.21)
4	(0.79, 1.50, 3.15, 4.35,)	price=$0.74t^{1.26}$	(5.62, 0.18)
5	(0.79, 1.50, 3.15, 4.35, 5.09)	price=$0.75t^{1.22}$	(6.71, 0.24)
6	(0.79, 1.50, 3.15, 4.35, 5.09, 5.74)	price=$0.77t^{1.17}$	(7.57, 0.34)
7	(0.79, 1.50, 3.15, 4.35, 5.09, 5.74, 7.40,)	price=$0.77t^{1.17}$	(8.77, 0.31)
8	(0.79, 1.50, 3.15, 4.35, 5.09, 5.74, 7.40, 7.94)	price=$0.79t^{1.15}$	(9.76, 0.35)
9	(0.79, 1.50, 3.15, 4.35, 5.09, 5.74, 7.40, 7.94, 8.55)	price=$0.80t^{1.12}$	(10.62, 0.42)
10	(0.79, 1.50, 3.15, 4.35, 5.09, 5.74, 7.40, 7.94, 8.55, 10.15)	price=$0.81t^{1.11}$	(11.68, 0.40)

5.3 Scenario 3

In the third experiment, a buyer wants to purchase a monitor from a seller. The suitable price for the buyer is in [\$0, \$250]. In this experiment, the buyer employs a boulware strategy in the negotiation. The deadline is still 11 cycles. According to Table 3, the estimated power regression function at the $10th$ cycle is $price = 0.64 \times t^{2.45}$. Also, the result estimated that it is more than 68.2% that the real price at the $11th$ cycle is in [\$222.17, \$236.07]. According to Figure 5, it can be seen that the estimated power function almost exactly fits all the real prices given by the buyer. Therefore, the seller could have enough reasons to trust and adopt the estimation result for the next cycle. Finally, the Figure 6 illustrates the comparison results with other two estimation functions. It can be seen that when the agent performs a boulware behavior, the proposed approach outperforms other two approaches.

From the experimental results in the above three experiments on three general kinds of agents, we can say that the estimated power function regression approach can estimate partners' potential behaviors successfully, and also the estimation results are accurate and reasonable enough to be adopted by agents to modify their strategies in negotiation. The comparison results among three types of agents' behaviors estimation approaches demonstrate the outstanding performance of our proposed approach.

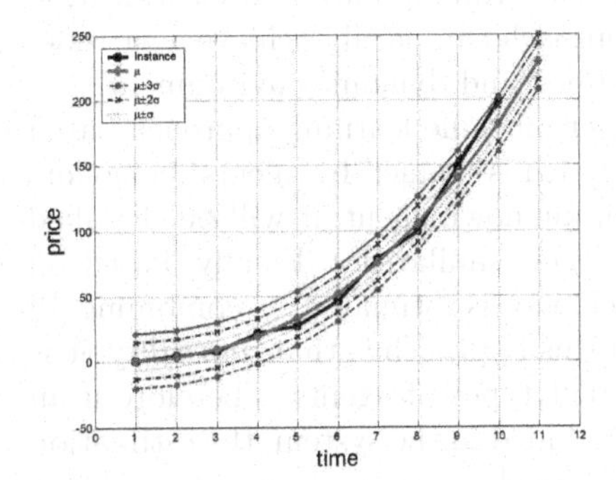

Fig. 5. Prediction results for scenario 3

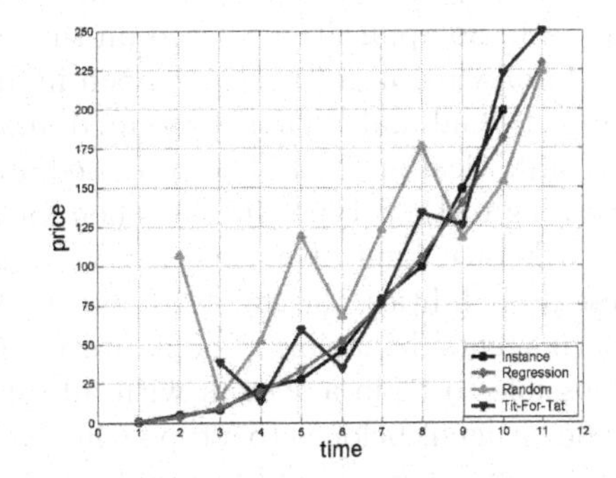

Fig. 6. Results comparison for scenario 3

Table 3. Prediction results for scenario 3

Cycle	Instance	Power regression function	Estimated (μ, σ)
2	(0.59, 4.71)	price=$0.59t^{3.01}$	(15.95, 0.00)
3	(0.59, 4.71, 8.06)	price=$0.65t^{2.45}$	(19.50, 1.11)
4	(0.59, 4.71, 8.06, 21.94)	price=$0.64t^{2.52}$	(36.60, 1.27)
5	(0.59, 4.71, 8.06, 21.94, 27.51)	price=$0.67t^{2.40}$	(49.70, 2.59)
6	(0.59, 4.71, 8.06, 21.94, 27.51, 46.12)	price=$0.68t^{2.38}$	(69.68, 2.36)
7	(0.59, 4.71, 8.06, 21.94, 27.51, 46.12, 78.45)	price=$0.67t^{2.41}$	(99.99, 3.50)
8	(0.59, 4.71, 8.06, 21.94, 27.51, 46.12, 78.45, 99.38)	price=$0.67t^{2.41}$	(132.56, 3.29)
9	(0.59, 4.71, 8.06, 21.94, 27.51, 46.12, 78.45, 99.38, 148.86)	price=$0.65t^{2.43}$	(176.42, 5.15)
10	(0.59, 4.71, 8.06,21.94, 27.51, 46.12, 78.45, 99.38, 148.86, 199.08)	price=$0.64t^{2.45}$	(229.12, 6.95)

6 Related Work

In this section, we introduce some related works and give discussions on the proposed approach. In [10], Schapire et. al. proposed a machine learning approach based on a boosting algorithm. In the first place, the estimation problem is reduced to a classification problem. All training data are arranged in ascending order and then partitioned into groups equally. For each of the breakpoints, a learning algorithm is employed to estimate the probability that a new bid at least should be greater than the breakpoint. The final result of this learning approach is a function which gives minimal error rate between the estimated bid and the real one. Based on this function, agents' behaviors can be estimated. However, the accuracy of this approach is limited by the training data and classification approach. So applications based on this approach can hardly achieve a satisfactory level when negotiations happen in an open and dynamic environment.

In [11], Gal and Pfeffer presented another machine learning approach based on a statistical method. The proposed approach is trained by agents' behaviors according to their types firstly. Then for an unknown agent, it will be classified into a known kind of agents according to their similarities. Finally, based on these probabilities, the unknown agent's behavior is estimated by combining all known agents' behaviors. The limitation of this approach is that, in reality, it is impossible to train a system with all different types of agents. Therefore if an unknown agent belongs to a type which is excluded in the system, the estimation result may not reach an acceptable accuracy level.

Chajewska et. al. [7] proposed a decision-tree approach to learn and estimate agent's utility function. The authors assumed that each agent is rational which looks for maximum expected utility in negotiation. Firstly, a decision tree is established which contains all possible endings for the negotiation. Each possible ending is assigned with a particular utility value and possibility. Based on the partner's previous decisions on the decision tree, a linear function can be generated to analogy the partner's utility function, and each item in the function comes from an internal node on the decision tree. The limitation of this approach is the requirement that all possible negotiation endings and the corresponding probabilities should be estimated in advance, which is impossible in some application domains when the variance of negotiation issues is discrete or the negotiation environment is open and dynamic.

Brzostowski and Kowalczyk [12] presented a way to estimate partners' behaviors based only on the historical offers in the current negotiation. In this first place, partners' types are estimated based on the given functions. For each type of agents, a distinct prediction function is given to estimate agents' behaviors. Therefore, based on the classification about partners' types and their individual estimation functions, the proposed approach can predict partners' behaviors in next negotiation cycle. However, a partner can only perform as a time-dependent agent or a behavior-dependent agent, which limits some applications. Also the accuracy of classification on partners' types may impact the accuracy of prediction result.

By comparing our approach with the above estimation strategies on agents' behaviors, our proposed approach has two attractive merits. (1) The proposed approach do not need any training or preparation in advance, and it can estimate partners' behaviors based only on the current historical records and generate reasonable and accurate estimation results quickly and timely. Therefore, agents can save both space and time resources by employing the proposed approach; and (2) the proposed approach estimates partners' possible behaviors in the form of interval, and the probability that each particular behavior will happen in the future is also represented by the proposed power regression function. Therefore, agents can adopt the estimation results by the proposed approach much easier and more convenient to administrate their own negotiation behaviors in future.

7 Conclusion and Future Work

In this paper, we proposed power regression function to estimate partners' behaviors in negotiation. We introduced the procedures to calculate the parameters in the regression function, and the method to predict partners possible behaviors. The experimental results demonstrate that the proposed approach is novel and valuable for the agents' behaviors estimation because (1) it is the first time that the regression analysis approach is applied on the agents' behaviors estimation; (2) the proposed approach does not need any training process in advance; (3) the representation format of the estimation results is easy to be further adopted by agents; and (4) the probability that each estimation behavior will happen in future on partners is also a significant criterion for agents to dominate their own behaviors in future.

The future works of this research will focus on two directions. Firstly, the multi-attribute negotiation is another promoting issue in recent years. Therefore, one of the emphases in our future works is to extend the proposed approach from the single-issue negotiation to the multi-issue negotiation. Secondly, as the negotiation environment becomes more open and dynamic, the proposed approach should be extended in order to predict not only agents' possible behaviors, but also impacts from potential changes on the negotiation environment.

Acknowledgement

This research is supported by the Australian Research Council Research Network in Enterprise Information Infrastructure (EII).

References

1. Kraus, S.: Strategic Negotiation in Multiagent Environments. The MIT Press, Cambridge, Massachusetts (2001)
2. Fatima, S., Wooldridge, M., Jennings, N.: Optimal Agendas for Multi-issue Negotiation. In: AAMAS 2003. Second International Joint Conference on Autonomous Agents and Multi-Agent Systems, pp. 129–136. ACM Press, New York (2003)

3. Brzostowski, J., Kowalczyk, R.: On Possibilistic Case-Based Reasoning for Selecting Partners in Multi-agent Negotiation. In: Webb, G.I., Yu, X. (eds.) AI 2004. LNCS (LNAI), vol. 3339, pp. 694–705. Springer, Heidelberg (2004)
4. Munroe, S., Luck, M., d'Inverno, M.: Motivation-Based Selection of Negotiation Partners. In: AAMAS 2004. 3rd International Joint Conference on Autonomous Agents and Multiagent Systems, pp. 1520–1521. IEEE Computer Society, Los Alamitos (2004)
5. Parsons, S., Sierra, C., Jennings, N.: Agents that Reason and Negotiate by Arguing. Journal of Logic and Computation 8(3), 261–292 (1998)
6. Zeng, D., Sycara, K.: Bayesian Learning in Negotiation. International Journal of Human-Computer Studies 48(1), 125–141 (1998)
7. Chajewska, U., Koller, D., Ormoneit, D.: Learning An Agent's Utility Function by Observing Behavior. In: Proc. 18th International Conf. on Machine Learning, pp. 35–42. Morgan Kaufmann, San Francisco, CA (2001)
8. Faratin, P., Sierra, C., Jennings, N.: Negotiation Decision Functions for Autonomous Agents. Journal of Robotics and Autonomous Systems 24(3-4), 159–182 (1998)
9. Fatima, S., Wooldridge, M., Jennings, N.: An Agenda-Based Framework for Multi-Issue Negotiation. Artificial Intelligence 152(1), 1–45 (2004)
10. Schapire, R.P., McAllester, D., Littman, M., Csirik, J.: Modeling Auction Price Uncertainty Using Boosting-based Conditional Density Estimation. In: ICML 2002. Machine Learning, Proceedings of the Nineteenth International Conference, pp. 546–553. Morgan Kaufmann, San Francisco (2002)
11. Gal, Y., Pfeffer, A.: Predicting Peoples Bidding Behavior in Negotiation. In: AAMAS 2006. 5th International Joint Conference on Autonomous Agents and Multiagent Systems (2006)
12. Brzostowski, J., Kowalczyk, R.: Predicting partner's behaviour in agent negotiation. In: AAMAS 2006. 5th International Joint Conference on Autonomous Agents and Multiagent Systems (2006)

Enhancing Web-Based Adaptive Learning with Colored Timed Petri Net

Shang Gao and Robert Dew

School of Engineering and Information Technology, Deakin University, Geelong VIC 3217, Australia
{shang,rad}@deakin.edu.au

Abstract. One of the issues for Web-based learning applications is to adaptively provide personalized instructions for different learning activities. This paper proposes a high level colored timed Petri Net based approach to providing some level of adaptation for different users and learning activities. Examples are given to demonstrate how to realize adaptive interfaces and personalization. Future directions are also discussed at the end of this paper.

1 Introduction

Supporting adaptation in Web-based learning applications is a challenging task. Before attempting a solution, we had better analyze the ground where a Web-based learning application is constructed. Besides different technologies ranging from database generated views to intelligent agents and user models, usually a Web-based learning application is made up of a group of hypertext knowledge nodes related to a specific knowledge field and connected to each other via hyperlinks. Students explore the knowledge space by following hyperlinks. Whilst enjoying flexibility of selection, students are suffering from disorientation and lack of personalized instructions in the hypertext knowledge space.

To prevent them from getting lost and to support some level of adaptation, researchers have been studying adaptation from many perspectives:

Brusilovsky [1] distinguished two different forms of adaptation for Web applications: adaptive presentation and adaptive navigation support. Patterno [2] proposed another adaptation, i.e. changes to the layout that do not affect the content, such as colors, font type or font size.

Nora and Gustavo [3] summarized the above techniques and adaptive patterns as follows:

- *adaptive content* consists of selecting different information, such as different text, images, videos, animation, etc. depending on the current state of the user profile. For example, an adaptive Web application provides an expert in a certain domain with more straightforward information than the step by step fundamental introductions for a novice.
- *adaptive navigation* consists of changing the anchors appearance, the link targets or the number of anchors presented to the users as well as the order

Z. Zhang and J. Siekmann (Eds.): KSEM 2007, LNAI 4798, pp. 177–185, 2007.

in which these anchors are presented. Adjusting links or anchors to make the application adaptive are also treated as link adaptation.

• *adaptive presentation* shows different layout of perceivable user interface elements, such as different type of media, different ordering or different colors, font size, font type or image size.

The above adaptive techniques focus on different delivery levels: adaptive content is at content-level; adaptive navigation support is at link-level and adaptive presentation is at presentation-level. We focus on realizing adaptive learning at the link-level as a hyperlink is the key point of accessing a knowledge node. It would be easier and more convenient to make a hyperlink/anchor visible or invisible based on a given learning strategy or control mechanism.

Inspired by the ideas presented in [4][5][6] and [7], a model of *hypertext learning state space* was proposed in [8]. The idea behind this model is categorizing closely related knowledge nodes into one single *learning state.* A group of *learning states*, connected by each other via *outside links*, builds up a *hypertext learning state space.* Accessing the *key knowledge node* where an *outside link* resides in is a key point of transiting from one *learning state* to another. A transition path of *learning states* is called a *learning path.* By defining the learning state space and manipulating learning-state-transition thresholds, students' learning paths are under control, while at the same time flexibility of the hypertext is maintained.

Given the similarity between *learning state space* and a Petri Net model [9] [10], a timed Petri Net based approach was adopted to interpret the browsing semantics of a learning state space in [11][12].In the future work of papers [11][12], the authors mentioned that a timed Petri Net could possibly be extended to integrate colored tokens. The extended model, called colored timed Petri Net, would be an ideal tool to describe multiuser behavior in an event-driven environment, where different users are granted different colored tokens. Transition enabling and firing is decided by classified colors held by a place and the color consuming function defined for each arc. In our case, if students at different levels are represented by different colored tokens, their access control becomes a matter of color allocation and consumption.

In this paper, the authors explain how colored tokens are allocated and consumed, and how multiuser's behavior is therefore controlled. The whole paper is organized as follows: a brief background introduction to timed Petri Nets and colored Petri Nets is in section two, followed by the description of the proposed high level colored timed Petri Net model and its content and link adaptive operations on browsing semantic structure of a hypertext learning state space. Future research directions are also discussed at the end of this paper.

2 High Level Petri Nets Background Knowledge

An original Petri Net (PN) [13] is invented by Carl Adam Petri, which consists of two kinds of nodes called places (represented by circles) and transitions (represented by bars) where arcs connect them.

Based on the original Petri Nets, extension has also been made over the years in many directions including time, data, color and hierarchy, which are all called high level Petri Nets.

Timed Petri Nets are to describe the temporal behavior of a system, time attributes can be associated with places (called P-timed), tokens (called age) and transitions (called T-timed).

If two time attributes are adopted, one is defined as the minimum delay d^{min} and the other as maximum delay d^{max}, the firing rules are that a transition is enabled after the minimum delay d^{min} elapses; it remains enabled in the interval (d^{min}, d^{max}) ; if after the maximum delay d^{max}, the enabled transition has not been fired, it is forced to do so, moving tokens from its input places to the output places. If the transition can not fire, the input tokens becomes unavailable. This "dead end" should be avoided by setting appropriate (d^{min}, d^{max}) and adjusting them dynamically.

In Figure 1(a), each place has a delay pair (d_i^{min}, d_i^{max}). Token in place p_1 is not available until 2 time units elapse. If t_1 is enabled but has not been fired after 10 time units, it is forced to fire with token flows from p_1 to p_2 and p_3. A ∞ value of d_4^{max} means no maximum delay constraint applied to p_4. If p_4 has an output transition, it can fire whenever it is chosen after 10 time units.

Colored Petri Nets can describe multiple resources(e.g. humans, goods, objects) of a system, using colors or types associated with tokens. In a colored Petri Nets (CPN), each token has a value often referred to as "color". Transitions use the values of the consumed tokens to determine the values of the produced tokens. The relationship between the values of the "input tokens" and the values of the "output tokens" is represented by a transition's color consuming function or arc function, denoted E. For instance, in Figure 1(b), $E_{f(p_1,t_1)} = a, E_{f(t_1,p_2)} = b$.

The firing rules of a CPN are that a transition is enabled after colored tokens deposited in all of its input places satisfy the input arc functions of that transition. If fired, colored tokens are removed from input places; and new colored tokens, produced by the output arc functions, are moved into output places.

In Figure 1(b), place p_1 is marked with two colored tokens a, b from a color set $\{a, b, c\}$, denoted $m(p_1) = \{a, b\}$. t_1 is enabled as its input arc only requires one colored token a which is available in place p_1. If fired, token a is removed from p_1, two new colored tokens b and c are generated by t_1's output arcs and deposited in p_2 and p_3, respectively, as shown in Figure 1(c).

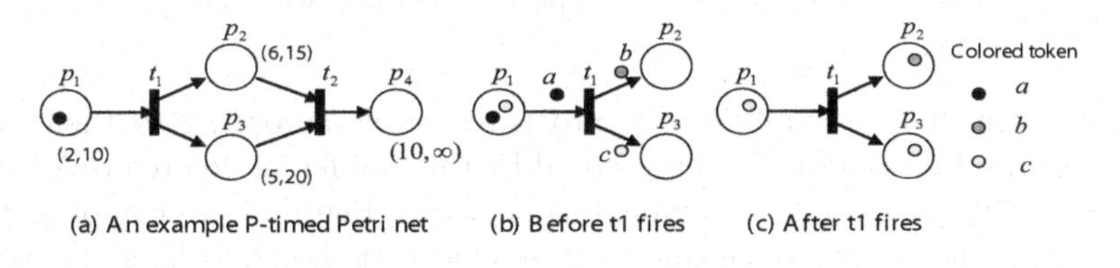

Fig. 1. An example of P-timed Petri Net and colored Petri Net

3 Colored P-Timed Petri Net Based Adaptation Model for Hypertext Learning State Space

As specified in [11], a timed Petri Net can be used to model temporal events in a discrete system. Students' learning activities in a Web-based environment are series of discrete events, such as reading hypertext, typing words and clicking hyperlinks. These activities help students move around in hyperspace.

A colored Petri Net can be used to model multiple resources or objects in a system. In a hypertext learning state space, each student is an individual object which has different behavior and should be treated individually. Some nodes contain advanced knowledge for "senior" students, while others are for "junior".

If all these factors, especially link-clicking and multiple user access control, are modeled with one colored timed Petri Net, we can not only benefit from dynamic executive semantics of a Petri Net, but also embed path and access control information in a Petri Net structure. Learning adaptation is, therefore, realized in some degree.

While providing adaptation, the underlying structure should be kept intact. In other words, students are only presented with a personalized interface and the original underlying structure is kept untouched.

In the following sections, a colored P-timed Petri Net (CPTPN) based adaptation model for hypertext learning state space is proposed, followed by a discussion about how to realize adaptive instruction with such a model.

3.1 CPTPN Based Adaptation Model

An example used in [11] is also adopted here to demonstrate how a *hypertext learning state space*(Figure 2) is converted to a CPTPN model(Figure 3).

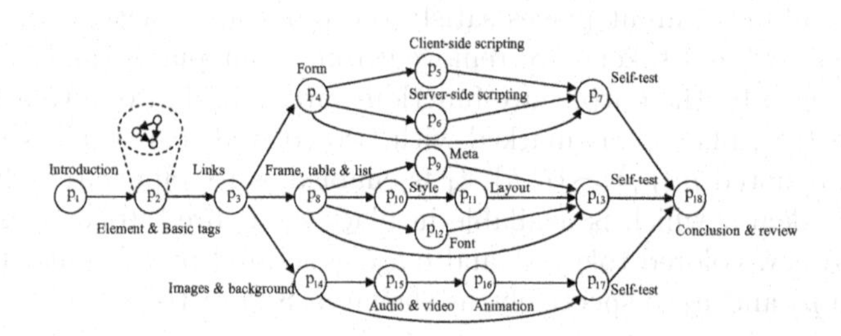

Fig. 2. An example of 18-hypertext learning state space

Assume students are categorized into three levels according to their entry records: low, middle and high, represented by three different colored token l, m, h, respectively. A CPTPN is obtained as shown in Figure 3 after applying the four typical structure mapping rules (e.g. sequence, merging, forking and Multi-path) discussed in [11] and allocating recommended learning time attributes and colored tokens access control.

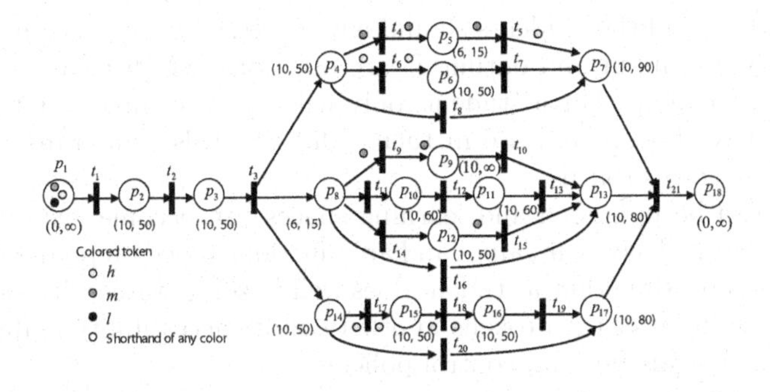

Fig. 3. A converted CPTPN model of the above hypertext learning state space

To simplify Figure 3, smaller colored circles are used to represent colored tokens instead of text notation. For a transition, only specific arc functions are identified with colored tokens allocated on its input and output arcs. Transitions with blank arc functions are assumed using identical arc function I_d, which means the output tokens are the same as input.

3.2 Firing Rules of CPTPN Based Model

In a CPTPN based model, when a place receives a colored token, its knowledge nodes become accessible to the student at that level. Before minimum delay d^{min}, the outside link is not click-able or visible. The student is required to concentrate on the current learning state and freely select knowledge nodes inside this state. A transition (which is an outside link) t is enabled only when each of its input places p_i has the colored tokens required by t's input arc function and is still in its (d_i^{min}, d_i^{max}) interval, where the (d_i^{min}, d_i^{max}) is measured relatively to the time the place becomes visible. In the (d_i^{min}, d_i^{max}) interval, the student is free to select visible links, transferring to other related states. If no link is selected by d_i^{max}, an enabled transition is chosen to fire by the system automatically, the current content becomes unavailable and the student is recommended to attempt other learning paths. The predefined firing priority of transitions and the colored tokens allocated usually reflect teaching preference.

For instance, in Figure 3, p_1 knowledge nodes are visible because of the three colored tokens deposited in place p_1. The link represented by transition t_1 is immediately activated due to the 0 value of d_1^{min} and the identical arc function I_d. A student can browse knowledge nodes of p_1 without any time constraint because d_1^{max} is set to ∞. If t_3 is fired in sequence, p_4, p_8 and p_{14} become visible simultaneously. A level m student can stay in p_4 for at most 50 time units before forced to leave for p_5. Any level students can go for p_7 without any constraint. Assuming a m level student chooses p_5 because he wants to learn more, he is required to spend at least 6 time units in it. After the 15 time units, t_5 is fired. Token m is consumed and a new colored token h is generated in p_7 due to the level upgrade of that student.

To activate t_{21}, all the students should have accessed p_7, p_{13} and p_{17} in whatever order. t_{21} has input and output I_d arc functions, which means students at any level can enable it. For a student, only when all the three input places p_7, p_{13} and p_{17} have tokens and are in their valid intervals, can transition t_{21} be fired and p_{18} receives tokens.

Students can be upgraded and learning spaces can become accessible adaptively by adjusting delay pair values and arc functions based on predefined teaching model. It is possible that a student does well in some states with higher level tokens while with lower level tokens in others. State accessibility really depends on the teaching strategies and control policies.

3.3 Access Control in CPTPN Based Model

Various access controls on place p_i can be easily achieved by allocating colored tokens and using appropriate arc functions, as shown in Table 1.

Table 1. Learning Access Control

Arc function	Definition	Notes
I_d	$I_d(c_i) = c_i$	Identical; student type kept intact
Dis	$Dis(c_i) = \bullet$	color removed; student types no longer distinguished
$Succ$	$Succ(c_i) = c_{i+1}$	student type upgraded
$Pred$	$Pred(c_i) = c_{i-1}$	student type degraded
$Succ_1$	$Succ_1(c_i, c_j) = (c_{i+1}, c_j)$	upgraded one component i of a compound color
$Succ_2$	$Succ_2(c_i, c_j) = (c_i, c_{j+1})$	upgraded one component j of a compound color
$Proj_1$	$Proj_1(c_i, c_j) = c_i$	removed one component j of a compound color
$Proj_2$	$Proj_2(c_i, c_j) = c_j$	removed one component i of a compound color

Compound color can be used if more than one resource is involved in access control. For instance, a student's entry level and assessment record can be represented using one token with two colors (i.e. the compound color). Different arc functions result in different learning paths. These functions provide user oriented content and link adaptation to some degree, in addition to the adaptations brought by delay pairs.

4 Adaptive Operations on the CPTPN Based Model

As discussed above, personalized learning paths can be produced by allocating colored tokens, adjusting arc functions, and updating the (d^{min}, d^{max}) pair. Simple content and link adaptation can be accomplished by this means.

4.1 Content Adaptation

Content adaptation corresponds to place visibility. A place is invisible if its preceding transitions never fire. A transition fires if its required tokens never arrive. If simply aiming at a particular group of students, content adaptation can be achieved by updating transitions' input arc functions. If aiming at all students, only adjust the target place's delay pair.

To transfer from user-oriented to non-user-oriented, arc function Dis can be used to remove colors, which means the succeeding states no longer differentiates leveled students.

Regardless of student levels, a place is skipped with $(0, 0)$ delay pair, which is called *learning skip.* When any colored token flows into this place, no delay is permitted and it enters its successors instantaneously. If no such a successor exists, students try other paths, or accept the system's advise. However, students are not able to see any content contained in that place.

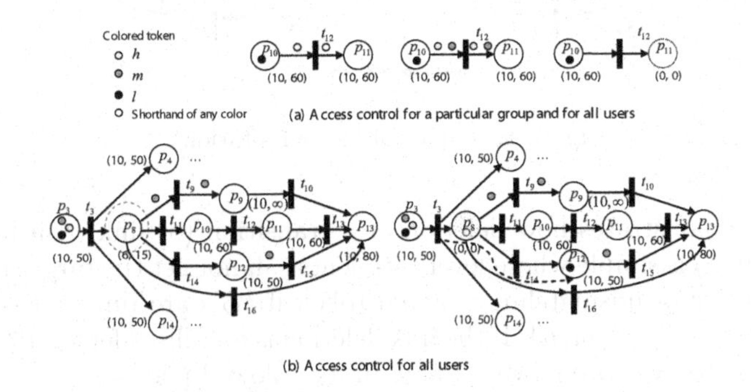

Fig. 4. Examples of content adaptation

Figure 4(a) illustrates a simply case where level l students are blocked from accessing place p_{11}. It can be easily achieved by changing the required input of arc function from I_d to $E_{f(t_{12},p_{11})} = \{m, h\}$.

Consider the complex situation like (b) where place p_8 is to be inaccessible. If (d_8^{min}, d_8^{min}) is set to $(0, 0)$, any tokens from p_3 bypass p_8 and flow into one of its following places. Which enabled transition to fire depends on the transitions' arc functions and teaching model applied. For instance, if faster learning is preferred, the place with the smallest d^{min} would be a better choice. If extra learning is required, p_{10} is another choice. If equivalent d^{min} values, d^{max} is the next variable to consider, like p_9, p_{10}, p_{12} and p_{13}.

4.2 Link Adaptation

Link adaptation corresponds to deletion or addition of transitions. A hyperlink becomes inactive if the associated transition never has the right colors to consume or the delay pairs of its output places are set to (∞, ∞). If a link is to be blocked only from a group of students, change its input arc function. If to be blocked from all students, a delay pair (∞, ∞) is the best choice. According to the executive semantics of CPTPN, if a token flows into a place with pair

(∞, ∞), the token can only leave after infinite delay. Such a "dead-end" transition would never be activated, as well as its corresponding links. An effect of link deletion is achieved.

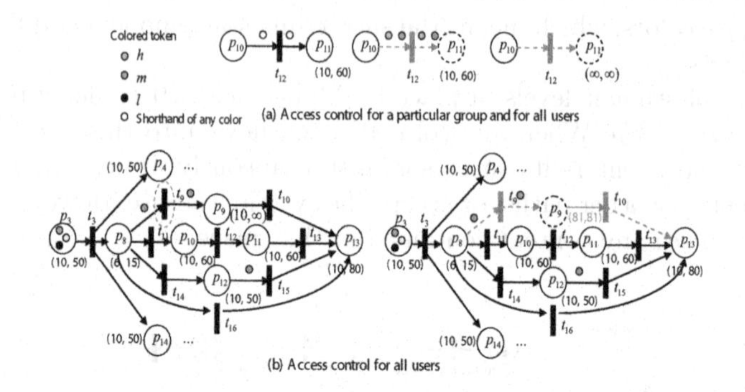

Fig. 5. Examples of link adaptation

Figure 5 demonstrates how deletion of a transition (or link) can be accomplished. Case (a) is simple where level l students can not get the target transition t_{12} fired because of mismatching colored token if t_{12}'s arc function is changed from I_d to $E_{f_{(p_{10}, t_{12})}} = \{m, h\}$. If the link deletion is for all students, $(d_{11}^{min}, d_{11}^{max})$ can be set to (∞, ∞) to remove t_{12} and all its followed places.

For a more complex situation like (b), target transition t_9 can be removed by updating its input arc function if aiming at particular group of users; otherwise, (∞, ∞) for p_9 definitely makes t_9 never be chosen. As it has three parallel transition t_{11}, t_{14} and t_{16}, however, $(81, 81)$ for p_9 also works given d_{13}^{max} is 80.

5 Discussion and Conclusion

Adaptive learning and personalized instruction is always one of the goals of technology-enhanced learning environments. This paper first introduces some background knowledge, such as the concept of timed Petri Nets and colored Petri Nets. Then a colored P-timed Petri Net model is proposed to simulate the learning state space and its link relationships. We benefit from the embedded executive semantics and dynamic timing and access control mechanism of CPTPN by mapping state to place, converting outside link to transition, using time delay pair (d^{min}, d^{max}) and colored tokens.

According to Nora and Gustavo's summary of adaptive pattern [3] , the adaptations proposed here are focusing at link level. By adjusting time delay pairs and upgrading colored tokens conditionally, content and link adaptations become possible.

There are many situations in which this adaptation model can be applied to, for instance, multiuser involved temporal activity coordination and multiple resources management in a time-sensitive system.

In our future research, we are interested in analyzing the reachable space of proposed CPTPN and applying colored timed Petri Net to adaptively coordinate tasks and activities in collaborative working environment.

References

1. Brusilovsky, P.: Methods and Techniques of Adaptive Hypermedia. International Journal of User Modeling and User-Adpated Interaction 6(2-3), 87–129 (1996)
2. Paterno, F., Mancini.: Designing WEb Interfaces Adaptable to Different Types of Use. In: Proceedings of Workshop Museumns and the Web (1999)
3. Koch, N., Rossi, G.: Patterns for Adaptive WEb Applications. In: Proceedings of 7th European Conference on Pattern Languages of Programs (2002), Available: citeseer: `http://citeseer.ist.psu.edu/koch02patterns.html`
4. Albert, D., et al.: Adaptive and Dynamic Hypertext Tutoring Systems Based on Knowledge Space Theory. In: Artificial Intelligence in Education: Knowledge and Media in Learning Systems. Frontiers in Artificial Intelligence and Applications, vol. 39, IOS Press, Amsterdam (1997)
5. Halasz, F., et al.: The Dexter hypertext reference model. Communications of the ACM 37(2), 30–39 (1994)
6. Rodrigo, A.B., et al.: Identifying Aggregates in Hypertext Structures. In: ACM Proceedings of the Hyptertext 1991 Conference, San Antonio, Texas, (December 1991)
7. Johnson, A., et al.: Adaptive Clustering of Scientific Data. In: Proceedings of the 13th IEEE International Phoenix Conference on Computers and Communication, Tempe, Arizona (1994)
8. Gao, S.: A Web-based Hypertext Learning State Space Model and Learning Control. In: Proceedings of the Sixth IASTED Int. Conf.: Internet and Multimedia Systems and Applications, Kaua'I, Hawaii, USA, August 12-14, 2002, pp. 92–96 (2002)
9. Stotts, P.D., Furuta, R.: Petri-net-based Hypertext: Document structure with browsing semantics. ACM Trans. on Information Systems 7(1), 3–29 (1989)
10. Furuta, R., Na, J.: Applying caTs Programmable Browsing Semantics to Specify World-Wide Web Documents that Reflect Place, Time, Reader, and Community. In: Proceedings of DocEng 2002, McLean, Virginia, USA, November 8-9, 2002, pp. 10–17 (2002)
11. Gao, S., Zhang, Z., Hawryszkiewycz, I.: Supporting Adaptive Learning in Hypertext Environment: a High Level Timed PetriNet based Approach. In: ICALT 2005. Proceedings of the 5th IEEE International Conference on Advanced Learning Technologies, Kaohsiung, Taiwan, July 5-8, 2005, pp. 735–739 (2005)
12. Gao, S., Zhang, Z., Hawryszkiewycz, I.: Supporting adaptive learning with high level timed Petri Nets. In: Khosla, R., Howlett, R.J., Jain, L.C. (eds.) KES 2005. LNCS (LNAI), vol. 3681, pp. 834–840. Springer, Heidelberg (2005)
13. Murata, T.: Petri nets: Properties, analysis and applications. Proceedings of the IEEE 77(4), 541–580 (1989)

Proof Explanation for the Semantic Web Using Defeasible Logic

Grigoris Antoniou[1], Antonis Bikakis[1], Nikos Dimaresis[1], Manolis Genetzakis[1], Giannis Georgalis[1], Guido Governatori[2], Efie Karouzaki[1], Nikolas Kazepis[1], Dimitris Kosmadakis[1], Manolis Kritsotakis[1], Giannis Lilis[1], Antonis Papadogiannakis[1], Panagiotis Pediaditis[1], Constantinos Terzakis[1], Rena Theodosaki[1], and Dimitris Zeginis[1]

[1] Institute of Computer Science, FORTH, Greece
[2] School of ITEE, The University of Queensland, Australia

Abstract. In this work we present the design and implementation of a system for proof explanation in the Semantic Web, based on defeasible reasoning. Trust is a vital feature for Semantic Web. If users (humans and agents) are to use and integrate system answers, they must trust them. Thus, systems should be able to explain their actions, sources, and beliefs. Our system produces automatically proof explanations using a popular logic programming system (XSB), by interpreting the output from the proof's trace and converting it into a meaningful representation. It also supports an XML representation (a RuleML language extension) for agent communication, which is a common scenario in the Semantic Web. The system in essence implements a proof layer for nonmonotonic rules on the Semantic Web.

1 Introduction

The next step in the development of the Semantic Web will be the logic and proof layers. The implementation of these two layers will allow the user to state any logical principles, and permit the computer to infer new knowledge by applying these principles on the existing data. Rule systems appear to lie in the mainstream of such activities.

Many recent studies have focused on the integration of rules and ontologies, including DLP [1], [2] and [3], TRIPLE [4] and SWRL [5]. Ongoing standardisation work includes the RuleML Markup Initiative [6], and the Rule Interchange Format (RIF) W3C Working Group.

Apart from classical rules that lead to monotonic logical systems, recently researchers started to study systems capable of handling conflicts among rules and reasoning with partial information. Recently developed nonmonotonic rule systems for the Semantic Web include DR-Prolog [7], DR-DEVICE [8], SweetJess [9] and dlvhex [10].

The upper levels of the Semantic Web have not been researched enough and contain critical issues, like accessibility, trust and credibility. The next step in the architecture of the Semantic Web is the proof layer and little has been written and done for this layer.

The main difference between a query posed to a "traditional" database system and a semantic web system is that the answer in the first case is returned from a given collection of data, while for the semantic web system the answer is the result of a reasoning process.

Z. Zhang and J. Siekmann (Eds.): KSEM 2007, LNAI 4798, pp. 186–197, 2007.

While in some cases the answer speaks for itself, in other cases the user will not be confident in the answer unless she can trust the reasons why the answer has been produced. In addition it is envisioned that the semantic web is a distibuted system with disparate sources of information. Thus a semantic web answering system, to gain the trust of a user must be able, if required, to provide an explanation or justification for an answer. Since the answer is the result of a reasoning process, the justification can be given as a derivation of the conclusion with the sources of information for the various steps.

In this work we implement a defeasible reasoning system for reasoning on the Web, which provides the additional capability of presenting explanations to users for the answers to their queries.

2 Basics of Defeasible Logics

2.1 Basic Characteristics

Defeasible reasoning is a simple rule-based approach to reasoning with incomplete and inconsistent information. It represents facts, rules, and priorities among rules. This reasoning family comprises defeasible logics [11] and Courteous Logic Programs [12]; the latter can be viewed as a special case of the former [13]. The main advantage of this approach is the combination of two desirable features: enhanced representational capabilities allowing one to reason with incomplete and contradictory information, coupled with low computational complexity compared to mainstream nonmonotonic reasoning. The basic characteristics of defeasible logics are: (i) they are rule-based, without disjunction; (ii) classical negation is used in the heads and bodies of rules, (negation-as-failure in the object language can easily be simulated, if necessary ; (iii) rules may support conflicting conclusions; (iv) the logics are skeptical in the sense that conflicting rules do not fire. Thus consistency is preserved; (v) priorities on rules may be used to resolve some conflicts among rules; and (vi) the logics take a pragmatic view and have low computational complexity.

2.2 Syntax

A *defeasible theory* is a triple $(F, R, >)$, where F is a set of literals (called *facts*), R a finite set of rules, and $>$ a superiority relation on R. In expressing the proof theory we consider only propositional rules. Rules containing free variables are interpreted as the set of their variable-free instances.

There are three kinds of rules: *Strict rules* are denoted by $A \rightarrow p$, where A is a finite set of literals and p is a literal, and are interpreted in the classical sense: whenever the premises are indisputable (e.g. facts) then so is the conclusion. Inferences from facts and strict rules only are called *definite inferences*. Facts and strict rules are intended to define relationships that are definitional in nature. Thus defeasible logics contain no mechanism for resolving inconsistencies in definite inference.

Defeasible rules are denoted by $A \Rightarrow p$, and can be defeated by contrary evidence.

A superiority relation is an acyclic relation $>$ on R (that is, the transitive closure of $>$ is irreflexive). Given two rules *r1* and *r2*, if we have that $r1 > r2$, then we will say that *r1* is *superior* to *r2*, and *r2 inferior* to *r1*. This expresses that *r1* may override *r2*.

The proof theory was studied in [11] and argumentation semantics in [14].

2.3 Defeasible Logic Metaprogram

In order to perform reasoning over a defeasible theory, we have adopted the approach proposed in [15], [16]. This approach is based on a translation of a defeasible theory D into a logic metaprogram P(D), and defines a framework for defining different versions of defeasible logics, following different intuitions. In particular, this framework is based on two "parameters": (a) the metaprogram P(D) and (b) the negation semantics adopted to interpret the negation appearences within P(D). For part (b), different semantics have been proposed and studied in the literature: the welll-founded semantics, answer set semantics, and the classical negation-as-failure operator of Prolog.

In this work we use a similar metaprogram which fits better our needs for representing explanations. The first two clauses define the class of rules used in a defeasible theory.

```
supportive_rule(Name,Head,Body):- strict(Name,Head,Body).
supportive_rule(Name,Head,Body):- defeasible(Name,Head,Body).
```

The following clauses define definite provability: a literal is definitely provable if it is a fact or is supported by a strict rule, the premises of which are definitely provable.

```
definitely(X):- fact(X).
definitely(X):-strict(R,X,L),definitely_provable(L).
definitely_provable([X1|X2]):-definitely_provable(X1),definitely_provable(X2).
definitely_provable(X):- definitely(X).
definitely_provable([]).
```

The next clauses define defeasible provability: a literal is defeasibly provable, either if it is definitely provable, or if its complementary is not definitely provable, and the literal is supported by a defeasible rule, the premises of which are defeasibly provable, and which is unblocked.

```
defeasibly(X):- definitely(X).
defeasibly(X):- negation(X,X1),supportive_rule(R,X,L),defeasibly_provable(L),
unblocked(R,X),xsb_meta_not(definitely(X1)).
defeasibly_provable([X1|X2]):-defeasibly_provable(X1),defeasibly_provable(X2).
defeasibly_provable(X):- defeasibly(X).
defeasibly_provable([]).
```

A rule is unblocked when there is not an undefeated conflicting rule:

```
unblocked(R,X):- negation(X,X1), xsb_meta_not(undefeated(X1)).
```

A literal is undefeated when it is supported by a defeasible rule which in not blocked:

```
undefeated(X):-supportive_rule(S,X,L),xsb_meta_not(blocked(S,X)).
```

A rule is blocked either if its premises are not defeasibly provable, or if it is defeated:

```
blocked(R,X):-supportive_rule(R,X,L),xsb_meta_not(defeasibly_provable(L)).
blocked(R,X):-supportive_rule(R,X,L),defeasibly_provable(L),defeated(R,X).
```

A rule is defeated when there is a conflicting defeasible rule, which is superior and its premises are defeasibly provable:

```
defeated(S,X):-negation(X,X1),supportive_rule(T,X1,V),
defeasibly_provable(V),sup(T,S).
```

We define the predicate *negation* to represent the negation of a predicate and evaluate the double negation of a predicate to the predicate itself. Furthermore, we define the predicate *xsb_meta_not* in order to represent the *not* predicate when executing a program in XSB trace.

Our system supports both positive and negative conclusions. Thus, it is able to give justification why a conclusion cannot be reached. We define when a literal is not definitely provable and when is not defeasibly provable. A literal is not definitely provable when it is not a fact and for every strict rule that supports it, its premises are not definitely provable. A literal is not defeasibly provable when it is not definitely provable and its complementary is definitely provable or for every defeasible rule that supports it, either its promises are not defeasibly provable, or it is not unblocked.

3 Explanation in Defeasible Logic

3.1 Proof Tree Construction

The foundation of the proof system lies in the Prolog metaprogram that implements defeasible logic, with some additional constructs to facilitate the extraction of traces from the XSB logic programming system. XSB is the logic programming engine used to run the metaprogram, but other Prolog systems could have been used. We use the trace of the invocation of the metaprogram to generate a defeasible logic proof tree, that subsequently will be transformed into a proof suitable to be presented to an end user.

The negation we use in conjunction with the metaprogram is the classical negation-as-failure in Prolog. Unfortunately, the XSB trace information for well-founded semantics does not provide the information we would require to produce meaningful explanations.

To enable the trace facility, the XSB process executes the command *trace*. After the loading of the metaprogram and the defeasible theory, the system is ready to accept any queries which are forwarded unmodified to the XSB process. During the evaluation of the given query/predicate the XSB trace system will print a message each time a predicate is:

1. Initially entered (**Call**),
2. Successfully returned from (**Exit**),
3. Failed back into (**Redo**), and
4. Completely failed out of (**Fail**).

The produced trace is incrementally parsed by the Java XSB invoker front-end and a tree whose nodes represent the traced predicates is constructed. Each node encapsulates all the information provided by the trace. Namely:

- A string representation of the predicates name
- The predicates arguments
- Whether it was found to be true (**Exit**) or false (**Fail**)
- Whether it was failed back into (**Redo**)

In addition to the above, the traced predicate representation node has a Boolean attribute that encodes whether the specific predicate is negated. That was necessary for overcoming the lack of trace information for the *not* predicate (see next section).

One remark is due at this stage: Of course our work relies on the trace provided by the underlying logic programming system (in our case XSB). If we had used an LP directly for knowledge representation, explanations on the basis of LP would have been appropriate. However, here we use defeasible reasoning for knowledge representation purposes (for reasons explained extensively in previous literature), thus explanations must also be at the level of defeasible logics.

3.2 Proof Tree Pruning

The pruning algorithm, that produces the final tree from the initial XSB trace, focuses on two major points. Firstly, the XSB trace produces a tree with information not relevant for the generation of the proof tree. One reason for this is that we use a particular metaprogram to translate the Defeasible Logic into logic programming. For the translation to be successful, we need some additional clauses which add additional information to the XSB trace. Another reason depends on the way Prolog evaluates the clauses, showing both successful and unsuccessful paths. Secondly, the tree produced by the XSB trace is built according to the metaprogram structure but the final tree needs to be in a complete different form, compliant with the XML schema described in section 6. We will take a closer look at the details of these issues.

A main issue of the pruning process was the way Prolog evaluates its rules. Specifically, upon rule evaluation the XSB trace returns all paths followed whether they evaluate to true or false. According to the truth value and the type of the root node, however, we may want to maintain only successful paths, only failed paths or combinations of them. Suppose we have the following defeasibly theory, translated in logic programming as:

```
fact(a).
fact(e).
defeasible(r1,b,a).
defeasible(r2,b,e).
defeasible(r3,~(b),d).
```

If we issue a query about the defeasible provability of literal b, XSB trace fails at first to prove that b is definitely provable and then finds a defeasible rule and proves that its premises are defeasible provable. It produces a proof tree, which begins with the following lines:

```
Proof
    defeasibly(b):True
        definitely(b):False
            fact(b):False
```

```
        strict(_h144,b,_h148):False
    negation(b,~(b)):True
    supportive_rule(r1,b,a):True
        strict(_h155,b,_h157):False
        defeasible(r1,b,a):True
    defeasibly_provable(a):True
        defeasibly(a):True
            definitely(a):True
                fact(a):True
    ......
```

In this type of proof, we are only interested in successful paths and the pruning algorithm removes the subtree with the false goal to prove that b is definitely provable and the false predicate to find a strict supportive rule for b. It also prunes the metaprogram additional negation clause. The complementary literal is used in next parts of the proof. The corresponding final pruning subtree for this query has the following form:

```
Proof
    defeasibly(b):True
        supportive_rule(r1,b,a):True
            defeasible(r1,b,a):True
        defeasibly_provable(a):True
            defeasibly(a):True
                definitely(a):True
                    fact(a):True
        ......
```

Suppose we have the following defeasibly theory:

```
fact(a).
defeasible(r1,b,a).
defeasible(r2,~(b),a).
```

If we issue a query about the defeasible provability of literal b, XSB trace fails to prove it and thus at first fails to prove that b is definitely provable. It produces a proof tree, which begins with the following lines:

```
Proof
    defeasibly(b):False
        definitely(b):False
            fact(b):False
            strict(_h144,b,_h148):False
        ......
```

In this proof, we are interested in unsuccessful paths and the pruning algorithm keeps the initial proof tree. Thus the pruning tree remains the same in the first lines.

The other heuristic rules deal with the recursive structure of Prolog lists and the XSB 's caching technique which shows only the first time the whole execution tree for a predicate, during trace execution of a goal. Our pruning algorithm keeps a copy of the initial trace so as to reconstruct the subtree for a predicate whenever is required.

Using these heuristic techniques, we end up with a version of the proof tree that is intuitive and readable. In other words, the tree is very close to an explanation derived by the use of pure Defeasible Logic.

4 Graphical User Interface to the Proof System

The graphical user interface to the proof system, offers an intuitive way to interact with the underlying system and visualize the requested proofs. The proofs are rendered as a tree structure in which each node represents a single predicate. A tree node may have child nodes that represent the simpler, lower level, predicates that are triggered by the evaluation of the parent predicate. Thus, the leaf nodes represent the lowest level predicates of the proof system, which correspond to the basic atoms of a defeasible theory (facts, rules and superiority relations) which can not be further explained. Additionally, if an atom has additional metadata attached to its definition, such as references for the *fact* and *rule* predicates, those are displayed as a tooltip to the corresponding tree node. This is an optional reference to a Web address that indicates the origin of the atom.

The interaction with the graphical user interface is broken down to three or four steps, depending on whether it is desirable to prune the resulting proof tree in order to eliminate the artifacts of the meta-program and simplify its structure (see section 2) or not. Thus, to extract a proof, the following steps must be carried out:

1. The Defeasible logic rules must be added to the system. Rules can be added by pressing either the *Add Rule* or *Add rules* from file button at the right part of the interface. The *Add Rule* button presents a text entry dialog where a single rule may be typed by the user. Besides that, pressing the *Add rules* from file button allows the user to select a file that contains multiple rules separated by newlines. The added rules are always visible at the bottom part of the graphical user interface.
2. As soon as the rules are loaded, the system is ready to be queried by the user. By typing a 'question' at the text entry field at the right part of the screen, just below the buttons, and pressing enter, the underlying proof system is invoked with the supplied input and the resulting proof is visualized to the tree view at the left part of the interface.
3. By pressing the *Prune* button the system runs the algorithms described in the previous section to eliminate redundant information and metaprogram artifacts and thus bring the visualized proof tree to a more human friendly form.

5 Agent Interface to the Proof System

The system makes use of two kinds of agents, the 'Agent' which issues queries and the 'Main Agent' which is responsible to answer the queries. Both agents are based on JADE (Java Agent DEvelopment Framework)[17]. JADE simplifies the implementation of multi-agent systems through a middle-ware that complies with the FIPA specifications. The agent platform can be distributed across machines and the configuration can be controlled via a remote GUI. The configuration can be even changed at run-time by moving agents from one machine to another one, as and when required.

Figure 1 shows the process followed by the Main Agent in order to answer a query. All the above steps are illustrated next.

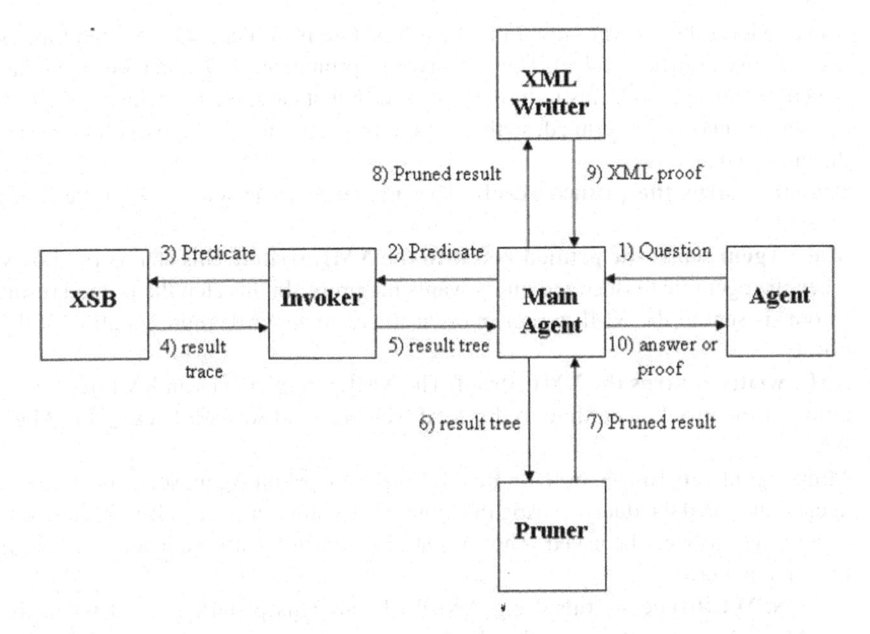

Fig. 1. The system architecture

1. **An agent issues a query to the Main Agent.** The query is of the form: **predicate::(proof/answer)**
   The predicate is a valid Prolog predicate, while the value in the parenthesis indicates the form of the answer that the Main Agent is expected to return. The 'answer' value is used to request for a single true/false answer to the agent's query, while the 'proof' value is used to request for the answer along with its proof explanation. Two examples of queries follow below:

   ```
   defeasibly(rich(antonis))::proof
   defeasibly(rich(antonis))::answer
   ```

2. **Main Agent sends the Predicate to the Invoker.** After receiving a query from an agent, the Main Agent has to execute the predicate. For this reason it extracts the predicate from the query and sends it to the Invoker who is responsible for the communication with the XSB (Prolog engine).
3. **Invoker executes the Predicate.** The Invoker receives the predicate from the Main Agent and sends it to XSB.
4. **XSB returns the result trace.** The XSB executes the predicate and then returns the full trace of the result to the Invoker.
5. **Invoker returns the result tree to Main Agent.** The Invoker receives the trace from the XSB and creates an internal tree representation of it. The result tree is then sent back to the Main Agent.
6. **Main Agent sends the result tree to the Pruner.** The Main Agent after receiving the result tree from the Invoker sends it to the Pruner in order to prune the tree. There exist two kinds of pruning. One is used when the agent that issued the query

wants to have the answer only. In that case the tree is pruned and the remaining is just the answer (true or false). The other type of pruning is used when the agent that issued the query also wants to have the proof. In that case, the brunches of the tree that are not needed are pruned, so the remaining is a pruned tree only with brunches that are needed.

7. **Pruner returns the pruned result.** The pruned result is sent back to the Main Agent.
8. **Main Agent sends the pruned result to the XML writer.** This step is used only when the agent that issued the query wants the proof. In this step the pruned result (proof) is sent to the XML writer in order to create an XML representation of the proof.
9. **XML writer returns the XML Proof.** The XML writer creates an XML representation of the proof, according to the XML schema, and sends it back to the Main Agent.
10. **Main Agent returns Answer or Proof.** Finally the Main Agent sends back to the agent that issued the query a string that contains the answer (true, false) or the proof accordingly to what he asked. The format of the string that is sent follows one of the three patterns:
    - **ANSWER(true — false)** e.g ANSWERtrue This pattern is used when the Main Agent wants to send only the answer. In this case it sends the string 'ANSWER' followed by the string representation of the answer (i.e. 'true' or 'false'). There is no space between the two words.
    - **PROOF:(proof string)** This pattern is used when the Main Agent wants to send the proof. In this case it sends the string 'PROOF:' followed by the string representation of the proof (written in XML)
    - **ERROR:(error message)** e.g. ERROR:invalid mode This pattern is used when an error occurs during the process. In this case the Main Agent sends the string 'ERROR:' followed by the string that contains the error message.

6 Extension of RuleML for Explanation Representation

The need for a formal, XML-based representation of an explanation in the Semantic Web led us to design an extension of the Rule Markup Language (RuleML) [6]. RuleML is an XML based language that supports rule representation for the Semantic Web.

In our XML schema, we use a similar syntax to RuleML to represent *Facts* and *Rules*. Specifically, we use the *Atom* element which refers to an atomic formula, and it consists of two elements, an operator element (`Op`) and a finite set of Variable (`Var`) or/and Individual constant elements (`Ind`), preceded optionally by a not statement (in case we represent a negative literal).

A *Fact* is consisted by an Atom that comprise a certain knowledge. The last primitive entity of our schema is *Rule*. In defeasible logic, we distinguish two kinds of Rules: *Strict Rules* and *Defeasible Rules*. In our schema we also note with a different element these two kind of rules. Both kinds consists of two parts, the `Head` element which is constituted of an Atom element, and the `Body` element which is constituted of a number of Atom elements.

The simplest proof explanation is in case of a definitely provable Atom. For that proof, we first have to denote the Atom, and then give the Definite Proof that explains why it is definitely provable. This explanation can come out in two ways: either a simple Fact for that Atom, or give a Strict Rule with Head this Atom and Body an Atom that should be also proved definitely with the same way. If the Body consists of multiple Atoms, then we state the definite provable explanation for every atom of the Body. The structure of a definite proof explanation is the following:

```
<Definitely_provable>
 <Atom>
  <Op>  rich </Op>
  <Ind> Bob</Ind>
 </Atom>
 <Definite_Proof>
 <Strict_rule Label="r1">
  <Head>
   <Atom>
    <Op>rich </Op>
    <Ind> Bob  </Ind>
   </Atom>
  </Head>
 <Body>
  <Atom>
   <Op>wins_lotto </Op>
    <Ind>Bob </Ind>
   </Atom>
  </Body>
 </Strict_rule>
 <Definitely_provable>
  <Definite_Proof>
   <Fact>
    <Atom>
     <Op> wins_lotto </Op>
     <Ind> Bob  </Ind>
    </Atom>
   </Fact>
  </Definite_Proof>
 </Definitely_provable>
</Definite_Proof>
</Definitely_provable>
```

Defeasible provability is represented in a similar fashion, but is more complicated to be presented in this paper. The full extension of RuleML will be presented in a companion paper.

7 Related Work

Besides teaching logic [18], not much work has been centered around explanation in reasoning systems so far. Rule-based expert systems have been very successful in applications of AI, and from the beginning, their designers and users have noted the need for explanations in their recommendations. In expert systems like [19] and *Explainable Expert System* [20], a simple trace of the program execution / rule firing appears to provide a sufficient basis on which to build an explanation facility and they generate explanations in a language understandable to its users. Work has also been done in explaining the reasoning in *description logics* [21]. This research presents a logical infrastructure for separating pieces of logical proofs and automatically generating follow-up queries based on the logical format.

The most prominent work on proofs in the Semantic Web context is *Inference Web* [22]. The Inference Web (IW) is a Semantic Web based knowledge provenance infrastructure that supports interoperable explanations of sources, assumptions, learned information, and answers as an enabler for trust. It supports provenance, by providing proof metadata about sources, and explanation, by providing manipulation trace information. It also supports trust, by rating the sources about their trustworthiness.

It is an interesting and open issue how our implemented proof system could be registered in the Inference Web, so as to produce PML proofs (an OWL-based specification for documents representing both proofs and proof meta-information). This would possibly require the registration of our inference engine, that is a defeasible logic reasoner, along with the corresponding inference rules, which are used in the defeasible logic proof theory and the explanations produced by our proof system. Extra work needs to be done in Inference Web in order to support why-not questions. Current IW infrastructure cannot support explanations in negative answers about predicates. This is the case that corresponds to our system 's explanations when an atom is not definitely or defeasibly provable.

8 Conclusion and Future Work

This work presented a new system that aims to increase the trust of the users for the Semantic Web applications.The system automatically generates an explanation for every answer to users queries, in a formal and useful representation. It can be used by individual users who want to get a more detailed explanation from a reasoning system in the Semantic Web, in a more human readable way. Also, an explanation could be fed into a proof checker to verify the validity of a conclusion; this is important in a multi-agent setting.

Our reasoning system is based on defeasible logic (a nonmonotonic rule system) and we used the related implemented meta-program and XSB as the reasoning engine. We developed a pruning algorithm that reads the XSB trace and removes the redundant information in order to formulate a sensible proof. Furthermore, the system can be used by agents, a common feature of many applications in the Semantic Web. Another contribution of our work is a RuleML extension for a formal representation of an explanation using defeasible logic. Additionally, we provide a web style representation for the facts, that is an optional reference to a URL. We expect that our system can be used by multiple applications, mainly in E-commerce and agent-based applications.

There are interesting ways of extending this work. The explanation can be improved to become more intuitive and human-friendly, to suit users unfamiliar with logic. The XML Schema should be made fully compatible with the latest version of RuleML. Finally, integration with the Inference Web infrastructure will be explored.

References

1. Grosof, B.N., Horrocks, I., Volz, R., Decker, S.: Description logic programs: combining logic programs with description logic. In: WWW, pp. 48–57 (2003)
2. Levy, A.Y., Rousset, M.C.: Combining Horn rules and description logics in CARIN. Artificial Intelligence 104(1-2), 165–209 (1998)
3. Rosati, R.: On the decidability and complexity of integrating ontologies and rules. WSJ 3(1), 41–60 (2005)
4. Sintek, M., Decker, S.: TRIPLE - A Query, Inference, and Transformation Language for the Semantic Web. In: International Semantic Web Conference, pp. 364–378 (2002)

5. Horrocks, I., Patel-Schneider, P.F.: A proposal for an OWL Rules Language. In: WWW 2004: Proceedings of the 13th international conference on World Wide Web, pp. 723–731. ACM Press, New York (2004)
6. RuleML: The RuleML Initiative website (2006), `http://www.ruleml.org/`
7. Antoniou, G., Bikakis, A.: DR-Prolog: A System for Defeasible Reasoning with Rules and Ontologies on the Semantic Web. IEEE Transactions on Knowledge and Data Engineering (accepted for publication)
8. Bassiliades, N., Antoniou, G., Vlahavas, I.P.: Dr-device: A defeasible logic system for the semantic web. In: Ohlbach, H.J., Schaffert, S. (eds.) PPSWR 2004. LNCS, vol. 3208, pp. 134–148. Springer, Heidelberg (2004)
9. Grosof, B.N., Gandhe, M.D., Finin, T.W.: SweetJess: Translating DAMLRuleML to JESS. In: RuleML (2002)
10. Eiter, T., Ianni, G., Schindlauer, R., Tompits, H.: dlvhex: A System for Integrating Multiple Semantics in an Answer-Set Programming Framework. In: WLP, pp. 206–210 (2006)
11. Antoniou, G., Billington, D., Governatori, G., Maher, M.J.: Representation results for defeasible logic. ACM Trans. Comput. Logic 2(2), 255–287 (2001)
12. Grosof, B.N.: Prioritized conflict handing for logic programs. In: ILPS 1997: Proceedings of the 1997 international symposium on Logic programming, pp. 197–211. MIT Press, Cambridge, MA (1997)
13. Antoniou, G., Maher, M.J., Billington, D.: Defeasible logic versus logic programming without negation as failure. J. Log. Program 42(1), 47–57 (2000)
14. Governatori, G., Maher, M.J., Antoniou, G., Billington, D.: Argumentation semantics for defeasible logic. J. Log. and Comput. 14(5), 675–702 (2004)
15. Antoniou, G., Billington, D., Governatori, G., Maher, M.J.: Embedding defeasible logic into logic programming. Theory Pract. Log. Program. 6(6), 703–735 (2006)
16. Maher, M.J., Rock, A., Antoniou, G., Billington, D., Miller, T.: Efficient defeasible reasoning systems. International Journal on Artificial Intelligence Tools 10(4), 483–501 (2001)
17. JADE: Java Agent Development Framework (2006), `http://jade.tilab.com/`
18. Barwise, J., Etchemendy, J.: The language of first-order logic. Center for the study of Language and Information (1993)
19. Shortliffe, E.: Computer-based medical consultations: MYCIN. American Elsevier, New York (1976)
20. Swartout, W., Paris, C., Moore, J.: Explanations in knowledge systems: Design for explainable expert systems. IEEE Expert: Intelligent Systems and Their Applications 06(3), 58–64 (1991)
21. McGuinness, D.L., Borgida, A.: Explaining subsumption in description logics. IJCAI (1), 816–821 (1995)
22. McGuinness, D.L., da Silva, P.P.: Explaining answers from the semantic web: the inference web approach. J. Web Sem. 1(4), 397–413 (2004)

Automatic Construction of a Lexical Attribute Knowledge Base*

Jinglei Zhao, Yanbo Gao, Hui Liu, and Ruzhan Lu

Department of Computer Science, Shanghai Jiao Tong University
800 Dongchuan Road Shanghai, China
{zjl,gaoyanbo,lh_charles,rzl}@sjtu.edu.cn

Abstract. This paper proposes a method to automatically construct a common-sense attribute knowledge base in Chinese. The method first makes use of word formation information to bootstrap an initial attribute set from a machine readable dictionary and then extending it iteratively on the World Wide Web. The solving of the defining concepts of the attributes is modeled as a resolution problem of selectional preference. The acquired attribute knowledge base is compared to HowNet, a hand-coded lexical knowledge source. Some experimental results about the performance of the method are provided.

1 Introduction

It is well known that lexical knowledge bases are very important to natural language processing. In general, a lexical concept can be represented as a set of attribute-value pairs. For instance, attributes such as *density*, *popularity* and *temperament* play very important roles in defining and distinguishing different concepts.

Attribute can be defined using Woods' linguistic test [1] : A is an attribute of C if we can say "*V is a/the A of C*", in which, C is a concept, A is an attribute of C and V is the value of the attribute A. For example, in "*brown is a color of dogs*", *color* is an attribute of concept *dog* with the value *brown*. For such a definition, several essentials need to be made clear. First, attributes are nouns. Second, attributes are defined on concepts and must allow the filling of values. If a C or V cannot be found to fit Woods' test, A cannot be an attribute. Third, different concepts can have the same attribute. For example, besides *dog*, concepts such as *elephant*, *furniture* can also have the attribute *color*.

In this paper, we propose a method to automatically construct a large common-sense attribute knowledge base in Chinese. The attribute base we built contains not only attributes, but also their associated defining concepts. Such an attribute centered knowledge source will be of great use, for example, for knowledge engineering and natural language processing. In knowledge engineering, because different concepts can have the same attribute, an attribute centered view will

*This work was supported by the NSFC under Grant No. 60496326 (Basic Theory and Core Techniques of Non Canonical Knowledge).

Z. Zhang and J. Siekmann (Eds.): KSEM 2007, LNAI 4798, pp. 198–209, 2007.

help very much for the engineers to select the most appropriate attributes for defining related concepts. Also, many natural language processing applications such as noun clustering, text categorization can benefit from such an knowledge source. [2] illustrated that enriching vector-based models of concepts with attribute information led to drastic improvements in noun clustering.

Our proposed method for constructing the attribute knowledge base contains two phases. The first is attribute set acquisition and the second is defining concept resolution. In the first phase, word formation information is used to first bootstrap a basic attribute set from a Machine Readable Dictionary (MRD) in Chinese. Then, the attribute set is extended and validated on the World Wide Web using lexical patterns. In the second phase, we model the solving of the corresponding defining concepts for the final extended attribute set acquired in the first phase as a resolution problem of selectional preference. The Web is again exploited as a large corpus to compute the associated concepts of the attributes. To evaluate our method, we compare the automatically acquired attribute base to a hand-coded lexical knowledge source HowNet [3]. The experimental results show that our method is very effective.

The remainder of the paper is organized as follows: Section 2 describes the related works. Section 3 introduces the method for the extraction of a set of general purpose attributes. Section 4 gives descriptions of the algorithm for solving the defining concepts for the attributes. Section 5 presents the experimental results. Finally, in Section 6, we give the conclusions and discuss future work.

2 Related Works

Two main sources have been exploited for the task of lexical resource acquisition in general. One is MRD, the other is large scale corpus. The definitional information in an MRD describes basic conceptual relations of words and is considered easier to process than general free texts. So, MRDs have been used as a main resource for deriving lexical knowledge from the beginning. [4, 5] constructed hyponymy hierarchies from MRDs. [6–8] extracted more semantic information from MRDs beyond hyponymy such as meronymy etc. The main difficulty of using MRDs as a knowledge source, as noted by [9], is that many definitional information of a word are inconsistent or missed.

Corpus is another important source for lexical knowledge acquisition. [10] used lexical-syntactic patterns like "NP such as List" to extract a hyponymy hierarchy and [11, 12] acquired *part-of* relations from the corpus. Data sparseness is the most notorious hinder for acquiring lexical knowledge from the corpus. However, the World Wide Web can be seen as a large corpus [13–15]. [16, 17] proposed Web-based bootstrapping methods which mainly employed the interaction of extracted instances and lexical patterns for knowledge acquisition.

Specific to attribute discovery, previous methods fall into two categories. One is to use mainly the layout information of web pages [18, 19], such as HTML tables and structural tags like TD, CAPTION etc., which can be clues of attributes for describing specific objects. However, such kind of methods suffers from the subjectivity of the web page writer. The other kind of method exploits lexico-syntactic patterns in which an object and its attributes often co-occur. [2] used the pattern "*the * of the C [is|was]*" to identify attributes from the Web in English. But as noted by [20], using such pattern matching directly results in too many false positives.

Overall, in attribute acquisition, all the previous works have concentrated on the acquisition of the attributes given a concept or an instantiated object. No work has ever attempted to automatically construct a common-sense attribute centered knowledge base.

3 Attribute Acquisition

3.1 Basic Attribute Set Extraction

Different from western languages, Chinese is a character-based language like Japanese and Thai. Several works have been done for exploring characters for processing such kind of languages [21–23]. In Chinese, a word is composed of one or several characters[1]. Almost each character is a morpheme with its own meaning. Usually, the constituent characters of a word can provide strong clues for the word's meaning. For example, the character 度(*du*) has the meaning *measure*. When used as the morpheme head (the rightmost character) in word formation, the resulted words such as 感光度(*photosensitivity*) are usually quality measure attributes. Such compositional information is very useful for identifying attributes.

In the phase of attribute acquisition of our algorithm, we first make use of the word formation information stated above to extract a basic attribute set. Such an algorithm, which is illustrated in Figure 1, uses both an MRD[2] and the Web. In which, a seed set of attribute indicating characters (morphemes) is needed for the algorithm to boot off, such that when they are used as head morpheme in word

```
Given: A  seed set of M attribute indicating morphemes.
       A MRD containing W words.
Begin:
       Find subset  S of W with a morpheme head in M.
       Get a validated set S' of  S using web-based  attribute filter.
       While |S'| can still increase do
           Find a new set X of W , such that every  x in X  have a hypernym  in S' .
           Get a validated set X' of  X using web-based attribute  filter.
           S'=S'∪X'.
End.
Output: An initial set S' of attributes from MRD.
```

Fig. 1. The Algorithm for Attribute Extraction

[1] In most cases, a morpheme in Chinese is a character. In the following, they are used interchangeably.

[2] The Standard Dictionary of Modern Chinese, Li Xingjian Ed., Foreign language teaching and research press, 2005.

formation, the resulting word will be very likely to be attributes. The initial seed set used in our experiment contains 15 attribute indicating morphemes selected manually according to Chinese linguistic knowledge which is listed as follows: 量(*liang*), 性(*xing*), 度(*du*), 程(*cheng*), 状(*zhuang*), 况(*kuang*), 态(*tai*), 势(*shi*), 级(*ji*), 力(*li*), 形(*xing*), 貌(*mao*), 能(*neng*), 质(*zhi*), 相(*xiang*). Such a seed set is not an exhausted set of head morphemes for forming attributes but just some prototypical attribute morphemes to boot off.

All the words in the MRD that have a morpheme head in the above set of seed morphemes are extracted as attribute candidates. After validation, all the hyponyms of the validated attribute set are extracted from the MRD iteratively using the simple heuristic pattern (1) for finding hyponyms from word definitions, where A is a known attribute and H is the identified hyponym of A.

(1) H: * 的(de) A ,|;|.

For example, 温度(*wen-du*, *temperature*) is an attribute with a seed morpheme 度(*du*) as the head. The word 冰点(*bing-dian*, *freezing-point*) has the definition in the MRD as (2). It matches the template (1), which suggests that 冰点(*bing-dian*) is also a possible hyponym of 温度(*wen-du*), and thus is a possible candidate of the attribute set.

(2) a. 冰点: 水开始凝结成冰时的温度.
 b. bing-dian: water start to freeze de wen-du.
 c. freezing-point: the temperature when water start to freeze.

3.2 Web-Based Attribute Filter

In every iteration of the above Algorithm, the newly identified attribute candidates must be verified before the start of next iteration. In some cases, even words with a morpheme in the initial morpheme set are not truly attributes. For example, while most nouns with a morpheme head 程(*cheng*) such as 病程(*course-of-disease*) are attributes, the word 工程(*gong-cheng*, *project*) is not an attribute word. So, a web-based classifier is used to filter out the false positives.

The classifier is based on the assumption that semantic information can be learned by matching certain associating lexico-syntactic patterns. Suppose T_a is the pattern set indicating that the word X is an attribute and T_c the pattern set indicating X as a defining concept. We use a likelihood ratio λ [24] for decision making:

$$\lambda = \frac{P(T_a, X)/P(T_a)}{P(T_c, X)/P(T_c)}$$

We will call such a likelihood ratio the attribute score. The pattern set T_a and T_c are some variants of Woods' linguistic definition of attribute, illustrated as (3) and (4) that express the syntactic behavior of attributes and defining concepts respectively.

(3) a1: * 的(*de, 's*) x 很(*hen ,very*)
a2: * 的(*de, 's*) x 为(*wei ,is*)

(4) b1: x 的(*de, 's*) * 很(*hen ,very*)
b2: x 的(*de, 's*) * 为(*wei ,is*)

The computation of the frequencies such as $f(T_a, X)$ is approximated by the web counts returned by the Google[3] search engine. The pattern set T_a and T_b contain two patterns a1 and a2, b1 and b2 respectively. The frequencies such as $f(T_a, X)$ is computed by a simple linear combination of $f(T_{a1}, X)$ and $f(T_{a2}, X)$. For example, if the word X is 冰点(*freezing-point*), $f(T_a, X)$ will be the web counts returned by Google for the query "的冰点很" (*'s freezing-point very*) combined with the web counts for query "的冰点为" (*'s freezing-point is*). Using a small set of labeled training data, we can get a threshold H of λ for filtering, that is, the attribute candidates with an attribute score lower than H will be filled out.

3.3 Extending the Basic Attribute Set on the Web

We don't think it possible to exhaust the attributes used in natural language. This is mainly because that attribute words can be compounds and newly compounded attributes are emerging everyday. For example, the attribute word *ability* can compound with *flying* and *language* to form new attributes *flying ability* and *language ability* respectively. However, in this paper, we make a closed world assumption for the attributes covered by a general MRD, because we think that such an attribute set can form a solid base for defining and differentiating various concepts.

The initial set of attributes extracted from the MRD doesn't contain all the attribute words in the MRD. Actually, it acts as a basic seed set for the second stage of extraction. The World Wide Web is used as a large corpus for extending the initial attribute set. The algorithm is illustrated in figure 2.

The method for extending the attribute set uses the conjunctive phrase pattern (5).

(5) 的(*de, 's*) x 和(*he, and*) NP.

The motivation for using such a phrase pattern is based on Resnik's assumption [25] that coordinated parts have strong semantic similarity. Given a known attribute x, we can assume that its coordinated part NP in (5) is also an attribute. For example, if x is 感光度(*photosensibility*), the conjunctive phrase (6) will be a good indication that 快门速度(*shutter speed*) is also an attribute.

(6) 的(*'s*) 感光度(*sensibility*) 和(and) 快门速度(*shutter-speed*).

[3] www.google.com

```
Given: A basic set S' of attributes extracted from MRD.
       A conjunctive template t for extending S'.
       A MRD containing W words.
Begin:
       Initialize known attribute set KS=S' and extended attribute set ES=empty.
       While iteration<N do
         For every s in KS
              Instantiate t by s and get a query q.
              Get the Google result set R of q.
              Use an NP chunker to find the candidate coordination set C of s in R.
              Use PES to find the candidate attribute set A of s in C.
              Get a validated set A' of A using Web Attribute Filter.
              A'=A'∩W and ES=ES∪A'.
         KS=ES and S'=S'∪ES.
End.
Output: A full set S' of attributes of the MRD.
```

Fig. 2. The Algorithm for Attribute Extending

The main difficulty for employing the conjunctive pattern above for extended attribute acquisition lies in determining the boundary of the coordination. Specific to the conjunctive phrase (6), for example, because 快门(*shutter*) and 速度(*speed*) are both nouns, there exist structural ambiguities in 感光度和快门速度(*sensibility and shutter speed*), as illustrated in (7).

(7) a1. [感光度和快门] 速度
    a2. [sensibility and shutter] speed
    b1. [感光度] 和[快门速度]
    b2. [sensibility] and [shutter speed]

In (a), the parallel part of 感光度(*sensibility*) is 快门(*shutter*), while in (b), the corresponding part is 快门速度(*shutter speed*).

We use a method called position exchanging search (PES) to solve the problem. PES assume that if A and B are coordinated part in coordination "A and B", then there would be enough pragmatic evidence for the structure "B and A". Then, given a known attribute a and a possible coordination b recognized by a NP chunker. We search for the phrase "的('s) b 和(and) a" on the Web to test the appropriateness of b to be an attribute candidate. If b passed the PES test, it will then be sent to the web filter introduced in Section 3.2 for further validation.

The algorithm for extended attribute acquisition is an iterative one. In the End of each iteration, the newly extended attribute set would be used as the known attribute set in the new iteration to further acquire more attributes.

4 Finding the Defining Concept for Attributes

Given a known attribute acquired by the first phase of our algorithm, we model the solving of its defining concept as a selectional preference resolution problem. Selectional preferences are limitations on the applicability of natural predicates to arguments [26]. Most works on selectional preference resolution are conducted for the verb predicates. For example, the verbal concept *eat* has the AGENT preferences for *human* and *animal*, and a PATIENT preference for *food*. Similarly, the defining concepts associated with a specific nominal attribute also have preference selection. For example, while the attribute *financial-power* is strongly related to *organization* involved in marketing activities, the attribute *vital-capacity* is restricted to *human* and *animal*.

The sub-algorithm for solving the defining concepts is illustrated in Figure 3. In the algorithm, first, a training set of <concept word , attribute word> pairs is extracted from the Web. For an attribute A, the heuristic pattern (8) is used to extract from the returning results of Google (www.google.com) for the possible concept words (the noun N in the pattern) associated with it. For example, if the attribute is 创造力(*creativity*), we can extract from the Web using (8) for the possible attribute words such as 青年(*young*), 员工(*staff*), 孩子(*children*), etc.

(8) N 的(*de, 's*) *A* 很(*he, and*).

From a large set of such <concept word , attribute word> pairs, then, we generalize for the concept classes using HowNet's IS-A concept taxonomy and get the space of candidate preference classes for specific attributes. The appropriateness of the candidates is evaluated using the selectional association measure [26].

Given: A training set of <concept word, attribute word> extracted from the Web.
An IS-A concept hierarchy.
Begin:
Create the space of candidate preference classes for each attribute.
Evaluate of the appropriateness of the candidates by a statistical measure.
Select the most appropriate subset in the candidate space to express the selectional preference.
End.
Output: The Domain of defining concepts for each attribute.

Fig. 3. The Algorithm for Defining Concept Resolution

Formally, let A be a random variable ranging over the set $a_1, ..., a_m$ of attributes under consideration. Let C be a random variable ranging over the set $c_1, ..., c_n$ of classes in the taxonomy. The selectional association between an attribute a_i and a class c can be defined as:

$$Assoc(a_i, c) = \frac{P(c|a_i) \log \frac{P(c|a_i)}{P(c)}}{S(a_i)}$$

In which, $S(a_i)$ is called the selectional preference strength of the attribute a_i that is modeled as the relative entropy between the probability distribution $P(c|a_i)$ and $P(c)$ as:

$$S(a_i) = \sum_{c} P(c|a_i) \log \frac{P(c|a_i)}{P(c)}$$

The joint probabilities in the model were estimated from the training set of <concept word, attribute word> pairs as follows:

$$P(a_i, c) = \frac{1}{N} \sum_{n \in words(c)} \frac{1}{|classes(n)|} freq(a_i, n)$$

where $words(c)$ denotes the set of nouns for which any sense is subsumed by c, classes(n) denotes the set of classes to which noun n belongs. $freq(a_i, n)$ is the number of times n appeared as the concept word of attribute a_i and N is the total number of instances in the extracted data set. After the evaluation of the association score of each candidate class, the most appropriate subset in the candidate space is selected for each attribute as its domain of defining concept.

5 Experimental Results

5.1 Results of Attribute Extraction

The MRD we used to extract the attribute set contains totally 35219 nouns. Using the 15 attribute indicating morphemes, 745 candidate attribute words whose morpheme head is in the seed morpheme set are first extracted from the MRD. From which, we randomly selected 300 words for training and testing the web attribute filter introduced in Section 3.2. Two PhD student majored in computational linguistics are employed for annotating the words as either attributes or non-attributes. The inter-annotator agreement measured by Kappa Score [27] is 91%. After discussion on a final annotation, 225 words are used to train the threshold of the log likelihood ratio and 75 words are used to test the classifier. A correct classification rate of 88.0% is achieved.

Using such a classifier to classify the above 745 candidate attribute words, we get a validated set S' of 521 attributes. The hyponyms of such a known attribute set are found from the MRD iteratively. Figure 4 shows the result of the numbers of newly extracted attributes before and after validation by the web-based filter in each iteration. After 5 iterations, we get a basic attribute set of totally 880 attributes from the MRD.

Using the conjunctive pattern and PES discussed in Section 3.3 to determine the coordination, we extend the 880 attributes using the web-based extending algorithm illustrated in Section 3.3. The number of attributes extracted before and after filtering by the Web-based attribute filter is illustrated by Figure 5. In which, for example, in the first iteration, the 880 attributes are used as seeds for the coordination based extending. Before web-based filtering, a number of 1235 attribute candidates is extracted and after filtering, a validated subset of 841 attributes is used as known attributes for the next iteration of extending. 1462

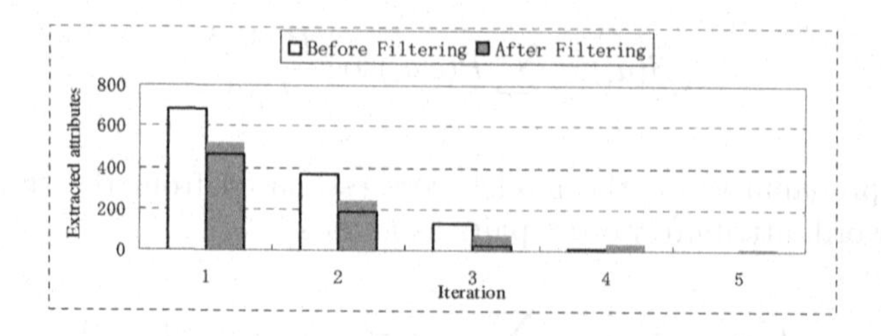

Fig. 4. Number of Attributes Extracted from MRD in Basic Attribute Extraction

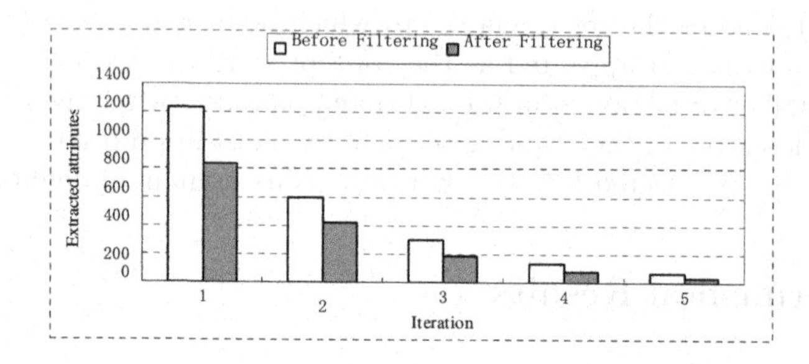

Fig. 5. Number of Attributes Extracted from MRD during Web-Extending

new attributes are got in the second stage of extraction and totally we acquire 2342 attributes. Among them, 1927 are correct after manually inspecting.

5.2 Results of Defining Concept Resolution

For the extracted attribute set, we totally acquired 220M <concept word, attribute word> training pairs from the Web to compute the defining concepts. When using such a training set and the concept class taxonomy to create the space of candidate classes, a threshold is used to ignore the possible noise introduced by the training set. Specifically, only those classes that have a higher number of the occurrences than the threshold are considered.

For an attribute, the domain of its defining concepts is explicitly computed by selecting from the candidate preference concept classes using the selectional association measure discussed in section 4. Table 1 illustrates some examples of the selectional preferences acquired. In which, for example, the domain of defining concept of attribute 气质(*temperament*) is 人(*human*). The quality of the acquired defining concepts of the attributes is compared to HowNet.

HowNet is a hand-coded common-sense lexical knowledge base describing inter-conceptual relations and inter-attribute relations of concepts as connoting in lexicons of the Chinese and their English equivalents. The top-most level of classification in HowNet includes entity, event, attribute and attribute value.

Table 1. Examples of Acquired Selectional Preferences of Attributes

Defining Concept	Selectional Association	Examples of Concept Word
	气质(temperament)	
人(human)	0.51	女孩(girl) ,男性(man)
	开本(book-size)	
读物(readings)	0.28	期刊(periodical) ,教材(teaching-material)
	题材(subject-matter)	
信息(information)	0.31	故事(story) ,艺术(art)
	处境(plight)	
人(human)	0.55	居民(resident), 乞丐(beggar)
场所(InstitutePlace)	0.1	诊所(clinic), 企业(enterprise)
团体(community)	0.05	家庭(family), 球队(team)

Under the attribute level, it contains 2093 attribute words each with a definition like follows:

(9) 寿命(life-span): attribute|属性, age|年龄, &animate|生物

which means that 寿命(*life-span*) is an attribute word with the defining concept 生物(*Animate*). The attribute base of HowNet provides a test-bed for the resulted selectional preferences of the attributes.

There are 1037 attributes in the set of extracted attributes overlapping with HowNet's attribute base. Among which, 110 attributes are randomly selected to test the appropriateness of the selectional preference computed above. The candidate preference class with the highest preference score are used to compare with HowNet's definition. Table 2 shows the number and the percentage of the defining concepts which are exactly matched, not matched at all, or matched by more general or more specific classes in the taxonomy.

The results of the comparison is encouraging. About 43.6% of the resulted selectional preferences exactly match with HowNet's definition and about 25.3% can match by 1 or 2 level of hyperonym or hyponym expansion. For example, our resulted defining concepts for 态度(*attitude*) is *human* while HowNet *animal*. Such an example is included in the case of "1 level of hyperonym" though we think that *human* is more appropriate than *animal*.

Table 2. Comparison of Acquired Defining Concepts with HowNet

exactly matched	48 (43.6%)
matched by 1 level hyperonym	12 (10.9%)
matched by 1 level hyponym	8 (7.2%)
matched by 2 level hyperonym	5 (4.5%)
matched by 2 level hyponym	3 (2.7%)
matched by >2 level hyperonym	13 (11.8%)
matched by >2 level hyponym	1 (0.91%)
not matched	20 (18.2%)

In some cases, HowNet tends to generalize too much the defining concepts of the attributes. For example, the defining concept for 处境(*plight*) in HowNet is entity, but as evidenced by the data of selectional association in Table 2, there is large difference between the selectional association for the concept class *human* and *InstitutePlace* which can't be reflected by HowNet at all. Such an example is included in the case of "matched by >2 level hyperonym" in Table 2.

6 Conclusions and Future Work

In this paper, we have presented a method for the automatic construction of a common-sense attribute knowledge base in Chinese. We first applied word-formation information to extract a basic set of attributes from an MRD and then extended it using a conjunctive pattern on the Web. During the extraction process, a web-based attribute filter is used to control the level of false positives. Then, by using an algorithm of selectional preference resolution, we computed the domain of defining concepts and the selectional preference strength for each extracted attributes, which provided a solid basis for applying such a source to various natural language tasks.

However, much work need to be done in the future. Currently, we don't differentiate senses of the attribute words. About 16.7% of the extracted attributes have more than one nominal senses in the MRD. To get a more precise computation of the defining concepts for each attribute, different senses of the attribute words should be disambiguated. Also, the constructed attribute knowledge base is a flat one by now. In the future, the attributes extracted should be clustered and a hierarchy of such attributes should be constructed.

References

1. Woods, W.: What's in a Link: Foundations for Semantic Networks. Bolt, Beranek and Newman (1975)
2. Almuhareb, A., Poesio, M.: Attribute-Based and Value-Based Clustering: An Evaluation. In: Proc. of EMNLP 2004, pp. 158–165 (2004)
3. Dong, Z., Dong, Q.: HowNet and the Computation of Meaning. World Scientific, Singapore (2006)
4. Amsler, R.: The Structure of the Merriam-Webster Pocket Dictionary (1980)
5. Chodorow, M., Byrd, R., Heidorn, G.: Extracting semantic hierarchies from a large on-line dictionary. Proceedings of the 23rd conference on Association for Computational Linguistics (1985) 299–304
6. Wilks, Y., Fass, D., Guo, C., McDonald, J., Plate, T., Slator, B.: A tractable machine dictionary as a resource for computational semantics. Longman Publishing Group White Plains, NY, USA (1989)
7. Alshawi, H.: Analysing the dictionary definitions. Computational lexicography for natural language processing table of contents (1989) 153–169
8. Richardson, S., Dolan, W., Vanderwende, L.: MindNet: acquiring and structuring semantic information from text. Proceedings of the 17th international conference on Computational linguistics (1998) 1098–1102

9. Ide, N., Veronis, J.: Extracting knowledge bases from machine-readable dictionaries: Have we wasted our time. Proceedings of KB&KS 93 (1993) 257–266
10. Hearst, M.: Automatic acquisition of hyponyms from large text corpora. Proceedings of the 14th conference on Computational linguistics-Volume 2 (1992) 539–545
11. Berland, M., Charniak, E.: Finding parts in very large corpora. Proceedings of the 37th Annual Meeting of the Association for Computational Linguistics (1999) 57–64
12. Poesio, M., Ishikawa, T., im Walde, S., Viera, R.: Acquiring lexical knowledge for anaphora resolution. Proceedings of the 3rd Conference on Language Resources and Evaluation (LREC) (2002)
13. Grefenstette, G., Nioche, J.: Estimation of English and non-English Language Use on the WWW. Arxiv preprint cs.CL/0006032 (2000)
14. Zhu, X., Rosenfeld, R.: Improving trigram language modeling with the World Wide Web. Proceedings of IEEE International Conference on Acoustics, Speech, and Signal Processing **1** (2001)
15. Keller, F., Lapata, M.: Using the web to obtain frequencies for unseen bigrams. Computational Linguistics **29**(3) (2003) 459–484
16. Brin, S.: Extracting patterns and relations from the world wide web. WebDB Workshop at 6th International Conference on Extending Database Technology, EDBT' 98 (1998) 172–183
17. Pennacchiotti, M., Pantel, P.: A Bootstrapping Algorithm for Automatically Harvesting Semantic Relations. Proceedings of Inference in Computational Semantics (ICoS-06), Buxton, England (2006)
18. Chen, H., Tsai, S., Tsai, J.: Mining tables from large scale html texts. 18th International Conference on Computational Linguistics (COLING) (2000) 166–172
19. Yoshida, M., Torisawa, K., Tsujii, J.: A method to integrate tables of the world wide web. in Proceedings of the International Workshop on Web Document Analysis (WDA 2001), Seattle, US (2001)
20. Poesio, M., Almuhareb, A.: Identifying Concept Attributes Using a Classifier. Ann Arbor 100 (2005)
21. Fujii, H., Croft, W.: A comparison of indexing techniques for Japanese text retrieval. In: Proceedings of the 16th annual international ACM SIGIR conference on Research and development in information retrieval, pp. 237–246. ACM Press, New York (1993)
22. Theeramunkong, T., Sornlertlamvanich, V., Tanhermhong, T., Chinnan, W.: Character cluster based Thai information retrieval. In: Proceedings of the fifth international workshop on on Information retrieval with Asian languages, pp. 75–80 (2000)
23. Baldwin, T., Tanaka, H.: Balancing up Efficiency and Accuracy in Translation Retrieval. Journal of Natural Language Processing 8(2), 19–37 (2001)
24. Mosteller, F., Wallace, D.: Inference and Disputed Authorship: The Federalist. Addison-Wesley, Reading (1964)
25. Resnik, P.: Semantic Similarity in a Taxonomy: An Information-Based Measure and its Application to Problems of Ambiguity in Natural Language. Journal of Artificial Intelligence 11(11), 95–130 (1999)
26. Resnik, P.: Selectional constraints: an information-theoretic model and its computational realization. Cognition 61(1-2), 127–159 (1996)
27. Siegel, S., Castellan, N.: Nonparametric statistics for the behavioral sciences. McGraw-HiU Book Company, New York (1988)

Populating *CRAB* Ontology Using Context-Profile Based Approaches

Lian Shi, Jigui Sun, and Haiyan Che

College of Computer Science and Technology, Jilin University,
Changchun 130012, China
Key Laboratory of Symbolic Computation and Knowledge Engineer
of Ministry of Education, ChangChun 130012, China
anningSL@163.com, JgSun@jlu.edu.cn, chehy@jlu.edu.cn

Abstract. Ontologies are widely used for capturing and organizing knowledge of a particular domain of interest, and they play a key role in the Semantic Web version, which adds a machine tractable, repurposeable layer to complement the existing web of natural language hypertext. Semantic annotation of information with respect to a underlying ontology makes it machine-processable and allows for exchanging these information between various communities. This paper investigated approaches for Ontology Population from the Web or some big corpus and proposed context-profile based approaches for Ontology Population. For each term extracted from web sites and web documents, we build a context profile of the term. The context profiles are represented as vectors such that we can calculate the similarity of two vectors. In our experiments we populate the *CRAB* Ontology with new instances extracted by presented approaches. Both theory and experimental results have shown that our methods are inspiring and efficient.

Keywords: Knowledge Extraction, Semantic Annotation Ontology Population, Context Profile.

1 Introduction

An ontology is an explicit formal conceptualization of a specific domain of interest. Ontologies play a key role in the Semantic Web version, which adds a machine tractable, repurposeable layer to compliment the existing web of natural language hypertext [1]. Therefore, ontologies are popular in a number of fields such as knowledge engineering and representation [2,3], information retrieval and extraction [4], knowledge management [5], agent system, and more [6]. The World Wide Web is probably the most and richest available repository as the source of information. The enrichment of the web with semantic annotations (metadata) is fundamental for the accomplishment of the Semantic Web [7]. Although there is no a univocally accepted definition for the Ontology Population task, a useful approximation has been suggested [8] as Ontology Driven Information Extraction, where, in place of a template to be filled, the

Z. Zhang and J. Siekmann (Eds.): KSEM 2007, LNAI 4798, pp. 210–220, 2007.

goal of the task is the extraction and classification of instances of concepts and relations defined in a Ontology. The task has been approached in a various of similar perspectives, including term clustering [9,10] and term categorization. Automatic Ontology Population(OP) with knowledge extracted from a web corpus is very beneficial. Ontology Population has recently emerged as a new field of application for knowledge acquisition techniques [11], however, manual ontology population is very labour intensive and time consuming. A number of semi-automatic approaches have investigated creating documents annotations and storing the results as ontology assertions. Our approaches which as many others are based on Harris' distributional hypothesis [12], i.e. that words that are semantically similar to the extent to which they share syntactic contexts.

The rest of the paper is structured as follows. Section 2 describes the state-of-the-art methods in Ontology Population. Section 3 gives information concerning the ontology of the Semantic-Based Intelligent Search Engine for Chinese documents *CRAB*[1]. Section 4 presents our approach. Section 5 reports on the experimental settings, results obtained, and gives a discuss. Section 6 concludes the paper and suggests directions for future work.

2 Related Works

Automated approaches for ontology population based on information from web documents are given by more and more attentions, since they are closely related with Semantic Web. There may be two classes to be distinguished for present approaches of Ontology Population. The first class is unsupervised approaches [13], which relies on no labeled training data but have low performance, while the another class is supervised approaches [14], which need to build a tagged training set manually. It is a time-consuming and burdened task that leads to a bottleneck for the latter in large scale applications. There are two main paradigms in state-of-art Ontology Population approaches [15]. The first one called Class-Pattern is performed using patterns [16] or relying on the structure of terms [17]. The second one called Class-Word is addressed using contextual features [13]. Our investigation focus on so-called Class-Example approach [15], which is based on Class-Word approach and adds an additional training set of training examples for each class, and then automatically learn s syntactic model- a set of weighted syntactic features. These training sets are produced by adding a few simple instances to each class. There are no restrictions on the training examples rather than except that they have to appear in Corpus.

3 The *CRAB* Ontology

The ontology used in our case study describes the financial domain, especially focuses on the stocks. The *CRAB* Ontology has been manually constructed using

[1] *CRAB* is a partial result of the project supported by NSFC Major Research Program 60496321; Basic Theory and Core Techniques of Non Canonical Knowledge.

the Protégé-based[2] management system developed in the context of the *CRAB* system. The *CRAB* ontology was implemented in OWL DL, a sublanguage of Web Ontology Language(OWL). Due to the complication of financial knowledge, we focus on the concepts and properties which are important and more interested by users, such as stocks, funds, bonds, companies, etc. and daily rates, prospect etc. Additionally, there is a "has attribute" property for each concept which links it either to another concept (e.g. concept "company" has attribute "launch") which its range could be a string (the name of a stock) or a numeric data-type (the code of a stock). Also, there are constraints on the range of numerical data-types which are defined by a minimum and maximum value. Moreover, we have populated a number of instances for the concepts. The *CRAB* ontology contains 200 concepts, 265 properties and 1000 instances as far as now. We enrich it by now and in future.

4 Our Approaches

We define Ontology Population as following process similar to Class-Word approach without more restrictions. Given a set of terms $T = t_1, t_2, \ldots, t_n$, a document collection D, where terms in T are supposed to appear, and a set of predefined classes $C = c_1, c_2, \ldots, c_m$, denoting concepts in a given ontology, each term t_i has to be assigned to a proper class in C. There a finite set I_i of sample instances for each class C_i. We build a context profile for each term to be classified using two kinds of context information, which are window profiles and syntactic profiles.

4.1 Window Profiles

We choosed n words to the left and right of a certian word of interest excluding so called stop-words and without trespassing sentencing boundaries, then the length of the window is n, and these words are called window profiles. For each sample instance in I_i, we calculate its feature weight vector $WinFeaVea(I_{ij})$, from which we get the feature weight vector $WinFeaVec(c_i)$ of the class by uniting all feature weight vectors of its sample instances. The weight of a feature $f_{c_i} \in F(c)$, is calculated [15] in three steps. $F(c)$ is a set of features of the class c.

1. The co-occurrence of with the training set is calculated:

$$weight_1(f_{c_i}) = \sum_{I_{ij} \in I_i} \alpha.log(\frac{P(f_{c_i}, I_{ij})}{P(f_{c_i}), P(I_{ij})}) \quad (1)$$

where $P(f_{c_i}, t)$ is the probability that feature f_{c_i} co-occurs with t, $P(f_{c_i})$ and $P(t)$ are the probability that f_{c_i} and t appear in the corpus, α=14 for

[2] Protégé Web Site: Http://protege.standord.edu

syntactic features with lexical element noun and α =1 for all the other syntactic features.

2. Normalize the feature weights: for each class c, to find the feature with maximal weight and denote its weight with $mxW(c)$;

$$mxW(c) = max_{f_{c_i} \in F(c)} weight_1(f_{c_i}) \tag{2}$$

Next, the weight of each feature is normalized by dividing it with $mxW(c)$;

$$weight_N(f_{c_i}) = \frac{weight_1(f_{c_i})}{mxW(c)} \tag{3}$$

3. Obtain the final weight of f_{c_i} by dividing $weight_N(f_{c_i})$ by the number of classes in which this feature appears.

$$weight(f_{c_i}) = \frac{weight_N(f_{c_i})}{|Classes(f_{c_i})|} \tag{4}$$

where $|Classes(f_{c_i})|$ is the number of classes for which f_{c_i} is present in the syntactic model.

We compute the feature weight vectors for each class denoted as

$$WinFeaVec(c) = [(f_{c1}, w_{c1}), \ldots, (f_{cn}, w_{cn})]$$

by the procedure described above. Next, we calculate the feature weight vector for a testing instance t.

1. To get the weight of testing instance in a new document, we use well-known formula of Term Frequency Inverse Document Frequency(TF-IDF).

$$w_t^{'} = tf \times idf = \frac{freq_t}{\Sigma_k(freq_k)} \times \log(\frac{N}{n_t}) \tag{5}$$

where N is the number of all documents in the Corpus, n_t is the number of documents in which the word t occurs, $freq_t$ is the number of occurrences of the word t.

2. To get the feature weight vector of testing instance t in the Corpus similar to the procedure calculating the weights of classes. This feature weight vector is denoted as

$$WinFeaVec(t)' = [(f_{t1}, w_{t1}), \ldots, (f_{tm}, w_{tm})]$$

3. Combining the two weights above, we modify each weight in $WinFeaVec(t)$. Since t occurs in a new document which is to be dealt with, for example to be annotated, that's why we load the weight of t in new document into the weights in Corpus in order to highlight the features . Then we get the final the feature weight vector of testing instance t denoted as $WinFeaVec(t)$.

$$WinFeaVec(t) = \left[(f_1, \frac{w_1}{w_1 + w_t'}), \ldots, (f_m, \frac{w_m}{w_m + w_t'})\right]$$

As far as now, we get the feature weight vectors for each class and the feature weight vector for t, we implement the classification by similarity formula:

$$sim(WinFeaVec(c_i), WinFeaVec(t)) = \cos\theta \frac{\sum_{k=1}^{n} w_{1k} \times w_{2k}}{\sqrt{\left((\sum_{k=1}^{n} w_{1k}^2)(\sum_{k=1}^{n} w_{2k}^2)\right)}} \quad (6)$$

We sketch the algorithm for window-profile based Ontology Population.

Algorithm 1: Window-Profile based Ontology Population(WPOP)

Input : Corpus, Ontological concept sets C , sample instance sets I , document d , testing instance t , window length l.
Output: class c_i to which belongs.
for (*each* $c_i \in C, 1 \le i \le |C|$) **do**
  for (*each* $I_{ij} \in I_i, 1 \le j \le |I_i|$) **do**
    $F(c_i) = (f_{c_{ij}} \mid \textit{the distance between } f_{c_{ij}} \textit{ and } I_{ij} \le l)$
  end
  for (*each* $f_c \in F(c_i)$) **do**
    calculate the weight of f_c ;
    $WinFeaVec(c_i) = [(f_{c_{i1}}, w_{c_{i1}}), \ldots, (f_{c_{in}}, w_{c_{im}})]$
  end
end
Calculate the feature weight of t in document d ;
$w_t^{'} = tf \times idf = \frac{freq_t}{\Sigma_k(freq_k)} \times \log(\frac{N}{n_t})$;
Calculate the feature weight vector of t in Corpus;
$F(t) = (f_{t_i} \mid \textit{the distance between } f_{t_i} \textit{ and } t \le l)$;
for (*each* $f_t \in F(t)$) **do**
  calculate the weight of f_t;
end
$WinFeaVec(t)' = [(f_{t1}, w_{t1}), \ldots, (f_{tn}, w_{tm})]$;
$WinFeaVec(t) = \left[(f_1, \frac{w_1}{w_1+w_t^{'}}), \ldots, (f_m, \frac{w_m}{w_m+w_t^{'}})\right]$;
$class(t) = argmax_{c_i \in C}\, sim(WinFeaVec(c_i), WinFeaVec(t))$;
Return class(t).

4.2 Syntactic Profiles

Syntactic Parse. In our approach, the Corpus is parsed by JAVA NLP API[3] developed by Stanford University. It supports Chinese national language processing well. We give an example. Here is a segmented sentence "中国人寿 于 2007年1月9日 正式 登陆 上海证券交易所 ， 股票代码 是 601628 。(China Life officially landed the Shanghai Stock Exchange in January 9, 2007, the number is 601628)".

The parsed result as follows, and Fig.1 shows the syntactic parse tree.

[3] http://www-nlp.stanford.edu/

```
(ROOT
  (IP
    (IP
      (NP (NR 中国人寿))
      (VP
        (PP (P 于)
          (NP (NT 2007年1月9日)))
        (ADVP (AD 正式))
        (VP (VV 登陆)
          (NP (NN 上海证券交易所)))))
      (PU ,)
      (IP
        (NP (NN 股票代码))
        (VP (VC 是)
          (QP (CD 601628))))
      (PU .)))
```

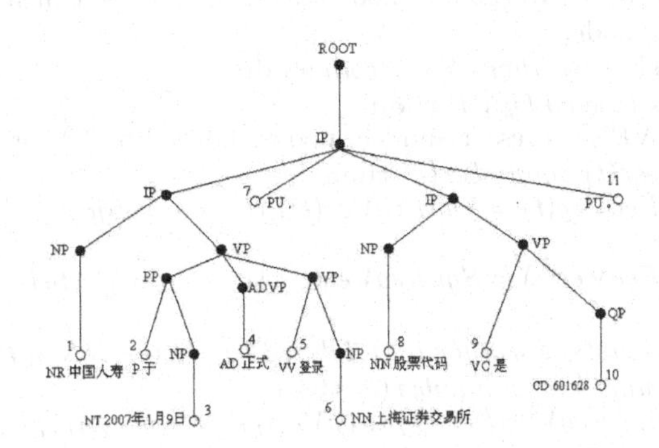

Fig. 1. The syntactic parse tree using Stanford JAVA NLP API

A parse tree contains kinds of information about syntactic relationship between words. In order to extract syntactic features from the parse tree, we first define three definitions and the process that how to extract syntactic features for a instance from a parse tree(see Alg2.).

Definition 1. *A node is a root inflexion, if it is the first IP node and it is a successor of root node.*

Definition 2. *A node is an inflexion, if it is a node labeled by IP and it is a successor of root inflexion. We call each branch of an inflexion a cluster, especially, a branch of root inflexion is root cluster. Two cluster having same inflexion are sibling clusters.*

Definition 3. *Two nodes are in a modificatory relationship, if the two node are in the same cluster or two sibling clusters. Accordingly, one node is a modifier of another node.*

Algorithm 2: Get Syntactic Features from the Parse tree(*GetSynFeature*)

```
Input  : PaserTree, an instance name t.
Output: SynFeaVec(t).
Traversal the parsing tree, find out the note where t occurs, and get the label and
number of t;
Get all modifiers of t;
for (each modifer m ∈ ModiferRelatedSetM) do
    if m is labeled DEG then
    |  ignore m node
    else
        if number(t) > number(m) then
        |  SynFeaVec(t) = SynFeaVec(t) ⋃ t − m − adj
        else
        |  SynFeaVec(t) = SynFeaVec(t) ⋃ t − m − adjof
        end
    end
end
Start from t, backtrack to the first node labeled by NP, and then get all brother
nodes of this NP node.
for (each brother ∈ Brothers Set Brother) do
    if brother is labeled by VP then
        find the VP's successor and leaf node v labeled by VV or VC;
        if number(t) > number(v) then
        |  SynFeaVec(t) = SynFeaVec(t) ⋃ t − v − subject
        else
        |  SynFeaVec(t) = SynFeaVec(t) ⋃ t − v − subjectof
        end
        if there exists a brother of v labeled by NP(or NN, QP, NT) then
            if number(t) > number(o) then
            |  SynFeaVec(t) = SynFeaVec(t) ⋃ t − o − objof + sub
            else
            |  SynFeaVec(t) = SynFeaVec(t) ⋃ t − o − subof + obj
            end
        end
    end
end
```

Feature Weight. Maximal Likelihood Ratio(MLHR) [18] is used for word co-occurrence mining in [19], while MLHR is used to calculate feature weight in our approach. That is because it is supposed to be outperforming in case of sparse data, a situation that may happen in case of questions with complex patterns that return small number of hits.

$$\begin{aligned} MLHR(t,i) &= -2\alpha \log \lambda \\ \lambda &= \frac{L(p,k_1,n_1)L(p,k_2,n_2)}{L(p_1,k_1,n_1)L(p_2,k_2,n_2)} \end{aligned} \tag{7}$$

where $L(p,k,n) = p^k(1-p)^{n-k}, p_1 = \frac{k_1}{n_1}, p_2 = \frac{k_2}{n_2}; p = \frac{k_1+k_2}{n_1+n_2}; k_1 = hits(i,t), k_2 = hits(i.-t); n_1 = hits(i), n_2 = hits(-t)$.

Here, $hits(i, -t)$ is the number of appearances of instance i when the feature t is not present and it is calculated as $hits(i) - hits(i, t)$. Similarly, $hits(-t)$ is the number of documents of the Corpus where t does not appear and it is calculated as $hits(-t) = Maxage - hits(t)(Marage$ is the number of documents in Corpus).

The α parameter reflects the linguistic intuition that nouns are most informative than all other, and verbs is more informative than adverbs or adjectives. $\alpha = 14$ in MLHR for syntactic features with lexical element noun and $\alpha = 1$ for all other syntactic features. The values of α is alterable according to different contexts.

Syntactic features with weights of a class are expressed in the form of a vector, and we share the same process in window-profile approach to obtain the similarity between a class and an instance to be classified.

5 Experimental Settings and Results

We evaluated the performance of context-profile based approaches for Ontology Population presented in section 4 on the stocks domain of the *CRAB* ontology. Our intention is to evaluate the proposed methods in populating the *CRAB* with new instances exploited from web documents. We conducted experiments concerning a given domain-specific corpus, which includes 8000 financial documents crawled from four portal web sites: Sina, Sohu, Yahoo, NetEase. The proposed method requires the pre-processing of the corpus by an Chinese word segmentation system which identifies text tokens in the Web pages. The corpus was parsed and syntactic profiles were extracted from the parse trees. We considered four classes: Company, Stock, Fund, Bond. For each class we created a training instance set.

Table 1. Performance of the Context-Profile approaches

Classes	Context-Profile	$P(\%)$	$R(\%)$	$F(\%)$
Company	Window-Profile	73	77	75
	Syntactic-Profile	78	84	81
Stock	Window-Profile	72	78	75
	Syntactic-Profile	76	82	79
Fund	Window-Profile	67	63	65
	Syntactic-Profile	72	68	70
Bond	Window-Profile	65	61	63
	Syntactic-Profile	70	66	68

The performance of the window-profile based approach and syntactic-profile based approach are evaluated separately using the precision and recall measures as well as their combination F measure. Table 1 shows results for 4 concepts chosen from the *CRAB* ontology. For each class, the first row shows the results of applying the window-profile based method, and the second row shows the results of the syntactic-profile based method. It is worth noting that the syntactic-profile based case performs better than the window-profile based case. For each

Table 2. Macro Average Performance of the Context-Profile approaches

Context-Profile	$MacroP(\%)$	$MacroR(\%)$	$MacroF(\%)$
Window-Profile	69	70	69
Syntactic-Profile	74	75	70

Table 3. Performance of the Context-Profile approaches

Classes	Context-Profile	$P(\%)$	$R(\%)$	$F(\%)$
Company	Window-Profile	73	77	75
	Syntactic-Profile	78	84	81
Stock	Window-Profile	72	78	75
	Syntactic-Profile	76	82	79
Fund	Window-Profile	67	63	65
	Syntactic-Profile	72	68	70
Bond	Window-Profile	65	61	63
	Syntactic-Profile	70	66	68

Table 4. The result with alternative α in the syntactic-profile approach

Classes	α	$P(\%)$	$R(\%)$	$F(\%)$
Company	6	62	68	65
	10	68	78	73
	14	78	84	81
Stock	6	63	70	66
	10	66	78	72
	14	76	82	79
Fund	6	56	52	54
	10	60	56	58
	14	72	68	70
Bond	6	54	50	52
	10	58	54	56
	14	70	66	68

approach we measured macro average precision, macro average recall, macro average F-measure and the result is in Table 2.

As we have already pointed out, that α could be referred to an heuristic parameter, which reflects the syntactic feature of a word and gives modificatory information to around words. We assigned different values to α, the comparative evaluation is shown in Table 3. The value of α is more higher, it is more easier to distinguish the word from its modifiers. Therefore, we get the best result when the value of α is 14.

6 Conclusion

In this paper we presented two kinds of context-profile based approaches for ontology population. Experimental results show that the context-profile based approaches enable as accurately and fast as to capture instances in application domains. It's is important to the *CRAB* system because *CRAB* Ontology plays a key role in the system and it is a core component. In our approaches, we aim to identify new instances for our *CRAB* Ontology by context-profiles of samples instances and testing instances. We populate the *CRAB* Ontology with these identified instances and put each instance into an appropriate place in the concept hiberarchy. The experimental results show our approaches are inspiring and gainful. Moreover, the *CRAB* will be enriched with more and more new instances, and it can provide more knowledge and information to users.

References

1. Bontcheva, K., Kiryakov, A., Cunningham, H., Popov, B., Dimitrov, M.: Semantic Web Enabled, Open Source Language Technology. In: Language Technology and the Semantic Web, Workshop on NLP and XML (NLPXML-2003), held in conjunction with EACL 2003, Budapest (2003)
2. Gruber, T.R.: A translation approach to portable ontology specifications. Knowledge Acquisition 5(2), 199–220 (1993)
3. Guarino, N.: Formal Ontology, Conceptual Analysis and Knowledge Representation. International Journal of Human and Computer Studies 43(5/6), 625–640 (1995)
4. Benjamins, V.R., Fensel, D.: The Ontological Engineering Initiative $(KA)^2$. In: Guarino, N. (ed.) Formal Ontology in Information Systems (this volume), IOS Press, Amsterdam (1998)
5. Poli, R.: Ontology and Knowledge Organization. In: ISKO 1996. Proceedings of 4th Conference of the International Society of Knowledge Organization, Washington (1996)
6. Guarino, N.: Formal Ontology and Information Systems. In: Guarino, N. (ed.) FOIS 1998. Proceedings of the 1st International Conference on Formal Ontologies in Information Systems, Trento, Italy, pp. 3–15. IOS Press, Amsterdam (1998)
7. Berners-Lee, T., Hendler, J., Lassila, O.: The semantic web. Scientific American 284(5), 34–43 (2001)
8. Bontcheva, K., Cunningham, H.: The Semantic Web: A New Opportunity and Challenge for HLT. In: ISWC. Proceedings of the Workshop HLT for the Semantic Web and Web Services (2003)
9. Lin, D.: Automatic Retrieval and Clustering of Similar Words. In: Proceedings of COLING-ACL 1998, Montreal, Canada (August 1998)
10. Almuhareb, A., Poesio, M.: Attributebase and value-based clustering: An evaluation. In: Proceedings of EMNLP 2004, Barcelona, Spain, pp. 158–165 (2004)
11. Buitelaar, P., Cimiano, P., Magnini, B. (eds.): Ontology Learning from Text: Methods, Evaluation and Applications. IOS Press, Amsterdam (2005)
12. Harris, Z.: Distributional structure. Word 10(23), 146–162 (1954)
13. Cimiano, P., Völker, J.: Towards large-scale, open-domain and ontology-based named entity classification. In: Proceedings of RANLP 2005, Borovets, Bulgaria, pp. 166–172 (2005)
14. Fleischman, M., Hovy, E.: Fine Grained Classification of Named Entities. In: Proceedings of COLING 2002, Taipei, Taiwan (August 2002)
15. Tanev, H., Magnini, B.: Weakly Supervised Approaches for Ontology Population. In: Proceedings of EACL-2006, Trento, Italy, 3-7 April (2006)
16. Hearst, M.: Automated Discovery of Word-Net Relations. In: WordNet: An Electronic Lexical Database, MIT Press, Cambridge (1998)

17. Velardi, P., Navigli, R., Cuchiarelli, A., Neri, F.: Evaluation of Ontolearn, a Methodology for Automatic Population of Domain Ontologies. In: Buitelaar, P., Cimiano, P., Magnini, B. (eds.) Ontology Learning from Text: Methods, Evaluation and Applications, IOS Press, Amsterdam (2005)
18. Magnini, B., Negri, M., Prevete, R., Tanev, H.: Is it the right answer? Exploiting web redundancy for answer validation. In: ACL- 2002. Proceedings of the 40th Annual Meeting of the Association for Computational Linguistics, Philadelphia, PA (July 2002)
19. Dunning, T.: Accurate Methods for the Statistics of Surprise and Coincidence. Computational Linguistics 19(1), 61–74 (1993)

Learning Dependency Model for AMP-Activated Protein Kinase Regulation

Yi-Ping Phoebe Chen[1,2], Qiumei Qin[1,3], and Qingfeng Chen[1]

[1] School of Engineering &Information Technology, Deakin University, Australia
[2] Australia Research Council (ARC) Centre in Bioinformatics, Australia
[3] Zhuhai State-owned Assets Supervision and Management Commission, China

Abstract. The AMP-activated protein kinase (AMPK) acts as a metabolic master switch regulating several intracellular systems. The effect of AMPK on muscle cellular energy status makes this protein a promising pharmacological target for disease treatment. With increasingly available AMPK regulation data, it is critical to develop an efficient way to analyze the data since this assists in further understanding AMPK pathways. Bayesian networks can play an important role in expressing the dependency and causality in the data. This paper aims to analyse the regulation data using B-Course, a powerful analysis tool to exploit several theoretically elaborate results in the fields of Bayesian and causal modelling, and discover a certain type of multivariate probabilistic dependencies. The identified dependency models are easier to understand in comparison with the traditional frequent patterns.

1 Introduction

AMP-activated protein kinase or AMPK plays an important role in cellular energy homeostasis [1, 2]. It consists of three subunits (α, β andγ) and regulates several intracellular systems. It has been proved that there is a close correlation between AMPK and mitochondria, glucose transport, lipid metabolism and adipocytokine [2]. Therefore, the investigation into the degree of activation and deactivation of subunit isoforms of AMPK can assist in better understanding of AMPK and the treatment of the type 2 diabetes, adiposity and other diseases [2, 3].

However, the efforts to develop methods to analyze AMPK regulation data lag behind the rate of data accumulation. Most of the current analysis rests on isolated discussion of single experimental results. Also, there has been a lack of systematic collection and interpretation of diverse AMPK regulation data. Besides, the existing approaches that seek to analyze biological data cannot cope with the AMPK regulation data that contains status messages of subunit isoforms and stimulus factors. This calls for the use of sophisticated computational techniques.

Recently, association rule mining was attempted to analyze the AMPK regulation data in [1]. It used Frequent Pattern Tree algorithm to identify the implicit, frequent patterns from the AMPK regulation data. However, not much work has been found to focus on the dependencies and causalities between the subunits and stimulus items of the AMPK regulation data. Hence, it is necessary to develop novel method to deal with the data.

Z. Zhang and J. Siekmann (Eds.): KSEM 2007, LNAI 4798, pp. 221–229, 2007.

Bayesian network is a model to present the dependencies and can effectively express causal relationship. Learning Bayesian networks is to find the best Bayesian network that models the observed correlations for giving data instances. WEKA [4] and B-Course [4, 11] are the popular Bayesian networks learning software. WEKA uses Hill Climbing algorithm, iteratively improve the best solution, Whereas B-Course uses Random and greedy heuristic robust learning algorithm. In comparison to WEKA, B-Course is an advanced analysis tool for interpretation and simple in use.

In this paper, we analyse the AMPK regulation data derived from the published experimental results by using the powerful tool, B-course, by which the dependency model, naive causal model and not-so-naive causal model are identified. The models express the dependencies and causalities between the subunits and stimulus items of the AMPK regulation data. They provide an easier way to understand the AMPK pathways in comparison with the traditional frequent patterns.

This paper is organized as follows. In section 2, the published experimental results of the AMPK regulation data was presented and the B-course approach was introduced as well. Section 3 shows the experimental results by using B-course. Conclusions and future work are discussed in section 4.

2 Data and Method

Bayesian networks are the directed acyclic graphs (DAGs) in which the nodes represent multi-valued variables and arcs signify the existence of direct causal influence between the linked variables [4,5]. For example, consider the dataset with 4 attributes: Cloudy(Yes/No), Sprinker(On/Off), Rain(Yes/No), WetGrass(Yes/No).

Table 1. A dataset with 4 attributes

	Cloudy	Rain	Sprinker	WetGrass
1	Yes	Yes	Off	Yes
2	No	No	On	Yes
3	No	No	Off	No
4	Yes	No	Off	Yes
5	Yes	No	On	Yes
6	Yes	No	Off	No
7	Yes	Yes	Off	Yes

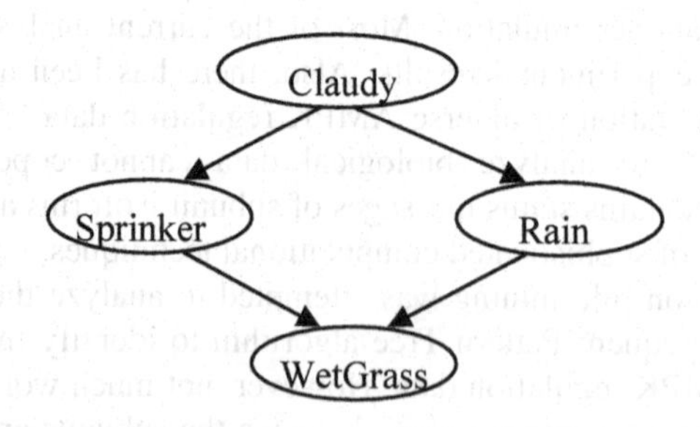

Fig. 1. A Bayesian networks representing the associations in Table 1

From the dataset, we can get the associations as follows, WetGrass correlates with Sprinkers On/Off, WetGrass correlates with Rain, Sprinker negatively correlates with Rain, etc. Hence, we get the Bayesian networks as in Figure 1.

Bayesian networks are attractive for their powerful ability to deal with the uncertainty. Recently they have been successfully applied to analyse expression data and predict the diseases [6-10].

B-Course is a free web-based online data analysis tool for Bayesian modelling (http://b-course.cs.helsinki.fi/obc/). It offers two kinds of trail, "C-trail" and "D-trail". "C-trail" data service offers a possibility for building classification models. "D-trail" offers a tool for building dependency models, it allows the users to analyse their data for multivariate probabilistic dependencies which are represented as Bayesian networks dependency model, naive causal model and not-so-naive causal model[11]. B-Course was to adopt the Bayesian framework which easier to understand than the classical frequent framework and has better performance than over the classical framework.

2.1 Data

In this paper we consider the AMPK regulation data which has been derived from the published experimental results, the same as in the paper *Mining frequent patterns for AMP- activated protein kinase regulation on skeletal muscle*[1]. AMPK contains catalytic subunit isoforms alpha1(α1) and alpha 2(α2) and regulatory subunit isoforms beta1(β1), beta2(β2), gamma1(γ1), gamma2(γ2) and gamma3(γ3). Plus the stimulus items and other parameters, there are seventeen variables in our database, namely, α1a, α1e, α1p, α2a, α2e, α2p, β1e, β2e, γ1e, γ2e, γ3e, Training, Glycogen, Diabetes, Nicotinic acid, Intensity, Duration. They are all the discrete variables. For α, β and γ subunits, there are four state items, "highly expressed", "expressed", "no change" and "no". For the stimulus items, the stimuli include "high intensity", "intense", "moderate" and "low".

2.2 Dependency Modeling Using Bayesian Networks

In B-Course, dependency modeling means finding the model of the probabilistic dependences of the variables. Let us look at a simple example to illustrate the type of models B-course searches for. There are four variables α1a, α1e, β1e and γ1e. The statements about dependencies were listed in Table 3.

Table 2. An Example of Experimental Database

EID	α1a	α1e	β1e	β2e	Intensity
E_1	no	expressed	No change	expressed	moderate
E_2	expressed	No	no	no	intense
E_3	Highly expressed	No	no	no	moderate
E_4	No change	Highly expressed	expressed	expressed	High intensity

In order to make the task of creating dependency model out of data computational feasible, B-Course makes two important restrictions to the set of dependency models it considers.

Table 3. An example of a dependency model

- ✧ α1a and α1e are dependent on each other if we know something about β1e or γ1e (or both).
- ✧ α1a and β1e are dependent on each other no matter what we know and what we don't know about α1e or γ1e (or both).
- ✧ α1e and β1e are dependent on each other on matter what we know and what we don't know about α1a or γ1e (or both).
- ✧ β1e and γ1e are dependent on each other no matter what we know and what we don't know about α1a or α1e (or both).

First, B-course only considers models for discrete data and it discretizes automatically all the variables that appear to be continuous. Secondly, B-Course only considers dependency models where the list of dependencies can be represented in a graphical format using Bayesian networks structures[11]. In this way, the list of dependencies in Table 3 can be represented as a Bayesian networks in Fig. 2.

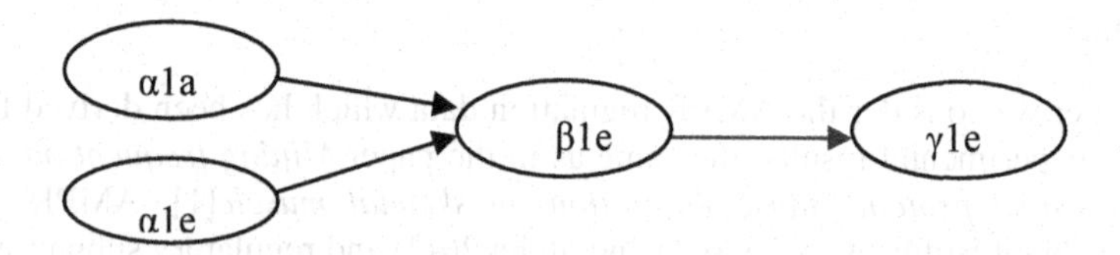

Fig. 2. A Bayesian network representing the list of dependencies in Table 2

2.3 Weighted Dependencies

It is natural to ask how probable or strong certain unconditional dependency statement represented by an arc in the dependency models. However, in non-liner models it is not an easy task to give "strengths" to arcs, since the dependencies between variables are determined by many arcs in a somewhat complicated manner. Furthermore, the strength of a dependency is conditional on what you know about the other variables.

Table 4. The different weights of dependencies in the model

Arcs	Arc weights(W)	explanation
α1e ➡ α2e	$W \geq 10^9$	The arc linked between α1e and α2e is so "strong" that removing it would cause the probability of the model go down to less than **one billionth** of the probability of original model. It means the arc is an important arc.
α1e ➡ β1e	$10^9 > W \geq 10^6$	The arc linked between α1e and β2e is an important arc that removing it would cause the probability of the model decrease to less than **one millionth** of that of the original model.
α1a ➡ α2a	35516	The model is 35516 times as probable as the model in which the arc has been removed.
α1a ➡ β1e	25	The arc linked between α1a and β1e is a "weaker" arc. The model is 25 times as probable as the model in which the arc has been removed.

Fortunately we can get a relatively simple measure of the importance of an arc by observing how much the probability of the model is changed by removing the arc. In the models, different arcs are used to describe the different probability. If the removal makes the model much worse (i.e., less probable), it can be considered as an important arc (dependency). If removing the arc does not affect the probability of the model much, it can be considered to be a weak dependency. Table 4 shows the different weights of dependencies in the model.

2.4 Inferring Causality

The current version of B-Course supports causal analysis of data and implements both "naive" and "not so naive" causal models. Naive causal models are built on assumptions that there are no latent (unmeasured) variables in the domain that causes the dependencies between variables. In naive causal model, there may be two kinds of connections between variables: undirected arcs and directed arcs. Directed arcs denote the causal influence from cause to effect and the undirected arcs denote the causal influence directionality of which cannot be automatically inferred from the data. Not so naive causal models restrict that every latent variable is a parent of exactly two observed variables, and none of the latent variables has parents. There are three kinds of arcs in not so naive causal models. Table 5 is the explanation.

Table 5. The different kinds of arcs in causal models

Arcs	Explanation
γ1e ⟶ β2e	γ1e has direct causal influence to β2e(direct meaning that causal influence is not mediated by any other variable that is included in the study)
α1e - - -➔ β1e	There are two possibilities, but we do not know which holds. Either α1e is cause of β1e or there is a latent cause for both α1e and β1e.
α1a - - - - α2a	There is a dependency but we do not know wether α1a causes α2a or if α2a causes α1a or if there is a latent cause of them both the dependency.

3 Discovering Dependencies Using B-Course

3.1 Experimental Results

"D-trail" has been chosen as our data analysis method. We uploaded our data file to the B-Course server and made the search last for 15 minutes (the system search time limit). Then we got the dependency model (Fig. 3) as follow.

In Fig. 3, different arcs represent the different weights of the arcs, details can be seen in Table 6.We also get the naive causal model (Fig. 4), which is built by the assumption that all the relevant variables of the domain have been included in the study, in other word, there are no unmeasured latent variables causing dependence. There may be two kinds of connections between variables: indirected arcs and directed arcs. Directed arcs denote the causal influence from cause to effect and the undirected arcs denote the causal influence directionality of which cannot be automatically inferred from the data. For example, under the situation of that there are no latent variables causing dependence, α2e was caused by α1a and α1e.In general all the dependencies between the variables might be caused by one latent variable.

Table 6. The arc weights of dependency model for AMPK

The kinds of the arcs	Arcs	Arc weights
So strong arcs	α1e → α2e	$W \geq 10^9$
	α2p → α1p	$W \geq 10^9$
Pretty strong arcs	α1e → β1e	$10^9 > W \geq 10^6$
	α1e → Training	$10^9 > W \geq 10^6$
	γ3e → γ2e	$10^9 > W \geq 10^6$
Less important arcs	Glycogen → Training	43365
	α1a → α2a	35516
	α1p → Training	7805
	β1e → Diabetes	3204
	β1e → Training	2294
	α1a → α2e	2110
Unimportant arcs	γ3e → γ1e	618
	α2p → Training	400
	γ1e → β2e	248
	β1e → γ3e	190
	α2p → Glycogen	70
	Training → Duration	69
	β1e → γ1e	68
	α2a → Diabetes	50
	α1a → β1e	25
	Training → γ3e	12
	α1e → γ3e	5.53
	Intensity → Glycogen	3.56
	γ3e → Duration	2.72
	Duration → Nicotinic acid	1.77
	γ3e → β2e	1.13
	γ2e → Diabetes	1.08
	β2e → Diabetes	1.05
	γ3e → Diabetes	1.05

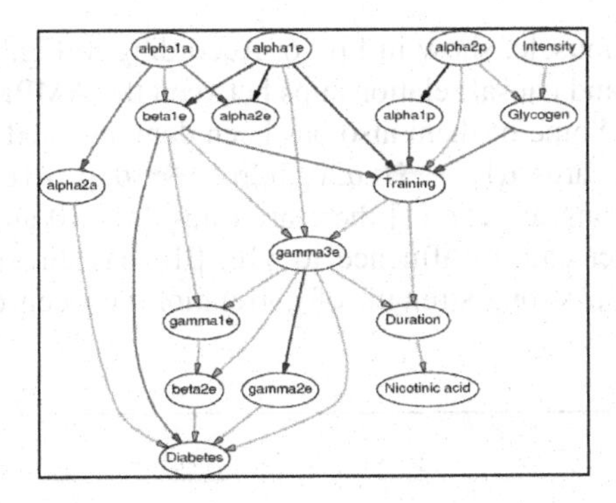

Fig. 3. The dependency model for AMPK

By restricting the latent variable that every latent variable is a parent of exactly two variables, and none of the latent variables has parents, we can get the "not-so naive causal models"(Fig.5). According to Fig. 5, we can get the following information:

- β1e has direct causal influence to diabetes, γ1e, γ3e, training.
- Training has direct causal influence to γ3e, duration.
- γ3e has direct causal influence to γ1e, β2e, γ2e, duration, diabetes.
- γ1e has direct causal influence to β2e.
- β2e has direct causal influence to diabetes.
- Duration has direct causal influence to nicotinic acid.

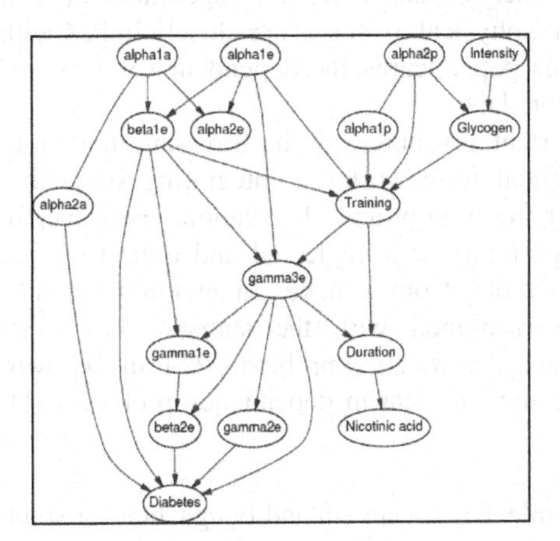

Fig. 4. The Bayesian networks naive causal model for AMPK

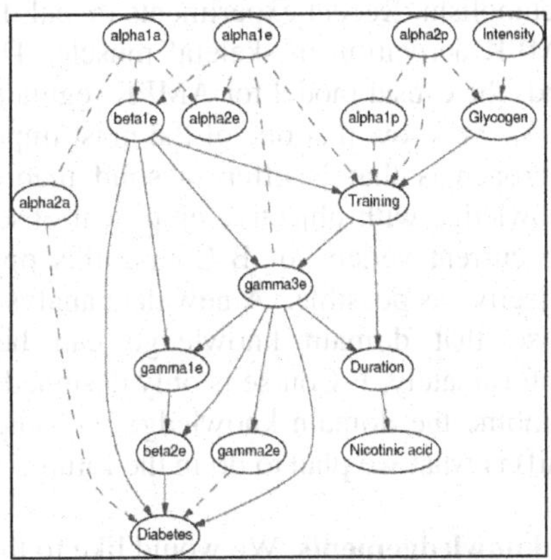

Fig. 5. The Bayesian causal model for AMPK

3.2 Interpretation

From the experimental results, it is obvious that the arcs linked with γ1e, β2e, duration and nicotinic acid respectively are unimportant arcs. Therefore, they can be moved from the networks to build the final Bayesian causal model to describe AMPK regulation in skeletal muscle, show in Fig. 6. According to Fig. 6 and Fig. 3, we can obtain the dependent and causal relationships between the AMPK isoform expression, subunits and activity. Some of them also has been demonstrated in recent experiment [1]. For example, the arcs $\alpha 1e \rightarrow \alpha 2e$, $\alpha 2p \rightarrow \alpha 1p$, $\alpha 1e \rightarrow \beta 1e$ and $\gamma 3e \rightarrow \gamma 2e$ are very strong arcs, in [1] there are rules 7, 8, 10 and 12. In addition, we see that $\gamma 3e$ has direct causal influence to $\gamma 1e$, $\beta 1e$ has direct causal influence to diabetes, and there is a strong strength of correlation between $\alpha 1a$ and training and so on.

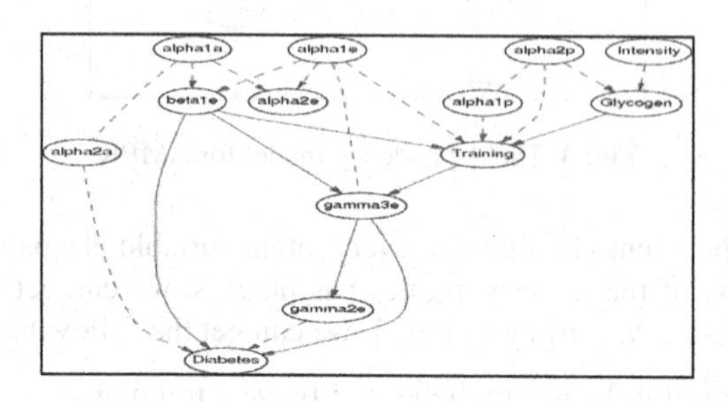

Fig. 6. The final Bayesian causal model for AMPK

4 Conclusions and Future Work

AMPK has emerged as an important energy sensor in the regulation of cell metabolism. Recent experiments reveal that physical exercises are closely linked with AMPK activation in skeletal muscle. This paper learns the dependency model and finds the causal model for AMPK regulation data.

As we know that one of the most important advantages of the Bayesian modelling approach is that is offers a solid theoretical framework for integrating subjective knowledge with objective empirical observation empirical observations. However in the current version of B-Course this opportunity was neglected and aims to be as objective as possible. A new data analysis tool P-Course makes an attempt to modify it so that domain knowledge can be combined with the statistic data [12]. Unfortunately, P-Course is only designed for classification problems (C-trail). How to combine the domain knowledge with the statistic data in dependency problems (D-trail) is what we plan to do in the future.

Acknowledgements. We would like to thanks Dr. Jiyuan An and Feng Chen for some useful suggestion on writing this paper.

References

1. Chen, Q., Chen, Y.P.P.: Mining Frequent Patterns for AMP-activated Protein Kinase Regulation on Skeletal Muscle. BMC Bioinformatics 7, 394 (2006)
2. AMP-activated protein kinase, Wikipedia, the free encyclopedia
3. Musi, N., Goodyear, L.J.: AMP-activated protein kinase and muscle glucose uptake. Acta Physiologica 178, 337–345 (2003)

4. Deforche, K.: Bayesian networks learning. In: The HIV Data Management and Data Mining Workshop (2004)
5. Pearl, J.: Probabilistic reasoning in intelligent systems: Networks of plausible inference. Morgan Kaufmann Publishers, San Mates (1988)
6. Friedman, N., Linial, M., Nachman, I., Pe'er, D.: Using Bayesian Networks to Analyze Expression Data
7. Fridman, N., Goldszmidt, M., Wyner, A.: Data Analysis with Bayesian networks A Bootstrap Approach
8. Voorbraak, F.: Uncertainty in AI and Bioinformatics. Linkoping Electronic Article in Computer and Information Science 3(7) (1998)
9. Helman, P., Veroff, R., Atlas, S.R., Willman, C.: A Bayesian networks classification methodology for gene expression data. Journal of computational biology 11(4) (2004)
10. Bottcher, S.G., Dethlefsen, C.: Prediction of the Insulin Sensitivity Index using Bayesian Networks
11. Myllymaki, P., Silander, T., Tirri, H., Uronen, P.: B-course: a web-based tool for Bayesian and causal data analysis. International Journal on Artificial Intelligence Tools 11(3), 369–387 (2002)
12. Lahtinen, J., Myllymaki, P., Ryynanen, O.-P.: P-Course: Medical Applications of Bayesian Classification with Informative Priors

Towards a Wrapper-Driven Ontology-Based Framework for Knowledge Extraction

Jigui Sun[1,2], Xi Bai[1,2], Zehai Li[1,2], Haiyan Che[1,2], and Huawen Liu[1,2]

[1] College of Computer Science and Technology, Jilin University, Changchun 130012, China
[2] Key Laboratory of Symbolic Computation and Knowledge Engineering of Ministry of Education, Changchun 130012, China
xibai@email.jlu.edu.cn

Abstract. Since Web resources are formatted in diverse ways for human viewing, the accuracy of extracting information is not satisfactory and, further, it is not convenient for users to query information extracted by traditional techniques. This paper proposes WebKER, a wrapper-driven system for extracting knowledge from Web pages in Chinese based on domain ontologies. Wrappers are first learned through suffix arrays. Based on HowNet, a novel approach is proposed to automatically align the raw data extracted by wrappers. Then knowledge is generated and described with Resource Description Framework (RDF) statements. After merged, knowledge is finally added to the Knowledge Base (KB). A prototype of WebKER is implemented and in the experiments, the performance of our system and the comparison between querying information stored in the KB and querying information extracted with traditional techniques are given, indicating the superiority of our system. In addition, the evaluation of the outstanding wrapper and the method for merging knowledge are also presented.

1 Introduction

With the emergence of the World Wide Web, making sense of large amount of information on the Web becomes increasingly important. Since most of the Web pages are written in the formatting language HTML nowadays, when users do some queries, the search engine cannot understand the semantics of the searching results or make the data sensible. Recently, Information Extraction (IE) techniques are more and more widely used for extracting relevant data from Web documents. The goal of an IE system is to find and link the relevant information while ignoring the extraneous and irrelevant one [2]. Among these techniques, wrappers are widely used for converting Web pages to structured data. However, the layouts of Web pages from different Websites are usually different. Moreover, the layout of a specific Web page may be updated from time to time. Therefore, to find out a generic method for automatically or semiautomatically generating wrappers becomes one of the hottest issues in information science community.

Several tools have already been applied in generating wrappers [3,6,7]. To the best of our knowledge, little work has been done to give complete frameworks or

Z. Zhang and J. Siekmann (Eds.): KSEM 2007, LNAI 4798, pp. 230–242, 2007.

systems for extracting knowledge form diverse Web pages in Chinese based on IE techniques and domain ontologies. Moreover, due to the lack of semantics, the traditional way of storing information is not convenient for users to query and sometimes the returned results are not ample. Motivated by the above analysis, in this paper, we expect to apply domain ontologies to assist users in extracting information from Web documents and then querying. We propose WebKER, a domain-ontology-based system for extracting knowledge from Chinese Web pages with wrappers generated based on pattern learning. Moreover, our system can be easily modified to deal with the Web documents in other languages if the corresponding ontologies and *Name Dictionary* are provided. The remainder of this paper is organized as follows. Section 2 describes the framework of WebKER. Section 3 describes the method of normalizing Web pages. Section 4 describes the raw-information-extraction process using wrappers based on pattern learning. Section 5 describes the knowledge-management process, including knowledge generation and knowledge merging. Section 6 gives the experiments in which the performance of our system and the comparison between querying information stored in our Knowledge Base (KB) and querying information stored in traditional database are presented. Section 7 describes the related works of recent information extraction techniques. Finally, some concluding remarks and our future research directions are described in section 8.

2 Framework

In this section, we describe the framework of WebKER. In Figure 1, *Retriever* block retrieves Web documents from the internet. Then these documents are transformed from ill-formed expression into well-formed expression with *Transformer* block. Two documents which are of the same layout are compared by

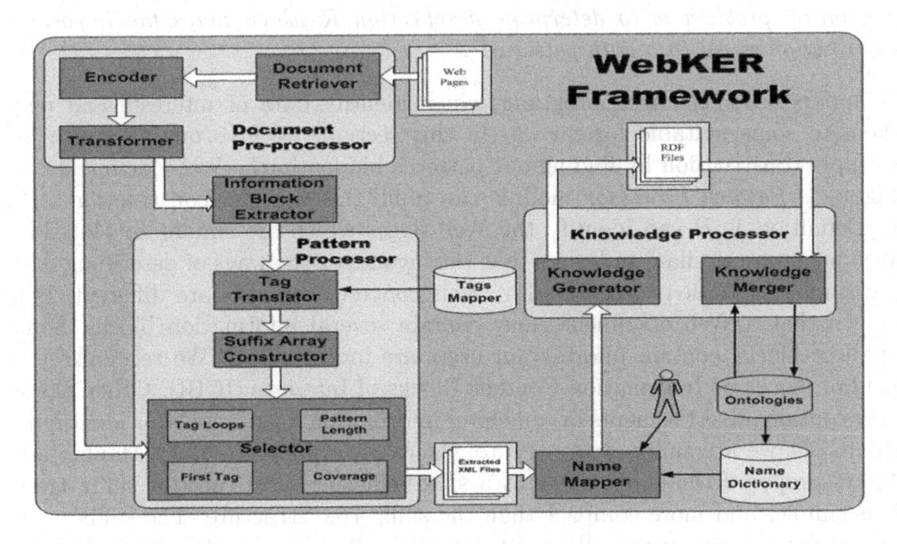

Fig. 1. Framework of WebKER

Information Block Extractor block and then the generated HTML snippet is sent to *Translator* block. According to the tag mapping table, tags are replaced with specific characters. *Suffix Array Constructor* block generates the repeated patterns hidden in the Web documents. According to the predefined criteria, *Selector* block finds out the most reasonable one from aforementioned repeated patterns. Then the properties and their values can be extracted and stored in XML files. Since the names of properties may not be identified by the domain ontologies, based on HowNet [19] and a *Name Dictionary*, these names are replaced with the predefined ones in domain ontologies by *Name Mapper* block. Finally, knowledge is generated and merged by *Knowledge Generator* block and *Knowledge Merger* block respectively before added to the KB.

3 Web Documents Preprocess

Most Web pages are written using HTML and much of their content is ill-formed. For instance, sometimes there is no close tag corresponding to an open tag. This will bring us troubles in the IE procedure when we need to determine the positions of interesting information. Therefore, we must first transform original Web documents from ill-formed expression into well-formed expression. XML expression is well-formed and XHTML is the extension of HTML based on XML. Thus we can transform the original HTML documents into XHTML documents to prepare for the extraction. Here, we use Tidy [18] to repair tags in original HTML documents. In Figure 1, *Document Pre-processor* module can fulfill this task.

4 Raw Information Extraction

Definition 1. *$\mathcal{W}$ is a Web page including information of interests. A wrappers generation problem is to determine a reflection $\mathcal{R}$ which maps the important information in $\mathcal{W}$ to a data pattern $\mathcal{P}$.*

Wrappers are specialized programs which identify data of interests and map them to some suitable format [8]. In this section, we give our approach for wrapper construction by finding out patterns hidden in the Web documents. In Figure 1, *Pattern Processor* module can fulfill this wrapper-construction task. In Definition 1, $\mathcal{R}$ can identify the Web pages which are similar to $\mathcal{W}$. "Two Web pages are similar" indicates that the layouts or the ways of describing data are similar in both pages, however, their concrete contents are different. It is known that a Web documents may contain several information blocks. Some of these blocks contain information users are interested in. We recognize this kind of blocks as Information Content Blocks of Interests (ICBI). Other blocks may contain advertisements or unhelpful information for users. After identifying the ICBIs, we use suffix arrays [17] to generate wrappers. In 1993, Manber and Myers proposed the suffix array as a space-efficient substitute for suffix trees. It is simpler and more compact than the suffix tree structure. The suffix array of a string is the array of all its suffixes sorted by lexicographically. Since each

Web document can be recognized as a string, we can use a suffix tree to store all the semi-infinite strings which is a substring of the text, defined by its starting position and continuing to the right as far as necessary to make the string unique. With the suffix array data structure to substitute for the suffix tree, we can analyze each sistring and discover helpful patterns. Actually, these patterns tell us how to construct wrappers to extract information from Web pages. Here, we give five steps to find out the most reasonable patterns as follows:

Step 1: Insignificant Block Elimination. There are some blocks in original HTML documents which do not contain any information. We first use regular expressions to match these blocks and then omit them. For instance, we can find out *Script* blocks using regular expression "$< SCRIPT[\hat{}>]*(> \mid \backslash n*)$".

Step 2: Information Block Extraction. In order to extract ICBIs, we compare each two lines of two documents generated from the same template (we can obtain these two documents by referring to their URLs). During the comparison, the line containing the same source codes is denoted by letter S(same) and the line containing different source codes is denoted by letter D(different). Then we can obtain a string composed of S and D. Through large amount of observation, we found that the number of S is usually much larger than that of D. Then the substring containing only one kind of characters is recognized as a lines group. If the number of S in a lines group exceeds a predefined threshold (20 characters), the corresponding source codes in the Web page will be omitted. Otherwise, the source codes will be retained.

Step 3: Suffix Array Construction. The distribution of tags can reflect the pattern of information in Web documents. We do not concern about the attributes within tags here. We omit them and only retain the tag names. Moreover, *Tag Translator* block replaces the literal content between two tags with a new tag </txt> and maps each tag to an identical letter. Now, we can get a string of letters representing tags in the Web document. Then based on algorithm described in [17], the suffix array of this string is constructed by *Suffix Array Constructor* block as shown in Figure 1.

Step 4: Pattern Discovery. Based on the generated suffix array, we expect to find out the maximal repeats hidden in its corresponding Web document. Suppose this array is denoted by SA and has n elements. We define an auxiliary array A with $n+1$ dimensions. $A[i]$ contains the length of the longest common prefix between adjacent suffixes $SA[i-1]$ and $SA[i]$, where $0 < i < n$. To simplify the algorithm, the first element $A[0]$ and the last element $A[n]$ are both equal to zero. We also define another array SP to store the start positions of the above common prefixes. The algorithm for generating array A and array SP are shown in Algorithm 1. According to array SA and array A, we can get the repeated substrings in the Web document string. By mapping them back to original tags we can obtain hidden patterns. For instance, a substring is a repeated patterns if it starts with the position $SP[i]$ and its length is $A[i+1]$, where $0 \leqslant i < n$.

Step 5: Pattern Selection. Since most of the extracted patterns are incomplete or redundant and some of them are even invalid, we should find out the

Algorithm 1. Patterns discovery algorithm

Input: The suffix array *sa* for a Web document string denoted by an array *d*
Output: The maximal common suffixes(patterns)

1. Initialize the auxiliary array a;
2. **for**(i = 0; i < the length of *sa*; i++) {
   2.1 Store the start position of sa_{i-1} of d in $left\_idx$;
   2.2 Store the start position of sa_i of d in $right\_idx$;
   2.3 Initialize a counter: $count = 0$;
   2.4 **while**($left\_idx <$ the length of d && $right\_idx <$ the length of d)
      2.4.1 **if**($d_{left\_idx} == d_{right\_idx}${
         2.4.1.1 $count = count + 1$; $left\_idx = left\_idx + 1$; $right\_idx = right\_idx + 1$;}
         else{2.4.1.2 $a_i = count$;**break**;}
   }//generating the auxiliary array A
3. Initialize the serial-number array $sp = b_1, b_2, ..., b_n$ and a temporary stack s;
4. Initialize the two integers: $t = 0, u = 0$;
5. **for**(int i = 0; i < the length of *sp*+1; i++) {
   5.1. **if**(*sp* is empty && $a_i > 0$)
      5.1.1. s.*push*(i);
      else{
         5.1.2. Store the top element of s in t;
         5.1.3. **if**($a_t < a_i$)5.1.3.1. Push i into s;
            else if($a_t == a_i$)5.1.3.2. **continue**;
            else{5.1.3.3. $sp_u = s.pop()$; $u = u + 1$; $i = i - 1$;}}
   }//generating the serial-number array *SP*

criteria to select an outstanding pattern. Here, "outstanding" means that with this pattern we can extract most interesting information from a specific kind of Web documents. In Figure 1, *Selector* block can fulfill this task. We give our patterns-selection method based on some of their features as follows:

- **F1: Tag Loop.** Based on the above pattern-discovery method, inevitably, some patterns contain tags loops. For instance, "<td><txt/></td><td>< txt/></td><td>" and "<td><txt/></td>" are equivalent patterns within the information-extraction process. This kind of longer patterns should be simplified firstly.
- **F2: Pattern Length.** Sometimes, the length of a pattern is too short (less than three or four tags). This kind of patterns can not match much of ICBI and the extracted information is usually incomplete, since they are incomplete. Therefore, if the length of a pattern dose not exceed the predefined threshold, we abandon it.
- **F3: First Tag.** A complete and valid pattern should start with an opening tag (like $< TagName >$) or a single tag (like $< TagName/ >$). Therefore, we omit the pattern starting with a closing tags (like $< /TagName >$).
- **F4: Coverage Percentage.** Some patterns can only cover small part of the string in the ICBI. Here, we give a function to calculate the Coverage percentage of patterns. Suppose there is a document string S and a candidate

pattern P. After matching, there are k break points $point_1, point_2, \ldots, point_k$. Each break point indicates a start position of the substring that this pattern matches. The Coverage percentage CP is defined as follows:

$$CP = \frac{k \times Length(P)}{Length(S)} \tag{1}$$

Here, $Length(X)$ denotes the length of string X. If the covering percentage of a pattern exceeds a predefined threshold, we retain it. Otherwise, we abandon it.

Through the above selections, we can get a more reasonable pattern for the ICBI. Of course, users can modify the pattern manually if necessary. We use a regular expression to describe this outstanding pattern and recognize it as the wrapper to extract raw data from its corresponding Web documents. The properties and their values of extracted individuals are saved in XML files finally.

5 Knowledge Generation and Merging

In this section, we give our approach for generating and merging knowledge. In Figure 1, *Knowledge Processor* module fulfills this task.

5.1 Names Mapping

The data newly extracted in Section 4 is raw since the names of extracted properties are the original names in Web pages and one property may have several synonymous names on different Websites. Therefore, before generating knowledge statements, we should map the original names of properties to the names predefined in domain ontologies. This process is named as *name mapping* and *Name Mapper* block fulfills this task.

HowNet is a bilingual general knowledge-base describing relations between concepts and relations between the attributes of concepts in Chinese and English [19]. Based on Hownet, we establish a *Names Dictionary* (*ND*) to fulfill the *name mapping*. *ND* is actually a table with two columns and each record appearing in the *ND* denotes a resource (class, property or individual). The resources' names predefined in the domain ontologies are listed in the left column (remarked as *OntName*). The corresponding synonymous names of resources from various Websites are listed in the right column. Figure 1 shows that synonymous names can be accumulated manually according to the users' specific needs if these names are not included in HowNet. Within the mapping process, the program will check the name of each resource stored in the XML files generated in section 4. If a specific resource's name is found in the right column of *ND*, it will be replaced with the corresponding name predefined in the domain ontologies. Otherwise, users will be invoked to decide whether the resource's corresponding ontology exists in the domain ontologies or not. If it is a known resource, the name will be recognized as a new synonymous name and replaced with its corresponding *OntName*. Otherwise, it will be abandoned finally. In order to get high efficiency of consulting *ND*, we use Binary Balance Tree (BBT) data structure to rearrange and store the resources' names.

5.2 Knowledge Generation

In this subsection, we use Resource Description Framework (RDF) [20] statements to represent knowledge. In RDF statements, things have properties which have values. The knowledge can be described as follows.

Definition 2. *The knowledge $\mathcal{K}$ is denoted by a 3 tuple $\mathcal{K}$=(subject, predicate, object):*

a. *subject is the part that identifies the thing that the statement is about;*
b. *predicate is the part that identifies the property or characteristic of the subject that the statement specifies;*
c. *object is the part that identifies the value of that property.*

Based on Definition 2, extracting information from Web pages is actually a process of extracting subjects, predicates and objects. Moreover, through large amount of observation, we find that predicates and their values often appear pairwise in structured content. Therefore, we can easily extract the names of properties and identify the subject and the object of each property by referring to domain ontologies. Ontologies can be expressed with RDF graphs and Jena [21] can query them easily. By querying the domain and the range of a specific property, we can identify the subject and the object of a specific property respectively. Then we can obtain the XML file which contains all the properties and their values using the wrapper generated in section 4. We also identify the type of each property before we get the final RDF statements. In the XML file, the properties can be extracted in the depth-first search manner. For each node in the XML tree structure, the domain and the range of its corresponding property are investigated respectively. Thereafter, we can get all the properties by referring to the domain ontologies and knowledge is generated by *Knowledge Generator* block.

Algorithm 2 describes the process for generating RDF statements. Noteworthy is that there may be some relationships between newly generated subjects. For instance, some subjects may be redundant resources which should be excluded or some subjects may be linked to other subjects through some properties predefined in the domain ontologies.

5.3 Merge Newly Generated Individuals in KB

When a new individual is generated, before adding it to the KB, we should check whether the individual has already been included in KB or not. This process is named as *individual aligning.* It is known that the same individual is usually described in different ways on different Websites. However, the RDF statement describing an specific individual should be unique in KB. Suppose ind_A denotes an individual which has already been stored in KB and ind_B denotes a newly generated individual. The set of ind_A's properties P_A is denoted by $\{Pind_{A1}, Pind_{A2}, ..., Pind_{An}\}$; the set of ind_B's properties P_B is denoted by

Algorithm 2. Knowledge generation algorithm

Input: The properties *pro* appearing in the mapped XML files and the domain ontologies *ont*.
Output: The RDF files which contain the extracted knowledge.

1. **for**(i = 0; i < the number of *pro*; i++){
   1.1. Generate a new individual sub_i;
   1.2. **if**(pro_i is an object property){
      1.2.1 Generate a new individual obj using the class of pro_i's range;
      1.2.2 Add the property pro_i and the value obj to sub_i;}
      else if(pro_i is a data property){
      1.2.3 Add the property pro_i and its value to sub_i;}
   1.3. Link the new individual sub_i to the individuals in *ont*;
   1.4. Merge the new individual sub_i and the individuals in *ont*;
   }
2. Return the RDF statements in *ont*;

$\{Pind_{B1}, Pind_{B2}, ..., Pind_{Bm}\}$. Here, we recognize the values of properties as strings. Since a string can be recognized as a set of characters or digits, we calculate the similarity of string α and string β by the function $simSTR(\alpha, \beta)$ defined as follows:

$$simSTR(\alpha, \beta) = \frac{\alpha \bigcap \beta}{min(|\alpha|, |\beta|)} \tag{2}$$

Here, $|\alpha|$ and $|\beta|$ denote the length of string α and string β respectively. If the similarity exceeds a predefined threshold 0.75, two strings will be recognized as similar strings. We give the following definition to decide whether two individuals are equivalent or not.

Definition 3. *ind_A and ind_B are equivalent iff their similarity simIND (ind_A, ind_B)$\geq \eta$, and*

$$simIND(ind_A, ind_B) = \frac{\sum_{k=1}^{|P_A \bigcap P_B|} simSTR(pv_A(P_k), pv_B(P_k))}{|P_A \bigcap P_B|} \tag{3}$$

Here, η denotes a predefined threshold; $pv_A(P_k)$ denotes the value of a specific property in ind_A and it is the kth property in set $P_A \bigcap P_B$; $pv_B(P_k)$ denotes the value of a specific property in ind_B and it is the kth property in set $P_A \bigcap P_B$. $|P_A \bigcap P_B|$ denotes the length of set $P_A \bigcap P_B$.

Based on Definition 3, we can decide whether two individuals are equivalent or not. If there is an individual stored in KB satisfying that the similarity between it and the newly generated individual exceeds the threshold η, they will be recognized as equivalent individuals and the new properties will be linked to the individual in KB. Otherwise, the newly generated individual and its properties will be added to KB. The above merging task can be fulfilled by *Knowledge Merger* block.

6 Experiments

The experiments for our approach are carried out on a PC with an Intel Pentium 1.8GHz CPU and 512MB main memory. In order to provide maximum transparency of our approach's performance, we gather a sample set of Web documents randomly from Sina, Sohu, Yahoo China, NetEase, Tencent which are large Websites in Chinese language covering a lot of information about finance, and HengXin, AllInOneNet, HeXun and GuTianXia which are professional Websites about negotiable securities. From each Website, we retrieve 20 documents and based on them, we expect to extract information about corporations which have come into the market.

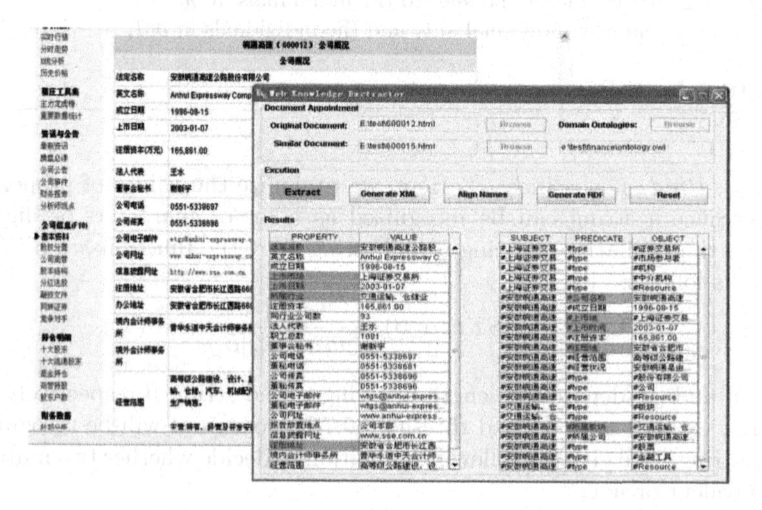

Fig. 2. Excerpt of a Web page and a snapshot of WebKER's extraction window

We developed a graphical user interface WebKER (**Web** **K**nowledge **E**xtract-o**R**) for extracting Web knowledge using the Java language. Figure 2 describes an excerpt of a Web page associated with the basic information for "安徽皖通高速公路股份有限公司" (Anhui Expressway Company Limited) from Yahoo China and shows a screen snapshot for our interface after knowledge extraction from this Web page. We evaluate the performance of our approach by calculating *Recall*(R), *Precision*(P) and *F-measure*(F). Assume that the individuals extracted by domain experts are sound and complete. *Recall* and *Precision* are defined as follows:

$$R = \frac{|S_{IES} \bigcap S_{DE}|}{|S_{DE}|} \times 100\%, \quad P = \frac{|S_{IES} \bigcap S_{DE}|}{|S_{IES}|} \times 100\% \tag{4}$$

Here, S_{IES} denotes the set of individuals extracted by our IE system and S_{DE} denotes the set of individuals extracted by domain experts manually. *F-measure* is another way of evaluating the extraction performance, defined as follows:

$$F = \frac{(\beta^2 + 1)RP}{R + \beta^2 P} \times 100\% \quad (5)$$

Here, β is a predefined value which adjusts the importance degrees of *Recall* and *Precision*. After normalized, all the "broken" tags in this document are repaired and then its corresponding suffix array is established. Then wrappers can be generated and used for extracting raw data. We can get the XML file describing raw information in this Web page. Based on *ND*, we map the resources' original names to *ONT names* and creates a new version of the XML file to describe this corporation. See the interface snapshot in Figure 2, the names with red background in the bottom left table are replaced with the ones with blue background in the bottom right table. Now, the mapped XML file can be used for generating knowledge based on the domain ontologies. According to Algorithm 2 described in subsection 5.2, the individuals can be extracted and finally described by RDF statements. After merged, these individuals can be added to the KB directly, since the knowledge in the KB are also described by RDF statements. Table 1 shows the performance of extracting the knowledge from the aforementioned Websites using our system.

Table 1. Performance of individual extraction

	Websites	URL	Individuals Extracted By DE	Performance				
				Cor.	Err.	Prec.	Rec.	F1
S1	Sina	www.sina.com	141	141	19	88.13%	100%	93.69%
S2	Sohu	www.sohu.com	100	100	20	80.00%	100%	88.89%
S3	Yahoo China	www.yahoo.com.cn	125	125	15	89.29%	100%	94.34%
S4	NetEase	www.163.com	120	0	0	0%	0%	0%
S5	Tencent	www.qq.com	58	58	22	65.00%	100%	78.79%
S6	HengXin	www.hengx.com	99	99	1	99.00%	100%	99.50%
S7	AllInOneNet	info.cmbchina.com	71	71	9	86.25%	100%	92.62%
S8	HeXun	www.hexun.com	100	100	0	100%	100%	100%
S9	GuTianXia	www.gutx.com	80	60	0	100%	75.00%	85.71%

From Table 1, we can see that for most Websites, *Precisions*, *Recalls* and *F-measures* are satisfactory. The extraction from NetEase fails, since the content of the ICBI in the documents on this Website is generated dynamically with JavaScript. So our system is helpless to the case that helpful information is not stored in the inputed HTML files. The recall for GuTianXia is lower than those for other Websites' since the values of a specific properties are missed in all pages. However, the Table 1 indicates that our approach can be applied to most of Websites well.

Traditional methods store information extracted by wrappers in a database for users's queries. This task can be easily fulfilled by saving the generated XML files into the relational database. Here, we compare the performance of querying the database with that of querying our KB. Focusing on five Websites, we ask 40 users to generate 120 queries information about these corporations.

Table 2. Querying database versus querying the KB

	Websites	Ques.	Performance For DB Case					Performance For KB Case				
			Cor.	Err.	Prec.	Rec.	F1	Cor.	Err.	Prec.	Rec.	F1
S1	Sina	120	67	9	88.16%	55.83%	68.37%	91	15	85.85%	75.83%	80.53%
S2	Sohu	120	72	14	83.72%	60.00%	69.9%	99	0	100%	82.50%	90.41%
S3	Yahoo China	120	57	7	89.06%	47.50%	61.96%	88	20	81.48%	73.33%	77.19%
S6	HengXin	120	49	1	98.00%	40.83%	57.64%	119	0	100%	99.17%	99.58%
S8	HeXun	120	52	1	98.11%	43.33%	60.11%	120	0	100%	100%	100%

Table 2 shows the results of this comparison. In this table, we can see that the *Recalls* of querying the KB are all higher comparing to those of querying the traditional database. The higher values of *F-measures* indicate the superiority of our domain-ontology-aided system.

7 Related Work

IE techniques are used for identifying and extracting specific resource fragments from corpus. Some related works have addressed the IE problem [1,3,6,7,10]. With the growth of the amount of information, the utilization of robust Web IE tools becomes necessary. Traditional IE concentrates on domains consisting of grammatical prose. Some researchers try to extract information using databases. It is known that the data in database are structured. However, most data on the Web are semi-structured or unstructured. So the conventional database cannot do the extraction work and some methods are proposed to extract data from Web sources to populate databases for further handling [4]. Some IE systems have also been proposed, which are usually based on wrappers and take a single approach or attack a particular kind of domain [3,6,7]. At the beginning, wrapper generation is tedious and time-consuming since it is usually done manually. Recently, some wrapper induction systems are proposed by researchers. Generally speaking, most of the recently developed wrappers are usually based on query languages [9], HTML structure analysis [11], grammar rules [12], machine learning [13], data modeling [14], ontologies [15], etc. The lack of homogeneity in the structure of the source data in the Websites becomes a most obvious problem in designing a system for Web IE. To the best of our knowledge, little work has been done to give complete frameworks to extract knowledge form diverse Web pages in Chinese.

8 Conclusions and Future Work

In this paper we propose an IE system WebKER based on suffix arrays and domain ontologies. It can automatically learn wrappers from Web documents and generate knowledge for users to easily query. Experiments on documents from large Chinese Websites yielded promising results. Our current-research goal focuses on improving our pattern processor by adding heuristics in the pattern discovery and selection. The performance of our system is based on the soundness and the completeness of the interrelated domain ontologies, and our long term goal is to automatically or semiautomatically extend the domain ontologies dynamically based on the raw information extracted with wrappers.

Acknowledgements

This work is supported by the National NSF of China under Grant No. 60496321, and Ministry of Education Program for New Century Excellent Talents in University (NECT).

References

1. Pinto, D., McCallum, A., Wei, X., Croft, W.B: Table Extraction Using Conditional Random Fields. In: Proceedings of the SIGIR 2003, pp. 235–242. ACM Press, New York (2003)
2. Cowie, J., Lehnert, W.: Information Extraction. Communications of the ACM 39, 80–91 (1996)
3. Cohen, W.W., Hurst, M., Jensen, L.S.: A Flexible Learning System for Wrapping Tables and Lists in HTML Documents. In: Proceedings of the WWW 2002, pp. 232–241. ACM Press, New York (2002)
4. Florescu, D., Levy, A.Y., Mendelzon, A.O.: Database Techniques for the World-Wide Web: A Survey. SIGMOD Record 27, 59–74 (1998)
5. Soderland, S.: Learning to Extract Text-based Information from the World Wide Web. In: Proceedings of the KDD 1997, pp. 251–254. Springer, Heidelberg (1997)
6. McDowell, L.K., Cafarella, M.: Ontology-driven Information Extraction with OntoSyphon. In: Cruz, I., Decker, S., Allemang, D., Preist, C., Schwabe, D., Mika, P., Uschold, M., Aroyo, L. (eds.) ISWC 2006. LNCS, vol. 4273, pp. 428–444. Springer, Heidelberg (2006)
7. Welty, C., Murdock, J.W.: Towards Knowledge Acquisition from Information Extraction. In: Cruz, I., Decker, S., Allemang, D., Preist, C., Schwabe, D., Mika, P., Uschold, M., Aroyo, L. (eds.) ISWC 2006. LNCS, vol. 4273, pp. 709–722. Springer, Heidelberg (2006)
8. Kushmerick, N.: Wrapper Induction for Information Extraction. Technical Report UW-CSE-97-11-04, University of Washington (1997)
9. Habegger, B., Quafafou, M.: WetDL: A Web Information Extraction Language. In: Yakhno, T. (ed.) ADVIS 2004. LNCS, vol. 3261, pp. 128–138. Springer, Heidelberg (2004)
10. Zhai, Y.H., Liu, B.: Web Data Extraction Based on Partial Tree Alignment. In: Proceedings of the WWW 2005, pp. 76–85. ACM Press, New York (2005)
11. Pek, E.H., Li, X., Liu, Y.Z.: Web Wrapper Validation. In: Goos, G., Hartmanis, J., Leeuwen, J.V. (eds.) APWeb 2003. LNCS, vol. 2642, pp. 388–393. Springer, Heidelberg (2003)
12. Chidlovskii, B., Ragetli, J., Rijke, M.D.: Wrapper Generation Via Grammar Induction. In: López de Mántaras, R., Plaza, E. (eds.) ECML 2000. LNCS (LNAI), vol. 1810, pp. 96–108. Springer, Heidelberg (2000)

13. Habegger, B., Debarbieux, D.: Integrating Data from the Web by Machine-Learning Tree-Pattern Queries. In: Meersman, R., Tari, Z. (eds.) OTM 2006. LNCS, vol. 4275, pp. 941–948. Springer, Heidelberg (2006)
14. Deng, X.B., Zhu, Y.Y.: L-Tree Match: A New Data Extraction Model and Algorithm for Huge Text Stream with Noises. Computer Science and Technology 20, 763–773 (2006)
15. Schindler, C., Arya, P., Rath, A., Slany, W.: HtmlButler–Wrapper Usability Enhancement Through Ontology Sharing and Large Scale Cooperation. Adaptive and Personalized Semantic Web 14, 85–94 (2006)
16. Lewis, D.D.: Naive Bayes at Forty: the Independence Assumption in Information Retrieval. In: Carbonell, J.G., Siekmann, J. (eds.) ECML 1998. LNCS, vol. 1398, pp. 4–5. Springer, Heidelberg (1998)
17. Manber, U., Myers, G.: Suffix Arrays: A New Method for On-Line Search. SIAM Journal on Computing 22, 935–948 (1993)
18. HTML Tidy Project, `http://www.w3.org/People/Raggett/tidy`
19. Gan, K.W., Wong, P.W.: Annotating Information Structures in Chinese Texts Using HowNet. In: Palmer, M., Marcus, M., Joshi, A., Xia, F. (eds.) Proceedings of the second workshop on Chinese language processing, pp. 85–92 (2000)
20. RDF Primer, `http://www.w3.org/TR/rdf-primer`
21. Jena Semantic Web Toolkit, `http://www.hpl.hp.com/semweb/jena.htm`

A Google-Based Statistical Acquisition Model of Chinese Lexical Concepts*

Jiayu Zhou[1], Shi Wang[2,3], and Cungen Cao[2]

[1] Department of Computer Information and Technology, Beijing Jiaotong University, Beijing China, 100044
zhoujiayu@computer.org
[2] Institute of Computing Technology, Chinese Academy of Sciences, Beijing China, 100080
wangshi_frock@hotmail.com,
cgcao@ict.ac.cn
[3] Graduate School of Chinese Academy of Sciences, Beijing China, 100049

Abstract. In this paper we propose a statistical model of Chinese lexical concepts based on the Google search engine and distinguish concepts from chunks using it. Firstly, we learn *concept boundary words* which can be seen as the statistical feature of concepts in large-scale corpora using Google. The instinctive linguistics hypothesis we believe is that if a chunk is a lexical concept, there must be some certain "intimate" words abut on its "head" and its "tail". Secondly, we construct a classifier according to the *concept boundary words* and then distinguish concepts from chunks by it. We consider the conditional probability, the frequency and the entropy of the *concept boundary words* and propose three attributes models to build the classifier. Experiments are designed to compare the three classifiers and show the best method can validate concepts with an accuracy rate of 90.661%.

Keywords: concept acquisition, knowledge discovery, text mining.

1 Introduction

Concepts play an essential role in human knowledge. In natural languages, concepts are expressed by lexical words called lexical concepts [1]. Extracting lexical concepts from large-scale unstructured text has become an important task in knowledge acquisition nowadays and is also helpful for information extraction, ontology learning, information retrieval, etc. It is an urgent task and great challenge to learn concepts from text automatically.

Recent related researches primarily center on term recognition. Terms are lexical concepts that are used in special domains, extraction of which is based on domain-specific corpora. Typical approaches are based on linguistic rules [2,3,4], statistical analysis [5,6,7] and a combination of both [8,9]. The accuracy of linguistic rules based approaches is very high, but they do not work well for Chinese corpora, because the

* This work is supported by the National Natural Science Foundation of China under Grant No. 60496326, 60573063, and 60573064; the National 863 Program under Grant No. 2007AA01Z325.

Z. Zhang and J. Siekmann (Eds.): KSEM 2007, LNAI 4798, pp. 243–254, 2007.

corpora need to be segmented, which is a real hard work. Comparing with linguistic rules, statistic methods are easier because of the independency of lexical structure and grammar. Frequency of occurrence, Mutual Information [5], Log-likelihood ration [6,7] are commonly used statistical devices. Besides, the methods are not restricted in a special domain. The combined methods are proven to be effective recently, especially in the field of Chinese term extraction [8,9].

Google has been used to extract similarities between words and phrases as normalized Google distance (NGD) based on information distance and Kolmogorov complexity in recent research[11,12,13], through which the World Wide Web (WWW) can be used as a large database with automatic semantics of useful quality. Because of the unique linguistic characteristics of Chinese language, methods of NGD cannot be directly applied to Chinese term. However, we can take advantage of the idea to serve the concept learning.

In our research, we realize that concepts are not just terms. Term is domain specific while concepts are general-purposed. Furthermore, terms are just restricted to several kinds of concepts such as named entities. So we cannot use the methods of term recognition directly to learn concepts from text even though from which we can benefit a lot.

In this paper, we propose a statistical acquisition model of Chinese lexical concepts based on Google. We achieve this by two statistical learning processes. Firstly, we learn *concept boundary words* which can be seen as the statistical features of concepts in large-scale corpora. The basic idea is that if a chunk is a lexical concept, there must be some certain "intimate" words in its context as its "head" (i.e. before the chunk) and its "tail" (i.e. after the chunk). Secondly, we construct a classifier according to the *concept boundary words* and then identify concepts from chunks using the classifier. The second step is to build a decision tree model which can finally tell concepts from other chunks based on the boundary words. This paper develops three different experiments to find the optimized classifier model considering the conditional probability, the frequency and the entropy of concept boundary words. The best method can validate concepts with an accuracy rate of 90.661%. Furthermore, our method does not require the corpora be segmented and pos-tagged, and therefore is domain independent.

The paper is organized as follows. Section 2 proposes our hypothesis concerning the Chinese lexical concept statistical feature and defines word associations and boundary words. In Sect.3 we describe the boundary words learning methods and an algorithm of building concept classifiers. Experimental results and comparisons are given in Sect.4. Section 5 concludes the paper.

2 Acquisition Model

Before presenting our method, we introduce a linguistics hypothesis which is the basis of our method:

Hypothesis 1. *If a chunk is a lexical concept, there must be some certain "intimate" words abut on its "head" and its "tail" in a corpus.*

Once such words are found, we can take advantage of Google to index the association of a certain chunk with patterns in Google and then determine whether the chunk is a concept or not. Figure 1 shows the acquisition process of concepts in our method.

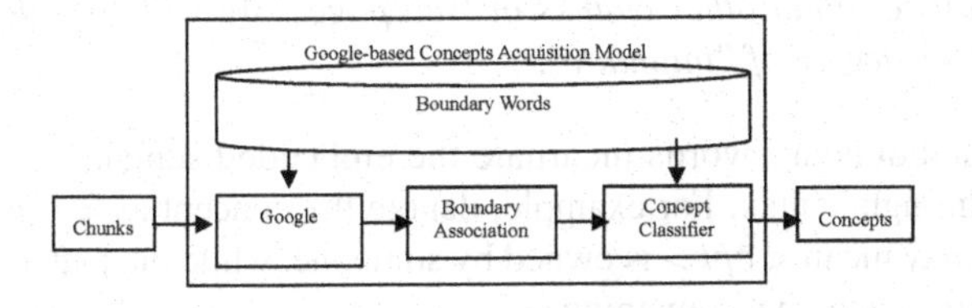

Fig. 1. Acquisition Model

We used selected Chinese lexical chunks as the input, Google then was used to run out the boundary association vector of the chunks, that is, the association metric between target word and boundary words set, which will be discussed in following sections. The vector was input to a classifier we trained and finally we were able to draw the conclusion whether the chunk is a concept or not.

Definition 1. *Denoting a web page as p and a keyword term as x, we write $x \in p$ if p is returned while querying x in Google. Let the page collection $X = \{p : x \in p\}$. The probability of a term x is defined as $p(x) = |X|/T$ where T means the total number of pages the Google indexes. Sometimes we need to index pages by combining two terms x,y immediately together, denoting by $x \oplus y$. The joint probability $p(x, y) = p(x \oplus y) = |\{p : x \oplus y \in p\}|/T$ means the number of web pages in which term y occurs immediately after x. Generally, $p(x, y)$ and $p(y, x)$ are not equal if terms x,y are not the same.*

For a word x and a chunk y, the probability that x occurs right before ck right before ck on the Web is defined as the head word association*:*

$$p_{head}(x|ck) = p(x, ck)/p(ck)$$

Similarly the probability that x occurs immediately after ck on the Web is defined as the tail word association*:*

$$p_{tail}(x|ck) = p(ck, x)/p(ck)$$

Note that in the calculation, constant T is divided out, which means we only need the number of pages returned by Google. Defined the frequency $f(x)$ as the number of pages returned by Google with the search term x, we have:

$$f(x) = p(x) \times T, f(x, y) = p(x, y) \times T.$$

Lemma 1. *Rewrite of Definition (1):*

$$p_{head}(x|y) = f(x, y)/f(y), p_{tail}(x|y) = f(y, x)/f(y)$$

The definitions and notions above are the fundamental statistical framework of our acquisition model.

Definition 2. *The Google boundary words (boundary words for short) are divided into two classes. The* head boundary words *are the words that usually occur right before a concept on Webpages, while* tail boundary words *are ones that usually appear immediately after concepts. According to our hypothesis, concepts and their boundary words are more "intimate" than other chunks in Web pages. We will consider three factors soon to reflect the degree of "intimacy".*

Essentially, the boundary words incarnate the embedded semantic features of lexical concepts through syntax. For example, denote the concept as < *cpt* >, the pattern <的>< *cpt* > may mean < *cpt* > is owned by someone, while the pattern < *cpt* ><的> may mean < *cpt* > can own something.

3 Recognizing Concepts Based on Boundary Words

3.1 Learning Concept Boundary Words with Google

The acquisition process of concept boundary words is a statistical learning process, as shown in Fig.2. We use manually extracted lexical concepts as our training set, in which Google is used to index such lexical concepts. Paragraphs that contain the lexical concepts are retrieved in Google results to extract and accumulate the frequency of words abut on concepts, which helped us to select candidate boundary words.

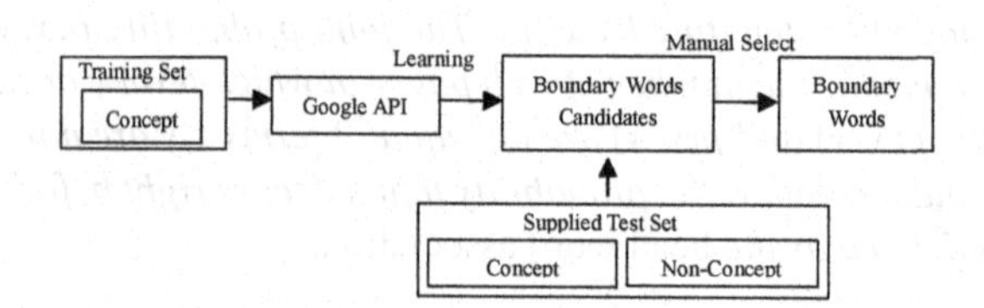

Fig. 2. Acquisition process of boundary words

The candidate boundary words need to be verified by a test set, which will be discuss later in this section. At the same time, some unpredictable interferences can disturb the test results, which makes it necessary to do a manual selection.

In our experiment, a training set of 1800 concept is used to extract candidate boundary words. Results of top eight candidates of each group are shown in Table 1, with each word assigned a unique ID. Note that most of such words are prepositions or conjunctions of a single Chinese character. In our experiment, some punctuation such as comma, semicolon and quotation marks also performed well. However, they are impossible to become boundary words in our system since Google will filter them from query terms.

The method above can find candidate words that are highly associated with lexical concept, yet it is hard to tell if the words are associated well with chunks which are not concepts. Thus, an extra selection process is required.

Table 1. Candidate Concept boundary words (Top 8)

Head Boundary Words			Tail Boundary Words		
ID	Words	Frequency	ID	Words	Frequency
h0	的 of	6141	t0	的 of	55128
h1	是 is	1224	t1	是 is	13025
h2	在 at	1101	t2	在 at	9035
h3	和 and	1187	t3	和 and	8068
h4	为 for	1033	t4	中 in	9509
h5	有 have	711	t5	有 have	6377
h6	与 and	746	t6	与and	4597
h7	了	748	t7	不 not	4485

Definition 3. *A lexical chunk is defined to be a Head Bad Non-Concept if its association of head boundary words is comparatively lower than that of common lexical concepts. Accordingly, we define a Tail Bad Non-Concept if its association of tail boundary words is comparatively lower. One chunk is a concept only if it has comparatively high association probability of both head boundary words and tail boundary words.*

Definition 4. *For a candidate head boundary word set H, a candidate tail boundary word set T, a concept set C and a non-concept set composed of head bad non-concept set NC_{head} and tail bad non-concept set NC_{tail}. The average association of concept AC and average association of non-concept ANC of a certain boundary word:*

$$AC_{hj} = \sum_i p_{head}(h_j|c_i)/|C|, \;\; ANC_{hj} = \sum_i p_{head}(h_j|nc_i)/|NC_{head}|$$

$$AC_{tj} = \sum_i p_{tail}(t_j|c_i)/|C|, \;\; ANC_{tj} = \sum_i p_{tail}(t_j|nc_i)/|NC_{tail}|$$

$$\text{where } h_j \in H, t_j \in T, c_i \in C, nc_i \in NC$$

The Lexical Concept Association (LCA) of a boundary candidate b is defined as:

$$LCA_b = AN_b - ANC_b$$

The LCA is a metric of measuring the quality of boundary words. The higher AC value, the more likely the word is genuinely a boundary word for a concept. We say a that word with better ability of signaling concepts only if the word has a higher AC value and a lower ANC value. Thus, using LAC as the metric, we are able to choose the real concept boundary words in the next step.

We use 9000 concepts and 12000 non-concepts (5500 labeled head bad and 6500 labeled tail bad) for experiment, and the result is shown in Fig.3.

As seen in Fig.3, it is still a little ambiguous to determine which candidate boundary words are real ones. However, we adopt a heuristic: the words with higher a LCA value.

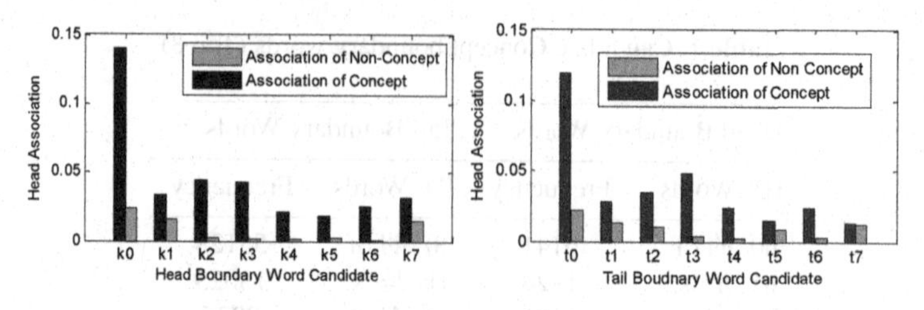

Fig. 3. Boundary word candidate association (AC and ANC value)

3.2 Building Concept Classifiers Using Boundary Words

Classifier Models Based on Statistics of Boundary Words. Once we select the boundary words, we are able to calculate the lexical chunks' conditional associations with such boundary words, or boundary word associations. In our experiment, several approaches based on different techniques are used to build the concept classifier based on concept boundary word associations. We build a decision tree to distinguish the lexical concepts from other chunks. In this method, the boundary word associations are used as attributes. Concept and non-concept sets are used to train the tree using J4.8 Algorithm, implemented by Weka [14,15]. In our first trial, we directly used the boundary word associations as the attributes. For a lexical chunk w and the probability of association $p(b|w)$ with a boundary word b, we have:

$$attrib_{[b]} = p(b|w) \tag{1}$$

We view (1) as a conditional probability attributes (*CP*) for b in building classifiers. Experimenting with such attributes gives pretty good results. But form the experiment result, we noticed that the boundary word associations only represent the conditional probability features. However, for some lexical concepts with very high frequency $f(w)$, $p(b|w)$ with some boundary word b could be quite small even though $f(b,w)$ is very large, which resulted in misleading results. This point led us to improve our conception of attributes:

$$attrib_{[b]} = p(b|w) \times log(f(w)) \tag{2}$$

However, we still find some problems that some concepts with a very low frequency are poorly recognized, shown as Fig.4(a). Such low frequency concepts are not associated well with all the boundary words. In fact, they may be only associated with one or two of them. This problem cannot be solved by only improving the expression of the attribute; according to our experiment, most common chunks have the frequency larger than 25600, shown in Fig.4(b). Hence, we separate them from our experiment.

From the results using *CPF*, we found that high frequency non-concept words may be well associated with some boundary words, which are considered to be concepts by the classifier. For this, we considered the entropy of the boundary association.

The Shannon's entropy is a good measure of randomness or uncertainty[16]. For a random variable X taking a finite number of possible values $x_1, x_2, \ldots, x_n$ with probabilities

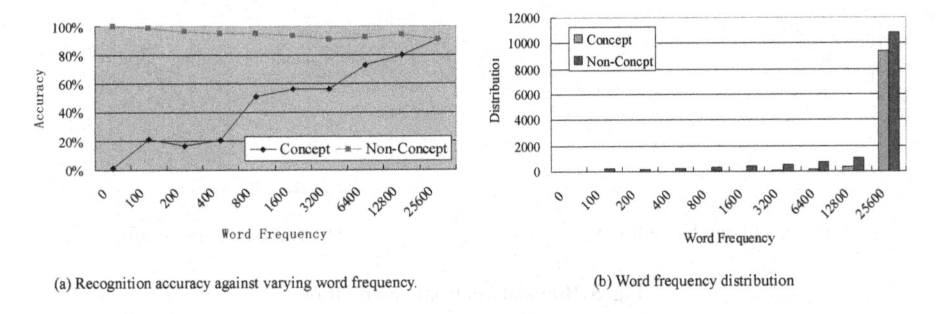

(a) Recognition accuracy against varying word frequency.

(b) Word frequency distribution

Fig. 4. Word Frequency Experiment Results

$p_1, p_2, \ldots, p_n$, where $p_i > 0, i = 1, 2, \ldots, n$ and $\sum p_i = 1$. The entropy of X can be defined as:

$$entropy(p_1, p_2, \ldots, p_n) = -\sum_{i=1}^{n} p_i log p_i$$

For a boundary word set $B = b1, b2, \ldots, bn$ and a chunk w, first we need to unify the probability as follow:

$$f_w(b_n) = p(b_n|w) / \sum_{i=1}^{n} p(b_i|w), \; \sum_{i=1}^{n} f_w(b_i) = 1$$

Accordingly, we have the *Boundary Entropy* of the chunk w:

$$E_w = entropy(b_1, b_2, \ldots, b_n) = -\sum_{i=1}^{n} f_w(b_i) log(f_w(b_i))$$

We divided our experiment into two parts, that is, the head boundary entropy and tail one. The result as shown in Fig.5, indicates that the boundary entropy of a lexical concept are generally higher than that of the non-concept ones, which provides us an important feature of concept recognition.

However, another question is raised: is the entropy used as a separate attribute or as a weight to other attributes. In our experiment, we chose the latter one, the reason is that boundary entropy for concepts is generally higher but is far from being a device to distinguish concepts from non-concepts, especially for the tail boundary entropy. Hence, we define the new attribute as conditional probability with frequency and entropy (*CPFE*):

$$attrib_{[b]} = p(b|w) \times log(f(w)) \times E_w.$$

Building Classifiers. During the training process of the decision tree, we first tried to use all the attributes of both head and tail boundary words at the same time; that is, totally with 16 boundary words, and 100,000 training samples. The output tree is enormous in size and has bad results, shown in Fig.6.

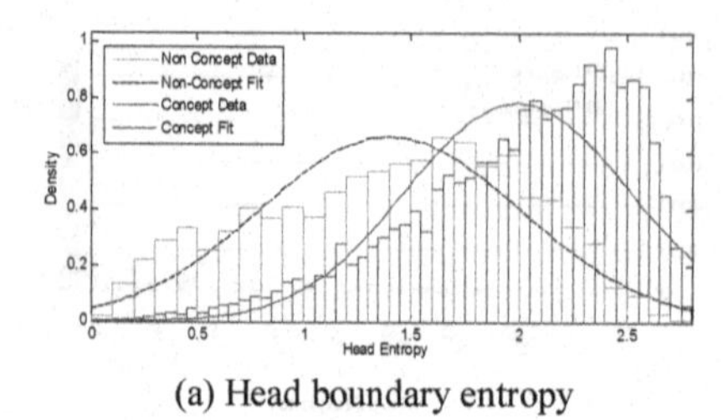

(a) Head boundary entropy

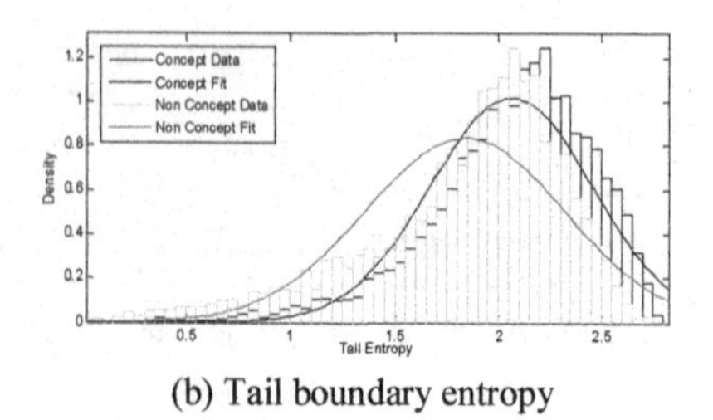

(b) Tail boundary entropy

Fig. 5. Boundary entropy distribution

```
h3 <= 0.015663
|   t3 <= 0.012789
|   |   h3 <= 0.003235
|   |   |   t7 <= 0.007519
|   |   |   |   t3 <= 0.009242: FALSE (1624.0/44.0)
|   |   |   |   t3 > 0.009242
|   |   |   |   |   t1 <= 0.002017: FALSE (40.0)
|   |   |   |   |   t1 > 0.002017
|   |   |   |   |   |   t4 <= 0.011384
|   |   |   |   |   |   |   h7 <= 0.012677
|   |   |   |   |   |   |   |   h2 <= 0.000195: FALSE (4.0/1.0)
|   |   |   |   |   |   |   |   h2 > 0.000195: TRUE (3.0)
|   |   |   |   |   |   |   h7 > 0.012677: FALSE (8.0)
|   |   |   |   |   |   t4 > 0.011384: TRUE (5.0)
                                ...500 lines omitted...
```

Fig. 6. Partial decision tree with 16 boundary words

The reason for the bad performance of the tree is that J4.8 Algorithm always gives a result with best precision of in-set test even if there is actually no any relation between input samples. To avoid this problem, there must be enough training samples according to the total attribute space. In our experiment, we considered the attribute space of each boundary association is 1000 (since the precision is 0.001). Suppose we use a boundary word set with this size, the size of samples should be 10^{3n}; that is, a boundary word set with size 16 should used a sample set size of 10^{48}, which is impossible to prepare such a training set because of the cost of time and space.

Additionally, we noticed that some leaves, as shown in Fig.6 with bold style, judge a non-concept by some attributes larger than some of leaves, which appear against our hypothesis. Through further analysis found that most of these leaves are over fitted; that is, with few samples support. Such leaves ostensibly increased the accuracy of the tree in the sample set, while in fact they lower the accuracy in open verification. Consequently, we need to manually split the tree in order to clear such mistaken leaves.

Arguments above led us to reduce the size of boundary word set, which means our training samples only allow us to use at most 2 boundary words according to the actual attribute space based on our precision. We divided our training process into two steps: building the head boundary tree and tail boundary tree independently. Therefore we could use 2 head boundary words and 2 tail boundary words to build the two trees. The former one is used to validate the head boundary, while the latter one is used in tail boundary. From the result, the sizes of trees are smaller than what we have built before, which gives us higher accuracy in the supplied data set. However, some over fitted leaves still exist in our boundary trees, which need to be manually split.

4 Experimental Results and Discussion

In the experiment, we used a data set with over 20,000 Chinese lexical concept and non-concept instances. Table 2 shows our data set. The estimated confidence of the data set is 90%. We run the data set in three classifiers, trained by *CP*, *CPF* and *CPFE*, shown as Fig. 7(a), Fig.7(b), Fig.7(c). Their accuracy results are shown as Table 3.

Table 2. Verification of data set

Type	Size
Concept	10103
Head bad non-concept	5968
Tail bad non-concept	8456
Total	24527

Table 3. Classifier accuracy comparison

Section	CP	CPF	CPFE
Concept Head	95.800%	97.816%	97.200%
Concept Tail	95.083%	94.355%	95.049%
Concept Total	87.668%	89.729%	89.819%
Non-Concept Head	95.552%	95.890%	95.111%
Non-Concept Tail	90.044%	88.805%	89.594%
Non-Concept Total	91.771%	91.292%	91.367%
Total	89.900%	90.580%	90.661%

The overall accuracy reflects the improvement in our attribute selection, which can be seen from Table 3, the accuracy of CPF and CPFE are over 90% and that of CP is very close to 90%. The high accuracy is expected since the cost of time in retrieving boundary associations is higher than other methods. From the results we can see that every improvement of the attribute has a comparably higher accuracy on the "concept total" section, although some lower the non-concept accuracy, resulting from the improving on constrains of our attributes. CPFE is a balanced choice between the precision and recall, considering its high accuracy in "concept" section and total accuracy, which is thought to be the best attribute.

From the results, we find that the accuracy of chunks with different numbers of characters varies obviously. Fig.8 reveals that the accuracy is high when the word is composed of 5 or 6 characters and decreases more or less. We also noticed that chunks with more than 6 characters tend to have complex structures which are generally have a relative lower frequency. On the other hand, some complex non-concept chunks may have good association with boundary words.

Carrying out the experiment several times via the same data set with several months, we find that the results are slightly fluctuating over an interval of $[-0.0001, +0.0001]$.

Head boundary tree

```
h3 <= 0.002764
|  h3 <= 0.001518: FALSE
|  h3 > 0.001518
|  |  h0 <= 0.022287: FALSE
|  |  h0 > 0.022287: TRUE
h3 > 0.002764
|  h0 <= 0.027647
|  |  h0 <= 0.006386
|  |  |  h3 <= 0.003571: FALSE
|  |  |  h3 > 0.003571
|  |  |  |  h0 <= 0.002565: FALSE
|  |  |  |  h0 > 0.002565: TRUE
|  |  h0 > 0.006386: TRUE
|  h0 > 0.027647: TRUE
```

Tail boundary tree

```
t0 <= 0.005501
|  t0 <= 0.002369: FALSE
|  t0 > 0.002369
|  |  t3 <= 0.012549: FALSE
|  |  t3 > 0.012549: TRUE
t0 > 0.005501
|  t3 <= 0.003769
|  |  t3 <= 0.001287: FALSE
|  |  t3 > 0.001287
|  |  |  t0 <= 0.047041: TRUE
|  |  |  t0 > 0.047041: FALSE
|  t3 > 0.003769: TRUE
```

(a) CP boundary decision tree

Head boundary tree

```
h3 <= 0.01437
|  h3 <= 0.004276: FALSE
|  h3 > 0.004276
|  |  h0 <= 0.113329: FALSE
|  |  h0 > 0.113329
|  |  |  h3 <= 0.007981: FALSE
|  |  |  h3 > 0.007981: TRUE
h3 > 0.01437
|  h0 <= 0.153173
|  |  h0 <= 0.033565
|  |  |  h3 <= 0.050315: FALSE
|  |  |  h3 > 0.050315: TRUE
|  |  h0 > 0.033565: TRUE
|  h0 > 0.153173: TRUE
```

Tail boundary tree

```
t0 <= 0.027553: FALSE
t0 > 0.027553
|  t3 <= 0.020365
|  |  t3 <= 0.005998: FALSE
|  |  t3 > 0.005998
|  |  |  t0 <= 0.296437: TRUE
|  |  |  t0 > 0.296437: FALSE
|  t3 > 0.020365: TRUE
```

(b) CPF boundary decision tree

Head boundary tree

```
h0 <= 0.053575
|  h3 <= 0.002641: FALSE
|  h3 > 0.002641
|  |  h3 <= 0.035137
|  |  |  h0 <= 0.033693: FALSE
|  |  |  h0 > 0.033693
|  |  |  |  h3 <= 0.010954: FALSE
|  |  |  |  h3 > 0.010954: TRUE
|  |  h3 > 0.035137: TRUE
h0 > 0.053575: TRUE
```

Tail boundary tree

```
t0 <= 0.024024
|  t3 <= 0.013975: FALSE
|  t3 > 0.013975
|  |  t3 <= 0.051881: FALSE
|  |  t3 > 0.051881: TRUE
t0 > 0.024024
|  t3 <= 0.02224
|  |  t3 <= 0.005768
|  |  |  t3 <= 0.002088: FALSE
|  |  |  t3 > 0.002088: TRUE
|  |  t3 > 0.005768: TRUE
|  t3 > 0.02224: TRUE
```

(c) CPFE boundary decision tree

Fig. 7. Boundary word decision trees based on our training data set (Dec.2006)

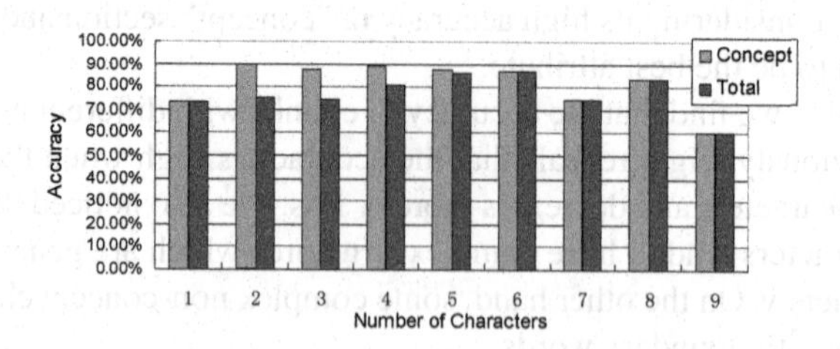

Fig. 8. Accuracy distribution of chunks composed of varying characters

Generally these changes are moving towards higher accuracy, because the pages indexed by Google can be thought as a dynamic corpus which is inevitable to fluctuate, and is however acceptable.

5 Conclusion

This paper presented a Google-based statistical acquisition model of Chinese lexical concepts. We constructed three decision trees based on different models and used the models to recognize concepts from chunks. Experiments are designed to compare the three models and show the best method should take (1) conditional probability, (2) the frequency and (3) the entropy of the concept boundary words into account, which can identify concepts with an accuracy rate of 90.661%.

Future research will be put into acquisition of chunks with low frequency and approaches such as structural analysis of chunks are needed to process complex chunks. Context of chunks extracted from Google can also be used in concept recognition.

References

1. Laurence, S., Margolis, E.: Concepts and cognitive science. MIT Press, Cambridge (1999)
2. Wu, S.H., Hsu, W.L.: SOAT: A Semi-Automatic Domain Ontology Acquisition Tool From Chinese Corpus. In: 19th International Conference on Computational Linguistics, Taipei, Taiwan, pp. 1313–1317 (2002)
3. Smeaton, A., Quigley, I.: Experiments on using semantic distances between words in image caption retrieval. In: 19th International Conference on Research and Development in Information Retrieval, pp. 174–180. ACM Press, Zurich, Switzerland (1996)
4. Maynard, D., Ananiadou, S.: Identifying Contextual Information for Multi-Word Term Extraction. In: 5th International Congress on Terminology and Knowledge Engineering, Innsbruck, Austria, pp. 212–221 (1999)
5. Church, K.W., Hanks, P.: Word Association Norms, Mutual Information, and Lexicography. Computational Linguistics 16(1), 22–29 (1990)
6. Dunning, T.: Accurate Methods for the Statistical of Surprise and Coincidence. Association for Computational Linguistics 19(1), 61–76 (1993)
7. Cohen, J.D.: Highlights: Language and Domain-Independent Automatic Indexing Terms for Abstracting. Journal of the American Society for Information Science 46(3), 162–174 (1995)
8. Kwong, O.Y., Tsou, B.K.: Automatic Corpus-Based Extraction of Chinese Legal Terms. In: 6th Natural Language Processing Pacific Rim Symposium, Tokyo, Japan, pp. 669–676 (2001)
9. Luo, S.F., Sun, M.S.: Two-Character Chinese Word Extraction Based on Hybrid of Internal and Contextual Measures. In: 2nd SIGHAN Work Shop on Chinese Language Processing, Sapporo, Japan, pp. 24–30 (2003)
10. Zhou, L.X.: Research of segmentation of Chinese texts in Chinese search engine. In: Systems, Man, and Cybernetics, 2001 IEEE International Conference, Tucson, USA, vol. 4, pp. 2627–2631 (2001)
11. Cilibrasi, R.L.: Statistical Inference Through Data Compression, PhD Thesis, ILLC DS-2007-01, University of Amsterdam (2007)
12. Cilibrasi, R.L., Vitanyi, P.: The Google Similarity Distance. Knowledge and Data Engineering. IEEE Transactions 19(3), 370–383 (2007)
13. Cilibrasi, R.L., Vitanyi, P.: Automatic Meaning Discovery Using Google (2004), http://xxx.lanl.gov/abs/cs.CL/0412098

14. Witten, I.H., Frank, E.: Data Mining, Practical Machine Learning Tools and Techniques, 2nd edn. Morgan Kaufmann, San Francisco (2005)
15. Weka, `http://www.cs.waikato.ac.nz/ml/weka/`
16. Taneja, I.J.: Generalized Information Measures and Their Applications, on-line book (2001), `http://www.mtm.ufsc.br/taneja/book/book.html`

Learning Concepts from Text Based on the Inner-Constructive Model*

Shi Wang[1,2], Yanan Cao[1,2], Xinyu Cao[1,2], and Cungen Cao[2]

[1] Graduate University of Chinese Academy of Sciences, Beijing China, 100049
wangshi_frock@hotmail.com, caoyanan01@163.com, cxy8202@163.com
[2] Key Laboratory of Intelligent Information Processing, Institute of Computing Technology, Chinese Academy of Sciences, Beijing China, 100080
cgcao@ict.ac.cn

Abstract. This paper presents a new model for automatic acquisition of lexical concepts from text, referred to as *Concept Inner-Constructive Model* (CICM). The CICM clarifies the rules when words construct concepts through four aspects including (1) parts of speech, (2) syllable, (3) senses and (4) attributes. Firstly, we extract a large number of candidate concepts using lexico-patterns and confirm a part of them to be concepts if they matched enough patterns for some times. Then we learn CICMs using the confirmed concepts automatically and distinguish more concepts with the model. Essentially, the CICM is an instances learning model but it differs from most existing models in that it takes into account a variety of linguistic features and statistical features of words as well. And for more effective analogy when learning new concepts using CICMs, we cluster similar words based on density. The effectiveness of our method has been evaluated on a 160G raw corpus and 5,344,982 concepts are extracted with a precision of 89.11% and a recall of 84.23%.

Keywords: concepts acquisition, *Concept Inner-Constructive Model*, knowledge discovery, text mining.

1 Introduction

From the cognitive point of view, knowing concepts is a fundamental ability when human being understands the world. Most concepts can be lexicalized via words in a natural language and are called *Lexical Concepts*. Currently, there is much interest in knowledge acquisition from text automatically and in which concept extraction is the crucial part[1]. There are a large range of other applications which can also be benefit from concept acquisition including information retrieval, text classification, and Web searching, etc.[2–4]

Most related efforts are centralized in term recognition. The common used approaches are mainly based on linguistic rules[5], statistics [6, 7] or a combination of both [8, 9]. In our research, we realize that concepts are not just terms. Terms are domain-specific

* This work is supported by the National Natural Science Foundation of China under Grant No.60496326, 60573063, and 60573064; the National 863 Program under Grant No. 2007AA01Z325.

Z. Zhang and J. Siekmann (Eds.): KSEM 2007, LNAI 4798, pp. 255–266, 2007.

while concepts are general-purpose. Furthermore, terms are just restricted to several kinds of concepts such as named entities. So even we can benefit a lot from term recognition we can not use it to learn concepts directly.

Other relevant works are focused on concepts extraction from documents. Gelfand has developed a method based on the Semantic Relation Graph to extract concepts from a whole document[10,11]. Nakata has described a method to index important concepts described in a collection of documents belonging to a group for sharing them[11]. A major difference between their works and ours is that we want to learn huge amount of concepts from a large-scale raw corpus efficiently rather than from one or several documents. So the analysis of documents will lead to a very higher time complexity and does not work for our purpose.

In this paper, we use both linguistic rules and statistical features to learn lexical concepts from raw texts. Firstly, we extract a mass of concept candidates from text using lexico-patterns, and confirm a part of them to be concepts according to their matched patterns . For the other candidates we induce an *Inner-Constructive Model* (CICM) of words which reveal the rules when several words construct concepts through four aspects: (1) parts of speech, (2) syllables, (3) senses, and (4) attributes.

The structure of the paper is as follows. Section 2 will show how to extract concept candidates from text using the lexico-patterns. The definition of CICM and CICM-based concepts learning algorithm will be discussed in Sect. 3. In Sect. 4, we will show experimental results. Conclusion and future works will be given in Sect. 5.

2 Extracting Concepts from Text Using Lexico-Patterns

In this research, our goal is to extract huge amount of domain-independent concept candidates. A possible solution is to process the text by Chinese NLU systems firstly and then identity some certain components of a sentence to be concepts. But this method is limited dut to the poor performance of the existing Chinese NLU systems, which still against many challenge at present for Chinese[12]. So we choose another solution based on lexico-patterns.

2.1 The Lexico-Patterns of Lexical Concepts

Enlightened Hearst's work[13], we adopt lexico-patterns to learning lexical concepts from texts. But first design a lot of lexico-patterns patterns manually, some of which are shown in Table 1.

Here is an example to show how to extract concepts from text using lexico-patterns:

Example 1. Lexico-Pattern_No_1 {

Pattern: <?C1><是><一 |><个|种><?C2>

Restrict Rules:

not_contain(<?C2>,<!标点>) ∧

lengh_greater_than(<?C1>,1) ∧

Table 1. The Lexico-Patterns for Extracting Concepts from Text

(Only 20 are listed and detail restrictions about the patterns are omitted for simplicity. <?C> stands for concepts, and <?X> represents any characters.)

<table>
<tr><th>ID</th><th>Lexico-Patterns</th></tr>
<tr><td>1</td><td><?C1><是><一|><个|种|><?C2></td></tr>
<tr><td>2</td><td><?C1><、><?C2><或者|或是|以及|或|等|及|和|与><其他|其它|其余></td></tr>
<tr><td>3</td><td><?C1><、><?C2><等等|等><?C3></td></tr>
<tr><td>4</td><td><?C1><如|象|像><?C2><或者|或是|或|及|和|与|、><?C3></td></tr>
<tr><td>5</td><td><?C1><、><?C2><是|为><?C3></td></tr>
<tr><td>6</td><td><?C1><、><?C2><各|每|之|这><种|类|些|样|流><?C3></td></tr>
<tr><td>7</td><td><?C1><或者|或是|或|等|及|和|与><其他|其它|其余><?C2></td></tr>
<tr><td>8</td><td><?C1><或者|或是|或|及|和|与><?C2><等等|等><?C3></td></tr>
<tr><td>9</td><td><?C1><中|里|内|><含|含有|包含|包括><?C2></td></tr>
<tr><td>10</td><td><?C1>由<?C2><组成|构成></td></tr>
<tr><td>11</td><td><?C1><包括|包含><?C2>在内的<?C3></td></tr>
<tr><td>12</td><td><?C1><是|作为|成为><?C2>部分<之一|></td></tr>
<tr><td>13</td><td><?C1>是<用|由><?C2><做|制|作|加工|炼|制造><而成></td></tr>
<tr><td>14</td><td><?C1><是|以|是以><?C2><为原料></td></tr>
<tr><td>15</td><td><?C1><作为|是><?C2>的<开始|开端|开头|中间过程|结束|结尾|结局></td></tr>
<tr><td>16</td><td><?C1><又称|简称|全称|俗称|旧称|今称|人称|史称|中文名称|英文名称><?C2></td></tr>
<tr><td>17</td><td><?C1><!左括号><?C2><!右括号><?X></td></tr>
<tr><td>18</td><td><?C1><的><?C2><是><?C3></td></tr>
<tr><td>19</td><td><?C1><称为><?C2>的<?C3></td></tr>
<tr><td>20</td><td><?C1><被视为|被称为|被誉为><?C2></td></tr>
</table>

lengh_greater_than(<?C2>,1) ∧
lengh_less_than(<?C1>,80) ∧
lengh_less_than(<?C2>,70) ∧
not_end_with(<?C1>,<这|那>) ∧
not_end_with(<?C2>,<的|而已|例子|罢了>) ∧
not_begin_with(<?C2>,<这|的|它|他|我|那|你|但>) ∧
not_contain(<?C2>,<这些|那些|他们|她们|你们|我们|他|它|她|你|谁>)

}

Sample sentences and the concepts extracted:

(1) 地球是一个行星，地球会爆炸吗？(The earth is a planet, will it blast?) →<?C1>=地球(The earch); <?C2>=行星(a planet)

(2) 很久很久以前地球是一个充满生机的星球.(Long long ago the Earth is a planet full of vitality.) →<?C1>=很久很久以前地球(Long long ago the Earth); <?C2>=充满生机的星球(a planet full of vitaligy)

How to devise good patterns to get as much concepts as possible? We summarized the following criteria through experiments:

(1) **High accuracy criterion.** Concepts distributing in sentences meet linguistics rules, so each pattern should reflect at least one of these rules properly. We believe that we should know linguistics well firstly if we want create to good patterns.

(2) **High coverage criterion.** We want to get as much concepts as possible. Classifying all concepts into three groups by their characteristics, (i.e. concepts which describe physical objects, concepts and the concepts which describe time) is a good methodology for designing good patterns to get more concepts.

2.2 Confirming Concepts Using Lexico-Patterns

Obviously, not all the chunks we got in section 2.1 are concepts, such as *<?C1>=很久很久以前地球(Long long ago the Earth)* in Example 1 above.

In order to identify concepts from the candidates, we introduce a hypothesis, called Hypothesis 1.

Hypothesis 1. *A chunk ck extracted using lexico-patterns in section 2.1 is a concept if (1) ck has been matched by sufficient lexico-patterns, or (2) ck has been matched sufficient times.*

To testify our hypothesis, we randomly draw 10,000 concept candidates from all the chunks and verify them manually. The association between the possibility of a chunk to be a concept and its matched patterns is shown as Fig. 1:

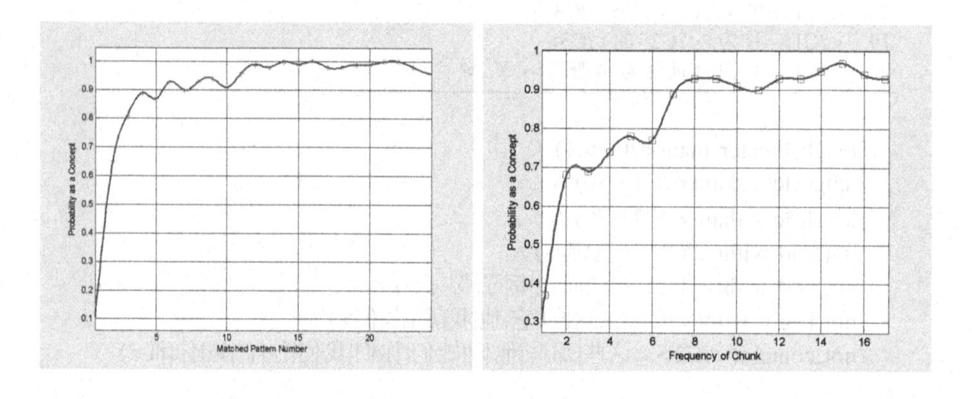

Fig. 1. Association between the lexico-patterns number / the times matched by all the patterns of chunks and their possibility of being concepts

The left chart indicates our hypothesis that the chunks which matched more patterns are more likely to be concepts and the right chart shows that the frequency of the chunk does work well to tell concepts from candidate chunks too. In our experiments, we take the number of patterns matchings to be 5 and threshold of matching frequency as 14, and single out about 1.22% concepts from all the candidate chunks with a precision rate of 98.5%. While we are satisfied with the accuracy, the recall rate is rather low. So in the next step, we develop CICMs to recognize more concepts from chunks.

3 Learning Concepts Using CICM

The CICM is founded on an instinctive hypothesis:

Hypothesis 2. *Most lexical concepts obey certain inner constructive rules.*

That means, when some words form a concept, each word must play a certain role and has certain features. We develop the hypothesis enlightened mainly from the knowledge of linguistics [14] and the cognitive process of human beings creating lexical concepts[15]. Some examples will be given to illuminate the Hypothesis 2 after present the definition of CICM.

3.1 Definition of CICM

According to Hypothesis 2, we can tell whether an unknown chunk is a concept or not by checking whether each word in it whether obeys the CICM. The problems are how to materialize these rules and how to get them. The POS models can reveal these rules using the parts of speech of words but is not precise enough and has many defections[16]. To get better performance we probe into the structure of concepts more deeply and find that besides POS, we must ensure each word's more definite role through at least other three aspects.

Definition 1. *The word model W=<PS, SY, SE, AT> of a word w is a 4-tuple where (1) PS is all the parts of speech of w; (2) SY is the number of w's syllable; (3) SE is the senses of w in HowNet; and (4) AT is the attributes of w.*

The *word models* are integrated information entities to model words. The reason of choosing these four elements listed above will be clarified when we construct CICMs.

Definition 2. *Given a concept cpt* $= w_1 \ldots w_{i-1} w_i w_{i+1}, w_n$ *with n words, the C-Vector of the word* w_i *towards cpt is a n-tuple:*

$$C - Vector(w_i) = <i, W_1, \ldots, W_{i-1}, W_{i+1}, \ldots, W_n> \quad (1)$$

The *C-Vector* of a word stands for one constructive rule when it forms concepts by linking other words and *i* is its position in the concept. A word can have same *C-Vectors* towards many different concepts. The *C-Vector* is the basis of CICM.

Definition 3. *The Concept Inner-Constructive Models (CICMs) of a word w is a bag of C-Vectors, in which each C-Vector is produced by a set of concepts contain w.*

Essentially, CICMs of words represent the constructive rules when they construct concepts. In the four elements of word models, PS and SY embody the syntactical information which have significant roles when conforming concepts in Chinese[14] and are universal for all types of words. SE and AT reveal the semantic information of words and are also indispensably. HowNet is an elaborate semantic lexicon attracted many attentions in many related works[17]. But there are still some words which are missing in it so we need to introduce attributes as a supplement. Attributes can tell the

Table 2. CICM of "生产"

ID	*C-Vectors*	Sample Concepts
1	<1, W(管理) >	生产 管理
2	<1, W(许可证) >	生产 许可证
3	<1,W(实习),W(报告)>	生产 实习 报告
4	<2,W(食品)>	食品 生产
5	<2,W(分布式)>	分布式 生产
6	<2,W(国民),W(总值)>	国民 生产 总值
7	<2,W(新疆),W(建设),W(兵团)>	新疆 生产 建设 兵团
8	<3,W(广东省),W(春耕)>	广东省 春耕 生产
9	<3,W(国家),W(安全),W(监督),W(管理局)>	国家 安全 生产 监督 管理局
…	…	…

semantic differences at the quantative level or qualitative level between concepts. Tian has developed a practicable approach to acquire attributes from large-scale corpora[18].

Table 2 displays the CICMs for the word *"生产(produce, production)"*:

Note that we omit the details of each word vector for simplicity.Taking "国民 生产 总值" for example, the full *C-Vector* is:

< 2,
<{n},2,{属性值,归属,国,人,国家},{有组成,有数量}>,
<{n},2,{数量,多少,实体},{有值域,是抽象概念}> >

3.2 Learning CICMs

Using CICMs as the inner constructive rules of concepts, our next problem is how to get these models. We use the confirmed concepts obtained in section 2.2 as a training set and learn CICMs hidden in them automatically. It is an instance learning process and the following procedure is implemented for this task:

Algorithm 1. *CICMs Instance Learning Algorithm:*

(1) Initializing the resources including (1.1) A words dictionary in which each one has fully parts of speech; (1.2) The HowNet dictionary; and (1.3) An attributes base of words[18].

(2) Constructing a model set MSet to accommodate all the words' models which is empty initially.

(3) For each concept cpt in the training set, segment it and create each word's C-Vector(w_i). Subsequently, if C-Vector(w_i) $\in$ MSet(w_i), then just accumulate the frequency; otherwise add C-Vector(w_i) to MSet(w_i).

(4) Removing the C-Vectors which have low frequency for each word's MSet.

Based on experiments, we choose 10% as the threshold of the number of the concepts containing the word in the training set. We exclude the vectors which have low frequency, that is, if a $C-Vector$ for a word is supported by just a few concepts, we look at it as an exception.

4 Clustering Words for More Efficient Analogy

Essentially, CICMs are models of instance analogy. We want to learn new concepts by "recalling" the old ones just as human beings. For example, we can build CICMs for the word *"生产(produce, production)"* like Table 2 and then identify that *"药品 生产(pharmaceutical production)"* is also a concept, because the latter has the same constructive rule as *"食品 生产(food production)"*.

But unluckily, even we know *"药品 生产"* is a concept, our system still can not tell whether *"药品 制造(pharmaceutical manufacture)"* is also a concept for there are no CICMs for the word *"制造"*. The reason for this is that the system still can not make use of word similarity. Therefore, we need to cluster words based on the similarity of CICMs and then learn more new concepts.

4.1 Similarity Measurement of Words

The similarity measurement of CICMs is the basis of clustering words in our task. Our measurement is founded on the intuitive distribute hypothesis that:

Hypothesis 3. *In concepts, similar words have similar CICMs.*

According to Hypothesis 3, the similarity of two words w_1, w_2 is defined as:

$$sim(w_1, w_2) = sim(CICM(w_1), CICM(w_2)) \tag{2}$$

The commonly used similarity measure for two sets includes *minimum distance*, *maximum distance*, and *average distance*. Considering that there are still some noises in our training set which would result in some wrong $C - Vectors$ in CICMs, we choose the average distance for it is more stable for noisy data, that is:

$$\begin{aligned} sim(w_1, w_2) &= \frac{1}{|CICM(w_1|} \sum_{vec_i \in CICM(w_1)} sim(vec_i, CICM(w_2) \\ &= \frac{1}{|CICM(w_1||CICM(w_2|} \sum_{vec_i \in CICM(w_1)} \sum_{vec_j \in CICM(w_2)} sim(vec_i, vec_j) \end{aligned} \tag{3}$$

Now the problem is how to calculate the similarity of two $C - Vectors$ of two words now. For two *C-Vectors*:

$$C - Vector_i =< i, W_1, \ldots, W_n >, C - Vector_j =< j, W_1, \ldots, W_m > \tag{4}$$

We standardize them to an *N-Vector* that is:

$$C - Vector_i =< W^i_{-N}, \ldots, W^i_N >, C - Vector_j =< W^j_{-N}, \ldots, W^j_N > \tag{5}$$

and $W_k = \emptyset$ if there is no word model in position k for both of them We adopt the cosine similarity when compare two vectors, that is:

$$sim(vec_i, vec_j) = \cos(\overrightarrow{vec_i}, \overrightarrow{vec_j}) = \frac{\overrightarrow{vec_i} \cdot \overrightarrow{vec_j}}{|\overrightarrow{vec_i}| \times |\overrightarrow{vec_j}|} \tag{6}$$

4.2 Clustering Words Based on the Density

Among all the clustering methods using density functions has prominent advantages–anti-noisiness and the capability of finding groups with different Inspired by DENCLUE[19], we define a influence function of a word w_0 over another word w:

$$f_B^w = f_B(w_0, w) \tag{7}$$

which is a function proportionately to the similarity of w_0 and w, and reveals the influence degree w_0 over w. Commonly used influence functions include *Square Wave Function* and *Gauss Function*. The former is suitable for the data which dissimilar distinctly while the later is more suitable for reflect the smooth influence of w_0. Because a word is related with many other words in different degrees but no simply 1 or 0 in corpus, it is more reasonable to choose *Gauss Influence Function*:

$$f_{Gauss}^w(w_0) = e^{\frac{-(1-sim(w_0,w_1))^2}{2\sigma^2}} \tag{8}$$

We call Equation (8) the *Gauss Mutual Influence* of w, w_0 for $f_{Gauss}^w(w_0) = f_{Gauss}^{w_0}(w)$. It makes each word linked with many other words to some extent. According to it, we can cluster words into groups. Before giving the definition of a word group, we develop some definitions first for further discussing:

Definition 4. *Given a parameter ξ, $\xi\_region(w_0) = \{w|fGuess(w, w_0) > \xi\}$ is called ξ_region of w_0. Given a parameter MinPts, w_0 is called a CoreWord if $|\xi\_region(w_0)| > MinPts$. The minimal ξ which makes w_0 to be a CoreWord is called the CoreDistance of w_0 and be marked as ξ^*.*

Definition 5. *We call w_0 is direct reachable to w' if w_0 is a CoreDistance and $w' \in \xi\_region(w_0)$ and marked as $d_reachable(w_0, w')$. For a set of words $w_0, w_1, \ldots, w_n = w'$, if $d\_reachable(w_i, w_{i+1})$ for all w_i, $1 \le i < n$, then w_0 is reachable to w', that is, $reachable(w_0, w')$.*

Based on the definitions above, a word group can be seen as the maximal words set based on the reachable property. The corresponding clustering algorithm is given below:

(1)Taking ξ=ξ^* and for all the words w perform the following operation:

$\xi_{cur} = \xi^*; cw\_cur = \{w\}$

$while(\xi_{cur} < 1)\{$

$cw\_pre = cw\_cur;$

$if(|\xi_{cur}\_region(w_0)| > MinPts)\{$

Build a word group cv_{cur} *which contains all the words in* $\xi_{cur}\_region(w)$ *and takes* w *as the* $CoreWord\ of\ it.$

$if(\frac{|cv_{cur} - cv_{pre}|}{|cv_{cur}|} < \alpha)\{break;\}$

$\}//if$

$else\{break;\}$

$\xi_{pre} = \xi_{cur}; \xi_{cur} += \Delta;$

$\}//while$

$\}cw = cw_{pre};$

(2)For each pair of $CoreWords$ w_i, w_j

$if(d\_reachable(w_i, w_j))$

Merge cw_i, cw_j *into a new group* cw_{i+j} *which has two* $CoreWords$ cw_i *and* cw_j

(3)Repeate (2) until no new groups are generated.

Many groups with different density will be generated in (2) for we set value for ξ not a single number but a large range of field. The groups with high density will be created firstly and be covered by the dilute groups. We escape choosing the parameter of ξ by doing this.

4.3 Identifying Concepts Using CICMs

Having the learned CICMs and word cluster, identifying method of new concepts is straightforward. Given a chunk, we just create its local $L - Vector$ and judge whether it satisfies one of its or its similar words' $C - Vector$ we have learned.

Definition 6. *For a chunk* $c_k = w_0 \ldots w_n$*, the Local C-Vector for a word* w_i *in it :* $L\_Vector(w_i, c_k) =< i, W_0, \ldots, W_{i-1}, W_{i+1}, \ldots, W_n, >$.

Theorem 1. *For a chunk* $c_k = w_0 \ldots w_n$ *, for each word* w_i *in it, there is* $L\_Vector(w_i, c_k) \in CICM(gw_i)$*, then* c_k *is a concept, where* gw_i *is the similar word group of* w_i.

5 Experimental Result and Discussion

5.1 Measurement and Result

Our system is called *Concept Extractor* (CptEx) and use the following formulae to evaluate its performance:

$$p = \frac{||m_a \bigcap m_m||}{||m_a||}, r = \frac{||m_a \bigcap m_m||}{||m_m||}, F - Measure = \frac{2 \times p \times r}{p + r} \quad (9)$$

where m_a are the concepts CptEx extracts and m_m are the ones built manually. To calculate the performance, we selected 1000 chunks from the raw corpus and label the concepts in them manually. We compare the results based on CICMs with those based the Syntax Models and the POS Models as shown in Table 3:

Table 3. Performance of CptEx

Measurement	Syntax Models	POS Models	CICM
p	98.5%	86.1%	89.1%
r	1.2%	87.8%	84.2%
F-measure	2.3%	86.9%	86.6%

Having adopted CICMs to distinguish concepts from the chunks extracted by lexico-patterns, the precision rate drops down to 89.1% while the recall rate flies to 84.2%. The precision rate reduces because there are still some improper CICMs which will confirm fake concepts. The samples below in Table 4 will show this.

Compared with POS Models, CICMs has a higher accuracy rate because we consider more factors to clarify the inner constructive rules rather than using part of speech only. On the other hand, our stricter models result in a lower recall rate.

5.2 Limitations and Analysis

Table 4 shows some chunks and their output result after introducing CICMs in CptEx.

Table 4. sample concepts extracted by CptEx

ID	Chunks	m_m	m_a
1	照相/ 器材/	照相器材	照相器材
2	打印/过程/	打印过程	打印过程
3	学生/ 思想/ 政治/ 工作/	学生思想政治工作	学生思想政治工作
4	国际/ 特/ 奥/ 会/ 董事会/	国际特奥会董事会	国际 董事会
5	网上/ 交易/ 可以/ 降低/ 经营/ 成本	网上交易 经营成本	网上交易 经营成本
6	在/ 整个/ 数码/ 处理/ 过程/	整个数码处理过程	整个数码处理过程
7	加盟/ 汉堡/	加盟汉堡	汉堡

We look into the chunk $ck1$ ="照相/ 器材/". The $L - Vectors$ of words "照相" and "器材" are as below:

+ L_Vector("照相", ck1) = <1, <{n},2,{器具},{有形状,有功能,是物质}>>
+ L_Vector("器材", ck1) = <2, <{v},2,{拍摄},{是行为}>>

and the CICMs of the two words are as below:

+ CICM(“照相”) ={
  - <2, <{b, n},2,{器具},{有形状,有功能}>, <{n},2,{器具},{有形状,有功能}>>
  - <1, <{n},2,{器具},{有形状,有功能,是物质}>>
  - <1, <{n},1,{场所，商，设施，机构},{有形状}>>
  - ...

  }
+ CICM (“器材”) ={
  - <3, <{n},2,{地点},{有功能,是物质}><{k, n},2,{事务,传播,信息},{有功能,是抽象}>, <{n},1,{场所,制造,厂房},{有形状,有功能}>>
  - <2, <{k, n},2,{事务,传播,信息},{有功能,是抽象}>, <{n},1,{场所,制造,厂房},{有形状,有功能}>>
  - <2, <{v},2,{保护},{是行为}>>
  - <2, <{v},2,{拍摄},{是行为}>>
  - <1, <{n},1,{树,木},{有形状,有高度,是植物}>>
  - ...

  }

Then we can see that (1) L_Vector(“照相”, ck1) ∈ CICM (“照相”) ,and (2) L_Vector (“器材”, ck1) ∈ CICM(“器材”). and can confirm “照相器材” is a concept. Chunks 2/3/5/7 are the same.

On the other side, for the chunk *ck*4= “国际特奥会董事会”, CptEx doesn’t work at present. The reason for this is that “特奥会” is a abbreviation but be wrongly parsed. CICMs of these three separate characters don’t be helpful for there are not much similar training concepts.

The chunk “加盟汉堡队” has been wrongly extracted for the CICMs of “加盟”, “汉堡队”. And for Chunk *ck*8=“加盟/v 汉堡/n”, the *L_Vectors* are all in the CICMs and then produce errors. We are going to cope with this problem through setting confidence of each $C - Vector$ of CICMs according to the frequency of training samples which support it. And the confidence of the concept confirmed by CICMs with the same confidence of the $C - Vector$ in *L_Vector* of it. The concept with a low confidence will be discarded or validated again in open corpus.

6 Conclusion and Future Works

We have described a new approach for automatic acquisition of concepts from text based on Syntax Models and CICMs of concepts. This method extracted a large number of candidate concepts using lexico-patterns firstly, and then learned CICMs to identify more concepts accordingly. Experiments have shown that our approach is efficient and effective. We test the method in a 160G free text corpus, and the outcome indicates the utility of our method.

There are still some more works be done to get better performance for there are some improper CICMs. We plan to validate concepts in an open corpus such as in the World Wide Web in the future.

References

1. Cao, C., et al.: Progress in the Development of National Knowledge Infrastructure. Journal of Computer Science & Technology 17(5, 1), C16 (2002)
2. Ramirez, P.M., Mattmann, C.A.: ACE: improving search engines via Automatic Concept Extraction. In: Proceedings of the 2004 IEEE International Conference, pp. 229–234. IEEE Computer Society Press, Los Alamitos (2004)
3. Zhang, Y.-T., Gong, L., Wang, Y.-C., Yin, Z.-H.: An Effective Concept Extraction Method for Improving Text Classification Performance. Geo-Spatial Information Science 6(4) (2003)
4. Acquemin, C., Bourigault, D.: Term Extraction and Automatic Indexing. Oxford University Press, Oxford (2000)
5. Chen, W.L., Zhu, J.B., Yao, T.: Automatic learning field words by bootstrapping. In: Proc. of the JSCL. Beijing: Tsinghua University Press, pp. 67–72 (2003)
6. Zheng, J.H., Lu, J.L.: Study of an improved keywords distillation method. Computer Engineering 31(194), C196 (2005)
7. Agirre, E., Ansa, O., Hovy, E., Martinez, D.: Enriching very large ontologies using the WWW. In: Proc. of the ECAI 2004 Workshop on Ontology Learning (2004)
8. Du, B., Tian, H.F., Wang, L., Lu, R.Z.: Design of domain-specific term extractor based on multi-strategy. Computer Engineering 31(14), 159–C160 (2005)
9. Velardi, P., Fabriani, P., Missikoff, M.: Using text processing techniques to automatically enrich a domain ontology. In: Proc. of the FOIS, pp. 270–284. ACM Press, New York (2001)
10. Gelfand, B., Wulfekuler, M., Punch, W.F.: Automated concept extraction from plain text. In: AAAI 1998 Workshop on Text Categorization, Madison, WI, pp. 13–17 (1998)
11. Nakata, K., Voss, A., Juhnke, M., Kreifelts, T.: Collaborative Concept Extraction from Documents. In: Reimer, U. (ed.) PAKM 1998. Proc. Second International Conference on Practical Aspects of Knowledge Management, Basel (1998)
12. Zhang, C., Hao, T.: The State of the Art and Difficulties in Automatic Chinese Word Segmentation. Journal of Chinese System Simulation 17(1), 138–C147 (2005)
13. Hearst, M.A.: Automatic acquisition of hyponyms from large text corpora. In: COLING 1992. Proceedings of the 14th International Conference on Computational Linguistics, pp. 539–545 (1992)
14. Lu, C., Liang, Z., Guo, A.: The semantic networks: a knowledge representation of Chinese information process. In: ICCIP 1992, pp. 50–57 (1992)
15. Laurence, S., Margolis, E.: Concepts: Core Readings. MIT Press, Cambridge, Mass (1999)
16. Yu, L.: A Research on Acquisition and Verification of Concepts from Large-Scale Chinese Corpora. A dissertation Submitted to Graduate School of the Chinese academy of Sciences for the degree of master. Beijing China (May 2006)
17. Dong, Z., Dong, Q.: HowNet and the computation of meaning. World Scientific Publishing Co., Inc., Singapore (2006)
18. Tian, G.: Research os Self-Supervised Knowledge Acquisition from Text based on Constrained Chinese Corpora. A dissertation submitted to Graduate University of the Chinese Academy of Sciences for the degree of Doctor of Philosophy. Beijing China (May 2007)
19. Hinneburg, A., Keim, D.: An efficient approach to clustering in large multimedia databases with noise. In: Proceedings of the 4th International Conference on Knowledge Discovery and Data Mining (1998)

Contextual Proximity Based Term-Weighting for Improved Web Information Retrieval

M.P.S. Bhatia[1] and Akshi Kumar Khalid[2]

Netaji Subhas Institute of Technology, University of Delhi, India
mpsbhatia@nsit.ac.in, akshi.kumar@gmail.com

Abstract. Despite its success as a preferred or de-facto source of information, the Web implicates two key challenges: To provide improved systems that retrieve the most relevant information available, and, secondly, how to target search on information that satisfies user's need with accurate balance of novelty and relevance. Nevertheless, Web content is not always easy to use. Due to the unstructured and semi-structured nature of the Web pages & design idiosyncrasy of Websites, it is a challenging task to organize & manage content from the Web. Web Mining tries to solve these issues that arise due to the WWW phenomenon. This paper proposes a novel context-based paradigm for improving Web Information Retrieval, given a multi-term query. The technique referred to as the Contextual Proximity Model (CPM), captures query context and matches it against term context in documents to determine term significance and topical relevance. It makes use of the co-information metric to detect the query context. This contextual evidence is used as an additional input to disambiguate and augment the user's explicit query and dynamically contribute to the term frequency metric to ensure a vital, positive impact on retrieval accuracy.

Keywords: Web Retrieval, Co-information, Query Context, Term Proximity, Contextual Similarity.

1 Introduction

The unabated growth of the Web and the increasing expectation placed by the user on the search engine to anticipate and infer his/her information needs and provide relevant results uphold the inevitable creation of intelligent server and client-side systems that can effectively mine for knowledge both across the Internet and in particular web localities. Despite its success as a preferred or de facto source of information, the Web implicates two key challenges: To provide improved systems that retrieve the most relevant information available, and, secondly, how to target search on information that satisfies user's need with accurate balance of novelty and relevance. Web Mining tries to solve these issues that arise due to the WWW phenomenon [1, 2]. It tries to overcome these problems by applying data mining techniques to content, (hyperlink) structure and usage of Web resources.

On the Web, the most commonly used tool for learning is the search engine [3]. The user first submits a query representing the 'subject of interest' to a search engine

Z. Zhang and J. Siekmann (Eds.): KSEM 2007, LNAI 4798, pp. 267–278, 2007.

system, which finds and returns the related Web pages. He/she then browses through the returned results to find those suitable Web pages. Search engines are critically important to help users find relevant information on the World Wide Web. In order to best serve the needs of users, a search engine must find and filter the most relevant information matching a user's query, and then present that information in a manner that makes the information most readily palatable to the user. However, the current search techniques are not designed for in-depth learning on the Web. The impetus for the work presented in the paper arises from the current state of development of the Web Search paradigm which is sufficient if one is looking for a specific piece of information, e.g., homepage of a person, a paper etc. But it is insufficient for open-ended research or exploration, for which more can be done. Moreover, user queries are often not best formulated to get optimal results. The goal is to drive the next generation of Web search with the key to support relevant search results by effectively and efficiently digging out user- centric information. This means retrieving as many relevant web documents in response to an inputted query that is typically limited to only containing a few terms expressive of the user's information need. The generic IR task can be specified as "Retrieve that amount of information which a user needs in a specific situation for solving his/her current problem" [4]. This definition implies two major research issues:

- IR should consider the specific user, the situation and the problem to be solved. This view leads to the notion of contextual retrieval.
- For retrieving the necessary information, all accessible sources should be exploited.

At their heart, most information retrieval models utilize some form of term frequency. The notion is that the more often a query term occurs in a document, the more likely it is that document meets an information need. We examine an alternative. We propose a model which assesses the presence of a term in a document not by looking at the actual occurrence of that term, but by a set of non-independent supporting terms, i.e. context. This yields a weighting for terms in documents which is different from and complementary to TF-based methods, and is beneficial for retrieval. In this paper we propose *Contextual Proximity Model (CPM),* novel context-based paradigm for improving Web Information Retrieval, given a multi-term query. CPM interprets and makes use of context for retrieval to determine term significance and relevancy. It makes use of the Co-information metric for formulating a context coverage list to finally extract the query context. The aim is to dynamically generate a measure for document term significance during retrieval that is used as an additional input to disambiguate and augment the user's explicit query. This contextual evidence contributes to the term frequency metric to ensure a significant and positive impact on retrieval accuracy.

2 Background Work

Traditionally, a term in a given document is considered to be significant if it occurs multiple times within that document. This observation, commonly referred to as Term Frequency (TF), was made by Luhn [5]. His study was based on the fact that the authors of documents typically emphasize a topic or concept by repeatedly using the same words. Since then, most information retrieval approaches [1, 6] have adopted TF (or variations of it) as a benchmark for indicating term significance or relevance within a given document. In particular, it is normally combined with inverse document frequency (IDF) to form the TFIDF measure [7]. Even with the emergence of Web Information Retrieval, TF still continues to be a standard measure of term significance within a document. There are several examples content-based web information retrieval systems [8, 9, 10, 11] that assess term significance using TF. But as is the case for many potentially relevant documents, TF is not always the best or most useful indicator of term significance or relevancy. Quite often, there are relevant documents that contain only a single or a few occurrences of a particular term. Consequently, through TF these terms will rarely be considered significant, and thus never contribute impressively to the rank score of the potentially relevant document they appear within. This is especially the case when infrequently occurring terms appear in large documents containing hundreds or even thousands of terms.

In the query-centric approaches to retrieval [15, 16], queries can be classified to aid in the choice of retrieval strategy. Kang et al. [15] classify queries as either pertaining to topic relevance, homepage finding or service task and use this classification as a basis of dynamically combining multiple evidences in different ways to improve retrieval. Plachouris et al. [17] use WordNet in a concept-based probabilistic approach to information retrieval where queries are biased according to their calculated scope. In their work, scope is an indication of generality or specificity of a query and is used as a factor of uncertainty in Dempster–Shafer's theory of evidence.

The use of context in information retrieval is not a new idea. Jing et al. [18] use context as a basis of measuring the semantic distances between words. During indexing, the context of terms in documents is generated and stored in vector form. During retrieval, the context of a term in a query is generated and is used to measure the semantic distance between itself and candidate morphological variants in documents. Mutual information of terms is used to match related terms during the calculation of context distance. Billhardt et al. [19] propose a context-based vector space model for information retrieval. After the term-document matrix has been constructed, it is used is a basis for generating a term context matrix where each column is considered a semantic description of a term. This term context matrix is then combined with the document vectors from the term-document matrix to transform it into the final document context vector used for retrieval. The WEBSOM [20] system is an example of another way in which context has been used for information retrieval. It uses a two level Kohonen's self-organizing map approach to group words and documents of contextual similarity. Context in WEBSOM is limited to the terms that occur direct either sides of the term in question. IntelliZap [21] is a context-based web search engine that requires the user to select a key word in the context of some text. The approach makes effective use of the contextual information in the immediate vicinity of the keywords selected, so that retrieval precision can be improved. Inquirus [22, 23] is another web search engine that uses contextual information to improve search results. A user must specify some contextual information, considered as preferences, pertaining to the query. This context (preferences) provides a high-level description of the users information need and ultimately control the search strategy used by the system. Hyperlink information can be a very valuable source of evidence for web information retrieval and it is either based on a set of retrieved documents during retrieval or on a global analysis of the entire document collection during indexing. Kleinberg [24] illustrates how hyperlink information in web pages can be used for web search when using a set of retrieved documents. An approach that also uses the characteristics of link information from a set of retrieved documents for topic distillation is presented by Amitay et al. [25].

PageRank, as proposed by Brin et al. [26], is hyperlink-based retrieval algorithm that calculates document scores by considering the entire hyperlink connected graph represented by all the links in the entire document collection. It uses link information to model user behavior by calculating the probability that a user will eventually visit a certain page. This probability or PageRank of a page is used to prioritize its ranking during retrieval. The model with the most similar form of ours is [32], though it uses traditional query expansion to determine context of query. Another closely related work [31], implicitly deduce context using three different algorithms. Finally [30], offers Term context model as a new tool for accessing term presence in a document.

A novel technique called *Contextual Proximity Model (CPM)* is presented in this paper. It is a technique that generates term support in a *fundamentally different way to TF.* With CPM, a term appearing infrequently in within a large document can potentially be given a high confidence. *It does not rely on the number of times a particular term appears in a document to determine whether it is significant or not.* This is a significant characteristic of CPM that makes it different to TF. Secondly, we *use the co-information metric to determine a context coverage list and form the query context.* This context is then used as a basis for matching contexts with terms in documents and ultimately calculating term significance.

3 Proposed Technique

Contextual search refers to proactively capturing the information need of a user by automatically augmenting the user query with information extracted from the search context; for example, by using terms from the web page the user is currently browsing or a file the user is currently editing. The notion of Topical Relevance is atomic, vague and not explicit. Its essence is reasoning from evidence to a conclusion of concern (an answer to the user's query). We intend to use the simple criterion called domain dependency (Contextual Similarity) of words to structure topics.

The context of both terms in documents and terms in queries is fundamental to Contextual Toning (CT). The aim is to use "Context as a query" and treat the context as a background for topic specific search. The technique is based on the notion that if a term occurs in a document in the same context as the query, then that term is deemed significant to that document. The result of the technique is the generation of a Term Support. *Contextual Support (CS)* is a measure that yields a weighting for terms in document that can potentially complement TF.

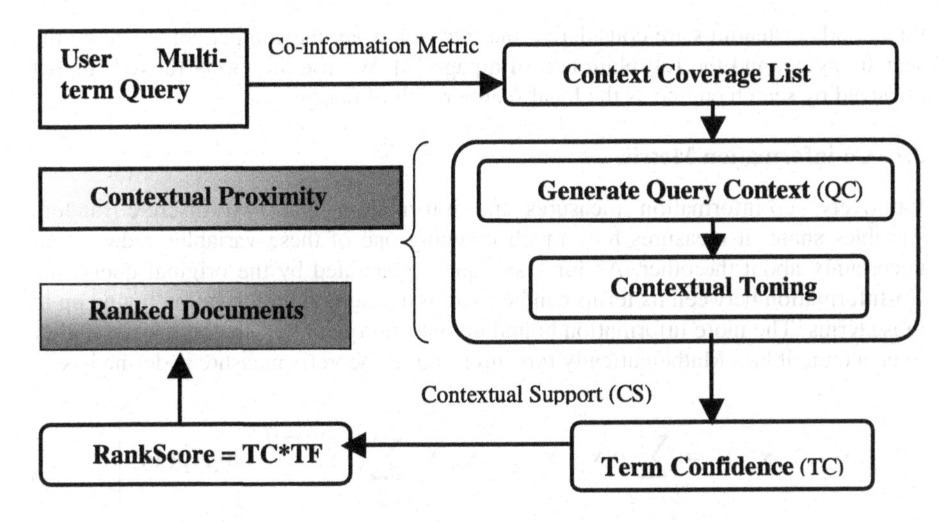

Fig. 1. Overview of the Contextual Proximity Model

3.1 Multi-term Query

Query, which entails a group of keywords, plays a quintessential role in the Web search paradigm. Recent studies claim that queries involving one or two keywords are most common in Web searches. While most Web Search engines perform very well for a single-keyword query, their precision is not good for query involving two or more keywords because the search results usually contain a large number of pages with weak relevance. Also, the users have a well-defined query re-formulation behavior, i.e., most multiple term queries include more than one context and users usually reformulate their queries by context instead of terms. A context is usually included as a sub-query in a user's query and it has strong impacts on the quality of search results. Thus we focus on how to detect the context of a user query.

A query typically contains only a few terms, which provide limited information. One straightforward method is to submit a query to a search engine to get the top ranked search pages. Those retrieved results provide some richer information about the query. In other words, we call the retrieved results of query as the local information of this query. Meanwhile, a query has its global information, based on the whole corpus, to provide more information. However, the global based approach can cause high computational complexity and it was shown in [2] that a local based approach outperforms the global based approach. In our approach, the top ranked search results are utilized to enrich the query.

Mathematically, we define a *Multi-term query* as:

A query $Q = t_1\ t_2\ \ldots\ t_n$ of n terms, where t_i ($1 <= i <= n$) is the ith term of Q.

Any subsets of $\{t_1, t_2, \ldots, t_n\}$, i.e., any combinations of $t_1\ t_2 \ldots t_n$ define the sub-queries. The *set of sub-queries* is defined as $SQ = \{sq_k\}$, $1 <= k <= 2^{n+1}-2$.

Our goal is to detect several sub-queries from SQ which could relate and contribute to the context. Given the search results for a query, we need to decide what features should be extracted from the search engine to construct the enrichment. Generally, three kinds of features are considered: the title of a page, the snippet generated by the search engine, and the full plain text of a page [3]. We use the top N ranked snippet retrieved by search engine as the local search result of query.

3.2 Co-information Metric

Intuitively, co-information measures the information that two discrete random variables share: it measures how much knowing one of these variables reduces our uncertainty about the other. As for a sub-query generated by the original query, the Co-Information between its terms can be used to measure the information bound up in those terms. The more information bound up one sub-query has, the higher possibility to be a topic it has. Mathematically this information-theoretic measure is defined as

$$I(x_{1i};x_{2j};\ldots;x_{nl}) = \sum_{i,j,\ldots,l} p(x_{1i},x_{2j},\ldots,x_{nl}) \left(\sum_{|v'|\subseteq|T'|} (-1)^{|v'|-|T'|} \log(p(T)) \right). \tag{1}$$

where $v' = (x_{1i}; x_{2j}; \ldots; x_{nl})$ and T' is any subset of v'. $|v'|$ means the number of elements in v', and the same with $|T'|$. In this way, for n events $x_{1i}, x_{2j}, \ldots, x_{nl}$, we define the co-information, CI, between them $I(x_{1i}; x_{2j}; \ldots; x_{nl})$ by

$$I(x_{1i}; x_{2j}; \ldots; x_{nl}) = \sum_{|v'|\subseteq|T'|} (-1)^{|v'|-|T'|} \log(p(T)). \tag{2}$$

We utilize the Co-Information metric to measure the degree of one sub-query being a topic. In our approach, the probability space of one query is built based on the local results of all sub-queries.

Given a query $Q = t_1 t_2 \ldots t_n$, where t_i ($1<= i <=n$) is the ith term of Q, we obtain the context coverage list as described in the following algorithm:

3.2.1 Coverage List Detection Algorithm

Step 1: Get the set of all sub-queries, $SQ = \{sq_k\}$, $1<=k<=2^{n+1}-2$, where $1<=k<=2^{n+1}-2$ is the number of all sub-queries and $2^n-2 = S^1{}_n + S^2{}_n + \ldots + S^{n-1}{}_n$

Step 2: Enrich each sub-query by submitting it into search engine and get the top N ranked snippets. In this way, we can get $(2^n-2).N^*$ snippets for query q. The snippet set is defined as $S(Q) = \{sn_i\}$, $1<=i<=(2^n-2).N^*$, where sn_i is the ith snippet. $N^* = \min(N, N')$, where N' is the actual number of retrieved snippets by search engine for each query.

Step 3: The probability of sub-query in S (Q) is defined as

$$p(sq_k) = \sum_{i=1}^{(2n-2).N^*} (sq_k \text{ can be found in } sn_i)/(2^n-2).N^* \;. \tag{3}$$

$p(sq_k)$ is the probability of occurrence of sq_k in collection.

Step 4: As for sq_k, we get its sub-query set, defined as $SQ_k = \{ sq_{k1}, sq_{k2}, \ldots sq_{k2^k-2} \}$. According to the theory of Co- information we mentioned in Section 3.2, CI value of sq_k is

$$I(sq_k) = \sum_{|sq_k| - |sq_{kj}|} (-1)^{|sq_k| - |sq_{kj}|} \log(p(sq_{kj})). \tag{4}$$

Step 5: Order all sub-queries in a descendant value of Co-information and split it into two parts by the threshold of zero. In other words, the first part includes the sub-queries with positive CI and the last part includes the sub-queries with negative CI.
Step 6: Finally, the list of context words we detected from SQ is defined as: Context Coverage List, CL = $\{ C_1, C_2, \ldots, C_i, \ldots C_M \}$, where C_i is one sub-query which has positive CI and M is the total number, and $I(C_i) >= 0$, $I(C_i) >= I(C_{i-1})$, $1 <= i <= M$.

3.3 Query Context

The context of a query consists of two sub-contexts, each of which is a set of terms with corresponding relatedness values:

- Set original query terms Q, and
- Set of context coverage terms CL.

These two sets of terms are sub-contexts and together they form the query context C= {Q, CL}.

3.4 Contextual Toning

The aim of Contextual toning is, using the query context determine the confidence that a term in a document is significant to that document. If a query term occurs in a document and it occurs in the context of the query, then it is considered to be important and given a high significance. The following sub-sections expound the contextual toning technique.

3.4.1 Proximity of Query Terms

Based on recent studies, the most common queries in Web searches involve one or two keywords. While most Web search engines perform very well for a single-keyword query, their precisions is not as good for queries involving two or more keywords. Search results often contain a large number of pages that are only weakly relevant to either of the keywords. One solution is to focus on the proximity of keywords in the search results. Proximity and semantic relationship between query terms can be used for web retrieval [28]. The relationship between query terms of a multi-term query can be mainly categorized into the following two types:

3.4.1.1 Topic Modifying. One query term represents a particular topic and the other modifies it. The query terms are subordinating.

3.4.1.2 Topic Collocating. Query terms represent individual topic.

3.4.2 Measures of Term Proximity

Proximity matters because terms that are close to each other in the text are more likely to be closely connected in the meaning structure of the text.

3.4.2.1 First-Appearance Term Distance (FTD). FTD means the term distance (TD) between the first appearances of X and Y in a document. The reason for using FTD is based on the hypothesis that important terms always appear in the forefront of a document, i.e, query terms emerge at the top of a document when they are contained in the subject of the document.

$$FTD(X, Y) = TD(\text{first}(X), \text{first}(Y)). \quad (5)$$

3.4.2.2 Minimum Term Distance (MTD). MTD means the smallest of all the term distances between X and Y in a document. The reason for using MTD as a proximity measure is based on the hypothesis that related terms appear in close proximity.

$$MTD(X, Y) = \min(\{TD(X, Y)\}). \quad (6)$$

3.4.3 Contextual Proximity

Given a term q and a set of terms that constitute a context C (i.e., Q or CL), then the Contextual Proximity (CP) of the occurrence of q in document D can be calculated

$$CP_{q,C,D} = (FTD_{q,c,D} + MTD_{q,c,D}) * R_c / \sum R_c. \quad (7)$$

where, c is a term in the context C, $FTD_{q,c,D}$ is the First-appearance term distance, i.e. is the term distance between the first appearances of q and c in D, $MTD_{q,c,D}$ is the Minimum Term Distance, i.e., the smallest of all distance between q and c in D and R_c is the relatedness of c to the original query Q. The Equation above effectively tones the contexts during retrieval. The best match is when terms in C occur directly next to occurrences of q in D. (During indexing, the position of the term in the document is recorded and stored in the index.)

For each query term, the technique separately calculates contextual proximity using both the original query Q and context coverage terms CL as contexts. The final measure is the *Contextual Support* (CS), which is a combination of the CP from both sub-contexts Q and CL. Given a query term q in the query Q, its CS is calculated by

$$CS_{q,D} = (CP_{q,Q,D} * w1) + (CP_{q,CL,D}) * (1 - w1)). \quad (8)$$

where, w1 is a weighting factor that is set to 0.5 by default. The resultant CS is a value in the range [0, 1] where a value close to 1 indicates a high confidence that the term q occurring in document D is a significant term and important indicator of relevance for D given Q. A value close to zero indicates insignificance and a low confidence of relevancy. The more related terms (terms in the context) that occur at a closer distance to the occurrence of the query term in the document, the higher the resultant confidence. On the other hand, the less related terms that occur a further distance from the occurrence of the query term, the lower confidence.

Unlike TF, that calculates term significance by counting the number of times a term occurs within a document, CPM relies on the context of the query and the context of term in the document to determine importance. Consequently, it has the significant advantage of potentially giving high confidence to terms that occur infrequently within documents. For example, consider a document D where term q occurs only once and each term in QC occurs only once. If q is close to the occurrences of terms in QC, then the resultant $CS_{q,D}$ will be high. The Toning technique heavily favors terms occurring very close to each other. The further they are apart or the closer to a distance of d, the less important the relationship of proximity is.

3.5 Combining CS and TF

As mentioned earlier, TF has long been used as a reliable indicator of term significance. We chose to incorporate it into CS by combining it with CS to give a final confidence measure of a term in a document. Given that contextual toning has been performed and we have a CS value for a term q in a document D, the final step of the technique is to combine CS with TF to give a final term confidence measure. TF is calculated by

$$TF_{q,D} = \log(count_{q,D} + 1) / \log(numWords_{D} + 1). \quad (9)$$

where $count_{q,D}$ is the number of times q occurred in D, and numWords is the number of terms in document D. $TF_{q,D}$ is calculated during indexing. Having both $TF_{q,D}$ and $CS_{q,D}$, the term confidence(TC) of q in D is calculated by

$$TC_{q,D} = (TF_{q,D} * w2) + (CS_{q,D} * (1 - w2)). \quad (10)$$

During retrieval the w2 is a weighting factor that is set to 0.5 by default.

There are a few important aspects of CPM that make it unique and different from existing techniques. Firstly, unlike TF that relies on a term to occur many times within a document to be considered significant or infrequently to be considered insignificant, CPM generates CS based on the notion that a term in a document is significant only if it occurs in the context of the query. This makes it frequency independent. Secondly, CPM is dynamic; the term significance is calculated at the retrieval time and not during indexing.

3.6 Efficiency

The added complexity of the CPM technique results in processing time overheads during retrieval. This is mainly because of the calculation of the closest distance between terms q and c that must be determined during matching so that chosen distance function can determine its importance as apart of CI calculation. Improved efficiency of CPM could be achieved by developing advanced functionality in the determination of closest distance between terms q and c. At the moment, the calculations iteratively to determine the closest distance between an occurrence of other term and the list of occurrences of the given term stored as a flat list of integers in memory. This can be thought as a type of linear search and is probably the most exhaustive and inefficient method of performing the determination of closest distance for context based matching. But on the other hand this type of data structure is easily created and managed during the reading of the term index from the file. Other methods such as binary tree base search may be introduced to improve efficiency, but was outside the scope of this research and not investigated.

4 Conclusion

The World Wide Web is undeniably "the place" for accessing information on any domain or subject. In this paper we discussed a novel context-based technique targeted to effectually mine user-centric information on Web. The proposed technique captures query context and matches it against term context in documents to determine term significance and relevance. It makes use of the Co-information metric for formulating a context coverage list to finally extract the query context. The model assesses the presence of a term in a document not by looking at the actual occurrence of that term, but by a set of non-independent supporting terms, i.e. context. This yields a weighting for terms in documents which is different from and complementary to TF-based methods, and is beneficial for retrieval. Given a multi-term query Q and a document collection DC, the technique can be used in a retrieval process in the following way:

I. *Generate query context QC for Q*
- ✓ Retrieve initial documents ID.
- ✓ Perform multi-term context detection on ID to obtain the context coverage list CL.
- ✓ Form Query context QC from Q & CL.

II. *Retrieve documents*
- ✓ For every query term q in Q
  - ➢ For every document D in DC containing q.
    - ❖ Tone QC with the term context of term q in D to calculate Contextual Support CS.
    - ❖ Combine CS with TF to give term confidence TC for q in D.
    - ❖ Add TC to rank score of D

References

1. Kosala, R., Blockeel, H.: Web Mining Research: A survey. SIGKDD Explorations 2(1), 1–15 (2000)
2. Bin, W., Zhijing, L.: Web Mining Research. In: ICCIMA 2003. Proc. 5th Int'l. Conf. on Computational Intelligence and Multimedia Applications (2003)
3. Kobayashi, M., Takeda, K.: Information Retrieval on the Web. ACM Computing Surveys 32(2) (2000)

4. Kuhlen, R.: Information and Pragmatic Value-adding: Language Games and Information Science. Computers and the Humanities 25, 93–101 (1991)
5. Luhn, H.: A statistical approach to mechanized encoding and searching of literary information. IBM J. Res. Develop. 1(4), 309–317 (1957)
6. Baeza-Yates, R., Ribeiro-Neto, B.: Modern Information Retrieval. Addison Wesley, New York (1999)
7. Salton, G., Yang, C.: On the specification of term values in automatic indexing. J. Doc. 29(4), 351–372 (1973)
8. Anh, V., Moffat, A.: Robust and web retrieval document-centric integral impacts. In: Proceedings of the 12th Text Retrieval Conference (TREC-12), Gaithersburg, USA, pp. 726–731 (2003)
9. Craswell, N., Hawking, D., Upstill, T., McLean, A., Wilkinson, R., Wu, M.: TREC 12 Web and interactive tracks at CSIRO. In: Proceedings of the 12th Text Retrieval Conference (TREC-12), Gaithersburg, USA, pp. 193–203 (2003)
10. Lawrence, S., Giles, C.: Context and page analysis for improved web search. IEEE Internet Computing 2(4), 38–46 (1998)
11. Robertson, S., Walker, S.: Okapi/Keenbow at TREC-8. In: Proceedings of the 8th Text Retrieval Conference (TREC-8), Gaithersburg, USA, pp. 151–161 (1999)
12. Yu, S., Cai, D., Wen, J., Ma, W.: Improving pseudo-relevance feedback in web information retrieval using web page segmentation. In: Proceedings of the 12th International Word Wide Web Conference (2003)
13. Voorhees, E.: Using WordNet for text retrieval. WordNet: An Electronic Lexical Database, pp. 285–303. MIT Press, Cambridge (1998)
14. Yu, S., Cai, D., Wen, J., Ma, W.: Improving pseudo-relevance feedback in web information retrieval using web page segmentation. In: Proceedings of the 12th International Word Wide Web Conference (2003)
15. Kang, I., Kim, G.: Query type classification for web document retrieval. In: Proceedings of the 26th Annual International ACM SIGIR Conference on Research and Development in Information Retrieval, pp. 64–71. ACM Press, New York (2003)
16. Plachouris, V., Cacheda, F., Ounis, L., van Rijsbergen, C.: University of Glasgow at the Web Track: Dynamic Application of Hyperlink Analysis using the Query Scope. In: Proceedings of the 12th Text Retrieval Conference (TREC-12), Gaithersburg, USA, pp. 636–642 (2003)
17. Plachouris, V., Ounis, I.: Query-biased combination of evidence on the web. In: Workshop on Mathematical/Formal Methods in Information Retrieval, ACM SIGIR Conference, pp. 105–121 (2002)
18. Jing, H., Tzoukermann, E.: Information retrieval based on context distance and morphology. In: Proceedings of the 22nd Annual International ACM SIGIR Conference on Research and Development in information Retrieval, pp. 90–96. ACM Press, New York (1999)
19. Billhardt, H., Borrajo, D., Maojo, V.: A context vector model for information retrieval. J. Am. Soc. Inf. Sci. Technol. 53(3), 236–249 (2002)
20. Honkela, T., Kaski, S., Lagus, K., Kohonen, T.: WEBSOM — self-organizing maps of document collections. In: Proceedings of WSOM_97 (Workshop on Self-Organizing Maps), Espoo, Finland, pp. 310–315 (1997)
21. Finkelstein, L., Gabrilovich, E., Matias, Y., Rivlin, E., Solan, Z., Wolfman, G., Ruppin, E.: Placing search in context: the concept revisited. In: Proceedings of the 10th International World Wide Web Conference, pp. 406–414 (2001)

22. Glover, E., Lawrence, S., Gordon, M., Birmingham, W., Lee Giles, C.: Web search — your way. Commun. ACM 44(12), 97–102 (2001)
23. Lawrence, S., Giles, C.: Context and page analysis for improved web search. IEEE Internet Computing 2(4), 38–46 (1998)
24. Kleinberg, J.: Authoritative sources in a hyperlinked environment. Journal of the ACM 46(5), 604–632 (1999)
25. Amitay, E., Carmel, D., Darlow, A., Lempel, R., Soffer, A.: Topic distillation with knowledge agents. In: Proceedings of the 11th Text Retrieval Conference (TREC-11), Gaithersburg, Maryland, USA (2002)
26. Brin, S., Page, L.: The anatomy of a large-scale hyper-textual web search engine. In: Proceedings of the 7th WWW Conference, Brisbane, Australia, pp. 107–117 (1998a)
27. Robertson, S.: On term selection for query expansion. J. Doc.
28. Tian, C., et al.: Web Search Improvement Based on Proximity and density of multiple keywords. In: ICDEW 2006. IEEE Proceedings of the 22nd International conference on Data engineering Workshops, IEEE Computer Society Press, Los Alamitos (2006)
29. Wen, J.R., et al.: Probabilistic Model for Contextual Retrieval, ACM SIGIR (2004)
30. Pickens, J., Farlance, A.M.: Term Context Models for Information Retrieval, ACM CIKM (2006)
31. Jonathan, S., et al.: Context Driven Ranking for the Web
32. Zakos, J., Verma, B.: A Novel Context-based Technique for Web Information Retrieval World Wide Web 9(4), 485–503 (December 2006)

Collection Profiling for Collection Fusion in Distributed Information Retrieval Systems

Chengye Lu, Yue Xu, and Shlomo Geva

School of Software Engineering and Data Communications
Queensland University of Technology
Brisbane, QLD 4001, Australia
(c.lu,yue.xu,s.geva)@qut.edu.au

Abstract. Discovering resource descriptions and merging results obtained from remote search engines are two key issues in distributed information retrieval studies. In uncooperative environments, query-based sampling and normalizing scores based merging strategies are well-known approaches to solve such problems. However, such approaches only consider the content of the remote database and do not consider the retrieval performance. In this paper, we address the problem that in peer to peer information systems and argue that the performance of search engine should also be considered. We also proposed a collection profiling strategy which can discover not only collection content but also retrieval performance. Web-based query classification and two collection fusion approaches based on the collection profiling are also introduced in this paper. Our experiments show that our merging strategies are effective in merging results on uncooperative environment.

Keywords: distributed information retrieval, peer to peer, collection fusion, collection profiling.

1 Introduction

As internet bandwidth become wider and wider and the network cost is continuing decrease, people are willing to share their own data collection with others over the internet. As a consequence peer to peer (P2P) systems become an important part of the cyber world. Since personal computers become more and more powerful and disk space is getting larger and cheaper, personal collection of data becomes larger and larger. Searching information in those large independent distributed collections can be treated as distributed information retrieval. According to the relationship between peers, distributed information retrieval systems can be divided into cooperative systems and uncooperative systems. Under cooperative environment, various information such as resource description, centralized index and collection statistics, etc., is hold in a central place. Clients can use such information to help their search. Under uncooperative environment, each client is independent and knows nothing about others. Clients can answer queries and return documents, but they do not provide other information such as collection statistics, collection description or retrieval model.

Z. Zhang and J. Siekmann (Eds.): KSEM 2007, LNAI 4798, pp. 279–288, 2007.

Collection fusion is one of key research area in distributed information retrieval systems. Collection fusion is referred to integrate the results from each individual distributed client. The final merged result list should include as more as relevant documents as possible and the relevant documents should have higher ranks. It is a difficult task because document scores returned by distributed collections are usually not comparable. In most case, collection statistics (e.g., size of the collection, inverse document/term frequency, etc.) are used to calculate document scores in most of the retrieval model such as Boolean model, probability model and vector space model. The use of collection statistics makes the document score quite different in different databases. Even the same document will have different score if it is in different databases. Therefore, merging results from different collections becomes a very complex task.

In this paper, we present our approach on collection fusion for uncooperative distributed IR systems. Specifically, we introduce a collection profiling technique for collection fusion which does not rely on sampling remote collections. The remainder of this paper is structured as follows: in section 2, we investigate on current collection fusion techniques; we present our web-based approach in section 3. Experimental evaluation and results discussion are presented in section 4 and we conclude in section 5.

2 Related Work

Under cooperative environment, global index is the most common technique for result merging. As the problem of result merging comes from the lack of collection statistic and the information of retrieval model, if the distributed collection can be treated as a logical centralized collection but documents are physically located distributed, there would be no merging problem. In global index architecture, usually there is a directory server that holds all the information of the distributed collections. Clients can get global collection statistics via the directory server then merge the distributed results together. As the document scores are calculated based on global collection statistics, the documents can be simply sorted by their scores. STARTS[3] is one of the best known protocols for the communication of peer to peer system. Clients can exchange their collection statistic via STARTS protocol. The global index architecture can achieve nearly 100% of the centralized information retrieval performance because a client will have all the information to calculate a document score as if they are in a centralized place. This architecture requires a deep collaboration between clients and fits well when all the clients are happy to share their entire collections such as libraries. However, this architecture is not practical in real word large scare distributed network because not all clients want to share their collection.

Normalized-score merging is a solution to real word large scare distributed information retrieval.[7, 8] The underlying idea is instead to keep all the information in a global index server, peers keep some information locally. Each collection is signed a rank according to their important. When merging, each document score is normalized by the rank of the corresponding collection, for example, increase document score if it comes from an import collection and decrease the document score if it comes from a less important collection. The collection rank can be calculated by collection ranking algorithms. CORI[2] and GlOSS[5] are two of the best known collection ranking algorithms. The collection information for collection ranking algorithms can be gathered by query-based sampling, user feedback or other training algorithms.

In GlOSS, collections are ranked by the similarity for a query. The similarity is calculated by the number of documents in the collection that are relevant to the query. This algorithm works very fine with large collection of heterogeneous data. Gravano and García-Molina[4] also suggest a variation of GlOSS known as gGlOOS which is based on vector space model. The similarity is calculated by the vector sum of the relevant documents instead of the number of relevant documents.

CORI is based on probabilistic inference network which is originally used for document selection. Callan [2] introduced this algorithm for collection selection. CORI uses document frequency (*df*, the number of documents containing the query term in a collection) and inverse collection frequency (*icf*, the number of collection not containing the query word) which is similar to term frequency (*tf*) and inverse document frequency (*idf*) in inference network. One of the advantages of CORI is it only use 0.4% the size of the original collection.

However, the merging strategies based on CORI and GlOSS are linear combination of the score of database and the score of the documents. It still requires the clients use the same indexing and retrieval model thus the document scores can be normalized. In today's p2p network environments, it is impossible to require peers to use the same software to manage their data. For example, in the popular BitTorrent and edonkey network, people use hundreds of different client softwares to share their files. It is reasonable to assume the trend of p2p network is to use various client softwares under the same communication protocol. Therefore, the document scores returned from peers may be based on different retrieval models. Thus the document scores cannot be normalized as the scores are not comparable. In some case, peers will only return an ordered list of documents without document score. Round-robin merging strategy is suggested to be use in the case that document scores are not comparable. Round-robin merging interleaves the results of each peer based on their original rank. It has been proved that the approach is simple but efficient when distributed collections have similar statistics and the retrieval performances are similar. However, Round-robin merging strategy will fail significantly when distributed collections have quite different collection statistics; for example, distributed collections are focus on different domains.

In summary, most of the result merging approaches requires information about the distributed collections' content, which is called resource descriptions. When processing a query, the collections will be signed ranks based on similarity between query and resource descriptions. In merging stage, the document score will be adjusted according to the collection rank. Query-based sampling by J. Callan and M. Connell (2001) [1]is the most famous technique to discover resource descriptions in uncooperative environment. It sends one-term query to distributed collections each time and learns the resource descriptions according to the returned top N documents. As the query terms are high frequency terms e.g. stop words, the sampling process can be treated as randomly selected files from distributed collections. It has been proved that this technique can produces accurate resource descriptions.

Voorhees et.al [9] present an merging approach based on the learning of past results of the past queries. Once training is complete, new queries are answered by matching the new query's content to that of the training queries and using the associated models to compute the number of documents to retrieve from each collection. Our work is based on Voorhees' work but differs in term of collection profiling and query clustering, which will be described in the rest of this paper in detail.

3 Web-Based Collection Profiling for Collection Fusion

In p2p networks, peer collections are managed by various IR systems. The retrieval performance of different IR systems could be quite different to a certain query. The retrieval performance should be taken into consideration when merging the results from different IR systems. In order to obtain both content quality and search engine retrieval quality, user feedback can be used together with collection ranking approaches such as CORI. This paper proposes a method that obtains resource descriptions and retrieval performances based on users' feedback.

Before we describe our approach, let us review how a user performs a query to distributed information systems. For the purpose of illustration, let's suppose that there are three remote collections: CA, CB and CC, and users have no prior knowledge of the content of the collections; each collection can be treated as a "Black Box". A user sends a query about art and computer science to these collections. Suppose that the user chooses 10 returned documents as relevant from each collection, if the 10 documents from CA are related to arts, the 10 documents from CC are related to computer sciences, and 5 each from CB are related to arts and computer sciences, it is reasonable to estimate that CA contains documents about arts but no computer sciences. CC contains documents about computer sciences but no arts. CB contains both computer sciences and arts. Therefore, according to query topics and user feedback, we could construct collection content profiles. This simple example also tells us that the remote IR systems will have different retrieval performance on different topics. An IR system that mainly contains computer science documents would not have good retrieval performance on art topics. User feedback can provide the information about how good a remote IR system's performance is on particular topics. Our idea is that the profiles of remote IR systems can be constructed based on user feedback. Based on the profile, the collection fusion can be improved by considering not only the content description but also the retrieval quality.

3.1 Query Classification

The internet provides huge amount of information. Some researchers have been studying on extracting topics classification from the internet.[6] However, their studies are based on western languages and their works are trying to build a complex tree structure for describing the relationship about various topics. Their works are effectiveness in homogeneous collections however, in a p2p environment, the documents are usually heterogeneous. Also, there are limited resources on text classification for Asian languages.

According to our investigation, many news websites classify their news into groups based on topics. For example, Yahoo news Taiwan (http://tw.news.yahoo.com/) groups their Chinese news into 12 topics. Google news Taiwan (http://news.google.com/news?ned=tw) groups their news into 9 topics. Those websites also provide very powerful search features on news. When searching in Yahoo news Taiwan, a list of news that contains the query term will be returned together with the source of the news, the catalogue and the news summary. Google news Taiwan would not return catalogue information of the news but it enables the user to search news within a specific catalogue. Mining such information may find the topics that a query term is related. For example, when searching for term "雅虎" (Yahoo) in Yahoo news, most of the returned news are about computer science. Therefore, we can determine that term "雅虎" (Yahoo) has strong relationship to computer science but has no relationship to arts, for example. As a result, with the help of such news sites, we can discover the relationship between a query term and a topic. Such information will then help identify query topics. For example, searching for "Linux" in Yahoo news Taiwan, out of 28 returned news, returns 20 news under topic "SCI/TECH", 1 news under topic "world news", 3 news under topic "financial" and 4 news under topic "education". This result indicates that, the term "Linux" will have a chance of 72% to be in topic "SCI/TECH", 4% to be in topic "world", 10% to be in topic "financial" and 14% to be in topic "education". By searching the term in all the catalogues in Google news Taiwan and calculating the number of results returned from each catalogues, we can also get the percentages that a term belongs to a particular catalogues.

Let $C=\{c_1,c_2,\cdots,c_{|C|}\}$ be a set of predefined classes, i.e., $c_j \in C$ is a class or catalogue, $N(w_i,c_j)$ be the number of news in class c_j returned from querying term w_i in a news web site. The probability that the term w_i belongs to a topic class c_j can be calculated by the following equation:

$$P(c_j \mid w_i)=\frac{N(w_i,c_j)}{\sum_{c_k\in C} N(w_i,c_k)} \tag{1}$$

Let $W=\{w_1,w_2,w_3,\ldots,w_{|W|}\}$ be a set of Chinese terms, $Q=\{w_1,w_2,\cdots,w_m\}$ represents a query that contains m terms. The probability that the Q belongs to a topic c_j can be calculated by equation:

$$P(c_j \mid Q)=P(c_j \mid w_1w_2\cdots w_m) \tag{2}$$

As the calculation of the probability of each term w_i belongs to a topic class c_j is independent event, we will have: $\forall w_i \in W, i=1,\ldots.m, c_j \in C$,

$$P(c_j \mid Q)=P(c_j \mid w_1w_2\cdots w_m)=\prod_{i=1}^{m} P(c_j \mid w_i) \tag{3}$$

As a result, we can calculate the probability that a query Q belongs to a topic c_j by using the equation above.

$\forall w_i \in W, i = 1,....m,$ the topic of a query $Q = \{w_1, w_2, \cdots, w_m\}$ is $c \in C$ which is determined by the following equation:

$$P(c \mid Q) = \max_{c_j \in C} \left(P(c_j \mid Q) \right) \tag{4}$$

3.2 Collection Profiling

For most of the existing work, collection profiles only contain resource description. Both collection selection and collection fusion are based on collection rank which is determined by the similarity of query and collection description. Our approach will consider both collection description and retrieval performance. In our system, a collection profile contains information about the contents and the performance of the search engine. A matrix $\{p_{i,j}\}$ is used to present the historical performance of collections, $p_{i,j}$ represents the average retrieval performance in catalogue c_j of collection i.

The performance of a search engine is usually measured by precision and recall. As most of the users only read top N results, the precision of top N results, denoted as PN, is a reasonable benchmark for the search engine performance. The average PN of a collection can measure how well a remote search engine preformed in the pass. However, using absolute value cannot tell if the performance is stable. For example, A= {0.6, 0.4, 0.5, 0.5} and B= {0.9, 0.1, 0.2, 0.8} are two historical performance for two collections. Their average performances are the same, however, obviously collection A is much stable than collection B. These results indicate that collection B is sensitive to some topics. For general use, collection A is a better choice than collection B. We introduce Overall Position (OP) as a measurement of the collections' performance.

Overall Position is a relativity measurement. Every time a collection returns results, it will be assigned a position according to its PN. For example, suppose tht we have totally M collections, the collection that has highest PN will be ranked 1; the one has second highest PN will have rank 2 etc. The collections that have no relevant documents will have rank M. The OPs indicate how well a collection performed comparing to other collections. The average position of pervious runs for collection i is denoted as OP_i.

3.3 Collection Fusion

In an uncooperative p2p environment, document scores may not be available in the result lists or the document scores are not comparable. In this section, we will introduce our merging strategy in an uncooperative p2p environment. Generally, the retrieval process is conducted with the following steps. Firstly the user's query Q is classified using equation (4) and then broadcasted to all the remote peer IR systems. When results are returned from peer IR systems, the results will be merged according to the query catalogue and collection profile. We propose a merging method called Sorted round robin strategy which incorporates collection profile into the standard round robin method to enhance the quality of result merging.

The basic idea of the round robin merging, in general, is to interleave the result list returned from each remote peer. Every time the top document in the list from each peer will be popped up and inserted into the final result list. The order of the peers to be visited is usually the order of the peer collection id. It is obvious that the basic round robin approach does not consider the performance of remote peers. In the worst case, the results from the worst peer will be popped up first and the results from the best peer will be popped up last. Therefore, the irrelevant documents will appear on the top of the merged result list. As a result, the distributed retrieval performance will be harmed significantly in the worst case. Further more, remote systems will have different retrieval performances on difference topics, as we described in previous sections. However, traditional round robin always sorts the results from remote systems in same order. Therefore, even the order of the remote systems have been optimized, the performance of the merged result cannot be guarantee.

For the above reasons, we proposed a modified round robin approach called sorted round robin margining strategy. Instead of using fixed order of remote system to merge results, we dynamic change the order of remote system based on query classes and pervious performance. In other words, the order of the visiting is determined based on the matrix $\{p_{i,j}\}$. The merging strategy can be described as following steps:

1. Determine query class c_j using equation 4.
2. Sort Collections by $\{p_{i,j}\}$.
3. Using round robin strategy to merge results, based on the collection order generated in step 2.
4. Repeat steps 1-3 for all input queries.

By using such merging strategy, we always ensure the order of the merge order be optimized no matter what type of query we are using. The more important collection will always be visited first and the quality of merged results can always be guarantee.

We also propose another merging strategy called Sorted Rank. The idea of round robin merging is one-by-one merging strategy. This strategy means if we have n remote systems, the second document in the most important system will be in the n+1 position in the merged result list. However, from our intuition, the more important system should have more documents in the top of the merged result list than the less important one. The simplest way to calculate a document score is to make the score linear to the collection rank (OP) and original rank in the remote system. Therefore, the designed the document score calculated by the following equation:

$$score(r_i, c_j) = r_i * cScore(i, c_j) = r_i * p_{i,j} \tag{5}$$

Where r_i is the document rank, i is collection that returns the document, c_j is the catalogue that the query belongs to and $p_{i,j}$ is the historical performance of collection i in collection j. Then the documents will be sorted by the calculated scores.

4 Evaluation

4.1 Test Set

We conducted the experiments with 30 databases from NTCIR6 CLIR track document collections (http://research.nii.ac.jp/ntcir/ntcir-ws6/). The documents in the collection are news articles published on United Daily News(udn), United Express(ude), MingHseng News(mhn), and Economic Daily News(edn) in 2000-2001, all together 901,446 articles. The articles are evenly separated into 30 databases which makes each database has around 30048 documents. In order to make the databases cover different topics, according to relevance judgments, relevant documents on different topics are manually put into different databases. 50 queries from NTCIR5 CLIR task are used as training set. That is, the collection profiles were created based on those 50 queries. P20 were used in profiling. 50 queries from NTCIR6 CLIR task are used in evaluation.

4.2 Retrieval System

The documents were indexed using a character-based inverted file index. In the inverted file, the indexer records each Chinese character, its position in the document, and the document ID. Chinese phrase is determined by each Chinese character position and document ID. A character sequence will be considered as a phrase in the document only when character positions are consecutive and have the same document ID. The English word and numbers in the document are also being recorded in the inverted file. The retrieval model that is used in the system is a Boolean model with *tf-idf* weighting schema. All retrieval results are initial search results without query expansion.

4.3 Experiment Design

Several runs were conducted in the experiments which are defined as follows:

- Centralize: all documents are located in a central database.
- Round robin (RR): results are merged using the standard round robin method. The order of visiting result lists is the order of collection id.
- Sorted round robin (SRR): results are merged using sorted round robin method described in section 3.3. The order of visiting result lists is the order of corresponding collection rank (OP).
- Sorted rank (SR): results are merged using sorted rank method described in section 3.3. Document scores are calculated by equation 6 and ascending sorted.

5 Results and Discussion

The average precision in table 1 show that SR is the most effective way of merging distributed results while the standard round robin method produced the lowest precision. If we use the centralized system's precision as the baseline, the precision produced by SR is 9.2% higher than SRR's and 15.2% higher than RR's, and the precision of SSR approach is about 6% higher than RR's. As the only difference between SRR and RR is the order of the collections to be visited, it is easy to conclude that sorting the returned results according to the importance of the collections can improve the precision. If we only at P5 and P10, SRR is 21% better than RR at P5 and about 9.1% better than RR at P10. The P-R curves clearly indicate that the precision of SRR is much higher than RR at the top of result list. The performance gain of SRR is mostly come from high precision up to P20.

Table 1. Average Precision

	RR	SRR	SR	Central
P5	0.1400	0.2323	0.2720	0.4360
P10	0.1680	0.2080	0.2340	0.4040
P15	0.1640	0.1987	0.2170	0.3787
P20	0.1560	0.1880	0.2000	0.3500
P30	0.1587	0.1580	0.1900	0.3167
All	0.0896	0.1029	0.1230	0.2202
% of central (all)	40.7	46.7	55.9	-

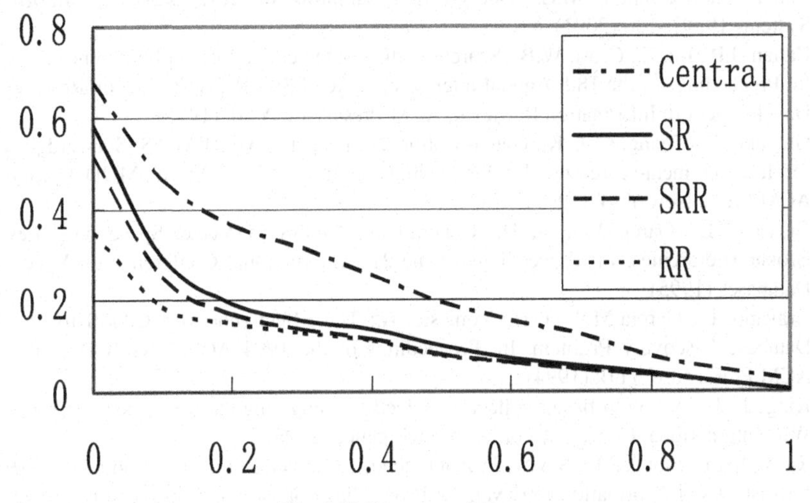

Fig. 1. P-R curves of 4 runs

SR has 9.2% improvement than SRR. This is because the SRR method sorts the collections simply according to their importance. The results from distributed collections are still evenly distributed in the merged result list. In SR, the more important the collection is the more documents from the collection will appear in the top of the merged result list. For example, if collection A has OP 5 and collection B has OP 20. According to equation (5), the first 3 documents from A will have higher rank than the first document in B. P-R curves clearly show that SR is much better than SRR in extracting relevant documents in the middle of the result list. Table 1 also shows the same result. At P10 and P15, SR is only about 5% better than SRR but from P20, SR is about 10% better than SRR.

Although centralized collection produces the highest precision, SRR and SR still provide reasonable better performance than the standard round robin method.

6 Conclusion

In this paper, we proposed and evaluated our approach on result merging that can be applied to uncooperative distributed information environments such as p2p systems. A Web based query classification method is also introduced in this paper. Learning user behaviors and using query classification can create collection profiles which contain not only collection content but also performance of remote information retrieval systems. Using the information in the profile can help merging results. Our experiments proved that our proposed SRR and SR approaches can provide much better results than the standard round robin method.

References

[1] Callan, J.P., Connell, M.E.: Query-Based Sampling of Text Databases. Information Systems 19, 97–130 (2001)

[2] Callan, J.P., Lu, Z., Croft, W.B.: Searching distributed collections with inference networks. In: Proceedings of the 18th Annual International ACM SIGIR Conference on Research and Development in Information Retrieval, ACM Press, New York (1995)

[3] Gravano, L., Chang, C.-C.K., Garca-Molina, H., Paepcke, A.: STARTS: Stanford proposal for Internet meta-searching. In: Proceedings of the 1997 ACM SIGMOD Conference, ACM Press, New York (1997)

[4] Gravano, L., García-Molina, H.: Generalizing GlOSS to Vector-Space databases and Broker Hierarchies. In: Proceedings of the 21st International Conference on Very Large Databases (1995)

[5] Gravano, L., García-Molina, H., Tomasic, A.: The Effectiveness of GlOSS for the Text Database Discovery Problem. In: Proceedings of the 1994 ACM SIGMOD conference, ACM Press, New York (1994)

[6] King, J., Li, Y.: Web based collection selection using singular value decomposition. In: WIC International Conference on Web Intelligence (2003)

[7] Rasolofo, Y., Abbaci, F., Savoy, J.: Approaches to collection selection and results merging for distributed information retrieval. In: Proceedings of the tenth international conference on Information and knowledge management, Atlanta, Georgia, USA (2001)

[8] Si, L., Callan, J.: Using sampled data and regression to merge search engine results. In: Proceedings of the 25th annual international ACM SIGIR conference on Research and development in information retrieval, Tampere, Finland (2002)

[9] Voorhees, E.M., Gupta, N.K., Johnson-Laird, B.: The Collection Fusion Problem, The Third Text REtrieval Conference (TREC-3), National Institute of Standards and Technology, Gaithersburg, M.D. (1994)

Integration of Descriptors for Software Component Retrieval

Yuhanis Yusof and Omer F. Rana

School of Computer Science, Cardiff University, Wales, UK
{y.yusof,o.f.rana}@cs.cardiff.ac.uk

Abstract. Software component retrieval is an important task in software reuse; after all, components must be found before they can be reused. In this paper, we propose the use of compound index to integrate two types of information (i.e functional and structural) that are used to represent a component in a component retrieval system. The proposed mechanism is flexible as it can be expanded to include additional information extracted from a software component. In order to retrieve components that are relevant to a given query, similarity measurement based on vector model and data distribution were performed. Experiments of program retrieval undertaken using existing approach (function-based) and the combination approach were also performed.

1 Introduction

Software component retrieval is an important task in software reuse [1]; after all, components must be found before they can be reused. Based on existing work in software component retrieval [2,3,4,5,6,7,8,9,5,10,11], there are two types of retrieval: function-based and structure-based.

Given a query that emphasizes what a component should do, function-based retrieval presents developers with software components that *act* similarly. This means that the retrieved components illustrate the same function as that defined in the query. Similarities between the query and components in the repository may have been performed using the textual analysis approach [6,12,7], which uses term occurrences to represent the function of a program. Nevertheless, various other methods have been used in representing functionality of a software component and these include textual description (e.g facet-based [2]), operational semantics (e.g execution-based [4]) and denotational semantics (e.g signature matching [13]).

In contrast to function-based retrieval which identifies components that *act* similarly, structure-based retrieval presents developers with components that *look* alike. A simple example would be two distinct programs that illustrate factorial function using different approaches, e.g. recursion and looping; even though they have the same function, they have a different structure. Among the existing studies that focus on structure information contained in a component are the work by Willis [14], and Santanul and Atul [15].

Z. Zhang and J. Siekmann (Eds.): KSEM 2007, LNAI 4798, pp. 289–300, 2007.

In this paper, we are proposing to combine the two types of retrieval by integrating the functional and structural descriptors into a single index. This index known as a compound index is used to represent a search query and components (i.e program source code) in a software repository. The work described in this paper is based on the following:

Definition 1. *Functional descriptors consist of information extracted from a program that represents the functionality of the program. This includes terms (identifier names) extracted from the code and comment statements written in a program.*

Definition 2. *Structural descriptors are information inferred from relationships between properties (e.g methods, objects) of a program. Such relationships includes class inheritance, interface hierarchies, method invocations and dependencies, parameters and return types, object creations, and variable access within a method. In this paper, the information inferred using the structural relationships are the design patterns and software metrics.*

Definition 3. *A compound index is an index containing several indices which includes functional and structural descriptors. Elements of this index are represented using continuous value (e.g 1, 4, 10, etc.) which indicates the importance of a term in the program or existence of a particular structural descriptor.*

Based on the work undertaken by Mili et al. [16], relevant components can be retrieved by identifying components that minimize some measure of distance to the user query. Such an approach expects that the outcome will either be an *exact match* [16] or (failing an exact match) one or more *approximate matches* [16]. As many studies [17,3,18] have demonstrated the effectiveness of using vector model in identifying similarities between two objects (e.g text documents, images), we investigate the use of this model in retrieving programs that are similar to a given query. Details of the formulas (i.e Euclidean Distance and Cosine Measure) which are based on this model are described in section 4.2. On the other hand, we also investigate the use of information on distribution of data (in a compound index) as the distance measure. The idea is to investigate whether programs with similar data distribution illustrate similar function and structure. An elaboration on similarity measurement based on a data distribution can be seen in section 4.1. In section 6, we present the results of program retrieval undertaken using existing approach (function-based) and the combination approach.

2 Related Work

Approaches undertaken in software component retrieval that employ functional descriptors as components representation are identified as using either information retrieval [6,12,7], descriptive [2,19,3], operational semantics [10,11,4] or denotational semantics methods [8,9,5]. The first method depends on a textual analysis of software components while the second relies on an abstract representation of the components, such as facet-based [2]. In the operational semantics methods, components are represented by how they function (e.g sample of input/output). On the other hand, the denotational semantics is an approach to formalizing the semantics of a component by constructing mathematical objects (called denotations or meanings) which express the semantics of these components (e.g using a formal language).

Function-based retrieval is very important, as it can provide effective and precise retrieval results [2,9,7]. Unfortunately the semantics of a software component identified using information retrieval and descriptive methods may be hard to determine if the software is not well-documented. Therefore there has been studies on using structure information as the basis of comparison between a query and components in a repository. Such information is based on structural relationships, thus it does not rely on the semantics of the natural language that exist in a component. Existing work on structure information includes the use of programming cliches [14,20] and pattern language [15]. A cliche is a pattern that appears frequently in many different programs (and possibly many different languages). For example, developers have probably already learnt the pattern for iterating through an array and can write such behaviour very quickly and reliably. Paul and Prakash [15] presented a framework in which pattern languages are used to specify interesting code features. In *SCRUPLE* [15], specifications for a software component are written using a pattern language, which is an extension of the programming language being used.

3 Model of Program Retrieval Using a Combination Approach

Figure 1 illustrates the model of program retrieval used in this work. There are three types of information extraction performed on a program. The first extractor, namely `Descriptor`, builds a functional index file of a program-term matrix (see Definition 4). This index file is updated whenever the repository receives new applications to be stored or when there is a request to withdraw any particular applications.

Definition 4. *A program-term matrix contains rows corresponding to the programs and columns corresponding to the weighted terms. Terms extracted from a program are given different weights to indicate their importance in determining the program's function. An example of such matrix as follow:*

The second extractor, which is known as the `Design Pattern`, is used to identify the existence of three design patterns (i.e Singleton, Composite and Observer)

Table 1. Program-term matrix

	database	connect	connectString	setDriver	driver
P1	3	2	2	0	0
P2	3	0	0	2	2

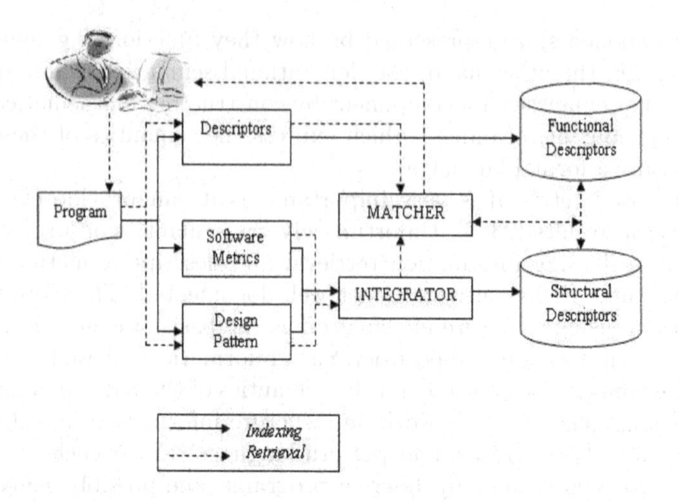

Fig. 1. Model of Program Indexing and Retrieval using a Combination Approach

in a program. Meanwhile, the `Software Metrics` extractor, extracts software metrics from a program and uses these metrics to classify the program into the appropriate application domain. Information on the program's application domain and relevant software metrics, that depict the reusability of the program [21,22], are later submitted to an `Integrator`. The `Integrator` will then combine this information with that obtained from the `Design Pattern` and generate the following:

1. confidence level of the existence of design patterns - (p_1, p_2, p_3). The higher the value of p_i in a program, the more the program is believed (through the existence of structural relationships) to employ a particular design pattern.
2. confidence of classification of a program into appropriate application domain - (d_1, d_2). The greater the value of d_i, the more a program is believed to be in a particular application domain.
3. software metrics used in determining program complexity and reusability - $(m_1, m_2, m_3, m_4, m_5, m_6, m_7)$. The fewer the value of m_i in a program, the more reusable the program is.

Variable p comprises data on the existence of three different design patterns, namely Singleton(p_1), Composite(p_2) and Observer(p_3). The second variable, d, is built upon information generated by topic classifiers that currently classify a given program into either database(d_1) or graphics(d_2). Finally, a total of seven software metrics is assembled in variable m that represents the complexity and reusability of a program. All of this information is mapped into a program-structural matrix which is later used to create the structural index file. In the context of this paper, such matrix is defined as follows:

Definition 5. *A program-structural matrix contains rows corresponding to the programs and columns corresponding to the structural descriptors. An example of such a matrix is as follows:*

Table 2. Program-structural matrix

	p_1	p_2	p_3	d_1	d_2	m_1	m_2	m_3	m_4	m_5	m_6	m_7
P3	2	0	3	1	0	0	8	7	0	0	7	1
P4	0	2	0	0	1	5	2	5	1	1	2	1

A program that is submitted to the retrieval system as a search query will undergo the same process: the `Descriptor` extracts relevant weighted terms from the program while the `Integrator` integrates information on design patterns, application domain and software metrics of the program. This information is then sent to the `Matcher` which will combine them into a compound index. This index is later known as a `Query Compound Index` (Qci).

For example: given a program, Q as the query, the Integrator produced the following compound index $Qci = 3, 2, 2, 2, 0, 3, 1, 0, 0, 8, 7, 0, 0, 7, 1$. This index is built upon the following:

Weighted terms (w): **{3, 2, 2 }** These values represent the three terms (e.g database, connect, connectString) extracted from program Q.

Design pattern (p): **{2, 0, 3 }** The three values in the tuple represent the existence of three design patterns in a program. For example, value 2 illustrates that two of Singleton's rules have been fulfilled while the value 3 indicates that there are three structural relationships in the program that illustrate existence of Observer design pattern.

Topic Classification (d): **{ 1, 0 }** The values in this tuple depict how sure the topic classifiers are in predicting the topic of a given program. Value 1 as index number 4 shows that all three classifiers have categorized the query as being in the database domain.

Program Reusability (m): **{ 0, 8, 7, 0, 0, 7, 1 }** The first two indices represent the complexity of a program while the rest of the values are used in determining the reusability of a given program. These metrics are based on the metrics defined by Chidamber and Kemerer [23].

Upon creating the Qci to represent a search query, the `Matcher` creates another program-term matrix. This matrix contains rows corresponding to the programs in the repository and columns corresponding to the weighted terms (e.g database, connect, connectString) in Qci. Nevertheless, the matrix only includes programs that contain term(s) as defined in the query. For each of the programs in the matrix, the `Matcher` combines the weighted terms, based on data in the functional index file (i.e generated by the `Descriptor`), with structural descriptors defined in the structural index file, into a Program Compound Index (Pci). Similarities between a Qci and each of the Pci, (Pci_n), are determined in order to present the user with relevant programs.

4 Similarity Measurement

In this section similarities between programs (represented as Qci and Pci) are identified using calculation undertaken based on vector model and data distribution.

4.1 Data Distribution

In the context of this paper, data distribution is the information on how data in a compound index are distributed. An example of data distribution measures is the skewness [24]. In order to determine similarities between two programs, the distance between the data distribution measurement (i.e skewness) is calculated.

Each of the programs stored in our repository is represented by its skewness value, which is determined based on the distribution of the indices in its compound index. Skewness characterizes the degree of asymmetry of a distribution around its mean [24]. For the data in a compound index, w_1, w_2, ..., w_n, the formula for skewness is:

$$skewness = \frac{\sum(w_i - \overline{w})^3}{(n-1)s^3} \tag{1}$$

where $\overline{w}$ is the mean, s is the standard deviation, and n is the number of indices in the compound index.

Our idea is for the `Matcher` to retrieve programs that illustrate similarities in both descriptors, functional and structural, when compared to a given query. Negative values for the skewness indicate data that are skewed left and positive values for the skewness indicate data that are skewed right. In the context of this paper, if data in a compound index is skewed left, it means that the program is represented more by its functional descriptors, and vice versa when it is skewed to the right. Based on a Qci, the `Matcher` identifies programs from the repository, with skewness values similar to Qci. These programs are then sorted based on their skewness values to be presented to the user. Programs having the same value of skewness as the Qci will be presented on the top of the hit list while others with the least value of difference will be sorted decreasingly.

4.2 Vector Model

Using the vector model as the similarities measurement, a program c_j and a user query q are represented as t-dimensional vectors. In this work, the vector model evaluates the degree of similarity of the component c_j with regard to the query q using two calculations: Cosine Measure and Euclidean Distance.

Cosine Measure proposes to evaluate the degree of similarity of the component c_j with regard to the query q as the correlation between the vectors $\mathbf{c_j}$ and $\boldsymbol{q}$. This correlation can be quantified, for instance, by the cosine of the angle between two vectors. That is,

$$sim(c_j, q) = \frac{\sum_{i=1}^{n} w_{i,c_j} * w_{i,q}}{\sqrt{\sum_{i=1}^{n}(w_{i,c_j})^2} * \sqrt{\sum_{i=1}^{n}(w_{i,q})^2}} \tag{2}$$

Ranking for Cosine Measure is done from highest value to the lowest value, i.e. highest Cosine Measure are placed first.

Euclidean Distance or simply *ED*, examines the root of square differences between indices of a pair of compound index. In mathematics, the Euclidean Distance or Euclidean metric is the distance between the two points that one would measure with a ruler, which can be proven by repeated application of the Pythagorean theorem. By using this formula, the distance between a program in a repository and a given query can be obtained using the following:

$$distance(c_j, q) = \sqrt{\sum_{i=1}^{n}(w_{i,c_j} - w_{i,q})^2} \quad (3)$$

For the ED, ranking is done from lowest distance to highest distance, i.e. the program with lowest ED is placed first.

5 Experiments

Based on the elaboration in the previous section, each of the programs in the repository are represented by: weighted terms, design patterns and software metrics. The search query which is in the form of a program is also represented by the same type of information. Experiments of program retrieval undertaken using distinct similarities measurements were then conducted to investigate whether one of the retrieval techniques outperforms the other for a given set of queries. A total of ten queries were used in the experiment. The relevance of programs retrieved by the system was determined by the researcher, (i.e identifying programs that the researcher thought to be relevant to the query). This is undertaken by determining similarities in terms of (1) function, (2) design patterns and (3) application domain. The sets of relevant programs determined by the above criteria are by no means extensive.

5.1 Retrieval Evaluation

Performance of the retrieval system using different similarity measurements was undertaken based on the average value of the measurement scores (i.e precision and recall) at a fixed document cut-off value (DCV), i.e $n = 10$. Precision is the ratio of the number of relevant programs retrieved to the total number of all (irrelevant and relevant) programs presented in a hit list, while recall is the ratio of the number of relevant programs retrieved to the total number of relevant programs in a repository.

Three similarity measurements are supported by our retrieval system, namely, Skewness, Euclidean Distance and Cosine Measure. The precision and recall values obtained from the experiments undertaken are depicted in Table 3 .

From the data depicted in Table 3, it can be seen that program retrieval for 9 out of 10 queries undertaken using ED as the similarity measurement generated

Table 3. Precision and Recall Scores for the Top 10 Programs

Query	# Relevant Programs	Skewness		Euclidean Distance		Cosine Measure	
		Precision	Recall	Precision	Recall	Precision	Recall
Q1	67	0.3	0.045	0.6	0.090	0.6	0.090
Q2	194	0.2	0.010	0.4	0.031	0.5	0.026
Q3	15	0.2	0.133	0.3	0.2	0.3	0.2
Q4	20	0.3	0.150	0.4	0.2	0.4	0.2
Q5	28	0.4	0.143	0.6	0.214	0.5	0.179
Q6	45	0.5	0.111	0.7	0.156	0.7	0.156
Q7	105	0.5	0.048	0.8	0.076	0.8	0.076
Q8	88	0.4	0.045	0.6	0.068	0.6	0.068
Q9	275	0.5	0.018	0.7	0.025	0.7	0.025
Q10	144	0.6	0.042	0.8	0.056	0.7	0.049

a precision greater or at least similar to using the Cosine Measure. The highest precision (i.e 0.8) for program retrieval was obtained when the ED and Cosine Measure were used as the similarity measurements for Q7. In addition, by using ED as the similarity measurement, we also obtained the highest recall (i.e 0.214) in the experiment, and this is revealed by the recall score for Q5. The low scores in recall can be accounted for the small value of DCV used in the experiment. If the DCV is smaller than the number of relevant programs, it is difficult to obtain a recall score of one. For example, if only five programs are examined and 50 relevant programs exist for a given query, then the recall is only 0.1 (10%) even if all the programs examined are relevant. Hence, this makes the search methods (i.e similarity measurements) appear much worse than they actually are.

We also reveal the average of precision and recall scores for all ten queries that were used in the experiment undertaken and this can be seen in Table 4. The results suggest that program retrieval is better undertaken using the Euclidean Distance rather than Cosine Measure and skewness as the similarity measurements. Therefore, it has been used as the default similarity measurements in our program retrieval system.

Table 4. Average values of Precision and Recall Scores

	Skewness		Euclidean Distance		Cosine Measure	
	Precision	Recall	Precision	Recall	Precision	Recall
Average	0.390	0.075	0.590	0.112	0.580	0.107

The decision of choosing ED rather than Cosine Measure or skewness as the similarity measurement in our program retrieval system is also supported by the statistical analysis. We performed the ANOVA test [24] (as the scores were normally distributed) to detect significant difference in the retrieval scores across multiple similarity measurements. In implementing the test, our null hypothesis, H_0, was that all the similarity measurements being tested were equivalent in terms of precision and recall. If the `p-value` obtained in the test is less then the identified significance level, α, we could conclude that the similarity measurements were significantly different. In this test, the α value was set to be 0.05 which is an acceptable value in any statistical test [24]. The ANOVA test reveals that there was a statistically significant difference in precision scores across the three similarity measurements as the obtained `p-value` was 0.004 (smaller than the α). The statistical significance of the differences between each pair of similarity measurements is later identified through Post-Hoc test [24]. The test reveals that there was a significant different in the scores between skewness and ED and between skewness and Cosine Measure.

6 Comparison with Other Tools

Our retrieval approach concerns the use of functional and structural descriptors in representing a program. It is similar to other work that is related to source code retrieval such as Google code search [25], Koders search engine [26] and SCRUPLE [15], nevertheless this work focuses on using functional or structural descriptors on their own. Comparing our approach to [25,26,15], which involves searching for code samples that match a specified regular expression [26], our approach include the identification of information that may not be explicitly available in a program (e.g design patterns and software metrics). Such an approach is required to fulfill users' search requirements that may include both types of descriptions; function and structure. For example, a developer may require programs that illustrate how to connect to a SQL database. Therefore the required programs need to determine whether or not the user has the appropriate JDBC driver loader. If so, then only the connection to the database can be made. Such functionality depicts dependencies between two objects – loader and database. Hence, illustrating existence of Observer design pattern. However, information on design patterns employed in a program are not usually included explicitly in the code or comment statements, for example, comment statement such as *This class implements the Observer pattern.* Therefore, including the keyword *Observer* into the search query may not be useful as relevant comparison could be made by the retrieval system. With this, program retrieval undertaken based on semantic terms, as employed by Koders search engine and Google code search, may present developers with programs that illustrate the required function only.

In order to determine whether the combination of structural and functional descriptors increase the effectiveness of program retrieval, we performed two experiments. The retrieval mechanism which is similar to the one used in Koders search engine [26] is employed as the functional approach, and our approach of combining functional and structural descriptors represents the combination approach. In Table 5, we present the processing times and precision scores that are calculated based on the top 10 programs presented in the functional and combination approach hit lists.

Table 5. Precision Scores and Processing Times

	Functional approach		*Combination approach*	
Query	Precision	Time(ms)	Precision	Time(ms)
Q1	0.1	835	0.4	1141
Q2	0.2	880	0.3	1188

Data in Table 5 shows that there is an improvement in precision when the retrieval system employed the combination approach. The precision has increased from 0.1 to 0.4 for Q1 and from 0.2 to 0.3 for Q2. Even though it took a longer time to generate a hit list when the combination approach was employed, the extra time was caused by the need of identifying structural descriptors in the query, and creating the appropriate compound index.

7 Conclusion

We have demonstrated the use of a compound index to integrate information on the function and structure of a program. As the functional descriptors represent what does a component do, structural descriptors symbolize structural relationships that exist in achieving the function. Therefore, an integrated index of functional and structural descriptors represents a component better, compared to using either the functional or structural descriptors on their own. The compound index is flexible as the number of functional and structural descriptors used to represent a query and/or a component is not fixed. It can be expanded to include other information that relates to a component, such as the software architecture (structural descriptor) and the sample of input/output for the component (functional descriptors). In addition, the index is generated automatically by the retrieval system and therefore is economical as it does not require the involvement of a human (expert) such as in a facet-based approach [2].

Based on the results depicted in Tables 3 and 4, it is suggested that it is better to use Euclidean Distance (ED) in identifying programs that are relevant to a given query. Because the evaluation showed that ED generates the best performance (compared to Cosine Measure and skewness), the default setting of the program retrieval system is to use ED in calculating similarities between a given query and programs in the repository. Even though there is work reporting that Cosine Measure works no worse than ED in a retrieval system [27], the experiment undertaken in this work shows that ED has outperformed the Cosine Measure. One possible explanation is that, as we are using program as the query, we intent to identify programs that show the least difference (i.e differences between elements of the Qci and Pci) as relevant (i.e similar function and structure) to the query. Using ED as the similarity measurement, we showed that program retrieval is better undertaken using a combination of functional and structural descriptors rather than using functional descriptors on their own. The experiment undertaken showed that precision scores of the combination approach have doubled the scores obtained using the functional approach.

In future, a similarity measurement that includes weighting schema can also be introduced. This is to help users of the software repository system to sort the importance of descriptors used in identifying the most relevant programs. If a user requires examples of programs that illustrate a particular design pattern and h/she does not emphasis on the programs' function, then larger weight can be given to the compound index element representing the design pattern. With such an approach, the retrieval system will be more flexible as it can easily be modified to represent users' search requirements and preferences.

References

1. Sommerville, I.: Software Reuse. In: Software Engineering, 7th edn., Addison-Wesley, Reading (2004)
2. Prieto-Diaz, R.: Domain analysis: An introduction. ACM SIGSOFT Software Engineering Notes 15(2), 47–54 (1990)
3. Yang, Y., Zhang, W., Zhang, X., Shi, J.: A weighted ranking algorithm for facet-based component retrieval system. In: ACST 2006: Proceedings of the 2nd IASTED international conference on Advances in computer science and technology, pp. 274–279. ACTA Press, Anaheim, CA (2006)
4. Niu, H., Park, Y.: An execution-based retrieval of object-oriented components. In: ACM-SE 37: Proceedings of the 37th annual Southeast regional conference (CD-ROM), p. 18. ACM Press, New York (1999)
5. Ye, Y., Fischer, G.: Supporting reuse by delivering task-relevant and personalized information. In: ICSE 2002: Proceedings of the 24th International Conference on Software Engineering, pp. 513–523. ACM Press, New York (2002)
6. Frakes, W.B., Nejmeh, B.A.: Software reuse through information retrieval. SIGIR Forum 21(1-2), 30–36 (1987)
7. Marcus, A., Sergeyev, A., Rajlich, V., Maletic, J.I.: An information retrieval approach to concept location in source code. In: WCRE 2004: Proceedings of the 11th Working Conference on Reverse Engineering, pp. 214–223. IEEE Computer Society, Washington, DC (2004)
8. Mili, A., Mili, R., Mittermeir, R.T.: Storing and retrieving software components: A refinement based system. IEEE Transactions on Software Engineering 23(7), 445–460 (1994)
9. Fischer, B.: Deduction-Based Software Component Retrieval. PhD thesis, Faculty of Mathematics and Informatics, University of Passau, Germany (June 2001)
10. Atkinson, S., Duke, R.: A methodology for behavioural retrieval from class libraries. Technical Report 94-28, Software Verification Research Centre, Department of Computer Science, The University of Queensland, Australia (September 1994)
11. Podgurski, A., Pierce, L.: Retrieving reusable software by sampling behavior. ACM Trans. Softw. Eng. Methodol. 2(3), 286–303 (1993)
12. Maarek, Y.S., Berry, D.M., Kaiser, G.E.: Guru: Information retrieval for reuse. In: Hall, P. (ed.) Landmark Contributions in Software Reuse and Reverse Engineering, Uni-com Seminars (1994)
13. Zaremski, A.M., Wing, J.M.: Signature matching: A key to reuse. In: Proceedings of SIGSOFT, Los Angeles, California, pp. 7–10 (1993)
14. Wills, L.M.: Automated program recognition by graph parsing. Master's thesis, Massachusetts Institute of Technology (1992)

15. Santanul, P., Atul, P.: A framework for source code search using program patterns. IEEE Transaction on Software Engineering 20(6), 463–475 (1994)
16. Mili, A., Mili, R., Mittermeir, R.T.: A survey of software reuse libraries. Annals of Software Engineering 5, 349–414 (1998)
17. Salton, G., Lesk, M.E.: Computer evaluation of indexing and text processing. J. ACM 15(1), 8–36 (1968)
18. Billhardt, H., Borrajo, D., Maojo, V.: A context vector model for information retrieval. J. Am. Soc. Inf. Sci. Technol. 53(3), 236–249 (2002)
19. Girardi, M.R., Ibrahim, B.: Automatic indexing of software artifacts. In: Proceedings of 3rd. International Conference on Software Reuse, Rio de Janeiro, Brazil, pp. 24–32 (1994)
20. Waters, R.C.: The programmer's apprentice: a session with kbemacs. IEEE Trans. Softw. Eng. 11(11), 1296–1320 (1985)
21. Kitchenham, B., Pfleeger, S.L.: Software quality: The elusive target. IEEE Software 13(1), 12–21 (1996)
22. Rotaru, O.P., Dobre, M.: Reusability metrics for software components. In: The third ACS/IEEE International Conference on Computer System and applications, Cairo, Egypt, pp. 24–31. IEEE Computer Society Press, Los Alamitos (2005)
23. Chidamber, S.R., Kemerer, C.F.: A metrics suite for object oriented design. IEEE Transactions on Software Engineering 20(6), 476–493 (1994)
24. Pallant, J.: 6. In: SPSS Survival Manual, pp. 53–54. Open University Press, Stony Stratford (2004)
25. Google: Google code search, `http://www.google.com/codesearch`
26. Koders: Koders search engine, `http://www.koders.com/`
27. Qian, G., Sural, S., Gu, Y., Pramanik, S.: Similarity between euclidean and cosine angle distance for nearest neighbor queries. In: SAC 2004 Proceedings of the 2004 ACM symposium on Applied computing, pp. 1232–1237. ACM Press, New York (2004)

Framework for Text-Based Conversational User-Interface for Business Applications

Shefali Bhat, C. Anantaram, and Hemant Jain

Tata Consultancy Services Limited, 249 D&E Udoyg Vihar Phase IV, Gurgaon, Haryana, India 122016
{shefali.bhat,c.anantaram,hemant.j}@tcs.com

Abstract. Business application systems traditionally have menu-driven interfaces (whether stand-alone or web-enabled etc.) that users operate on. However, such an interface can become rather cumbersome for users who want some data from the system, but do not know how to get it. Natural language based user-interface to business applications is one alternative. Further, as email and SMS based interactions becomes more ubiquitous, future business application systems may enable email and SMS based interfaces to their systems. This would entail a natural language interface to business applications. We describe a framework for text-based natural language conversational user-interface, for business applications. Our framework permits the user to carry out a dialog with the system in order to fetch relevant data and carry out various tasks of the system. The framework uses semantic web based ontology of the domain, to aid in the retrieval of the relevant data and concepts from the system.

Keywords: Conversational systems, Natural Language Interface, semantic web, ontology, email.

1 Introduction

Business applications are designed to be used by end-users for a variety of possible actions in the business domain. Most of the user-interface of such systems is menu-driven in order to cater to a wide variety of users. In a sense, menu-driven interfaces have become a standard interface for most of the business application systems, for example ERP packages, online banking systems, e-commerce applications etc. However, while the advantages of menu-driven interfaces are significant, such interfaces can become cumbersome for applications with large number of choices and options. Further, a user needs to be logged into the system to use the business application. Natural language interfaces are designed to provide a mechanism to allow the user of an application to interact with the system in a natural human language, such as English. Natural language is one of many "interface styles" (or "interaction modalities") that can be used in the dialog between a user and a computer. There is a significant appeal in being able to address a system and direct its operations by using the same language as

Z. Zhang and J. Siekmann (Eds.): KSEM 2007, LNAI 4798, pp. 301–312, 2007.

we use in everyday human-to-human interaction. By implementing natural language interface to business applications, the user interaction can move from just pushing buttons and dragging options, to specifying operations and assessing their effects through the use of language.

This paper describes a framework for natural language conversational user interface to a business application system, wherein inputs to the system are described in text-based natural language, either through an email to the system or through a text interface to the system. The idea is to carry out a conversation with the user, in order to drill down to what the user actually wants and then identify the application task(s) that would carry out the user's requirement. Thus, instead of the user going through the process of translating his requirements into the exact set of system commands, in a conversational interface a user specifies the task that needs to be done at a high-level and then the system carries out a dialog to map it to the exact set of commands of the system. The natural language interface interprets the text and calls appropriate APIs of the application or generates the appropriate query to accomplish the requested tasks. The main advantage of such a system is that the user is free to enter any information that he has, in the raw unstructured form. It is the job of the system to process that and get whatever else is required.

2 State of Art

A number of attempts have been made to build natural language interfaces to business applications. Sybase Inc. has built a system called "Answers Anywhere" [11], to provide a natural language interface to a business application through a wireless phone, a handheld PDA, a customized console, or a desktop computer. Their method is based on agents and networks. An agent processes requests either directly or by combining its processing with results produced by other agents. While their system shows promise, their approach does not involve ontology based querying or retrieval. Further, they do not handle semantic description of web resources, or traversal of the ontology graph. PRECISE NLI system [1] is designed for a broad class of semantically tractable natural language questions, and guarantees to map each question to the corresponding SQL. The problem of finding a mapping from a complete tokenization of a question to a set of database elements such that the semantic constraints are satisfied is reduced to graph matching problem. PRECISE uses the max-flow algorithm in order to solve the problem. While their work seems quite interesting, they restrict each question to start with "wh" token. Further, there does not seem to be any provision for firing domain based rules. Though a lot of other related work in Natural Language Interface to databases has been done, none of them use ontology based methodologies to answer the queries posed. NaLIX system [12] discusses about construction of generic Natural Language query interface to an XML database. On the other hand TRIPS [3] enforces strict turn taking between the user and the system and process each utterance sequentially through three stages – interpretation, dialogue management and generation. These restrictions make the interaction unnatural [6]. Some efforts for carrying out conversations in common-sense knowledge domains, such as ELIZA [7], Cyc project [2] and Open Mind Common Sense project [10] have been carried out. However, none of these address natural language interface with business applications.

3 Issues in Natural Language Interface to Business Applications

In our framework we have focused on the following core issues which we believe should be addressed in a NL interface to business systems:

a) **Ontology of the domain:** How can the domain ontology be created from the application data? How can it be seeded automatically?
b) **Concepts of the domain:** what should be the *concepts* of the domain? How should they be represented?
c) **Handling queries of the domain:** what should be the mechanism for resolving the queries posed on the domain?
d) **Application specific tasks:** how should the NL interface handle application specific tasks – especially tasks that require detailed application logic?
e) **Context resolution and clarification:** Often, the system cannot fully identify user's intent, in such a scenario what should be the ways and means of clarifying the user's intent.

4 Framework for NL Interface

In our framework we discuss some mechanisms for addressing the core issues discussed above.

4.1 Ontology Creation

The ontology of the domain describes the domain terms and their relationships. We use semantic web technology to create the ontology. Semantic Web technology allows seamless integration of different resource definitions that semantically mean the same thing. This permits easier integration of domain knowledge, which in turn makes the natural language system more robust in answering queries posed by the user. The application data (i.e. the database of the business application) forms a part of the domain terms and their relationships in the ontology. This helps forms the main concepts of the domain and their relationships with a <subject-predicate-object> structure for each of the concepts. Figure 1 shows the ontology creation process and the levels of ontology.

The Seed Ontology describes the basic relations that are applicable in the domain, for example in a Project Management domain where details about all the employees, projects and the relation between the employees and the projects are handled; facts like "project has a project number", "an employee is a person", etc populate the Seed Ontology. The Application Data (also termed as static facts) provides the actual data that is present in the system. The Ontology Generator takes in the Seed Ontology and Application Data and creates an instance of the Seed Ontology populated with the application data. This is called the Application Ontology. Next, the rules of the application domain are then evaluated together with the Application Ontology by a Rule Engine, such as Closed World Machine (CWM) [5], to create the Domain Ontology. Domain Ontology is used by the Natural language Interface system to answer questions on / carry out the tasks of the domain.

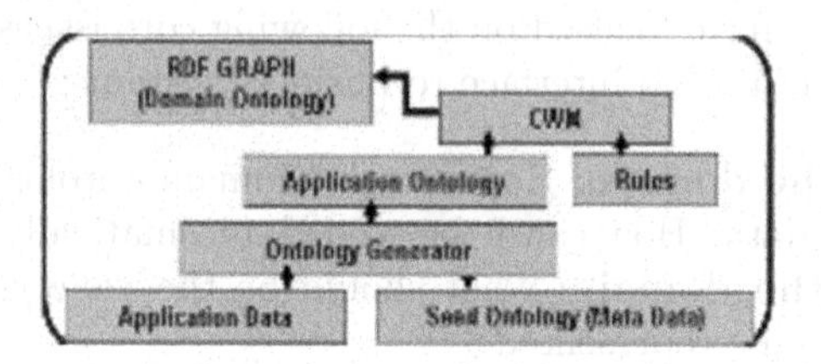

Fig. 1. Ontology Creation

In our work, domain-specific rules (defined manually for each specific domain) are defined on the ontology in W3C's N3 format to state possible derivable facts about data. For example, "has_boss" can be inferred from the rule as shown below:

```
{?x ds:NAME ?a. ?x ds:Role ds:ProjectLeader. ?y ds:NAME ?b.
?y ds: Role ds:TeamMember.} => {?b ds:has_boss ?a}.
```

Thus, If x=1, a = Ritesh, **y**=2, **b** = Rajat. Then **Rajat has_boss Ritesh** The ontology is used in conjunction with the domain rules on the data and new facts are derived based on the data of the domain. These new facts are called derived facts. The static and derived facts are then converted into Resource Description Framework (RDF) format. The RDF format has each fact (both static and derived) as a set of URLs in XML form.

4.2 Concepts of the Domain

The RDF file is read and a <subject-predicate-object> graph structure in created in the memory. Once we have the domain ontology in memory, we can traverse it using the graph traversal functions to get the subject, predicate or object (or a combination of these). A set of class objects is created in memory to represent each subject, predicate and object of the <subject-predicate-object> structure as *concept* of the domain. Each *concept* has the concept name and its synonyms to help identify the concept from the natural language sentence that the user inputs. The synonyms are derived by posing the concept name to WordNet - an open source lexical reference system [8]. In WordNet, English nouns, verbs, adjectives and adverbs are organized into synonym sets, each representing one underlying lexical concept. Different relations link the synonym sets. The relevant synonyms are loaded with each concept in memory.

4.3 Handling Queries of the Domain

As mentioned earlier, the system drills down from the concepts to the answers in three possible ways:

a) Firing of SPARQL queries
b) Ontology Traversal
c) Retrieving answers through API

The queries that are executed are written in a generic form in SPARQL (which is a query language for RDF) [4]. Since the domain ontology is in RDF format, the general structure of the query is (subject, predicate, object). We have identified seven types of queries for the subject-predicate-object (hence forth referred to as <s-p-o>) structure of our ontology; these are: s (only subject); p (only predicate); o (only object); s-p (subject and predicate); s-o (subject and object); p-o (predicate and object); s-p-o (subject, predicate and object specified). The actual query is formulated by binding the value of the concept raised in the input sentence to the generic SPARQL query of one of the above seven types, in order to formulate the precise query and retrieve the answer. For example, let the question be "In which project is Ritesh a group leader?". The concepts raised for this particular query are, "*Project*", "*Ritesh*" and "Group Leader". Post concept identification, the system will try to fire one of the above seven mentioned SPARQL query. For this example, since s, p, and o are known, therefore following SPARQL query is fired (shown below).

```
Select = (''?f'')
where=GraphPatttern([(''?a'',ds[prd1],ds[val1]),(''?b'',ds[prd2],
ds[val2]), (''?a'', ''?c'', ''?a'') , (''?b'', ''?c'' ,''?a'') ,
(''?b'' , ''?d'' , ''?e'') , (''?e'', ''?d'', ''?e'') ,
(''?e'', ds[prd], ''?f'')])
result = sparqlGr.query (select, where)
```

Where val1=Ritesh and prd1= ename, val2 = Group Leader and prd2 = role and prd = pname This query states that for an unknown subject "*a*" having "*ename*" as predicate and "*Ritesh*" as object , there is a "*b*" having "*role*" as predicate and "*Group leader*" as object, *also subject "b" is bound to objects "a" and "e"* with *predicates* "c" and "d" respectively, also there exists some "*e*" such that this "*e*" having "pname" as a predicate gives us the value for the project "*f*". In case the query generation does not fetch an answer then the system traverses the RDF graph. Ontology traversal takes in concepts identified from the input sentence and determines which part of the ontology these concepts satisfy. That is, the concepts could be leaf nodes or some intermediate nodes in the ontology graph. Once this is established, the traversal tries to determine the relationship (direct or inherited) between the concepts identified in the graph structure. When the application data is transformed into the Domain Ontology, a graph structure is created in the memory. It is this graph structure that is traversed in order to obtain an answer for the query. The graph traversal figures out the node which is directly or indirectly connecting two or more different nodes. Thus questions like "what is common between X and Y?" which are quite cumbersome to be answered by a query or through an API function, are answered easily through ontology traversal. For example, if a user wants to know "*what is common between Anubha and Shefali*", the query generation mechanism is not going to give an answer; whereas an ontology traversal answers "*Anubha and Shefali are in the VirginAtlantic project*".

4.4 Handling Application Specific Tasks

In case both SPARQL as well as ontology traversal does not lead to an answer, the system tries to pose the concepts to the application. This requires tasks to be identified that can be carried out by the application system. For each possible task of the application, a task description file describes the concepts that are required to perform the task. From the concepts that are identified from the input sentence, the task that has the maximum number of raised concepts is carried out. The answers generated by the application system are then fed to a response manager, which formulates the appropriate response and sends it to the user in natural language.

4.5 Context Resolution and Clarification

One of the issues is the need for context resolution and clarification. This need arises in conversations, especially in case of input sentences such as "what is his allocation percentage?" The framework tries to relate the pronoun/determinant in the input sentence with the corresponding context of the conversation with the user. In our framework, the set of interactions carried out with a user in a session is used to identify the context of that session. The system resolves gender-specific pronouns based on the concepts identified from the context of a session However, in case the question posed by the user is ambiguous or the system is unable to resolve the pronoun, the system responds back to the user through e-mail and seeks clarification. To understand this better, consider the following conversation between a user and a natural language enabled application.

```
User: Who all are allocated to Technology Program?
System: Employees in TechnologyProgram are.
                Puneet Shefali Ritesh Rajat
User: What is his allocation percentage?
System: We are talking about 3 men -- Puneet, Ritesh and Rajat
                when you say ' his ' whom do you mean? Your answer
User: Puneet
System: Puneet's allocation percentage is 25 percent.
```

The moment the user clarifies the content, Puneet in this case, the system fetches the answer and replies back.

5 Our Architecture

We have implemented an email-based Natural language interface to business applications, called NATAS, based on the framework discussed in Section 4, wherein the explicit domain ontology is described using semantic web technology. The natural language interface uses the ontology and the data, to derive facts and reason on them in order to answer the user's input. In this section, we describe the architecture of our natural language interface system. Figure 2 shows the overall architecture of the interface system, where the arrows depict the information flow in the system. A conversation is initiated when the user types in the input that s(he) wants to convey to the system.

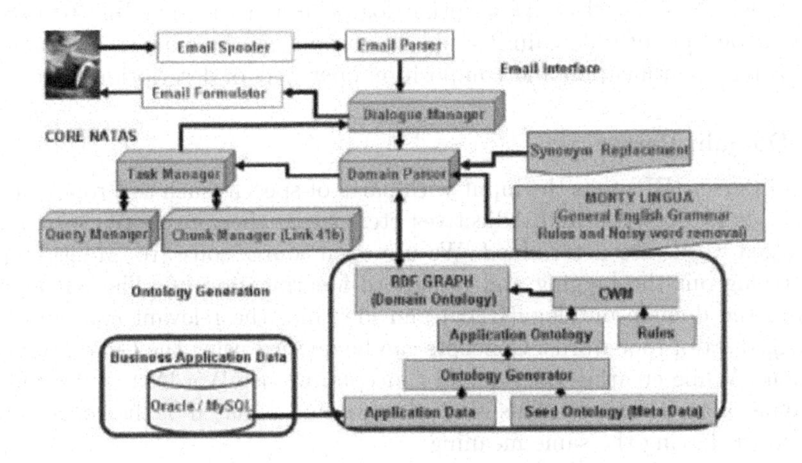

Fig. 2. Broad Architecture

5.1 Dialog Manager

The input is received by a Dialog Manager (DM) which manages the user interaction. Since we have implemented an email-based interaction, our architecture has an email parser. Alternatively, it could also be a direct input into a text-box in a business application system. The DM is the main dialoguer in the system and keeps track of the user preferences, the current context, the concepts referred in the previous interactions (if any) etc.

5.2 Domain Ontology

The application ontology is described in N3 notation, using OWL schema. The rules on the application ontology that describe the possible derivable facts are also described in N3. For example, in a "Project Allocation and Monitoring system, the basic application ontology is shown below.

```
ds:WON rdfs:subClassOf ds:project.
ds:SWON rdfs:subClassOf ds:project.
ds:project ds:has_status "created, submitted, approved, active,closed".
ds:employees ds:has_param "empno, name".
```

The rules on the application ontology describe inferences that can be carried out on the ontology. For example, a "manager" can be inferred from the project allocations as shown below:

```
{?x ds:role ds:GroupLeader. ?z ds:role ds:TeamMember.
?x ds:projectno ?y. ?z ds:projectno ?y }=>{?x ds:has_manager ?z }.
```

Combining these rules with the basic application ontology and the actual data, gives the complete domain ontology for the application. Since we are using OWL schema, which is based on Description logics, it is easily possible to express complex concepts of a domain. Complex concepts can be expressed as a set of terms using quantification and composition operators of description logics.

5.3 Domain Parser

Domain Parser (DP) tags the input with parts-of-speech (such as Proper Nouns, Nouns, Verbs, Adverbs and Adjectives etc.) Next, the root words for each of the tagged words are determined. We use open source software, MontyLingua, for carrying out the tagging and root-word identification [9]. The parser next reads in the domain ontology(RDF) and identifies the relevant concepts from the tagged set of root-words. Concepts can be queried using the Graph traversal functions. While creating the domain concepts, we use WordNet [8] to find the synonyms of each concept. This is required to find a match of the concept with other words having the same meaning.

5.4 Task Manager

The concepts are used to identify the task which the user wants to carry out. The task may require just a traversal of the domain ontology which is loaded in memory, or may correspond to an API call in the application system. The Task Manager first tries to carry out the task by traversing the ontology with reference to all the concepts referred by the Concept Manager. If the task cannot be carried out by ontology traversal, the Task Manager tries to formulate a query on the RDF graph using one of the seven types of queries discussed in section 4.3. If such a query cannot come up with an answer, the relevant API of the application is invoked. The answers retrieved are formulated and sent to the user.

6 Examples of Implementation

6.1 Project Management System

We consider a logical subset of the Project Management System for an organization. In this example the system consists of a number of tables containing data about the projects, the various costs associated with the projects, employees and their allocations in different projects. The important tables are named ProjectDetails, Employees, Cost Details and Allocations.

Let us assume that we want to find out, "*What is the revenue of the project in which Ritesh is a Group Leader?*" Clearly from the above give data this query involves joins over four tables in order to extract answer. NATAS fetches answer for this query in the following manner. The domain ontology is loaded into the memory as RDF triples <subject, predicate, object>. An example of the domain ontology follows:

Table 1. ProjectDetails

Project No	**Name**	**...**	**Type**	**Status**	**Startdate**	**Endate**
100582	TechProg	...	SWON	Active	01-01-05	31-12-05
.	.	.	.	.	.	.
.	.	.	.	.	.	.
.	.	.	.	.	.	.
200315	Bechtel	...	WON	Created	10-10-05	31-12-06

Table 2. CostDetails

Cost No	**Project No**	**...**	**Costing**	**Billing**	**Revenue**	**Realized**
12	200234	...	25	25	75	20
.	.	.	.	.	.	.
.	.	.	.	.	.	.
.	.	.	.	.	.	.
10	100582	...	70	56	1000	60

Table 3. Allocations

Alloc No	**Project No**	**...**	**Emp No**	**Start Date**	**End Date**	**Alloc Per**	**Role**
100	100582	...	160784	01-09-05	31-12-05	100	Team Member
.	.	.	.	.	.	.	.
.	.	.	.	.	.	.	.
.	.	.	.	.	.	.	.
1000	100582	...	180180	01-01-05	01-10-05	50	Project Leader

Table 4. Employees

Employee No	**Name**	**Age**	**...**	**Gender**
160982	Shefali	24	...	Female
.	.	.	.	.
.	.	.	.	.
.	.	.	.	.
180180	Ritesh	28	...	Male

```
<rdf:Description rdf:about=
        "http://Demo:Demo@172.21.107.145:8080/SfleXProd/#180180">
<age rdf:resource=
        "http://Demo:Demo@172.21.107.145:8080/SfleXProd/#27"/>
<dob rdf:resource=
        "http://Demo:Demo@172.21.107.145:8080/SfleXProd/#08-01-1979"/>
<eName rdf:resource=
        "http://Demo:Demo@172.21.107.145:8080/SfleXProd/#Ritesh"/>
<empNo rdf:resource=
        "http://Demo:Demo@172.21.107.145:8080/SfleXProd/#180180"/>
<sex rdf:resource=
        "http://Demo:Demo@172.21.107.145:8080/SfleXProd/#Male"/>
</rdf:Description>
```

The primary key forms the subject, fieldname forms the predicate whereas the values of the fields forms the object in the ontology file.

SPARQL Query One of the seven query templates is chosen for this example it is:

```
select = ("?g")
        Where = GraphPattern([("?a", ds[prd1], ds[val1]), ("?a", "?c", "?a"),
        ("?b", "?c" ,"?a"),("?b", ds[prd2], ds[val2]),("b", "?d" ,"?e"),
        ("?f", "?d" ,"?e"),("?f", ds[prd], "?g" )])
        result = sparqlGr.query(select, where)
```

This query when fired fetches the appropriate answer. The deduction is as follows:

If empno **180180** (*a*) has **name** (*prd1*) **Ritesh** *(val1)* & **180180** (*a*) has **empno** (*c*) **180180** (*a*) &Some allocation **100**(*b*) has **empno** (*c*) **180180** (*a*) & Same allocation **100**(*b*) has **projectno**(*d*) **100582** (*e*) & Same allocation **100** (*b*) has **role**(*prd2*) **Projectleader** (*val2*) & Some cost No (*f*) has **projectno** (d) **100582** (e) & Same cost No (*f*) has **revenue** (*prd*) **1000** (g) Hence deducing: **The revenue is 1000.**

Domain Rules Domain specific rules are written over the ontology to describe the facts that can be derived from the given set of static facts. The rules operate on the ontology provided and generate an RDF structure containing facts as well as derived facts. With the help of these rules all the queries regarding the data can be answered. A typical rule is as follows:

```
{?s ds:name ?o . ?x ds:empno ?s . ?x ds:role ds:TeamMember .
?x ds:projectno ?y . ?z ds:projectno ?y . ?z ds:role ds:ProjectLeader .
?z ds:empno ?p . ?p ds:name ?q } => {?o ds: has_manager ?q }.
```

This rule when instantiated for a data item would derive the appropriate answer. **Anubha has manager Ritesh.**

6.2 Retail Management System

In case of retail outlet, having a number of products, promotion offers and catering to a large customer needs. Consider the following query posed by the customer.

```
User: Which camcorders have more than 20% discount?
Concepts Identified: Camcorders, more than, 20%, discount.
SPARQL Fired:
            Select= ("? f")
            where.addPatterns([("?a","?c","?a"),("?a",ds[prd],"?f"),
            ("?b","?c","?a"),("?b","?d","?e"),("?b",ds[prd2],ds[val2]),
            ("?e","?d","?e"),("?e", ds[prd1], ds[val1])])
            result = self.sparqlGr.query(select,where)
System: The Camcorders are
DXG 3MP Digital Camcorder - DXG-301V
Panasonic Mini DV Camcorder
Aiptek IS-DV2 Digital Camcorder
Panasonic 2.8" LCD Digital Camcorder with 3CCD Technology
                - Silver (SDR-S150).
```

We show a screen shot of the interaction for the above query.

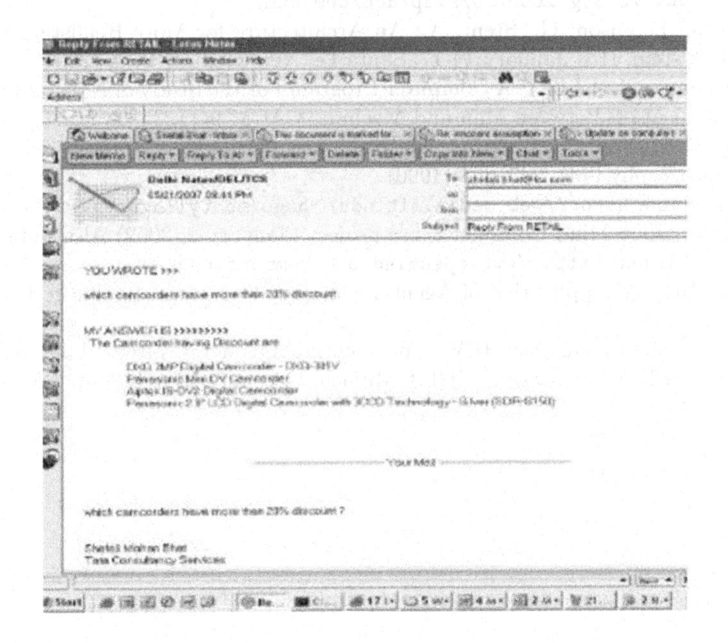

Fig. 3. Sample Output

7 Conclusion

We have described a text-based conversational user-interface as an alternate mechanism of interaction with business application systems. This not only enables the user to interact with a business application in a language that one uses in their day-to-day conversation but also enables the user to carry out a dialog with the application. The use of semantic web ontology allows the natural language system to interact "intelligently" with the user by traversing the domain ontology for both static as well as derived facts. This permits easier integration of domain knowledge, which in turn makes the natural language system more robust in answering queries posed by the user. We have implemented an email-based interaction with business applications using the above framework.

References

1. Popescu, A.M., Etzioni, O., Kautz, H.: Towards a Theory of Natural Language Interfaces to Databases. IUI, USA, pp. 149–157 (2003)
2. Lenat, D.B.: Cyc: A large-scale investment in knowledge infrastructure. CACM 38(11) (1995)
3. Ferguson, G., Allen, J.F.: TRIPS: An integrated intelligent problem-solving assistant. In: Proc. AAI, pp. 567–573 (1998)
4. `http://dev.w3.org/cvsweb/2004/PythonLib-IH/Doc/sparqlDesc.html?`
5. `http://www.w3.org/2000/10/swap/doc/cwm.html`
6. Allen, J., Ferguson, G., Stent, A.: An Architecture for More Realistic Conversational System. IUI, January 14-17, Santa Fe, New Mexico, USA (2001)
7. Weizenbaum, J.: ELIZA–A Computer Program For the Study of Natural Language Communication Between Man and Machine. CACM 9(1), 35–36 (1966)
8. Miller, G.A.: Nouns in WordNet: a lexical inheritance system. International Journal of Lexicography 3(4), 245–264 (1990)
9. Montylingua, `http://web.media.mit.edu/~hugo/montylingua`
10. Push, S.: Open Mind Common Sense project (January 2, 2002) Also published on KurzweilAI.net, `http://www.openmind.org/commonsense`
11. Sybase Inc. An Application of Agent Technology to Natural Language User Interface
12. Li, Y., yang, H., Jagadish, H.V.: Constructing a Generic Natural Language Interface for an XML Databases. EDBT Munich, Germany, March 26-31 (2006)

Ontology Mining for Semantic Interpretation of Information Needs

Xiaohui Tao, Yuefeng Li, and Richi Nayak

FIT, Queensland University of Technology, Australia
{x.tao,y2.li,r.nayak}@qut.edu.au

Abstract. Ontology is an important technique for semantic interpretation. However, the most existing ontologies are simple computational models based on only "super-" and "sub-class" relationships. In this paper, a computational model is presented for ontology mining, which analyzes the semantic relations of "part-of", "kind-of" and "related-to", and interprets the semantics of individual information need. The model is evaluated by comparing the knowledge mined by it, against the knowledge generated manually by linguists. The proposed model enhances Web information gathering from keyword-based to subject(concept)-based. It is a new contribution to knowledge engineering and management.

1 Introduction

The semantic interpretation of a user's information need is a great challenge in Web intelligence. Ontology, as a formal description and specification of knowledge, provides a common understanding of topics to be communicated between users and systems [2,11]. By using an ontology, information systems are expected to be able to understand the semantic meanings of words and phrases, and to compare information needs by concepts instead of keywords [8]. Thus, ontology is deemed by the Web intelligence community as one of the most useful techniques for the semantic interpretation.

Over the last decade, many attempts have been suggested to use ontology to describe and specify the knowledge possessed by humans. A great work conducted by Maedche & Staab [9] is a comprehensive ontology learning framework consisting of four phases: Import, Extract, Prune, and Refine. Li & Zhong [6] proposed a method aimed to discover the top backbone of an ontology based on the interesting patterns found in documents. Gauch et al. [3] proposed to learn personalized ontology based on the online portals. King et al. [4] learn ontology from the Dewey Decimal Classification system[1]. However, these existing works are only based on "super-" and "sub-class" relationships, and do not specify the semantic relations of "part-of" and "kind-of" well. In previous works, the semantic relations existing between the classes in the ontology are either pre-defined constantly, or mined from the related text without a well-defined structure. However, a user may not really follow the pre-defined semantic relations that exist

[1] Dewey Decimal Classification, http://www.oclc.org/dewey/

Z. Zhang and J. Siekmann (Eds.): KSEM 2007, LNAI 4798, pp. 313–324, 2007.

in the concepts, while he/she is going through the process of generating a topic for the information need. Web users do possess a concept model in the process of information gathering. Usually, users can easily determine if a Web page interests them or not while they read through the content. The rationale behind this is that users implicitly possess a concept model based on their knowledge, although they may not be able to express it [6]. Consequently, these ontology mining methods are neither comprehensive nor realistic in the real world knowledge acquisition.

In this paper, a computational model is presented, along with two novel concepts of *Exhaustivity* and *Specificity*, for mining knowledge from a subject ontology. The semantic relations existing between the subjects in the ontology are dynamically analyzed, instead of being constantly defined by the traditional means. The proposed method is evaluated by applying to a system that gathers information from a large corpus. The proposed model can improve the semantic interpretation of information needs and thus the performance of the Web information gathering systems. It is a significant contribution to knowledge engineering and knowledge management in the Semantic Web.

The paper is organized as follows. Section 2 presents the formal definitions of the subject ontology discussed in this paper. Section 3 introduces the proposed ontology mining method for the semantic interpretation. Section 4 discusses the experiments and the results. Finally, Section 5 presents the related work, and Section 6 makes the conclusions.

2 Formalization of the Subject Ontology

The subject ontology is constructed based on a taxonomic knowledge base, which is built by using the Library of Congress Subject Headings[2] (LCSH) system as the backbone and each subject heading as a class node. The taxonomic knowledge base is formalized as follows.

Definition 1. *Let OntoBASE be a taxonomic ontology base. An ontology base is formally defined as a 2-tuple* $OntoBASE :=< \mathbb{S}, \mathbb{R} >$, *where*

- $\mathbb{S}$ *is a set of subjects* $\mathbb{S} := \{s_1, s_2, \cdots, s_m\}$;
- $\mathbb{R}$ *is a set of relations* $\mathbb{R} := \{r_1, r_2, \cdots, r_n\}$.

Definition 2. *A subject* $s \in \mathbb{S}$ *is formalized as a 3-tuple* $s :=<$ *label,instanceSet,* $\sigma >$, *where*

- *label is a label assigned by experts to a subject s in the LCSH system;*
- *instanceSet is a set of objects associated to the subject s, in which each element is an instance denoted by inst and specifying a semantic meaning referring by s;*
- σ *is a signature mapping* ($\sigma : s \rightarrow 2^s$) *that defines a set of relevant subjects to the given s.*

[2] The Library of Congress, http://www.loc.gov/

Definition 3. *A relation* $r \in \mathbb{R}$ *is a 2-tuple* $r := < type, r_\nu >$, *where*

- *type is a set of relationships,* $type = \{kindOf, partOf, relatedTo\}$;
- $r_\nu \subseteq \mathbb{S} \times \mathbb{S}$. *Each* $(x, y) \in r_\nu$ *may be written as* $r_\nu(x, y)$, *means that* y *is the subject who holds the type of relation to* x, *e.g.* s_x *is* $kindOf$ s_y.

$KindOf$ is a directed relation in which one subject is in different form of another subject. *PartOf* is also a directed relation but used to describe the relationship held by a compound subject class and its component classes. $KindOf$ and $partOf$ hold properties of transitivity and asymmetry. Transitivity means if s_1 is a kind (part) of s_2 and s_2 is a kind (part) of s_3, then s_1 is a kind (part) of s_3 as well. Asymmetry means if s_1 is a kind (part) of s_2 and $s_1 \neq s_2$, s_2 may not be a kind (part) of s_1 necessarily. *RelatedTo* is a non-taxonomic relation describing the relationship held by two subjects that overlap in their semantic spaces. *RelatedTo* holds the property of symmetry but not transitivity. If s_1 is related to s_2, s_2 is also related to s_1. However, if s_1 is related to s_2, s_2 is related to s_3, s_1 may not be related to s_3 necessarily.

The subject ontology facilitates a user's concept model. It is dynamically constructed based on a user's individual information need. The user interacts with the system and identifies the positive and negative (ambiguous) subjects according to a topic (in this paper a user's information need is called as a *topic*) and the possessed concept model. A subject ontology is built based on the user feedbacks and the taxonomic knowledge base. Fig. 1 presents an incomplete subject ontology constructed for the topic "Economic espionage", where the white nodes are positive subjects, the dark gray are the negative, and the light gray are the unlabelled subjects. The unlabelled subjects are those in the volume of a positive subject but not being identified by the user as either positive or negative. The semantic relations existing between the subjects are addressed by different type of lines. A subject ontology is formalized by the following definition.

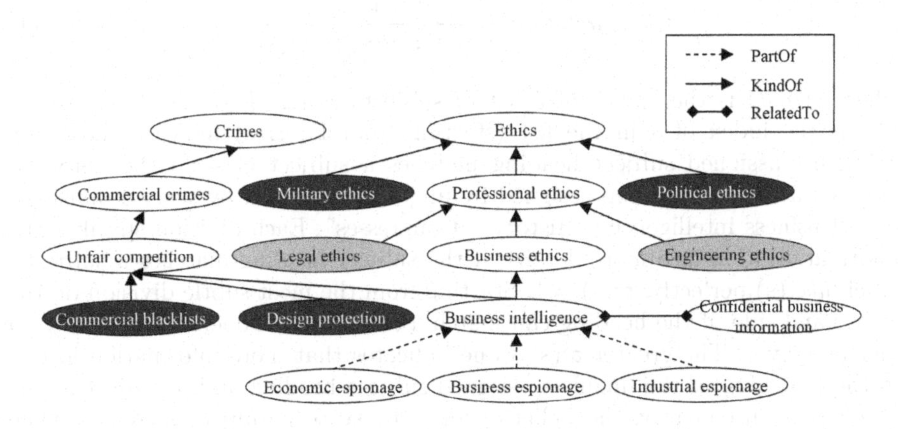

Fig. 1. Ontology Constructed for Topic "*Economic Espionage*"

Definition 4. *The structure of an ontology that formally describes and specifies topic $\mathcal{T}$ is a 5-tuple $\mathcal{O}(\mathcal{T}) := \{\mathcal{S}, \mathcal{R}, tax^{\mathcal{S}}, rel, \mathcal{A}^{\mathcal{O}}\}$, where*

- *$\mathcal{S}$ is a set of subjects and $\mathcal{S} \subseteq \mathbb{S}$. $\mathcal{S}$ has three subsets, where $\mathcal{S}^{+} \subseteq \mathcal{S}$ is a set of positive subjects to $\mathcal{T}$, $\mathcal{S}^{-} \subseteq \mathcal{S}$ is a set of negative subject to $\mathcal{T}$, and $\mathcal{S}^{\diamond} \subseteq \mathcal{S}$ is a set of unlabelled subjects to $\mathcal{T}$;*
- *$\mathcal{R}$ is a set of relations and $\mathcal{R} \subseteq \mathbb{R}$;*
- *$tax^{\mathcal{S}}$: $tax^{\mathcal{S}} \subseteq \mathcal{S} \times \mathcal{S}$ is a taxonomic backbone of the ontology, which consists of two directed relations $kindOf$ and $partOf$;*
- *rel is a function defining non-taxonomic relations;*
- *$\mathcal{A}^{\mathcal{O}}$ is a set of rules mined from $\mathcal{O}$.*

Given a subject s, its $vol(s)$ refers to the union of all subjects in its volume, $dom(s)$ refers to the set of its directed child subjects and $dom(s) \subseteq vol(s)$. Where $dom(s) = \{s_1\}$, for $partOf(tax^{\mathcal{S}}) = r_{\nu}(s_1, s)$ one may also write $partOf(s_1, s)$ meaning that s_1 is a part of s, for $kindOf(tax^{\mathcal{S}}) = r_{\nu}(s_1, s)$ one may also write $kindOf(s_1, s)$ meaning that s_1 is a kind of s.

3 Ontology Mining Model

The expert knowledge can be mined from the ontology backbone and the instances associated to the subjects. An ontology requires expert knowledge to fill the taxonomic backbone with instances [1]. Usually each information item in a library is briefly described by the information provided by the author, e.g. title, table of content, and linguists, e.g. summary, a list of subject headings. An information item specified and summarized forms an instance *inst* in the ontology. The belief *bel* of an instance *inst* to a subject s can be determined by:

$$bel(inst, s) = \frac{1}{\iota(s) \times \xi(inst) \times \varpi(s)} \tag{1}$$

where $\xi(inst)$ is the length of a list of subjects assigned to *inst* by linguists; $\iota(s)$ is the index of s in the list (starting from 1); $\varpi(s)$ indicates how well a linguist assigned subject heading matches a subject class in the ontology. Usually, a subject heading assigned by linguists is formed by several divisions, e.g. "Business intelligence – History – Congresses". Each division specifies the information item into more details. If the subject heading matches a subject's label $label(s)$ perfectly, $\varpi(s) = 1$. Starting from the most subtle division on the right hand side, if the heading can match a $label(s)$ with one division off, $\varpi(s)$ increases by 1. The greater $\varpi(s)$ value indicates that more information lost in in the matching of the linguists assigned subject heading and a subject in the ontology. Consequently, the belief of an instance to a subject increases while there are less subjects in the associated list, the higher position the subject is indexed at, and the less divisions are cut off from the subject heading in order to match a subject in the ontology.

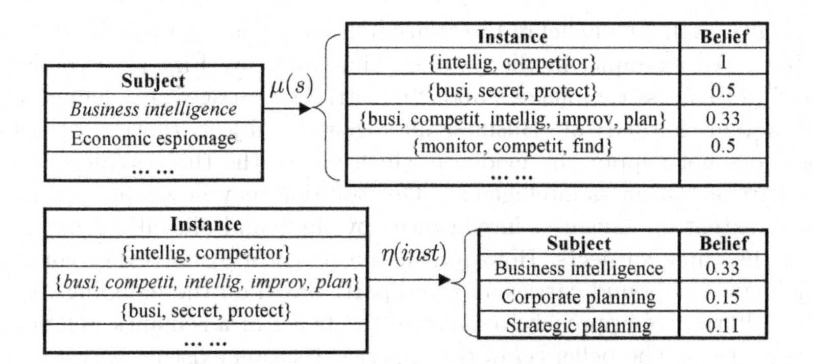

Fig. 2. Mappings of Subjects and Instances Related to "*Economic Espionage*"

3.1 Semantic Matrix Between Subjects and Instances

Let $\Omega = \{inst_1, inst_2, \ldots, inst_i\}$ be a finite and nonempty set of instances in $\mathcal{O}(\mathcal{T})$, $inst \leftrightarrow s$ be the binary relation between $inst$ and s. Based on an $inst \in \Omega$, we may have the mapping:

$$\eta : \Omega \to 2^{\mathcal{S}}, \quad \eta(inst) = \{s \in \mathcal{S} | inst \leftrightarrow s\} \subseteq \mathcal{S}. \tag{2}$$

The mapping function $\eta(inst)$ is used to describe the subjects to which the instance is relevant. In order to classify the list of instances, we may have the reverse mapping μ of η:

$$\mu : \mathcal{S} \to 2^{\Omega}, \quad \mu(s) = \{inst \in \Omega | inst \leftrightarrow s\} \subseteq \Omega. \tag{3}$$

The mapping $\mu(s)$ divides the list of instances and assigns them to a group of subjects, and reveals the relationship between instances and subjects. Each $inst$ is relevant to one or more subjects in $\mathcal{S}$, and each s refers to one or more instances in Ω. Each pair of $(inst, s)$ is assigned with a belief value defined by Eq. (1). Fig. 2 presents an incomplete mapping related to the topic "Economic espionage", where the top is for $\mu(s)$ and the bottom for $\eta(inst)$. These mappings aim to explore the semantic matrix existing between the subjects and instances. Based on the mappings, the binary relation $(inst \leftrightarrow s)$ can be defined by:

$$(inst \leftrightarrow s) = \{(inst, s)\} \subseteq \Omega \times \mathcal{S}. \tag{4}$$

And the complement ^{c} of $(inst \leftrightarrow s)$ can be defined by:

$$(inst \leftrightarrow s)^c = \{(inst, s) | \neg(inst \leftrightarrow s)\} = \Omega \times \mathcal{S} - (inst \leftrightarrow s). \tag{5}$$

An instance $inst$ does not refer to the set of subjects $\{s \in \mathcal{S} | (inst \leftrightarrow s)^c\} = (\eta(inst))^c \subseteq \mathcal{S}$. A subject s is not referred by the set of instances $\{inst \in \Omega | (inst \leftrightarrow s)^c\} = (\mu(s))^c \subseteq \Omega$.

Let $s_1 \in \mathcal{S}, s_2 \in \mathcal{S}$. The semantic relation held by s_1 and s_2 is defined by the taxonomic knowledge base, e.g. $kindOf(s_1, s_2)$, $partOf(s_1, s_2)$, or

$relatedTo(s_1, s_2)$. The belief of the relation, however, is not given by the knowledge base. For example, in the ontology illustrated by Fig. 1, "Economic espionage", "Business espionage", and "Industrial espionage" are defined by the LCSH system as a part of "Business intelligence" subject. However, the LCSH system does not supply any evidence which one of the three subjects forms a bigger part of "Business intelligence". One solution may be simply dividing the semantic extent of "Business intelligence" by three and allocating one third to each of the three subjects. However, in the real world, not every component assembling a compound carries an equal part. Based on the mappings, we may have an alternative approach to measure the belief of a semantic relation held by two subjects. The belief *bel* of the semantic relation existing between s_1 and s_2 can be measured by the following equation, in respect to s_1:

$$bel(s_1|s_2) = \frac{\sum_{inst \in (\mu(s_1) \cap \mu(s_2))} bel(inst, s_1) + \sum_{inst \in (\mu(s_1) \cap \mu(s_2))} bel(inst, s_2)}{2 \times \sum_{inst \in \mu(s_1)} bel(inst, s_1)} \tag{6}$$

Since $(\mu(s_1) \cap \mu(s_2)) \subseteq \mu(s_1)$, the $bel(s_1, s_2)$ value is between [0,1], where 0 indicates that the relationship held by s_1 and s_2 is not supported at all, although it may be stated existing according to the LCSH system, and 1 indicates that the semantic space referred by s_1 is completely contained by the semantic extent of s_2, and the relationship held by s_1 and s_2 is fully supported by the expert knowledge.

Based on the mappings, the belief of an instance *inst* to a topic $\mathcal{T}$ in the ontology $\mathcal{O}(\mathcal{T})$ can be determined by:

$$bel(inst, \mathcal{T}) = \sum_{s \in \eta(inst) \wedge s \in \mathcal{S}^+} bel(inst, s) - \sum_{s \in \eta(inst) \wedge s \in \mathcal{S}^-} bel(inst, s). \tag{7}$$

If $bel(inst, \mathcal{T}) > 0$, the instance supports the topic. Otherwise, it is against the topic or makes the topic more confusing. The instances associated to an unlabelled subject count nothing to the topic because there is no evidence that they appreciate any site of positiveness or negativeness. Similarly, the belief of a subject to the topic is determined by:

$$bel(s, \mathcal{T}) = \sum_{inst \in \mu(s)} bel(inst, \mathcal{T}). \tag{8}$$

If $bel(s, \mathcal{T}) > 0$, the subject supports the topic. Otherwise, it is against the topic or makes it more confusing. The greater $bel(s, \mathcal{T})$ value makes the support (or confusion) stronger. Again, the unlabelled subjects hold belief value of 0 to the topic because their beliefs can not be clarified. Comparing to the positive and negative subjects in the feedback from a user, these subjects may be called "confirmed" positive and negative subjects, and those subjects in the user feedback may be called "candidate" subjects.

3.2 Exhaustivity and Specificity of Subjects

Mining an ontology means discovering knowledge from the backbone and the concepts that construct and populate an ontology. Two concepts are introduced here for mining an ontology: *specificity* (*spe* for short) describing the focus of the semantic meanings of a subject according to a topic, whereas *exhaustivity* (*exh* for short) restricting the extent of the semantic meanings covered by a subject. They aim to assess the certainty belief of a subject in the ontology constructed to facilitate a user's knowledge model about a given topic. A subject in the ontology may be highly exhaustive, although it may not be specific to the topic. Similarly, a subject may be highly specific, although it may deal with only a few aspects of the topic.

Input: $\mathcal{O}(\mathcal{T})$: the constructed ontology; $s \in \mathbb{S}$: a subject in $\mathcal{O}(\mathcal{T})$;
θ: a parameter between (0,1);
Output: $\delta(s)$: the parameter applied to specificity assignment;

---

1. $\delta(s) = 1$ //*initialize* $\delta(s)$
2. Let $S = dom(s), (S \subset \mathbb{S})$;//*determine direct child subjects*
   $S_1 \subseteq S$ such that $\forall s_1 \in S_1 \Rightarrow r_\nu(s_1, s) = kindOf(s_1, s)$;
   $S_2 \subseteq S$ such that $\forall s_2 \in S_2 \Rightarrow r_\nu(s_2, s) = partOf(s_2, s)$;
3. Let $\delta_1 = 1, \delta_2 = 1$;
4. If $S_1 \neq \emptyset$, calculate $\delta_1 = \theta \times min\{\delta(s_1) | s_1 \in S_1\}$;
5. If $S_2 \neq \emptyset$, calculate $\delta_2 = \frac{\sum_{s_2 \in S_2} \delta(s_2)}{|S_2|}$;
7. $\delta(s) = min\{\delta_1, \delta_2\}$.

Algorithm 3.2. Analyzing Semantic Relations for Specificity Assignment

The specificity of a subject refers to the semantic extent of the subject according to a topic. There are a few rules applied to the determination of a subject's specificity. The specificity of a subject increases if more instances refer to it, and if greater belief of these instances are to the topic. Secondly, the specificity decreases if a subject is at a higher level in the taxonomy as its description becomes more abstractive and less focused, e.g. from "Economic espionage" to "Business intelligence" in Fig. 1. Based on these, the specificity of a subject is defined by:

$$spe(s, \mathcal{T}) = bel(s, \mathcal{T}) \times \delta(s), \tag{9}$$

where $\delta(s)$ is a relative parameter applied by the semantic relation held by s and its peers. Algorithm 3.2 presents the technique of how $\delta(s)$ is determined, based on the analysis of semantic relations. In the algorithm, θ is a parameter indicating the graduate decrease of specificity from the leaves to the root in the ontology. While in the experiments we set θ as 0.9. The specificity is used to determine the strength of a subject in the personalized ontology supporting or against a topic.

The exhaustivity of a subject refers to the extent of the semantic meanings dealt by the subject. The semantic extent spreads if more subjects related and more details these subjects hold. Two different kinds of exhaustivity are proposed: the lower-bound $^{\perp}$ exhaustivity of a subject s refers to the semantic extent covered by s, and is compounded by all the subjects appeared in $vol(s)$:

$$exh^{\perp}(s,\mathcal{T}) = \sum_{s' \in vol(s)} bel(s',\mathcal{T}). \tag{10}$$

The upper-bound exhaustivity refers to the semantic extent dealt by the s, and is compounded by all the subjects appeared in its $lib(s)$. It is defined by the same technique as Eq. (10), but $s' \in lib(s)$. According to Eq. (8), $bel(s,\mathcal{T})$ may be positive, negative, or zero, in respect to that s is a confirmed positive, negative, or unlabelled subject. Consequently, $exh^{\perp}(s,\mathcal{T})$ and $exh^{\top}(s,\mathcal{T})$ may be resulted in positive or negative as well. A subject with the positive lower-bound exhaustivity values makes the semantic meaning of the topic clearer. In contrast, a subject with the negative lower-bound exhaustivity values makes the topic more confusing. Thus, the lower-bound exhaustivity is used to extract expert knowledge for a topic, e.g. the positive lower-bound exhaustive subjects for the extraction of positive training set, and the negative lower-bound exhaustive subjects for the negative training set. The upper-bound exhaustivity indicates the importance of a subject in $\mathcal{O}(\mathcal{T})$. A subject with a low upper-bound exhaustivity would not have significant impact to the given topic $\mathcal{T}$.

4 Experiments and Discussions

The Reuters Corpus Volume 1 (RCV1) is used as the testbed in the experiments. RCV1 is an archive of 806,791 documents produced by the Reuters journalists. It is the official testbed used in TREC-11 2002. TREC-11 has topics designed by linguists and associated with the training sets and testing sets. These topics (R101-120) are used in the experiments.

A system is implemented for Web information gathering by employing different models that attempt to interpret the semantic meanings of a user's information need from a training set (see [6] for more details). The performance of the system relies on the semantic interpretation of the information need, where the information gathering method remains the same. In the experiments, three models, TREC, Web, and Ontology model, are developed for training set:

TREC model. The training sets are manually generated by the TREC linguists who read each document and mark it either positive or negative according to a topic [12]. These training sets reflect a user's concept model perfectly, and may be deemed as the "perfect" sets;

Web model. The training sets are automatically generated from the Web (see [14] for technical details). The model analyzes a given topic and identifies the relevant subjects, then uses the subjects to gather a set of Web documents by using a selected Web search engine (Google is chosen for the experiments as it has become the most popular search engine nowadays). The model then measures the certainty of each document supporting/against the topic and assigns a float type of positive (or negative) judgment to the document. These documents then become the input training set to the Web information gathering system;

Ontology mode.l The training sets are generated by the instances associating to the subjects and representing the expert knowledge. Each document d is generated by an instance *inst* in the training set, and holds a specificity value $spe(d_{inst})$ of supporting, against, or unlabelled to the topic:

$$spe(d_{inst}) = bel(inst, \mathcal{T}) \times \sum_{s \in \eta(inst)} spe(s, \mathcal{T}). \tag{11}$$

The documents with positive *spe* value go to the positive set, with negative value go to the negative set, and with zero go to the unlabelled set. The world knowledge base contains 394,070 topical subjects. A large volume (138MB) of information stored in the catalogue of a library[3] is used in the experiments for knowledge extraction. The text pre-processing of the information volume includes stopword removal, word stemming, and word grouping. As a result, there are total of 448,590 documents and 162,751 unique terms in the volume.

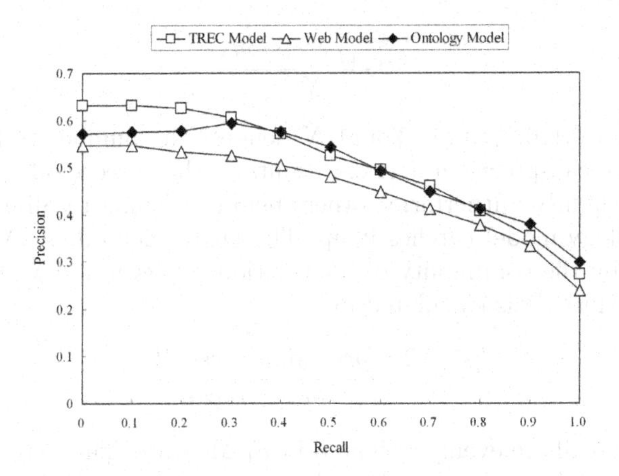

Fig. 3. The Recall-Precision Average Results

The performances of the system by applying the three models are compared and analyzed in order to find out if the Ontology model outperforms the TREC and Web models. The performance is assessed by two methods: the precision averages at 11 standard recall levels, and F_1 Measure. The former is used in TREC evaluation as the standard to compare the performance of different information filtering models [15]. A recall-precision average is computed by summing the

[3] The Queensland University of Technology Library, http://www.library.qut.edu.au/

Table 1. The Detailed Experiment Results

	Macro-F_1 Measure			Micro-F_1 Measure		
Topic	TREC Model	Web Model	Onto Model	TREC Model	Web Model	Onto Model
R101	0.733318	0.655451	0.597792	0.666026	0.598174	0.542757
R102	0.728482	0.558831	0.575412	0.671198	0.517944	0.532683
R103	0.359997	0.334705	0.385867	0.324216	0.305887	0.344459
R104	0.644069	0.616248	0.628044	0.585115	0.566150	0.578606
R105	0.554753	0.566169	0.578187	0.509154	0.516342	0.529305
R106	0.232358	0.243348	0.279385	0.222259	0.227022	0.258641
R107	0.229675	0.202800	0.205733	0.206138	0.186638	0.193576
R108	0.179407	0.152000	0.138759	0.167580	0.142408	0.129480
R109	0.450758	0.656368	0.665913	0.420498	0.602643	0.611943
R110	0.217572	0.156042	0.280089	0.201898	0.146614	0.256817
R111	0.108190	0.090479	0.126678	0.101694	0.086330	0.121804
R112	0.193970	0.174479	0.198679	0.179967	0.163100	0.181294
R113	0.315167	0.212564	0.351950	0.286667	0.197450	0.325234
R114	0.412804	0.424706	0.419160	0.373158	0.389225	0.384021
R115	0.506326	0.539497	0.507864	0.452275	0.483148	0.455121
R116	0.632010	0.555766	0.574003	0.577878	0.511178	0.525036
R117	0.361151	0.370189	0.290755	0.330699	0.339865	0.270744
R118	0.111399	0.166860	0.220595	0.107719	0.159185	0.208374
R119	0.409717	0.289761	0.290104	0.380344	0.273500	0.272663
R120	0.672907	0.649673	0.632343	0.614788	0.587924	0.569371
Average	0.402701	0.380797	0.397366	0.368964	0.350036	0.364596

interpolated precisions at the specified recall cutoff at first, and then dividing by the number of experimental topics:

$$\frac{\sum_{i=1}^{N} precision_{\lambda}}{N} \tag{12}$$

where $\lambda = \{0.0, 0.1, 0.2, \ldots, 1.0\}$ and N denotes the number of topics. Fig. 3 illustrates the recall-precision average results of the three models. The perfect TREC model slightly outperforms others before reaching recall cutoff 0.3, and then the Ontology model catches it up. The latter method, F_1 Measure [5], is well accepted by the community of information retrieval and Web information gathering. F_1 Measure is calculated by:

$$F_1 = \frac{2 \times precision \times recall}{precision + recall}. \tag{13}$$

Precision and recall are evenly weighted in F_1 Measure. The *macro-*F_1 Measure averages each topic's precision and recall values and then calculates F_1 Measure, whereas the *micro-*F_1 Measure calculates the F_1 Measure for each returned result in a topic and then averages the F_1 Measure values. The greater F_1 values indicate the better performance. The detailed F_1 Measure results are presented in Table. 1. The TREC model performs best, followed by the Ontology model, and then the Web model. In 11 out of 20 topics (the highlighted rows), the Ontology model outperforms the TREC model. The number of documents included in a TREC training set is limited (about 61 documents per topic in average). Consequently, some semantic meanings contained by the topic are not fully covered by the TREC training set. In contrast, the expert knowledge extracted by the Ontology model is from a large volume of expert classified information stored in library. The broad semantic coverage is the Ontology model's strength. The Ontology model has about 1568 documents per topic in average, covering much broader semantic extent than the TREC training set. Although in average, the perfect TREC model outperforms the Ontology model slightly, considering the TREC model employs the manpower of linguists to read every single document in the training set, which reflects a user's concept model perfectly, it is not realistic to expect that it can be beaten. Bear in mind of these, the close performance to the TREC model is a great achievement. Therefore, the experiments are evaluating the proposed method and confirm the success of it.

5 Related Work

Much effort has been invested in ontology learning or mining for semantic interpretation. Staab & Studer [13] formally define an ontology as a 4-tuple of a set of concepts, a set of relations, a set of instances and a set of axioms. Slightly different, Maedche & Staab [9] have another slightly different definition which differentiates the relations to hierarchical and plain relations. Zhong [16] proposed a learning approach for task (or domain-specific) ontology, which employs various mining techniques and natural-language understanding methods. Li & Zhong [6] proposed an semi-automatic ontology learning method, in which a class is called compound concept assembled by primitive classes that are the smallest concepts and can not be divided any further. Navigli et al. built an ontology called *OntoLearn* [10] to mine the semantic relations among the concepts from Web documents. Gauch et al. [3] used *reference ontology* and *personalized user profile* built based on the categorization of online portals and proposed to learn personalized ontology for users. However, their work does not specify the semantic relationships of "part-of" and "kind-of" existing in the concepts but only "super-class" and "sub-class". Singh et al. [7] developed *ConceptNet* ontology and tried to specify common sense knowledge. However, *ConceptNet* does not count expert knowledge. Developed by King et al. [4], *IntelliOnto* is the one closest to the goal of facilitating the human user's concept model. It is built based on DDC system, and tries to describe the world knowledge. Unfortunately, *IntelliOnto* covers only a limited number of concepts, which limits the coverage of the world knowledge described.

6 Conclusions

In this paper, a computational model is proposed for ontology mining. Two novel concepts, specificity and exhaustivity, are introduced for analyzing the semantic relations in the ontology. The model aims (i) to discover knowledge from the ontology, and (ii) to help interpret the semantic meanings underlying from a user's information need. In this paper, the semantic relationships of "kind-Of", "part-of", and "related-to" existing between the subjects in the ontology are investigated in respect to the information need, and the strength of relations is analyzed dynamically according to the related expert knowledge. The semantic extent of a subject in the ontology is formally restricted by using the concepts of exhaustivity and specificity according to the information need. The proposed model improves the semantic interpretation of information needs for the Web information gathering systems, and enhances Web information gathering from keyword-based to subject(concept)-based as shown by the experiments. The ontology mining model is a significant contribution to the knowledge discovery and knowledge management by using ontology.

References

1. Antoniou, G., van Harmelen, F.: A Semantic Web Primer. The MIT Press, Cambridge, Massachusetts (2004)
2. Curran, K., Murphy, C., Annesley, S.: Web intelligence in information retrieval. In: Proceedings of IEEE/WIC International Conference on Web Intelligence, pp. 409–412. ACM Press, New York (2003)
3. Gauch, S., Chaffee, J., Pretschner, A.: Ontology-based personalized search and browsing. Web Intelli. and Agent Sys. 1, 219–234 (2003)
4. King, J.D., Li, Y., Tao, X., Nayak, R.: Mining World Knowledge for Analysis of Search Engine Content. Web Intelli. and Agent Sys. 5, 1–21 (to appear, 2007)
5. Lewis, D.D.: Evaluating and optimizing autonomous text classification systems. In: Proceedings of the 18th annual international ACM SIGIR conference on Research and development in information retrieval, pp. 246–254. ACM Press, New York (1995)
6. Li, Y., Zhong, N.: Mining Ontology for Automatically Acquiring Web User Information Needs. IEEE Trans. on Knowledge and Data Engineering 18(4), 554–568 (2006)
7. Liu, H., Singh, P.: ConceptNet: A Practical Commonsense Reasoning Toolkit. BT Technology Journal 22, 211–226 (2004)
8. Liu, J.: New Challenges in the World Wide Wisdom Web (W4) Research. In: Konstantas, D., Léonard, M., Pigneur, Y., Patel, S. (eds.) OOIS 2003. LNCS, vol. 2817, pp. 1–6. Springer, Heidelberg (2003)
9. Maedche, A., Staab, S.: Ontology learning for the Semantic Web. IEEE Trans. on Intelligent Systems 16(2), 72–79 (2001)
10. Navigli, R., Velardi, P., Gangemi, A.: Ontology learning and its application to automated terminology translation. IEEE Trans. on Intelligent Systems. 18, 22–31 (2003)
11. Noy, N.F.: Semantic integration: a survey of ontology-based approaches. SIGMOD Rec. 33(4), 65–70 (2004)
12. Robertson, S.E., Soboroff, I.: The TREC 2001 Filtering Track Report. In: Text REtrieval Conference (2001), `http://citeseer.ist.psu.edu/750837.html`
13. Staab, S., Studer, R. (eds.): Handbook on Ontologies. Springer, Heidelberg (2004)
14. Tao, X., Li, Y., Zhong, N., Nayak, R.: Automatic Acquiring Training Sets for Web Information Gathering. In: Proceedings of the 2006 IEEE/WIC/ACM International Conference on Web Intelligence, pp. 532–535. ACM Press, New York (2006)
15. Voorhees, E.M.: Overview of TREC 2002. In: The Text REtrieval Conference (TREC) 2002 Proceedings (2002), `http://trec.nist.gov/pubs/trec11/papers/OVERVIEW.11.pdf`
16. Zhong, N.: Representation and construction of ontologies for Web intelligence. International Journal of Foundation of Computer Science 13(4), 555–570 (2002)

Knowledge Flow-Based Document Recommendation for Knowledge Sharing

Chin-Hui Lai and Duen-Ren Liu

Institute of Information Management, National Chiao Tung University, Taiwan
{chlai,dliu}@iim.nctu.edu.tw

Abstract. Knowledge is a critical property that organizations use to gain and maintain competitive advantages. In the constantly changing business environment, organizations have to exploit effective and efficient approaches to help knowledge workers find task-relevant knowledge, as well as to preserve, share and reuse such knowledge. Hence, an important issue is how to discover knowledge flow (KF) from the historical work records of knowledge workers in order to understand their task-needs and the ways they reference documents, and actively provide adaptive knowledge support. This work proposes a KF-based document recommendation method that integrates KF mining and collaborative filtering recommendation mechanisms to recommend codified knowledge. The approach consists of two phases: the KF mining phase and the recommendation phase. The KF mining phase can identify each worker's knowledge flow by considering the referencing time and citation relations of knowledge resources. Then, based on the discovered KF, the recommendation phase applies sequential rule mining and the CF method to recommend relevant documents to the target worker. Experiments are conducted to evaluate the performance of the proposed method and compare it with the traditional CF method using data collected from a research institute laboratory. The experiment results show that the proposed method can improve the quality of recommendation.

Keywords: Knowledge Flow, Clustering, Sequential Rule, CF Recommendation.

1 Introduction

Knowledge and expertise are generally codified in textual documents, e.g., papers, manuals and reports, and preserved in the knowledge base. This codified knowledge can be circulated in an organization to support workers engaged in management and operational activities. Because most of these activities are knowledge-intensive tasks, knowledge management performs a key role in preserving and sharing organizational knowledge. Consequently, the effectiveness of knowledge management depends on providing task-relevant documents to suit the information needs of knowledge workers for tasks.

The *KnowMore* system [1] provides context-aware knowledge retrieval and delivery to support the procedural activities of workers. The task-based K-support

Z. Zhang and J. Siekmann (Eds.): KSEM 2007, LNAI 4798, pp. 325–335, 2007.

system [2, 3] adaptively provides knowledge support to suit a worker's dynamic information needs. However, previous researches on task-based knowledge support do not analyze and utilize the flow of knowledge among various codified knowledge (documents) to provide effective recommendations of task-relevant documents.

Knowledge flow (KF) research focuses on how KF can carry, share, and accumulate knowledge when it passes from one team member/process to another. Working knowledge may flow among workers in an organization, while process knowledge in a workflow may flow among various tasks [4, 5]. However, there is no systematic method that can flexibly identify KF in order to understand the information needs of workers. Furthermore, conventional KF approaches did not focus on analyzing KF from the perspective of information needs and recommending relevant documents based on the discovered KF.

This work proposes a KF-based document recommendation method based on the KF model. The proposed method identifies the flow of knowledge and the information needs of knowledge workers by analyzing their historical work records. Workers who are similar to the target worker are identified based on KF and similarity analysis. Then, sequential rules discovered from the neighbors of the target worker are applied in order to recommend topics and documents. As a result, our method keeps track of the KF by analyzing a worker's knowledge referencing behavior for a task over time, and by proactively providing relevant topics and support documents to the worker.

2 Background

In this section, we introduce the background of this research, including knowledge flow, information retrieval, document clustering, and collaborative filtering system.

Knowledge Flow: The concept of knowledge flow has been applied in various domains, e.g., scientific research, communities of practice, teamwork, industry, and organizations [6-8]. Scientific articles represent the major medium for disseminating knowledge among scientists to inspire new ideas. A knowledge flow model combined with a process-oriented approach has been proposed to capture, store, and transfer knowledge. KF in weblogs (blogs) is regarded as a communication pattern whereby the post of one blogger links to the post of another blogger to exchange knowledge.

Information Retrieval: Information retrieval (IR) deals with the representation, organization, storage, and access to information items [9]. The vector space model [10] is typically used to represent documents as vectors of index terms and the weights of the terms are measured by the *tf-idf* approach.

Association Rule Mining: Association rule mining [11] is a data mining technique that is widely used to generate recommendations in recommender systems. An association rule can describe the relationships among items, such as products, documents or movies, based on patterns of co-occurrence across transactions. To identify association rules in transactions, the *Apriori* algorithm is usually employed.

Collaborative Filtering Recommendation: Collaborative filtering (CF), the most successful recommendation approach available, is used in many applications [12, 13].

CF is based on the concept that if like-minded users like an item then the target user will probably like it as well [14]. Therefore, CF predicts and recommends items, e.g., products, movies, and documents, based on the preferences of people who have the same or similar interests to those of the target user. The CF approach involves two steps: neighborhood formation and prediction. The neighborhood of a target user is selected according to user similarity, which is computed by Pearson's correlation or the cosine measure. Either the k-NN (nearest neighbor) approach or a threshold-based approach is used to choose the most similar *n* users to the target user. In this study, we use the k-NN approach. In the prediction step, the predicted rating is calculated from the aggregate weights of the selected *k* nearest neighbors' ratings, as shown in Eq. 1:

$$p_{u,j} = \overline{r_u} + \frac{\sum_{i=1}^{n} w(u,i)\left(r_{j,i} - \overline{r_i}\right)}{\sum_{i=1}^{n} \left|w(u,i)\right|} \qquad (1)$$

where $P_{u,j}$ denotes the prediction rating of item *j* for the target user *u;* $\overline{r_u}$ and $\overline{r_i}$ are the average ratings of user *u* and user *i*, respectively; *w(u,i)* is the similarity between target user *u* and user *i*; $r_{i,j}$ is the rating of user *i* for item *j;* and *n* is the number of users in the neighborhood.

3 Knowledge Flow-Based Document Recommendation

An overview of the propose method is illustrated in Fig. 1. KFs are identified from a large number of documents of tasks and workers' log. A knowledge flow mining technique is used to determine the KF of each worker. According to the different level of knowledge abstraction, a topic-level KF and a codified-level KF are derived from the referenced documents with access time recorded in worker's work logs on a specific task. Workers with similar KF to that of the target worker are considered as neighbors of the target worker. The behavioral patterns of these neighbors are identified by a sequential rule mining method. According to the discovered sequential rules and the KF of the neighbors, relevant topics and documents are recommended to the target worker to support the execution of the task at hand and enhance work efficiency.

The proposed document recommendation method consists of two phases: a knowledge flow mining phase and a recommendation phase. The knowledge flow mining phase identifies the knowledge flow of a knowledge worker. The recommendation phase recommends documents to the target worker based on his knowledge flow. The knowledge flow mining phase involves four steps: document profiling, document clustering, knowledge flow mining, and knowledge flow adjustment. Each document is represented as a document profile which is an *n*-dimensional vector. Then, the document profiles are used to group documents with similar content into clusters according to the similarity of profiles. Next, the topic-level and codified-level KFs are generated. Finally, the codified-level KFs are adjusted by the number of times documents are cited.

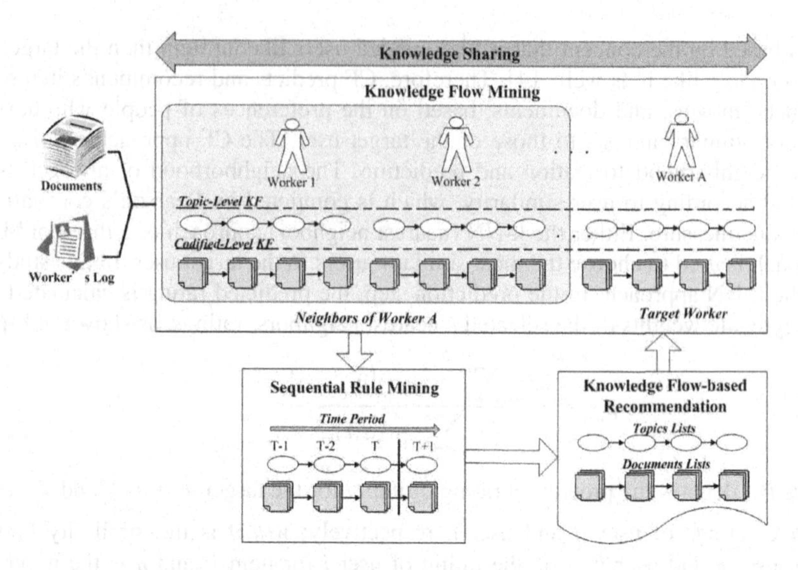

Fig. 1. An overview of the KF-based recommendation

The recommendation phase also comprises four steps: flow comparison, sequential rule mining, rule matching, and recommendation. In the flow comparison step, knowledge workers' KFs are compared to derive the similarity of the workers. Next, to identify the referencing behavior of workers, sequential rules are mined in the topic-level KF. The rules are used to match topic-level KF and then recommend topics. Moreover, the CF recommendation method is used to recommend documents relevant to the recommended topic.

3.1 Knowledge Flow Model

Knowledge Flow Definition. A knowledge flow (KF) is defined as the evolution of a worker's knowledge requirements on conducting a task, including information needs, preferences and referencing behavior for codified knowledge; KF is identified from the worker's work logs on historical task executions.

A KF consists of two levels: the codified level and the topic level. The knowledge in the codified-level indicates the knowledge flows among documents based on the access time. In most situations, the knowledge obtained from a document prompts knowledge workers to access the next relevant document (codified knowledge). The document sequence contains several task-related documents. Moreover, documents with similar subjects of interests can be grouped into the same topic to form a topic-level abstraction of knowledge. The codified-level KF can be abstracted to form a topic-level KF, which represents the transition among various topics of codified knowledge. The knowledge in the topic-level indicates that task knowledge may flow among topics and prompt knowledge workers to obtain knowledge from the next related topic.

3.2 Knowledge Flow Mining Phase

KF mining consists of document profiling, document clustering and KF mining.

Document Profiling and Clustering: Two profiles, a document profile and a topic profile are needed to identify a worker's KF. The document profile describes the characteristics of a document based on its content. A document can be represented as an *n*-dimensional vector composed of terms with respective weights derived by the *tf-idf* approach [9, 15]. The approach derives the term weights by measuring the term frequency and the inverse document frequency to estimate the degree of importance of terms in a document [10]. Based on the term weights, the terms with higher values are selected as discriminative terms to describe the characteristics of a document. The document profile of d_j is composed of these discriminative terms and represented by an *n*-dimensional vector. Let the document profile be $DP_j =< dt_{1j} : dtw_{1j}, dt_{2j} : dtw_{2j}, \cdots, dt_{nj} : dtw_{nj} >$, where dt_{ij} is the term *i* in d_j. To cluster documents, the document profiles are used to measure the similarities of the documents; then, the documents with similar profiles are grouped together.

Documents are clustered to form a topic set based on the similarity among documents. We use a single-link hierarchical clustering method [16] to automatically group documents into clusters of documents. Each cluster is defined as a topic and a topic profile is used to describe the key features of the cluster. A topic profile can be derived from the profiles of the documents within a cluster. Let $TP_x =< tt_{1x} : ttw_{1x}, tt_{2x} : ttw_{2x}, \cdots, tt_{nx} : dtw_{nx} >$ be the profile of topic (cluster) *x*, where tt_{ix} is a topic term and ttw_{ix} is the weight of the topic term. Let D_x be the set of documents in cluster *x*. The weight of a topic term is determined by Eq. 2.

$$ttw_{ix} = \frac{\sum_{j \in D_x} dtw_{ij}}{|D_x|} \times tf_{ix} \tag{2}$$

where dtw_{ij} is the weight of term *i* in document *j*, $|D_x|$ is the number of documents in cluster *x*, and tf_{ix} is the frequency of term *i* in cluster *x*. The terms with higher average term weights and frequency are selected as topic terms.

Knowledge Flow Mining: The KF mining approach considers the time factor to keep track of the accumulation of knowledge and determine each knowledge worker's KF. This mining approach identifies two kinds of KF: codified-level KF and topic-level KF. The codified-level KF is extracted from the documents recorded in the worker's work log. The documents are arranged according to the time they were accessed and then a document sequence as a codified-level KF is obtained. Each worker has his/her own codified-level KF, which represents his/her process of knowledge accumulation.

The topic-level KF is derived by mapping documents into corresponding clusters. We use the document clustering results to map the documents in the codified-level KF into topics (clusters) to build the topic-level KF. The knowledge in the topic-level KF is an abstraction of the codified-level KF, and indicates how knowledge flows among various topics.

3.3 Recommendation Phase

The recommendation phase consists of four steps: flow comparison, sequential rule mining, pattern matching, and CF recommendation. To determine the similarity of various topic-level KF, the KF of the target worker is compared with those of other workers in the flow comparison step. Workers with similar KF to that of the target worker are regarded as the latter's neighbors and their topic-level KF are used to identify sequential rules for the target user. Then, we set a knowledge window on the target user's KF and compare it with the discovered sequential rules by computing the matching degree of rules. The rules with higher matching degrees are selected to recommend topics. Moreover, documents belong to the recommended topics will be recommended to the target user based on the CF approach. Next, we describe recommendation phase in detail.

Flow Comparison

To find the target worker's neighbors, the topic-level KF of the target user is compared with those of other workers to compute the flow similarities. Since the KFs are sequences, we need to compute the similarity of two sequences. The sequence alignment method [17, 18], which basically computes the cost of aligning one sequence to another sequence, can be used to compute the similarity of two sequences. Moreover, the similarity of two workers can also be derived based on the similarity of their referenced documents. Accordingly, we propose a hybrid similarity measure, which is composed of the KF alignment similarity and the aggregated profile similarity, as shown in Eq. 3.

$$sim(TKF_i,TKF_j) = \alpha \times sim_a(TKF_i,TKF_j) + (1-\alpha) \times sim_P(AP_i, AP_j) \tag{3}$$

where $sim_a(TKF_i,TKF_j)$ represents the KF alignment similarity, $sim_p(AP_i,AP_j)$ represents the aggregated profile similarity, and α is a parameter used to adjust the relative importance of these two types of similarity.

We adopt the sequence alignment method [17] to derive the KF alignment similarity, $sim_a(TKF_i,TKF_j)$, by adding the estimation of the overlapping of topics in the two compared topic-level KF, as shown in Eq.4.

$$sim_a(TKF_i,TKF_j) = \delta \times \frac{2 \times |T_i \cap T_j|}{|T_i| + |T_j|} \tag{4}$$

where TKF_i and TKF_j are the topic-level KF of workers *i* and *j* respectively; δ is the normalized maximal flow alignment score derived by the dynamic programming approach [17]; T_i and T_j are the set of topics in topic-level KF TKF_i and TKF_j respectively. $T_i \cap T_j$ is the intersection of topics common to TKF_i and TKF_j; and $|T_i|$ and $|T_j|$ are the number of topics in topic-level KF TKF_i and TKF_j respectively.

The aggregated profile similarity, defined as $sim_p(AP_i,AP_j)$, computes the similarity of two workers based on their aggregated profiles that are derived from the profiles of documents referenced by them; AP_i and AP_j are two vectors of the aggregated profile of workers i and j respectively. The cosine formula is used to calculate the similarity between two aggregated profiles. The aggregated profile of a worker i is defined as $AP_i = \sum_{t=1}^{T} tw_{t,T} \times DP_t$, where $tw_{t,T}$ is the time weight of the document x referenced at time t in KF; T is the index of the time the worker referenced the latest documents in his KF; and DP_t is the profile of the document referenced by worker i at time t. The aggregation considers the time decay effect of the documents. Each document is assigned a time weight according to the time the document was referenced. If a document was referenced in the recent past, it is given a higher time weight; otherwise, a lower time weight is given. The time weight of each document profile is defined as $tw_{t,T} = \frac{t - St}{T - St}$, where St is the start time of the worker's KF.

Mining the referencing behavior

Based on the similarities mentioned in the previous sub-section, knowledge workers with similar referencing behavior (high similarities) are grouped together and regarded as neighbors of the target worker. Using the topic-level KF of each group member, we apply time-based association rule (sequential rule) mining [19, 20] on the KF to identify topic-level sequential rules for the target worker. Let R_y be an association rule with time constraints, as defined in Eq. 5.

$$R_y\text{: } r_{y,T-l}, \ldots, r_{y,T} \Rightarrow r_{y,T+1} \; (Support_y,\ Confidence_y)$$
$$\text{where } r_{y,T-s} \in TKF \text{ or } \phi \text{ and } r_{y,T+1} \in TKF;\ s=0 \text{ to } l \qquad (5)$$

The conditional part of the sequential rule is $<r_{y,T-l}, \ldots, r_{y,T}>$, and the consequent part is $r_{y,\ T+1}$. TKF is the set of all topics. The values of support and confidence are determined in order to evaluate the importance of a rule. Based on the values, the significant rules can be chosen for recommending topics.

Sequential Rule Matching

The sequential rules obtained in the previous step are matched with the topic-level KF of the target worker to predict the topics required at time T+1. We set a knowledge window on the KF before time T+1. Let the knowledge window $KW_u = < TP_u^{T-l}, TP_u^{T-l+1} \ldots, TP_u^{T} >$ be on the topic-level KF of the target worker u before time T+1. Note that TP_u^{T-s} is the topic referenced by user u at time T-s, s=0...l. The matching method compares the topic subsequences derived from the conditional part of a sequential rule with a given knowledge window. The matching degree is calculated based on the similarity of topics to identify the sequential rules qualified to recommend topics at time T+1.

CF Recommendation

There are two phases in the CF recommendation step: recommending topics and recommending documents. In our recommendation method, topics are recommended first based on the sequential rules, and then documents belonging to those topics are recommended. According to the sequential rule matching results, we sort the rules in descending order of their degree of matching, and then take the top-N rules with high-matching degrees to compile a topic recommendation list for the target worker at time *T+1*. After recommending topics, our method recommends the top-N documents with the highest predicted ratings that belong to the recommended topics. User similarity is calculated according to the similarity of user profiles. A user profile UP_u of a worker u can be derived from the profiles of the documents referenced by u on conducting a task, which is similar to the derivation of topic profile (Eq. 2). Let NB_a be the set of target worker a's neighbors that are selected according to the similarity derived using Eq. 3. The predicted ratings of documents are measured by Eq. 6:

$$\hat{p}_{a,d,t} = \beta \times \left(\overline{R}_a + \frac{\sum_{x \in NB_a} sim(UP_a, UP_x) \times (R_{x,d,t} - \overline{R}_x)}{\sum_{x \in NB_a} sim(UP_a, UP_x)} \right) + (1-\beta) \times ((c_d \times df_d) \times \lambda) \quad (6)$$

where $\overline{R}_a / \overline{R}_x$ is the average rating given by the target worker a / worker x, UP_a/UP_x is the user profile of the target worker/ worker x; $R_{x,d,t}$ is the rating of document d that belongs to topic t and worker x; c_d is the number of times a document d is cited by other documents in the document set Z; df_d is the document frequency, i.e., the number of times that document d was referenced by worker a's neighbors; λ is the normalized factor to normalize the value of $(c_d \times df_d)$ to the range from 0 to 5; and β is the weighting determined by experiments. We note that the rating score is from 0 to 5, thus λ is equal to 5 divided by the maximum value of $(c_d \times df_d)$ over all $d \in Z$. In Eq. 6, we modify the predicted formula in the traditional CF method and add two parameters, namely, the number of times a document is cited and the document frequency, to improve the recommendation quality. If a document has been cited many times and referenced by many neighbors, it has the first priority for recommendation. To recommend documents, the top-N documents with highest predicted ratings are chosen and recommended to the target worker.

4 Experiments and Evaluation

We use 35 knowledge workers from a research laboratory and a real-world dataset consisting of 424 research papers to evaluate the proposed method. For each worker, the documents and the time they were referenced are used to identify the worker's referencing behavior on conducting a task. The dataset is divided as follows: 80% for training and 20% for testing. The training set is used to generate recommendation lists, while the test set is used to verify the quality of the recommendations.

We compare three recommendation methods: KF-based recommendation, KF-based recommendation without topics, and the traditional CF method. The KF-based recommendation (with topic) method recommends topics based on sequential rules and then recommends documents belonging to the recommended topics (Eq. 6). Similar to the KF-based recommendation with topics, the KF-based recommendation without topics uses Eq 6 to derive the predicted ratings of documents. However, it recommends relevant documents without restricting the documents belonging to the topics recommended by the sequential rules. The traditional CF method uses Eq. 1 to derive the predicted rating, where user similarity is computed based on user profiles.

To evaluate the quality of recommendations, we use recall and precision measures, which are widely used in recommender systems [21]. *Recall* represents the percentage of total known helpful documents that are recommended. *Precision* represents the percentage of recommended documents that the user finds helpful. The F1-metric [21, 22] is used to balance the trade-off between precision and recall. The F1-metric assigns equal weight to both recall and precision and is used in our evaluation, as shown in Eq. 7.

$$F1 = \frac{2 \times recall \times precision}{recall + precision} \tag{7}$$

4.1 Experiment Results

We conduct several experiments to measure the recommendation quality of our proposed method. First, documents in the data set are grouped into eight clusters using a single-link clustering method. Based on the clustering results, topic-level KF are generated by mapping documents from the codified-level KF into these clusters for each knowledge worker. Then, the topic-level and codified-level KF are used to enhance the quality of document recommendation. We set α=0.3 and β ranges from 0 to 1 in order to obtain the average F1 values for KF-based recommendation methods.

From the results shown in Fig 2, we observe that the F1 value decreases as k increases, and KF-based recommendation performs better than the other methods. Both KF-based recommendation methods obtain better F1 scores when k=2, whereas the CF method performs better when k= 4.

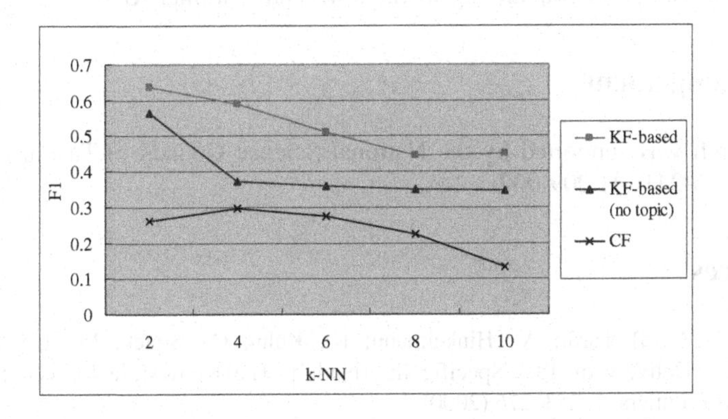

Fig. 2. Compare the KF-based recommendation method with CF under different k

Fig. 3 shows the comparison of the three recommendation methods under various numbers of top-N documents. The F1 scores of both KF-based recommendation methods increase as the number of top-N increases. When recommending the top-25 documents, both methods have higher F1 scores. By comparison, CF performs well if it recommends the top-10 documents. The KF-based recommendation method performs better than other methods.

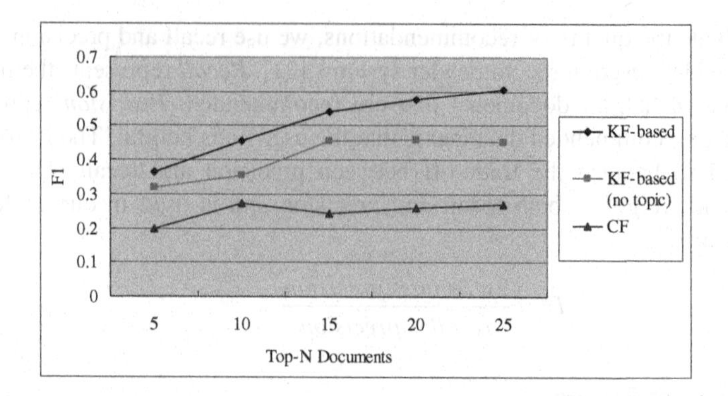

Fig. 3. Compare the KF-based recommendation method with CF under various Top-*N*

5 Conclusions

We have proposed a KF-based document recommendation method that integrates KF mining and collaborative filtering techniques to proactively recommend codified knowledge to knowledge workers. We evaluated the recommendation quality of the proposed method under various parameters and compared it with the traditional CF method. The experiment results show that the method improves the quality of document recommendation and performs better than the traditional CF method. The recommended documents not only meet the information needs of workers, but also facilitate knowledge sharing among workers who have similar KFs.

Acknowledgement

This research was supported by the National Science Council of Taiwan under the grant NSC 96-2416-H-009-007.

References

1. Abecker, A., Bernardi, A., Hinkelmann, K., Kuhn, O., Sintek, M.: Context-Aware, Proactive Delivery of Task-Specific Information: The KnowMore Project. Information Systems Frontiers. 2, 253–276 (2000)
2. Liu, D.R., Wu, I.C., Yang, K.S.: Task-based K-Support system: disseminating and sharing task-relevant knowledge. Expert Systems With Applications. 29, 408–423 (2005)
3. Wu, I.C., Liu, D.R., Chen, W.H.: Task-stage knowledge support: coupling user information needs with stage identification. In: IEEE International Conference on Information Reuse and Integration, IRI-2005, pp. 19–24 (2005)
4. Zhuge, H.: A knowledge flow model for peer-to-peer team knowledge sharing and management. Expert Systems with Applications 23, 23–30 (2002)
5. Zhuge, H.: Knowledge flow network planning and simulation. Decision Support Systems 42, 571–592 (2006)

6. Kim, S., Hwang, H., Suh, E.: A process-based approach to knowledge-flow analysis: a case study of a manufacturing firm. Knowledge and Process Managment. 10, 260–276 (2003)
7. Zhuge, H.: Discovery of knowledge flow in science. In: Communications of the ACM, vol. 49, pp. 101–107. ACM Press, New York (2006)
8. Anjewierden, A., de Hoog, R., Brussee, R., Efimova, L.: Detecting knowledge flows in Weblogs. In: Sunderam, V.S., van Albada, G.D., Sloot, P.M.A., Dongarra, J.J. (eds.) ICCS 2005. LNCS, vol. 3514, pp. 1–12. Springer, Heidelberg (2005)
9. Baeza-Yates, R., Ribeiro-Neto, B.: Modern Inofrmation Retrieval. Addison-Wesley, Boston (1999)
10. Salton, G., Buckley, C.: Term-weighting approaches in automatic text retrieval. Information Processing & Management 24, 513–523 (1988)
11. Agrawal, R., Imielinski, T., Swami, A.: Mining association rules between sets of items in large databases. ACM SIGMOD Record 22, 207–216 (1993)
12. Konstan, J.A., Miller, B.N., Maltz, D., Herlocker, J.L., Gordon, L.R., Riedl, J.: GroupLens: applying collaborative filtering to Usenet news. Communications of the ACM 40, 77–87 (1997)
13. Rucker, J., Polanco, M.J.: Siteseer: personalized navigation for the web. Communications of the ACM 40, 73–76 (1997)
14. Breese, J.S., Heckerman, D., Kadie, C.: Empirical Analysis of Predictive Algorithms for Collaborative Filtering. In: Proceedings of the Fourteenth Annual Conference on Uncertainty in Artificial Intelligence, pp. 43–52 (1998)
15. Porter, M.F.: An algorithm for suffix stripping. Program. 14, 130–137 (1980)
16. Jain, A.K., Murty, M.N., Flynn, P.J.: Data clustering: a review. ACM Computing Surveys (CSUR) 31, 264–323 (1999)
17. Charter, K., Schaeffer, J., Szafron, D.: Sequence alignment using FastLSA. In: METMBS 2000. International Conference on Mathematics and Engineering Techniques in Medicine and Biological Sciences, pp. 239–245 (2000)
18. Oguducu, S.G., Ozsu, M.T.: Incremental click-stream tree model: Learning from new users for web page prediction. Distributed and Parallel Databases. 19, 5–27 (2006)
19. Agrawal, R., Srikant, R.: Mining sequential patterns. In: Proceedings of the Eleventh International Conference on Data Engineering, pp. 3–14 (1995)
20. Cho, Y.B., Cho, Y.H., Kim, S.H.: Mining changes in customer buying behavior for collaborative recommendations. Expert Systems with Applications 28, 359–369 (2005)
21. Sarwar, B., Karypis, G., Konstan, J., Riedl, J.: Analysis of recommendation algorithms for e-commerce. In: Proceedings of the ACM Conference on Electronic Commerce, pp. 158–167. ACM Press, New York (2000)
22. Van RijsBergen, C.J.: Information retrieval. Butterworths, London (1979)

Finding Similar RSS News Articles Using Correlation-Based Phrase Matching

Maria Soledad Pera and Yiu-Kai Ng

Computer Science Dept., Brigham Young University, Provo, Utah 84602, U.S.A.

Abstract. Traditional phrase matching approaches, which can discover documents containing exactly the same phrases, fail to detect documents including phrases that are semantically relevant, but not exact matches. We propose a *correlation-based phrase matching* (CPM) model that can detect RSS news articles which contain not only phrases that are exactly the same but also semantically relevant, which dictate the degrees of similarity of any two articles. As the number of RSS news feeds continue to increase over the Internet, our CPM approach becomes more significant, since it minimizes the workload of the user who is otherwise required to scan through huge number of news articles to find related articles of interest, which is a tedious and often an impossible task. Experimental results show that our CPM model on matching bigrams and trigrams outperforms other phrase, including keyword, matching approaches.

1 Introduction

Phrase queries are frequently used to retrieve documents from the Web. A phrase, which is often defined as a *sequence of words* [3], can be represented in two folds: (i) the syntactic structure that the words are organized in, and (ii) the semantic content it delivers. Changing either one of the two representations may result in a phrase with a different meaning. Traditional phrase matching techniques aim to retrieve documents including phrases that match exactly with the query phrase, although some advanced approaches tolerate errors to some extent (e.g., proximity of words, word order, and missing words in a phrase). These inherent characteristics draw restrictions on their potential usages, i.e., they may fail to detect potentially relevant phrases and hence documents. For example, the phrase "heterogeneous node" (on wireless networks) is semantically relevant to "heterogeneous device" and "heterogeneous transport," which could be used along with "heterogeneous node" in retrieving closely related documents.

Neither keyword matching (nor traditional phrase matching as mentioned earlier) can solve the inexact phrase matching problem. Using keywords "heterogeneous" and "node" individually in keyword search could match documents that include either the word "heterogeneous" or "node," but not necessarily both, and thus the content of retrieved documents might be totally unrelated to "heterogeneous node." Some of these documents may address "heterogeneous alloys," whereas others may discuss "homogeneous node." Even though the "matched" documents include both words, they are not necessarily in the same order, which

Z. Zhang and J. Siekmann (Eds.): KSEM 2007, LNAI 4798, pp. 336–348, 2007.

might run into the same "content mismatched" problem. The more sophisticated similarity matching approaches, such as [15], can detect documents that include similar (not necessarily the same) words; they, however, cannot resolve the word-ordering problem. For example, consider the sentences "They *jog* for thirty minutes and *walk* for an hour" and "They *run* for an hour and *stroll* for thirty minutes." Ignoring the word order and simply considering the degrees of (single-)word similarity, i.e., *jog* versus *run* and *walk* versus *stroll*, causes these sentences to be treated as closely related, even though they are semantically different, and filtering out mismatched documents manually is a waste of time.

We propose a *correlation-based phrase matching* (CPM) model that can detect RSS news articles which contain phrases that are semantically relevant, in addition to exact matches. We are interested in RSS news articles, since there is no precedent in the amazing amount of online news that can be accessed by Internet users these days. Thus, the problem of seeking information in online news articles is not the lack of them but being overwhelmed by them. This brings a huge challenge in finding related online news with distinct information automatically, instead of manually, which is a labor-intensive and impractical process. The proposed CPM model measures the degrees of similarity among different RSS news articles using phrase similarity to detect *redundant* and discover *similar* news articles. We call the proposed model *correlation-based*, since we adapt the correlation factors in fuzzy sets to model the similarity relationships among different phrases. For each phrase p, its fuzzy set S is constructed that captures the *degrees of memberships*, i.e., closeness, of p to all the other phrases in S, which are called *phrase correlation factors*.

The rest of the paper is organized as follows. In section 2, we discuss research work in phrase matching. In section 3, we present the design of CPM. In section 4, we verify the accuracy of CPM in detecting related documents using various test cases. In section 5, we draw a conclusion and include future work on CPM.

2 Related Work

Phrase matching has been applied in solving different problems, such as ranking relevant documents, document clustering [3], and Web document retrieval [1]. In [1], a system for matching phrases in XML documents, called PIX, is presented. PIX allows users to specify both (i) tags and annotations in an XML document to ignore and (ii) phrases in the document to be matched. This technique relies on *exact* and *proximity* phrase matching (i.e., words in a phrase that are within a distance of $k(\geq 1)$-words in a document) in retrieving relevant documents.

[3] cluster Web documents based on matched phrases and their levels of significance (e.g., the title and the body) in the documents. This method uses *exact* phrase matching to determine the degrees of overlap among documents, which yield their degrees of similarity.

[14] use phrase matching for ranking medical documents. The similarity of any two phrases in [14] is detected by the number of consecutive three-*letter* triples in common, with various scores assigned to different triples of letters,

e.g., uncommon three-letter triples are given a *higher weight*, three-letter triples at the beginning of a word are *more important* than the ones at the end of the word, and long phrases are discounted to avoid bias on their lengths. In [14] phrases are treated as documents and tri-grams of letters are treated as words.

[8], who use phrase and *proximity* terms for Web document retrieval and treat every word in a query as a phrase, show that the usage of phrases and proximity terms is highly beneficial. However, their experimental results show that even though phrases and proximity terms have a positive impact on 2- or 3-word queries, they have less, or even negative, effects on other types of queries.

[2] present a compression method that searches for words and phrases on natural-language text. This method performs an *exact* search for words and phrases on compressed text directly using any sequential pattern-matching algorithm, in addition to a word-based *approximate* for extended search. Thus, searches can be conducted for approximated occurrences of a phrase pattern.

[9] emphasize the importance of phrase extraction, representation, and weighting and claim that phrases obtained by syntactic (instead of statistical) processing often increase the effectiveness of retrieval when proximity and weighting information are adequately attached to a query phrase representation. [12], however, determine the degree of similarity between any two documents by computing the number of common phrases in the documents, and dividing the number of common phrases by the total number of phrases in both, which is intuitively another *exact* phrase matching approach.

3 Correlation-Based Phrase Matching

Semantically relevant phrases detected by our CPM model hold the same syntactic features as in other phrase matching approaches, i.e., a phrase is treated as a *sequence* of words and the *order* of words is significant. Unlike existing phrase matching approaches, we develop novel *phrase correlation factors* for the n-gram ($1 \leq n \leq 5$) phrases. Using one of these chosen sets of n-gram phrase correlation factors, the n-gram phrases in an RSS news article are matched against the n-gram phrases in another RSS news article to determine their degrees of similarity. We detail the design of our n-gram CPM approach on RSS news articles below.

3.1 Content Descriptors of RSS News Articles

Two of the essential elements in an RSS (XML) news feed file, in which RSS news articles are posted, are the *title* and *description* of an *item* (i.e., a news article), since the former contains the headline and the latter includes the first few sentences of the article. Furthermore, several items can appear in the same RSS feed file. (See, as an example of, an RSS news feed file as shown in Figure 1.) We treat the title and description of each item as the *content descriptor* of the corresponding article and determine its degree of similarity with the *content descriptor* of another item (in the same or a different RSS news feed file) according to the correlation factors of phrases in the two content descriptors.

<?xml version="1.0" encoding="utf-8"?> <rss version="2.0">
...
Includes the name and link of the Website where the RSS file is located
<title>english.people.com.cn</title> ... <link>http://english.people.com.cn</link> ... ...
<item>
Several items (i.e., articles) can be specified in an RSS news feed
<title>Ugandan rebels withdraw from Juba peace talks</title>
<link>http://english.people.com.cn/200701/12/eng20070112_340796.html</link>
<description>The peace talks between the Ugandan government and rebels of the Lord's Resistance Army (LRA) are faltering following a walkout of the LRA delegation which cited frequent attacks by the Ugandan army as the reason. </description>
</item>
Contains the headline of the article
URL where the article in full can be retrieved
<item>
Contains a few lines about the article
<title>4 ASEAN countries agree to expand air linkages</title>
<link>http://english.people.com.cn/200701/12/eng20070112_340782.html</link>
<description>Brunei, Indonesia, Malaysia and the Philippines agreed Friday to expand their air linkages in a move to further boost trade and tourism among the four countries. </description>
</item>
...

Fig. 1. Portion of an RSS news feed file

3.2 Computing the Phrase Correlation Factors

Prior to computing n-gram ($1 \leq n \leq 5$) phrase correlation factors, we first decide at what level the correlation factors are to be calculated, which dictates how the subsequent process of phrase comparison should be conducted.

The major drawback of the phrase-level granularity is its excessive overhead. A phrase may start at any position in a document, and the lengths of phrases vary in practical usage. Thus, the number of possible phrases to be considered could be huge. For example, consider a portion of the paragraph that is randomly chosen from www.cnn.com: "... the organ's unwrinkled surface resembled **that of the** brain of an idiot. ... Researchers contend **that if the** plant-eating beasts ... ." Even only considering trigram phrases, there are 71 trigram phrases in the entire paragraph. However, not all of them, such as the phrases "that of the" and "that if the," are useful in determining the content of the paragraph, or its corresponding document in general. Thus, our CPM model, which pre-computes the correlation factors of any two n-gram phrases, considers only *non-stop, stemmed words*[1] in an RSS news article to form phrases to be matched.

The Unigram Correlation Factors. We construct the unigram (i.e., single-word) correlation factors using the documents in the Wikipedia Database Dump (http://en.wikipedia.org/wiki/Wikipedia:Database_download). We chose the Wikipedia documents for constructing each of the n-gram ($1 \leq n \leq 5$) correlation factors, since the 850,000 Wikipedia documents were written by more than 89,000 authors on various topics. The diversity of the authorships leads to a representative group of documents with different writing styles and a variety of subject areas. Thus, the set of Wikipedia documents is an effective representative set of documents that is appropriate for computing the general correlation factors among unigram, as well as other n-gram ($2 \leq n \leq 5$), phrases. The

[1] *Stopwords* are words that appear very frequently (e.g., "him," "with," "a," etc.), which include articles, conjunctions, prepositions, punctuation marks, numbers, non-alphabetic characters, etc., and are typically not useful for analyzing the informational content of a document. *Stemmed words* are words with the same meaning.

correlation factors of the unigrams are computed according to the *distance* and *frequency of occurrence* of the unigrams in each Wikipedia document.

Prior to constructing the unigram correlation factors, we first removed all the words in the Wikipedia documents that are *stopwords.* Eliminating stopwords is a common practice, since the process (i) filters the noise within both a query and a document [4,7] and (ii) enhances the retrieval performance [13], which enriches the quality of our unigram phrase correlation factors. After stopwords were removed, we stemmed the remaining words by using the Porter Stemmer [11], which stems each word to its grammatical root but retains the semantic meaning of the words, e.g., "driven" and "drove" are reduced to their stemmed word "drive." The final count of non-stop, stemmed unigrams is 57,926. The *unigram correlation value* of word w_i with respect to word w_j, which is constructed by using the Wikipedia documents without stop- or non-stemmed words, is defined as

$$c_{i,j} = \sum_{w_i \in V(S_i)} \sum_{w_j \in V(S_j)} \frac{1}{d(w_i, w_j)} \tag{1}$$

where $d(w_i, w_j) = |Position(w_i)$ - $Position(w_j)|$ is the distance, i.e., the number of words, between w_i and w_j in a Wikipedia document, and $V(S_i)$ ($V(S_j)$, respectively) denotes the set of (non-)stemmed words of w_i (w_j, respectively).

Correlation factors among unigrams that co-occur more frequently than others in a document are assigned higher values. To avoid the bias on the frequency of occurrences in a "long" Wikipedia document, we *normalize* $c_{i,j}$ as

$$nc_{i,j} = \frac{c_{i,j}}{|V(S_i)| \times |V(S_j)|} \tag{2}$$

where $|V(S_i)|$ ($|V(S_j)|$, respectively) is the number of words in $V(S_i)$ ($V(S_j)$, respectively).

Given k different $nc_{i,j}$ values, one from each of the Wikipedia documents in which both w_i and w_j occur, the *unigram correlation factor* $cf_{i,j}$ of w_i and w_j is

$$cf_{i,j} = \frac{\sum_{m=1}^{k} nc_{i,j}^{m}}{m} \tag{3}$$

where $nc_{i,j}^{m}$ is the normalized correlation value $nc_{i,j}$ (as defined in Equation 2) of w_i and w_j in the mth ($1 \leq m \leq k$) document in which both w_i and w_j occur, and k is the total number of Wikipedia documents in which w_i and w_j co-occur.

Example 1. Consider the following two RSS news articles represented by their content descriptors:

Article 1. Ice storm threatens chaos: A *brutal ice storm* is *threatening* to *turn Oklahoma* into a sheet of ice and cause dangerous conditions from Texas to New York. "This is a one-in-maybe-15-to 25-year event," CNN severe weather expert Chad Myers said.

Article 2. Ice storm threatens chaos: Freezing rain hit Oklahoma today, the start of what forecasters say could be a brutal ice storm. Millions of people from Texas through Oklahoma to Missouri are being *warned*

Table 1. Correlation factors of some unigrams in two different RSS news articles

Article 1/Article 2	warned	conditions	deteriorate	afternoon	severe	weather	expert
brutal	1.9E-7	2.3E-7	2.5E-7	3.3E-8	2.1E-7	8.7E-8	6.3E-8
ice	1.4E-7	1.6E-7	9.5E-8	7.9E-8	1.5E-7	3.2E-7	5.8E-8
storm	1.4E-6	1.7E-7	9.5E-8	5.0E-7	7.9E-7	1.3E-6	7.9E-4
threatening	3.8E-7	2.4E-5	2.6E-7	1.4E-7	3.8E-7	1.6E-7	6.6E-8
turn	1.5E-7	1.1E-7	1.2E-7	1.2E-7	1.1E-7	1.3E-7	8.5E-8
Oklahoma	3.7E-8	1.5E-8	9.7E-9	4.1E-8	4.0E-8	1.0E-7	3.1E-8

that *conditions* will *deteriorate* this *afternoon.* "This is a one-in-15- to 25-year event," CNN *severe weather expert* Chad Myers said."

Table 1 shows some of the unigrams in the two news articles and their correlation factors generated by Equation 3 using the Wikipedia documents. □

The N-gram Phrase Correlation Factors. The phrase correlation factors of any two n-grams ($2 \leq n \leq 5$) are calculated according to the correlation factors of their corresponding unigrams, since unigram correlation factors are reliable in detecting similar words. (See experimental results in Section 4.) We compute the bigram, trigram, 4-gram, and 5-gram phrase correlation factors using the (prior) Odds (*Odds* for short) [5] that measures the predictive or prospective support according to a hypothesis H by the prior knowledge $p(H)$ alone to determine the strength of a belief, which is the unigram correlation factor in our case.

$$O(H) = \frac{p(H)}{1 - p(H)} \tag{4}$$

Based on the computed unigram correlation factors, $cf_{i,j}$, in Equation 3, we generate the n-gram ($2 \leq n \leq 5$) phrase correlation factors between any n-gram phrases p_1 and p_2 using Equation 4 such that $p(H)$ is defined as the *product of the unigram correlation factors* of the corresponding unigrams in p_1 and p_2, i.e.,

$$pcf_{p_1,p_2} = \frac{\prod_{i=1}^{n} cf_{p_{1_i},p_{2_i}}}{1 - \prod_{i=1}^{n} cf_{p_{1_i},p_{2_i}}} \tag{5}$$

where p_{1_i} and p_{2_i} ($1 \leq i \leq n$) are the ith word in p_1 and p_2, respectively.

Tables 3, 2, 4, and 5 show the phrase correlation factors of some bigrams, trigrams, 4-grams, and 5-grams, respectively, in the two articles in Example 1.

3.3 Phrase Comparison

Phrases in the content descriptor of an RSS news article are compared against their counterparts in another RSS news article. CPM can detect phrases in RSS news articles that are semantically relevant (or the same) to phrases in other articles. To accomplish this, the phrase of a chosen length k ($1 \leq k \leq 5$) in a news article A_1 is compared with each phrase of the same length in another article

Table 2. Correlation factors of some *bigrams* in two different RSS news articles

Article 1/Article 2	warned conditions	conditions deteriorate	deteriorate afternoon	weather expert
brutal ice	3.0E-14	2.2E-14	2.0E-14	5.1E-15
ice storm	2.3E-14	1.5E-14	4.7E-14	2.5E-10
storm threatening	3.2E-11	4.3E-14	1.3E-14	8.2E-14
threatening turn	4.2E-14	2.8E-12	3.1E-14	1.3E-14

Table 3. Correlation factors of some *trigrams* in two different RSS news articles

Article 1/Article 2	warned conditions deteriorate	conditions deteriorate afternoon	severe weather expert
brutal ice storm	2.8E-21	1.1E-20	5.4E-17
ice storm threatening	5.9E-21	2.1E-21	1.2E-20
storm threatening turn	3.8E-18	5.3E-21	1.0E-20
threatening turn Oklahoma	4.0E-22	1.1E-19	1.5E-21

Table 4. Correlation factors of some *4-grams* in two different RSS news articles

Article 1/Article 2	warned conditions deteriorate afternoon	severe weather expert Chad
brutal ice storm threatening	4.0E-28	4.4E-24
ice storm threatening turn	7.2E-28	7.2E-28
storm threatening turn Oklahoma	1.5E-25	1.6E-27

Table 5. Correlation factors of some *5-grams* in two different RSS news articles

Article 1/Article 2	Missouri warned conditions deteriorate afternoon	severe weather expert chad myers
brutal ice storm threatening turn	1.2E-35	2.1E-31
ice storm threatening turn Oklahoma	5.0E-33	7.2E-35
storm threatening turn Oklahoma sheet	4.5E-37	5.1E-35

A_2. If there are m (n, respectively) different words in A_1 (A_2, respectively), then there are m-k+1 different phrases in A_1 to be compared with n-k+1 different phrases in A_2, which include overlapped phrases. In computing the degree of similarity of A_1 and A_2, the correlation factors of phrases of the chosen length, i.e., in between 1 and 5, in A_1 and A_2 are used. Since the average content descriptor of an RSS new article is 25 words in length, the computation time for matching phrases in two news articles is negligible.

3.4 Similarity Ranking of RSS News Articles

In CPM, we use the correlation factors of n-gram ($1 \leq n \leq 5$) phrases of a chosen length to define the degrees of similarity of two RSS news articles A_1 and A_2.

The *degree of similarity* of A_1 with respect to A_2 is not necessary the same as the *degree of similarity* of A_2 with respect to A_1, since A_1 and A_2 may share common information but also include information that are unique of their own.

Using the n-gram ($1 \leq n \leq 5$) phrase correlation factors, we define a fuzzy association of each n-gram phrase in A_1 with respect to all the n-gram phrases in A_2. The degree of correlation between a phrase p_i in A_1 and all the phrases in A_2, denoted $\mu_{p_i,2}$, is calculated as the complement of a negated algebraic product of all the correlation factors of p_i and each distinct phrase p_k in A_2, i.e.,

$$\mu_{p_i,2} = 1 - \prod_{p_k \in A_2} (1 - pcf_{i,k}) \text{ or } \mu_{p_i,2} = 1 - \prod_{p_k \in A_2} (1 - cf_{i,k}) \tag{6}$$

which is adapted from the fuzzy word-document correlation factor in [10], and the 1st (2nd, respectively) formula in Equation 6 is used for n-gram ($2 \leq n \leq 5$) phrases (unigram phrases, respectively). The correlation value $\mu_{p_i,2}$ falls in the interval [0, 1] and reaches its maximum when $pcf_{i,k}$ ($cf_{i,k}$, respectively) = 1, i.e., when p_i ($\in A_1$) = p_k ($\in A_2$).

The *degree of similarity* of A_1 with respect to A_2, denoted $Sim(A_1, A_2)$, using the chosen n-gram phrase correlation factors is calculated as the average of all the values $\mu_{p_i,2}$ for each $p_i \in A_1$ ($1 \leq i \leq m$), and m is the total number of n-gram phrases in A_1.

$$Sim(A_1, A_2) = \frac{\mu_{p_1,2} + \mu_{p_2,2} + \cdots + \mu_{p_m,2}}{m} \tag{7}$$

$Sim(A_1, A_2) \in [0, 1]$. When $Sim(A_1, A_2) = 0$, it indicates that there is no n-gram phrase in A_1 that can be considered similar to any n-gram phrase in A_2. If $Sim(A_1, A_2) = 1$, then either A_1 is (semantically) identical to A_2, or A_1 is *subsumed* by A_2, i.e., all the n-gram phrases in A_1 are (semantically) the same as (some of) the n-gram phrases in A_2, and in this case A_1 is treated as a *redundant* article and can be ignored. $Sim(A_2, A_1)$ can be defined accordingly.

Given any pair of A_1 and A_2, we use the Stanford Certainty Factor (SCF) [6] to combine $Sim(A_1, A_2)$ and $Sim(A_2, A_1)$, which yields the *relative degree of similarity* that should reflect how closely related A_1 and A_2 are. We adapt SCF, since it is a monotonically increasing (decreasing) function on combined assumptions for creating confidence measures and is easy to compute.

$$SCF(A_1, A_2) = \frac{Sim(A_1, A_2) + Sim(A_2, A_1)}{1 - MIN(Sim(A_1, A_2), Sim(A_2, A_1))} \tag{8}$$

The following table shows that when the degrees of similarity of two articles are both *high* (see row 1), their SCF is also *high*, and the same occurs when one of the degrees of similarity is *high* (e.g., one of the two articles is subsumed by another), their SCF yields a comparatively *high* relative degree of similarity (see row 2). Furthermore, if both degrees of similarity are *low*, their SCF is also *low*, indicating that the articles are not related.

	$Sim(A_1, A_2)$	$Sim(A_2, A_1)$	SFC
1	0.60	0.76	3.40
2	0.52	0.23	0.97
3	2.5E-6	3.26-6	5.76E-6

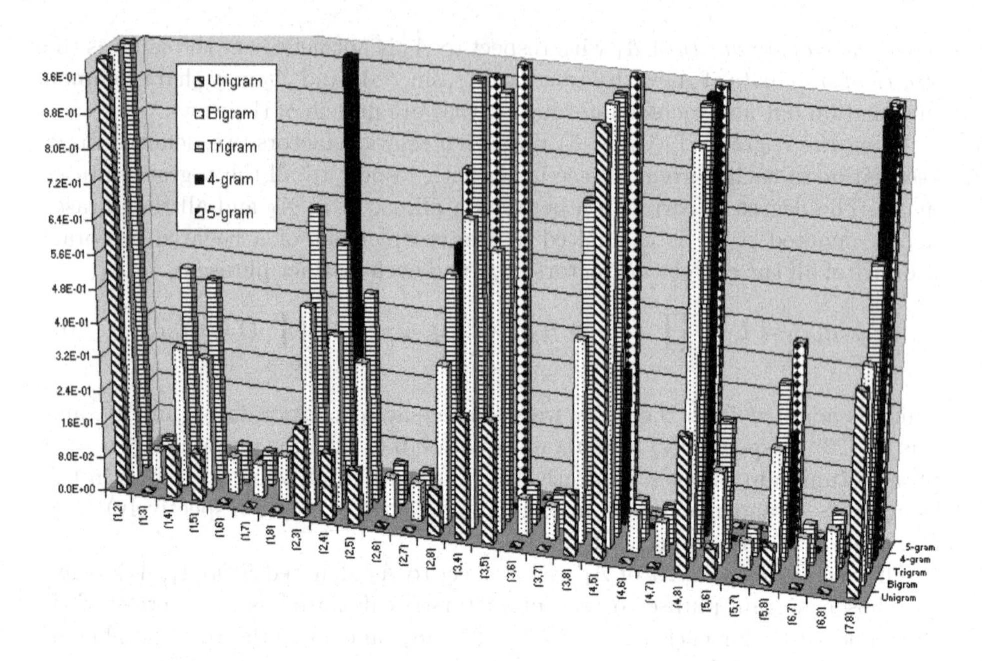

Fig. 2. *SCF*s of the eight different RSS news articles in Table 6

Example 2. Consider Figure 2, which includes the *SCF*s computed for different pairs of the articles shown in Table 6, demonstrates the applicability of our CPM on bigrams and trigrams. According to the content of the RSS news articles

- Articles 1 and 2, as also shown in Example 1, are very closely related, since both give an account of a *storm that took place in Oklahoma*, and the *SCF*s of bigrams and trigrams are higher than the SCFs of other n-grams.
- Articles 4 and 5 (7 and 8, respectively) resort last year's *bird flu outbreak* (the new *iPhone*, respectively), and again the SCFs of bigrams and trigrams correctly identify their similarity.
- Even if articles 1, 2, and 6 report the *weather* in different parts of the world, they are not all related. Unigrams in this case provide a more accurate measure to detect that article pairs (1, 6) and (2,6) are not related.
- The rest of the pairs are *unrelated*, and in most cases the use of unigrams is more effective than any other n-grams. However, the *SCF*s computed by using bigrams and trigrams are also *low* when the pairs of articles are *unrelated*, but are *higher* than unigrams if the article pairs are *related*, and thus are useful in *detecting* and *ranking* related news articles.
- Many *SCF*s computed by using 4-grams and 5-grams are not consistent compared with the *SCF*s obtained using unigrams, bigrams, or trigrams, and as a result, they are not dependable in detecting (un-)related articles. □

Table 6. Sample articles (Art) used for demonstrating their *SCF*s as a result of *n*-gram ($1 \leq n \leq 5$) phrase matching

Art	Title and Description
1	*Ice Storm Threatens Chaos.* Freezing rain hit Oklahoma today, the start of what ...
2	*Ice Storm Threatens Chaos.* A brutal ice storm is threatening to turn Oklahoma ...
3	*Fashion Designers Issue Model Guidelines.* The American fashion industry says ...
4	*New Outbreak of Bird Flu Hits Nigeria.* A new outbreak of H5N1 bird flu has hit ...
5	*Indonesian Woman 59th Bird Flu Death.* Indonesian woman died from bird flu ...
6	*EU proposes ambitious climate target.* ... "the most ambitious policy ever" to ...
7	*The iPhone: Revolution? Gamble? Flop?.* The hype around Apple Inc.'s upcoming ...
8	*The iPhone Phenomenon.* There's a new cell phone that has had consumers ...

4 Experimental Results

In evaluating the accuracy of using our *n*-gram ($1 \leq n \leq 5$) CPM approach to determine the degree of similarity of any two RSS news articles, we collected thousands of articles from different sources as partially shown in Table 7.

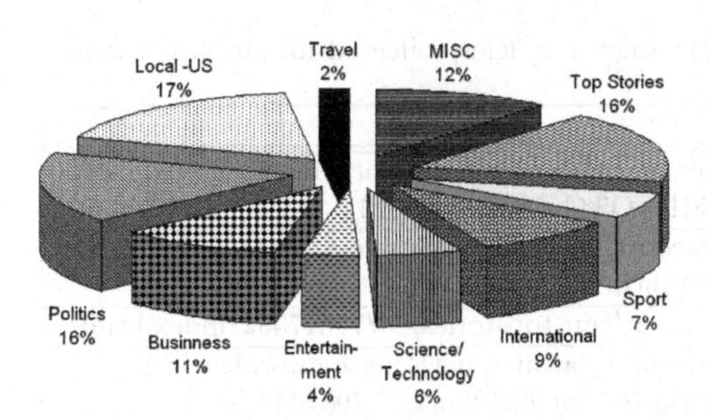

In order to guarantee the impartiality of our experiments, the articles were randomly selected from hundreds of different news feeds, collected between July 2006 and July 2007. The chart on the left shows the variety of subject areas covered by the chosen RSS news feeds which demonstrate the suitability of our CPM model to news articles independent of their content, and Table 8 shows a few collected news articles.

We determined the relative degree of similarity of each pair of the 1059 news articles extracted from 200 RSS news feeds using each *n*-gram ($1 \leq n \leq 5$) phrase matching approach. In this empirical study, we (i) randomly selected 410 pairs, (ii) manually examined each pair to determine their relative degree of similarity, and (iii) compared the manually determined similarity with the automatically computed *SCF*s based on each one of the five *n*-gram phrase matching approaches. To verify which *n*-gram CPM is the most accurate in determining related pairs of news articles, we consider the number of *False Positives* (*FP*s), i.e., *unrelated* pairs with *high SCF*s, and *False Negatives* (*FN*s), i.e., *related* pairs with *low SCF*s, generated by using CPM on each type of *n*-grams as follows:

$$Accuracy = \frac{Total\_Number\_of\_Examined\_Pairs - Misclassified\_Pairs}{Total\_Number\_of\_Examined\_Pairs} \quad (9)$$

where $Misclassified\_Pairs$ is the sum of FPs and FNs encountered.

Figure 3(a) shows the number of *correctly* classified pairs, as well as *incorrectly* identified pairs, i.e., the sum of the FPs and FNs, on the randomly chosen 410

Table 7. Sources of RSS news feeds, the number of feeds (Fd) of each source (200 total), and the number of news articles (Art) from each source (1059 total)

Sources	Fd	Art	Sources	Fd	Art	Sources	Fd	Art
1115.org	3	19	abcnews.go.com	10	135	adn.com	2	5
blogs.zdnet.com	1	6	boston.com	8	26	businessweek.com	8	41
cbsnews.com	8	24	chron.com	7	21	cnn.com	8	29
dailymail.co.uk	1	4	english.people.com	6	54	forbes.com	2	8
foxnews.com	10	35	guardian.co.uk	4	12	health.telegraph.co.uk	1	4
hosted.ap.org	11	39	iht.com	9	27	latimes.com	3	9
microsoftwatch.com	1	4	money.cnn.com	4	17	money.telegraph.co.uk	1	4
msnbc.com	1	6	news.bbc.co.uk	2	13	news.ft.com	9	47
news.telegraph.co.uk	3	25	news.yahoo.com	3	37	nytimes.com	10	60
online.wsj.com	6	70	politics.guardian.co.uk	1	3	portal.telegraph.	1	3
primezone.com co.uk	2	8	prnewswire.com	8	24	seattletimes.nwsource. com	8	50
slashdot.org	1	5	sltrib.com	2	6	sportsillustrated.cnn.com	3	9
timesonline.com	3	10	today.reuters.com	12	112	usatoday.com	10	27
washingtonpost.com	2	6	wired.com	3	9	worldpress.org	2	6

Table 8. Portion of the thousands of RSS news articles collected for empirical study

Source	Date	URL
abcnews.com	7/3/2007	abcnews.go.com/TheLaw/Politics/story?id=3339302& page=1&CMP=OTC-RSSFeeds0312
cbsnews.com	7/2/2007	feeds.cbsnews.com/~r/CBSNewsBusiness/~3/130066324/ main3010949.shtm
cnn.com	1/10/2007	rss.cnn.com/~r/rss/cnn_topstories/~3/74617352/index.html
latimes.com	1/12/2007	feeds.latimes.com/~r/latimes/news/nationworld/world/ ~3/74277710/la-fg-somalia12jan12,1,2987342.story
news.bbc.co.uk	1/16/2007	news.bbc.co.uk/go/rss/-/1/hi/world/south_asia/6268487.stm

pairs of news articles, which is a subset of the news articles listed in Table 7, whereas Figure 3(b) shows the accuracy computed by using Equation 9 for each of the *n*-gram CPM approaches. Clearly, *bigrams* and *trigrams* yield the *lowest* number of *misclassified* pairs of articles, while achieve the *highest* count ($\geq 90\%$) of *correctly* detected pairs among all the *n*-grams. The use of *4-* and *5-grams* reduces the accuracy to as low as 60%, whereas *unigrams* has an accuracy of 86%.

When using bigrams and trigrams, the misclassified pairs occur, since if they have at least one common bigram or trigram, then their *odds* increase. Also, the degrees of similarity for unigram, bigram, and trigram are relatively higher, and thus their SCFs are comparatively higher, whereas the degrees of similarity generated by using 4-grams and 5-grams tend to be much lower, and thus their SCFs are often extremely low, which might explain their *low* accuracy in detecting similar pairs of RSS news articles. In general, the *unrelated* pairs of articles detected by using *bigrams* and *trigrams* have a SCF close to the power of E-6,

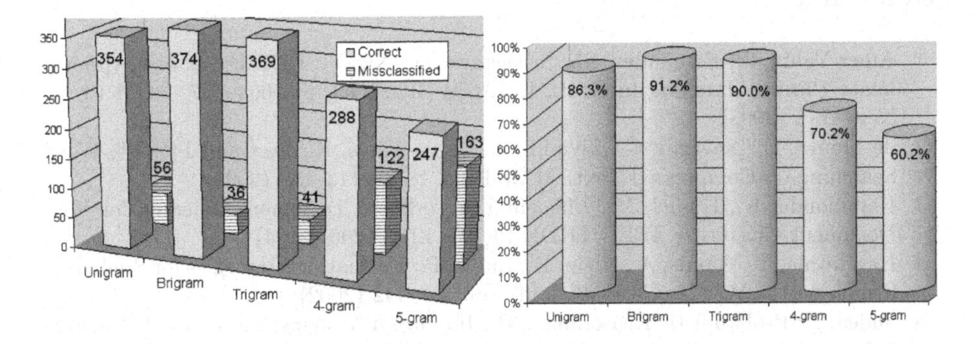

(a) (In)Correctly classified, related articles using our n-gram CPM.

(b) Accuracy of using our n-gram CPM to detect related articles.

Fig. 3. Classified 410 pairs of news articles and their accuracy

whereas *related* pairs have a SCF above 0.1. Neither FPs nor FNs are desired, and in this study they contribute only 10% of the bigram (trigram) pairs, out of which close to 90% are FP pairs. In fact, FPs are less harmful in our similarity detection approach, since we do not lose many similar pairs.

Based on the empirical study, we conclude that (i) *bigram* and *trigram* outperform others in detecting similar RSS news articles. In most cases, the SCFs computed by using bigrams and trigrams on similar RSS news articles are higher than the ones computed by using *unigrams*. (ii) *4-grams* and *5-grams* are not reliable in determining the relevance between any two RSS news articles as explained earlier. Our empirical study further verify the claims made by [8,9], which state that the use of bigrams and trigrams is often more effective than the use of other n-gram phrases in retrieving information.

5 Conclusions

We have presented a novel approach for finding similar RSS news articles using n-gram ($1 \leq n \leq 5$) phrase matching and shown that bigrams and trigrams outperform other n-grams in detecting similar articles. We have also verified the accuracy of our correlation phrase matching (CPM) approach by analyzing hundreds of pairs of randomly selected RSS news articles from multiple sources and concluded that CPM on bigrams and trigrams is highly accurate ($\geq$ 90%) and requires little overhead (using predefined correlation factors) in finding related articles. Our CPM can also be used for (i) detecting (similar) junk emails and spam Web pages, (ii) clustering (Web) documents with similar content, and (iii) discovering plagiarization, which form the core future work for our CPM model.

References

1. Amer-Yahia, S., Fernandez, M., Srivastava, D., Xu, Y.: PIX: Exact and Approximate Phrase Matching in XML. In: ACM SIGMOD, pp. 664–667. ACM Press, New York (2003)
2. de Moura, E., Navarro, G., Ziviani, N., Baeza-Yates, R.: Fast and Flexible Word Searching on Compressed Text. ACM TOIS 18(2), 113–139 (2000)
3. Hammouda, K., Kamel, M.: Efficient Phrase-Based Document Indexing for Web Document Clustering. IEEE TKDE 16(10), 1279–1296 (2004)
4. Haveliwala, T., Gionis, A., Klein, D., Indyk, P.: Evaluating Strategies for Similarity Search on the Web. In: WWW Conf., pp. 432–442 (2002)
5. Judea, P.: Probabilistic Reasoning in the Intelligent Systems: Networks of Plausible Inference. Morgan Kaufmann, San Francisco (1988)
6. Luger, G.: Artificial Intelligence: Structures and Strategies for Complex Problem Solving, 5th edn. Addison-Wesley, Reading (2005)
7. Luo, G., Tang, C., Tian, Y.: Answering Relationship Queries on the Web. In: WWW Conf., pp. 561–570 (2007)
8. Mishne, G., de Rijke, M.: Boosting Web Retrieval through Query Operations. In: European Conf. on Information Retrieval, pp. 502–516 (2005)
9. Narita, M., Ogawa, Y.: The Use of Phrases from Query Texts in Information Retrieval. In: ACM SIGIR, pp. 318–320. ACM Press, New York (2000)
10. Ogawa, Y., Morita, T., Kobayashi, K.: A Fuzzy Document Retrieval System Using the Keyword Connection Matrix and a Learning Method. Fuzzy Sets and Systems 39, 163–179 (1991)
11. Porter, M.: An Algorithm for Suffix Stripping. Program 14(3), 130–137 (1980)
12. Toud, S.: Creating a Custom Metrics Tool. MSDN Magazine (April 2005), `http://msdn.microsoft.com/msdnmag/issues/05/04/EndBracket/`
13. Tzong-Han, T., Chia-Wei, W.: Enhance Genomic IR with Term Variation and Expansion: Experience of the IASL Group. In: Text Retrieval Conf. (2005)
14. Wilbur, W., Kim, W.: Flexible Phrase Based Query Handling Algorithms. In: American Society for Information Science and Technology, pp. 438–449 (2001)
15. Yerra, R., Ng, Y.-K.: Detecting Similar HTML Documents Using a Fuzzy Set Information Retrieval Approach. In: IEEE GrC 2005, pp. 693–699. IEEE, Los Alamitos (2005)

Automatic Data Record Detection in Web Pages

Xiaoying Gao, Le Phong Bao Vuong, and Mengjie Zhang

School of Mathematics, Statistics and Computer Science
Victoria University of Wellington, PO Box 600, Wellington, New Zealand
{xgao,Phong.Le,mengjie}@mcs.vuw.ac.nz

Abstract. Wrapper induction is currently the main technology for data extraction from semi-structured web pages. However, wrapper induction has the limitation of requiring training Web pages, and the information extraction process is quite complex involving pattern induction, data extraction and data transformation. This paper introduces a new approach that achieves automatic data extraction by applying clustering to detecting similar text tokens, developing a new method to label text tokens to capture the hierarchical structure of HTML pages, and developing an algorithm for transforming labelled text tokens to XML. The approach is examined and compared with a number of existing wrapper induction systems on three different sets of web pages. The results suggest that the new approach is effective for data extraction and that it outperforms existing approaches on these web sites. This approach has the advantages of requiring no training and has no explicit processes for pattern induction or data extraction, therefore the whole process has been simplified.

1 Introduction

Semi-structured Web pages such as foreign currency exchange tables and online shopping lists often contain important and useful data. Information Extraction (IE) systems (wrappers) automatically extract important data from these pages so that the data can be used by software programs for data mining, knowledge discovery and accurate search, etc. Wrapper induction is a machine learning technology that learns IE patterns from training example pages to be used for extracting data from web pages. It has been successfully used in extracting useful information from different web sites in many tasks [1,2,3,4,5].

However, information exaction using wrapper induction technology has two main limitations. Firstly, most existing wrapper induction systems [1,2,3,4,5] require a number of training Web pages from each site. These systems assume that there are multiple Web pages with similar data records in similar data format available on every targeted Web site. Human effects (or other programs) are needed to find the training pages [3,4] and/or to manually label the pages [1,2,5].

Secondly, information extraction based on wrapper induction technology has a complex process involving three separate stages: pattern induction from training pages, data extraction by matching the pattern on testing pages, and data transformation by representing extracted data to structured form such as databases

Z. Zhang and J. Siekmann (Eds.): KSEM 2007, LNAI 4798, pp. 349–361, 2007.

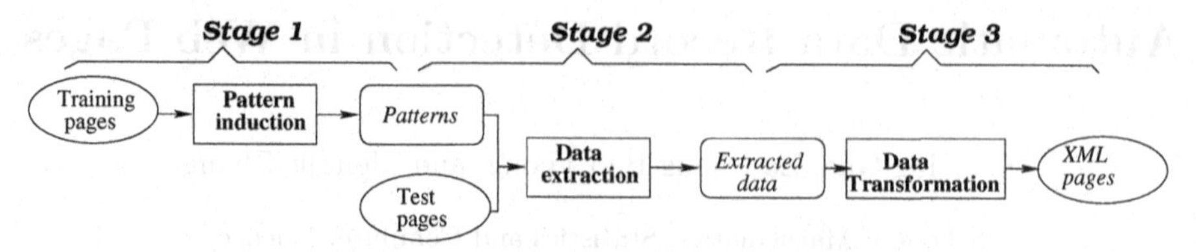

Fig. 1. Information extraction based on wrapper induction

or XML pages. The process is shown in Figure 1. The main problem of separating the three stages is that a lot of structural information (e.g. the hierarchical structure of HTML page) is lost during the data extraction. As the extracted data is a collection of raw data without any structural information, it is quite complex to transform the extracted data into a structured form such as XML, especially for complex Web pages with nested tables or lists. Figure 2 shows an example page with nested tables where the second cell in each row is another table. Because the nested data structure is lost during data extraction, it is quite difficult to reconstruct the parent-child relationship between data in the data transformation stage.

Lecturers	**David**	16/06/1952	New Zealand
	Daniel	13/05/1969	Australia
Students	**John**	06/12/1979	New Zealand
	Peter	12/05/1980	France

Fig. 2. Example of data with nested structures

1.1 Goals

To solve these problems, this research aims to build a system that directly transforms data format from semi-structured HTML to structured XML without explicit training, pattern learning or data extraction. Given any single semi-structured HTML page, the data fields are expected to be automatically identified, labelled and detected (rather than extracted), then the whole page is rewritten to XML. This approach will be examined and compared with similar existing systems on three groups of web sites of varying difficulty. Specifically, this paper will investigate the following research issues.

- How the text tokens containing the data fields to be extracted in a semi-structured HTML web page can be properly represented for data field identification;
- How the available information on a single HTML page can be used to categorise the text tokens into different clusters so that the text tokens in the same cluster are identified as candidate data fields of the same attributes;

- How the text tokens can be labelled using the structural information in the HMTL page;
- How the data records can be detected using the labels of the text tokens;
- How the detected data records can be represented in XML without losing any important structural information so that the data records can be directly used by software programs for other purposes; and
- Whether this approach is more effective for processing pages with nested data structures.

The remainder of the paper is organized as follows. Section 2 briefly summarizes the related work. Section 3 outlines this approach describe the text token representation, clustering and labelling methods. Section 4 details the algorithm for data record detection. Section 5 demonstrates the experimental results on three different groups of Web sites. Finally, Section 6 concludes the paper.

2 Related Work

As shown in a survey paper [6], many wrapper induction systems have been built for learning information extraction patterns to extract data from semi-structured Web pages. Most earlier wrapper induction systems such as WIEN [1] and STalker [2] require manually labelled training pages, which need considerable amount of human efforts. One recent wrapper induction system RoadRunner [3] can learn from unlabelled web pages but it requires at least two training pages generated with the same template and with different number of data records, which may not be available for many Web sites. AutoWrapper [4] is a relatively recent system that uses a single unlabelled web page to learn patterns but it is restricted to tabular pages.

Regarding the motivation of this research, the closest related work is MDR (Mining data records in Web pages) [7]. MDR does not require training and does not have explicit pattern induction stage. The main technology of MDR is based on a string matching algorithm. MDR can handle non-contiguous data records, but it can not effectively handle nested data.

Regarding the learning technologies used in existing systems, the closest related work to our approach is the one in [8]. It uses a clustering algorithm called AutoClass [9] to exploit data format and data content to classify data. It uses grammar induction for identifying data records. This system requires multiple pages from each site, and is restricted to pages with tables and lists. There is no evidence to show that this system can handle pages with nested structure.

Our preliminary research on clustering algorithm introduces a variant Hierarchical Agglomerative Clustering (HAC) algorithm, *K-neighbors-HAC* [10]. Its main characteristic is that it introduces a *Location ID* to specify the location of the objects, and compares with K neighbour objects in each cycle for similarity comparison. Since similar text tokens are often located close to each other, this algorithm is more efficient than the standard HAC algorithm.

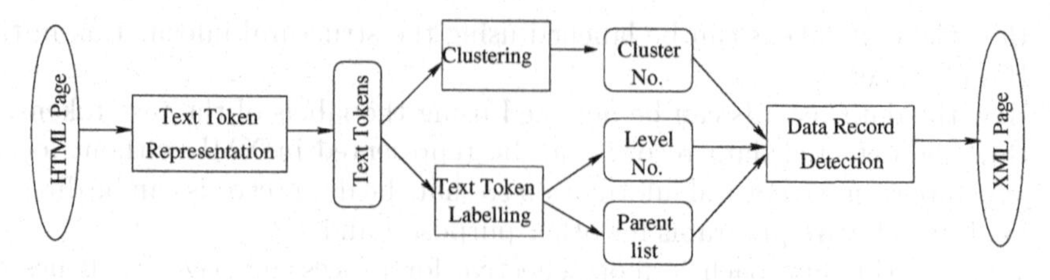

Fig. 3. Overview of our approach

3 The New Approach/System

The target source of this approach is semi-structured Web pages containing a number of "data records", each of which has a number of "data fields". Each data field (e.g. "David") is the value of an attribute (e.g. NAME). The main task of this approach is to identify these data fields and data records. Figure 3 shows an overview of this approach.

As shown in the figure, the approach first represents a single web site as a list of text tokens. A clustering algorithm is developed for obtaining the cluster of the text tokens and a method is introduced for labelling the text tokens and producing the level number and the parent list of the text tokens. According to the cluster number, level number and the parent list of the text tokens, a data record detection method is developed to find all the possible data records in the web page and put them into the XML format.

In the rest of this section, we will describe the process of this approach, including the text token representation, clustering, and text token labelling. Due to the length of the data record detection, we will leave it to the next section.

3.1 Text Token Representation

Due to the nature of the semi-structured web pages, the data fields of the same attribute are similar in two ways. Firstly, they are often constructed in the same HTML format. For example, "David", "Daniel" and "John" for the NAME attribute are presented as cells in a table.. Secondly, they share some common textual patterns, for instance, prices are usually represented by a number preceded by a dollar sign($). As neither of the format or content feature is less important than the other, we exploit both of them to compute the similarities of the data fields.

The format of a data field is mainly defined by its HTML tag markup and its content is usually a text string. We use the term "text token" to capture both the format and the content of a data field. We represent a web page as a set of text tokens. A text token is a text string wrapped in an HTML block consisting of a text string and the HTML tags preceding and succeeding the text string.

In order to meet the requirements of the text token clustering and text token labelling modules in this approach, a *text token* is represented with four features namely a *string value*, *left* and *right delimiters*, and a *location ID*. The

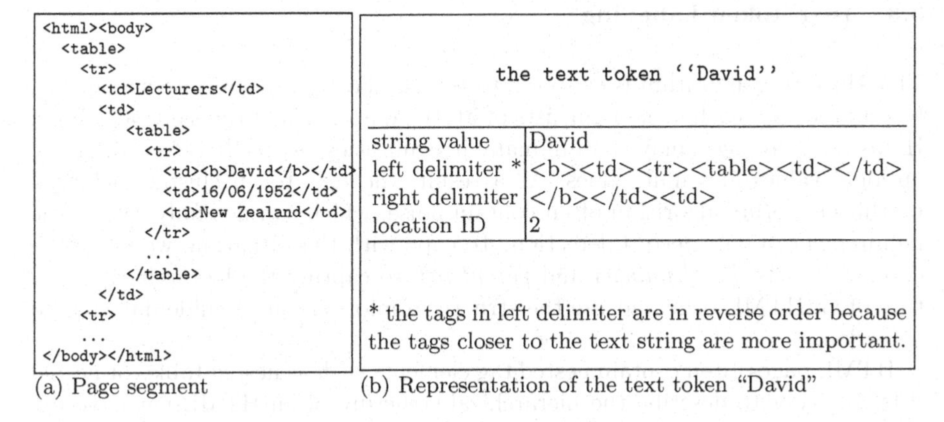

(a) Page segment (b) Representation of the text token "David"

Fig. 4. Text token representation

string value of a text token represents a string of characters between HTML tags captured by the text token. The *left* and *right delimiters* are the sequences of consecutive HTML tags preceding and succeeding the string value, respectively. Each text token is associated with a *location ID* which corresponds to the ordinal appearance of the text token in the HTML page. As an example, for the page given in Figure 2, the HTML fragment and the representation of the text token centred at "David" are shown in Figure 4.

3.2 Clustering Algorithm

Based on the representation of the text tokens, we expect similar data fields to be detected and clustered into a single category. To achieve this objective, we apply a variant Hierarchical Agglomerative Clustering (HAC) algorithm K-neighbours-HAC [10] to group similar text tokens into clusters. The algorithm is briefly outlined as follows. Details of similarity measures for the text strings and tags delimiters can be seen from [10].

Step 1: Initialize every text token as a single cluster.

Step 2: Each cluster (a single object or an object representative) is compared to K nearby (by the *location IDs*) surrounded clusters. Similar clusters are merged.

Step 3: A representative of each merged cluster is created by selecting the text token with the median *location ID.*

Step 4: K is increased by a certain step *delta.*

Step 5: Repeat steps 2-4, until all clusters have been compared and no similar clusters can be merged any more, or alternatively after D iterations.

The parameters K and *delta* are determined based on the total number of text tokens on the Web page t. In our prototype system, we used k = $t/20$ and *delta* = $t/10$. These values were found quite reasonable to obtain good results.

3.3 Text Token Labelling

This clustering algorithm is expected to put similar text tokens into single clusters and ideally each cluster consists of all the data fields of the same attribute. However, Web pages may contain many irregularities, especially when there are optional values and null values. As a result, the text token clusters achieved by this clustering algorithm often contain noise, which can not fully satisfy the requirement of data record detection. To cope with this situation, we introduce two new labels, *Level number* and *parent list*, to capture the hierarchical structure of an HTML page and use the structural information to guide data record detection.

HTML pages often contain nested tag elements such as nested tables or nested lists. In order to describe the hierarchical structure of an HTML page, we use HTML DOM (Document Object Model) [11] to construct a tree for each page. In our case, however, only the block tags that add structure to the document are employed to build the DOM tree. Non-block tags such as <B>, <I> and <BR> do not have any contribution into building the hierarchy and are therefore ignored. The partial DOM tree for the example page given in Figure 2 are shown in Figure 5. Each tag node is labelled by its name and an ID number such as TR0.

We associate a *level number* and a *parent list* to each text token to capture the HTML structural information. The *level number* corresponds to the hierarchical level of a text token in the tree representation of the HTML document. As shown in Figure 5, the *level number* for the text token centered at "David" is 9. For each text token, the sequence of tags along the path from the current node that contains the text token to the root node of the tree called its *parent.* For the above example, the *parent list* for the text token "David" is [TD4, TR2, TABLE1, TD1, TR0, TABLE0, BODY0, HTML0].

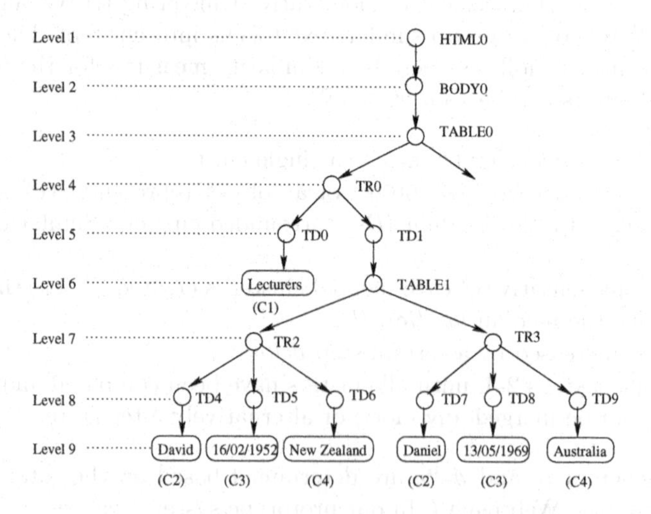

Fig. 5. An example DOM tree

```
<xml>
  <level level_no='6'>
    <element cluster='1' value='Lecturers'>
    <level level_no='9'>
      <tuple>
        <element cluster='2' value='David'>
        <element cluster='3' value='16/06/1952'>
        <element cluster='4' value='New Zealand'>
      </tuple>
      <tuple>
        <element cluster='2' value='Daniel'>
        <element cluster='3' value='13/05/1969'>
        <element cluster='4' value='Australia'>
      </tuple>
    </level>
  </level>
  <level level_no='6'>
    <element cluster='1' value='Students'>
    <level level_no='9'>
      <tuple>
        <element cluster='2' value='John'>
        ...
  </level>
</xml>
```

Fig. 6. Example output in XML

4 Data Record Detection

The three labels *cluster number*, *level number* and *parent list* are used for marking up candidate data fields and grouping them into data records. In this section, we first introduce how the detected data records are represented in XML as our final output, and then detail our method for detecting/generating the data records using the three text token labels.

4.1 System Output Format in XML

Our system automatically detects data records and saves them into XML. As an example, partial XML output for the page given in Figure 2 is shown in Figure 6. The proposed XML representation consists of three basic components: *element*, *tuple* and *level*. Each *element* is a candidate data field. A *tuple* component contains a collection of *element* components at the same level which constitute to a data tuple (a candidate data record). A *level* is used to describe data with nested structure and a *level* consists of elements and tuples at the same level and lower levels (higher level numbers).

4.2 XML Data Tuple Generation from HTML Using Token Labels

For each labelled text token in the HTML page, a new *element* component is created. An *element* component captures the string value of the text token that is the candidate data field of users interests. It also captures the cluster number within which the text token belongs to. We include the cluster number here as the metadata to allow data of the same type, which are data fields of the same attribute and hence belong to the same cluster, to be easily queried by composing a filter on the cluster number.

A *tuple* is generated by grouping text tokens at a single level, and this is detailed in two rules shown in R1 and R2. R1 is based on the observation that the majority of the data fields in a data record are located in the same sub-tree and also located at the same level, so they are often presented as text tokens with a common nearest parent. The text tokens with common nearest parent constitute the core of a *tuple*, called a *base tuple*.

R1: A group of *element* components are considered in the same *base tuple* if they are at the same level and all share the same nearest parent node. For instance, "David", "16/06/1952" and "New Zealand" in Figure 5 constitute a *base tuple* because they all are located at level 9 and they have the same nearest parent node which is <TR2>.

Ideally, all data records are correctly identified as base tuples using the rule above. In our example shown in Figure 6, a set of *base tuples* are successfully identified which correspond to a set of data records. The cluster labels of the elements of each *base tuple* are the same, which are (c2, c3, c4), and this means this page has a set of data records and each data record has three data fields corresponding to three attributes.

However, a data record may be split into two (or more) different *base tuples* if the text tokens share different common nearest parent nodes which form more than one sub-trees as a result. An example of this type is shown in Figure 7. The first two text tokens with cluster labels (c1, c2) are grouped in a *base tuple* and the other three text tokens with cluster labels (c3, c4, c5) are grouped in another. Since the five text tokens belong to different clusters, they are more likely to be values of different attributes, and accordingly, they are more likely to be data fields of the same data record. However, they are grouped into two *base tuples* because the first two tokens have a slightly different HTML format (so with different common parents). These two *base tuples* should be merged together into one with the cluster labels (c1, c2, c3, c4, c5) using rule R2. R2 is based on the observation that each data record usually has a set of different kinds of attributes, that is, each data tuple consists of a sequence of text tokens belonging to different clusters.

R2: We define two *base tuples* are mergable if they are at the same level and the majority of the cluster labels of the elements of the two *base tuples* are

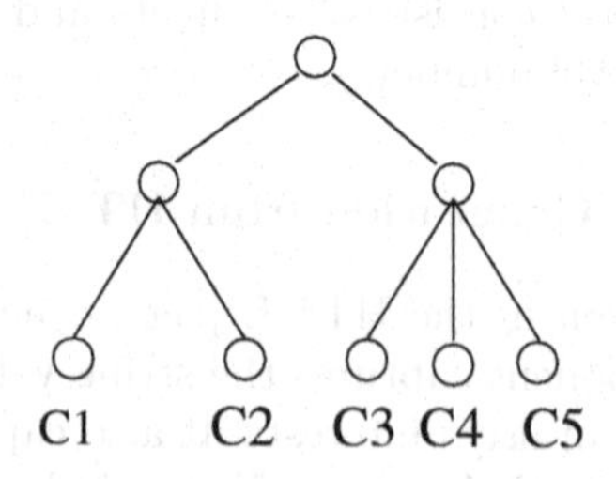

Fig. 7. Example of mergable base tuples

different. Two base tuples are merged into one by finding their common parent node. The merging is repeated until all mergable *base tuples* are grouped into a single tuple.

A *level* component contains *element* components and/or *tuple* components which belong to the same level. It may contain other lower level components in the DOM tree. Each *level* component has a level number (*level_no*) which equals to the values of the *level* attributes of all text tokens it contains.

4.3 Data Records Represented in XML

Given the output format, we consider a data *tuple* with a lot of candidate data fields as a candidate data record, that is, a data tuple with a number of text tokens that belong to different clusters (e.g. "David, 16/06/1952, New Zealand"), is considered a candidate data record. We consider a *level* with a number of similar candidate data records a data region (e.g. level 6 is a data region since it has two similar tuples with the same cluster labels). For the example page given in Figure 2, two data regions are detected (two *levels* and each level has two data records (two *tuples*) as partially shown in Figure 6.

Our system does not explicitly extract the data, but data extraction from the resulted XML page becomes a trivial task because queries can be written easily using XML query languages such as XQuery. Our XML representation also allows the users to write queries that retrieve any subsets of data. For example, for the output shown in Figure 6, XML query can be written to retrieve all data with the cluster number 4 to get a country list or to retrieve all data on a particular level[1].

The main characteristic of our system is that it automatically maps HTML to XML by utilising the hierarchical structure of HTML to transform data to structured form. For the example page shown in Figure 2, the text token labelling retains the hierarchical tree structure of the data and hence the parent-child relationship between nested structured data are retained in the final XML page.

5 Experimental Results

Our prototype system ADRD (Automatic Data Record Detection) was examined on three groups of Web sites where the first one was collected by ourselves and the other two were publicly available. Our system was compared with five other similar systems. *MDR* [7] is a relatively new system and is available online. We downloaded MDR from the web and tested it on all our Web sites. The other four systems are relatively well known and we use their results found in the

[1] Please note that our system can detect/group data but do not label the attributes(semantics, contents, meaning) of the data, for example, our system identify the data record with three data fields "David, 16/06/1952, New Zealand", but it does not know "David" is a name, "16/06/1952" is the date of birth and "New Zealand" is the country. The attributes labelling should be done a human user, which is the same as other automatic wrapper induction systems.

literature. The experiments and results with comparisons are discussed in the rest of the section.

For each HTML example page in the experiments, we identify whether it has any of these following characteristics: (1) N: containing nested data, (2) M: containing missing data, meaning that most data records have values for a particular field but some do not; and (3) O: containing optional data, meaning that only a few records have some extra fields for special notes.

If more than 90% of the data records are correctly constructed as tuples, the result is considered *correct*, represented by $\surd$. If 75% to 90% of data records are correctly constructed as tuples, the result is defined as *not perfect*, represented by $\sim$. Otherwise, the extracted results are considered *wrong*, represented by $\times$.

5.1 Experiment 1

The first group of experiments was run on 10 Web pages which were selected to represent different HTML pages with some of the three characteristics. The pages were downloaded directly from the Internet by the time the experiments were conducted. The Web page details with the results of our system *ADRD* and another system *MDR* are shown in Table 1. Row one shows that the first example web page has nested data structure and optional data but does not have missing data; while our ADRD successfully built all the correct tuples in the page, the MDR wrapper did not (not perfect).

Table 1. Comparison of two approaches on our Web site collection

#	Description	URL	N	M	O	ADRD	MDR
1	AutoWeb car	autoweb.drive.com.au	Y	N	Y	$\surd$	$\sim$
2	Yahoo hot job	www.hotjobs.com	N	Y	N	$\surd$	$\surd$
3	Over stock Product	www.overstock.com	N	N	Y	$\surd$	$\surd$
4	VUW Grad Courses	www.mcs.vuw.ac.nz/courses	N	N	Y	$\surd$	$\surd$
5	VUW Staff Directory	www.mcs.vuw.ac.nz/people	Y	N	N	$\surd$	$\times$
6	VUW software group	www.mcs.vuw.ac.nz/research/se-vuw	N	N	Y	$\surd$	$\sim$
7	WorldCup 2006	fifaworldcup.yahoo.com	Y	N	N	$\surd$	$\times$
8	DickSmith Support	www.dse.co.nz	Y	N	N	$\times$	$\times$
9	Froogle Product	www.froogle.com	N	N	Y	$\surd$	$\surd$
10	Yahoo Auction	auctions.yahoo.com	N	N	N	$\surd$	$\surd$

As can be seen from Table 1, our system successfully built correct tuples for 90% of the example pages, suggesting that our system can successfully identify majority of data records in these pages with nested data structures, missing data or optional data. The MDR system, however, only achieved successful results on 50% of these pages, suggesting that our system outperforms the MDR system on these Web pages. In particular, for the four pages containing nested data structures, our system performed well for three of them. The MDR system, however, did not achieve correct results for any pages with nested structures.

The only failure of our system on the first experiment is on the example #8. In this page, our system cannot recognize the nested-structure as the data in the first level is in image format.

5.2 Experiment 2

Our second group of experiments was run on the 30 Web sites collected by Kushmerick which were available on RISE (Repository of Online Information sources Used in Information Extraction Tasks)[12]. The results for the existing AutoWrapper approach [4] and the WIEN approach with six sub-systems [1] were known and we compared them with our ADRD system and MDR.

Table 2. Comparison of the four approaches on 30 Web sites

	ADRD	MDR	AutoWrapper	WIEN
# sites for correct result (√)	23	11	14	4-17*
# sites for non-perfect result (∼)	4	3	0	0
# sites for wrong result (×)	3	16	16	26–13*

* WIEN has six sub-systems, so the result is a range

As shown in Table 2, our approach presented in this paper achieved correct results on 76.7% of the pages (23 out of 30) and the results suggest that our approach outperforms the MDR, AutoWrapper, and the WIEN approaches on these pages.

Further inspection of the results reveals that for the seven pages with nested structures, three pages with missing data and five pages with optional data, the new ADRD system performed better than all other systems investigated here, suggesting that this approach is more effective than other systems in processing Web pages with nested structures, missing or optional data.

5.3 Experiment 3

Our third group of experiments was designed based on the literature [4,3,2] which showed comparisons of existing systems on five typical Web sites from RISE [12]. We tested our system and MDR on these five sites to further investigate the effectiveness of our approach. Table 3 shows the experimental results of ADRD

Table 3. Comparison of six approaches on five Web sites

#	Site	N	M	O	ADRD	RR	WIEN	STalker	AutoWrapper	MDR
1	Okra	N	N	N	√	√	√	√	√	√
2	BigBook	N	N	N	√	√	√	√	√	√
3	LA Weekly	N	N	Y	√	√	×	√	∼	×
4	Address Finder	N	Y	N	×	×	×	√	×	∼
5	PharmWeb	Y	N	Y	√	√	×	×	×	×

and those existing systems RoadRunner (RR)[3], WIEN [1], STalker [2], the AutoWrapper [4], and the MDR [7] on the five typical Web sites.

According to the results, our system resulted in correct performance on four out of five web sites. The only one that our system fails is the page with a very special situation: the data field values are presented in different order and the formats of these disordered data fields are very similar.

Our system achieved clearly better results than the WIEN approach, the AutoWrapper and the MDR systems and produced comparable performance with the RoadRunner(RR) and the STalker systems. However, RoadRunner requires at least two example pages for training, while STalker requires a number of manually labelled training example pages. Our system, on the other hand, only needs one unlabelled Web page. Please note that only RoadRunner and our system can handle #5 with nested data.

6 Conclusions

This paper investigated an approach to automatic information extraction by clustering and labelling text tokens. This approach does not require any training and the information extraction process does not have explicit pattern learning or data extraction. The main contributions include new labels to categorize text tokens to capture the hierarchical structure of a Web page, and a new method that transforms labelled text tokens into XML format for data record detection.

The approach was examined and compared with a number of typical existing approaches on three different groups of web sites. These web site groups contain both some typical pages with missing, optional and nested data and some typical Web pages recently used in well known approaches which can be found publicly. The results suggest that our new approach is effective for web information extraction from the pages with missing data, optional data, or data with nested structures, and that it outperforms many existing approaches on these web sites.

The future work is to explore ways to extend this work to information extraction from more flexible pages such as pages where data are not separated by tags. Further research is need to further analyse and to improve the clustering performance.

References

1. Kushmerick, N., Weld, D.S., Doorenbos, R.B.: Wrapper induction for information extraction. In: IJCAI 1997. Intl. Joint Conference on Artificial Intelligence, pp. 729–737 (1997)
2. Muslea, I., Minton, S., Knoblock, C.: A hierarchical approach to wrapper induction. In: Etzioni, O., Müller, J.P., Bradshaw, J.M. (eds.) Agents 1999. Proceedings of the Third International Conference on Autonomous Agents, pp. 190–197. ACM Press, Seattle, WA (1999)
3. Crescenzi, V., Mecca, G., Merialdo, P.: Roadrunner: Towards automatic data extraction from large web sites. In: Proceedings of 27th International Conference on Very Large Data Bases, pp. 109–118 (2001)

4. Gao, X., Andreae, P., Collins, R.: Approximately repetitive structure detection for wrapper induction. In: Zhang, C., W. Guesgen, H., Yeap, W.-K. (eds.) PRICAI 2004. LNCS (LNAI), vol. 3157, pp. 585–594. Springer, Heidelberg (2004)
5. Carme, J., Ceresna, M., Frlich, O., Gottlob, G., Hassan, T., Herzog, M., Holzinger, W., Krpl, B.: The lixto project: Exploring new frontiers of web data extraction. In: Bell, D., Hong, J. (eds.) Flexible and Efficient Information Handling. LNCS, vol. 4042, pp. 1–15. Springer, Heidelberg (2006)
6. Muslea, I.: Extraction patterns for information extraction tasks: A survey. In: Proceedings of AAAI Workshop on Machine Learning for Information Extraction, Orlando, Florida (July 1999)
7. Liu, B., Grossman, R., Zhai, Y.: Mining data records in web pages. In: Proceedings of the ninth ACM SIGKDD international conference on Knowledge discovery and data mining, pp. 601–606. ACM Press, New York (2003)
8. Lerman, K., Knoblock, C., Minton, S.: Automatic data extraction from lists and tables in web sources. In: IJCAI 2001. Proceedings of the workshop on Advances in Text Extraction and Mining (2001)
9. Cheeseman, P., Stutz, J.: Bayesian classification (autoclass): Theory and results. In: Advances in Knowledge Discovery and Data Mining. American Association for Artificial Intelligence USA, pp. 153–180 (1996)
10. Vuong, L.P.B., Gao, X., Zhang, M.: Data extraction from semi-structured web pages by clustering. In: proceedings of the 2006 IEEE/WIC/ACM International Conference on Web Intelligence, pp. 374–377. The IEEE Computer Society Press, Los Alamitos (2006)
11. Object Management Group: W3c document object model (2005), `http://www.w3.org/dom/`
12. Muslea, I.: Repository of online information sources used in information extraction tasks (2005), `http://www.isi.edu/info-agents/rise/repository.html`

Visualizing Trends in Knowledge Management

Maria R. Lee[1] and Tsung Teng Chen[2]

[1] Shih Chien University, Department of Information Management
104 Taipei, Taiwan
maria.lee@mail.usc.edu.tw
[2] National Taipei University, Graduate School of Information Management
104 Taipei, Taiwan
misttc@mail.ntpu.edu.tw

Abstract. Knowledge visualization is a creative process, but difficult to formalize. This paper presents a system that is capable of analyzing voluminous citation data and visualizing the result. The system offers visualizations of trends by clustering scientific papers taken from the web (CiteSeer papers). Two methods are implemented: factor analysis and PFNET. An experiment has been carried out with the literature in knowledge management. A deep analysis of current trends in KM is then performed to check the relevance of these results. While the topical content is specific to knowledge engineering, semantic web, and related sub-areas, the approach could be applied to any general topic area in AI.

Keywords: Knowledge Management, Knowledge Visualization, Factor Analysis, PFNET.

1 Introduction

Knowledge visualization is a creative process, but difficult to formalize. Massive scientific papers publish every year. It would be useful to comprehend the entire body of scientific knowledge and track the latest developments in specific science and technology fields. However, effective and efficient comprehension of vast knowledge is a challenging task.

Knowledge Management (KM) is a fast growing field with great potential. However, researchers have disagreeing opinions about what constitutes the content and context of the KM research area [11]. It will be instrumental if the intellectual structure of KM domain could be constructed using Knowledge Domain Visualization (KDV) techniques. Previous studies of intellectual structure of the KM domain has been constructed by researchers with predominantly information systems and management oriented factors [18, 19]. In contrast, our study drew primarily on voluminous science and engineering literature that has given us some interesting results.

This paper is organized as follows. Section 2 introduces the start-of-the-art of knowledge domain visualization. Section 3 introduces a scientific analysis system, which offers visualizations of trends by clustering scientific papers taken from the web (CiteSeer papers). Two methods are implemented: factor analysis and Pearson correlation coefficients in PFNET. An experiment has been carried out with the literature in knowledge management in section 3. A deep analysis of current trends in KM is then carried out to check the relevance of these results in section 4.

Z. Zhang and J. Siekmann (Eds.): KSEM 2007, LNAI 4798, pp. 362–371, 2007.

2 Knowledge Domain Visualization

Visual exploration of large data sets provides insight by the visualizing of data [16]. However, lacking the ability to adequately explore the large amounts being collected, and despite its potential usefulness, the data become useless.

Fortunately, the task of knowledge comprehension could be facilitated by an emerging field of study - Knowledge Domain Visualization (KDV), which tries to depict the structure and evolution of scientific fields [3]. A knowledge domain is represented collectively by research papers and their inter-relationships in this research area. A knowledge domain's intellectual structure can be discerned by studying the citation relationships and analyzing seminal literatures of that knowledge domain.

Researchers in information science studied the intellectual structure of a discipline in the early eighties [21]. One of the pioneering studies, Author Co-citation Analysis (ACA), is used to present the intellectual structure of knowledge domain. Recent studies in knowledge visualization adopt this ACA approach as its underlying methodology and outfitted the intellectual structure with visual cues and effects [4,5,6,7]. In addition, some recent work in knowledge discovery systems and data mining systems carry out analyses and visualizations a scientific domain [29, 30, 31].

We proposed an approach, which is comparative to ACA analysis to derive knowledge visualization [8]. Figure 1 shows the process of the proposed approach. We propose an approach to construct a full citation graph from the data drawn from the online citation database CiteSeer [2]. The proposed procedure leverages the CiteSeer citation index by using key phrases to query the index and retrieve all matching documents from it. The documents retrieved by the query are then used as the initial seed set to retrieve papers that are citing or cited by literatures in the initial seed set [9, 19]. The full citation graph is built by linking all articles retrieved, which includes more documents than the other schemes reviewed earlier. Factor analysis and pathfinder network (PFNET) [21, 22, 23] are used to analyze the citation network.

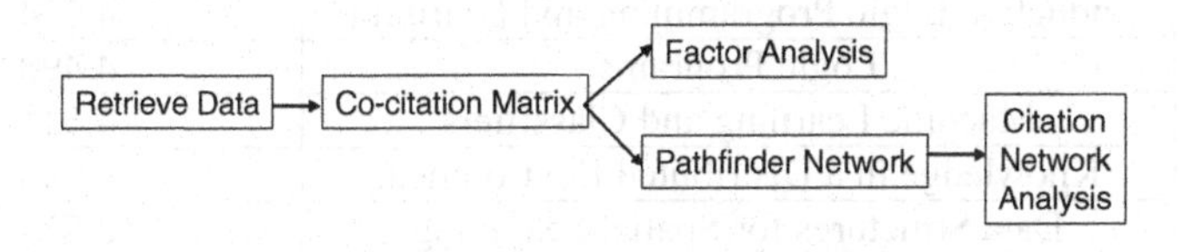

Fig. 1. Citation Analysis Process

3 Research Trends in Knowledge Management

Many knowledge visualization studies drew their citation data by using a key phrase to query citation indexes. However, it is rather limited to retrieve the citation data by a simple query of citation indexes.

Factor analysis and pathfinder network (PFNET) are amalgamated to perform citation analysis. The resulted citation graph was built from the literatures and citation information retrieved by querying the term "Knowledge Management" from CiteSeer on March, 2006. The complete citation graph contains 599,692 document nodes and 1,701,081 citation arcs. In order to keep the highly cited papers and keep the literature to a manageable size, we pruned out papers that were cited less than 150 times. The resultant citation graph contains 199 papers and 640 citation arcs.

In addition, we search literatures published during the eight years between 1998 and 2005 in KM. We divided these eight years into five consecutive overlapping time slots of four years each. Following the same citation processing procedure described above, the citation graph in each time slot is tabulated in table 1.

The threshold value is applied to prune out papers that were cited less than the threshold value. The information of the resulting citation graphs are listed on the far right-hand two columns.

Table 1. Citation Graphs Data

	Time span	No. Papers	No. Links	Thres-hold	No. Papers	No. Links
1	98-01	33,836	78,176	20	666	2,182
2	99-02	27,115	55,819	15	621	1,978
3	00-03	20,178	36,303	10	548	1,680
4	01-04	13,852	21,520	10	265	680
5	02-05	8,506	11,003	5	305	658

3.1 Factor Analysis

The factors and the variances of these factors are listed in Table 2. Eighteen factors were extracted which collectively explained 56.028 cumulative variances.

Table 2. KM Main Factors and Variances Explained

Factor	Descriptive Name	Variance Explained
1	Semi-structured/Object Databases	6.406
2	Inductive Logic Programming and Learning	4.694
3	Logic Programs	4.498
4	Machine Learning and Classifiers	4.312
5	Knowledge in a Distributed Environment	4.046
6	Data Structures for Spatial Searching	3.591
7	AI Concept Symbols	3.303
8	Data Mining	2.871
9	Information Integration of Varying Sources	2.771
10	Functional Languages and Development Environment	2.568
11	Planning and Problem Solving	2.477
12	Distributed Agents Cooperation	2.407
13	Modal and Temporal Logic	2.251
14	Inductive Logic Programming	2.077
15	Views Maintenance	2.042
16	Probabilistic Reasoning	2.005
17	Computational Geometry	1.930
18	Search Engines	1.781

We identified eight to ten top research themes of each time period and consolidated them into table 3. We took the ten themes found in the earliest time span (1998-2001) as the basis and merged new themes or their variants uncovered in the latter periods. Forty-eight themes found in the five periods are merged into eighteen main trends.

Table 3. Top New KM Research Trends 1998-2001

Theme (Trend)	Descriptive Name	Found Period	Average Ranking
1 (1)	Search Engine and Web Information Categorization	3	4.7
2	Materialized Views and Queries Using Views	1	2
3	Information Extraction Using Machine Learning Approach	1	3
4 (2)	P2P Issues	4	3.9
5	Agent-Oriented System Analysis and Design	4	3
6 (3)	Auctions Protocols and Algorithms	5	4.6
7	Web Usage and Web Mining	2	5
8	Semi-Structured Data Query and Schema Integration	3	4.3
9	Collaborative Filtering	1	9
10 (4)	Time Series Data Mining	2	10
11 (5)	Web Annotation and Knowledge Embedding	4	5.3
12 (6)	Continuous Query and Query Over Streaming Data	4	4
13 (7)	Schema Reconciliation and Ontology Merging	2	8.5
14	Privacy Preserving Data Mining	2	9
15	Dialogs Modeling of Agents	3	9
16 (8)	Automatic Programs Spec. Generation and Checking	3	7
17	Automated Trust Mgmt.	4	9
18	Context and Semantic Rich HCI	5	7

From table 3 listed above, eight themes consistently appear in these periods. Theme number 1, 4, 6, 10, 11, 12, 13, and 16 appeared in the earlier period and lasted well into the last period. We therefore regard these eight themes as research trends that will continue to gather interest, evolving, and lasting for sometime. Theme numbers 2, 3, and 9 appear only once in period one and this may imply their transient nature. Or, they may just converge with other themes; for example, theme 3—information extraction using machine learning approach—may converge with data mining or web mining. The study of agent-oriented software engineering, web usage mining, XML data processing, and privacy preserving data mining (themes 5, 7, 8, 14) appear to be insignificant in the last period implying that these studies may be maturing and so are gathering less interest. Theme numbers 15, 17, and 18 appear once in the later period and this may indicate these are ephemeral or emerging studies [1,10, 12, 13, 17, 18, 20, 24, 25].

4 Relationships Between Trends and KM Studies

The trends and themes found in the last section seem predominantly related to the Internet. However, we can see the links between them and the KM studies discussed in section three by further analysis. The first trend deals with search engine related issues, which also includes Web information categorization and ranking as well as focused searches of particular information. A focused search engine uses reinforcement learning to optimize its sequential decision making.

The second trend is characterized by studies in P2P infrastructure and semantic-based P2P systems. These researches try to provide the conceptual modeling, mechanism, and data model to integrate heterogonous data models or semantics that is local to each peer. The P2P paradigm is very useful when it comes to sharing files over the Internet; however, the more powerful knowledge sharing paradigm and semantic-based retrieval in P2P are hindered by the lack of common semantics shared by the peers. The research in this trend tries to lay down the infrastructure that facilitates effective knowledge sharing and exchange in the P2P paradigm.

The third trend includes studies in auctions protocols and algorithm, combinatorial auctions, and preference elicitation using learning algorithms. The parallels between the preference elicitation problem in combinatorial auctions and the problem of learning an unknown function from learning theory were discussed [26]. The fifth trend deals with the issue of semantic annotation [27], knowledge embedding, and data mapping for semantic web [28]. Semantic web promises to provide a common framework that allows data to be shared and reused across applications. However, most of the existing data is in the form of XML not in RDF; a mapping mechanism is required to render the data sources interoperable.

The seventh trend encompasses research of ontology merging, schema reconciliation, and knowledge sources integration. The separate ontology development has led to a large number of ontologies covering overlapping domains. In order for these ontologies to be reused, tools and methods that facilitate the merging or alignment of them need to be developed [17]. A key hurdle in building a data-integration system that uniformly accesses multiple sources of data is the acquisition of semantic mappings. A system that uses current machine learning techniques to find such mappings semi-automatically was proposed by researchers.

4.1 Pathfinder Network

The Pearson correlation coefficients between items (papers) were calculated when factors analysis was applied. The correlation coefficients are used as the basis for PFNET scaling [23]. The value of Pearson correlation coefficient falls between the range -1 and 1. The coefficient approaches to one when two items correlate completely. Items that closely relate, i.e., are highly correlated, should be placed closely together spatially.

The distance between nodes is normalized by taking $d = 1/(1 + r)$, whereas r is the correlation coefficient. The distance between items is inversely proportional to the correlation coefficient, which maps less correlated items apart and highly correlated items spatially adjacent. As we have mentioned earlier that the nodes located close to the center of a PFNET graph represents papers contributed to a fundamental concept, which are frequently referred by other peripheral literatures that are positioned in outer branches. Figure 2 shows PFNET scaling with papers under same factors close to each other.

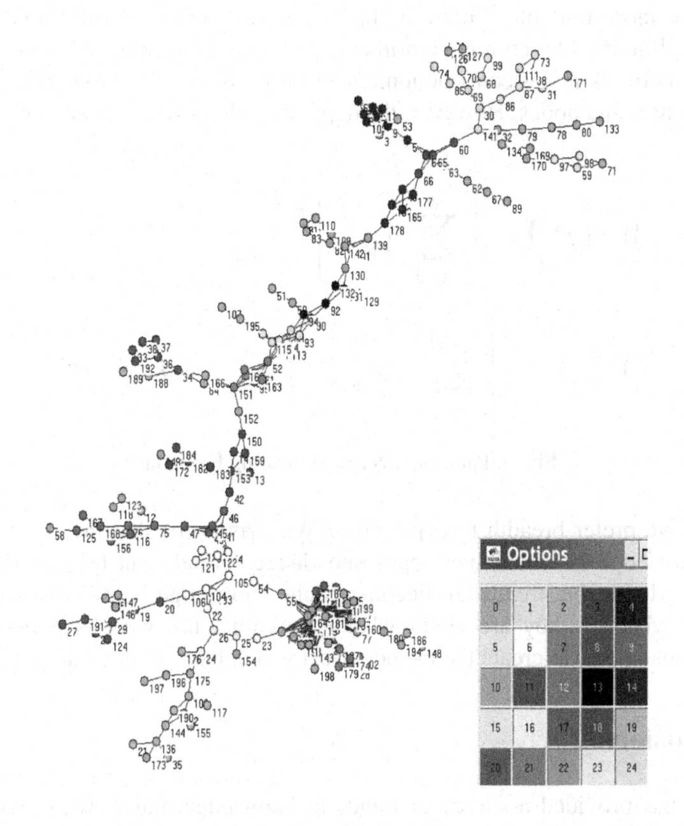

Fig. 2. PFNET Scaling with Papers under Same Factors Close to Each Other. Nodes belong to the same factor are painted with corresponding color of the factor. Nodes that do not belonged to any factor are painted with color palette numbered 0.

The cluster that corresponds to logic programs and AI concept symbols papers (colored in 3 and 7 palette) seems to play a key role in the KM intellectual structure as shown in the PFNET. AI concept symbols include a paper discussing the philosophical problems of AI, a paper addresses the "knowledge level" of computer system, and a book discusses the logical foundations of Artificial Intelligence. A paper in factor seven is dated back as early as 1969.

5 Discussions

Based on the research trends and changes in knowledge management communities over the past 20 years, the proposed trends in knowledge management are rivaling [14]. Of particular interest has been the proposed approach that could be re-used in a number of ways and could possibly be shared across different domains.

One of the potential future works of the proposed approach is to develop niche knowledge. The current approach provides a broad view of the trends, but we would like to work more in-depth. Figure 3 shows the mathematic function of Pathfinder Network Scaling [5]. Our current algorithm is to increase the value of parameter q or r, which limit network nodes construction. One of the advantages is to retrieve the most significant paths and nodes. However, the depth of nodes will be missed out.

$$W(P) = \left[\sum_{i=1}^{k} w_i^r \right]^{\frac{1}{r}}$$

$$W_{n_i n_k} = \left[\sum_{i=1}^{k-1} w_{n_i n_k}^r \right]^{\frac{1}{r}} \qquad \forall k \leq q$$

Fig. 3. Pathfinder Network Scaling Formula

Why do we prefer breadth to depth now? We may argue that the construction of breadth knowledge is easier than depth knowledge. Trends can be provided by the breadth knowledge, but trends are fleeting. Niche knowledge is less about trends and more about vision - they are about what is possible, not what is popular. Niche knowledge also helps to create the medici effect which leads to innovation [15].

6 Conclusions

This paper has provided a survey of trends in knowledge management, which have been developed through analysis of citations and relationships among citations in CiteSeer. Factor analysis and Pearson correlation coefficients in PFNET have been implemented and the results obtained in the domain of knowledge management.

The research themes that play the central role, according to the layout of PFNET, are knowledge representation, information integration and query related studies, modal and temporal logic, data mining, text categorization and constraint logic programming. The studies of inductive logic programming, machine learning, planning, and active network and mobile agents seem to fall to the side line or just play a peripheral role in KM related studies.

Ten of the most important current research trends of KM were summarized. Semantic web, semantic-based P2P and agent systems, Ecommerce related topics, Ontology, and human computer/robot interaction researches are recent popular research trends in the KM domain area. Distributed knowledge representation and reason systems are also research interests due to World Wide Web proliferation. In addition, classifiers and patterns learning, especially in the area of Webs and hidden databases with Web front end, are active research areas too. The fusion of disparate research areas such as computer science, economics, and law is an interested new trend; the implication of other research areas, leveraging Internet as a standard platform, may follow this cross discipline trend is expected. Privacy and security related issues are getting more attention due to the burgeoning Ecommerce activities.

Research that intertwines World Wide Web, P2P systems, and intelligent agents with classical AI topics seems to be the new direction in large. However, we have only seen limited new research that tries to leverage the rich AI tradition of the past to pursue Web and Internet related fields.

One of the future works is to emerge and apply the proposed approach to other disciplines. We also would like to extend the approach to work in-depth to find out niche knowledge of a domain. Niche knowledge adds more value. Niche knowledge creates the medici effect that when you step into an intersection of fields, disciplines, or cultures, you can combine existing concepts into a large number of extraordinary new ideas. More and more, innovation is arising not from particular industries or disciplines, but rather across them [15].

References

1. Boella, G., Torre, L.: Permissions and Obligations in Hierarchical Normative Systems. In: Proceedings of the 9th international conference on Artificial intelligence and law Scotland, ACM Press, United Kingdom (2003)
2. Bollacker, K.D., Lawrence, S., Giles, C.L.: CiteSeer: an Autonomous Web Agent for Automatic Retrieval and Identification of Interesting Publications. In: Proceedings of the second international conference on Autonomous agents Minneapolis, ACM Press, Minnesota, United States (1998)
3. Börner, K., Chen, C., Boyack, K.: Visualizing Knowledge Domains. In: Cronin, B. (ed.) Annual Review of Information Science and Technology. American Society for Information Science and Technology, Medford, New Jersey, vol. 37, pp. 179–255 (2002)
4. Chen, C.: Visualization of Knowledge Structures. In: Chang, S.K. (ed.) Handbook Of Software Engineering And Knowledge Engineering, vol. 2, p. 700. World Scientific Publishing Co, River Edge, NJ (2002)
5. Chen, C.: Searching for Intellectual Turning Points: Progressive Knowledge Domain Visualization. PNAS 101, 5303–5310 (2004)
6. Chen, C., Kuljis, J., Paul, R.J: Visualizing Latent Domain knowledge. Systems, Man and Cybernetics, Part C, IEEE Transactions 31, 518–529 (2001)
7. Chen, C., Paul, R.J.: Visualizing a Knowledge Domain's Intellectual Structure. Computer 34, 65–71 (2001)
8. Chen, C., Steven, M.: Visualizing Evolving Networks: Minimum Spanning Trees versus Pathfinder Networks. In: IEEE Symposium on Information Visualization, pp. 67–74 (2003)
9. Chen, T.T., Lee, M.: Revealing Themes and Trends in the Knowledge Domain's Intellectual Structure. In: Hoffmann, A., Kang, B.-H., Richards, D., Tsumoto, S. (eds.) PKAW 2006. LNCS (LNAI), vol. 4303, pp. 99–107. Springer, Heidelberg (2006)
10. Chen, T.T., Xie, L.Q.: Identifying Critical Focuses in Research Domains. In: IV 2005. Proceedings of the Information Visualisation, Ninth International Conference, London, pp. 135–142 (2005)

11. Dahl, T.S., Mataric, M.J., Sukhatme, G.S.: Adaptive Spatio-Temporal Organization in Groups of Robots. In: Intelligent Robots and System, 2002. IEEE/RSJ International Conference, vol. 1, pp. 1044–1049 (2002)
12. Earl, M.: Knowledge Management Strategies: Toward a Taxonomy. Journal of Management Information Systems 18, 215–242 (2001)
13. Feigenbaum, J., Shenker, S.: Distributed Algorithmic Mechanism Design: Recent Results and Future Directions. In: Proceedings of the 6th international workshop on Discrete algorithms and methods for mobile computing and communications Atlanta, ACM Press, Georgia (2002)
14. Guarino, N., Welty, C.: Evaluating Ontological Decisions with OntoClean. Communications of the ACM 45, 61–65 (2002)
15. Hoffmann, A., Kang, B.-H., Richards, D., Tsumoto, S. (eds.): PKAW 2006. LNCS (LNAI), vol. 4303. Springer, Heidelberg (2006)
16. Johansson, F.: The Medici Effect: Breakthrough Insights at the Intersection of Ideas, Concepts & Cultures. Harvard Business School Publishing, Boston (2004)
17. Keim, D.: Visual Exploration of Large Data Sets. Communication of the ACM 44(8), 39–44 (2001)
18. Noy, N.F., Musen, M.A.: The PROMPT Suite: Interactive Tools for Ontology Merging and Mapping. Int. J. Hum.-Comput. Stud. 59, 983–1024 (2003)
19. Ponzi, L.J.: The Intellectual Structure and Interdisciplinary Breadth of Knowledge Management: a Bibliometric Study of Its Early Stage of Development. Scientometrics 55, 259–272 (2002)
20. Subramani, M., Nerur, S.P., Mahapatra, R.: Examining the Intellectual Structure of Knowledge Management, -2002 - An Author Co-citation Analysis, Management Information Systems Research Center, Carlson School of Management, University of Minnesota, 2003, p. 23 (1990)
21. Tews, A.D., Mataric, M.J., Sukhatme, G.S.: A Scalable Approach to Human-Robot Interaction. In: ICRA 2003. Robotics and Automation, 2003. Proceedings. IEEE International Conference, pp. 1665–1670 (2003)
22. White, H.D., Griffith, B.C.: Author Cocitation: A Literature Measure of Intellectual Structure. Journal of the American Society for Information Science 32, 163–171 (1981)
23. White, H.D.: Pathfinder Networks and Author Cocitation Analysis: A Remapping of Paradigmatic Information Scientists. Journal of the American Society for Information Science & Technology 54, 423–434 (2003)
24. White, H.D.: Author Cocitation Analysis and Pearson's r. Journal of the American Society for Information Science & Technology 54, 1250–1259 (2003)
25. Vaidya, J., Clifton, C.: Privacy Preserving Association Rule Mining in Vertically Partitioned Data. In: Proceedings of the eighth ACM SIGKDD international conference on Knowledge discovery and data mining, ACM Press, Edmonton, Alberta, Canada (2002)
26. Vaughan, R.T., Stoy, K., Sukhatme, G.S., Mataric, M.J., LOST,: Localization-Space Trails for Robot Teams. Robotics and Automation, IEEE Transactions 18, 796–812 (2002)
27. Lahaie, S., Parkes, D.: Applying Learing Algorithms to Preference Elicitation. In: Proceesings pf the 5th ACM Conference on Electronic Commerce, ACM Press, New York (2004)
28. Erdmann, M., Maedche, A., et al.: From manual to semi-automation tools. In: Proceedings of the COLING 2000 Workshop on Semantic Annotation and Intelligent Content (2000)

29. Doan, A.: Madhavan et al Learning to map between ontologies on the semantic web. In: Proceedings of the 11th International Conference on World Wide Web, ACM Press, New York (2002)
30. Mothe, J., Chrisment, C., Dkaki, T., Dousset, B., Karouach, S.: Combining Mining AND Visualization Tools TO Discover THE Geographic Structure OF A Domain. Computers, environment and urban systems 30, 460–484 (2006)
31. Crimmins, F., Smeaton, A., Dkaki, T., Mothe, J.: TetraFusion: information discovery on the Internet. Intelligent Systems and Their Applications, IEEE Intelligent Systems 14, 55–62 (1999)
32. Mothe, J., Dousset., B.: Mining document contents in order to analyze a scientific domain. In: Sixth International Conference on Social Science Methodology, Amsterdam, The Netherlands (2004)

Constructing an Ontology for a Research Program from a Knowledge Science Perspective

Jing Tian[1], Andrzej P. Wierzbicki[1,2], Hongtao Ren[3], and Yoshiteru Nakamori[1,3]

[1] Center for Strategic Development of Science and Technology,
Japan Advanced Institute of Science and Technology (JAIST),
Asahidai 1-1, Nomi, Ishikawa 923-1292, Japan
{jtian, andrzej, nakamori}@jaist.ac.jp
[2] National Institute of Telecommunications, Szachowa 11, 04-894 Warsaw, Poland
[3] School of Knowledge Science, JAIST, Asahidai 1-1, Nomi, Ishikawa 923-1292, Japan
hongtao@jaist.ac.jp

Abstract. Although ontologies as an important component are widely used for different purpose in different communities and a number of approaches have been reported for developing ontologies, few works have been done to clarify the concept of knowledge science as far as we know. This paper presents a novel attempt to create an ontology characterizing a research program *"Technology Creation Based on Knowledge Science"* from a *Knowledge Science* perspective. We address a combination of bottom-up and top-down approaches to ontology creation, which is a first time to put forward a perspective of combining explicit knowledge with tacit, intuitive and experiential knowledge for constructing an ontology. An example of application of this ontology, related to a software tool named *adaptive hermeneutic agent* (AHA), is also given in the paper.

Keywords: ontology, knowledge science, knowledge engineering, knowledge management, technology management.

1 Introduction

The word ontology was taken from philosophy, where it means a systematic explanation of being [1]. The term of ontology was borrowed by computer scientists in the middle of 1980s as a means to represent information and knowledge. It got significant development in the 1990s and the emergence of the Semantic Web has marked an important step in the evolution of ontologies. Regarded as a means for a shared knowledge understanding and a way to represent real world domains, in the last decades, ontologies are expected to play a crucial role in data and application integration at public and corporate level, for example, development of information systems, organization of content in web sites, categorization of products in e-commerce, structured and comparative searches of digital content, standard vocabularies in expert domains, product configuration in manufacturing, among many others [2][3][4][5].

Z. Zhang and J. Siekmann (Eds.): KSEM 2007, LNAI 4798, pp. 372–383, 2007.

This paper presents a novel attempt to create an ontology characterizing a research program *"Technology Creation Based on Knowledge Science"* at Japan Advanced Institute of Science and Technology (JAIST) from a *Knowledge Science* perspective. Knowledge Science is a new academic field which relates to the philosophy, methodology and techniques for creating knowledge, modeling knowledge creation process and conducting research on knowledge management. The School of Knowledge Science at JAIST is the first school established in the world to make knowledge creation as the core of its scientific research. From the second half of 2003, the 21st-century COE (Center of Excellence) Program "Technology Creation Based on Knowledge Science" of JAIST sponsored by the Ministry of Education, Culture, Sports, Science and Technology (MEXT, Japan) has been initiated. This program aims to establish an interdisciplinary research field focusing on research and education exploring issues related to "knowledge science," including how to 1) create knowledge that can help spark innovation in a variety of situations, 2) develop individuals capable of coordinating knowledge creation processes, and 3) ensure ethical behavior in a knowledge-based society. In order to fuse theoretical research and practical research, a series of projects have been promoting at the Center of Strategic Development of Science and Technology of JAIST.

In this research, we tried to construct an ontology for COE program at JAIST with a new understanding of *knowledge science* and make explicit (at least, as much as possible) assumptions about this concept that are often tacitly made and never defined well. This work also will help the development of some projects of this program, clarify basic concepts for COE program itself, and help researchers in this program with vocabularies of keywords, with literature searches, and so on.

There are many methods for developing ontologies, we can distinguish among those i) from scratch, ii) by reuse or iii) with the help of (automatic) knowledge acquisition techniques [4]. In our case, we combined the bottom-up classification and specification and the top-down reflection on concept of knowledge science to build the ontology from scratch. The paper emphasized one of the most difficult aspects of constructing ontologies, namely, combining explicit knowledge, which typically used in bottom-up approaches, with tacit, intuitive and experiential knowledge, which typically used in top-down approaches. Some knowledge acquisition techniques were also taken into account in this work.

The rest of this paper is organized as follows. Section 2 introduces our research goals and methods. Section 3 and Section 4 respectively explain the bottom-up keywords analysis and top-down reflection in knowledge science. Detailed description of the ontology we built for COE program is given in Section 5, and one possible application of the ontology is also included. To understand our result better, we reflect also about other possible views on such ontology in Section 6. Section 7 summarizes this paper.

2 The Goals and Ways of Constructing Ontology of the COE Program

In addition to the philosophical origin, ontology is also given diverse other meanings. In contemporary computer science, ontology is defined as a formal language-like specification of a domain knowledge – actually equivalent to a taxonomy of concepts in a given field of knowledge, enhanced by a structure of hierarchical dependences and other links between concepts constituting the taxonomy, see, e.g., Dieng and Corby [6]. *Ideally, an ontology should provide* [7]*:*

1) *a common vocabulary,*
2) *explication of what has been often left implicit,*
3) *systematization of knowledge,*
4) *standardization of terms,*
5) *meta-model functionality* (providing a metalanguage for specific models in the domain).

Actually, these goals are not attainable: in order to have a formal meta-model [3], we need a meta-meta-model and so on, therefore we have to stop at some level of explication of basic assumptions and rely on an *hermeneutical horizon* – an intuitive perception what concepts and assumptions are basic and true and how we understand them. Thus, any ontology will achieve the ideal goals mentioned above only to a certain degree. Note, however, that this implies that any ontology can be re-engineered, corrected according to changes in the hermeneutical horizon.

In the paper, we tried to construct the ontology of 21st Century COE Program *Technology Creation Based on Knowledge Science* at JAIST as a case study, with the following goals:

(1) To clarify the use of the concept of Knowledge Science in this Program and make explicit (at least, as much as possible) assumptions about this concept that are often tacitly made (ideal goals 2, 5);

(2) To represent a vocabulary of terms used in this COE Program, together with a systematization of terms used (ideal goals 1, 3);

(3) To help in the development of a software system designed to support hermeneutic search of literature, and possibly in other projects related to the COE Program.

The ideal goal 4) – standardization – is addressed only to limited degree, because of the heterogeneity of the interdisciplinary projects in the COE Program. Thus, we design ontology for COE program at JAIST not only for helping in the development of some projects of this program, but also make to clarify basic concepts for COE program itself.

Known ways of constructing ontologies can be treated not as absolute recipes, but hints how to proceed. There is a distinction of a top-down approach - actually, starting with an intuitive perception of the basic concepts in hermeneutical horizon and specifying them in detail subsequently – and a bottom-up approach - starting, say, with the concepts actually used in a given field of knowledge and trying to interpret them and their structural relations. The top-down approach starts with issues related to meta-model functionality (idea goals, 5); the bottom-up approach starts with issues related to systematization (ideal goals, 3) and standardization (ideal goals, 4). Obviously, we need a combination of both approaches in order to construct a useful ontology.

To create ontology, we proceeded along several lines. First, we checked the terms and concepts used by the program leader in a paper presenting an introduction to the

COE program, thus providing an outline of COE ontology. Then, we collected 43 papers composed by COE project members, which have appeared either at an international conference or journal. We extracted the keywords from the papers and counted the frequency of keywords in the full paper by using a computer program. We chose the keywords with high frequency to supplement the outline of COE ontology. We chose also pairs of keywords occurring with non-zero frequency to make a simple QT clustering of them [8] and compared the ontology emergent bottom-up from such clustering with the top-down outline of COE ontology. Finally, we took into account the reflection on knowledge sciences [9] and used this reflection for corrections of the supplemented outline; this way, we finally created the ontology for COE program.

3 Bottom-Up Classification and Specification: Keyword Analysis

To build an outline of the ontology of COE program, we started with the paper presenting an introduction to this program authored by the program leader [6]. After analyzing the purpose and sub-projects of the program, we selected the key terms and concepts mentioned in the paper and organized an ontology outline with three levels of branches. The first level included five main topics:

- Knowledge science,
- Systems science and methodology,
- Education in knowledge science,
- Knowledge creation,
- Management of technology.

In addition, we also referred to the program reports presented by the program leader in later periods to check and revise the outline.

Furthermore, we collected the papers authored by COE project members - as many as were available. Since we had to limit this search to electronic files, we finally considered only 43 papers, which were either included in Proceedings of International Symposium on Knowledge and Systems Sciences (JAIST, 2004), or Proceedings of the First World Congress of the International Federation for Systems Research (Kobe, 2005), or in the International Journal of Knowledge and Systems Sciences (Issues 1 to 6). We extracted keywords (including keyphrases) from all papers and counted the frequency of their occurrences in the full body of papers by using a computer program designed by a member of our group.

The keywords were specified by authors, but we thought they were not enough for our research to compare the contents of all papers. Thus, the additional keyphrase extraction was taken into account. This may increase the correlation and improve the clustering. Keyphrase extraction techniques for the English language have been well developed. Keyphrase frequency associated with keyphrase significationce was proposed in Luhn [10]. Based on the knowledge bases of "*stop words*"[1], there are three steps to the keyphrases extractor:

[1] We used a list of stop words gotten from http://en.wikipedia.org/wiki/Stop_words

(1) *Find Keyphrases:* Extract keyphrases from the text file and make a list of all phrases. A phrase is defined as a sequence of one, two, or three words that appear consecutively in the text, with no intervening stop words or punctuation. In our case, phrases of four or more words are relatively rare.

(2) *Score Keyphrases:* For each keyphrase, count how often the keyphrase appears in the text. Assign a score to each phrase. The score is the number of times the keyphrase appears in the file.

(3) *Final Output:* We now have an ordered list of mixed-case phrases (upper and lower case, if appropriate). The list is ordered by the scores calculated in step 2.

Another attempt was a clustering of keywords based on their joint occurrence. We selected a simple QT (quality threshold) clustering algorithm [4]. The goal of QT clustering is go form large clusters of genes with similar expression pattern, and to ensure a quality guarantee for each cluster. Quality is defined by the cluster diameter and the minimum number of genes contained in each cluster. In our case, if the frequency of occurrence of a pair of keywords equals or exceeds an assumed threshold t, the pair might be counted to belong to a candidate cluster; the largest of such candidate clusters is counted as an actual cluster, it is subtracted from the entire set of keywords, and the procedure is repeated on the remaining keywords. It turned out that the joint occurrence of keywords is not common, most frequencies of such co-occurrence are zero, thus the clustering was done at the threshold level $t = 1$. Because of the space, we only list the outputs of following two clusters as examples.

Example of Keywords Clustering

```
Cluster1:

Papers:
{
09_1_Minh.txt, 09_2_Nagai-kss04.txt, 12_2_phan.txt,
12_3_Tran.txt, 15_1_Zhang.txt, 15_2_huang-wei.txt,
20055.pdf.txt, 20057.pdf.txt, 20073.pdf.txt,
06_1_Hao.txt, 20177.pdf.txt
}
Keywords
{
Text summarization, Natural language processing, Text
mining, Association rule mining, Coreference
resolution, Anaphora resolution, Clustering algorithm,
Information extraction, Natural language processing,
Data mining, Knowledge discovery, Clustering, Genetic
algorithm, K-means algorithm, Text clustering, Ant-
based Clustering, Semantic similarity measure,
Ontology, Phrase indexing, Sentence extraction,
Ensemble learning, SVM ensemble, Direct space method,
Rough sets
}
Cluster2:

Papers:
{
```

```
05_1_ma.txt, 05_3_ JieYAN.txt, 20060.pdf.txt
}
Keywords:
{
Transportation fuel cell forecast, Technology
roadmapping, Technology creation, Roadmapping process,
Technology forecasting, Roadmapping, Interactive
planning
}
```

With respect to the outline of COE ontology with three levels branch that we summarized from the project introduction and reports, the key phrases included in cluster one belong to the topic of "Knowledge Representation and Acquisition", we have not listed a independent branch for it yet. The key phrases included in cluster two belong to the topic of "Management of Technology". And we found the researchers were very interested in "Technology Roadmaps". It could be one of the subtopics which belongs the "Management of Technology". Drawing inferences about other cases from above instance, the clusters give us the hints to categorize the keywords, rethink the ontology outline of COE Program and finally merge the results with the top-down reflection introduced in next section.

4 Top-Down Reflection on the Concept of Knowledge Science

Knowledge science (KS) is often confused with or tacitly assumed to be subordinated to *knowledge management* (KM), thus we first reflect on the origins and meaning of the second term. *Knowledge management* has much popularity in management science, but its technological origins are often forgotten. It was first introduced by computer technology firms in early 1980-ies – first in IBM, then Digital Equipment Corporation who probably was the first to use the term *knowledge management* – as a computer software technology in order to record the current work on software projects. This started the tradition of treating knowledge management as a system of computer technologies. Later this term was adopted by management science, and made a big career. This has led to two opposite views how to interpret this term [11][12]:

- As *management of information relevant for knowledge-intensive activities*, with stress on information technology: databases, data warehouses, data mining, groupware, information systems, etc.
- As *management of knowledge related processes,* with stress on organizational theory, learning, types of knowledge and knowledge creation processes.

The first view is naturally represented by information technologists and hard scientists; the second by social scientists, philosophers, psychologists and is clearly dominating in management science. Representatives of the second view often accuse the first view of perceiving *knowledge to be an object* while it should be seen as *knowledge related to processes;* they stress that knowledge management should be *management of people.* For example, in an excellent book on the dangers of postponing action *The Knowing-Doing Gap* [9], say that "[an] article asserted that 'knowledge management starts with technology'. We believe that this is precisely wrong. ...Dumping technology on a problem is rarely an effective solution."

However, while it is correct that knowledge management cannot be reduced to management of information, such a correct assessment tends to overlook both the complexity and the essence of the controversy. The complexity is that, historically, knowledge management has started with technology and cannot continue without technology; thus, both interpretations should be combined in adequate proportions. The essence of the controversy is that *management of people* should be also understood as *management of knowledge workers;* and knowledge workers are today often mostly information technologists, who should be well understood by managers. Thus, we believe that the two views listed above should be combined. Moreover, they incompletely describe what knowledge management is; there is a third, essential view, seeing knowledge management as the *management of human resources in knowledge civilization era,* concentrating on knowledge workers, their education and qualities, assuming a proper understanding of their diverse character, including a proper understanding of technologists and technology.

This is particularly visible concerning the concepts of *technology management* versus *knowledge management.* Management science specialists in knowledge management often tend to assume that *technology management* is just a branch of *knowledge management;* technologists specializing in *technology management* stress two aspects. First, an essential meaning of the word *technology* is *the art of designing and constructing tools or technological artefacts* (thus, *technology* does not mean *technological artefacts,* although such a meaning is often implied by a disdainful use of the word *technology*, e.g., in the quoted above phrase *dumping technology*). In this essential meaning sense, the term is used in the phrase *technology management.* Secondly, *technology management* might be counted as a kind of special *knowledge management,* but it is an older discipline, using well developed concepts and processes, such as *technology assessment, technology foresight* [13] and *technology roadmapping* [14][15]. Only recently, some of these processes have been also adapted to knowledge management [16].

All the above discussion implies that we are observing now an emergence process of a new understanding of *knowledge sciences* – an interdisciplinary field that goes beyond the classical epistemology, includes also some aspects of *knowledge engineering* from information technology, some aspects of *knowledge management* from management and social science, some aspects of *interdisciplinary synthesis* and other techniques (such as decision analysis and support, multiple criteria analysis, etc.) from systems science. This emergence process is motivated primarily by the needs of an adequate education of *knowledge workers* and *knowledge managers and coordinators;* however, also the research on knowledge and technology management and creation needs such interdisciplinary support.

The classical understanding of the words *knowledge science* might imply that it is epistemology enhanced by elements of knowledge engineering, knowledge management and systems science. However, the strong disciplinary and historical focus of epistemology suggests an opposite interpretation: knowledge science must be interdisciplinary, thus it should not start with epistemology, although it must be enhanced by elements of epistemology. The field closest to knowledge science seems to be systems science – at least, if it adheres to its interdisciplinary origins and does not suffer too much from the unfortunate (but unavoidable today) disciplinary division into *soft* and *hard systems science.* The noticeable tension between *soft* and *hard* systems science is just an older version of the tension between understanding *knowledge management* either from the perspective of knowledge engineering, or from the perspective of social and management science, mentioned above.

To summarize, we should thus require that *knowledge sciences* gives home to several disciplines (quoted here in an alphabetic order):

- *Epistemology and philosophy of science,*
- *Knowledge engineering,*
- *Management science and knowledge management,*
- *Sociological and soft systems science,*
- *Technological and hard systems science,*

On equal footing, with a requirement of mutual information and understanding, this basic classification should be also reflected in the proposed ontology of the COE Program.

To our knowledge, only one university in the world, the Japan Advanced Institute of Science and Technology, founded – already in 1998 – the School of Knowledge Science, while the field is understood similarly as described above. The university supports only graduate education, for master and doctoral degrees; in knowledge science, three types of graduates are typical:

- Specialists in management, with understanding of knowledge engineering and systems science;
- Specialists in systemic knowledge coordination, with understanding of knowledge engineering and management;
- Specialists in knowledge engineering, with understanding of management and systems science.

The ontology of the COE Program might be also treated as a first step towards constructing an ontology for the School of Knowledge Science in JAIST, providing a better understanding of what is (or are) knowledge science (or sciences).

5 Final Proposal of the Ontology and Its Application

Based both on the bottom-up classification and on the above reflection as a basis of top-down approach, the ontology of the COE Program can be proposed. It is organized as an inverted tree, with fourth-level branches corresponding to keywords found in the papers of COE Program members. The general category of the domain of Knowledge Science includes the following eight sub-domains as the first lever of ontology of the COE Program:

- *Knowledge Creation and Transformation*
- *Knowledge Representation, Systematization, Acquisition*
- *Knowledge Management*
- *Systems Science*
- *Education and Knowledge Science*
- *Management of Technology*
- *Technology Creation*
- *Diverse Related Themes*

Each sub-domain is consisted of several topics (Second lever); the different topics include particular sub-topics (Third Lever). All keywords was summarized as and categorized into the sub-topics (Fourth lever). In addition, the clustering of the keywords gave us the hints to find the relations between the subtopics and the further relations between topics as well as sub-domains. Because of the limitation of pages, we can not list the proposed ontology here. Our classification is naturally not absolute nor the ultimately final; it might be further enhanced and corrected as new data will become available.

On the basis of requirements of researchers [17][18] and the phenomenon of *Hermeneutics*[19], a software tool for information and knowledge retrieval was designed [20][21], in order to help researchers in gathering and interpreting relevant knowledge or research materials; this software tool is called *Adaptive Hermeneutic Agent (AHA)*. The AHA is equipped with a simple and intuitive search interface and uses familiar search syntax, such as used by popular search engines (like Google, Yahoo). The search support can be extended to the definition of queries that will be automatically executed by the system with a fixed period of time. The definition of a query by the user is helped by ontological information; actually, the ontology described above is used in AHA as a basis of defining queries that can be selected from this ontology, supplemented or modified, for example, by adding new keywords that are relevant to the searched topic. After the query is executed, the AHA can also filter the obtained results by using a reinforcement learning approach that relies on a profile of the user's interests. The AHA could also use a visual interface for the clustering and graphical presentation of search results.

Therefore, the COE ontology as described earlier is an important element and first step in developing the software tool of AHA. The second step is the creation of user profile. It is to say, the user could extract the knowledge from COE ontology to formulate the outline of user profile. For instance, the user (e.g. COE member) can select the domains (keywords) he are most interested in and give the weights for different keywords. Then, the user could gather relevant knowledge and information based on his profile by using search engines connected to AHA. The AHA will do adaptive selection automatically as following steps: text extraction (from MS-word file to text or from PDF file to text); keyword extraction and frequents calculation (extracting keywords from the search results by statistics method); measurement of the similarity of each file and user profile; giving a ranking list including top N results. The fig. 1 shows an interface of creating user profile based on ontology of COE Program.

Other possible applications of the work on ontology formation described here include, for instance, the development of an ontology of Knowledge Science in JAIST, an ongoing project that will include the lessons from the work described here; or a construction of a Knowledge Map or a research network for professionals interested in related domain, etc.

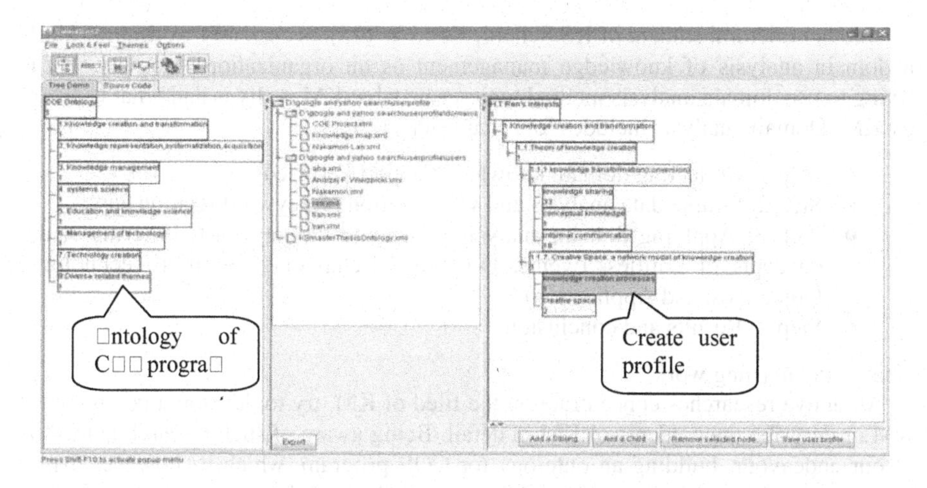

Fig. 1. The main interface of creating user profile based on ontology of COE Program

6 Other Perspectives and Ontological Approaches

Either before our work or at the same time, there are some related works in other perspectives for building an ontology or taxonomy, mostly in terms of knowledge management.

Dr. Totok H. Wibowo [22][23], a postdoctoral research fellow of COE Center, worked with other two colleagues to construct a Knowledge Map for the faculty members of School of Knowledge Science (KS) in order to provide a critical mechanism in creating a research network of professionals interested in knowledge creation and knowledge management. They conducted the questionnaire survey and interview with the professors in KS School, and also referred many related journal papers, information from websites and other publications. However, they concentrated on the perspective of knowledge management (KM) which, as discussed above, can be counted only as a part of knowledge sciences. Finally, they concluded with a taxonomy of knowledge management, which consists of eight disciplines: Business of KM, Technologies of KM, KM Processes, KM Systems, Sociology of KM, Creativity, Psychology of KM, and Philosophy of KM. This work was based on the classification of existing research fields and topics of KM by taxonomical method and it clearly represented a KM perspective.

In order to distinguish and describe KM technologies according to their support for strategy, Saito et al. [24] employed an ontology development method to describe the relations between technology, KM and strategy, and to categorize available KM technologies according to those relations. This study focused particularly on two sub-domains of the KM field: KM strategies and KM technologies. The processes of developing the ontology in this work included four steps:

- Definition of the domain and scope
- Identification of key terms and concepts, and their relationships
- Definition of the structure of the ontology as a hierarchy of categories
- Survey of KM technologies according to the ontology.

Another doctoral student of KS School, Kun Nie [25], is currently working towards a domain analysis of knowledge management as an organizational activity. He is trying to use domain analysis method to describe what KM really is and what is meant by KM. Domain analysis includes four main steps [26]:

- Step 1: Collecting domain knowledge/expert knowledge
- Step 2: Simple data analysis and visualization of keyword relationships
- Step 3: Applying domain analysis (distribute the keywords in terms of the concepts of Entities, Events, Functions, Behaviours, Support Technology, Objectives, and Application)
- Step 4: Results and conclusion.

This is an on-going work.

All above researches concentrate on the filed of KM, try to develop a taxonomy of KM or describe the contents of KM in detail. Being aware of such research helped us in our endeavour, building an ontology for COE program, which is, however, based on the assumption that Knowledge Science (KS) has much more rich meaning than Knowledge Management (KM).

7 Conclusion

We presented a process of constructing ontology of the 21st Century COE Program *Technology Creation Based on Knowledge Science* together with one of possible applications – helping in the development of an adaptive hermeneutic agent (AHA). The construction of ontology is a complex, multidimensional process; we must combine bottom-up approaches (from recorded documents) with top-down processes (from intuitive hermeneutical horizon), also look from diverse perspectives to improve the final product. Nevertheless, the effort spent on ontology construction is profitable in terms of diverse possible applications and of a creative illumination and enlightenment.

References

1. Heidegger, M.: Sein und Zeit. Niemayer, Halle (1927)
2. McGuinness, D.L.: Ontologies Come of Age. In: Fensel, D., Hendler, J., Lieberman, H., Wahlster, W. (eds.) Spinning the Semantic Web: Bringing the World Wide Web to its Full Potential, pp. 171–192. MIT Press, Cambridge, MA (2002)
3. Pinto, H.S., Martins, J.P.: Ontologies: How can They Built? Knowledge and Information Systems 6, 441–464 (2004)
4. Bontas, E.P., Tempich, C.: Ontology Engineering: A Reality Check. In: Meersman, R., Tari, Z. (eds.) OTM 2006. LNCS, vol. 4275, pp. 836–854. Springer, Heidelberg (2006)
5. Corcho, O., Fernández-López, M., Gómez-Pérez, A.: Methodologies, Tools and Languages for Building Ontologies. Where is their meeting Point? Data & Knowledge Engineering 46, 41–64 (2003)
6. Dieng, R., Corby, O.: Knowledge Engineering and Knowledge Management: Methods, Modesl and Tools. Springer, Heidelberg (2000)

7. Mizoguchi, R., Kozaki, K., Sano, T., Kitamura, Y.: Construction and Deployment of a Plant Ontology. In: Dieng, R., Corby, O. (eds.) EKAW 2000. LNCS (LNAI), vol. 1937, pp. 113–128. Springer, Heidelberg (2000)
8. Heyer, L.J., Kruglyak, S., Yooseph, S.: Exploring Expression Data: Identification and Analysis of Coexpressed Genes. Genome Res. 9, 1106–1115 (1999)
9. Wierzbicki A.P., Nakamori Y.: Knowledge Sciences - Some New Developments. Zeitschrift für Betriebswirtschaft (in print)
10. Luhn, H.P.: The Automatic Creation of Literature Abstracts. I.B.M. Journal of Research and Development 2(2), 159–165 (1958)
11. Wiig, K.M.: Knowledge management: an introduction and perspective. Journal of Knowledge Management 1(1), 145–156 (1997)
12. Davenport, T., Prusak, L.: Working Knowledge: How Organizations Manage What They Know. Harvard Business School Press, Boston Ma (1998)
13. Salo, A., Cuhls, K.: Technology Foresight - Past and Future. Journal of Forecasting 22(2-3), 79–82 (2003)
14. Willyard, C.H., McClees, C.W: Motorola's Technology Roadmap Process. Research Management 30(5), 13–19 (1987)
15. Phaal, R., Farrukh, C., Probert, D.: Technology Roadmappingg - a Planning Framework for Evolution and Revolution. Technological Forecasting and Social Change. 71, 5–26 (2004)
16. Ma, T., Liu, S., Nakamori, Y.: Roadmapping for Supporting Scientific Research. In: 17th International Conference on Multiple Criteria Decision Making, Whistler, Canada (2004)
17. Tian, J., Nakamori, Y.: Knowledge Management in Scientific Laboratories: A Survey-based Study of a Research Institute. In: The Second International Symposium on Knowledge Management for Strategic Creation of Technology, Koba, Japan, pp. 19–26 (2005)
18. Tian, J., Wierzbicki, A.P., Ren, H., Nakamori, Y.: A Study of Knowledge Creation Support in a Japanese Research Institute. International Journal of Knowledge and System Science 3(1), 7–17 (2006)
19. Wierzbicki, A.P., Nakamori, Y.: Creative Space: Models of Creative Processes for the Knowledge Civilization Age. Springer, Heidelberg (2006)
20. Ren, H., Wierzbicki, A.W: Implementing Creative Environments for Scientifc Research. Journal of Systems Science and Systems Engineering (to be published)
21. Ren, H., Tian, J., Nakamori, Y., Wierzbicki, A.P: Electronic Support for Knowledge Creation in a Research Institute. Journal of Systems Science and Systems Engineering (to be published)
22. Totok, H.W.: Towards Knowledge Mapping of Advanced Education in Knowledge Creation. Presentation in group seminar at COE Center of JAIST (2006)
23. Totok, H.W., Nie, K., Ji, Z.: An Implementation of Knowledge Maps for Sharing Explicit Knowledge. Presentation in group seminar at COE Center of JAIST (2006)
24. Saito, A., Umemoto, K., Ikeda, M.: A Strategy-based Ontology of Knowledge Management Technologies. Journal of Knowledge Management 11(1), 97–114 (2007)
25. Kun, N.: Towards a Domain Analysis of Knowledge Management as an Organizational Activity. Presentation in group seminar at COE Center of JAIST (2007)
26. Bjørner, D.: Software Engineering?: Domains, Requirements, and Software Design. Springer, Heidelberg (2006)

An Ontology of Problem Frames for Guiding Problem Frame Specification

Xiaohong Chen[1,3], Zhi Jin[1,2], and Lijun Yi[1,3]

[1] Academy of Mathematics and System Science, CAS, Beijing, China
[2] Institute of Computing Technology, CAS, Beijing, China
[3] Graduate University of Chinese Academy of Sciences, Beijing, China
{chenxh,zhijin,yilijun}@amss.ac.cn

Abstract. Problem Frames approach is a new and prospective tool for classifying, analyzing and structuring software development problems. However, it has not yet been widely used mainly because lacking of CASE tools for guiding the problem frame specification development. This paper proposes an ontology based solution for this kind of CASE tools. An ontology of Problem Frames approach has been developed for this purpose. It specifies the basic terms elicited from Problem Frames approach and gives a concept model of this approach. This ontology can serve as the guidance of specifying the application problems. A case study has been given for illustration.

Keywords: Ontology, Problem Frames Approach, Problem Frame Specification, Requirement Engineering.

1 Introduction

Requirement engineering is receiving increasing attention as it is generally acknowledged that the early stages of the system development life cycle are crucial to the successful development and subsequent deployment and ongoing evolution of the system. At present the prevalent approaches of RE include KAOS [1], the i* framework [2], the scenario-based approach [3], etc. KAOS is a goal-oriented approach. It starts with high-level, composite goals of the stakeholders, and then refines those goals into hierarchical goal structures. The i* framework models social dependency relationships that the future system actors have to take over and handle. Such social analysis helps developer form a clearer understanding to one of the fundamental aspects of the design problem-the social and organizational aspect. The scenario-based approach tries to use scenarios that the domain users are familiar with to guide them to input system information step by step, thereby helping to identify requirements. It collects a set of running scenarios of real system as the guidance for eliciting the goals, and builds the requirement goal tree of application software.

Instead of modeling requirements by using those subjective concepts such as goals, dependency, ect., Problem Frames approach [4] follows another way. It is a problem domain oriented analysis approach which addresses the real world

Z. Zhang and J. Siekmann (Eds.): KSEM 2007, LNAI 4798, pp. 384–395, 2007.

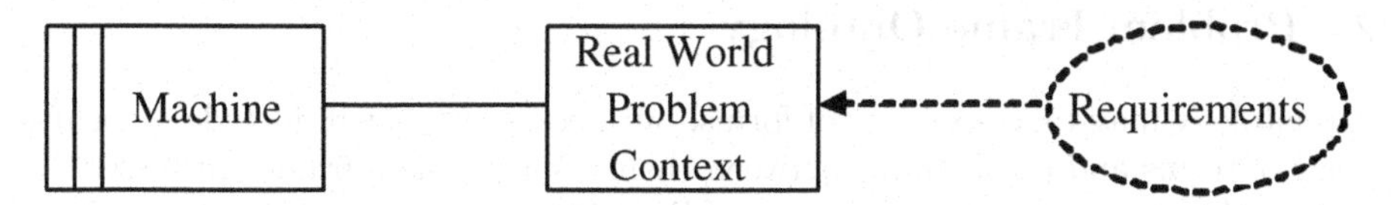

Fig. 1. Elements of of Problem Frames Approach

problems, and provides a means of analyzing and structuring problems. The starting point of analysis is a problem which has to be solved by software development. The approach makes the most of recurring class of problems which is defined to be a problem frame in software development. Some diagrams, such as the context diagram, the problem diagram and the problem frame diagram etc., have been used to serve as the regulator of requirements elicitation and denote the intermediate results in requirements specification development. Figure 1 illustrates some essential elements of the Problem Frames approach. Problem Frames approach assumes the software development as the action of building a machine(Machine) to solve a problem in an identified part of the world(Real World Problem Context) to meet a customer's need(Requirements). To begin with, the context diagram structures the world into problem domains and machine domain. Then by adding the requirements to extend the context diagram, the problem diagram is obtained. After that, Problem Frames approach goes on to find the solution of the problem by decomposing it into sub-problems based on predefined heuristic rules, or matching it with those existing problem frames. Finally when each of the sub-problems can fit a problem frame whose solution is known, the original problem gets its solution.

Compared with the approaches mentioned above, Problem Frames approach has the following advantages. Firstly, it helps users to focus on the problem, instead of drifting into inventing solutions. Secondly, it helps identify problem domain types. Thirdly, it provides rational principals for problem analysis, which is not appeared in other analysis approaches. Fourthly, it provides a specific guidance on how to deal with different types of problems. Finally, it admits the relations of problem domains that really constitute the requirements.

However, although many tools exist to support other approaches(UML tools are notable examples), no such tools exist for problem frames [6]. It has been argued in [7] that a tool for problem frames is 'urged and justified', and an essential ingredient in the uptake of the approach by practitioners. We propose that such a kind of tool must [5]: (a) contain knowledge about Problem Frames approach; (b) accommodate a certain degree of formalization; and (c) possess the ability of reasoning. Building an ontology[8] [9] of this approach is the first step for this kind of CASE tool.

This paper intends to define an ontology of Problem Frames approach for guiding the problem frame specification and present the potential ability of this ontology for guiding the requirements specification development. The structure of this paper is as follows. Section 2 provides problem frame ontology. Section 3 presents the guided process for developing the application specification by using this ontology with a small case study. Section 4 concludes the paper.

2 Problem Frame Ontology

An ontology may take a variety of forms, but necessarily it will include a vocabulary of terms and associations between them. Our problem frame ontology will include three part. They are concepts of Problem Frames approach, associations between the concepts and constraints on the concepts and the associations.

2.1 Concept Hierarchy

According to modeling process, we categorize the concepts in Problem Frames approach into five groups. The first group named as Problem contains all the concepts about the problem. The second group named as ProblemModel contains all the diagram models of Problem. The third group named as Domain contains all the concepts in the form of a vertex in the ProblemModel. The fourth group named as Interaction contains all the concepts in the form of a edge in the ProblemModel. The last group named as BasicTerm contains all the basic concepts in Problem Frames.

Figure 2 illustrates the partial hierarchy of the Problem Frame concepts. T is the universal type without particular meaning. The partial ordering, existing on these concepts, represents the "IsA" relationship between two concepts. The following will give the definitions of these concepts.

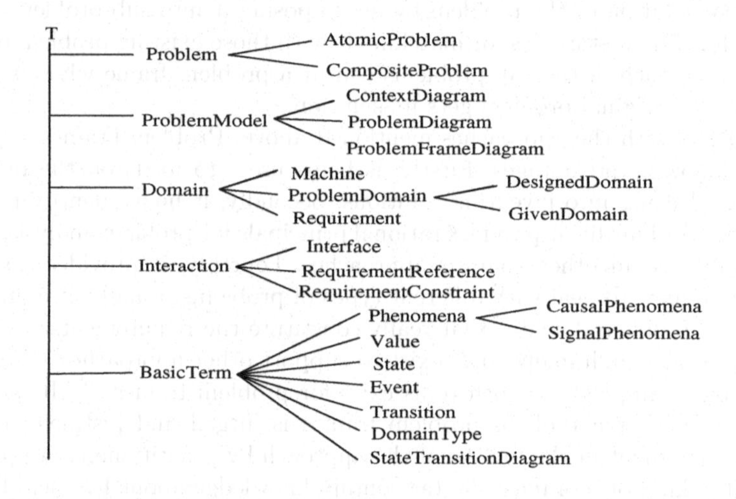

Fig. 2. A Concept Hierarchy of Problem Frames Approach

About the BasicTerm. BasicTerm concepts are the basic concepts of Problem frames approach including Phenomenon, Domain etc.

- A **Phenomenon**(phenomenon) is an element of what we can observe in the world. Here, we'll recognize two categories of phenomena:
  - **CausalPhenomena** are phenomena that can cause other phenomena and can be controlled.

  - **SymbolicPhenomena** are phenomena that are used to symbolize other phenomena and relationships between them.
- **Event** is an individual that is an occurrence at some point in time, regard as atomic and instantaneous. It is a kind of phenomenon.
- **State** is a relationship among two or more individuals that can be true at one time and false at another. It is a kind of phenomenon, also an element of a StateTransitionDiagram.
- **Value** is an individual that can not undergo change over time. It is a kind of phenomenon. The values we are interested in are such thing as numbers and characters, represented by symbols.
- **DomainType** is the type of domains. Domains have three types, which are Causal Domain, Anonymous Domain, and Lexical Domain.
  - A **Causal Domain** is one whose properties include predictable causal relationship among its causal phenomena.
  - An **Autonomous Domain** usually consists of people. The most important characteristic of an Anonymous Domain is that it's physical but lacks positive predictable internal causality.
  - A **Lexical Domain** is a physical representation of data-that is, of SymbolicPhenomena.
- **Transition** is a connection among two states consisting of events that they share.
- **StateTransitionDiagram** is a diagram showing behaviors in the form of a set of states with transitions between them.

About the Domain. Domain is a set of related phenomena that are usefully treated as a unit in problem analysis. Domain concepts are Machine, ProblemDomain and Requirement.

- **Machine** is the machine to be built by the developer in a software development problem, and in diagrams is represented by rectangle with a double vertical stripe.
- **ProblemDomain** is a domain in a problem other than the machine domain. ProblemDomains can be classified into:
  - **GivenDomain** is a domain that is given in a particular problem, and in diagrams generally represented by a rectangle.
  - **DesignedDomain** is the physical realization of a description or model that the developer is free to design, and in diagrams generally represented by a rectangle with a vertical stripe.
- **Requirement** is a condition on one or more domains of the problem context that the machine must bring about, and in diagrams generally represented by a dashed oval.

About the Interaction. Interaction concepts are Interface, RequirementReference and RequirementConstraint.

- **Interface** is a connection among two or more domains consisting of phenomena that they share, and in diagrams is represented by a solid line.

- **RequirementReference** is a reference by the requirement to some phenomena of a domain, and in diagrams is represented by a dashed line.
- **RequirementConstraint** is a requirement reference that constrains the domain to which it offers, and in diagrams is represented by a dashed arrow.

About the ProblemModel. ProblemModel concepts in Problem Frames are ContextDiagram, ProblemDiagram and ProblemFrameDiagram.

- **ContextDiagram** is a diagram showing the structure of the problem context in terms of domains and connections between them.
- **ProblemDiagram** is a diagram describing a particular problem. It shows the problem parts: the requirement, the domains, and the interfaces and references among them.
- **ProblemFrameDiagram** is a diagram describing a particular class of problem. It takes the form of a generic problem diagram, with some special naming and annotation conventions.

About the Problem. Finally, we come to the starting point of the approach-problem. Here we specify two different kinds of problems.

- **AtomicProblem** is a problem fitting a problem frame in its simplest form, with the smallest possible number of domains, that is to say, the solution is known.
- **CompositeProblem** is a problem that can not be matched with a problem frame and need to be decomposed into sub-problems to find its solution.

2.2 Associations

Besides classifying things, an ontology provides the associations between the ontological categories. These associations form a general conceptualization of

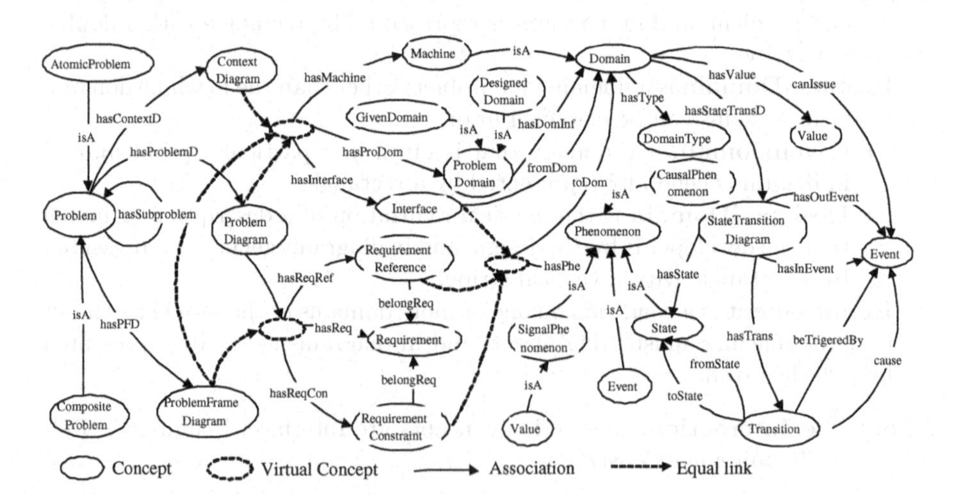

Fig. 3. Concept Model of Problem Frames Ontology. The equal link links a concept to a virtual concept. The concept has all the associations of corresconding virtual concept.

Table 1. Associations of Problem Frames ontology, where PFDiagram is short for ProblemFrameDiagram, STD for StateTransitionDiagram, ReqRef for RequirementReference, and ReqCon for RequirementConstraint

Association	formation	description
hasSubproblem	Problem → Problem	a set of subProblems of Problem
hasContextD	Problem → ContextDiagram	ContextDiagram of Problem
hasProblemD	Problem → ProblemDiagram	a set of ProblemDiagrams of Problem
hasPFD	Problem → PFDiagram	a set of PFDiagrams of Problem
hasMachine	ContextDiagram → Machine	ContextDiagram, ProblemDiagram
	ProblemDiagram → Machine	and PFDiagram have
	PFDiagram → Machine	its corresponding Machine
hasProDom	ContextDiagram → ProblemDomain	ContextDiagram, ProblemDiagram
	ProblemDiagram → ProlemDomain	and PFDiagram have
	PFDiagram → ProlemDomain	its corresponding ProblemDomain
hasReq	ProblemDiagram → Requirement	ProblemDiagram and PFDiagram
	PFDiagram → Requirement	have Requirement
hasInterface	ContextDiagram → Interface	ContextDiagram, ProblemDiagram
	ProblemDiagram → Interface	and PFDiagram have its
	PFDiagram → Interface	corresponding Interface set
hasReqRef	ProblemDiagram → ReqRef	ProblemDiagram and PFDiagram
	PFDiagram → ReqRef	have ReqRef
hasReqCon	ProblemDiagram → ReqCon	ProblemDiagram and PFDiagram
	PFDiagram → ReqCon	have ReqCon
belongReq	ReqRef→Requirement	a ReqRef or ReqCon
	ReqCon→Requirement	belongs to a Requirement
hasPhe	Interface→ Phenomenon	observed Phenomena from
	ReqRef→ Phenomenon	Interaction
	ReqCon → Phenomenon	
conPhe	Phenomenon → Domain	phenomena constitute Domain
hasValue	Domain → Value	Domain has value
canIssue	Domain → Event	Domain can issue Event
hasStateTransD	Domain → STD	Domain has STD
hasDomInf	ProblemDomain → Domain	ProblemDomain has Domain information
fromDom	Interface→ Domain	a interface controlled by a Domain
toDom	Interface→Domain	controlled Domain by Interface
	ReqRef→Domain	a domain connecting ReqRef
	ReqCon→Domain	a domain connecting ReqCon
fromState	Transtion → State	Transition's source state
toState	Transition → State	Transition's sink state
hasTrans	STD → Transition	Transition Set of a STD
beTriggeredBy	Transition → Event	Events triggered by Transition
hasInEvent	STD → Event	a set of event in STD
hasOutEvent	STD → Event	Events out of STD
cause	Transition → Event	a Transition cause a Event

Problem Frames approach. Some of this kind of associations are listed in table 1. For each formation of an association, the left side is called the source concept and the right side is called the sink concept. According to the five concept groups, associations can be classified into four groups. In table 1, the second row contains the associations between the Problem and ProblemModel concepts, the third row contains the associations between ProblemModel and Domain concepts, the fourth row contains the associations between Domain and Interaction concepts, and the fifth row contains the associations between Domain and BasicTerm concepts, Interaction and BasicTerm concepts, and BasicTerm itself. According to the concepts and associations above, the conceptual model of the Problem Frames approach can be formed by the ontological categories as shown in Figure 3.

2.3 Constraints

The last part of the ontology is a set of constraints which represent the semantic constraints among the concept categories and support reasoning. The general form of a rule definition is

$$\text{declaration} \bullet \text{predicate.}$$

In which, declaration represents the condition, and predicate represents the conclusion under the condition. These constraints are usually used in the consistency and completeness checking of the application problem statements. The following gives some examples of the constraints.

ProblemModel constraints:

- $\forall pm \in ProblemModel \bullet \exists! pm.hasMachine$
- $\forall pm \in ProblemModel \bullet pm.hasProDom \neq \emptyset$
- $\forall pm \in ProblemModel \bullet pm.hasInterface \neq \emptyset$
- $\forall pd \in ProblemDiagram \bullet pd.hasReqRef \cup pd.hasReqCon \neq \emptyset$
- $\forall pd \in ProblemDiagram \bullet pd.hasReq \neq \emptyset$

Interaction constraints

- $\forall in \in Interaction \bullet in.hasPhe \neq \emptyset$
- $\forall in \in Interaction \bullet in.toDom \neq \emptyset$
- $\forall rr \in RequirementReference \bullet rr.belongReq \neq \emptyset$
- $\forall rc \in RequirementConstraint \bullet rc.belongReq \neq \emptyset$
- $\forall i \in Interface \bullet i.fromDom \neq \emptyset$
- $\forall i \in Interface \bullet i.fromDom \neq i.toDom$

3 Application Specification Development

3.1 Process of Ontology-Guided Specification Development

This section presents the process of eliciting the application description. This process is guided by the problem frame ontology developed in the above section. In terms of the Problem Frames approach, the process mainly contains the following four steps. Because the requirements are mainly captured by a variety of diagrams, we first define some notations for these diagrams. '?' means that users input is requested. '$\uparrow$' means that the information may be inherited and may need some modifications. '$A \stackrel{X}{\Longrightarrow} B$' has different meanings according to different X in '$\Rightarrow$'. If 'X' is 'correspond', it means B is a corresponding to A. If 'X' is 'instance', it means B is an instance of A. If 'X' is 'add', it means B can be obtained by adding more information to A. If 'X' is 'partof', it means B is a part of A.

Step 1. Develop the ContextDiagram. Generate the context diagram template according to problem frame ontology as shown in the second part of figure 4. The constraints make sure that there is only one Machine in the ContextDiagram, and a interface must connect two different domains. The users are requested to follow the template to fulfill the following tasks: 1) identify the machine domain and the problem domains, 2) show how they are connected in interface, and 3) name the shared phenomena in each interface.

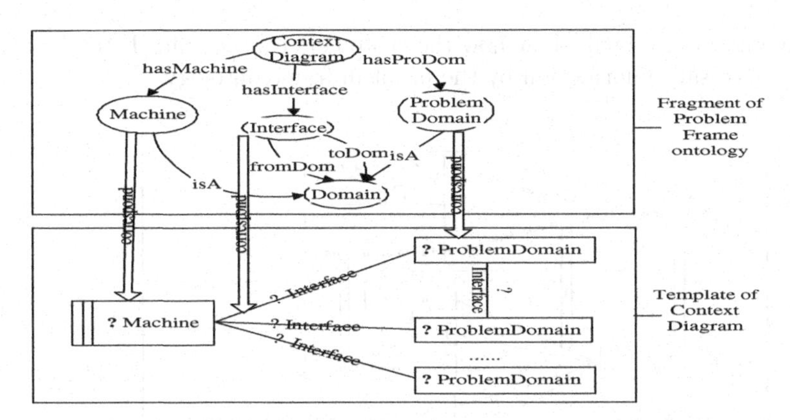

Fig. 4. Context Diagram Template Generation

Step 2. Develop the ProblemDiagram. On the basis of the context diagram obtained in step 1, we could get a problem diagram template of the application problem using the problem frame ontology. The users need to instantiate the Requirements, RequirementReferences and RequirementConstraints, and add more information to the interfaces (about which domain controls(!) the share phenomena in interfaces).

Step 3. Decompose the ComplexProblems and develop ContextDiagrams and ProblemDiagrams of sub-problems. Complex problem needs to be decomposed into several sub-problems. For each sub-problem, we need develop its context diagram as well as its problem diagram. Different from step 1, ContextDiagram template is generated by inheriting relative domains and Interfaces from context diagram from step 1. ProblemDiagram template is generated by inheriting the domains and interfaces from the sub-problem context diagram, and inheriting Requirements, RequirementReferences and RequirementConstraints from the ProblemDiagram obtained in step 2.

Step 4. Group the specifications of (sub)problems to compose the specification of original problem. If all the (sub)problems can be fit with basic problem frames or variants whose specifications are known, group these specifications together. Then, get the specification of the original problem.

3.2 Case Study

In this section, we use an example to illustrate the guided process of problem frame elicitation. This application case [10] is stated as follows:

> A society needs a telemarketing system for selling lottery tickets to its supporters. This society has different kind of lottery campaigns. The available campaigns may be updated by the campaign Designer. The telemarker communicate with the supporters via telephones. All of these connections are made by this system. If one supporter is willing to buy a special kind of lottery, the ticket placement should be recorded for producing a ticket order. The ticket orders will be sent to order processor.

In the following, we will show how the customers can be guided step by step to supply necessary information by the problem frame ontology.

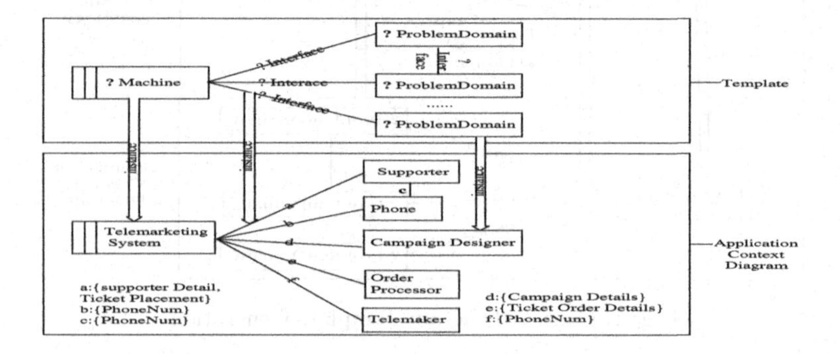

Fig. 5. ContextDiagram Template and Application Context Diagram

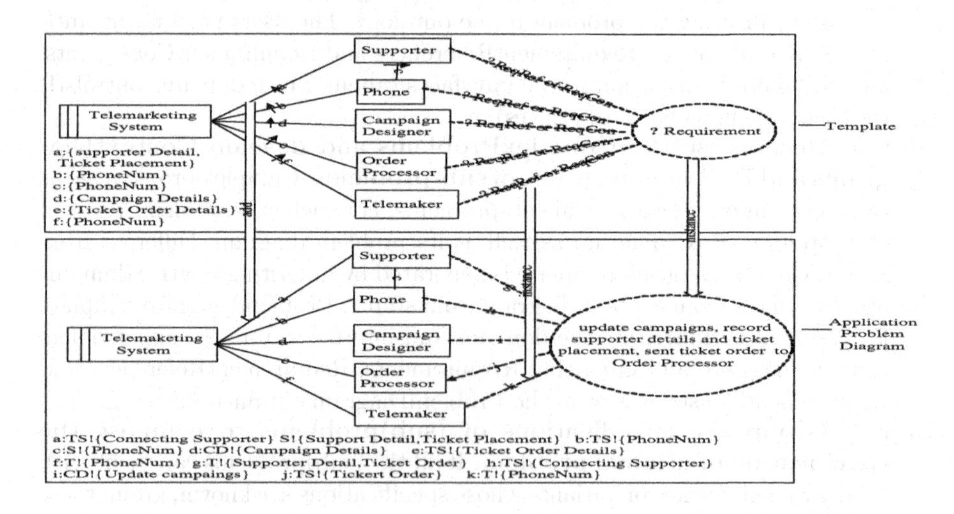

Fig. 6. Template of ProblemDiagram and Application Problem Diagram

Step 1. *Develop the ContextDiagram.* Follow the ContextDiagram template as shown in the first part of figure 5, then get a ContextDiagram of the application case as shown in the second part of figure 5.

Step 2. *Develop the ProblemDiagram.* According to the guidance of step 2, generate the template of the ProblemDiagram of the application case as shown in figure 6. Then follow the template,thus get a ProblemDiagram of the case in the second part of figure 6.

Step 3. *Decompose the ComplexProblems and develop ContextDiagrams and ProblemDiagrams of sub-problems.* The case is a complex problem. We decompose the problem in [10]. First, we get a sub-problem-Supporter Details Editor, then analyze it. Same with this, we go on to decompose the case, until every sub-problem can be fitted to a basic problem frame or a variant.

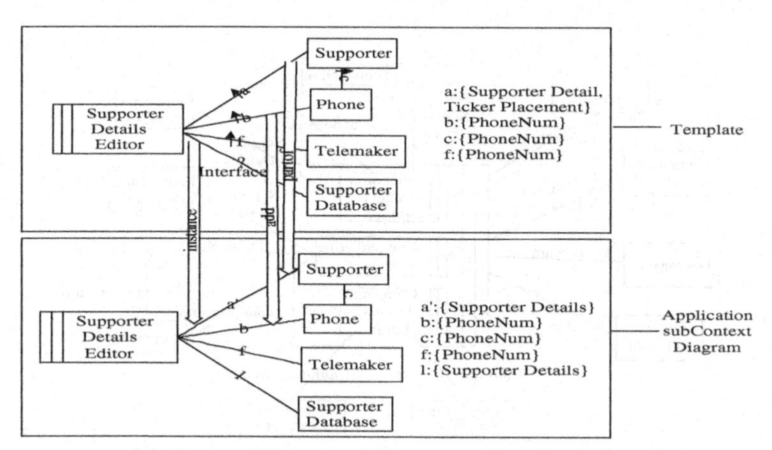

(a) ContextDiagram of Supporter Details Editor

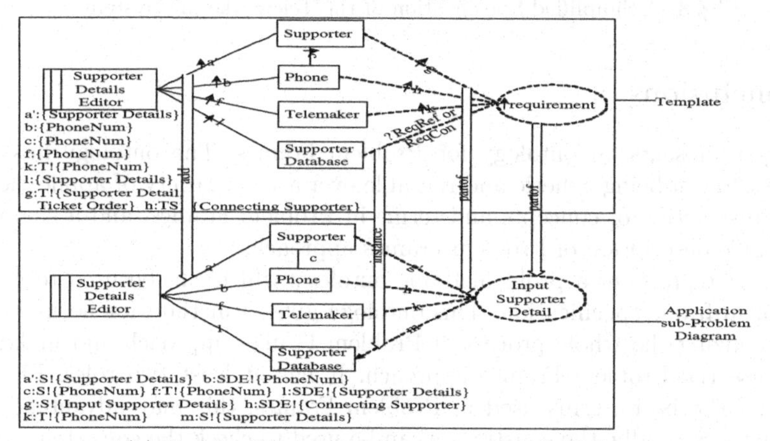

(b) ProblemDiagram of Supporter Details Editor

Fig. 7. ContextDiagram and ProblemDiagram of Supporter Details Editor

(a) *Develop its ContextDiagram.* According to the guidance of step 3, generate a template for this sub-problem as shown in the first part of figure 7(a). Follow the template. At last we get a ContextDiagram of the sub-problem as shown in the second part of figure 7(a).

(b) *Develop its ProblemDiagram.* According to the guidance of step 3, generate a template for this sub-problem as shown in the first part of figure 7(b). Follow the template. At last get a ProblemDiagram of the subproblem as shown in the second part of figure 7(b).

Step 4. *Group the (sub)problems to compose the specification of the case.* This case can be decomposed into 3 sub-problems(details see [10]), where Supporter Details Editor and Campaign Details Editor can be fitted to Commanded behavior frame [4] or frame variant, and Telemarketing Scheduler can be fitted to workpieces frame [4]. Group specifications of all these subproblems, we get a simplified specification in figure 8.

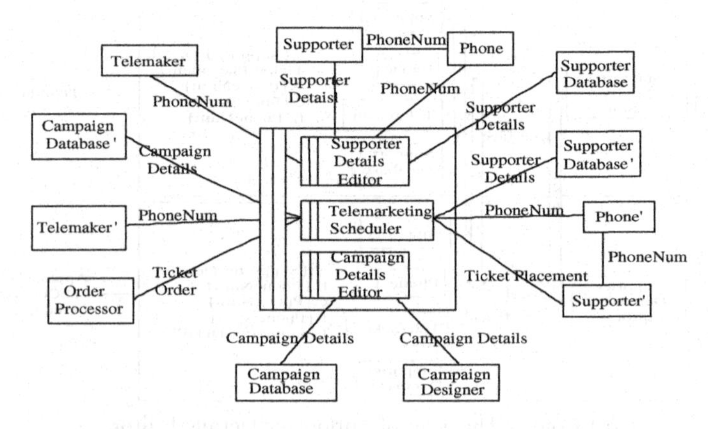

Fig. 8. A Simplified Specification of the Telemarketing System

4 Conclusions

This paper presents an ontology for Problem Frames. The ontology has the characteristics of being generic and reusable over a wide variety of applications. It is an abstraction of concepts and terms in Problem Frames, and a sharable and reusable description of Problem Frames approach.

This paper proposes a problem frame ontology which can be used to guide the problem frame specification. This ontology gives a method for normalizing and systemizing the whole process of Problem Frames approach and makes it easy to use the Problem Frames approach. It has at least two roles. Firstly, it provides the basic terms used in Problem Frames, this may advantage the applications. Secondly, the constraints can be used to check the consistency and completeness of application requirements, and support reasoning. This work lays the foundation for further research in CASE tool building for Problem Frames approach.

Acknowledgments. The authors wish to thank the anonymous reviewers for their constructive comments and suggestions. This work was supported by the National Natural Science Fund for Distinguished Young Scholars of China under Grant No.60625204, the Key Project of National Natural Science Foundation of China under Grant No. 60496324, the National 973 Fundamental Research and Development Program of China under Grant No. 2002CB312004, the National 863 Hight-tech Project of China under Grant No. 2006AA01Z155, the Knowledge Innovation Program of the Chinese Academy of Sciences and MADIS.

References

1. van Lamsweerde, A.: Goal-Oriented Requirements Engieering: A Guided Tour. In: RE2001. Proceedings of the 5th IEEE International Symposium on Requirements Engineering, Toronto, Canada, pp. 27–31 (August 2001)
2. Yu, E.: Towards Modeling and Reasoning Support for Early-Phase Requirements Engineering. In: RE 1997. Proceedings of the 3rd IEEE International Symposium on Requirements Engineering, Washington D.C., USA, Jan 6-8, 1997, pp. 226–235 (1997)
3. Sutcliffe, A.: Scenario-based Requirements Engineering. In: RE 2003. Proc. of 11th IEEE International Requirements Engineering Conference, pp. 320–329 (2003)
4. Jackson, M.: Problem Frames: Analyzing and Structuring software development problems. Addison-Wesley, Reading (2001)
5. Yi, L.: An Ontology-Based Approach for Problem Frame Oriented Requirements Modeling, PhD Thesis, Academy of Mathematics and System Science (2007)
6. Cox, K., Hall, J.G., Rapanotti, L.: Editorial: A roadmap of Problem Frames research, Information and Software Technology (2005)
7. Ourusoff, N.: Towards a CASE tool for Jackson's JSP, JSD and Problem Frames. In: Cox, K., Hall, J.G., Rapanotti, L. (eds.) Proceedings of the 1st International Workshop on Applications and Advances of Problem Frames, IEE, Edinburgh, pp. 69–73 (May 24, 2004)
8. Borst, W.: Construction of Engineering Ontologies for Knowledge Sharing and Reuse, PhD Thesis, University of Twente, Enschede (1997)
9. Studer, R., Benjamins, V.R., Fensel, D.: Knowledge engineering principles and methods. Data and Knowledge Engineering 25(1-2), 161–197 (1998)
10. Jin, Z., Liu, L.: Towards Automatic Problem Decomposition: An Ontology-based Approach. In: Proceedings of the 2006 international workshop on Advances and applications of problem frames, pp. 41–48 (2006)

A Speaker Based Unsupervised Speech Segmentation Algorithm Used in Conversational Speech

Yanxiang Chen[1,2] and Qiong Wang[1]

[1] College of Computer Science & Information,
Hefei University of Technology, Hefei, Anhui 230009, China
[2] Department of Electrical Computer Engineering,
University of Illinois at Urbana-Champaign, Urbana, IL 61801, USA
chenyx@uiuc.edu

Abstract. Difference between acoustic characteristics of speakers can be applied to segment conversational speech. In this paper, an unsupervised speech segmentation algorithm is emphasized while Euclidean distance measure and the distance measure based on GLR (Generalized Likelihood Ratio) and duration model are compared. The latter measure makes use of the likelihood ratio to describe the similarity and text-independent two-speaker verification system shows it is effective in verifying segment points as the result of being sensitive to speaker changes.

Keywords: Conversational speech, GLR distance measure, unsupervised speech segmentation.

1 Introduction

The ability to segment conversational speech based on changes in speakers is useful [1] [2]. Speaker change makers in the audio stream provide the capability to skip to the next speaker when reviewing audio data, or to playback only those portions of the audio corresponding to a particular speaker. In this paper, we describe techniques for segmentation of speech based on the identity of speaker.

Speech segmentation algorithms can be mainly divided into two portions: one is supervised training when data is available for each speaker and therefore such model as GMM (Gaussian Mixture Model) can be made in advance by using the data [3]. However, speakers are not always the same, and it is not practical to assume that speech samples for each speaker are available beforehand for many tasks. As a consequence, no speaker models can be built in advance [4].Therefore; unsupervised approaches without prior speaker models are desirable in many applications. One method is based on Bayesian Information Criterion (BIC) [5], which utilizes a maximum likelihood criterion penalized by the model complexity. But it is difficult to apply to various data, since it is necessary to determine a penalty threshold in order to control the balance between the likelihood and model complexity.

To the problem, an unsupervised agglomerative clustering with no need of threshold is proposed. The distance measure for the clustering is likelihood ratio while merging

Z. Zhang and J. Siekmann (Eds.): KSEM 2007, LNAI 4798, pp. 396–402, 2007.

the two clusters with the minimum distance at each stage of clustering results in no need of threshold. A duration model is used to bias the likelihood ratio by taking advantage of the temporal information. The paper is organized as follows: Section 2 presents the framework for an unsupervised speech segmentation based on speakers. Then two different distance measures are described in section 3. In section 4, the experiments and results are reported. The conclusion and future work are in section 5.

2 Unsupervised Speech Segmentation Based on Speakers

The unsupervised segmentation is widely used because of no need for the training data. The motivation of the approach is to use hierarchical agglomerative clustering to segment the data into approximate speaker clusters.

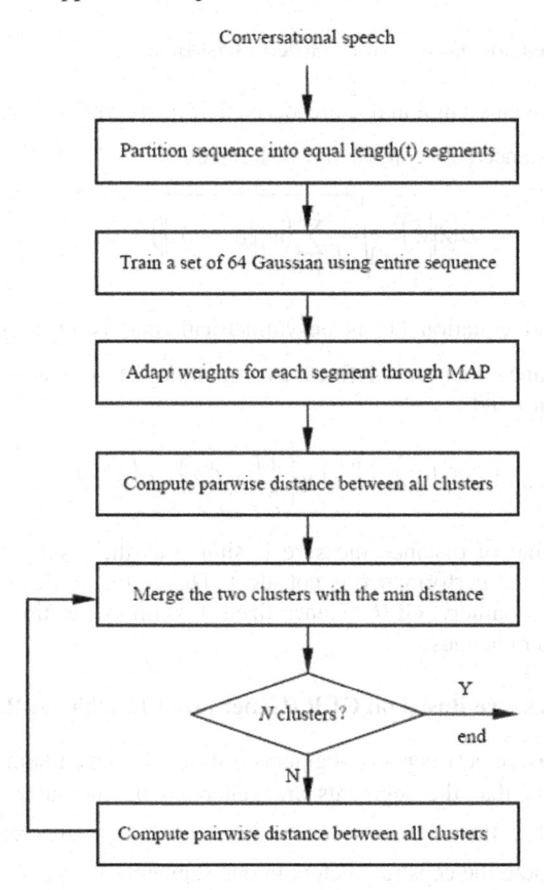

Fig. 1. Diagram of unsupervised speech segmentation

Figure 1 shows the diagram. The sequence is first partitioned into equal length segments. These segments form the initial set of clusters, each containing only one segment. We use a set of 64 Gaussians, which are trained by the entire sequence of feature vectors from the file being segmented. For each segment, adapt mixture weights from the common set of Gaussians through MAP (Maximum A Posterior).

Agglomerative clustering then proceeds by computing the pairwise distance between all clusters and merging the two clusters with the minimum distance. When two clusters are merged, new mixture weights using the union of segments in both clusters are estimated and distances to the remaining clusters are computed. This is repeated until the desired number of clusters is obtained.

3 Distance Measure

The key point in clustering on how to measure distance between clusters is formulated in detail.

3.1 Distance Measure Based on Euclidean Distance

Mixture of multi-Gaussian λ and $\tilde{\lambda}$ are derived from segments x and y respectively. The Euclidean distance between λ and $\tilde{\lambda}$ is defined as

$$D\left(\lambda \| \tilde{\lambda}\right)=\sqrt{\frac{1}{M}\sum_{j=1}^{M}\left(w_j\left\|\mu_j-\tilde{\mu}_j\right\|\right)} \tag{1}$$

The distance of equation (1) is unsymmetrical, that is, $D\left(\lambda\|\tilde{\lambda}\right)\neq D\left(\tilde{\lambda}\|\lambda\right)$. So a symmetrical distance measure defined as equation (2) is given to represent the similarity between x and y ,

$$d(x,y)=\log\left(d\left(\lambda,\tilde{\lambda}\right)\right)=\log\left(\frac{1}{2}\left(D\left(\lambda\|\tilde{\lambda}\right)+D\left(\tilde{\lambda}\|\lambda\right)\right)\right) \tag{2}$$

Though this kind of distance measure is simple as the result of non-probability distance measure, the performance is not ideal. Therefore, we describe a probability distance measure, namely GLR (Generalized Likelihood Ratio), which is more sensitive to speaker changes.

3.2 Distance Measure Based on GLR (Generalized Likelihood Ratio)

This distance measure between two segments is derived from a likelihood ratio test for the hypothesis H_0 that the segments are generated by the same speaker and the hypothesis H_1 that the segments are generated by two different speakers. Let $x=\{v_1,\cdots,v_r\}$ denote the cepstral vectors in one segment, $y=\{v_{r+1},\cdots,v_n\}$ denote the vectors in another segment, and $z=\{v_1,\cdots,v_n\}$ denote the combined collection of vectors. Furthermore, x and y vectors are assumed to be independently distributed, and are not necessarily time adjacent. Let $L(x:\theta_x)$ be the likelihood of the x segment, and

θ_χ denotes the parameters based on maximum likelihood estimation for samples in the x segment. Rather than computing the likelihood of a segment of speech assuming a single Gaussian, the likelihood is based on a mixture of M-Gaussian. Therefore,

$$L(x:\theta_x)=\prod_{j=1}^{r}\sum_{k=1}^{M}w_k(x)N_k(v_j) \tag{3}$$

$L(y:\theta_y)$ is similarly defined. Let $L(z:\theta_z)$ be the likelihood of the merged segments of x and y. In fact, we estimate the means and covariance matrices for components of the Gaussian mixture by using the entire utterance, and then they are fixed. Mixture weights for each segment, the only free parameters to be estimated, are adapted through MAP.

Let $\Pr(H_1)=L(x:\theta_x)L(y:\theta_y)$ be the likelihood that the segments are generated by different speakers. Let $\Pr(H_0)=L(z:\theta_z)$ be the likelihood that the two segments are generated by the same speaker. Then GLR (Generalized Likelihood Ratio) is defined as

$$\lambda_L=\frac{\Pr(H_0)}{\Pr(H_1)}=\frac{L(z:\theta_z)}{L(x:\theta_x)L(y:\theta_y)} \tag{4}$$

The distance measure used in the hierarchical clustering is

$$d(x,y)=-\log\lambda_L \tag{5}$$

3.3 Bias by Duration Model

In the experiments described below, the median duration for a given talker is about 18 seconds. Therefore, temporally close segments are more likely to be from the same speaker than from different speakers. In order to take advantage of this information at the level of hierarchical clustering, the likelihood ratio is biased using a duration model based on speaker changes over the original equal length segments.

Define a two-state Markov chain, where state 1 of the chain represents speaker a, and state 2 represents speaker b. Then 1-step transition probabilities for the same speaker and for the speaker change are

$$\Pr[S_{i+1}=a\mid S_i=a]=p \qquad \Pr[S_{i+1}=b\mid S_i=a]=1-p \tag{6}$$

The n-step transition probability that the speaker for segment i is also speaking for segment $i+n$ is

$$f(n)=\Pr[S_{i+n}=S_i]=\frac{1+(2p-1)^n}{2} \tag{7}$$

Using this equation we can compute the prior probabilities that two given clusters X and Y are produced either by the same speaker or by two different speakers. Let Z be the cluster formed by merging X and Y. There will be $Z=z_1,\cdots,z_{m+n}$, so that

$z_j \in X$ and $z_{j+1} \in Y$ or vice versa correspond to intervals in which the beginning and ending speakers are different. Let C be the number of such intervals in Z and then $(n_i + 1)$ be the difference between time indices of the first and the last segments of the i interval. A duration bias is then defined as

$$\lambda_D = \frac{\Pr[H_0]}{\Pr[H_1]} = \frac{\prod_i^C f(n_i)}{\prod_i^C (1 - f(n_i))} \tag{8}$$

The duration biased distance between two clusters is defined as

$$d(x, y) = -\log \lambda_L - \log \lambda_D . \tag{9}$$

4 Experiments

The data set used is subset of the two-speaker verification task in the NIST (National Institute of Standard and Technology) evaluation [4]. The verification task is, given an audio file containing conversational speech and given a hypothesized speaker, to determine if the hypothesized speaker is talking in the audio file. The training data contains 160 speakers (80 men and 80 women) and for each speaker there are three whole conversations. The testing data contains 1000 utterances, each of which is approximately 1 min in duration and also with two sides of the conversation summed together. The proposed unsupervised clustering is used to partition the speech file into speaker homogenous regions. After clustering, the multi-speaker speech has been turned into a set of single-speaker speech. Then as in the single-speaker case, each is scored. The maximum score is selected as the detection score for the entire utterance.

All speech files are processed into mel-frequency cepstrum coefficients (MFCC), which includes 23 cepstral coefficients. The frame size is 20 ms and the shift is 10 ms.

4.1 Comparison of Different Distance Measures

The performance based on Euclidean distance was compared with the performance based on GLR and duration model on the same task.

Table 1. EER(%) of different distsnce measures

Euclidean distance	GLR + duration
36.97	18.85

Table 1 shows the EER (Equal Error Rate) and Figure 2 is DET (Detection Error Tradeoff) curve. Since the smaller the EER is and the closer the curve is to the origin, the better the performance is, so unsupervised clustering based on GLR and duration model is better than based on Euclidean distance.

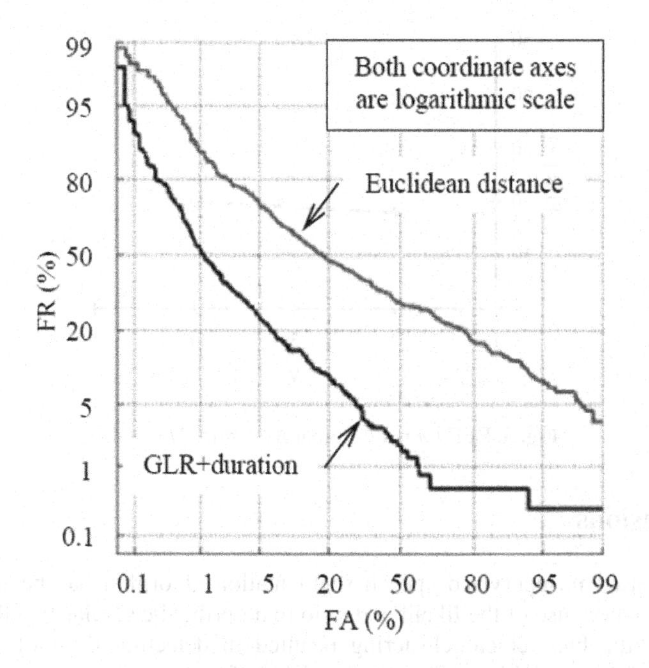

Fig. 2. DET of different distance measures

4.2 Influence of Different Parameters

Three important parameters that must be set for the segmentation are the initial segment size t, the mixture M and the number of clusters into which the segments are grouped N.

Table 2. EER(%) of different t and M ($N = 2$)

M \ $t(s)$	0.5	1	2
32	23.35	24.21	30.72
64	19.03	18.85	26.67

From Table 2, we find that the performance is not very sensitive to the initial segment sizes ranging from 0.5s to 1s and for 2s the performance is reduced because some of the initial segments contain more than one speaker. Mixture 32 is worse than mixture 64. From Figure 3, we find that the performance varies negligibly when varying the number of clusters from 2 to 4, and although for 5 clusters the performance is slightly reduced. The use of more than 2 clusters on two-speaker speech is based on helping with overlapped speech.

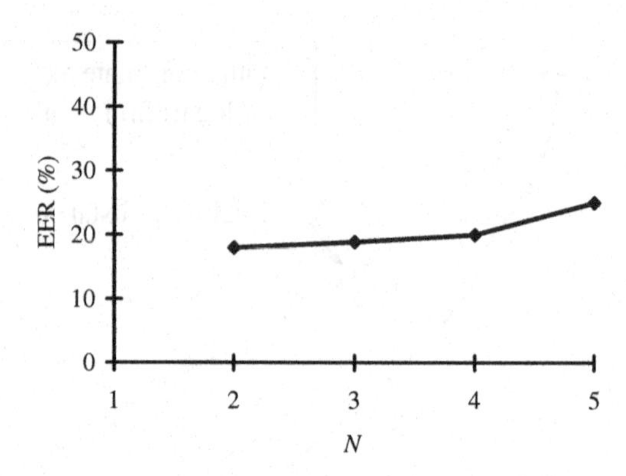

Fig. 3. EER (%) of different N ($t = 1s$ $M = 64$)

5 Conclusions

A speaker based unsupervised speech segmentation algorithm has been described. Because of making use of the likelihood ratio to describe the similarity, GLR distance measure for the hierarchical clustering resulted in detecting most of the speaker changes, as did the addition of a durational bias for the distance between clusters. Text-independent two-speaker verification system showed to have significant success in verifying segment points. Future work will focus on resolving short segments, people speaking simultaneously in conversation, because of the inherent instability of short analysis windows.

References

1. Gish, H., Siu, M-H., Rohlicek, R.: Segregation of Speakers for Speech Recognition and Speaker Identification. In: ICASSP. Proceeding of the International Conference on Acoustics, Speech and Signal Processing, Toronto, pp. 873–876 (1991)
2. Meignier, S., Bonastre, J.F., Chagnolleau, I.M.: Speaker Utterances Tying Among Speaker Segmented Audio Documents Using Hierarchical Classification: Towards Speaker Indexing of Audio Databases. In: ICSLP. Proceeding of the International Conference on Speech Language Processing, Denver, pp. 577–580 (2002)
3. Jin, H., Kubala, F., Schwartz, R.: Automatic Speaker Clustering. In: Jin, H., Kubala, F., Schwartz, R. (eds.) Proceeding of the DARPA Speech Recognition Workshop, Chantilly, pp. 108–111 (1997)
4. Reynolds, D.A., Singer, E.: Blind Clustering of Speech Utterances Based on Speaker and Language characteristics. In: ICSLP. Proceeding of the International Conference on Speech and Language Processing, Sydney, pp. 3193–3196 (1998)
5. Delacourt, P., Wellekens, C.: DISTBIC: A speaker-based segmentation for audio data indexing. Speech Communications 32, 111–126 (2000)
6. Bonastre, J.F., Delacourt, P., Fredouille, C.: A Speaker Tracking System Based on Speaker Turn Detection for NIST Evaluation. In: ICASSP. Proceeding of the International Conference on Acoustics, Speech and Signal Processing, Istanbul, pp. 1177–1180 (2000)

Distributed Knowledge Management Based on Ontological Engineering and Multi-Agent System Towards Semantic Interoperation

Yonggui Wang, Jiangning Wu, Zhongtuo Wang, and Yanzhong Dang

Institute of Systems Engineering, Dalian University of Technology, Linggong Road 2, Dalian 116024, Liaoning, P.R. China
ygwang@student.dlut.edu.cn, {jnwu, wangzt, yzhdang}@dlut.edu.cn

Abstract. Currently, the available architectures for knowledge management are mainly centralized and focus on basic string processing in essence. They tend to ignore that knowledge is distributed and full of semantics in and among complex knowledge-based organizations. Distributed knowledge management appears to solve the issues of distributed knowledge resources. Meanwhile, ontological engineering, which aims at describing the semantics for the knowledge or information, has been increasingly used in artificial intelligence and business in the past several years. Taking this into consideration, an architecture of distributed knowledge management, towards semantic interoperation (SiDKM) by introducing ontological engineering theory and multi-agent system (MAS) for distributed knowledge resources, is proposed in this paper. This architecture involves almost all the core modules in the knowledge management process, and can process large amounts of heterogeneous and distributed knowledge, which enables efficient knowledge sharing, reuse and evolution.

1 Introduction

The World Wide Web has dramatically increased the availability of electronically available information and the relevant numbers such as the number of documents are expected to grow exponentially. Besides, with the increasingly global competition and co-operation trends, the organization tends to be more dispersed, which, meanwhile, results in more different departments and more different types of staff. How to effectively manage the large number of information and provide corresponding personalization services for different types of staff are vital and knotty issues, especially with the circumstance that knowledge has been one of the key competitive factors for organizations. However, existing information or knowledge management systems are mainly central oriented, which can only be accessed by users in a uniform way. These inherently centralized approaches tend to ignore that the knowledge is usually distributed in and among complex knowledge-based organizations. Plus, the classical knowledge management projects mainly use keywords matching as a search method or managing approach which in essence belongs to the string processing level, that is, they can't embody the semantics of the knowledge.

Z. Zhang and J. Siekmann (Eds.): KSEM 2007, LNAI 4798, pp. 403–411, 2007.

Nowadays, ontologies' applications have been applied in many different areas such as e-commerce, knowledge engineering and artificial intelligence. Ontology has the strong potential to facilitate the knowledge creation of semantic relationships between various pieces of relevant and useful information to realize the information processing at the semantic level. Moreover, ontology is also effective in the context of integration and representation of various knowledge resources in organizations [1]. Besides, ontology is the backbone of semantic web which facilitates sharing and reuse of knowledge not only between software agents and computers but also between individuals [2].

Based upon this background, we introduce ontological engineering theory and multi-agent system (MAS) technology and propose an architecture of distributed knowledge management aiming at efficiently processing the large numbers of heterogeneous, distributed, and unstructured or semi-structured documents at the semantic level. This architecture enables the semantic interoperation among the different departments and different types of staff in the large distributed organizations and organization alliance.

The remainder of this paper starts with a clarification of ontological engineering theory, MAS and DKM in Section 2. The characteristics of DKM and the architecture of SiDKM by introducing ontological engineering theory and MAS are presented in Section 4. After a related case study is given in Section 4, Section 5 arrives at the conclusions and suggests some possible future works at last.

2 Backgrounds

2.1 Ontological Engineering

Ontological engineering refers to the set of activities that concern the ontology development process, the ontology life cycle, and the methodologies, tools and languages for building ontologies [3]. Its focus mainly includes three aspects: ontology/ontology building, ontology mapping and ontology evolution.

One popular ontology definition is given by Gruber [4], "An ontology is a formal and explicit specification of a shared conceptualization". Ontology can be regarded as a vocabulary of concepts and relationships between those concepts in a given domain. It can facilitate the acquisition and building of domain knowledge and enables semantic integration of heterogeneous and distributed knowledge.

Ontology mapping is defined as that given two semantically related ontologies O_1 and O_2, for each entity (concept, instance or relation) in the source ontology O_1, we try to find a corresponding entity (concept, instance or relation) in the target ontology O_2, which has the same intended meaning. It enables the existences of different approaches of knowledge representation and makes it possible for realizing personalization services for different types of users. There have been many such ontology mapping systems, such as GLUE [5], IF-Map [6], MAFRA [7], SF [8] and OMEN [9]. The specific ontology mapping approaches include language-based, structure-based, instance-based, semantic-based and reasoning-based.

Ontology evolution as one of the basic theories of ontology is described as the process of modifying an ontology in response to a certain change, such as adding a new concept, a revised attribute, a new instance etc. in a given domain. There have been some approaches or tools for ontology evolution [10-14]. In particular, ontology evolution is the renewal mechanism for ontology's change.

2.2 Multi-Agent Systems

Multi-Agent system (MAS) is an ad hoc topic in the domain of distributed artificial intelligence. Its purpose is to enable more than one agent to interrelate and interact for making one common task finish. The communication between agents is considered as a sequence of communication and computation steps [15, 16]. The referred communication capabilities naturally allude to twofold, namely, the process of the mediator agents generating and sending messages to the local agents and the other mediator agents as well as the process of receiving and decoding messages from the local agents and the other mediator agents. Generally, the agents used in distributed artificial intelligence appreciate the following several properties [17, 18]:

- **Autonomy.** Each agent works by itself, in other words, it doesn't need the direct intervention of humans.
- **Reactivity.** Each agent can perceive and respond to the relevant changes timely.
- **Pro-activity.** Agents are able to exhibit goal-directed behaviors by taking the initiative, but not simply act in response to their environments.
- **Co-operability.** Each agent can communicate with other relevant agents in order to attain common objectives.
- **Mobility.** Agents are able to travel through computer networks. An agent in one host may create another agent or transport from host to host during execution.

2.3 Distributed Knowledge Management

Traditionally, the knowledge management projects are centralized knowledge management, which can only be accessed by users in a uniform way. These inherently centralized approaches tend to ignore that the knowledge is usually distributed in and among complex knowledge-based organizations. Peter F. Drucker predicated in 1998 that the pervasive adoption of information technologies in every day business correlated with changes in demographics and economics, will induce a major reorganization of companies [19]. In order to cope with the distributed knowledge resources, distributed knowledge management system (DKMS) is presented to try to deal with these issues. It was introduced by Bonifacio, Bouquet and Cuel [20], in which the knowledge node (KN) is an abstraction of formal (e.g. division) or informal (e.g. community of practice) organizational units which are parts of social networks. The knowledge flow (KF) takes place between these nodes.

Schmücker & Müller released their experiences in the application of DKMS [21]. Compared with the centralized solutions in terms of knowledge sharing, they generalized some main advantages. DKMS ensures instant access to up-to-date knowledge. Knowledge providers can follow a self-determined organization of knowledge, and users have the direct access to other persons' knowledge through direct communication. Besides, the DKMS supports indirectly the socialization knowledge transfer process. Also, DKMS solution is less costly than those centralized counterparts, as they require less infrastructure investment and do not need large maintenance costs. However, they also pointed out a number of disadvantages of

DKMS. Searching for needed information becomes more difficult as there is no shared structure to organize it. Moreover, the same document can be found in different places, and it can be available in many versions, which implies that there is no guarantee for the quality of the found information.

3 The SiDKM Architecture

Besides taking the aforementioned information about knowledge management, for better designing the architecture of DKMS towards semantic interoperation, we still need to analyze the basic characteristics of DKMS. After the investigation, we find that the DKMS is prone to including the following several basic characteristics:

- DKMS is a collaboration group of inter-linked and network-based knowledge management system. All sub-KMSs (Knowledge Management Systems) are dispersed and belong to different departments, communities, and cities.
- Each sub-KMS is an autonomous and self-governed entity with all the necessary functionalities for the corresponding community. On one hand, all sub-KMSs can work effectively in their own community without intervention of the other sub-KMSs. On the other hand, each sub-KMS is also active as a component in the relative community league under the control of network agreement and collaboration mechanism.
- Interoperation or collaboration is another characteristic of the DKMS. All sub-KMSs are subordinate to one collaboration league and abide by the common communication infrastructure and interaction middleware mechanism. With the support of inter-linked network environment, all sub-KMSs can realize the information transformation and sharing, which will make collaboration and application integration among relative communities come true.

Considering the characteristics of DKM and the existing knotty issues in traditional knowledge management, we propose an architecture of DKM towards semantic interoperation by introducing ontological engineering theory and MAS, which is short

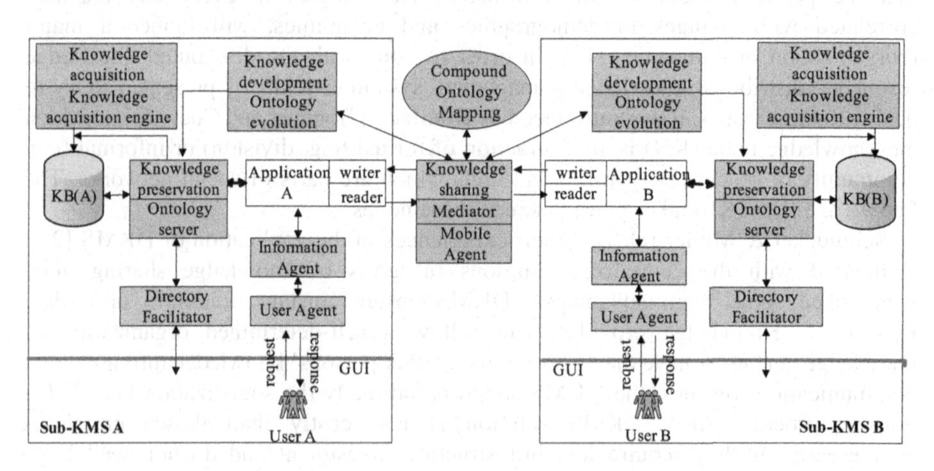

Fig. 1. The SiDKM architecture

for SiDKM. The architecture is shown in Fig.1. Probst et al. [22] identified that the knowledge kernel modules should include knowledge acquisition, knowledge preservation, knowledge development, knowledge use and knowledge sharing. Our architecture involves the entire above-mentioned modules.

3.1 Knowledge Acquisition

With the dramatically increased the availability of information for the organization, knowledge acquisition is an evitable process for maintaining the organization's knowledge base. This module is designed for knowledge seizing, especially for the large amount of electronic documents etc. It transforms all the unstructured or semi-structured electronic information into structured information under the supervision of domain ontology, which is located in ontology server. In particular, it contains three processes, namely externalization (conversion of tacit knowledge to explicit knowledge), internalization (conversion of explicit knowledge to tacit knowledge) and intermediation (conversion of explicit knowledge to explicit knowledge). The three processes all need the supervision of the domain ontology.

3.2 Knowledge Preservation

Ontology server is designed to organize the knowledge of organization at the conceptual level and enable the generation of the corresponding database. Besides, for user's each query, ontology server can be used to provide navigational view and integrate information view of knowledge items. It also enables the relevant modules to put knowledge in the right perspective. Moreover, directory facilitator, a service for providing yellow pages, is introduced here. It describes the services that the corresponding node provides.

3.3 Knowledge Sharing

Knowledge communicating and sharing is the most important module in our architecture. Most distributed tasks depend on it. It aims at realizing the communication and sharing at the semantic level between two sub-KMSs. Each sub-KMS has its own ontology and interface of application programming with readers and writers. Its workflow can be described as follows: When sub-KMS A needs to visit sub-KMS B, its request must be processed by the mediator. The mediator encapsulates a set of mapping rules between two sub-KMSs' ontologies. Then the transformed request is sent to sub-KMS B. After some operations on sub-KMS B, sub-KMS B generates the response represented by its style and returns the response to sub-KMS A. Again, the response also needs to be transformed by mediator into the style of sub-KMS A. Then two sub-KMSs finish one time knowledge communication.

Especially, we take the implementation of both approach and technology into consideration. We adopt the compound ontology mapping approach as the communication approach, which is proposed in our last paper [23]. It integrates the language-based, structure-based and instance-based semantic similarities from the dimensionality of commonsense semantic and domain semantic. Besides, mobile agent technology is adopted for the transportation between source sub-KMS and target sub-KMS. It takes the ontology mapping approach, the relevant data and state. It is especially effective in distributed knowledge retrieval system [24].

3.4 Knowledge Development

The organizations' knowledge is not like the dictionary with the vocabularies constant for a long time, but changes ceaselessly. The knowledge management process must seize the changing of organizations knowledge, such as the new concepts, attributes, or instances. So we introduce the module of ontology evolution to crack this hard nut. In our proposed architecture, the ontology evolution is based on the ontology mapping results and WordNet (http://www.cogsci.princeton.edu/~wn). For example, when we acquire a mapping concept pair, such as "Personal Computer" and "PC", we will change the concept node "Personal Computer" in the source ontology into a form of synonym set, that is, "(Personal Computer, PC)". This kind of format would be much more flexible for future applications.

3.5 Knowledge Retrieval

Users can access the knowledge through a GUI by submitting their requests to a search engine. The ontology in the source sub-KMS will provide navigation services and standardize the query requests. In fact, the knowledge retrieval here is a kind of distributed information retrieval. The retrieval flow can be described as follows:

Step 1: Users submit their request expression in their own words by GUI.

Step 2: The user agent and information agent transform the users' requests into the standard formats according to the local ontology in the source sub-KMS.

Step 3: Put the standard requests into the mobile agent. According to the metadata information provided by the directory facilitator, find all the relevant target sub-KMSs and establish PRI (priority) list for them.

Step 4: Mobile agent visits the target sub-KMS with the first priority. According to the mapping rule set by means of ontology mapping in advance, transform the requests into the standard formats of the target sub-KMS, then search the relevant information in terms of the ontology in the target sub-KMS and return the search results to the mobile agent.

Step 5: Mobile agent visits next target sub-KMS in the PRI list and meanwhile repeats the same work as Step 4 until the last target sub-KMS in the PRI list.

Step 6: Mobile agent returns to the source sub-KMS with all search results acquired from all relevant target sub-KMSs. Then the source sub-KMS outputs the search results to the users.

4 A Case Study

For better understanding the proposed SiDKM architecture, we describe a case study in this section and demonstrate how the SiDKM architecture works.

Case description

The staff of the production department in the Lenovo Group Ltd. of China needs to acquire the sales information of the retailers all over the world, in order to make the production plan for the next phase. How could they acquire the relevant information?

The solution provided by the SiDKM

It sets out the whole process of the solution with the start of distributed knowledge retrieval. The corresponding flow can be expatiated as follows:

Step 1: The staff of the production department in the *Lenovo* Group Ltd. of China inputs his query keywords "*The sales volume of PC*" to the search engine in the distributed knowledge management system.

Step 2: After submitting this request expression, the user agent and information agent get into work to preprocess the submitted request according to the local ontology located in the domain of knowledge management in China. They find that the standard description of this expression is "*PC_Sales_Volume*" in the local ontology. So the request expression "*The sales volume of PC*" is transformed into "*PC_Sales_Volume*". Meanwhile, acquire its hypernym concept "*PC_Sales_Department*" and its hyponym concepts "*Jiayue_Sales_Volume*, *Fengxing_Sales_Volume*, *Yangtian Sales_Volume*, *Tianjiao_Sales_Volume*" (*Jiayue*, *Fengxing*, *Yangtian* and *Tianjiao* are different types of personal computers in *Lenovo*). Then put the standard format "*PC_Sales_Volume*" into the local mobile agent.

Step 3: For the staff of the production department in the *Lenovo* Group Ltd. of China, the distribution of retailers are known in advance. So the PRI list has existed.

Step 4: Here just take the retailer of Hong Kong in the PRI list as an example. According to the mapping rule set generated by the compound ontology mapping approach, we find that the concept "*Personal_Computer_Sales_Quantity*" in the local ontology of Hong Kong retailer has the same intended meaning as the requested concept "*PC_Sales_Volume*". Then find the hypernym concept "*Personal_Computer_Sales_Department*" and the hyponym concepts "*Jiayue_Sales_Quantity*, *Fengxing_Sales_Quantity*, *Yangtian_Sales_Quantity*, *Tianjiao_Sales_Quantity*" of the concept "*Personal_Computer_Sales_Quantity*" in the retailer's local ontology of Hong Kong. Now search all the information about the aforementioned concepts in the retailer's local ontology of Hong Kong from the corresponding knowledge base. Then store the search results in the mobile agent. The visit of Hong Kong retailers is over now.

Step 5: The mobile agent turns around the next retailer until the last retailer in the PRI list.

Step 6: The mobile agent returns to the distributed knowledge management in China. Then it responds the staff with the sales information all over the world. Especially, the information is more specific than expected. In other words, the staff can not only gain the overall sales volume, but also the sales volume of each type of personal computers.

Besides, the mapping rule set generated in the query process is vital for knowledge development. Also, take the mapping rule set between the local ontology in China and that of Hong Kong as an example. The mapping rule set is shown as follows:

PC_Sales_Department **is equivalence to** *Personal_Computer_Sales_Department*
PC_Sales_Volume **is equivalence to** *Personal_Computer_Sales_Quantity*
Jiayue_Sales_Volume **is equivalence to** *Jiayue_Sales_Quantity*
Fengxing_Sales_Volume **is equivalence to** *Fengxing_Sales_Quantity*
Yangtian Sales_Volume **is equivalence to** *Yangtian Sales_Quantity*
Tianjiao_Sales_Volume **is equivalence to** *Tianjiao_Sales_Quantity*

According to above mapping rule set, the database administrater can reorganize the relevant concepts in the local ontology in forms of synonym set. For instance, the concept "*PC_Sales_Volume*" is reorganized as (*PC_Sales_Volume, Personal_Computer_Sales_Department*). This will be more convenient for other services in the future.

5 Conclusion and Future Work

In this paper, we present an architecture of DKM towards semantic interoperation (SiDKM) based upon ontological engineering theory and MAS technology. This architecture can facilitate knowledge sharing, reuse and evolution in the heterogeneous and distributed knowledge environments. Especially, ontological theory is introduced to realize the semantic interoperation among distributed organizations, while MAS technology is used as an intelligent communication tool in and among all sub-KMSs.

Implementation of a prototype based on SiDKM architecture is our main future work. Actually, we have implemented several modules, such as compound ontology mapping [23], semi-construction of Chinese ontology [25]. In general, our future work will focus on threefold: (1) Design and implementation of MAS; (2) Ontology evolution algorithm; (3) Integration of all algorithms and technologies.

Acknowledgements

This work is sponsored by the National Natural Science Foundation of China (NSFC) under Grant Nos. 70431001 and 70620140115.

References

1. Berners-Lee, T., Hendler, J., Lassila, O.: The Semantic Web. Scientific American 284(5), 34–43 (2001)
2. Fensel, D.: A Silver Bullet for Knowledge Management and Electronic Commerce. Springer, Heidelberg (2001)
3. Asunción, G.-P., Fernandez-Lopez, M., Corcho, O.: Ontological Engineering. Springer, Heidelberg (2004)
4. Gruber, T.: Towards Principles for the design of Ontologies used for Knowledge Sharing. International Journal of Human Computer Studies 43, 907–928 (1995)
5. Doan, A., Madhavan, J., Dongmingos, P., Halevy, A.: Learning to Map between Ontologies on the Semantic Web. In: Proceedings of the Eleventh International World Wide Web Conference, Honolulu, Hawaii, USA, pp. 662–673 (2002)
6. Kalfoglou, Y., Schorlemmer, M.: IF-Map: An Ontology Mapping Method Based on Information Flow Theory. Int. J. on Data Semantics 1, 98–127 (2003)
7. Macedche, A., Motik, B.: MAFRA-A Mapping Framework for Distributed Ontologies. Web Intelligence and Agent System 1, 235–248 (2003)
8. Melnik, S., Garcia-Molina, H., Rahm, E.: Similarity Flooding: A Versatile Graph Matching Algorithm. In: ICDE 2002. Proc. of the 18th International Conference on Data Engineering, San Jose.CA, pp. 117–128 (2002)

9. Mitra, P., Noy, N.F., Jaiswal, A.R., OMEN,: A Probabilistic Ontology Mapping Tool. In: McIlraith, S.A., Plexousakis, D., van Harmelen, F. (eds.) ISWC 2004. LNCS, vol. 3298, Springer, Heidelberg (2004)
10. Maedche, A., Volz, R.: The ontology extraction and maintenance framework text-to-onto. In: Proceedings of the ICDM 2001 Workshop on Integrating Data Mining and Knowledge Management (2001)
11. Klein, M., Fensel, D., Kiryakov, A., Ognyanov, D.: Ontology versioning and change detection on the web. In: Gómez-Pérez, A., Benjamins, V.R. (eds.) EKAW 2002. LNCS (LNAI), vol. 2473, pp. 1–4. Springer, Heidelberg (2002)
12. Gabel, T., Sure, Y., Voelker, J.: KAON- ontology management infrastructure. SEKT informal deliverable 3.1.1.a, Institute AIFB, University of Karlsruhe (2004)
13. Stojanovic, N., Hartmann, J., Gonzalez, J.: Ontomanager- a system for usage-based ontology management. In: Proceedings of FGML Workshop. Special Interest Group of German Information Society (FGML- Fach Gruppe Maschinelles Lernen der GI e.V.) (2003)
14. Cederqvist, P., et al.: The CVS manual- Version Management with CVS. Network Theory Limited (2003)
15. Luck, M.: From definition to deployment: what next for agent-based system? The Knowledge Engineering Review 14(2) (1999)
16. Barbuceanu, M., Fox, M.S.: Coordinating multiple agents in the supply chain. In: Proceedings of the 5th Workshop on Enabling Technologies, pp. 134–141 (June 19-21, 1996)
17. Wang, H.: Intelligent agent assisted decision support systems: integration of knowledge discovery, knowledge analysis, and group decision support. Expert Systems with Applications 12(3), 323–335 (1997)
18. Wooldridge, J.N.: Intelligent agents: theory and practice. The Knowledge Engineering Review 10(2), 115–152 (1995)
19. Drucker, P.: The coming of the new organization. Harvard Business Review 66(1), 45–53 (1988)
20. Bonifacio, M., Bouquet, P., Cuel, R.: Knowledge-Nodes: the Building Blocks of a Distributed Approach to Knowledge Management. In: Procedings of I-KNOW 2002, Graz, pp. 191–200 (2002)
21. Schmücker, J., Müller, W.: Praxiserfahrungen bei der Einfürung dezentraler Wissensmanagement Losungen. Wirtschaftsinformatik 45(3), 307–311 (2003)
22. Probst, G., Raub, S., Romhardt, K.: Wissen managen. Gabler Verlag, Wierbaden (1998)
23. Jiangning, W., Yonggui, W.: IAOM: An Integrated Automatic Ontology Mapping Approach towards Knowledge Integration. In: LSMS 2007. The 2007 International Conference on Life System Modeling and Simulation, Shanghai. LNCS, Springer, Heidelberg (2007)
24. Nottelmann, H., Fuhr, N.: Evaluating Different Methods of Estimating Retrieval Quality for Resource Selection. In: Proceedings of the 26th Annual International ACM SIGIR Conference on Research and Development in Information Retrieval, Toronto, Canada (2003)
25. Zhang, X., Yanzhong, D.: A Methodology for Automatic Concept Acquisition in Domain Ontology Construction. In: Submitted to the 3rd Asia-Pacific International Conference on Knowledge Management (2006)

WTPMiner: Efficient Mining of Weighted Frequent Patterns Based on Graph Traversals*

Runian Geng[1,2], Wenbo Xu[1], and Xiangjun Dong[2]

[1] School of Information Technology, Jiangnan University, Wuxi, Jiangsu 214122, China
gengrnn@163.com
[2] School of Information Science and Technology, Shandong Institute of Light Industry, Jinan, Shandong 250358, China

Abstract. Data mining for traversal patterns has been found useful in several applications. Traditional model of traversal patterns mining only considered un-weighted traversals. In this paper, a transformable model of *EWDG*(*E*dge–*W*eighted *D*irected *G*raph) and *VWDG*(*V*ertex–*W*eighted *D*irected *G*raph) is proposed. Based on the model and the notion of support bound, a new algorithm, called *WTPMiner*(*W*eighted *T*raversal *P*atterns *Miner*), and two methods for the estimations of algorithm are developed to discover weighted frequent patterns from the traversals on graph.Experimental results show the effect of different estimation method of support bound.

Keywords: Data mining, WDG, traversal patterns, frequent patterns.

1 Introduction

Data mining on graph traversals have been an active research during recent years. Graph and traversal on it are widely used to model several classes of data in real world, especially in a distributed information providing environment where objects are linked together to facilitate interactive access.Example for environments include *WWW* and on–line services,where users, when seeking for information of interest, travel from one object to another via corresponding facilities (i.e., hyperlinks) provided. Clearly, understanding user access patterns in such environments will not only help improve the system design (e.g., provide efficient access between highly correlated objects, better authororing design for pages, etc.), but also be able to lead to better marketing decisions (e.g., putting advertisements in proper places,better customer/user classification and behavior analysis, etc.). Capturing user access patterns in such environments is referred to as *mining traversal patterns* [1].

The structure of Web site can be modeled as a graph in which the vertices represent Web pages, and the edges represent hyperlinks between the pages.

* This work was supported by the Excellent Young Scientist Foundation of Shandong Province, China (No. 2006BS01017) and the Scientific Research Development Project of Shandong Provincial Education Department, China (No. J06N06).

Z. Zhang and J. Siekmann (Eds.): KSEM 2007, LNAI 4798, pp. 412–424, 2007.

Furthermore, user navigations on the Web site can be modeled as traversals on the graph. Each traversal can be represented as a sequence of vertices, or equivalently a sequence of edges. Once a graph and its traversals are given, valuable information can be discovered. Most common form of the information may be frequent patterns, i.e., the sub-traversals that are contained in a large ratio of traversals. In current information world, capturing this information has more important actual significance. However, traditional model of traversal patterns mining hardly considered weighted traversals on the graph [1][2].

This paper extends previous works by generalizing this to consider attaching weights to the traversals. Such traversals weight may reflect some importance. For example, each Web page may have different importance which reflects the value of its content. Each edge, which represents a transition between Web pages, can be assigned with a weight standing for the user stay time. With the weight setting, the mining algorithm *can not* be relied on the well-known Apriori paradigm any more. Therefore, we adopt the notion of *support bound* [4]. On top of the notion, we propose a mining algorithm called *WTPMiner* for the discovery of weighed frequent patterns from traversals on graph.

The rest of this paper is organized as follows. Section 2 reviews previous works related with the traversal pattern mining and weighted mining. The related definitions of problem are given in Section 3. Section 4 proposes an algorithm called *WTPMiner*. Section 5 describes two methods for the estimation of weight and *support bound* used in this paper. Empirical research and the analyses of algorithm are reported in Section 6. Finally, Section 7 gives the conclusion as well as future research works.

2 Related Works

The main stream of data mining, which is related to our work, can be divided into two categories, i.e. the traversal pattern mining and the weighted mining. For the traversal pattern mining, there have been few works. Chen et al. [1] proposed two algorithms – FS and SS of the problem about traversal pattern mining. However, they did not consider graph structure, on which the traversals occur. Nanopoulos et al. [2] proposed three algorithms with a trie structure of the problem of mining patterns from graph traversals, one which is level-wise w.r.t. the lengths of the pattern and two which are not. They considered the graph, on which traversals occur. The above works dealt with the mining of un-weighted traversal patterns.

For the weighted mining, most of previous works are related to the mining of association rules and its sub-problem, the discovery of frequent itemsets. Cai et al. [3] generalized the discovery of frequent itemsets to the case where each item is given an associated weight. Wang et al. [4] extended the problem by allowing weights to be associated with items in each transaction. Their approach ignores the weights when finding frequent itemsets, but considers during the association rule generation. Tao et al. [5] proposed an improved model of weighted support measurement and the weighted downward closure property. Seno et al.

[6] and Yun et al. [7][8] also considered weighted items in the process of frequent itemsets, the length-decreasing support constraints for a new measurement of support, and closed frequent patterns with weight constraints. Although the above works take the notion of weight into account as examined in this paper, they only concerned on the mining from items, but not from traversals.

3 Correlative Definitions and Notions

In order to commendably describe the problem, i.e., mining weighted frequent patterns from the traversals on graph, we give some correlative definitions and notions as follows.

Definition 1. *A WDG(Weighted Directed Graph) is a finite set of vertices and edges, in which each edge joins one ordered pair of vertices, and each vertex or edge is associated with a weight value. A base graph is a weighted directed graph,on which traversals occur.*

By Definition 1, we know that there should be two kinds of *WDG*s. One is *VWDG*(Vertex-*WDG*) which assigns weights to each vertex in base graph, and the other is *EWDG*(Edge-*WDG*) which assigns weights to each edge. Nextly, we will know they are essentially equivalent. So, we only explore the former in this paper. For example, Fig.1(a) is a base graph G which has 6 vertices and 8 edges, in which each vertex is associated with a weight.

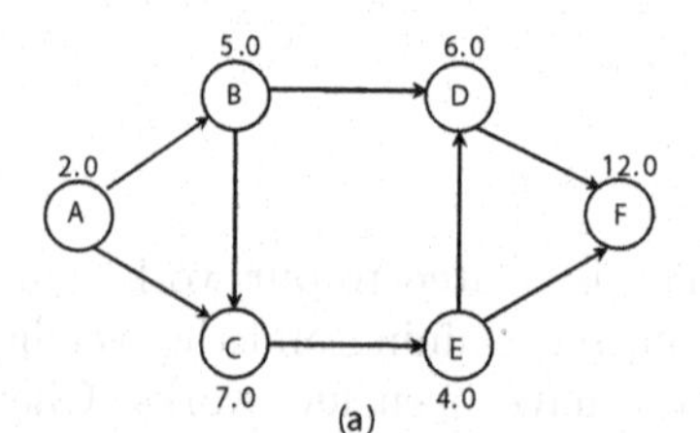

TID	Traversal
1	<A, B>
2	<B, C, E, F>
3	<A, C>
4	<B, C, E>
5	<A>
6	<A, C, E, D>

(b)

Fig. 1. (a) Example of base graph G (b) Traversal database T from G

Definition 2. *A traversal is a sequence of consecutive vertices along a sequence of edges on a base graph. Essentially, a traversal is a pattern. We assume that every traversal is path, which has no repeated vertices and edges. The length of a traversal is the number of vertices in the traversal. The weight of a traversal is the sum of vertex weights in the traversal. A traversal database is a set of traversals.*

Figure 1(b) is a traversal database T from G which has totally 6 traversals, each of which has an identifier and a sequence of consecutive vertices.

By Definition 1 and 2, obviously, there should be two corresponding traversal cases — VWDG traversal and EWDG traversal. Figure 2(a) and (b) respectively describe them, and (c) is the combination of two cases.

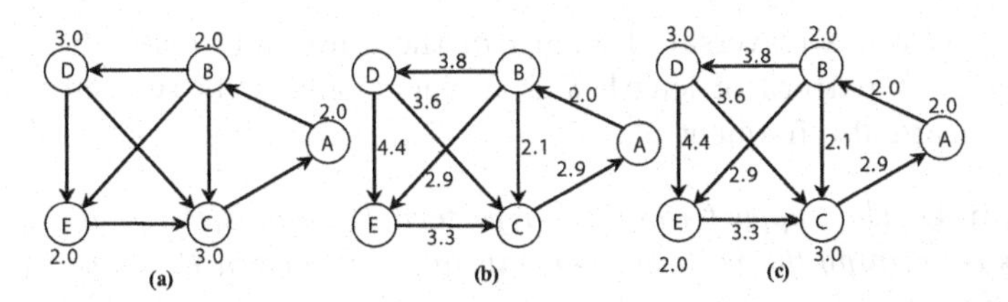

Fig. 2. Three cases of assigning weights to traversals

Essentially, the two *WDG* cases can be reduced to one case, i.e., assigning weights to vertices or edges. There we reduce the two cases to one case — assigning weights to vertices. The reason why we can reduce is that two nodes with a weighted edge in a *EWDG* can be thought as a node with same weight value in the corresponding *VWDG*, and the edges between vertices in *VWDG* have no weight value, their linking directions refer to the source *EWDG*. Figure 3 describes this change method, each vertex's weight in Fig.3(a) or (c) is 0, and each edge's weight in Fig.3(b) or (d) is also 0. Each node in Fig.3(b) or (d) represents the corresponding two nodes with a directed weighted edge in Fig.3(a) or (c), e.g., the nodes named *'bd'*,*'de'*,*'ec'* and *'bc'* respectively represents the corresponding nodes with directed edges named *<B,D>*,*<D,E>*,*<E,C>* and *<B,C>*. Figure 3(e), (f), (g) and (h) truly describe how to change *EWDG* into *VWDG*. Each edge's direction in new birth *VWDG* is gotten as follows, e.g., for node *'de'* in (h), because it's corresponding field in (e) is joined by two directed edges named *<B,D>* and *<E,C>*, so there are two edges (*'bd'*→*'de'* and *'de'*→*'ec'*), the other edges's direction is similar to the above. By the above change method, we can transform *EWDG* to *VWDG*. This union is convenient to solve pattern mining problem on weighted graph.

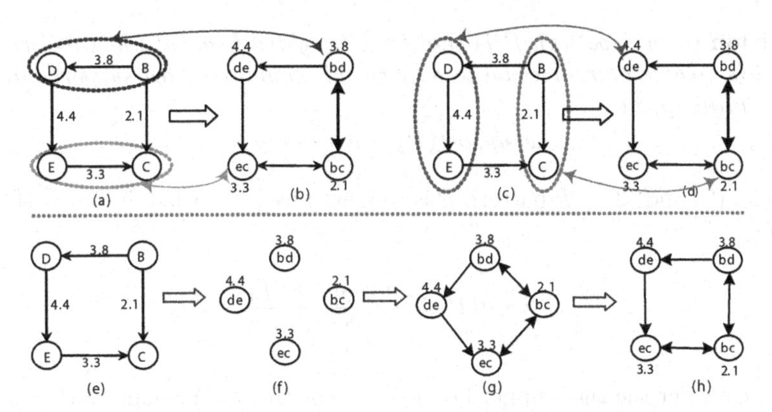

Fig. 3. The change process from *EWDG* to *VWDG*

Definition 3. *A subtraversal is any subsequence of consecutive vertices in a traversal. If a pattern P is a subtraversal of a traversal T, then we say that P is contained in T, and vice versa T contains P.*

Property 1. Given a traversal of length k, there are only two subtraversals of length $k-1$. A pattern of length k is frequent only if its two subpatterns of length $k-1$ are also frequent [2].

Definition 4. *The support count of a pattern P, scount(P), is the number of traversals containing the pattern. The support of a pattern P, support(P), is the fraction of traversals containing the pattern P. Given a traversal database T, let $|T|$ be the number of traversals.*

$$support(P) = \frac{scount(P)}{|T|}. \tag{1}$$

From practice, we can easily draw a property as follows.

Property 2. The support count and the support of a pattern decrease monotonically as the length of the pattern increases. In other word, given a k-pattern P and any l-pattern containing P, respectively denoted by P_k and P_l, where $l > k$, then $scount(P_k) \geq scount(P_l)$ and $support(P_k) \geq support(P_l)$.

Given a base graph G with a set of vertices V = $\{v_1, v_2, \ldots, v_n\}$, in which each vertex v_j is assigned with a weight $w_j \geq 0$, k-pattern $P \subset V$, we will define the weighted support of a pattern.

Definition 5. *The weighted support of a pattern P, called wsupport(P), is*

$$wsupport(P) = \left(\sum_{v_j \in P} W_j\right)(support(P)). \tag{2}$$

Definition 6. *A k-pattern P is said to be weighted frequent when the weighted support is greater than or equal to a given minimum weighted support threshold called wminsup, i.e.,*

$$wsupport(P) \geq wminsup\ . \tag{3}$$

From (1), (2) and (3), a k-pattern P is weighted frequent when its support count satisfies:

$$scount(P) \geq \frac{minsup \times |T|}{\sum_{v_j \in P} W_j}\ . \tag{4}$$

We can consider the right upper bound of (4) as the lower bound of the support count for a pattern P to be weighted frequent. Such lower bound, called k-support bound, is given by

$$sbount(P) = \left\lceil \frac{minsup \times |T|}{\sum_{v_j \in P} W_j} \right\rceil\ . \tag{5}$$

We take the ceiling of the value since the function *sbound*(P) is an integer. From (4) and (5), we can say a pattern P is weighted frequent when the support count is greater than or equal to the support bound.

$$scount(P) \geq sbound(P). \quad (6)$$

Note 1. *sbound*(P) can be calculated from base graph without referring traversal database. On the contrary, *scount*(P) can be obtained by referring traversal database.

The problem concerned in this paper is stated as follows. Given a weighted directed graph (base graph G) and a set of path traversals on the graph (traversal database T), find all weighted frequent patterns.

4 A Framework for Mining Weighted Frequent Patterns

We propose a framework for the mining of weighted frequent patterns from traversals on the graph. An efficient algorithm for mining large itemsets has been Apriori algorithm [9]. The reason why Apriori algorithm works is due to the downward closure property, which says all the subsets of a large itemset must be also large. For the weighted setting, however, it is not necessarily true that all the subpatterns of a weighted frequent pattern are weighted frequent. Therefore, we can not directly adopt Apriori algorithm. Instead, we will extend the notion of *support bound*, which can be applied to the pruning and candidate generation.

4.1 Pruning by Support Bound

One of key technologies to enhance the mining performance is to devise a pruning method which can reduce the number of candidates as many as possible. We must prune such candidates which have no possibility to become weighted frequent and keep those candidates that have a possibility to become weighted frequent in the future. The main concern point is how to decide such possibility.

Definition 7. *A pattern P is said to be a scalable pattern when it has a possibility to become weighted frequent in the future if extended to longer patterns, i.e., when some future patterns containing P will be possibly weighted frequent.*

Now, the pruning problem is converted to the scalability problem. For the decision of such scalability, we will first devise the weight bound of a pattern. Let the maximum possible length of weighted frequent patterns be u, which may be the length of longest traversal in the traversal database. Given a k-pattern P, suppose l-pattern containing P, denoted by P_l, where $k < l \leq u$. For the additional $(l-k)$ vertices, if we can estimate upper bounds of their weights as $w_{r1}, w_{r2}, \ldots, w_{rl-k}$, then the upper bound of the weight of the l-pattern P_l, called l-weight bound of P, is given by

$$wbound(P_l) = \sum_{v_j \in P} w_j + \sum_{j=1}^{l-k} w_{rj} \ . \tag{7}$$

The first sum, in (7), is the sum of the weights for the k-pattern P. The second one is the sum of the $(l-k)$ estimated weights, which can be estimated in several ways. We will propose two estimation methods in the following section.

From (5) and (7), we can derive the lower bound of the support count for l-pattern containing P to be weighted frequent. Such lower bound, called $l-$ *support bound* of P, is given by

$$sbount(P_l) = \left\lceil \frac{minwsup \times |T|}{wbound(P_l)} \right\rceil . \tag{8}$$

Lemma 1. *A pattern P is scalable if $scount(P) \geq sbound(P_l)$ for some $k < l \leq u$, otherwise is not scalable.*

Proof. Let l_i is any expression of l. If $scount(P) \geq sbound(P_{l_i}) \Rightarrow scount(P) \geq scount(P_{l_i})$ (by property 2)$\Rightarrow$there is a possibility: $scount(P_{l_i}) \geq sbound(P_{l_i})$ $\Rightarrow P_{l_i}$ will possibly be weighted frequent (by (6)). $\Rightarrow P$ is scalable (by defintion 7) Otherwise, if $scount(P) < sbound(P_{l_i})$, $\because$ by property 2, $scount(P) \geq scount(P_{l_i})$ $\therefore$ $scount(P_{l_i}) < sbound(P_{l_i}) \Rightarrow P_{l_i}$ will definitely not be weighted frequent$\Rightarrow$ P is not scalable. □

Corollary 1. *A pattern P is scalable if $scount(P) \geq sbound(P)$.*

Proof. From (5), (7) and (8)$\Rightarrow sbound(P) \geq sbound(P_l)$ for all $k < l \leq u \Rightarrow$ $scount(P) \geq sbound(P_l)$ for all $k < l \leq u \Rightarrow P$ is scalable by Lemma 1. □

Obviously, Corollary 1 states that a weighted frequent pattern must be scalable. On the contrary, in the case of $scount(P) < sbound(P)$, and because $sbound(P) \geq sbound(P_l)$, i.e., $scount(P) < sbound(P) \geq sbound(P_l)$, so we can not decide the relation of $scount(P)$ and $sbound(P_l)$, i.e., can not decide if P is scalable, therefore we need to estimate $sbound(P_l)$ to decide the scalability by Lemma 1.

According to Lemma 1 and Corollary 1, we can devise a pruning algorithm called *Pruning-SB* (*P*runing by *S*upport *B*ounds). The details of this are in Alg.1 of Table(1).

Table 1. Alg.1: Pruning-SB & Alg.2: Pruning-MSB

Alg. 1: *Pruning-SB* (*Pruning* by *S*upport *B*ounds)	Alg. 2: *Pruning-MSB*(*Pruning* by *MinS*upport *B*ound)
1: for each pattern P in candidates set C_k {	1: for each pattern P in candidates set C_k {
2: if ($scount(P) \geq sbound(P)$) then	2: if ($scount(P) \geq sbound(P)$) then
3: continue; // P is scalable then keep	3: continue; // P is scalable then keep
4: for each l from $k+1$ to u {	4: estimate $sbound(P,+)$;
5: estimate $sbound(P_l)$;	5: if ($scount(P) \geq sbound(P,+)$) then
6: if ($scount(P) \geq sbound(P_l)$) then	6: continue; // P is scalable then keep
7: break; } // P is scalable then keep	7: $C_k = C_k - \{P\}$;
8: if ($l > u$) then	// P is unscalable then prune
9: $C_k = C_k - \{P\}$;} // P is unscalable prune	8: }

Definition 8. *The max l-weight bound, $wbound(P, +)$, and the min l-support bound of a pattern P, $sbound(P,+)$, are defined as follows.*

$wbound(P,+) = max(wbound(P_l)); sbound(P,+) = min(sbound(P_l)), k < l \leq u.$

Corollary 2. *A pattern P is scalable if* $scount(P) \geq sbound(P,+)$*, but not scalable if* $scount(P) < sbound(P,+)$.

Proof. If $scount(P) \geq sbound(P,+)$, $\exists l_i, sbound(P_{l_i}) = sbound(P,+) \Rightarrow scount(P) \geq sbound(P_{l_i}) \Rightarrow P$ is scalable (by Lemma 1). Otherwise, if $scount(P) < sbound(P,+) \Rightarrow scount(P) < sbound(P_l)$, for all $k < l \leq u \Rightarrow P$ is not scalable. □

According to Corollary 2 along with Corollary 1, we can devise another pruning algorithm called *Pruning-MSB P*runing by *M*in *S*upport *B*ounds). The details of this are in Alg.2 of Table(1).

4.2 Candidate Generation

We will devise candidate generation algorithms by defining downward closure properties between scalable patterns. If there is a downward closure property between scalable patterns, new candidates can be generated from current scalable patterns.

Definition 9. *We say that there is partial downward closure property when the* $(k-1)-$ *subpattern* $<p_1, p_2, \ldots, p_{k-1}>$ *of a scalable* $k-$ *pattern* $<p_1, p_2, \ldots, p_k>$ *is also scalable, and there is full downward closure property when two* $(k-1)-$ *subpatterns* $<p_1, p_2, \ldots, p_{k-1}>$ *and* $<p_2, p_3, \ldots, p_k>$ *of a scalable* $k-$ *pattern* $<p_1, p_2, \ldots, p_k>$ *are also scalable.*

Note 2. There are only two $(k-1)-$ *subpatterns* of a k-pattern by Property 1. When there is the partial downward closure property, we can generate candidate $(k+1)$-patterns, C_{k+1}, from scalable k-patterns, C. The details of this are in Alg.3 of Table(2). When there is the full downward closure property, we can generate C_{k+1} in a similar way. The details of this are in Alg.4 of Table(2).

Table 2. Alg.3: Gen-PDC & Alg.4: Gen-FDC

Alg. 3: *Gen-PDC*(*Gen-*Partial Downward Closure)	Alg. 4: Gen- FDC (Gen-Full Downward Closure)
1: for each $P = <p_1, p_2, \ldots, p_k>$ in C_k {	1: for each $P = <p_1, p_2, \ldots, p_k>$ in C_k {
2: for each edge $<p_k, v> \in G$	2: for each edge $<p_k, v> \in G$
3: if $(v \notin P)$ then // not repeated vertex	3: if $(v \notin P)\&(Q = <p_2, \ldots, p_k, v> \in C_k)$ then
4: P is extended to $P' = <p_1, p_2, \ldots, p_k, v>$;	4: P is extended to $P' = <p_1, p_2, \ldots, p_k, v>$;
5:}	5:}

4.3 WTPMiner(Weighted Traversal Patterns Minier) Algorithm

By combing the pruning and candidate generation algorithms into a whole, we can get an algorithm, called *WTPMiner (Weighted Traversal Patterns Minier)*, for mining weighted frequent patterns from traversals on graph. Table(3) shows the algorithm proposed in this paper, which performs in a level-wise manner.

From Table(3), we can know that the framework of our proposed algorithm, comparing with the Apriori Gen Algorithm [9], contains some significant difference in the detailed steps. In the algorithm, each step is outlined as follows.

Table 3. Alg. 5: WTPMiner Algorithm for mining weighted frequent patterns

```
Alg. 5: Mining weighted frequent patterns-WTPMiner
Inputs: Base graph G, Traversal database T,Minimum weighted support wminsup
Output: List of weighted frequent patterns L_k
1: u = max(length(t)), t ∈ T;  //step1. maximum length of weighted frequent patterns
2: C_1 = V(G);  // step2. initialize candidate patterns of length 1
3: for (k = 1; k ≤ u and C_k ≠ Q; k++){
4:   for each traversal t ∈ D   //step3.obtain support counts
5:     for each pattern P ∈ C_k
6:       If ( P is contained in t) then P.scount++;}
   //step4. determine weighted frequent patterns
7:   L_k = {P|P ∈ C_k, wsupport(P) ≥ wminsup or scount(P) ≥ sbound(P)}
8:   if (k<u) then {   //step5. prune candidates
9:     C_k= pruneCandidates(C_k, G, u);
    //step6. generate new candidates for next pass
10:    C_{k+1} = genCandidates(C_k, G);}}
```

Step 1 is to find out the maximum possible length of weighted frequent patterns, which is limited by the maximum length of traversals. Step 2 initializes candidate patterns of length 1 with the vertices of base graph. In step 3, traversal database is scanned to obtain the support counts of candidate patterns. Step 4 is to determine weighted frequent patterns if the weighted support is greater than or equal to *wminsup* or the support count is greater or equal to the *support bound.* In step 5, the subroutine *pruneCandidates*(C_k,G,u) is to prune candidate patterns by checking their scalability. The algorithm *Pruning-SB* or *Pruning-MSB* can be used according to their efficiency. The remaining patterns are scalable patterns. In step 6, the subroutine *genCandidates*(C_k,G) generates new candidate patterns of length k+ 1 from the scalable patterns of length k for the next pass. The algorithm *Gen-PDC* or *Gen-FDC* can be used according to its applicability and efficiency.

5 Estimations of Support Bound

Inspired by [3], we propose two methods for the estimation of l-support bound of P.

5.1 Estimated by All Remaining Vertices in Graph

Given a k-pattern P, suppose l-pattern containing P, where $k < l \le u$. Let V be the set of all vertices in the base graph. Among the remaining vertices $(V-P)$, let the vertices with the $(l-k)$ greatest weights be $v_{r_1}, v_{r_2}, \ldots, v_{r_{l-k}}$, and, let $wbound(P_l)$ and $sbound(P_l)$ be defined same as (7) and (8), respectively. For example, in Fig. 1, the 3-support bound for the pattern $P = <A>$ is $sbound(P_3) = \left\lceil \frac{5.0\times 6}{2.0+(12.0+7.0)} \right\rceil = 2$.

Similar to Property 2, we can get the corollary as follows.

Corollary 3. *wbound*(P_l) *increases monotonically, and accordingly sbound*(P_l) *decreases monotonically as* l *increases.*

Let the upper limit of the length of possible weighted frequent patterns be known as u. By Corollary 3, the min support bound of P is the u-support bound of P,

$$sbount(P, +) = sbound(P_u) \ . \tag{9}$$

By (9) along with Corollary 2, if $scount(P) \geq sbound(P_u)$, then P is scalable. On the contrary, if $scount(P) < sbound(P_u)$, then P is not scalable. This means that we do not need to calculate l-support bounds of P for $k < l < u$. Therefore, the pruning algorithm *Pruning-MSB* is more efficient than $Pruning-SB$ under this station.

Corollary 4. *For* $\forall p_i \in \{P =< p_1, p_2, \ldots, p_k >, wbound((P - \{p_i\})_l) \geq wbound(P_l)$, *and accordingly* $sbound((P - \{p_i\})_l) \leq sbound(P_l)$.

Proof

$$By\ (7), wbound\Big((P - \{p_i\})_l\Big) = \sum_{v_j \in (P - \{p_i\})} w_j + \sum_{j=1}^{l-k+1} greatest\Big(w_{rj}\Big);$$

$$\sum_{v_j \in \left(P - \{p_i\}\right)} w_j + \sum_{j=1}^{l-k+1} greatest\Big(w_{rj}\Big) \geq \sum_{v_j \in P} w_j + \sum_{j=1}^{l-k} greatest\Big(w_{rj}\Big) = wbound(P_l);$$

Furthermore, by Corollary 3, $sbound((P - \{p_i\})_l) \leq sbound(P_l)$. □

Lemma 2. *There is the full downward closure property among scalable patterns. That is, if a k-pattern* $P =< p_1, p_2, \ldots, p_k >$ *is scalable, then the two* $(k-1)$-*subpatterns* $P_a =< p_1, p_2, \ldots, p_{k-1} >$ *and* $P_b =< p_2, p_3, \ldots, p_k >$ *are also scalable.*

Proof. The if condition means $scount(P) \geq sbound(P_u)$. By Property 2, for $P_a, scount(P_a) \geq scount(P)$, and $sbound((P_a)_u) \leq sbound(P_u)$ by Corollary 4. Therefore $scount(P_a) \geq sbound\,((P_a)_u)$, which implies P_a is scalable. This is similar to P_b. □

5.2 Estimated by Reachable Vertices

To prune unnecessary candidates as many as possible, the support bounds need to be estimated as high as possible. It means that we must estimate the weight bounds as low as possible. The previous method, however, has a tendency to over-estimate the weight bounds. This tendency is mainly due to the non-consideration of the topology of base graph. Specifically, the vertices with greatest weights are chosen one after one, even though they can not be reached from the corresponding pattern.

Definition 10. *Given a base graph G, k-reachable vertices from a vertex v is all the vertices reachable from v within the distance k. The distance is truly equal to the arc numbers between v and object vertex.*

Table 4. Alg. 6: Generating R(P,l)

```
Alg. 6: R(P, l)
1: T = {head vertex of P}
2: X = T;  //X is a temp set of vertices in path for v to objects
3: for (t = 1; t <= l - k; t++){
4:   for each vertex v ∈ X
5:     for each edge < v, w > in G
6:       if w ∉ X then insert w to X;}
7:         R(P, l) = X - P;
```

Such k-reachable vertices can be regarded as the vertices within the radius k from v. Obviously, k-reachable vertices include all the $(k-1)$-reachable vertices.

Given a k-pattern P, let $R(P,l)$, $k < l \leq u$, be the $(l-k)$-reachable vertices from the head vertex of P, but not in P and not through the vertices in P. The details of generating $R(P,l)$ are in Alg. 6 of Table(4).

For example, from Fig. 1(a), $R(< A >, 2)$ is $\{B, C\}$, and $R(< A >, 3)$ is $\{B, C, D, E\}$.

Among the vertices in $R(P,l)$, let the vertices with the $(l-k)$ greatest weights be $v_{r_1}, v_{r_2}, \ldots, v_{r_{l-k}}$, and, let $wbound(P_l)$ and $sbound(P_l)$ be defined same as (7) and (8), respectively. For example, refer to Fig. 1, the 3-support bound for the pattern $P =< A >$ is $sbound(P_3) = \left\lceil \frac{5.0 \times 6}{2.0+(6.0+7.0)} \right\rceil = 2$.

Similar to corollary 3, we can get the corollary as follow.

Corollary 5. *$wbound(P_l)$ increases monotonically, and accordingly $sbound(P_l)$ decreases monotonically as l increases.*

By corollary 5, the min support bound of P is the u-support bound of P, the corresponding expression is $sbound(P, +) = sbound(P_u)$. It is same as (9).

Corollary 6. *For $\forall p_i \in \{P=< p_1, p_2, \ldots, p_k>\}$, $wbound((P - \{p_i\})_l) \geq wbound(P_l)$, and accordingly $sbound((P - \{p_i\})_l) \leq sbound(P_l)$.*

It's proof can refer to Corollary 4.

Lemma 3. *There is the partial downward closure property among scalable patterns. That is, if a k-pattern $P =< p_1, p_2, \ldots, p_k >$ is scalable, then the $(k-1)$-subpattern $P_a =< p_1, p_2, \ldots, p_{k-1} >$ is also scalable.*

It's proof can refer to Lemma 2. Obviously, the candidate generation algorithm 3 can be only applied.

6 Experiments

This section presents experimental results of the mining algorithm, and compares two estimation algorithms, All vertices and Reachable vertices, using synthetic dataset. We implemented our algorithm with C++ language, running under Microsoft VC++ 6.0. The experiments are performed on 2.93GHz Pentium IV

PC machine whose operating system is Windows XP Professional and database is Microsoft SQL Server 2000 for managing base graphs and traversals on them.

During the experiment, base graph is generated synthetically according to the parameters, i.e., number of vertices and average number of edges per vertex. And then, we assigned distinctive random weight to each vertex of the base graph. All the experiments use a base graph with 100 vertices and 300 edges, i.e., 3 average edges per vertex. The number of traversals is 10,000 and the minimum weighted support is 2.0. We generated six sets of traversals, in each of which the maximum length of traversals varies from 5 to 10.

Figure 4(a) shows the trend of the number of scalable patterns with respect to the max length of traversals. We measured the number of scalable patterns when the length of candidate patterns is (max length of traversals − 1). As shown in the figure, the number of scalable patterns for Reachable vertices is smaller than that of All vertices. The difference of the number of scalable patterns between two estimation algorithms becomes smaller as the max length of traversals increases.

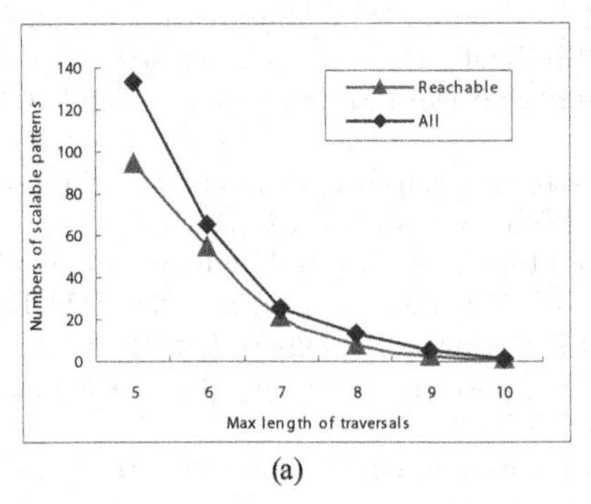

(a)

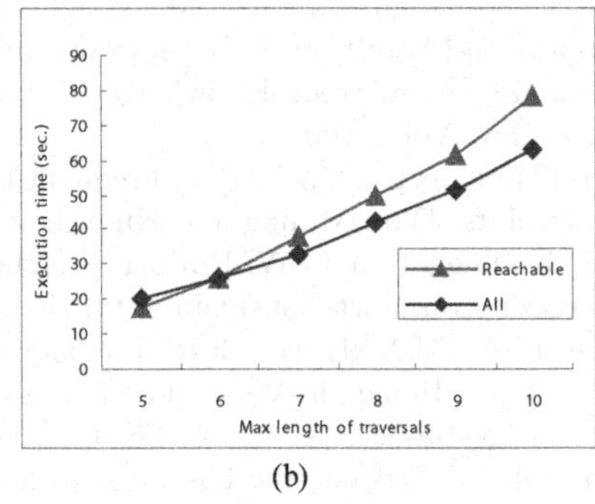

(b)

Fig. 4. The trends of number of scalable patterns (in a) & execution time (in b) w.r.t different max length of patterns

Figure 4(b) shows the trend of the execution time with respect to the max length of traversals. As shown in Fig. 4(b), when the max length of traversals is short, Reachable vertices is more efficient than All vertices. When the max length of traversals increases, however Reachable vertices is less efficient. The performance difference becomes larger when the max length of traversals becomes longer. This is because Reachable vertices spends more time to find reachable vertices as the max length of traversals increases.

7 Conclusions and Future Works

This paper explored the problem of discovering frequent patterns from weighted traversals on graph. A changing model between *EWDG* and *VWDG* is proposed. Based on the model, we presented the mining algorithm named *WTPMiner*. In this algorithm, we use the the notion of support bound. We also proposed two methods for the estimation of support bound, and then experimented on them. How to scale the model and algorithms to a more larger scale, e.g., Web-size scale and other large relation moldels etc., and how to efficiently put it into practice are still be worthy of further researches.

References

1. Chen, M.S., Park, J.S., Yu, P.S.: Efficient Data Mining for Path Traversal Patterns. IEEE Trans. on Knowledge and Data Engineering 10, 209–221 (1998)
2. Nanopoulos, A., Manolopoulos, Y.: Mining Patterns from Graph Traversals. Data and Knowledge Engineering 37, 243–266 (2001)
3. Cai, C.H., Ada, W.C., Fu, W.C., Cheng, C.H., Kwong, W.W.: Mining Association Rules with Weighted Items. In: IDEAS. Proc. of International Database Engineering and Applications Symposium, pp. 68–77. IEEE Press, New York (2001)
4. Wang, W., Yang, J., Yu, P.S.: In: SIGKDD. Proc. of ACM SIGKDD International Conference on Knowledge Discovery and Data Mining, pp. 270–274. ACM Press, New York (2000)
5. Tao, F., Murtagh, F., Farid, M.: Weighted Association Rule Mining using Weighted Support and Significance Framework. In: SIGKDD. Proc. of ACM SIGKDD International Conference on Knowledge Discovery and Data Mining, pp. 661–666. ACM Press, New York (2003)
6. Seno, M., Karypis, G.: Finding Frequent Patterns Using Length-Decreasing Support Constraints. Data Mining and Knowledge Discovery 10, 197–228 (2005)
7. Yun, U., Leggett, J.J.: WLPMiner: Weighted Frequent Pattern Mining with Length-Decreasing Support Constraints. In: Ho, T.-B., Cheung, D., Liu, H. (eds.) PAKDD 2005. LNCS (LNAI), vol. 3518, pp. 555–567. Springer, Heidelberg (2005)
8. Yun, U.: Mining lossless closed frequent patterns with weight constraints. Knowledge-Based Systems 20, 86–97 (2007)
9. Agrawal, R., Srikant, R.: Fast algorithms for mining association rules. In: Pro. of the 20th LDB Conference, Santiago, pp. 487–499. Morgan Kaufmann, San Francisco (1994)

Development of Enhanced Data Mining System to Approximate Empirical Formula for Ship Design

Kyung Ho Lee[1], Kyung Su Kim[1], Jang Hyun Lee[1], Jong Hoon Park[2], Dong Geun Kim[2], and Dae Suk Kim[2]

[1] INHA University, Department of Naval Architect & Ocean Engineering, 253 Yonghyun-dong, Nam-gu, Inchon, Korea
{kyungho,ksukim,jh_lee}@inha.ac.kr
[2] INHA University, Graduate School, Department of Naval Architect & Ocean Engineering, 253 Yonghyun-dong, Nam-gu, Inchon, Korea

Abstract. Companies must have tools to manage effectively their huge engineering data. So we developed a data mining system based on GP which can be one of the components for the realization of the utilization of engineering data. The System can derive and extract necessary empirical formulas to predict design parameters in ship design. But we don't have enough data to carry out the learning process of genetic programming. When the learning data is not enough, not good result such like overfitting can be obtained. Therefore we have to reduce the number of input parameters or increase the number of learning and training data. In this paper we developed the improved data mining system by Genetic Programming combined with Self Organizing Map (SOM) to solve these problems. By using this system, we can find and reduce the input parameters which do not influence on output less, as a result of this study we can solve these problems.

1 Introduction

Recently the importance of the utilization of engineering data is gradually increasing because that can secure the productivity. Engineering data contains the experiences and know-how of experts. In intelligent system for ship design, it have been slowly changing from those developed by the knowledge-based approach, whose knowledge is difficult to extract and represent, to those developed by the data-driven approach, which is relatively easy to handle.

As mentioned before, the engineering data contains many meaningful information so the development of data analysis and utilization method is very important concept. This paper focuses on Data Mining and Knowledge Discovery in Database (KDD) technique which is one of useful method to support this concept.

In ship design process, the utilization of existing data is one of the important issues because most of ship design is performed by modification of the previous ship design data. Especially in preliminary ship design state, most of design parameters are determined by using empirical formulas, which are generated from existing ship data. But these empirical formulas had been made on the basis of traditional ships, such as

Z. Zhang and J. Siekmann (Eds.): KSEM 2007, LNAI 4798, pp. 425–436, 2007.

bulk carrier, crude oil tanker and so on. So they are too old-fashioned to apply to build new typed ships, such as LNG carrier, drill ship and so on. Generation of new-fashioned empirical formulas to adapt to high value added ships is very important issue. Although Korean shipyards have accumulated a great amount of data from much ship building experience. But they do not have appropriate tools to utilize the data in practical works. We developed data utilization tool to adapt to ship design has been developed by using genetic programming (GP) since last several years. [1] That focuses on how to adapt for ship designing, especially on the generation of empirical formulas to predict design parameters in preliminary ship design stage. Generally, a field of study for the prediction, artificial neural network (ANN) is used as training system in most engineering fields. But if the characteristic of the training data is nonlinear and discontinuous, the performance of the training result is deteriorative. And the ANN shows a black-box system its own process so user cannot perceive the predicted formula. On the other hand, GP has excellent ability to approximate with non-linear and discontinued data. Above all, GP can show the trained results as a function tree form. Generally, a lot of accumulated data are needed for the training by using ANN or GP. But actually, in a real situation in the ship designing field, we don't have enough data to utilize for the training procedure. Therefore we introduced enhanced GP techniques to fit an approximated function from accumulated data with small learning data which used Polynomial GP (PGP), Linear Model GP (LM-GP) and combining GP (PLM-GP) in previous papers.

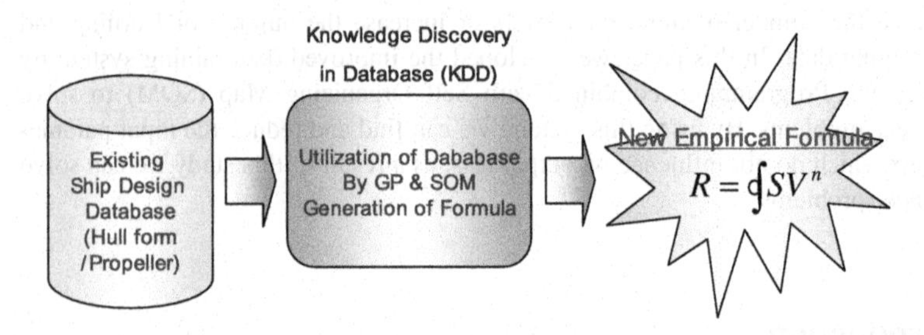

Fig. 1. The process of preliminary ship design

As mentioned before, a lot of data are needed for generating empirical formulas to predict the relations of input parameters and output design parameters. But in practical cases, we cannot get enough learning samples, in addition there are many design input parameters. Therefore if we reduce the number of input parameters, we can get good results. In this paper, we adopted Self Organizing Map (SOM) to reduce the number of input parameters. By applying SOM to accumulated data, the influence of input parameters on design outputs can be found. The developed data mining system by combining GP with SOM can be powerful tool for the generation of empirical formulas to predict design parameters in ship design. Fig.1 is concept of the developed data mining system in this study.

In fore part of this paper, the concept of GP and the developed data mining system based on GP will be introduced briefly. And then the enhanced data mining system combining GP with SOM will be introduced.

2 Genetic Programming for Data Mining

Polynomials are widely used in many engineering applications such as Response Surface Modeling [3, 4, 5], since the mathematical form of a polynomial is simple, and very easy to handle. In many engineering field, the response surface modeling method is adopted to reduce the computational cost required for analysis and simulation during the optimization design process. Thus, it is desirable to use the minimal size of samples to construct response surfaces. The classical method for attaining good polynomials is to use "all-possible-regressions" and "stepwise regression" methods [7,8], but there are limitations in obtaining polynomials with a desired accuracy.

In the paper [2], we tried to use Genetic Programming for generating optimal polynomials that approximate very highly nonlinear response surfaces using only minimal or very small size of learning samples. Major issues regarding finding such polynomials using GP have advantages that are addressed below.

First, the GP tree can easily represent the polynomial if a function set contains only "+", "-", "*" operators, and a terminal set includes only variables and constants, but it is difficult to expect for GP to generate polynomials enabling to model a nonlinear function using only such a function and terminal set. We tackled this problem by the use of low order Taylor series of various mathematical functions in a function set. That effectively makes GP produce very high order polynomials. But the generated polynomial tends to become too complex, and it is necessary to control the size of polynomials.

Second, we have not enough learning samples so the overfitting problem can be very serious, but there are no other kinds of additional samples. So, we used the EDS (Extended Data Set) method with the FNS (Function Node Stabilization) method to generate additional samples.[10]

2.1 Function Set with Taylor Series for GP

In this paper, we use the Taylor Series of mathematic functions in the Function set to generate a high order polynomial easily. By the way, if we take the high order Taylor Series, the GP tree produces a very complex polynomial. So, we determined to take only two or third order polynomials from the Taylor Series.

We use the following function set and terminal set.

$$F = \{+, -, *, g_i (i = 1, \ldots, 18)\}$$

$$T = \{one, rand, x_i (i = 1, \ldots, n)\}$$

"one" returns 1, and "rand" is a random number whose size is less than 1. x_i represents a variable. All function and terminals have their weights. The weights are estimated using Hooke & Jeeves method towards further minimizing its fitness function defined in the next section. Since we adopt Taylor Series, the ordinary least squared method [8] cannot be used easily.

Where, g_i is a low order Taylor Series as followings.

$$g_1 : \sin(x) = x - \frac{x^3}{3!},\ g_2 : \cos(x) = 1 - \frac{x^2}{2!},\ g_3 : \tan(x) = x + \frac{x^3}{3!},$$

$$g_4 : \log(1+x) = x - \frac{x^2}{2} + \frac{x^3}{3},\ g_5 : \exp(x) = 1 + \frac{x}{1!} + \frac{x^2}{2!} + \frac{x^3}{3!},$$

$$g_6 : \sinh(x) = x + \frac{x^3}{3!},\ g_7 : \cosh(x) = 1 + \frac{x^2}{2!} \cdots$$

2.2 Fitness Function for Overfitting Avoidance

In previous page, we introduced the matter of overfitting, so we must solve that. But overfitting is very complicated problems. Therefore in this chapter, we will introduce the overfitting avoidance method.

Without considering overfitting, the fitness function can be defined by (1).

$$\vartheta = \vartheta_{MSE} \tag{1}$$

Where, $\vartheta_{MSE} = 1/m \sum_{i=1}^{m} [f_{GP}(\overline{X}_i) - y_i]^2$

The learning set takes the form of $L\{(\overline{X}_i, y_i), 0 \le y_i \le 1\}_{i=1,\ldots,n}$.

Here, $\overline{X}_i(x_{i,1},\ldots,x_{i,n}, 0 \le x_{i,j} \le 1)$ is a n-dimensional vector, and y_i is a desired output of the GP tree (f_{GP}) at $\overline{X}_i$. Since m is very small, the GP tree is overfitted if only $\vartheta = \vartheta_{MSE}$ is used. The EDS method [10] can be included in the fitness function for smooth fitting.

$$\vartheta_E = \vartheta_{MSE} + \lambda \hat{\vartheta}_{MSE} \tag{2}$$

Where, $\hat{\vartheta}_{MSE} = 1/p \sum_{i=1}^{p} [f_{GP}(\overline{X}_i^E) - y_i^E]^2$, and λ is a constant that determines the contribution of $\hat{\vartheta}_{MSE}$.

The extended data set $L^E\{(\overline{X}_i^E, y_i^E)\}_{i=1,\ldots,p}$ can be constructed by simple linear interpolations of closest learning samples. In this paper, we simply take 0.1 as the value of λ. That is a small value, and certainly the GP tree is overfitted. To improve this situation, we introduced the FNS method. If the GP tree contains several function nodes $(g_i\ T)$, where T is the subtree, and if the value of $(g_i\ T)$ is very large compared with others, then the slight change of T's value might cause the very large change of the GP tree's output. As shown in (3), the FNS method penalizes the GP tree if the tree contains such nodes.

$$\vartheta_{EF} = \vartheta_{MSE} + 0.1\hat{\vartheta}_{MSE} + \vartheta_{FNS} \tag{3}$$

Where, $\vartheta_{FNS} = 0$ if for all g_i in the GP tree are

$|(g_i\ T) - f_i(T)| \le \delta$, otherwise $\vartheta_{FNS} = \alpha$

f_i is a mathematic function corresponding to f_i, α is very large positive number such as 1.0E10, and $\delta(0.1)$ is a tolerance. If that is large, then is very different from f_i, and ϑ_{FNS} becomes α. In this case, the Hooke & Jeeves method tries to make approach to f_i within the tolerance δ by estimating weights of the GP tree through minimizing ϑ_{EF}. If this effort fails, the GP tree has a very large fitness value, and will be excluded in the next generation. Note that in FNS, the only way of $(g_i T)$ approximating f_i is that T has a small value because g_i is a low order polynomial.

2.3 Controlling the Size of Polynomials

During the evolving process, many GP trees produce too complex polynomials although low order Taylor Series is used. There is need for reducing the complexity of a polynomial. One way of doing this is to give the allowable maximum number of term ($n_{\max}$) of polynomials to the GP system so as not to generate GP trees representing a too complex polynomial. The algorithm for computing $n_{\max}$ of polynomial from the GP tree can be implemented using the stack structure. Its description is rather lengthy. So, we will not discuss due to the space limitation.

3 Numerical Results for the GP as a Data Mining

Table 1 is shown the parameters used in PGP.

Table 1. The Parameters used in PGP

Maximum generation	**40**
Selection method	**Tournament with 30 trees**
Maximum terms of a polynomial	**100**
Reproduction probability	**0.15**
Crossover probability	**0.7**

The function below shows the Goldstein-price function typically used as a benchmark problem for testing the performance of the optimization algorithms.

$$f(x_1,x_2)=(1+(x_1+x_2+1)^2\cdot(19-14x_1+3x_1{}^2-14x_2+6x_1x_2+3x_2{}^2))$$
$$\cdot(30+(2x_1-3x_2)^2\cdot(18-32x_1+12x_1{}^2+48x_2-36x_1x_2+27x_2{}^2))$$

$$-2\le x_i\le 2 \qquad i=1,2$$

It is nearly impossible to approximate the Goldstein-price function with a good accuracy because the function value is changed from 0 to 1.0E6, and this large value is too dominant to others. So, we take the logarithm scale ($f_{\log}(x_1,x_2)=\log(1+f(x_1,x_2))$).

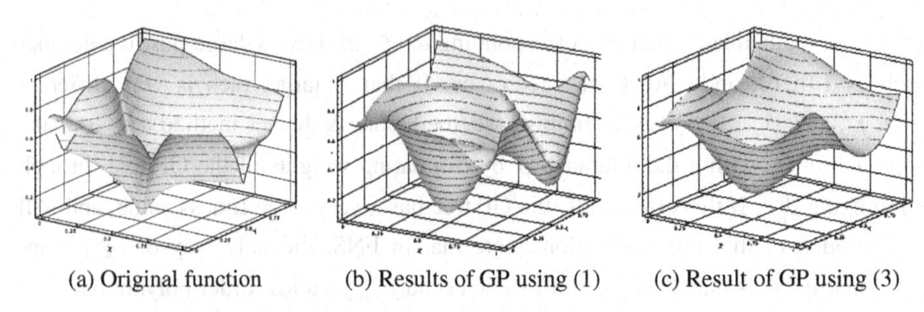

(a) Original function (b) Results of GP using (1) (c) Result of GP using (3)

Fig. 2. The logarithm-scaled Goldstein-price function

The learning set is prepared as 5x5 grid type, and in the same manner the test set is made by 200x200 grid type. The size of the population is 5000. The learning and test set are normalized. Fig.1.a shows $f_{\log}(x_1, x_2)$. Since this function is highly nonlinear and the size of the learning set is only 25, do not expect for GP to produce the good GP tree. Fig.2.b shows the results of GP when (1) is used for the fitness function. The GP tree is severely overfitted. The mean square error (MSE) of learning and test set are 30386E-6 and 0.0712, respectively. On the other hand, as shown in Fig.2.c, when (3) used as the fitness function, the results give the smooth surface, and roughly picture the overall feature of $f_{\log}(x_1, x_2)$. The MSE of learning and test set are 3.645 E-3 and 4.365E-3, respectively. We can see the results are enhanced.

Using the translation program, the GP tree is transformed into the normal polynomial form, and this polynomial is simplified by Mathematica. The result is shown in Fig.3. The size of polynomial is manageable.

```
0.693707 (-1.94697 (-2.56609 + 4.154 x2)
    (1. - 0.253758 (1.76982 - 3.07988 x1 - 1.45075 x2)^2
      (1. - 0.520818 x2^2)^2)
    (0.598926 (1. + 0.556292 (1.66548 x1 + 0.751999
              (-1.0079 x1 + 0.892572 x2)^3 - 1.47491 x2) ^ 2) -
      0.97981 (1. - 0.439762 (-1.20691 x1^3 + x1^2 (3.11693 -
                2.61313 x2) - 1.88593 x1 (-2.20564 + x2)
              (-0.179953 + x2) - 0.4537 (-2.94709 + x2)
              (-1.1928 + x2) (0.561499 + x2)) ^ 2)) +
  1.542 (1.14388 (-1.08873 + x2) (0.0949399 + x2) +
      1.28987 (-0.327209
          (1. - 0.0880898 (1.66792 - 4.38709 x1 - 0.462422 x2)^2
            (1. - 0.0417391 (1.10438 x1 - 0.286908 x2)^2
              (-2.54243 + x2)^2 (1.16337 + x2)^2) ^ 2) +
        0.846979 (1. + 0.341689 (1. - 0.292192
              (1.64987 - 1.24615 x1 - 3.537 x2)^2
              (1. - 0.473873 x2^2)^2) ^ 2))))
```

Fig. 3. The polynomial obtained by GP

4 Integrated Data Mining System for Ship Design

We developed the data mining system for a data analysis and utilization by using enhanced genetic programming with integrated model. That system is contrived to apply to ship design under the case that the accumulated data is not enough to make learning process.

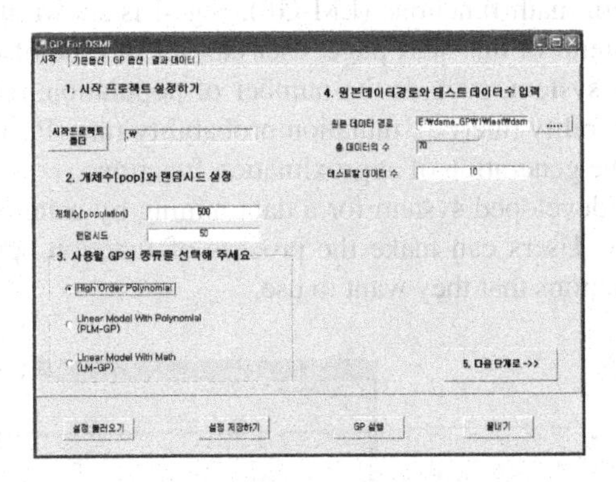

Fig. 4. The Start page of the developed Data mining system

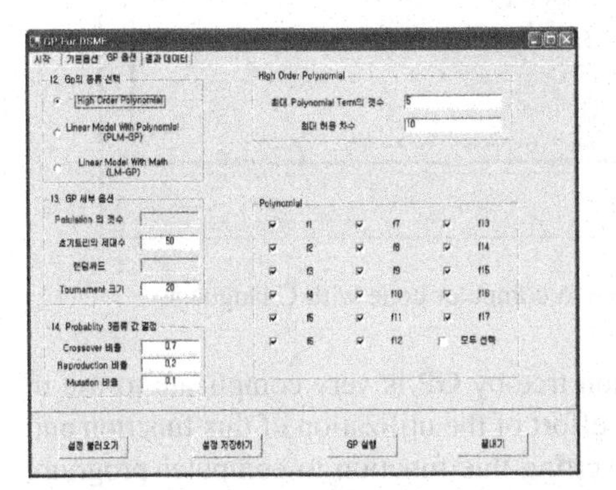

(a) High Order Polynomial GP

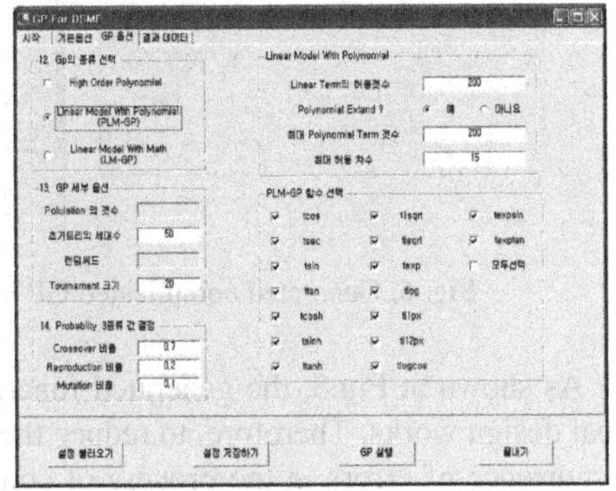

(b) Linear model with Polynomial (PLM-GP)

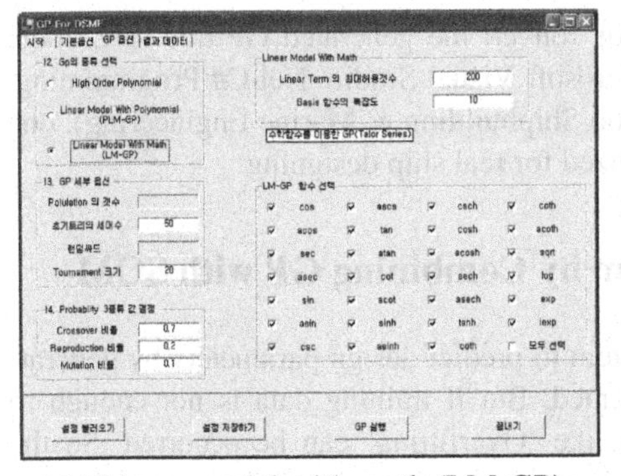

(c) Linear model with math (LM-GP)

Fig. 5. 3 kinds of GP and their optional functions

The integrated system can make fitting functions with 3 types of GP, such as GP with high order polynomial (PGP), linear model GP with polynomial (PLM-GP) and linear model GP with math functions (LM-GP). Fig. 4 is shown the start page of the GP data mining system. In this start page, user can select type of GP, all sorts of options to execute the system, such as the number of population, the number of initial tree, crossover probability rate(Pc), mutation probability rate (Pm) and optional function to be used for the generation of approximation function.

Fig. 5 shows the developed system for a data mining by using PGP, PLM-GP and LM-GP respectively. Users can make the process of function approximation by selecting arbitrary functions that they want to use.

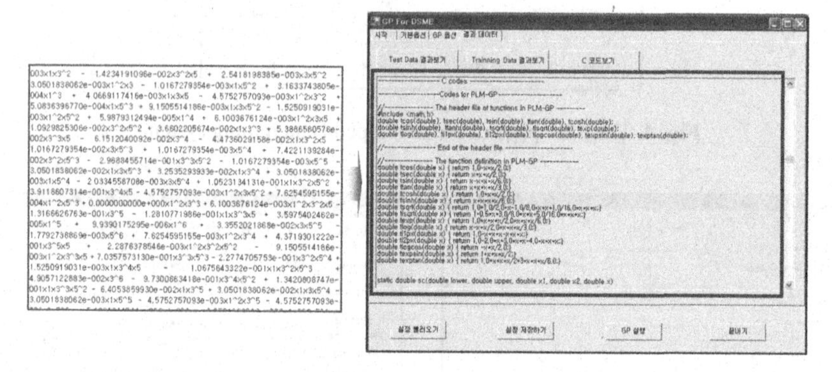

Fig. 6. Generated complicated GP tree to computer code with C language

As shown at Fig.3, the generated function tree by GP is very complicate to use in real design works. Therefore, to reduce the effort of the utilization of this function and occurrence of errors in the process of converting this function to computer program, the data mining system can convert the generated GP tree to C language code that can interface with other program.

Fig.6 is shown an example of result of convert the generated GP tree to C code. This system is implemented by using Microsoft Visual Studio .Net C# Programming. And the system is used in DSME (Daewoo Shipbuilding & Marine Engineering), one of world best shipyards in Korea, and utilized for real ship designing.

5 Enhanced Data Mining System by Combining GP with SOM

In the previous chapters, data mining system to predict design parameters by generating empirical formulas with GP is presented. But if training data is not enough to execute this system, fatal problem such like 'Overfitting' can be occurred. So the avoidance of overfitting problem is the most important issue in regression domain. Generally, the number of training data is closely related to the number of input parameters. If the number of input parameters is increased, we need more and more training data. As mentioned before, we don't have enough data in ship designing process. So reducing input parameters prior to training process is required. In order to reduce input parameters, we have to know the influence of the input parameters. Self Organizing Map (SOM) is appropriate tool to find the influence of input parameters for output parameters. SOM is a well-known model of ANN. It is devised by Kohonen, and is a typical method for competitive learning.

In our approach, influence analysis for input parameters by SOM is performed firstly. According to the results, input parameters with low influence are removed in training data for GP. By combining GP with SOM, the performance and accuracy of training can be enhanced.

5.1 Experiment of Enhanced Data Mining System for Ship Designing Process

In this experiment, the prediction of block coefficient (Cb) for crude oil tanker is carried out. The Cb is one of the most important design parameter to be estimated.

In order to validate the data mining system, we gathered 70 training data, and 60 of them are used for training process and the remaining 10 data are used for test. As mentioned before, 70 training data are not enough to perform training. But we could not get more training data, because in Korean shipyard, most of those real data are forbidden to open to the public. Fig. 7 is a part of training sample data of this experiment. As shown at the figure, the learning samples are consist of 7 input parameters, such as dead weight (DWT), length between perpendiculars (LBP), ship speed, breadth, draft, depth and Froude number (Fn) and the output parameter, Cb.

	A	B	C	D	E	F	G	H
1	DWT	LBP	Speed	Breadth	Draft	Depth	Fn	Cb
2	103146.00	243,86	15,00	42,03	14,61	22,50	0,16	0,83
3	103214.00	244,00	15,50	42,00	14,61	23,84	0,16	0,82
4	159800.00	274,00	15,20	48,00	16,00	23,20	0,15	0,85
5	127000.00	272,00	14,80	46,00	14,80	22,60	0,15	0,86
6	146300.00	267,00	14,90	44,40	16,50	24,10	0,15	0,85
7	165000.00	274,19	15,50	50,00	16,00	23,10	0,15	0,84
8	159000.00	274,00	15,20	48,00	15,50	23,70	0,15	0,84
9	108721.00	267,00	14,70	44,40	16,50	24,10	0,15	0,86
10	124600.00	263,00	14,50	46,00	14,50	23,70	0,15	0,86
11	150000.00	270,45	15,00	44,60	16,10	24,20	0,15	0,85
12	144348.00	270,04	14,80	43,06	17,33	23,80	0,15	0,86
13	147916.00	274,00	15,00	47,80	16,02	22,80	0,15	0,85
14	146270.00	274,30	15,20	43,20	16,71	23,30	0,15	0,85
15	147500.00	268,00	14,50	43,24	17,53	23,50	0,15	0,86
16	146184.00	277,00	15,50	44,40	16,55	24,10	0,15	0,85
17	155982.00	281,00	15,20	53,70	15,77	22,30	0,15	0,85
18	155250.00	285,00	15,00	44,20	17,15	22,50	0,15	0,86
19	149999.00	267,00	14,50	46,20	16,81	22,00	0,15	0,86
20	38200.00	221,00	15,50	30,00	11,55	16,80	0,17	0,80
[illegible]	[illegible]	[illegible]	[illegible]	[illegible]	[illegible]	[illegible]	[illegible]	[illegible]

Fig. 7. Learning samples to predict block coefficient (Cb)

In this experiment, the performance of prediction which can be compared by the value of RMSE (root mean square error) is carried out for the following 3 approaches.

(1) Yamagata Formula (Prediction by traditional empirical formula)

$$Cb = 1.036 - 1.46 \cdot Fn \qquad Fn < 0.24$$

= 3.116-10.15·Fn Fn>0.24
= 10.40 Fn>0.267

(2) Prediction by PLM-GP
(3) Prediction by PLM-GP combining with SOM

Although there are 7 input parameters, we have just 70 training data. That means, input parameters are too many comparing with the number of training data. So we have to reduce input parameters by the evaluation of influence through SOM.

The cross dashed line is exact values of Cb that means estimated value is equal to real value. Fig. 8 represents the prediction result by Yamagata's empirical formula. The trend-line is far apart on the real value line. But they have their own trend. Fig. 9 shows the prediction result by PLM-GP, the trend-line is still far apart on the real value line. The result of the evaluation using SOM is shown at Fig. 10. By the result of the evaluation of influence for the input parameters, Froude number (Fn) is the lowest influence factor among input parameters. So we can remove the Fn parameter from 7 input parameters. Fig. 11 represents the prediction result by PLM-GP combining with SOM. There is no the Fn parameter in this prediction result. We can show the trend-line is closed to the real value line, so we get the improved result.

5.2 Evaluation of the Experiments

As we can see at Fig. 8, the prediction result by Yamagata empirical Formula is not good. But have their trend. The reason why the bad result is the formula was generated by using old-fashioned ship data. On the other hand, the result by GP trained by 60 learning data is considerably improved. Their RMSE is 0.00801. The developed data mining system using PLM-GP can predict the design parameters very well with small learning samples. Finally, the result by PLM-GP combining with SOM is better than other results as we expected. Their RMSE is 0.0049 so the result is better than previous experiment's RMSE 0.00801.

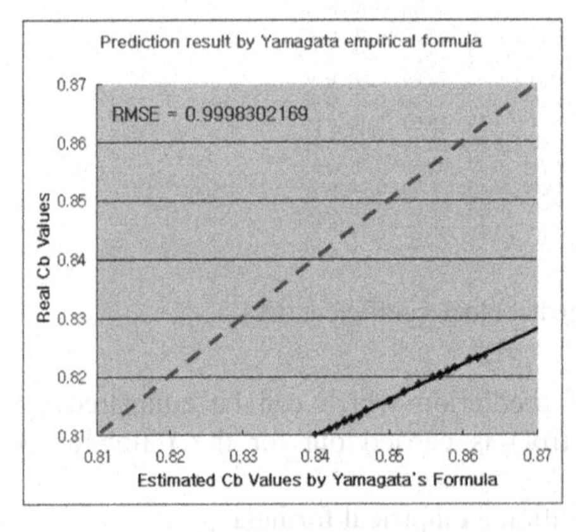

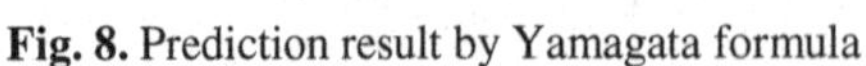
Fig. 8. Prediction result by Yamagata formula

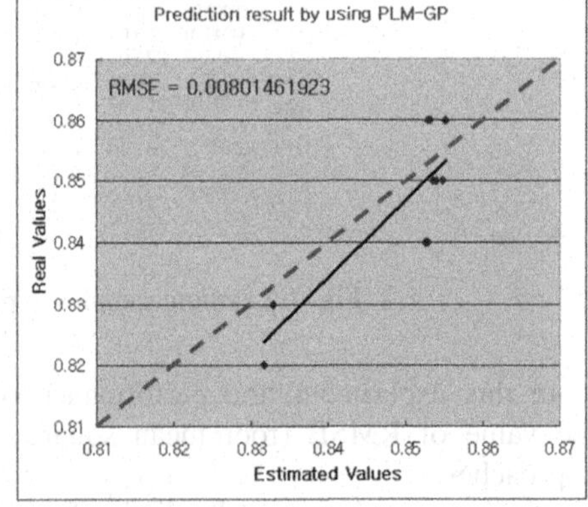

Fig. 9. Prediction result by using PLM-GP

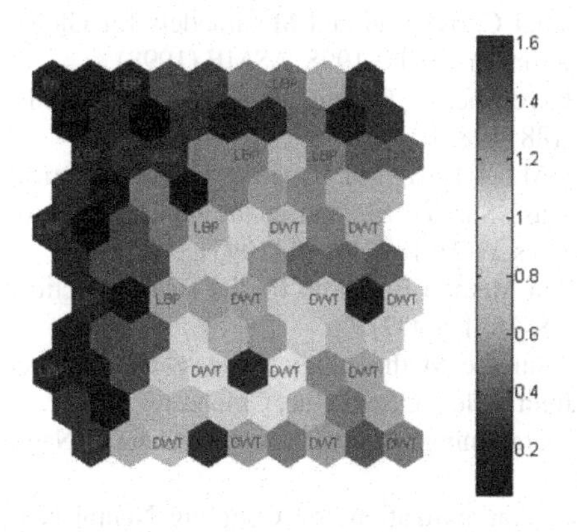

Fig. 10. Visualization of the influence value

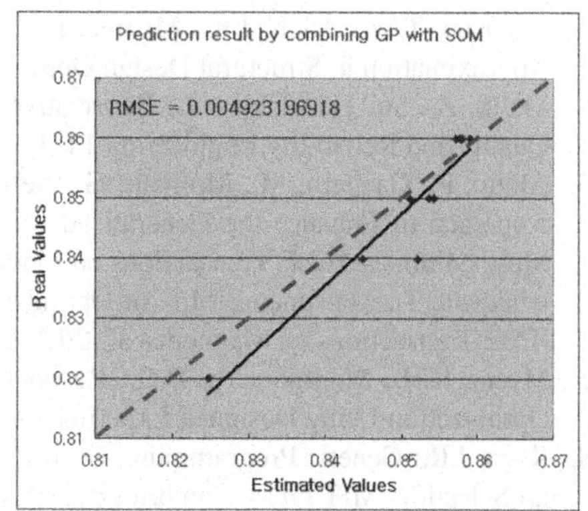

Fig. 11. Prediction result by GP with SOM for each parameter in SOM

6 Conclusions

In this paper, data mining system for the prediction of design parameters to assist the ship designing process with insufficient learning samples is developed. The integrated system is presented with PGP, LM-GP and PLM-GP. Designer can generate empirical formulas by using this system easily with small learning samples. In order to improve the performance of prediction, SOM is combined with GP program. Through the evaluation of influence for the input parameters by SOM, the performance of prediction was considerably enhanced. The validation test and the adoption of the developed method in the ship designing process are presented. As a result, the system is good for non-linear function approximation with limited amount of learning data, without overfitting. The developed data mining system can be used in real preliminary ship design process to generate new empirical formulas for the next generation high value added ships, such as LNG carrier, FPSO, and so on.

Acknowledgement

This work is supported by Advanced Ship Engineering Research Center (R11-2002-104-08002-0).

References

1. Lee, K.H., Yeun, Y.S., Yang, Y.S., Lee, J.H., Oh, J.: Data Analysis and Utilization Method based on Genetic Programming in Ship Design. In: Gavrilova, M., Gervasi, O., Kumar, V., Tan, C.J.K., Taniar, D., Laganà, A., Mun, Y., Choo, H. (eds.) ICCSA 2006. LNCS, vol. 3981, pp. 1199–1209. Springer, Heidelberg (2006)
2. Lee, K.H., Oh, J., Park, J.H.: Development of Data Miner for the Ship Design based on Polynomial Genetic Programming. In: Sattar, A., Kang, B.-H. (eds.) AI 2006. LNCS (LNAI), vol. 4304, pp. 981–985. Springer, Heidelberg (2006)

3. Simpson, T.W., Allen, J.K., Mistree, F.: Spatial Correlation and Metamodels for Global Approximation in Structural Design Optimization. In: DETC 1998. ASME (1998)
4. Malik, Z., Su, H., Nelder, J.: Informative Experimental Design for Electronic Circuits. Quality and Reliability Engineering 14, 177–188 (1986)
5. Alotto, P., Gaggero, M., Molinari, G., Nervi, M.: A Design of Experiment and Statistical Approach to Enhance the Generalized Response Surface Method in the Optimization of Multi-Minimas. IEEE Transactions on Magnetics 33(2), 1896–1899 (1997)
6. Ishikawa, T., Matsunami, M.: An Optimization Method Based on Radial Basis Function. IEEE Transactions on Magnetics 33(2/II), 1868–1871 (1997)
7. Myers, R.H., Montgomery, D.C.: Response Surface Methodology: Process and Product Optimization Using Designed Experiments. John Wiley & Sons, Inc, Chichester (1995)
8. Koza, J.R.: Genetic Programming: On the Programming of Computers by Means of Natural Selection. MIT Press, Cambridge (1992)
9. Yeun, Y.S., Lee, K.H., Yang, Y.S: Function Approximations by Coupling Neural Networks and Genetic Programming Trees with Oblique Design Trees. Artificial Intelligence in Engineering 13(3), 223–239 (1999)
10. Lee, K.H., Yeun, Y.S., Ruy, W.S., Yang, Y.S: Polynomial genetic programming for response surface modeling. In: FEA 2002. 4th International Workshop on Frontiers in Evolutionary Algorithms in conjunction with 6th Joint Conference on Information Sciences (2002)
11. Yeun, Y.S., Suh, J.C., Yang, Y.S: Function approximation by superimposing genetic programming trees: with application to engineering problems. Information Sciences 122, 2–4 (2000)
12. Barron, A., Rissanen, J., Yu, B.: The minimum description length principle in coding and modeling. IEEE Trans. Information Theory 44(6), 2743–2760 (1998)

Research on a Novel Word Co-occurrence Model and Its Application

Dequan Zheng, Tiejun Zhao, Sheng Li, and Hao Yu

MOE-MS Key Laboratory of Natural Language Processing and Speech
Harbin Institute of Technology, Harbin, 150001
dqzheng@mtlab.hit.edu.cn

Abstract. This paper presented a novel word co-occurrence model, which was based on an ontology representation of word sense. In this study, word sense ontology is firstly constructed by context multi-elements, and then, the usage of word co-occurrence in content was gotten in using part of speech, semantic, location, average co-occurrence transition probabilities, and was expressed as word co-occurrence feature; final, word cohesion is calculated to judge the co-occurrence degree by the same co-occurrence feature. The relation experiments in natural language processing acquire better results.

1 Introduction

A word co-occurrence refers to words often used together. The word co-occurrence statistical information is one of the key questions of natural language processing. The idea that at least some aspects of word meaning can be induced from patterns of word co-occurrence statistics is becoming increasingly popular and is applied to information retrieval, information extraction, data mining, text clustering, text categorization, etc.

In natural language processing, mutual information (MI)[1] was often used to describe word co-occurrence and measure word cohesion of word x and y. Word co-occurrence statistics were often computed for words appearing together in a sentence or within a window of *n* words. Finally, all sets were ranked by their co-occurrence statistics and the one receiving the highest rank was considered the best choice.

Yutaka Matsuo, et al studied keyword extraction from a single document using word co-occurrence statistical information[2], et al extracted semantic representations based on word co-occurrence approach[3], et al found User Semantics on the Web using Word Co-occurrence Information[4]. For information retrieval, this is important because if a simple word co-occurrence model is used with no lexical expansion[5], NIST automatically evaluate machine translation quality using n-gram co-occurrence statistical information[6]. In NTCIR1, Lin, et al also studied word co-occurrence information for translation disambiguation[7]. In their study, mutual information was used to measure word cohesion of two translations x and y within a text window of 3 for two query terms. Gao, et al designed a word co-occurrence model that considers the distance between two words in computing their cohesion score[8].

Z. Zhang and J. Siekmann (Eds.): KSEM 2007, LNAI 4798, pp. 437–446, 2007.

However, previous researches have failed to consider the semantic information and have ignored the changeful environment of topics. In order to overcome above shortcomings, in this paper, a novel Word Co-Occurrence Model was presented for natural language processing.

In this paper, a sentence is considered to be a set of words separated by a stop mark (".", "?" or "!"). We also include document titles, section titles, and captions as sentences. Two terms in a sentence are considered to co-occur once. That is, we see each sentence as a "basket," ignoring term order and grammatical information except when extracting word sequences.

2 Word Co-occurrence Model

Ontology was recognized as a conceptual modeling tool, which can descript an information system in the semantics and knowledge[9]. After ontology was introduced in the field of Artificial Intelligence[10], it was combined with natural language processing and applied in many field, such as knowledge engineering[11], information retrieval, and semantic Web[12]. It provides with theory to construct the word sense ontology and word co-occurrence statistical information.

In this study, we determined the structure of such a linguistic ontology knowledge, which is comprised of a word sense ontology description and a representation of word co-occurrence. In this study, firstly, word sense ontology is firstly constructed by context multi-elements. And then, word co-occurrence representation will be respectively acquired by determining, for each word, including its co-occurrence with semantics, pragmatics, and syntactic information. Subsequently, word cohesion is calculated to judge the co-occurrence degree by the word co-occurrence representtation, final, we will compare the semantic evaluation value in word co-occurrence.

2.1 Word Sense Ontology Construction

In practical application, ontology can be described in natural languages, framework structure, semantic web, logical language, etc[13]. At present, some popular methods, such as Ontolingua[14], CycL, Loom[15], are all based on logical language.

Despite the strong logical expression, it is not easy for logical language to deduce the process. In this study, we provided a framework structure of such a Chinese word. This structure is a readable format by computer and comprises of Chinese Pinyin, part of speech (POS), English translation, semantic information, relationship, synonymy, etc. Figure 1 shows the word sense ontology description framework for Chinese keywords.

In this study, we automatically construct the Chinese word sense ontology based on a combination of HowNet[16], Chinese Thesaurus, Chinese-English bilingual dictionary and other information.

Where, the Semantic information is mainly from the semantic definition of Chinese word in HowNet. There is only a number to denote the semantic definition in "HowNet-Definitions"[16] and the number will replace the semantic information in this paper. For example, '爱好(hobby)' is defined as 'DEF={fact|事情:{FondOf|喜欢:target={~}}}' and its number is 10086, then its semantic information is replaced by number 10086.

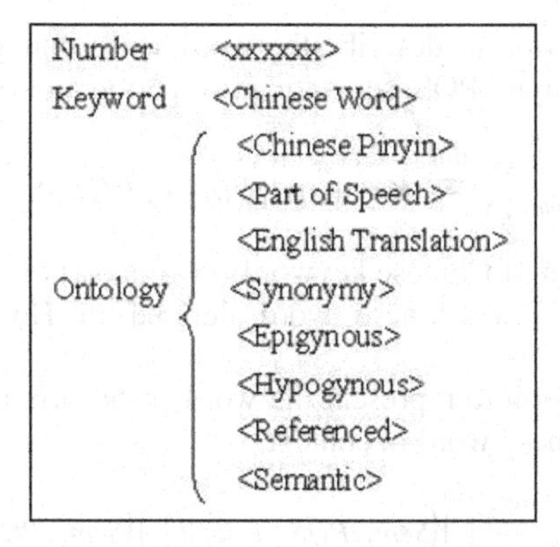

Fig. 1. Word sense ontology description

The POS information is from the Chinese POS tagging set developed by Mitlab in Harbin Institute of Technology. It has 52 labels including 10 punctuations and the label is from 0 to 51 and we define the POS of a Chinese word as label 52.

The English translation is from a Chinese-English bilingual dictionary developed by Mitlab and the dictionary contains basic Chinese-English word 102,615 pairs and auxiliary Chinese-English word 23,067 pairs.

We referred to the Chinese thesaurus developed by IR laboratory of Harbin Institute of Technology. The thesaurus got rid of some useless words and is expanded to 77,343 words.

2.2 Word Co-occurrence Description

The multi-elements of a Chinese word in context, including co-occurrence, the co-occurrence distance, position, will act as the composition of word co-occurrence representation. In figure 2, the characteristic string W_1, W_2, …, W_i represents POS and semantic string, *Keyword* is a Chinese word, *l* or *r* is the position of word that is left or right co-occurrence with *keyword*.

Fig. 2. Word co-occurrence description

2.3 Word Co-occurrence Representation

Word Co-occurrence description plays an important role in the language modeling. In this paper, we will consider multi-elements of a Chinese word and its co-occurrence such as POS, semantic, position and transition probabilities to describe the word co-occurrence.

We define an expression to describe the word sense ontology of a Chinese word including its Chinese Pinyin, POS, semantic information and English translation.

$$Keyword_{Ontology} \overset{def}{=} Keyword(Pinyin, POS, Sem, Tran) \quad (1)$$

Where, *Pinyin* is the sign of Chinese keyword, *POS* is part of speech, *Sem* denotes the semantic information of a Chinese word defined in Hownet, *Tran* is English translation.

We define an expression to represent the word co-occurrence, which considers the multi-elements of a Chinese word in context.

$$Keyword_{co-occur} \overset{def}{=} \left(\bigcup_{l=1}^{m} (Sem_l, POS_l, L, \overline{C_l}) / \bigcup_{r=1}^{n} (Sem_r, POS_r, L, \overline{C_r}) \right) \quad (2)$$

Where, $keyword_{co\text{-}occur}$ denotes the Chinese word that co-occurs with other word. The right side is regarded as word co-occurrence statistics that acquired from training data. $(Sem_l, POS_l, L, \overline{C_l})$ denotes the left side *l-th* co-occurrence of the keyword, which comprises of semantic, POS, the position and the weighed average co-occurrence distance, $(Sem_r, POS_r, L, \overline{C_r})$ denotes the right side *r-th* co-occurrence. The symbol "/" separates the left side and right side of the keyword. The symbol "∪" denotes the aggregate of word co-occurrence statistical information of the keyword.

We define the keyword and (Sem_i, POS_i, L) as a semantic pair to mark with <*Keyword*, (Sem_i, POS_i, L)> and uniquely denotes the keyword and its co-occurrence.

$\overline{C_l}$ denotes the weighed average co-occurrence distance of the semantic pair <*Keyword*, (Sem_i, POS_i, L)> in all training data.

$$\overline{C_l} = (1 + w_l) \frac{1}{all} \sum_{i=1}^{all} C_i \quad \text{wl=(0~1)} \quad (3)$$

Where, C_i denotes the *i-th* co-occurrence distance of the semantic pair <*Keyword*, (Sem_i, POS_i, L)>; all is the times in all training data; w_l denotes the weighing of the semantic pair. If the semantic pair appears several times, the weighing is bigger.

Formula 2 is word co-occurrence representations of a Chinese word, all word co-occurrence construct a knowledge bank in a specific field. Formula 4 is the word co-occurrence statistics.

$$Co-Occurrence_{Field} \overset{def}{=} \sum_{all} Keyword_{co-occur} \quad (4)$$

2.4 Word Co-occurrence Acquisition

To acquire word co-occurrence statistics, we first need to know their *POS* and semantic information in a sentence, and then, get the ***Characteristic String*** to replace this sentence for acquiring word co-occurrence information. An example is shown in table 1.

Table 1. Characteristic string acquisition

Items	Results ("游客" acts as keyword)
Chinese sentence	外国游客来北京游玩。
Segmentation and POS tagging	外国 nd/ 游客 ***Keyword***/来 vg/ 北京 nd/ 游玩 vg/ 。wj/
Semantic tagging	外国 nd/021243 游客 ***Keyword***/070366 来 vg/017545 北京 nd/021243 游玩 vg/092317 。wj/-1
Characteristic String	nd/021243***Keyword***/070366 vg/017545 nd/021243 vg/092317

In this study, we will get word co-occurrence statistics by learning the usage of a Chinese word and their co-occurrence in semantics, pragmatics and syntactic information in every document. Algorithm 1 is the processing of word co-occurrence information acquisition for a document.

Algorithm 1

Step1, extract the keywords of document.

For every document D_i, we firstly get Chinese word segmentation, subsequently, extract *i* keywords by *tf**idf* strategy for every document.

Step2, get characteristic string.

We will use the word sense ontology description to make all synonyms into the same one. Subsequently, extract the sentence that includes the keywords to construct a temporary file and regard a sentence as processing units. And then, word segmentation, *POS* tagging, semantic tagging will be gone and get rid of the auxiliary word likes "*的、地、得、了*" etc to get a ***Characteristic String***.

Step3, Calculate the co-occurrence distance.

We respectively get the co-occurrence distant between a semantic pair of word co-occurrence by formula 5. Where, α is a variable for different training data.

$$C_{li} = \left(\frac{1}{\alpha}\right)^{i-1} \qquad C_{rj} = \left(\frac{1}{\alpha}\right)^{j-1} \tag{5}$$

Step4: Calculate the weighed average co-occurrence distance between a semantic pair of word co-occurrence.

In a document, keywords and their co-occurrence information $\left(Sem_i, POS_i, L, \overline{C_i}\right)$ construct the word co-occurrence for this document.

And here, we get word co-occurrence of a semantic pare in a document, and all training data is learned, we will get word co-occurrence statistics in a field.

To avoid data sparseness, the weighed average co-occurrence distance is defined with β if word co-occurrence of a semantic pair does not appear in knowledge bank.

2.5 Word Cohesion Calculating

Mutual information was often used to measure word cohesion of word x and y. Word co-occurrence statistical information were often computed for words appearing together in a sentence or within a window. Finally, all sets were ranked by their co-occurrence statistics and the one receiving the highest rank was considered the best choice.

In this paper, we consider word co-occurrence representation to calculate word cohesion. We firstly define two symbols, Sem_Word_i and Sem_Pair_i. Sem_Word_i denotes the relation information appeared in document, including semantic label, POS, location, and Sem_Pair_i denotes the semantic pair constructed by the *Keyword* and its co-occurrence.

$$Sem\_Word_i \overset{Def}{=} (Sem_i, POS_i, i)$$

$$Sem\_Pair_i \overset{Def}{=} < Keyword, Sem\_Word_i >$$

In actual processing, we will first acquire the ***Characteristic String*** of every sentence, and then, regard the *Keyword* as the center to define the left side word cohesion of Keyword *Cohesion_Left*.

$$Cohesion\_Left = \prod_{i=1}^{m} P(Sem\_Word_{li} \mid Sem\_Pair_{li-1})$$

Where, $P(Sem\_Word_{li} \mid Sem\_Pair_{li-1})$ is the conditional probability, Sem_Word_0 and Sem_Pair_0 denotes the *Keyword*. And then,

$$\log Cohesion\_Left = \sum_{i=1}^{m} \log P(Sem\_Word_{li} \mid Sem\_Pair_{li-1})$$

According to word co-occurrence presentation and the monotony of logarithm, we replaced the conditional probability with the weighed average co-occurrence distance of the semantic pair, so, *Cohesion_Left* is as follows,

$$Cohesion\_Left = \sum_{i=1}^{m} \overline{C_{li}} \tag{6}$$

In a similar way, the right side word cohesion is defined.

And then, the word cohesion of word co-occurrence in a sentence is the left side word cohesion plus the right side. It is defined as follows,

$$Cohesion\_Keyword_S = Cohesion\_Left + Cohesion\_Right \tag{7}$$

The word cohesion of word co-occurrence in a document is defined as follows (t is the number of sentences in a document),

$$Cohesion\_Keyword_D = \sum_{S=1}^{t} Cohesion\_Keyword_S \tag{8}$$

So, the word cohesion of word co-occurrence in a document is defined as follows (k is the number of the keywords in a document),

$$Cohesion\_Docment = \sum_{D=1}^{k} Cohesion\_Keyword_D \tag{9}$$

3 Application Strategy

We applied the novel word co-occurrence model to information retrieval, Text similarity computing, document re-ranking, etc. There are various applications of word co-occurrence and the document will get different word cohesion from word co-occurrence statistics for different application.

In this study, we use NTCIR-3[17] formal Chinese test collection. For every topic, we will extract its keywords and acquire word co-occurrence statistics from its description type by algorithm 1. For example,

Title: 复制小牛之诞生 (The birth of a cloned calf)

Description: 与使用被称为体细胞核移植的技术创造复制牛相关的文章 (Articles relating to the birth of cloned calves using the technique called somatic cell nuclear transfer)

After Chinese words are segmented, we select "体细胞，移植，技术，创造，复制，牛" to act as the keyword of this topic and acquire word co-occurrence statistics of every topic from its description type run (D-Run) by algorithm 1. Figure 3 is the presentation of word co-occurrence.

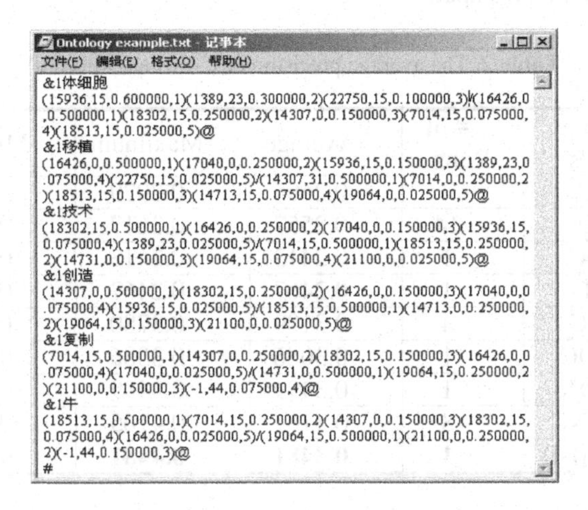

Fig. 3. Presentation of word co-occurrence

For description, we defined a similarity ratio to measure the text similarity between two documents by word cohesion of word co-occurrence. It is defined by formula 10.

$$Similarity\_Ratio = \frac{Cohesion\_Docment_i}{Cohesion\_Docment_j} \tag{10}$$

4 Experiment and Discussion

4.1 Information Retrieval

In this study, we use Chinese word as indexing units, we firstly do an initial retrieval according to user query, and we expand user query based on word sense ontology description of keywords. Subsequently, we respectively acquire word co-occurrence statistics from user query and retrieval documents, and then, compare the similarity between user query and retrieval documents according to the word cohesion of word co-occurrence to reorder the initial retrieval document set.

$$Similarity_{ratio-Doc_i} = \frac{Cohesion - Doc_i}{Cohesion - Topic} \tag{11}$$

We use NTCIR-3 Formal Chinese Test Collection, which contains 381,681 Chinese documents and 42 topics, and select 42 D-runs type query to evaluate our method[17]. We use the two kinds of relevant measures, i.e. relax-relevant and rigid-relevant[18] and compare the results with other group submission.

In NTCIR-3 workshop, in total, 8 different combinations of topic fields and 34 runs are used in Chinese-Chinese (C-C) single language information retrieval (SLIR) track. Table 2 shows the distribution and the corresponding average, maximum, minimum of average precision[17] and HLM-D is the result of the D-runs that is proposed method in this paper.

Table 2. The average precision of C-C runs (Relaxed)

Topic Fields	# of Runs	Average	Maximum	Minimum
C	4	0.2605	0.2929	0.2403
D	14	0.2557	0.3617	0.0443
DC	1	0.2413	0.2413	0.2413
T	1	0.2467	0.2467	0.2467
TC	4	0.3109	0.3780	0.2389
TDC	1	0.3086	0.3086	0.3086
TDN	1	0.3499	0.3499	0.3499
TDNC	8	0.3161	0.4165	0.0862
HLM-D	**1**	**0.4481**	**0.4481**	**0.4481**

Generally speaking, the *TDNC*-runs show good performance than other runs. The highest precision run of *C-C* runs is a *TNDC*-run, but the lowest is also a *TNDC* for all submissions in NTCIR-3 workshop[17]. In table 2, HLM-D-runs improve the performance than other runs, and there are about an average 9.34% increase in precision can be achieved for the relaxed relevance C-C run compared with rigid relevance C-C run according to the proposed method.

4.2 Text Similarity Calculation

Text similarity is a measure for the matching degree between two or more texts, the more high the similarity degree is, the more the meaning of text expressing is closer, vice versa. Some proposal methods included Vector Space Model[19], Ontology-based[20], Distributional Semantics model[21], etc.

In this paper, first, for two Chinese texts D_i and D_j, we respectively extract k same feature words, if the same feature words in the two texts is less than k, we don't compare their similarity. Second, acquire word co-occurrence statistics of every text by their feature words. Third, get the word cohesion of every text by the proposal method. Final, compute the similarity ratio of every two text D_i and D_j. The similarity ratio equals to the ratio of the similarity evaluation value of text D_i and D_j by formula 10, if the ratio is in the threshold γ, we think that text D_i is similar to text D_j.

We download four classes of text for testing from Sina, Yahoo, Sohu and Tom, which include 71 current affairs news, 68 sports news, 69 IT news, 74 education news.

For the test of current affairs texts, according to the strategy of similarity calculation, we choose five words as feature word, i.e. 贸易(trade), 协议(protocol), 谈判(negotiation), 中国(China), 美国(America). In the test, the word "经贸(economic and trade), 商贸(business)" are all replaced by "贸易(trade)" and other classes are similar. The testing result is shown in table 3.

Table 3. Testing results for text similarity

Items	0.95<α<1.05			0.85<α<1.15		
	Precision	Recall	F_1-measure	Precision	Recall	F_1-measure
Current affairs	97.14%	97.14%	97.14%	94.60%	100%	97.23%
Sports	88.57%	91.18%	89.86%	84.62%	97.06%	90.41%
IT	93.75%	96.77%	95.24%	91.18%	100%	95.39%
Education	94.74%	97.30%	96.00%	90.24	100%	94.87%
General	93.57%	95.62%	94.58%	90.07%	99.27%	94.42%

We analyzed all the experimental results to find that the results for current affairs texts are the best, while the sports texts are lower than others. We think it is mainly because some sports terms are unprofessional for the lower sports texts recognition, such as 汉家军(han jia jun), 救主(savior), 郝董(hao dong), etc. Other feature words are more fixed and more concentrated.

5 Conclusion

In this paper, we gave a novel word co-occurrence model based on ontology for natural language processing. Our relation experimental results show a good performance than other runs.

In the proposal approach, we only use a part of word sense ontology description and combined with some statistical information. Further work include: (1) use more semantic relation, (2) combine with some NLP technologies, (3) improve the model, (4) apply this method to other field.

Acknowledgements

This work is supported by the national natural science foundation of China (No.60435020) and the National High-Tech Development 863 Program of China (No.2006AA01Z150, 2006AA010108).

References

1. Kenneth, W.: Church and Patrick Hanks. Word association norms, mutual information and lexicography. In: Proceedings of the 27th Annual Conference of the Association of Computational Linguistics, pp. 76–82 (1989)
2. Matsuo, Y., Ishizuka, M.: Keyword extraction from a single document using word co-occurrence statistical information. International Journal on Artificial Intelligence Tools 13(1), 157–169 (2004)
3. Bullinaria, J.A., Levy, J.P.: Extracting Semantic Representations from Word Co-occurrence Statistics: A Computational Study, Behavior Research Methods (2006)
4. Mori, J., Matsuo, Y., Ishizuka, M.: Finding user semantics on the web using word co-occurrence information. In: PersWeb 2005. Proc. Int'l. Workshop on Personalization on the Semantic Web (2005)
5. Papka, R., Callan, J.P., Barto, A.G.: Text-Based Information Retrieval Using Exponentiated Gradient Descent. In: NIPS, pp. 3–9 (1996)
6. NIST: Automatic Evaluation of Machine Translation Quality Using N-gram Co-Occurrence Statistics (2002), http://www.nist.gov/speech/tests/mt/resources/scoring.htm
7. Lin, C.J., Lin, W.C., Bian, G.W., Chen, H.H.: Description of the NTU japanese-english cross-lingual information retrieval system. In: Proceedings of the NTCIR2 Workshop (1999)
8. Gao, J.F., Nie, J.Y., Zhang, J., Xun, E.D., Su, Y., Zhou, M., Huang, C.N.: TREC-9 CLIR experiments at MSRCN. In: The Nineth Text REtrieval Conference (November 2000)
9. Neches, R., Fikes, R., Finin, T., et al.: Enabling Technology for Knowledge Sharing. AI Magazine 12(3), 16–36 (1991)
10. Gruber, T.R.: Toward principles for the design of ontologies used for knowledge sharing. In: International Workshop on Formal Ontology (1993)
11. CycL, http://www.cyc.com
12. W3C Semantic Web, http://www.w3.org/2001/sw
13. Uschold, M.: Building Ontologies-Towards A Unified Methodology. In: expert systems (1996)
14. Ontolingua, http://www.ksl.stanford.edu/software/
15. Loom, http://www.isi.Edu/isd/LOOM/
16. Dong, Zh.D.: http://www.keenage.com/
17. NTCIR, http://research.nii.ac.jp/ntcir-ws3/work-en.html
18. Lingpeng, Y., Donghong, J., Li, T.: Document Re-ranking Based on Automatically Acquired Key Terms in Chinese Information Retrieval[A]. In: Proceedings of the COLING, Sydney, pp. 480–486 (2004)
19. Salton, G., Buckley, C.: Term weighting approaches in automatic text retrieval. Information Processing and Management 24(5), 513–523 (1988)
20. Oleshchuk, V., Pedersen, A.: Ontology Based Semantic Similarity Comparison of Documents. In: 14th International Workshop on Database and Expert Systems Applications, pp. 735–738 (September 2003)
21. Besancon, R., Rajman, M., Chappelier, J.C.: Textual similarities based on a distributional approach. In: Tenth International Workshop on Database and Expert Systems Applications, pp. 180–184 (September 1-3, 1999)

Cost-Time Sensitive Decision Tree with Missing Values*

Shichao Zhang[1,2], Xiaofeng Zhu[1], Jilian Zhang[3], and Chengqi Zhang[2]

[1] Department of Computer Science, Guangxi Normal University, Guilin, China
[2] Faculty of Information Technology, University of Technology Sydney
P.O.Box 123, broadway, NSW2007, Australia
[3] School of Information Systems, Singapore Management University
{zhangsc,chengqi}@it.uts.edu.au, xfzhu_dm@163.com,
jilian.z.2007@phdis.smu.edu.sg

Abstract. Cost-sensitive decision tree learning is very important and popular in machine learning and data mining community. There are many literatures focusing on misclassification cost and test cost at present. In real world application, however, the issue of time-sensitive should be considered in cost-sensitive learning. In this paper, we regard the cost of time-sensitive in cost-sensitive learning as waiting cost (referred to WC), a novelty splitting criterion is proposed for constructing cost-time sensitive (denoted as CTS) decision tree for maximal decrease the intangible cost. And then, a hybrid test strategy that combines the sequential test with the batch test strategies is adopted in CTS learning. Finally, extensive experiments show that our algorithm outperforms the other ones with respect to decrease in misclassification cost.

1 Introduction

Inductive learning techniques have met great success in building models that assign testing cases to classes in [1, 2]. Traditionally, inductive learning built classifiers to minimize the expected number of errors (also known as the 0/1 loss). Cost Sensitive Learning (CSL) is an extension of classic inductive learning for the unbalance in misclassification errors. Much previous work of CSL has focused on how to minimize classification errors. However, there are different types of classification errors, and the costs of different types of errors are often very different. For example, for a binary classification task in medical domain, the cost of false positive (FP) and the cost of false negative (FN) are often quite different. In addition, misclassification cost is not the only cost to be considered when applying the model to new cases. In fact, there are a variety of costs are often referred in cost-sensitive learning in [3], such as, misclassification cost, test cost, waiting cost, teacher cost, and so on.

* This work is partially supported by Australian large ARC grants (DP0559536 and DP0667060), a China NSF major research Program (60496327), China NSF grant for Distinguished Young Scholars (60625204), China NSF grants (60463003), an Overseas Outstanding Talent Research Program of Chinese Academy of Sciences (06S3011S01), an Overseas-Returning High-level Talent Research Program of China Ministry of Personnel, a Guangxi NSF grant, and an Innovation Project of Guangxi Graduate Education (2006106020812M35).

Z. Zhang and J. Siekmann (Eds.): KSEM 2007, LNAI 4798, pp. 447–459, 2007.

Much precious research focuses on test cost and misclassification cost in [4-7] and select some attributes to test in order to minimize the total cost. They considered not only the misclassification cost but also test cost instead of the right ratio of the classification. But in these work, the test cost and the misclassification cost have been defined on the same cost scale, such as dollar, which is incurred in a medical diagnosis. For example, the test cost of attribute A_1 is 50 dollars, A_2 is 20 dollars, the cost of false positive is 600 dollars, and the cost of false negative is 800 dollars. But in fact, a same cost scale is not always reasonable. For sometimes we may encounter difficulty in defining the multiple costs on the same cost scale. It is not only a technological issue, but also a social issue [20, 21]. For example, in medical diagnosis, how much money we should assign for a misclassification cost? So we need to involve the costs with different scales in cost-sensitive learning.

Next, it is necessary for considering waiting cost, which is defined as "the cost of performing a certain test may depend on the timing of the test" [3] in medical analysis. Time-sensitive costs are relatively common. For example, in any classification problem that requires multiple experts, one of the experts might not be immediately available. That will be paid the cost of waiting. The research on waiting time has broadly been studied in economics [8], medical service [9], and so on, but a little research [6, 10] can be found in cost-sensitive learning. Hence, it is necessary to take the waiting time into account in cost-sensitive learning.

Thirdly, the data is often incomplete in real-world applications, i.e. the data may contain missing values. There are many methods [11, 12] to handle the missing data and there are only a little literatures [4-7] focused on dealing with missing values in cost-sensitive learning. How to deal with the missing values in training sets and in test sets is still a hot issue in cost-sensitive learning.

In this paper, waiting cost (WC) will be defined firstly and talked about with misclassification cost and test cost in cost-sensitive learning later. Then cost-time sensitive decision tree (referred to CTS decision tree) will be constructed with missing values that are appearing in training sets as well in test sets. Next, a hybrid test strategy in which the usually sequential test strategy and single batch strategy will be combined to consider is proposed after constructing the cost-time sensitive decision tree. Different from the existing methods, our method presents some characteristics as fellows: 1) Introducing a new cost concept—waiting time cost—in details in cost-sensitive learning based on the research in medical application and economics. 2) In training set examples, the CTS decision tree is built based on a new attribute splitting method which is a trade-off among classification ability, misclassification cost and tangible cost (including test cost and waiting cost). 3) A hybrid test strategy which can receive the most decrease for misclassification cost or intangible cost in limited tangible costs is proposed for the test examples with missing values in order to test the constructed CTS decision tree.

The rest of the paper is organized as follows. In section 2 we review the related work. Then we present our methods in section 3 and experimental results will be presented in section 4. Finally, in section 5 we will conclude our work with a discussion of future work.

2 Review of Pervious Work

Cost-sensitive learning was first applied to the medical diagnosis for solving various problems. In real-world applications, there are many other costs besides the misclassification cost in [3]. We follow the categorization as mentioned in [13] and give a refined review in category 4 on classifiers sensitive to both attribute costs and misclassification costs as follows:

- **Classifiers minimizing 0/1 loss**
  This has been the main focus of machine learning, from which we mention only CART and C4.5. They are standard top-down decision tree algorithms. C4.5 introduced the information gain as a heuristic for choosing which attribute to measure in each node. CART uses the GINI criterion. Weiss et al. [14] proposed an algorithm for learning decision rules of a fixed length for classifications in a medical application; there are no costs (the goal is to maximize prediction accuracy).
- **Classifiers sensitive only to attribute costs**
  The splitting criterion of these decision trees combines information gain and attribute costs in [15, 16]. These policies are learned from data, and their objective is to maximize accuracy (equivalently, to minimize the expected number of classification errors) and to minimize expected costs of attributes.
- **Classifiers sensitive only to misclassification costs**
  This problem setting assumes that all data is provided at once [17], therefore there are no costs for measuring attributes and only misclassification costs matter. The objective is to minimize the expected misclassification costs.
- **Classifiers sensitive to both attribute costs and misclassification costs**
  More recently, researchers have begun to consider both test and misclassification costs: [4-7, 18] proposed a new method for building and testing decision trees that minimizes the sum of the misclassification cost and the test cost. The objective is to minimize the expected total cost of tests and misclassifications. Both algorithms learn from data as well.

As far as we know, some researchers believe that it is necessary to consider the misclassification cost and test cost simultaneously. Under this assumption, the doctors must make their optimal decisions concerning the trade-offs between the less test cost and lower misclassification cost. Some researchers use a simple strategy to quantify the misclassification cost to be the same as the test cost [4, 6, 19] (for example, use *dollar* as cost unit). However, the unit scales for different costs are usually various in real-world applications, and it is always difficult for people to quantify these various unit scales of costs into a sole unit [20, 21]. The literature [20, 21] proposed a cost-sensitive decision tree that considers two different unit scales for costs. However, this method did not treat the tangible cost (such as test cost and waiting cost etc) and intangible cost (such as misclassification cost) as two different resources essentially, instead it divides all the costs into two groups of cost, that is the target cost and the source cost.

It often happens that the result of a test is not available immediately. For example, a medical doctor typically sends a blood test to a laboratory and gets the result in the next day. In [6, 10], they regarded the cost of waiting time as waiting cost or delay cost. In this paper, we give our definition of waiting time and combine it with test cost and common cost as tangible cost in our proposed algorithm.

There exists missing values during the process of cost-sensitive learning. Sometimes, values are missing due to unknown reasons or errors and omissions when data are recorded and transferred. As many statistical and learning methods cannot deal with missing values directly, examples with missing values are often deleted. However, deleting cases can result in the loss of a large amount of valuable data. Thus, much previous research has focused on filling or imputing the missing values before learning and testing. However, under cost-sensitive learning, there is no need to impute any missing data and the learning algorithms should make the best use of only known values and that "missing is useful" to minimize the total cost of tests and misclassifications. There is a little research that has paid attention to this, such as, [22]. In this paper, we will talk about the case in which missing values can be encountered both in training sets and test sets.

In summary, waiting cost which will be defined in advanced is combined the others costs in order to cost-sensitive learning, a novelty CTS decision tree can be built and a hybrid test strategy will be proposed to test the constructed decision tree with missing values in the dataset. Extensive experiments will show the efficiency of our method.

3 Building CTS Decision Tree

We assume that the training data and test data contain some missing values. In our proposed CTS decision tree, these processes will be considered: 1) selecting the splitting attributes. 2) building the CTS decision tree. 3) testing the constructed CTS decision tree.

3.1 Some Definition of Costs

Turney [10] has created taxonomy of the different types of cost in inductive concept learning in [3]. According to this taxonomy there are nine major types of costs. In this paper, we will take three of them into account.

3.1.1 Misclassification Cost

Misclassification cost: costs incurred by misclassification errors. This type of errors is the most crucial one and most of the cost-sensitive learning research has investigated the ways to manipulate such costs. These error costs can either be constant or conditional depending on the nature of the domain. Conditional misclassification costs may depend on the characteristics of a particular case, on time of classification, on feature values or on classification of other cases.

3.1.2 Test Cost

Test cost: costs incurred for obtaining attribute values. The necessity for test cost is proportional to the cost of misclassification. If the cost of misclassification surpasses the costs of tests greatly, then all tests of predictive value should not be taken into consideration. Similar to error costs, test costs can be constant or conditional on various issues such as prior test selection and results, true class of instance, side effects of the test or time of the test.

3.1.3 Waiting Cost

It often happens that the result of a test is not available immediately. For example, a doctor typically sends a blood test to a laboratory and gets the result in the next day. A person waiting for the blooding test results will undertake pains, pressures. That is to say, a person must pay some cost for waiting the results of test. In this paper, the waiting cost only reflects the cost of waiting of results after a test is performed. And waiting cost will be confined to the assumption as follows:

1. Waiting cost relates to the waiting time. Different waiting time presents different waiting cost. However, the longer waiting time, the smaller waiting cost maybe. For instance, waiting time for 10 minutes will present more serious effect for the person with heart attack than the one only with catching a cold. The waiting cost of the former will be much more larger than the latter in the same waiting time.
2. Waiting cost varies from person to person, for place to place. For two persons with heart attack, to waiting 10 minutes for the old man is more danger than the twenties, so the waiting cost for the former will higher than the latter.
3. Waiting cost relates the resources of the test. That is to say, a higher income will have a higher waiting cost with the same amount of waiting time [8]. It is obvious that a rich maybe more care the waiting time than the poor in medical test.

In this paper, we assume all the waiting cost for each attribute is a constant that will be decided by the specialist in the domain, denoted as $WC_i, i=1,...,n$, where n is the number of attributes. For satisfied the above three assumptions, each will be multiplied by a coefficient α which is the ratio of a waiting cost to the corresponding assumptions.

In particular, $WC_i = 0$ if the result of test can be got at once or means the waiting have little impact for the result of later test. $WC_i = \infty$, if the next test cannot be performed until the result of the former is presented even if the real waiting time for the result of the test can be got in a finite time, such as several seconds. So the value of WC_i is between 0 and ∞. For ease of calculation, we will regard the scale of waiting cost as the same unit of the test cost. Furthermore, in a batch test, for example, the attributes $A_1,...A_i$ are tested at the same time, we will get $\underset{t=1,...i}{WC} = \max\{WC_t\}$.

3.1.4 Tangible Cost and Intangible Cost

In this paper, we regard Misclassification Cost as Intangible Cost (we refer it to IC) which means the cost exists but cannot be explained with some unit, such as dollar. On the contrary, Tangible Cost (noted as T_C) means the cost exists and can be explained with some unite, such as dollar, time and so on. The value of T_C for attribute i can be defined as follows:

$T\_C_i = TC_i + \alpha WC_i$, $i=1,...,n$, where n is the number of attributes.

In particular, in our test strategy shown in section3.3, a number of attributes will be test at the same time at the aim to receiving the best system performance. For example, a set of blood tests shares the common cost of collecting blood from the patient. This common cost (referred to CC) is charged only once, when the decision is made to do the first blood test. There is no charge for collecting blood for the second blood test, since we may use the blood that was collected for the first blood test. It is practical for considering common cost during the process of test and it is also helpful to improve the test accuracy in same test cost. Hence, in batch test, the value of T_C can be calculated as follows:

$$T\_C_{1,\ldots,i} = \sum_{t=1}^{i} TC_t - CC_t + \max_{t=1,\ldots i}\{\alpha WC_t\} \tag{1}$$

3.2 Building CTS Tree

3.2.1 Selecting the Attributes for Splitting

There exist many kinds of splitting criterion to construct decision trees. Such as, the Gain (in ID3), the Gain Ratio (in C4.5) and minimal total cost. However, the methods (for instance, in ID3 or C4.5) consider the classification ability for the attribute without taking the costs into account. On the contrary, the method with minimal total cost [19] concerns the cost without paying much attention to the classification ability of the attribute. In our view, we hope to obtain optimal results both on the classification ability and on cost under the assumption of the limited resources. So in our strategy, the term 'Performance' which is equal to the return (the Gain Ratio multiplied by the total misclassification cost reduction) divided by the investment (tangible cost) is employed to meet our demand. That is to say, we select the attribute with larger Gain Ratio, lower test cost and decrease the misclassification cost. The 'Performance' is defined as follows:

$$\text{Performance}(A_i) = (2^{\text{GainRatio}(A_i,T)} - 1) * \text{Redu\_Mc}(A_i) / (\text{TC}(A_i) + 1) \tag{2}$$

Where, GainRatio(A_i,T) is the Gain Ratio of attribute A_i, TC(A_i) is the test cost of attribute A_i. Redu_Mc(A_i) is the decrease of misclassification cost brought by the attribute A_i, and we get:

$$\text{Redu\_Mc}(A_i) = \text{Mc} - \sum_{i=0}^{n} \text{Mc}(A_i) \tag{3}$$

Where Mc is the misclassification cost before testing the attribute A_i. If an attribute A_i has n branches, then $\sum_{i=0}^{n} Mc(A_i)$ is the total misclassification cost after splitting on A_i.

For example, in a positive node, the Mc = fp * FP, fp is the number of negative examples in the node, FP is the misclassification cost for false positive. On the contrary, for examples, in a negative node, the Mc = fn * FN, where fn is the number of positive examples in the node, FN is the misclassification cost for false negative.

We select the attribute with maximal values of Performance(A_i) (i=1,…,n; n is the number of attributes) as the current splitting attribute during the process of constructing the CTS decision tree.

3.2.2 Building Tree

Our algorithm chooses an attribute with maximal Performance based on formula (2) to generate a node. Then, similar to C4.5, our algorithm chooses a locally optimal attribute without backtracking. Thus the resulting tree may not be globally optimal.

However, the efficiency of the tree-building algorithm is generally high. Some notes will be presented for constructing as follows:

Firstly, according to the formula (2), we select the attribute with maximal Performance as current splitting attribute to generate a node or a root node. If a number of the attributes' Performance is equal, the criterion to select test attribute should follow in priority order:

1) the bigger Redu_Mc;
2) the bigger test cost.

This is because that our goal is to minimize the misclassification cost.

Secondly, there are missing values in the training set while building the CTS decision tree. Zhang et al. [22] experimentally demonstrated all kinds of methods for dealing with missing values for constructing cost-sensitive decision tree, and made a conclusion that the best method would be the internal node strategy in [19] in which the missing values will be handled by internal nodes without being imputed. Hence, we will employ the internal nodes method to dealing with missing values in the training examples in our cost-time sensitive decision tree.

Another important point is how to stop building tree. In the process of classify a case using a decision tree, the layer that will be visited may be different if the resources provided is different. The more resources a case has, the more layers it will be visited. For allowing a decision tree that can fit for all kinds of needs, the condition of stopping building tree is similar to the C4.5. That is to say, when one of the following two conditions is satisfied, the process of building tree will be stopped.

(a) all the cases in one node are positive or negative;
(b) all the attributes are used up.

The next problem is how to label a node when it has both positive and negative examples. In traditional decision tree algorithms, the majority class is used to label the leaf node. In our case, as the decision tree is used to make predictions to minimize the misclassification cost or intangible cost under given resources. That is, at each leaf, the algorithm labels the leaf as either positive or negative (in a binary decision case) by minimizing the misclassification cost or intangible cost. Suppose that there is a node, P denotes that the node is positive and N denotes that the node is negative. The criterion is as follows:

$$\begin{matrix} P & \text{if} \\ N & \text{if} \end{matrix} \begin{cases} p*FN > n*FP \\ p*FN < n*FP \end{cases}$$

Where, p is the number of the positive case in the node, while n is the number of negative cases. FN is the cost of false negative, and FP is the cost of false positive. If a node including positive cases and negative cases, we must pay the cost of misclassification no matter what we conclude. But the two costs are different, so we will choice the smaller one. Here we consider that the cost of the right judgment is 0. For example, there is a node including 20 positive cases and 24 negative cases, and if the cost of the false positive and false negative is the same, the node will be considered negative. But in our view, they are different, such as FN is 800 and FP is 600, the node will be considered positive because 16000 is larger than 14400. In reality, considering the situation that a patient (positive) is classified as health (false negative), the result may be that he will lose of his life. On the contrary (false positive), he may only need to pay the cost of medicine fee, etc. So the cost is different. And in general, we think that the cost of false negative (FN) is larger than the cost of false positive (FP).

Finally, the attribute at the top or root of the tree is likely the attribute which is with zero (or small) tangible cost, more decrease of the intangible cost and strongly classification ability without missing values base on the principle of our constructing cost-time sensitive decision tree.

3.3 Performing Tests on Test Examples

After the decision tree is built and all the nodes are labeled, the next interesting question is how this tree can be used to deal with test examples and pay the minimal misclassification cost or intangible cost confined to any tangible costs. There exist many kinds of test strategies in a number of literatures in [4-7, 19]. We can classify all these test strategies into three catalogues: sequential test strategies, batch test strategies and the hybrid test strategies combined the sequential method with the batch one. In our CTS decision tree, it isn't practical for either the single sequential strategies or the batch strategies to be employed after considering waiting cost and the test examples with different resources. For example, a test example with enough test resources would not care about the test cost but consider the waiting time. In a medical test, the rich maybe test all the attributes under the unknown attribute (or node) which will be got the result costing more waiting time. This changes the sequential strategies to batch one. On the contrary, facing a number of attributes to be tested, the poor maybe only select one attribute to test due to the limited resources. Hence, in this paper, we will propose a hybrid method in which either sequential strategy or batch strategy will be chosen by some principles. The hybrid strategy is constructed as follows:

At first, calculating the utility for each attribute based on the below definition:

$$Utility = \frac{IC}{T\_C} \tag{4}$$

Secondly, computing the utility for some batch attributes which will satisfy some rules below:

1) This batch attributes will be recommended by the specialist in these domains. These attributes will be combined mainly by some common cost.
2) All the T_C of these attributes won't exceed the resource for this test example.

Different from the principle in [6] in which more attributes would be added into the batch until ROI does not increase, the time complexity of our method for computing utility is only linear because the batch is limited due to introduction by the specialist. And the number of the combination will be 2^n in the method in [6]. It is feasible for the specialists to decide the batch attributes as the researchers in this domains acquaint all the tests which can be grouped better than the single test. This can decrease the tangible costs as well as the time complexity of our algorithm. On the other hand, the limited resources for the test examples will be considered different from existing methods in which the limited resources have not been taken into account, such as, in [4-7, 19].

Thirdly, the attribute(s) with the maximal utility for the result coming from the former two processes will be tested firstly. And the test strategy will be regarded as the sequential strategy while the attribute tested firstly is a single one. Otherwise, it is a batch strategy. So we call our method as a hybrid strategy as there exists both sequential method and batch strategy in our method.

This process for testing will be continued until all tests are completed or the resources of the test example are used up.

4 Experimental Analyses

In this section, we empirically evaluate our algorithm with real-world datasets in order to show its effectiveness. We choose two real-world datasets, listed in Table 1, from the UCI machine learning repository [23]. Each dataset is split into two parts: the training set (60%) and the test set (40%). These datasets are chosen because they have some discrete attributes, popular in use and have a reasonable number of examples. The numerical attributes in datasets are discretized at first using a minimal entropy method [24]. Because there is no missing values in these four datasets, we artificially apply MCAR mechanism on these datasets under different missing rates. And the missing values in the conditional attributes are under missing rates of 10%, 20%, 30%, 40%, 50% and 60%, respectively. To assign Test Cost, we randomly select a cost value that satisfied the constraint of falling into [1, 100] for each attribute. The costs for test are generated at the beginning of each imputation. The misclassification cost is set to 600/800(600 for false positive and 800 for false negative). Note that here the misclassification cost is only a relative value. It has different scales with test costs.

Table 1. Datasets used in Experiments

	Instances	Attributes	Classes(N/P)
Tic-tac-toe	958	9	332/626
Mushroom	8124	21	4208/3916

For comparison, in section 4.1, we construct three CTS decision trees with different splitting criterion to demonstrate the efficiency of our splitting criteria without concerning about the waiting cost (i.e., the value of waiting cost for each attribute in each dataset is equal and constant). One CTS decision tree with the splitting criteria of gain ration is denoted as GR, the method with the minimal total cost is referred to as MTC and our method is regarded as PM. The experimental results of the effect for waiting cost will be presented in section 4.2.

4.1 Experiments for Algorithm's Efficiency

In this section, we evaluate the performances of different algorithms on datasets under different levels of missing rates. For each experiment, we do not consider the resources of test example, i.e., each test example has enough test cost to be tested. The test cost is 900 and 2100, respectively for dataset Tic-tac-toe and Mushroom. Figures 1 and 2 provide detailed results from these two datasets, where the x-axis represents the missing rate and the y-axis is the Misclassification Cost.

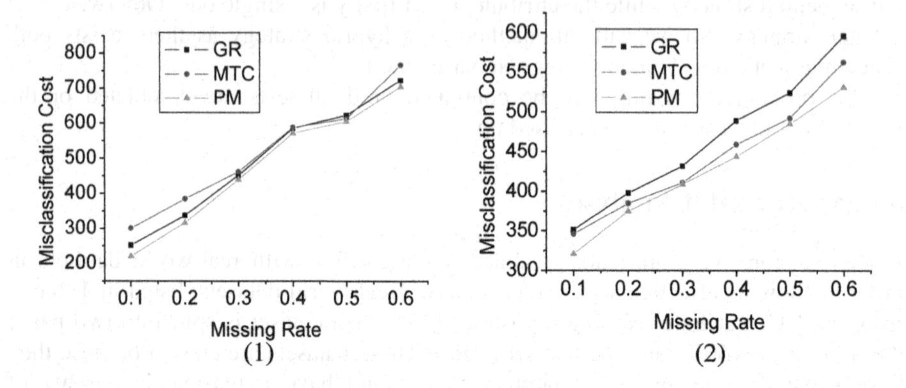

Fig. 1. The misclassification cost VS the missing rate

From Figures 1, we can see that with the increase of the missing rate, all methods suffer from the increase of misclassification cost, because more missing values have been introduced and the model generated from the data has been corrupted. When comparing with the different algorithms, we find that PM is the best among these methods in terms of misclassification cost at different missing rate. The experimental results demonstrate that the method with Formula (2) for spitting attributes is best than the naive methods, such as maximal grain ratio method or the minimal total cost method under the assumption of without limited resources.

In the above paragraph, we did not fix the test costs and we assumed each attribute will be tested with the enough test resources. However, in real applications, there exists a limited resources, such test cost. In this subsection, we will use all kinds of test cost levels to test the efficiency of our method comparing with the other two

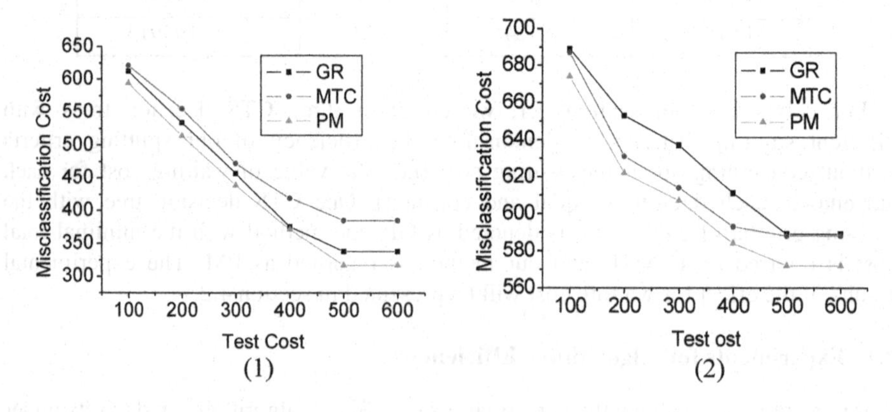

Fig. 2. The misclassification cost VS the test cost

methods. The experimental results of dataset Tic-tac-toe are shown in Figure 3 and 4, for missing rate 20% and 40% respectively. The domain of test cost is between 100 and 600 in Figure 3 and 4, the results of MC is presented in y-axis. The results show that our PM algorithm performs better than the other two algorithms in the terms of limited test costs.

From Fig. 1 to 2, we can make a conclusion that the splitting criterion of PM algorithm is best than the maximal grain ratio and minimal total cost method with all kinds of resources, such as test cost.

4.2 Experiments for Waiting Cost

In section 3, we analyze it is practical for talking about waiting cost into the CTS learning. In this section, we will experimental test this method. In order to investigate the effect of various waiting costs, we set up three levels for waiting costs: (a) all waiting costs are 0 (denoted as Max0 in Figure 3); (b) all delay costs are randomly assigned from 0 to 50; and (c) all delay costs are randomly assigned from 0 to 100. We also design some common cost in our experiments. In our opinion, the higher waiting cost, the less test cost is. This is because there are more common costs which can efficient decrease the test cost for the test examples. From Figure 3, we can see this meets our expectations.

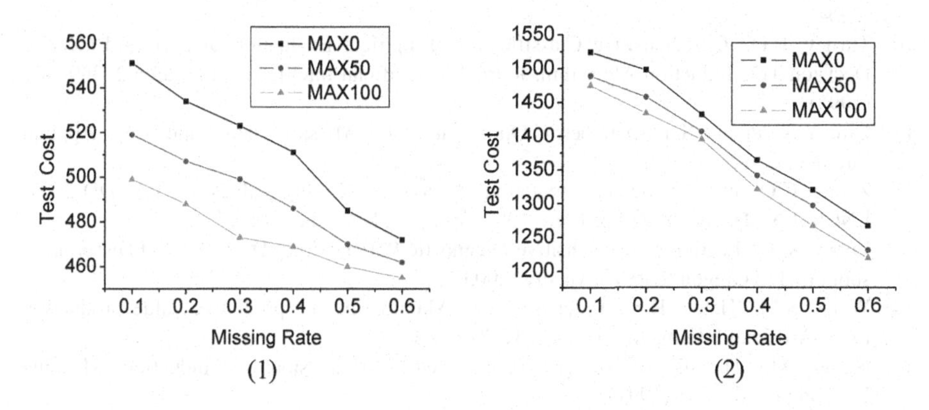

Fig. 3. The test cost VS the missing rate

5 Conclusions

In this paper, we introduce waiting cost into cost-sensitive learning with misclassification cost and test cost to construct CTS decision tree with a new splitting criterion. Then a hybrid test strategy is proposed to test the constructed CTS decision tree with missing values and limited resources. Extensive experiments show our method outperforms the existed methods and it is feasible for talking about waiting cost in CTS learning. In the future, we will further demonstrate the advantages of CTS decision tree.

References

1. Quinlan, J.R.: C4.5: Programs for Machine Learning. Morgan Kaufmann Publishers, San Francisco (1993)
2. Mitchell, T.M.: Machine Learning. McGraw Hill, New York (1997)
3. Turney, P.D.: Types of cost in inductive concept learning. In: Workshop on Cost-Sensitive Learning at the Seventeenth International Conference on Machine Learning, Stanford University, California (2000)
4. Ling, C.X., Sheng, S., Yang, Q.: Test Strategies for Cost-Sensitive Decision Trees. TKDE 18(8), 1055–1067 (2006)
5. Yang, Q., Ling, C.X., Chai, X., Pan, R.: Test-Cost Sensitive Classification on Data with Missing Values. TKDE 18(5), 626–638 (2006)
6. Sheng, V.S., Ling, C.X.: Feature Value Acquisition in Testing: A Sequential Batch Test Algorithm. In: ICML'2006, pp. 809–816 (2006)
7. Sheng, V.S., Ling, C.X., Ni, A., Zhang, S.: Cost-Sensitive Test Strategies. In: AAAI 2006. Proceedings of the Twenty-first National Conference on Artificial Intelligence (2006)
8. Koopmanschap, M.A., et al.: Influence of waiting time on cost-effectiveness. Social Science & Medicine 60, 2501–2504 (2005)
9. Cromwell, D.A.: Waiting time information services: An evaluation of how well clearance time statistics can forecast a patient's wait. Social Science & Medicine 59, 1937–1948 (2004)
10. Turney, P.D.: Cost-Sensitive Classification: Empirical Evaluation of a Hybrid Genetic Decision Tree Induction Algorithm. Journal of Artificial Intelligence Research 2, 369–409 (1995)
11. Qin, Y.S., et al.: Semi-parametric Optimization for Missing Data Imputation. Applied Intelligence (2007)
12. Zhang, C.Q., et al.: Efficient Imputation Method for Missing Values. In: PAKDD 2007. LNCS (LNAI), vol. 4426, pp. 1080–1087. Springer, Heidelberg (2007)
13. Zubek, V.B.: Learning Cost-Sensitive Diagnostic Policies from Data, A Ph.D Dissertation submitted to Oregon State University (2003)
14. Weiss, S.M., Galen, R.S., Tadepalli, P.V.: Maximizing the predictive value production rules. Artificial Intelligence 45(1-2), 47–71 (1990)
15. Nunez, M.: The use of background knowledge in decision tree induction. Machine Learning 6, 231–250 (1991)
16. Tan, M.: Cost-sensitive learning of classification knowledge and its applications in robotics. Machine. Learning Journal 13, 7–33 (1993)
17. Domingos, P.: MetaCost: A General Method for Making Classifiers Cost-Sensitive. Knowledge Discovery and Data Mining, 155–164 (1999)
18. Greiner, R., Grove, A., Roth, D.: Learning Cost-Sensitive Active Classifiers. Artificial Intelligence Journal 139(2), 137–174 (2002)
19. Ling, C., et al.: Decision Trees with Minimal Costs. In: Proceedings of 21st International Conference on Machine Learning, Banff, Alberta, Canada, July 4-8 (2004)
20. Qin, Z., Zhang, C., Zhang, S.: Cost-sensitive Decision Trees with Multiple Cost Scales. In: Webb, G.I., Yu, X. (eds.) AI 2004. LNCS (LNAI), vol. 3339, Springer, Heidelberg (2004)

21. Ni, A.L., Zhu, X.F., Zhang, C.Q.: Any-Cost Discovery: Learning Optimal Classification Rules? In: Zhang, S., Jarvis, R. (eds.) AI 2005. LNCS (LNAI), vol. 3809, pp. 123–132. Springer, Heidelberg (2005)
22. Zhang, S., Qin, Z., Ling, C., Sheng, S.: "Missing is Useful": Missing Values in Cost-sensitive Decision Trees. IEEE Transactions on Knowledge and Data Engineering 17(12), 1689–1693 (2005)
23. Blake, C.L., Merz, C.J.: UCI Repository of machine learning databases (1998), http://www.ics.uci.edu/ mlearn/MLRepository.html
24. Fayyad, U.M., Irani, K.B.: Multi-interval discretization of continuous-valued attributes for classification learning. In: Proceedings of the 13th International Joint Conference on Artificial Intelligence, pp. 1022–1027. Morgan Kaufmann, San Francisco (1993)

Knowledge in Product and Performance Support

Rossitza Setchi and Nikolaos Lagos

School of Engineering, Cardiff University, The Parade, Cardiff CF24 3AA, UK
{Setchi,LagosN}@cf.ac.uk

Abstract. The emphasis nowadays on individual information needs and personalization of content as well as the level of complexity involved in some domains necessitates the use of Knowledge Engineering approaches. In particular, product and performance support is a complex and knowledge-intensive domain which despite the advances in the last few years, still suffers from the lack of information and knowledge models. This paper addresses this problem by offering definitions the for the product and performance support domain suggesting a problem-solving approach which uses the proposed knowledge models to adapt existing and construct entirely new solutions.

Keywords: Performance support, product support, knowledge engineering, knowledge types, knowledge models.

1 Introduction

Performance support is defined as any assistance that a person may need while performing a certain job. The variety of user activities may include tasks such as maintenance, troubleshooting, servicing, operating, installing, etc. In performance support systems, the emphasis is on users and their tasks.

Product support systems are similar to performance support systems as they assist users in the effective use, maintenance and disposal of a certain product. The focus is on the effective use of the product rather than the efficient user performance (although these two aspects are always closely related). The product in the centre of such a system could be a consumer product (e.g. used for communication or transportation), a machine/system designed and manufactured for a client, or the manufacturing facilities on the shop floor of a company. This paper assumes that advanced working environments should support people as well as the products used by them.

Product support is often associated with the traditional paper-based technical manual, and performance support is perceived as training on the job. Indeed, these views have dominated the area for several decades. The technological advances in the 1990s in multimedia, hypermedia and information exchange, however, led to the development of the Interactive Electronic Technical Manual (IETM) [1] and Electronic Performance Support System [2]. Both were initially thought of as functional equivalents of the traditional instructions manuals. It soon became obvious that electronic documents specially formatted for viewing and user interaction on an electronic display provide much more benefits than those initially expected [1, 3].

Z. Zhang and J. Siekmann (Eds.): KSEM 2007, LNAI 4798, pp. 460–471, 2007.

Nowadays, the US Navy and Army utilize an integrated product support system for condition-based monitoring of helicopters [3]. US Air Force use an electronic technical manual for their F-16 test stations [2]. US Navy utilize a performance support and training system for electronic technicians [4], while the Royal Navy employ a performance support system for training maintenance technicians to use Radar 1007 [5]. Large companies such as Rolls-Royce [6], Oracle [7] and Boeing [8] have experienced teams of technical writers, engineers and computer scientists who produce high quality technical documentation.

Currently, most companies develop their product or performance support systems in-house. As a result, due to the lack of expertise in AI and knowledge engineering (KE), the vast majority of these applications do not employ KE approaches. This is rather restrictive considering the emphasis nowadays on individual information needs and personalization of content. Moreover, the level of complexity in some products today is such that the use of structured approaches to modeling data and knowledge is essential.

The aim of this paper is to outline a systematic way of structuring and modeling knowledge in the product and performance support domain.

The rest of the paper is organized as follows. Section 2 reviews the state-of-the-art in on-the-job training and work-based learning. Section 3 presents a classification of different types of performance support and relevant definitions. Section 4 outlines a problem solving approach to product and performance support. Finally, section 5 concludes the paper. The rest of the paper uses the term 'performance support' as connotation to both product and performance support.

2 Literature Review

Since the early 1970s corporative training has been based on delivering classroom instructions and treating trainees as a homogeneous group. In the 1990s, however, researchers started to question the effectiveness of this type of training. Bezanson [9], for example, believed that traditional training methods were not responsive to individual training needs while Wireman [10] felt that training in many corporate training programs was too theoretical, the students were not motivated, and there was not enough flexibility in terms of presentation styles and learning content. Other studies [11-12] suggested that the demand for highly skilled and trained persons in manufacturing, maintenance and diagnosis in particular, could be reduced by involving less skilled persons in more challenging tasks and providing them with in-house training and adequate instructions and information support.

These ideas and the advances in computer-based learning made possible in the mid 1990s the integration of learning into the workplace. The concept of the 'on-the-job training' was introduced, and the first electronic performance support systems (EPSSs) were built. EPSSs are computer-based systems that provide on-the-job access to integrated information, advice and learning experience [1, 9, 13]. The motivation was not only to reduce training costs and increase productivity but also to allow workers to control their own learning by giving them the ability to retrieve information at the workplace at the moment they need it.

The key principles of that philosophy were: adopt the perspective of the performers of the job; facilitate incremental learning, or learning while carrying out a task; allow performers to control their own training; and give them the ability to access information and knowledge at the time they need it. Translated in technical terms, these key principles required task-specific access to information and knowledge; just-in-time, on-the-job training; learning controlled by the worker, and accommodating different levels of knowledge and learning styles.

Subsequently, the idea "to provide the right information to the right person at the right time" [14] set the research agenda in this area in the following years. This included research into user cantered [15-16] and performance-cantered [17] design, pedagogical models [18-19], personalization [20-21] and the use of knowledge engineering techniques for representing product support knowledge [4, 22-24]. Nowadays, EPSSs increasingly use Intranets and especially the Internet as a technology platform, and are seen as the electronic infrastructure that preserves and distributes individual and corporate knowledge throughout an organization.

Recent studies suggest that there is a noticeable shift in responsibility for learning from organizations to individuals [25]. Nowadays, employees are expected to take greater personal responsibility to ensure that their skills are current or marketable. This trend has made it more difficult for individuals to base their careers on established paths [26], and has led to the development of a work-based learning philosophy which embraces independent self-managed study.

The Internet changed the way lifelong and distance learning courses are delivered [27]. Tailored course materials for individual students, a greater degree of interaction between students and instructors and creating a network of support make such courses successful. Nowadays, companies such as Dell, CISCO and HP adopt e-learning solutions for their corporate training [28]. Academics and practitioners alike agree that e-learning systems are a valuable knowledge sharing and transfer tool due to features like synchronous learning elements, peer collaboration, and performance support.

This review shows that there is a demand for flexible, tailored and just-in-time learning that can be accessed quickly, widely and cost-effectively by employees. Organizations have to tailor their approaches to training based on the subject and the individual.

Existing knowledge-based approaches to addressing these demands include:

- integrating product life-cycle information, expert knowledge and hypermedia [29],
- integrating users' and task models [21],
- developing information models [7, 30],
- developing knowledge models [4, 31-34],
- considering context [35], and
- developing a KE framework for performance support [36].

Sections 3 and 4 of this paper contribute to the research in this field by offering definitions, classification and a technical approach to structuring, modeling and reasoning with knowledge in the performance support domain.

3 Performance Support and Knowledge

3.1 Types of Performance Support

Performance support is a highly knowledge-intensive domain. In reality, each performance support system is unique as it needs to reflect the following variables:

- the product and its life cycle, volume, complexity, and variability;
- end users and their diversity in terms of qualification, expertise, experience, cognitive and information processing abilities;
- the tasks performed by the end-users and the level of their complexity;
- the company and its type, structure, technological level, operational practices and information infrastructure such as, for example, the use of Computer-Aided Design (CAD), Computer-Aided Manufacture (CAM), Product Data Management (PDM), Enterprise Resource Planning (ERP) or Electronic Document Management (EDM) Systems;
- the knowledge and skills for system development available.

Table 1 provides an insight into the variety of performance support solutions available. The ten different types illustrated in the table differ in terms of their structure, content, presentation, navigation, and most importantly, the benefits to their end-users. Starting from the traditional paper-based manual (#1), these applications show progression in terms of complexity and functionality although their numbering is not meant to be used for ranking purposes. Each subsequent type builds upon one, two or more other types. For example, multimedia applications (#3) incorporate the features and functionality of both paper-based (#1) and electronic systems (#2). The features of the first four types (#1)-(#4) and their associated functionality are the basic stones of the more complex applications. For example, a performance support system, which is completely integrated via a PDM system with the collaborative product development environment of the company (#8), would certainly be electronic (#2), multimedia (#3), hypermedia (#4) and have a paper version (#1). It does not necessarily need to be integrated with the product (#6) although such a possibility exists. Another example is provided by modern photocopiers, which have electronic performance support (#2) integrated with the product itself (#6). As Table 1 indicates, such applications are instruction-based and task-oriented. Due to their integration with the product and the availability of real time data, they provide good-quality diagnostics and efficient condition-based maintenance. The navigation in these systems is predefined by the underlying logic of the tasks performed as well as the operational state of the system, which depends on real-time signals from gauges.

Clearly, all ten types depicted in Table 1 require some structuring of knowledge. In the basic types (#1-#4), this structuring is led by the technical authors and is implicitly built in the system. Each subsequent type (#5-#10) becomes more knowledge intensive and hence, requiring more formal approaches. The most advanced type from a KE point of view, adaptive performance support (#10) uses knowledge models to provide user-tailored and task-specific support. These knowledge models are explicitly defined, formalized and used to reason the type of support the system offers at run time.

Table 1. Types of performance support

Type	Characteristics	Example
1. Paper-based	Hierarchical structure Content: technical specifications, descriptions, instructions, troubleshooting, warnings, etc. Presentation using text, diagrams, tables Benefits: well structured information, access to declarative and procedural knowledge, corrective maintenance	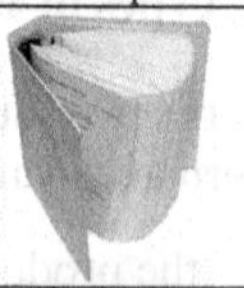
2. Electronic	Linear structure + Task-oriented and dominated by instructions + Task driven navigation (logical 'NEXT') + Improved display of data, often step-by-step instructions following the task(s)	
3. Multimedia	+ Multimedia-rich content + Coordinated multimedia, multimedia manipulation, hot spot referencing + Improved visualization due to multimodalities and cross-referencing	
4. Hypermedia	+ Associations between information elements +Alternative pathways through information + Improved information consumption due to the non-linear use of information	
5. Interactive	+ Decision trees integrated with the hypermedia structure + Dialogue-driven interaction + Simple support in diagnostics and troubleshooting	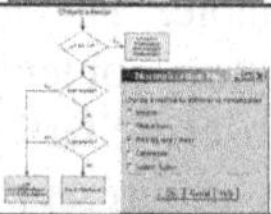
6. Integrated with the product	+ Real time monitoring + Simultaneous access to multiple information sources, NEXT function based on real-time data + Improved condition-based maintenance, diagnostics	
7. Integrated with a DB	+ Collecting real-time data about maintenance/operation + Access to historic data + Improved preventive maintenance	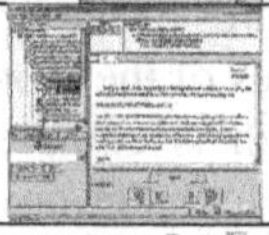
8. Integrated with a PDM system	+ Information models (e.g. product models, metadata) + Sharing of data, collaborative environments + Up-to-date product support, handling product modifications, version control	
9. Integrated with an ERP system	+ Standardized business processes + Sharing of information, content delivery driven by workflows + Improved planning of maintenance tasks, inventory of spare parts, equipment maintenance management	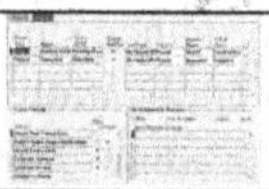
10. Adaptive	+ Knowledge models (e.g. users and task models) + Providing user-tailored and task-specific support in the most useful format, generation of adapted and new content + Personalized product and performance support	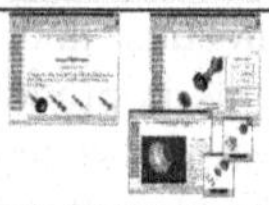

3.2 Types of Knowledge in Performance Support

The goal of any performance support system is to deliver knowledge that is accurate, applicable, reliable and user-tailored. Therefore, the knowledge base should contain relevant knowledge about the products, users and their tasks (Fig. 1), as well as ways of linking them with each other.

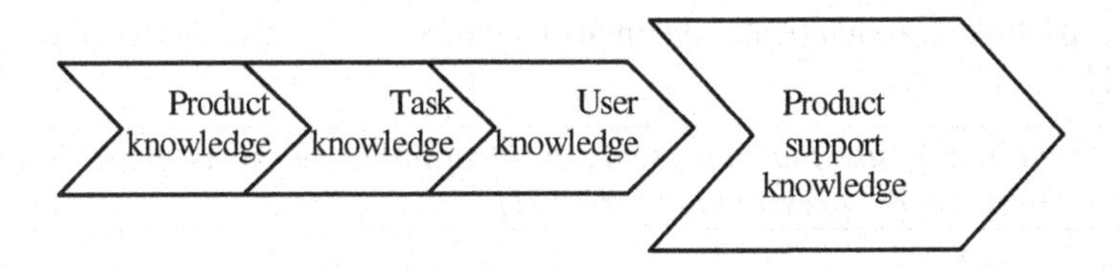

Fig. 1. Performance support knowledge

For the purpose of this research, the following definitions have been adopted [36]:

Knowledge is a specific semantic interpretation of information.
Product knowledge is a formal, temporal representation of knowledge related to the product.
Task knowledge is a formal, temporal representation of knowledge related to a task.
User knowledge is a formal, temporal representation of knowledge related to a user.

If the knowledge available in a performance support system is K_{PSS}, product knowledge is K_p, user knowledge is K_u and task knowledge is K_t, then for the performance support system to be able to deliver optimal support, the following formal requirement must be satisfied.

$$K_p \cup K_u \cup K_t \subseteq K_{PSS} . \quad (1)$$

4 Problem Solving Approach to Performance Support

4.1 Problems in Performance Support

One of the central questions when developing a performance support system is whether the system should be able to:

- assist with well-known and well-defined problems only, by retrieving existing solutions (pre-composed documents),
- assist with well understood although not known in advance problems by adapting existing solutions, or
- help in situations, which require entirely new solutions, by composing completely new documents.

A system able to offer support by adapting existing and creating new solutions should have built-in in its core an appropriate problem solving cycle.

The problem solving process starts by defining the performance support problem (PSP) and continues with determining the PSP type and the course of actions that should be followed for reaching a performance support solution (PSS). A PSS can take several forms, including explanations, instructions, warnings, descriptions, which are delivered through documents.

Ideally, to facilitate the process of adapting existing and composing new solutions, the performance support documents should be constructed using finer granularity, e.g. logically linked, semantically distinguishable and classifiable information objects (IOs) [31].

An IO is a data structure that represents an identifiable and meaningful instance of information in a specific presentation form [31].

An IO could be a picture that illustrates a part of a product or a textual description of its function. The notion of an Information Object Cluster (IOC) is introduced in this research as a means of organizing IOs.

Information Object Cluster is a 2-tuple IOC:=({IO}, S) where {IO} is the set of IOs sharing a common property that are arranged in a structure S.

For example, clutches can be classified in different types according to the number of disks they have. A single-disk clutch has a different structure and operation compared to a double-disk clutch. Assume that four IOs are available, as follows.

IO1: A textual description of the operation of a single-disk clutch,
IO2: A textual description of the operation of a double-disk clutch,
IO3: Image of a single-disk clutch, and
IO4: Image of a double-disk clutch.

Since all four IOs share the common property of describing the concept of a "clutch", they all belong to the same IOC. Within the IOC, the IOs are structured according to whether they describe a single-disk or a double-disk clutch. IO1 therefore is linked with IO3, and IO2 is linked with IO4. The links are in the form of directed arcs, defining the presentation order. In the case that a user needs a description of the car, other IOCs that correspond to the other subsystems of the car (e.g. brakes) are also needed to produce a different document.

Document (D) is a 2-tuple D := ({IOC}, S) where {IOC} is a set of logically structured IOCs.

There are two main factors that define the structure, content, and presentation of a document generated by a performance support system. These are the domain and the environment within which the system is deployed. Domain-specific elements include the products and the tasks that are supported, as well as the documentation resources that describe them. Environmental-specific elements consist of user, location, time and purpose dependent features. A performance support system should be able to extract observations from users' queries, related to both types of factors. Therefore, a performance support problem could be defined as follows.

A Performance support Problem (PSP) is a 4-tuple PSP := (MOD, HYP, CON, OBS)

where: MOD is a finite set that includes knowledge about the domain (i.e. supported products and tasks) that is represented with corresponding models. MOD also contains the links between these models and relevant IOCs and IOs.

HYP is a finite set of combinations of elements of MOD. Each combination corresponds to a possible documentation configuration. The arrangement of MOD combinations and documentation configurations will be referred to as hypotheses.

CON is a finite set that includes knowledge about the application environment. It is represented by related models (e.g. user model (UM) or system's specifications model). CON will be referred to as the system's context.

OBS is a finite set that consists of the observations acquired by the current query. OBS should correspond to elements of MOD and CON.

A performance support solution is defined as follows.

Given a performance support problem PSP := (MOD, HYP, CON, OBS) and a finite set of observations OB such that $OB \subseteq OBS$*, then valid hypotheses are represented by VH where* $VH \subseteq HYP$ *if and only if,* $MOD \cup CON \cup VH \vdash OB$*. Performance support solution (PSS) can be every hypothesis 'h' that* $h \in VH$*. The best solutions are considered the ones for which OB=OBS.*

Performance support solutions can be distinguished in terms of their content and presentation. The content can include general information (e.g. "any clutch is an assembly") or specific information (e.g. "this clutch has two disks"). It can be generated at run-time, be adapted, or pre-composed. The presentation can be text- or text and multimedia-based, as well as adaptive, adaptable, and predefined.

Previously solved/same PSPs are problems that have already been solved by the system. The reasoning mechanism used should be able to retrieve solutions without repeating the reasoning process, as well as be able to compare the specifics of different problems. Naturally, case-based reasoning (CBR) is considered as the most appropriate technique. The cases are represented as problem-solution pairs, where the attributes of each case correspond to elements of the domain- and environmental-specific models.

Similar PSPs are problems for which the current set of observations is not the same as any of the previous ones but which shares similarities with them. The new PSSs can be produced by removing, adding, replacing, and adapting IOs and IOCs used in previous situations. Assume, for instance, that a previous problem concerns a single-disk clutch. If the current query requires the description of a double-disk clutch then the new solution can be generated by replacing the IOs that describe the single-disk clutch with the IOs that correspond to the double-disk clutch. CBR provides means for adapting solutions according to previous experiences. The adapted cases form new problem-solution pairs, which are retained in the case-base.

New PSPs occur when there are no similarities between earlier observations and new ones. The PSS in this case requires additional IOCs that are not contained in the document base. For example, assume that a query relates to the description of a transaxle and that all previous PSSs use only an IOC that describes a clutch. If the transaxle IOC does not exist in the document base then means to automatically create it are needed. The existence of a new problem that has little or no similarity with previous experiences cannot be easily managed by CBR alone. In this case a multi-modal reasoning strategy is needed. General knowledge about the application domains (e.g. product domain), contained in the models, can be utilized to support the case-based reasoning component. In the example discussed, model-based reasoning can be used to identify that a new IOC is required to describe the transaxle. The transaxle concept can then be detected in the product model, and its structure and characteristics acquired. Consequently, a new IOC that describes a transaxle can be generated.

4.2 Illustrative Example

Fig. 2 shows an approach, which enables:

- using existing solutions by retrieving pre-composed documents (Fig. 2A),
- adapting existing solutions (Fig. 2B), and
- creating new solutions by composing completely new documents (Fig. 2C).

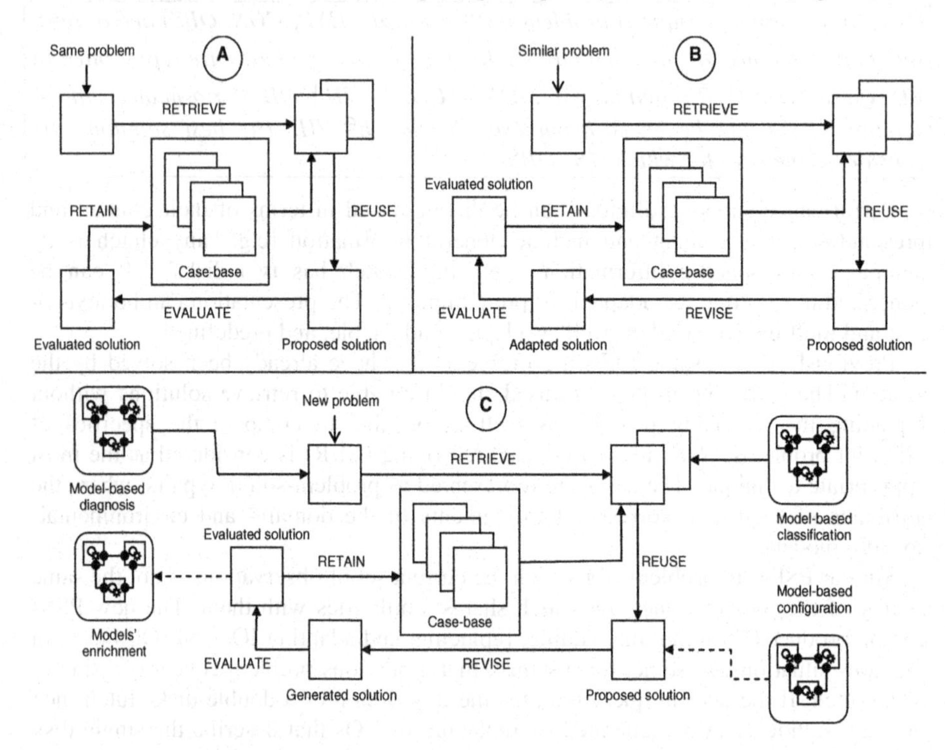

Fig. 2. Reasoning for different types of performance support problems

In those cases when a previously solved problem is identified (Fig. 2A), the system operates by retrieving all relevant cases, proposing the best match to the user as a solution, revising this solution, and retaining it in the case base. If the problem to be solved is not identical with any of the problems contained in the case base (Fig. 2B), an extra stage (i.e. solution adaptation) is needed, which is provided by case-based reasoning. Model-based reasoning is employed in Fig. 1C to deal with entirely new cases. This involves identifying a new problem, classifying the cases' attributes according to a model, and generating new documents. Enrichment of the model(s) employed is also possible.

A prototype system based on the approach proposed is developed in [36] (Fig. 3). The system uses product, task, and user models as described in [34]. The models are created with the Protégé ontology editor. The case-based reasoning tool utilizes FreeCBR. The system is implemented in Java; it uses Apache's Tomcat as a standalone web server.

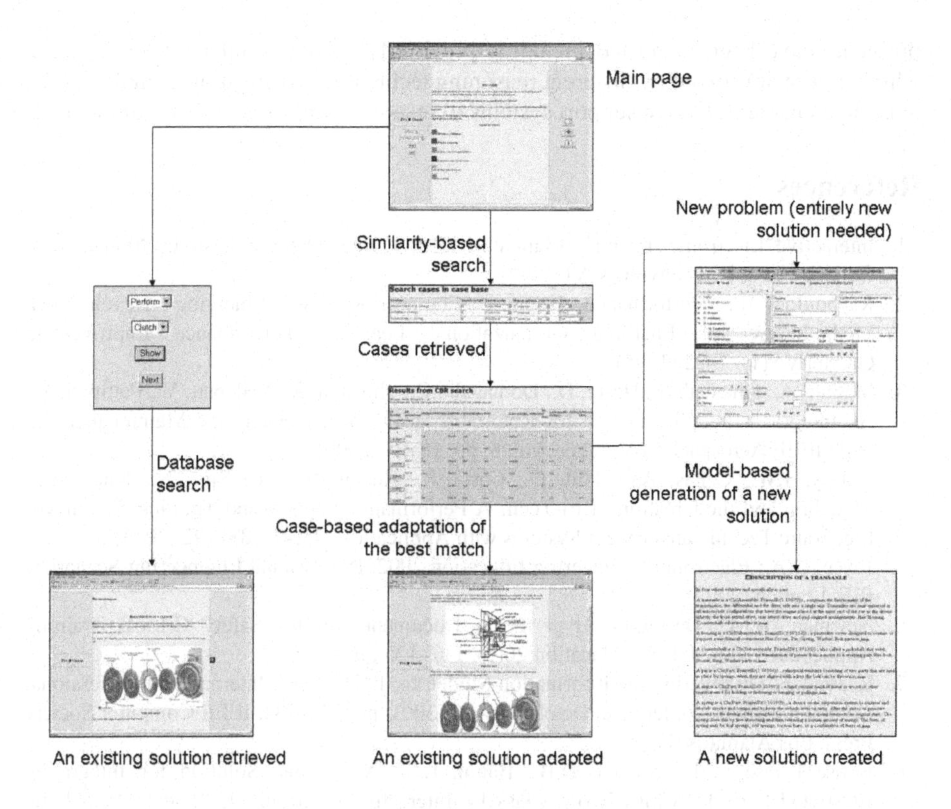

Fig. 3. Illustrative example

Fig. 3 shows that the solution provided changes according to the mode used. An existing solution is retrieved using a simple database search. If a case-based reasoning tool is employed, the user is given a list of ranked hypotheses and the choice to select those that best fit his/her information interests. He/she is then presented with a solution that is adapted to the attributes of the corresponding case. For example in Fig. 3, the adapted solution includes more details on the internal structure of a clutch than the existing solution illustrated on the same figure. The third possibility, when an entirely new solution is needed, involves the use of a model-based tool to generate a completely new document. In this study ontology relations (i.e. aggregation, abstraction, and referential relations) between clutch components are used to generate a new virtual document.

5 Conclusions

Both product and performance support systems assist users in the effective use of a certain product. Such systems should incorporate knowledge about the product, the task and the context of use and should be able to respond to different types of situations. Each situation can be represented as a problem that involves the use of virtual documents into developing an appropriate solution. Three different classes of problems have been identified including previously solved, similar, and new ones, which in turn are solved by different reasoning techniques. As illustrated, multi-modal reasoning enables adaptive support systems to respond to different user requirements.

References

1. Interactive Electronic Training Manual (IETM) Guide. Defense Systems Management College Press, Fort Belvoir, VA (1999)
2. Raybould, B.: Performance Support Engineering: An Emerging Development Methodology for Enabling Organizational Learning. Performance Improvement Quarterly 8(1), 7–22 (1995)
3. Mathur, A., Ghoshal, S., Haste, D., Domagala, C., Shrestha, R., Malepati, V., Pattipati, K.: An Integrated Support System for Rotorcraft Health Management and Maintenance. In: 2000 IEEE Aerospace Conference, vol. 6, pp. 18–25 (2000)
4. Coffey, J.W., Cañas, A.J., Hill, G., Carff, R., Reichherzer, T., Suri, N.: Knowledge Modeling and the Creation of El-Tech: A Performance Support and Training System for Electronic Technicians. Expert Systems with Applications 25(4), 483–492 (2003)
5. Joyce, D.A.: Electronic Engineering Education 2002: Professional Engineering Scenarios, IEE, London (2002)
6. Newman, R.M.: Designing Hypermedia Documentation for Safety Critical Training Applications. Eur. J. Eng. Education 26(2), 117–125 (2001)
7. Russell, J.: Discovering the Information Model. In: IPCC. IEEE International Professional Communications Conference, October 24–27, 2001, pp. 121–134. IEEE Computer Society Press, Los Alamitos (2001)
8. Boose Boose, M.L., Shema, D.B., Baum, L.S.: A Scalable Solution for Integrating Illustrated Parts Drawings into a Class IV Interactive Electronic Technical Manual. In: ICDAR 2003. Proc. 7th Int. Conf. on Document Analysis and Recognition, August 3–6, 2003, vol. 1, pp. 309–313 (2003)
9. Bezanson, W.R.: Performance Support: Online, Integrated Documentation and Training. In: ACM Proceedings, Savannah, Georgia, USA, pp. 1–10. ACM Press, New York (1995)
10. Wireman, T.: Developing Performance Indicators for Managing Maintenance. Industrial Press, New York (1998)
11. Harris, P.J.: An Expert Systems Technology Approach to Maintenance Proficiency. Robotics And Computer-Integrated Manufacturing 11, 195–199 (1994)
12. Pitblado, G.: Maintenance. In: Snow, D.A. (ed.) Plant Engineer's Reference Book, pp. 32/1 – 32/16. Butterworth Heinemann, Oxford (1996)
13. Cantando, M.M.: Vision 2000: Multimedia Electronic Performance Support Systems. In: 44th Annual Conf. of the Society for Technical Communication, Toronto, Canada, pp. 162–165 (1997)
14. Lindsey, J., Flowers, S.E.: Beyond Maintenance Training: Implementing Performance Support Software on the Plant Floor. In: Proc. Industrial Computing Conference, New Orleans, LA, USA, pp. 1–9 (1995)
15. Mackenzie, C.: The Need for a Design Lexicon: Examining Minimalist, Performance-Centered And User-Centered Design. Technical Communication 49(4), 405–410 (2002)
16. Jokela, T.: Evaluating The User-Centeredness of Development Organizations: Conclusions and Implications from Empirical Usability Capability Maturity Assessments. Interacting with Computers 16(6), 1095–1132 (2004)

17. Marion, C.: Attributes of Performance-Centered Systems: What Can We Learn From Five Years Of EPSS/PCD Competition Award Winners? Technical Communication 49(4), 428–443 (2002)
18. Bradley, C., Oliver, M.: The Evolution of Pedagogical Models for Work-Based Learning within a Virtual University. Computers & Education 38, 37–52 (2002)
19. Heijke, H., Meng, C., Ris, C.: Fitting to the Job: The Role of Genetic and Vocational Competencies in Adjustment and Performance. Labor Economics 10, 215–229 (2003)
20. Chandler, T.N.: Keeping Current in a Changing Work Environment: Design Issues in Repurposing Computer-Based Training for On-the-job Training. Int J Of Industrial Ergonomics 26, 285–299 (2000)
21. Pham, D.T., Setchi, R.M.: Case-Based Generation of Adaptive Product Manuals. Proceedings of the Institution of Mechanical Engineers 217(B), 313–322 (2003)
22. Setchi, R.M., Lagos, N.: Semantic-Based Authoring of Technical Documentation. In: World Automation Congress 2006 (WAC2006) - 10th International Symposium on Manufacturing and Applications (ISOMA 2006), Budapest, Hungary (2006)
23. Setchi, R.M., Lagos, N.: Context Modeling for Product Support Systems. ACTA Mechanica Slovaca 2(A), 425–434 (2006)
24. Setchi, R.M., Huneiti, A.M., Pasantonopoulos, C.: The Evolution of Intelligent Product Support. In: CIRP ICME 2006. 5th CIRP Int Seminar on Intelligent Computation in Manufacturing Engineering, Ischia, Italy, vol. 5, pp. 639–644 (2006)
25. Garofano, C.M., Salas, E.: What Influences Continuous Employee Development Decisions. Human Resource Management Review 15, 281–304 (2005)
26. Bosley, S., Arnold, J., Cohen, L.: The Anatomy of Credibility: A Conceptual Framework of Valued Career Helper Attributes. Journal of Vocational Behavior 70(1), 116–134 (2007)
27. Harper, K.C., Chen, K., Yen, D.C.: Distance Learning, Virtual Classrooms, and Teaching Pedagogy in the Internet Environment. Technology in Society 26, 585–598 (2004)
28. Wang, Y.-S., Wang, H.-Y., Shee, D.Y.: Measuring E-Learning Systems Success in an Organizational Context: Scale Development and Validation. Computers in Human Behavior 23(4), 1792–1808 (2007)
29. Pham, D.T., Dimov, S.S., Setchi, R.M.: Intelligent Product Manuals. Proc. Instn. Mech. Engrs. 213(I), 65–76 (1999)
30. Ghobashy, D.M., Ammar, H.H.: Interactive Electronic Technical Manual Object Model (IETMOM). In: ACS/IEEE Int. Conf. on Computer Systems and Applications, June 25-29, 2001, pp. 445–447 (2001)
31. Pham, D.T., Setchi, R.M.: Authoring Environment for Documentation Development. Proc. Instn. Mech. Engrs. 215(B), 877–882 (2001)
32. Szczepkowski, M.A., Wischusen, D., Hicinbothom, J.: Work-centered Support Development Methodology In Tactical Aviation. In: DASC 2003. The 22nd Digital Avionics Systems Conference, October 12-16, 2003, pp. 2.E.1 - 21-12 (2003)
33. Chang, S.E., Changchien, S.W., Huang, R.-H.: Assessing Users' Product-specific Knowledge for Personalization in Electronic Commerce. Expert Systems with Applications 30(4), 682–693 (2006)
34. Setchi, R.M., Pham, D.T., Dimov, S.S.: A Methodology for Developing Intelligent Product Manuals. Engineering Applications of AI 19(6), 657–669 (2006)
35. Lagos, N., Setchi, R.M., Dimov, S.S.: Towards the Integration of Performance Support and E-Learning: Context-Aware Product Support Systems. In: Meersman, R., Tari, Z., Herrero, P. (eds.) OTM 2005. LNCS, vol. 3762, pp. 1149–1158. Springer, Heidelberg (2005)
36. Lagos, N.: Knowledge-based Product Support Systems. PhD Thesis, Cardiff University, Cardiff, UK (2007)

What Drives Members to Continue Sharing Knowledge in a Virtual Professional Community? The Role of Knowledge Self-efficacy and Satisfaction

Christy M.K. Cheung[1] and Matthew K.O. Lee[2]

[1] Department of Finance and Decision Sciences, Hong Kong Baptist University, Hong Kong, China
ccheung@hkbu.edu.hk
[2] Department of Information Systems, City University of Hong Kong, Hong Kong, China
ismatlee@cityu.edu.hk

Abstract. The present research explains members' intention to continue sharing knowledge in a virtual community in terms of knowledge self-efficacy and satisfaction. The research model was tested with the current users of a virtual professional community (Hong Kong Education City) and was accounted for 32% of the variance. Both knowledge self-efficacy and satisfaction play an important role in explaining members' intention to continue sharing knowledge. The findings contribute to the foundation for future research aimed at improving our understanding of user continuance behavior in virtual communities.

Keywords: Knowledge sharing, knowledge management, virtual community, community of practice, satisfaction, knowledge self-efficacy, information systems continuance, e-business.

1 Introduction

In contemporary information systems literature, there exists a general agreement about the value and importance of knowledge management [2], [18], [30], [31]. Traditional literature on knowledge management focuses mostly on knowledge creation and dissemination within organizations. However, with the growth of the Internet and the high penetration rate of Internet usage, considerable attention is being focused on the role of online social spaces (e.g., virtual communities) in knowledge management [25].

Conversational systems (e.g., online discussion forums in virtual communities) are a useful medium for knowledge extraction, exchange, and creation. For instance, conversations are captured so as to accommodate contextualization, search, and community [34]. Without the physical or temporal constraints in virtual communities, members with diverse organizational, national, and cultural backgrounds can contribute, discuss, learn from the community's explicit knowledge, and share their implicit knowledge with other members [7]. Specifically, they can share knowledge by helping each other to solve problems, telling stories of personal experiences, and

Z. Zhang and J. Siekmann (Eds.): KSEM 2007, LNAI 4798, pp. 472–484, 2007.

debating issues based on shared interests [36]. Virtual communities support new combinations of existing knowledge and the creation of new knowledge. Since messages in virtual communities are openly available, members can easily identify the synergistic possibilities that arise from the potential combinations of information from multiple resources [36].

Many professional bodies realize the potential of information and communication technologies in connecting individuals with common interests. Some researchers [1], [29], [37], however argue that the creation of an online social space does not guarantee that knowledge exchange will actually take place. This concern basically pertains to user acceptance of online social structures for knowledge exchange. In recent years, some information systems research in knowledge management has been conducted to address this concern [23], [37]. The success of a virtual community however depends primarily on whether members are willing to continue to participate and share their knowledge with others. Obviously, if there are a lot of participants who are willing to stay and contribute their knowledge in the community, this will improve the likelihood of connecting individuals who are able and willing to help. Therefore, it is important to identify what affects members' decisions to continue to stay and share in a virtual community.

A Virtual Professional Community (VPC) is a distinct type of virtual community in which people with common interests, backgrounds, and goals participate and collectively contribute to a database of professional knowledge. Basically, there are three different types of VPC:

- Intra-firm professional communities: Many virtual professional communities operate inside large firms, where they are often called "community of practice" (CoP). The main motivation for a CoP is to improve knowledge sharing among employees and to foster a creative and innovative enterprise culture. Intra-firm VPCs depend on the infrastructure and administration from the management of the firm.
- Inter-firm professional communities: Some virtual professional communities are established to improve and strengthen relationships with customers and partners. Given the increasingly competitive business environment, more and more companies are forming enterprise networks.
- Public professional communities: This type of community is often organized by third-party organizations, and its memberships are many times larger than those of traditional professional societies. The aims of these VPCs are to bring together audiences on specific topics, and to provide a platform for professionals with common interests and similar working culture to freely exchange their experience, to share information, and to foster social relationships.

In the current study, we focus on knowledge sharing behavior in public professional communities. Specifically, the empirical research in this study is conducted in a well-established virtual professional community in Hong Kong, Hong Kong Education City (www.hkedcity.net). We examine the characteristics and usage behaviors of knowledge contributors, as well as the motivations that drive them to continue sharing knowledge in the virtual community.

2 Theoretical Background

Past research on knowledge management primarily focused on the initial adoption of knowledge sharing behavior. This study investigates the confirmation stage where users evaluate their knowledge sharing behavior and make the decision to continue or discontinue the behavior. The theoretical foundation of the present study is reviewed. Specifically, the concepts of knowledge sharing, user satisfaction, as well as knowledge self-efficacy are addressed.

2.1 Knowledge Sharing

Virtual communities provide people with common interests, backgrounds, and goals to participate and collectively contribute to a set of professional knowledge. Knowledge is commonly conceived as a public good. A public good is characterized as *"a shared resource from which every member of a group may benefit, regardless of whether or not they personally contribute to its provision, and whose availability does not diminish with use"* [10]. The fundamental problem of public goods is that individuals merely consume the public good without contributing to the group or the institution, resulting in a social dilemma situation. Social dilemmas occur whenever an individual attempts to maximize its self-interest and makes rational decisions. Applying the public good concept to the knowledge sharing in virtual communities, there is a tendency for individuals to refuse to contribute and enjoy a free-ride. Particularly, electronic networks of practice allow everyone to access and consumer knowledge without making any contribution. Wasko and Tiegland [36] however urged that though public goods are subject to social dilemmas, they are often created and maintained through collective action.

Considerable attention has focused on factors that drive people to share knowledge in electronic networks of practice. Cheung and Lee [12] built on Batson public good framework [5] and classified the key factors determining user intention to share knowledge into four categories. Table 1 summarizes the key factors determining user intention to share knowledge. Among all these key factors, reciprocity and enjoyment of helping are the most widely studied factors.

Table 1. Key Factors of Knowledge Sharing in Previous Studies

Category	Factors	References
Egoism	Extrinsic rewards	[illegible]
	Wage	[illegible]
	Organizational reward	[illegible]
	Reciprocity	[illegible]
	Reputation	[illegible]
	Self-interest	[illegible]
Collectivism	Social identity	[illegible]
Altruism	Enjoyment of helping others	[illegible]
Principlism	Normative Commitment (Moral obligation)	[illegible]

2.2 User Satisfaction

Early user satisfaction research tended to focus primarily on the operationalization of satisfaction construct and ignored the theoretical bases. According to Melone [27], "this lack of agreement on the conceptual definition of the user-satisfaction construct has lead to a situation in which there are many operationalizations and an equal number of conceptual definitions, for the most part lacking theoretical foundation" (p.80). In response to the call for a rigorous theoretical support in the study of user satisfaction, recent studies are more grounded with theories. For instance, Devaraj et al. [16] examined consumer-based channel satisfaction using technology acceptance model, transaction cost analysis, and service quality. Bhattacherjee [6], McKinney et al. [26] and Susarla et al. [33] adopted the expectation confirmation theory to examine satisfaction. Among diverse theoretical frameworks, expectation confirmation theory has been receiving a great deal of attention in recent IS research. These studies provided insights to user psychology and explained user satisfaction formation processes [11].

Bhattacherjee [6] proposed an IS continuance model that relates satisfaction and perceived usefulness to the degree in which users' expectations about an information system are confirmed. Expectation provides a baseline level to evaluate the actual performance of an IS and confirmation (disconfirmation) in turn determines satisfaction. This line of research is consistent with the expectancy value theory, where people form expectations and then evaluates their experiences.

2.3 Knowledge Self-Efficacy

Research on knowledge management has already suggested the importance of knowledge self-efficacy on people intention to share knowledge [9], [10], [23]. Knowledge self-efficacy refers to people believing their knowledge can help other members in virtual communities. This definition is built upon the social cognitive theory [4]. Bandura [4] suggested that the motivations of performing a behavior do not stem from the goals themselves, but from the self-evaluation that is made conditional on their fulfillment. He even suggested that "mastery experience" is the most important factor determining self efficacy. In other words, success raises self-efficacy, failure lowers it. When applying this concept in the public good context, knowledge self-efficacy refers to the perception of the criticality of the contributions to the provision of a public good. In general, knowledge self-efficacy promotes the sharing of knowledge.

3 Research Model and Hypotheses

Fig. 1 depicts a research model explaining user intention to continue using a virtual professional community for knowledge sharing.

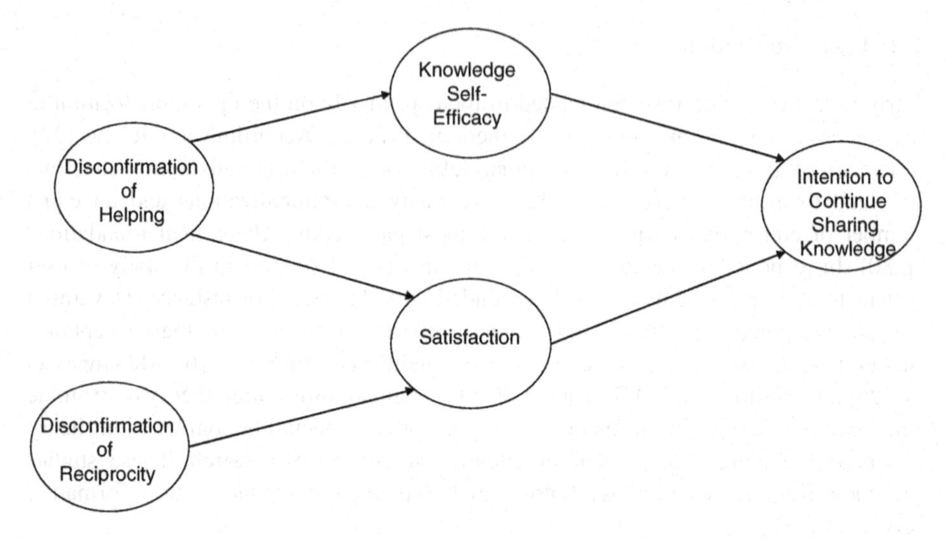

Fig. 1. Research Model

3.1 Intention to Continue Sharing Knowledge

Knowledge sharing is a necessary component of knowledge management. It embeds the notion of "*willingness to share*" or "*voluntary act of making information available to others...*" [15]. Since this study focuses on continuance behaviour, *"Intention to continue sharing knowledge"* is defined as *"the likelihood that a user will continue sharing knowledge in a virtual community"*. The concept is built on the IS continuance model.

3.2 Reciprocity

"A motive is egoistic if the ultimate goal is to increase the actor's own welfare" [5]. Explanations of action for the public good in terms of egoism can be linked to personal motivational theories, as well as social exchange theory [8]. Reciprocity is conceived as a benefit for individuals to engage in social exchange. People have an expectation that their contribution will result in returns in the future. In knowledge sharing literature, researchers found that knowledge sharing is facilitated when people who share knowledge in virtual communities (any electronic network of practices) believe in reciprocity. There is a positive relationship between reciprocity and knowledge contribution intention [23], [37].

The current study goes beyond the initial adoption and focuses on continuance behavior in electronic networks of practice, particularly continuous knowledge sharing in virtual communities. It is believed that after several interactions with other users in a virtual community, users are able to compare their expectations with the actual experiences of using the virtual community. Specifically, they can evaluate whether reciprocity has actually occurred. Based on expectation confirmation theory [28], confirmation (disconfirmation) will result in satisfaction (dissatisfaction), and satisfaction will lead to continuance intention. Applying this argument to the current investigation, it is believed that if the gap between a user's expectation and actual reciprocal experience is large, the user will feel satisfied (dissatisfied) with their experience with the virtual community, and in turn he/she will have a higher likelihood to continue (discontinue) to share knowledge. In other words, if the users find that they can receive the reciprocity as they expected, they will feel satisfied and in turn they will have a higher chance to continue sharing knowledge in the virtual community. Thus, in this study, the hypothesis is:

H1: Users with a higher confirmation (positive disconfirmation) of reciprocity in a virtual community relates positively to satisfaction with the virtual community.
H2: Satisfaction with the virtual community relates positively to intention to continue sharing knowledge in the virtual community.

3.3 Enjoyment of Helping and Helping Behaviour

"Altruism is motivation with the ultimate goal of increasing the welfare of one or more individuals other than oneself" [5]. Explanations of action for the public good in terms of altruism can be linked to empathic emotion [5]. Enjoyment in helping has been frequently cited as an important factor that determines user willingness to share or contribute knowledge in electronic networks of practice or online social spaces [21], [23], [37]. People are willing to help others to solve challenging problems because answering questions provides them with feelings of pleasure [24].

In this case, the goal is to help others and it is the motivation for them to contribute. After users have had several interactions with other users, they are able to judge whether their contributions are helpful to others. Users first form an expectation about the outcomes of their helping behaviours, for instance, they expect their messages would be helpful to others. After their interactions with other members in the virtual community, they will compare their expectation with their actual experience, that is, to evaluate whether their messages are really helpful to others. If there is a positive disconfirmation (their messages are more helpful than expected), users will feel satisfied. On the other hand, if there is a negative disconfirmation (their messages are less helpful than expected), users will feel dissatisfied. Thus, the hypothesis is:

H3: Users with a higher confirmation (positive disconfirmation) of helping others in a virtual community relates positively to satisfaction within the virtual community.

It is also believed that if users have a positive disconfirmation with helping in the virtual community, their knowledge self-efficacy will increase. Bandura [4] suggested that the motivations of performing a behaviour do not stem from the goals themselves, but from the self-evaluation that is made conditional on their fulfilment. If the users found their knowledge to be helpful to other members in the virtual community, it will enhance their confidence that their knowledge is able to help other people. Therefore, the hypothesis is:

H4: A higher confirmation (positive disconfirmation) of helping others in a virtual community relates positively to knowledge self-efficacy.

Research on knowledge management has already suggested the importance of knowledge self-efficacy on people's intention to share knowledge [9], [10], [23]. It is believed that knowledge self-efficacy will have an important impact on user intention to continue sharing knowledge in a virtual community. The hypothesis is:

H5: Knowledge self-efficacy relates positively to intention to continue sharing knowledge in the virtual community.

4 Research Design

The research model was empirically tested in a real virtual community, Hong Kong Education City (www.hkedcity.net). Hong Kong Education City (HKed City) is a leading and one-stop education portal with a vision to build Hong Kong into a learning city. With the vision of *"Bridging the Learning Communities. Building the Learning City"*, Hong Kong Education City aims at taking a leading role in promoting quality education and the use of IT in education to schools, teachers, students, parents, and the public. In the current study, the unit of analysis are teachers or educators who use the *"Teachers' Channel"* of the Hong Kong Education City (www.hkedcity.net). "Teachers' Channel" is a virtual professional community that provides teachers and educators with resources on professional development and updated news on educational related issues.

4.1 Data Collection and Responses

The target respondents of this study were the teachers who have used the "Teachers' Channel" in HKed City. In order to reach the respondents, an invitation email with the URL of the online questionnaire was sent to both primary and secondary school teachers. The participation of this study was voluntary. To increase the response rate, an incentive of three USB flash drives and thirty book coupons were offered as lucky draw prizes. Reminder emails were also sent a few weeks after the first invitation email.

A total of 315 responses were collected in this study and 60 of them have contributed in the virtual community before. Among the 60 contributors, 72% were male and 28% were female. About 22% were aged 21-30 and only 8% were aged 51 or above. 72% were secondary school teachers and 28% were primary school teachers, and 22% had more than 20 years teaching experiences. In terms of the usage behaviour in the virtual community (HKed City), about 40% had less than 2-year experience with the virtual community, but over 40% of them used it every week. Non-response error is estimated using the comparison of differences between the early and late respondents. This is the most commonly used method for non-response error estimation among IS researchers [30]. We did not find the error exists in this study.

4.2 Measures

Empirical research on continuance behaviour in knowledge management is still in its infancy. Most measures were developed and modified based on some established scales. The measure of *"Intention to continue sharing knowledge"* was adapted from Bagozzi and Dholakia [3]. Items measuring "Disconfirmation of reciprocity", "Disconfirmation of Helping", and "Knowledge self-efficacy" were adapted and modified from Kankanhalli et al. [23] to fit the specific context of virtual community. The measures of *"Satisfaction"* were borrowed from Bhattacherjee [6]. All the measures of the constructs in the current study are listed in Appendix A.

A multi-item approach was used. That means each construct was measured by a few items for construct validity and reliability. A slider scale was used in this study and provided a continuous scale from 0 to 100 or -50 to 50 (See Fig. 2). Respondents either clicked or dragged the slider to indicate their preference point.

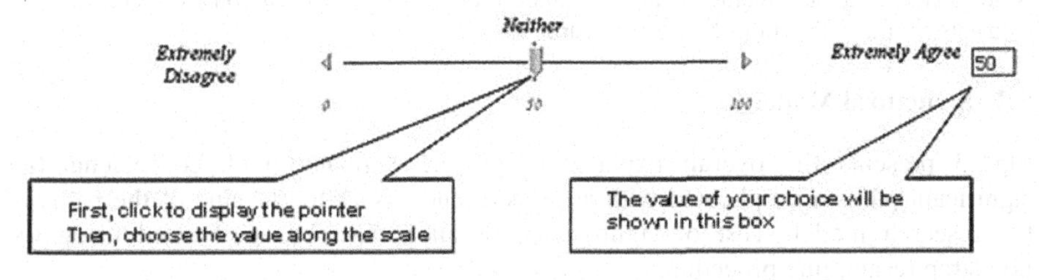

Fig. 2. The Slider Scale

5 Results

Following the two-step analytical procedures [20], the measurement model is first examined and then the structural model is assessed.

5.1 Measurement Model

Convergent validity, which indicates the extent to which the items of a scale that are theoretically related to each other should be related in reality, was examined using the composite reliability (CR) and the average variance extracted (AVE). The critical values for CR and AVE are 0.7 and 0.5 respectively [17]. As shown in Table 2, all CR and AVE values meet the recommended thresholds.

Discriminant validity is the extent to which the measure is not a reflection of some other variable. It is indicated by low correlations between the measure of interest and the measure of other constructs that is not theoretically related to Fornell and Larcker [17]. Evidence about discriminant validity can be demonstrated when the square root of the average variance extracted for each construct is higher than the correlations between it and all other constructs. Table 2 shows that the squared root of average variance extracted for each construct is greater than the correlations between the constructs and all other constructs. The results suggest that an adequate discriminant validity of the measures.

Table 2. Correlation Matrix and Psychometric Properties of Key Constructs

	CR	AVE	CI	DRECIP	DHELP	SAT	KSE
Continuance Intention (CI)	**0.99**	**0.98**	**0.99**				
Disconfirmation of Reciprocity (DRECIP)	**0.95**	**0.85**	0.77	**0.92**			
Disconfirmation of Helping (DHELP)	**0.97**	**0.92**	0.49	0.68	**0.96**		
Satisfaction (SAT)	**0.97**	**0.88**	0.55	0.69	0.71	**0.94**	
Knowledge Self-Efficacy (KSE)	**0.88**	**0.65**	0.52	0.65	0.86	0.77	**0.81**

Note: Shaded diagonal elements are the square root of AVE for each construct Off-diagonal elements are the correlations between constructs.

5.2 Structural Model

Fig. 3 presents the overall explanatory power, estimated path coefficients (all significant paths are indicated with asterisks), and associated t-value of the paths of the research model. Test of significance of all paths were performed using the bootstrap resampling procedure.

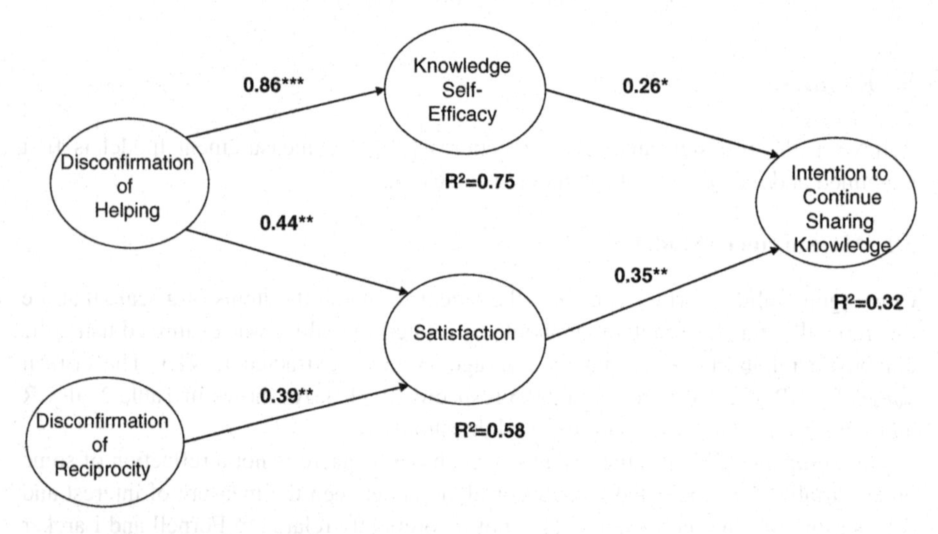

Fig. 3. Result of the Research Model (Note: *p<0.10, **p<0.05, ***p<0.01)

The results show that the exogenous variables explain 32% of the variation in *"Intention to Continue sharing Knowledge in a virtual community"*, 35% of the variance in *"Satisfaction"* and 75% of the variation in *"Self-Efficacy in Knowledge Sharing"*. All the structural paths are found to be statistically significant in the research model. Both 'disconfirmation of reciprocity' and 'disconfirmation of helping' have a significant impact on satisfaction with the virtual community. Their path coefficients are 0.39 and 0.44 respectively. Disconfirmation of helping is also strongly related to user's knowledge self-efficacy, with a path coefficient at 0.86. Finally, knowledge self-efficacy and satisfaction affect intention to continue sharing knowledge significantly, with path coefficients of 0.26 and 0.35 respectively.

6 Discussion and Conclusion

This study attempts to test a research model of knowledge-sharing behaviors in virtual communities. In particular, this study goes beyond initial adoption (why people share) to continuance (why people continue to share). The finding suggests that an important way to promote the continuance of knowledge sharing is to increase knowledge self-efficacy and enhance user satisfaction. When members find that their shared knowledge can help others, they will be satisfied and will gain higher knowledge self-efficacy, and thus they will tend to continue sharing knowledge in the virtual community.

This study contributes to existing virtual community and knowledge management research in several ways. First, this study adds to the limited research done on knowledge sharing behaviours in virtual communities of professional groups and allows future research to build upon it. This study also allows operationalization and validation of instruments in the research model. Finally, this study goes beyond initial adoption and examines continuance in the context of knowledge sharing in virtual communities.

Apart from the theoretical contributions, the results of this study also provide some insights to community designers for knowledge management. Specifically, it is important for community designers to use some guidelines and tools to encourage members to continue sharing knowledge.

- ***Providing members with a recognition mechanism:*** Community designers should propose the use of some recognition mechanisms where members who have provided useful suggestions to other members are identified and informed that they have helped others.
- ***Creating members' social network:*** Community designers should try to integrate member-produced content, as well as content from a member's connections, into the member profiles. This can help connect knowledge contributors and adopters so that adopters can show their appreciation for the knowledge received.

In interpreting the results of this study, one must pay attention to a number of limitations. First, the theoretical model accounts for 32% of the variance in continuance intention and suggests that some important predictors may be missing. Second, the online survey involves self-reported measures, which may be subject to the influence of common method bias. Finally, this study represents one type of professional group where the participants usually share some common interests, background, and goals to participate and collectively contribute to the professional knowledge. It is desirable to replicate the results with other types of virtual communities. Obviously, future research should examine these speculations.

References

1. Alavi, M., Leidner, D.E.: Knowledge Management Systems: Issues, Challenges, and Benefits. Communications of Association of Information Systems (1999)
2. Alavi, M., Leidner, D.E.: Review: Knowledge management and knowledge management systems: Conceptual foundations and research issues. MIS Quarterly 25(1), 107–136 (2001)
3. Bagozzi, R.P., Dholakia, U.M.: Intentional Social Action in Virtual Communities. Journal of Interactive Marketing 16(2), 2–21 (2002)
4. Bandura, A.: Social foundations of thought and action: A social cognitive theory. Prentice-Hall, Englewood NJ (1986)
5. Batson, C.D.: Why act for the public good? Four answers. Personality and Social Psychology 20(5), 603–610 (1994)
6. Bhattacherjee, A.: Understanding information systems continuance: An expectation confirmation model. MIS Quarterly 25(3), 351–370 (2001)
7. Bieber, M., Engelbart, D., Furuta, R., Hiltz, S.R., Noll, J., Preece, J.A., Stohr, E., Turoff, M., Walle, B.V.D.: Toward Virtual Community Knowledge Evolution. Journal of Management Information Systems 18(4), 11–35 (2002)
8. Blau, P.M.: Exchange and power in social life. John Wiley, New York (1964)
9. Bock, G.-W., Zmud, R.W., Kim, T.-G., Lee, J.-N.: Behavioural Intention Formation In Knowledge Sharing: Examining The Roles Of Extrinsic Motivators, Social-Psychological Forces, And Organizational Climate. MIS Quarterly 29(1), 87–111 (2005)
10. Cabrera, A., Cabrera, F.E.: Knowledge-sharing dilemmas. Organizational Studies 23(5), 687–710 (2002)
11. Cheung, C.M.K., Lee, M.K.O.: The Asymmetric Impact of Website Attribute Performance on User Satisfaction: An Empirical Study. e-Service Journal 3(3), 65–89 (2005)
12. Cheung, C.M.K., Lee, M.K.O.: Understanding Intention to Continue Sharing Knowledge in Virtual Communities. In: The Proceedings of the 15th European Conference on Information Systems, St. Gallen, Switzerland (2007)
13. Constant, D., Kiesler, S., Sproull, L.: What's mine is ours, or is it? A study of attitudes about information sharing. Information Systems Research 5(4), 400–421 (1994)
14. Constant, D., Sproull, L., Kiesler, S.: The kindness of strangers: The usefulness of electronic weak ties for technical advice. Organization Science 7(2), 119–135 (1996)
15. Davenport, T.H., Prusak, L.: Working knowledge: How organizations manage what they know. Harvard Business School Press, Boston (1998)
16. Devaraj, S., Fan, M., Kohli, R.: Antecedents of B2C channel satisfaction and preference: Validating e-commerce metrics. Information Systems Research 13(3), 316–333 (2002)
17. Fornell, C., Larcker, D.F.: Evaluating structural equation models with unobservable variables and measurement error. Journal of Marketing Research 18, 39–50 (1981)
18. Grover, V., Davenport, T.H.: General Perspectives on Knowledge Management: Fostering a Research Agenda. Journal of Management Information Systems 18(1), 5–21 (2001)
19. Gu, B., Jarvenpaa, S.: Online discussion boards for technical support: The effect of token recognition on customer contributions. In: Twenty-fourth international conference on information system, pp. 110–120 (2003)
20. Hair, J.F., Anderson, R.E., Tatham, R.L., Black, W.C.: Multivariate data analysis, 5th edn. (1998)
21. Hennig-Thurau, T., Gwinner, K.P., Walsh, G., Gremler, D.D.: Electronic word of mouth via consumer-opinion platforms: What motivates consumers to articulate themselves on the internet. Journal of Interactive Marketing 18(1), 38–52 (2004)

22. Kane, A.E.A., Argote, L., Levine, J.M.: Knowledge transfer between groups via personnel rotation: Effects of social identity and knowledge quality. Organizational Behavior and Human Decision Processes 96, 56–71 (2005)
23. Kankanhalli, A., Tan, B.C.Y., Wei, K.-K.: Contributing knowledge to electronic knowledge repositories: An empirical investigation? MIS Quarterly 29(1), 113–143 (2005)
24. Lakhani, K.R., Hippel, E.V.: How open source software works: "free" user-to-user assistance. Research Policy 1451, 1–21 (2002)
25. Lee, M.K.O., Cheung, C.M.K., Lim, K.H., Sia, C.L.: Understanding Customer Knowledge Sharing in Web-based Discussion Boards: An Exploratory Study. Internet Research 16(3), 289–303 (2006)
26. McKinney, V., Yoon, K., Zahedi, F.M.: The measurement of web-customer satisfaction: An expectation and disconfirmation approach. Information Systems Research 13(3), 296–315 (2002)
27. Melone, N.P.: A theoretical assessment of the user-satisfaction construct in information systems research. Management Science 36(1), 76–91 (1990)
28. Oliver, R.L.: Effect of examination and disconfirmation on post-exposure product evaluations: an alternative interpretation. Journal of Applied Psychology 62, 486–490 (1976)
29. Orlikowski, W.J.: Learning from Note: Organizational Issues in Groupware Implementation, in Computerization and Controversy, pp. 173–189. Academic Press, New York (1996)
30. Sambamurthy, V., Subramani, M.: Special Issue on Information Technologies and Knowledge Management. MIS Quarterly 29(2), 193–195 (2005)
31. Schultz, U., Leidner, D.E.: Studying knowledge management in information systems research: Discourses and theoretical assumptions. MIS Quarterly 26(3), 213–242 (2002)
32. Sivo, S.A., Saunders, C., Chang, Q., Jiang, J.J.: How Low Should You Go? Low Responses Rates and the Validity of Inference in IS Questionnaire Research. Journal of the Association for Information Systems 7(6), 351–414 (2006)
33. Susarla, A., Barua, A., Whinston, A.B.: Understanding the service component of application service provision: An empirical analysis of satisfaction with ASP services. MIS Quarterly 27(1), 91–123 (2003)
34. Wagner, C., Bolloju, N.: Supporting Knowledge Management in Organizations with Conversational Technologies: Discussion Forums, Weblogs, and Wikis. Journal of Database Management 16(2), i-viii (2005)
35. Wasko, M.M., Faraj, S.: It is what one does: Why people participate and help others in electronic communities of practice. Journal of Strategic Information Systems 9, 155–173 (2000)
36. Wasko, M.M., Tiegland, R.: Public goods or virtual commons? Applying theories of public goods, social dilemmas, and collective action to electronic networks of practice. Journal of Information Technology Theory and Application 6(1), 25–41 (2004)
37. Wasko, M.M., Faraj, S.: Why should I share? Examining social capital and knowledge contribution in electronic networks of practice? MIS Quarterly 29(1), 35–57 (2005)

Appendix A

Continuance Intention (Modified from Bagozzi and Dholakia [3])	
C□□	□lease e□press the degree to □hich you □ight intend to continue sharing in the □eachers□ Channel in the ne□t fe□ □eeks□□□tre□ely □nlikely□□□tre□ely □ikely□
C□□	□intend to continue sharing in the □eachers□Channel in the ne□t fe□ □eeks□□□tre□ely □isagree□□□tre□ely Agree□
Disconfirmation of Reciprocity (Modified from Kankanhalli et al. [23])	
□□□C□□	Co□pared to □y initial e□pectations□the le□el of reciprocity □i□e□□get back help □hen □need□ in the □eachers□Channel is □Much □orse than e□pected□Much better than e□pected□
□□□C□□	□o □hat e□tent does the degree of reciprocity □i□e□□so□ebody responds □hen □a□ in need□ occurring in the □eachers□Channel □eet your original e□pectations□□□ar belo□ □y e□pectation□□ar abo□e □y e□pectation□
□□□C□□	□o□ big is the difference bet□een □hat you e□pected □hen you are gi□ing an ans□er to others and □hat the reciprocity actually occurred in the □eachers□Channel□□□ar belo□ □y e□pectation□□ar abo□e □y e□pectation□
Disconfirmation of Helping (Modified from Kankanhalli et al. [23])	
□□□□□□	Co□pared to □y initial e□pectations□the helpfulness of □y ans□ers in the □eachers□Channel is □Much □orse than e□pected□Much better than e□pected□
□□□□□□	Co□pared to □y initial e□pectations□the helpfulness of □y response on helping other people to sol□e proble□s in the □eachers□Channel is □□ar belo□ □y e□pectation□□ar abo□e □y e□pectation□
□□□□□□	□o□ big is the difference bet□een □hat you percei□ed the helpfulness of your ans□ers to be and ho□ they actually helped others in the □eachers□Channel□□□ar belo□ □y e□pectation□□ar abo□e □y e□pectation□
Knowledge Self Efficacy (Modified from Kankanhalli et al. [23])	
□□□	□ha□e confidence in □y ability to pro□ide kno□ledge that others in the □eachers□Channel consider □aluable□□□tre□ely □isagree□□□tre□ely Agree□
□□□	□ha□e the e□pertise needed to pro□ide □aluable kno□ledge for □eachers□Channel□□□tre□ely □isagree□□□tre□ely Agree□
Satisfaction (Bhattacherjee [6])	
	□o□ do you feel the kno□ledge sharing e□perience □ith □eachers□Channel□
□A□□	□□tre□ely □issatisfied□□□tre□ely □atisfied□
□A□□	□□tre□ely □ispleased□□□tre□ely □leased□
□A□□	□□tre□ely □rustrated□□□tre□ely Contented□
□A□□	□Absolutely □errible□Absolutely □elighted□

MMFI_DSSW – A New Method to Incrementally Mine Maximal Frequent Itemsets in Transaction Sensitive Sliding Window

Jiayin Feng and Jiadong Ren

College of Information Science and Engineering,
Yanshan University, QinHuangdao 066004, China
jdren@ysu.edu.cn,
feng_ada2000@163.com

Abstract. Due to streaming data are infinite in length and fast changing with time, it is very significant to limit the memory usage in the process of mining data streams. Maximal frequent itemset is a subset of frequent itemsets; it can represent the important information of frequent itemsets with low computational cost. In this paper, we propose an algorithm MMFI_DSSW (Mining Maximal Frequent Itemsets in Data Streams Sliding Window) to mine maximal frequent itemsets with a novel MFI_BVT (Maximal Frequent Itemsets Binary Vector Table) summary data structure in sliding window. MFI_BVT builds a binary vector for each itemsets first. Then algorithm MMFI_DSSW performs logical AND operation to mine all the maximal frequent itemsets in MFI_BVT with a single-pass scan incoming data. Finally, the mining result can be updated incrementally. Experiment shows that algorithm MMFI_DSSW is efficient and scalable in memory usage and running time of CPU.

Keywords: data streams, maximal frequent itemsets, algorithm.

1 Introduction

Frequent itemsets mining in data streams is an important research field in mining of online data streams. It has been studied in recent years. Due to data stream is continuous, massive and infinite, it is essential to design an efficient summary data structure and single-pass algorithm in mining data streams. So far, three summary data structures are applied in mining data streams: landmark windows model[1,2], tilted-time windows model[3] and sliding windows model[4]. In sliding window model, user pays more attention to the analysis of most recent data stream, thus we only analyze the most recent data items in detail, summarize the old one and discard the expired data. There are mainly two methods in using sliding window model, time sensitive sliding window[4,5] and transaction sensitive sliding window[6-8]. In time sensitive sliding window method, a fixed time period is defined for the sliding window, and the incoming transactions will be processed once in each time period. In transaction sensitive sliding window, a fixed numbers of transactions are given to the basic window by user. The incoming transactions will be processed once when there are enough transactions.

Z. Zhang and J. Siekmann (Eds.): KSEM 2007, LNAI 4798, pp. 485–495, 2007.

However, the application of the summary data structure in memory can decrease amount of frequent itemsets, but the total numbers of frequent itemsets in data streams are also greatness, it consumes large numbers of resource in memory to store completely information of frequent itemsets all the same. Maximal frequent itemsets include all the items of frequent itemsets; it can represent all the frequent itemsets with small numbers of itemsets. Thus, mining maximal frequent itemsets from data streams will save memory resource. Now, the algorithms for mining maximal frequent itemsets in data streams include DSM-FMI[7]proposed by Hua Fu Li and estDec+[8] presented by Daesu Lee. Algorithm DSM-FMI[7]mines the set of all maximal frequent itemsets in landmark windows over data streams. A summary data structure SFI-forest(summary frequent itemset forest) is proposed to incrementally maintain the essential information about maximal frequent itemsets in the stream. SFI-forest is the improvement of FP-tree structure[9], it can single-pass and real-time record sequence information in each node. Algorithm estDec+[8] is derived from estDec[6] for mining frequent itemsets in data streams. A CP-tree(Compressed-Prefix tree) structure is proposed to compress several nodes which have the same counts into one node. Thus, it can saves memory space. However, DSM-FMI and estDec+ have to store the nodes of the tree for the maximal frequent itemsets, so the total numbers of nodes are huge. The memory for storing the nodes and its related information will not be sufficient when there are a huge number of maximal frequent itemsets in data streams. It is necessary to design an algorithm which can use less memory and the cost of CPU.

In this paper, we propose an algorithm MMFI_DSSW for mining maximal frequent itemsets in data streams with a small memory usage. First, a novel data structure MFI_BVT is used to maintain the incoming streaming data with binary vector. Then, algorithm MMFI_DSSW can mine all then maximal frequent itemsets in data streams sliding window with a single pass scan the incoming data. Finally, the mining result can be update incrementally.

The rest of the paper is organized as follows. Problem definitions are given in section 2. The novel summary data structure MFI_BVT and algorithm MMFI_DSSW are presented in detail in section 3. We will show our experiment result in section 4. Finally, we conclude our works in section 5.

2 Problem Definition

Let I=$\{i_1, i_2, ...i_n\}$ be a set of literals, called items. A itemset X is a non-empty subset of I. A k-itemset is a set with k items and is denoted by $(i_1, i_2, ...i_k)$. A transaction T with m items is denoted by T=$\{i_1, i_2, ...i_m\}$, such that T⊆I. Let DS denote the infinite sequence of data streams and the size of DS are expressed as |DS|. Let DS=$\{B_1, B_2, ...B_n\}$, n is the number of basic window, and there are the same number of transactions in each B_i, we called it transaction sensitive sliding window.

Definition 1. Suffix subsets: For each newly arrived transaction T=$\{i_1, i_2,...i_m\}$, it can be mapped into a suffix sets$\{(t_1, (x_1, x_2, ...x_m)), (t_2, (x_2, x_3, ...x_m)),$ ...

$(t_{m-1}, (x_{m-1}, x_m))$, $(t_m, (x_m))$}. All of these sets are the suffix subsets of transaction T, and these subsets of T can be considered as the independent transactions that are stored in MFI_BVT.

Definition 2. Maximal frequent itemsets: Let itemsets X be a subset of I, $X._{esup}$ expresses the estimated support of itemsets X. A itemsets is a maximal frequent itemsets, if (1) $X._{esup} \geq minsup \mid B_i \mid, minsup \subset [0,1]$ is the minimum support threshold given by user and $|B_i|$ is the length of B_i so far. (2) It is not the subsets of any other itemsets.

3 MFI_BVT Structure and MMFI_DSSW

3.1 Construct MFI_BVT (Maximal Frequent Itemsets Binary Vector Table)

In this section, we will discuss how to construct the summary data structure MFI_BVT in detail. First, the definition of MFI_BVT is given, and then the algorithm how to construct the MFI_BVT is described. Finally, an example is showed to explain to algorithm MFI_BVT.

Definition 3. MFI_BVT (Maximal Frequent Itemsets Binary Vector Table): A MFI_BVT consists of two tables, HT(head table) and BVT(Binary Vector Table), and they can be defined as follows.

(1) It contains a HT which has 2 fields: HT_data and HT_count. HT_data logs the items and HT_count records the number of the data so far.
(2) It includes a BVT which has 3 fields: BVT_flag, BVT_count and BVT_code. BVT_flag records the first item which is not 0 in BVT_code. BVT_count logs the number of the itemsets. BVT_code records the itemsets with binary vector representation.

Now, we will describe the process how to construct the MFI_BVT in detail. First, algorithm MFI_BVT maps the itemsets into suffix subsets, and then inserted these suffix subsets into BVT and HT. The algorithm for constructing MFI_BVT is described as follows.

```
Algorithm1 construct MFI_BVT
Input: the incoming data in B_i
Output: HT, BVT
(1) Initialize HT and BVT, set HT_data, HT_count, BVT_flag,
BVT_count and BVT_code to be empty.
(2) For each itemsets X_j in B_i, map X_j into suffix subsets
(t_1,(x_1, x_2, ...x_m))(t_1 is the first item in (x_1, x_2, ...x_m),
and then if t_1 has been in HT, increase the count by one; if not,
add t_1 into HT.
(3) For each suffix subsets, if there is the binary vector of
(x_1, x_2, ...x_m) in BVT, the BVT_count will be added by one;
if not, it will be added into BVT.
```

`(4) Check out each column in BVT, if the binary vector |BVT_code`$_i$`|`
`equals to 1, i.e. there is only one item in (`x_1`, `x_2`, ...`x_m`),`
`it can be expressed by HT, so we will not record it into BVT.`

Example 1. Let the size of sliding window is 6, the transactions in the first transaction sensitive window block B_1 of the data stream DS are $(acdef)$, $(abcdf)$, (cef), $(acdf)$, (cef), (df), When the transaction (def) comes, the transaction sensitive window appends the new transaction (def) and deletes the oldest transaction $(acdef)$ from the current window. The transactions in the second window B_2 include $(abcdf), (cef), (acdf), (cef), (df), (def)$, minsup=4 and a, b, c, d, e, f are items in the stream.

Above all, the first itemsets $(acdef)$ can be mapped into suffix subsets $(a, acdef)$, $(c, cdef)$, (d, def), (e, ef) *and* (f, f) . For $(a, acdef)$, algorithm MFI_BVT inserts a and $acdef$ into HT and BVT, HT_data=a, HT_count=1, BVT_flag$_1$=a, BVT_count$_1$=1 and BVT_code$_1$ = 101111. $(c, cdef)$, (d, def) , (e, ef) and (f, f) do the same as $(a, acdef)$. However, because $|BVT\_code5|=$ $|000001|$=1, we only insert (f, f) in HT. Then algorithm maps the next itemsets $(abcdf)$ into suffix subsets. Because b is a new item, algorithm adds b into HT and BVT with lexicographic order. Next, let the BVT_code of the new suffix subsets compare with the existed BVT_code in BVT, if the new one is different from the existed BVT_codes, the new suffix subset is inserted into BVT; if not, the relevant BVT_count value is added by one. $(acdef)$ and $(abcdf)$ are mapped into BVT as show in Table 1.

All of the six itemsets are added into the head table HT and BVT as show in table 2 and table 3.

Table 1. The first two itemsets in BVT

Flag	a	c	d	e	a	b	c	d
Count	1	1	1	1	1	1	1	1
a	1	0	0	0	1	0	0	0
b	0	0	0	0	1	1	0	0
c	1	1	0	0	1	1	1	0
d	1	1	1	0	1	1	1	1
e	1	1	1	1	0	0	0	0
f	1	1	1	1	1	1	1	1

3.2 Pruning and Merging Strategy in MFI_BVT

According to the apriori[10]principle,only the frequent itemsets are used to construct candidate itemsets in the next pass. So, (1) If the HT_count of a item in head table is less than the minimal support, it will be deleted from HT, then all the row including this item will be set to zero and all the columns including this items will be deleted in BVT. (2) Merge the columns which have same BVT_code and BVT_flag and check out whether there is a column whose |BVT_code| is 1.

Table 2. Head Table (HT)

Date	Count
a	3
b	1
c	5
d	4
e	3
f	6

Table 3. All the itemsets in BVT

Flag	a	c	d	e	a	b	c	d	c	a
Count	1	1	1	3	1	1	2	3	2	1
a	1	0	0	0	1	0	0	0	0	1
b	0	0	0	0	1	1	0	0	0	0
c	1	1	0	0	1	1	1	0	1	1
d	1	1	1	0	1	1	1	1	0	1
e	1	1	1	1	0	0	0	0	1	0
f	1	1	1	1	1	1	1	1	1	1

For example, in table 2 and table 3, (1) The count of b is less than minsup $\times$ $|B_1|$, so deleted b from HT. (2) Due to the second row in BVT_code is b, it is set to be zero. And we delete column 6 in table 3, because its BVT_flag is b. Then merge column 5 and column 10 into column 5. Finally check out whether there is a column which |BVT_code|=1. Table 4 shows the pruned and merged result.

Table 4. Pruned and merged result of BVT

Flag	a	c	d	e	a	c	d	c
Count	1	1	1	3	2	2	3	2
a	1	0	0	0	1	0	0	0
b	0	0	0	0	0	0	0	0
c	1	1	0	0	1	1	0	1
d	1	1	1	0	1	1	1	0
e	1	1	1	1	0	0	0	1
f	1	1	1	1	1	1	1	1

3.3 MMFI_DSSW Algorithm for Mining Maximal Frequent Itemsets

Algorithm MMFI_DSSW groups the BVT into different teams based on BVT_flag, i.e., if those different suffix subsets in BVT have the same BVT_flag, they will be grouped into the same team. These different suffix subsets perform logical AND operation with other BVT_code in the same team. Then, we store the maximal frequent itemsets in the temporary table (TT) based on definition2. Next, algorithm MMFI_DSSW outputs all the maximal frequent itemsets when it is needed. Finally, the Temporary Table(TT) is set to be empty.

The algorithm for mining maximal frequent itemsets in MFI_BVT is given as follows.

```
Algorithm2 MMFI_DSSW
Input: MFI_BVT, |Bi|, minsup
Output: maximal frequent itemsets
```

```
{
(1) Initialize temporary table TT = null; TTrow=0;
(2)  For each HT_data do
(3)  {For each BVT_flag= HT_data_i do
       {Group into team T (T= BVT_flag_i);
(4)     For each T_i do
         {J=i+1;
(5)      Call LgAnd(BVT_code_i, BVT_code_{i+1}, j);}
(6)     For each row in TT
(7)       {For each row'=row+1 in TT
          {TTrow._code =TTrow._code AND TTrow'._code;
             TTrow._count= TTrow._count+ TTrow'._count;
             Delete TTrow' ;} }}
(8)  Delete TTrow._count minsup*|B_i | in TT; }
}

LgAnd(BVT_code_i, BVT_code_{i+1},j)
{
     j=j+1;
(1)  If BVT_code_i=BVT_code_i AND BVT_code_{i+1} then
        BVT_count_i=BVT_count_i+BVT+count_{i+1};
        LgAnd(BVT_code_i, BVT_code_j);
        TTrow= TTrow +1;
        TTrow._codei= BVT_code_i;
        TTrow.count_i= BVT_count_i;
(2)  Else
        TTrow= TTrow +1;
        TTrow._codei= BVT_code_i;
        TTrow._counti=BVT_count_i;
        LgAnd(BVT_code_i, BVT_code_j);
(3)  End if
}
```

For the example showed in table 4, algorithm MMFI_DSSW initializes the temporary table (TT) first. Next, according to lexicographic order of BVT_flag, it groups column 1 and column 5 into one team, because the BVT_flag of column 1 and column 5 is a, then $(a, 3, 101101)$ is acquired, we can also acquire $(c, 3, 001011)$, $(c, 3, 001101)$, $(d, 4, 000101)$ and $(e, 3, 000011)$ from teams c, d, e. Because all the BVT_count of $(a, 3, 101101), (c, 3, 001011), (c, 3, 001101)$, $(d, 4, 000101)$ and $(e, 3, 000011)$ are $3 > 0.4 * 6$, and the logical AND operation for the different teams are $\{101101\ AND\ 001101\ AND\ 000101\} = 101101$, $\{001011\ AND\ 000011\} = 001011$, so, we consider $(c, 3, 001101)$ and $(d, 4, 000101)$ are included into $(a, 3, 101101)$, and $(e, 3, 000011)$ is included into $(c, 3, 001101)$. Next, due to $(a, 3, 101101)(c, 3, 001101)$ and $(c, 3, 001101) \nsubseteq (a, 3, 101101)$ too, we record $(a, 3, 101101)$ and $(c, 3, 001101)$ into TT. So, the maximal frequent itemsets are $(acdf)$ and (cef).

3.4 Append a New Transaction into MFI_BVT

The new transactions have to be appended into MFI_BVT when the sliding window slides forward. We will describe the appending process of a new transaction as follows.

```
Algorithm3 AppendNew
Input: the new transaction T_i, MFI_BVT
Output: the new MFI_BVT
(1) For transaction T_i in B_i, mapping T_i into suffix subsets
{(t_1,(x_1, x_2, ...x_m)), (t_2,(x_2, x_3,...x_m),...(t_{m-1}, (x_{m-1}, x_m)),
(t_m,x_m)}, and then if t_1 has been in HT, increase the count by
one; if not, add t_1 into HT.
(2) For each suffix subsets, if there is the binary vector of
(x_1, x_2, ...x_m)) in BVT, the BVT_count will be added by one;
if not, it will be added into BVT.
(3) Check out each column in BVT, if the binary vector
|BVT_code_i| equals to 1, we will not record it into BVT.
```

For example, when the new transaction (def) comes , algorithm AppendNew maps (def) into suffix subsets $(d, def), (e, ef), (f, f)$ first. And then, it maps those suffix subsets $(d, def), (e, ef), (f, f)$ into binary vector.$(d, 000111)$, $(e, 000011)$. Because the binary vector $|BVT\_code_i|$ of (f, f) equals to 1, so we only record it into HT. Next, due to $(d, 000111)$ and $(e, 000011)$ have been in MFI_BVT, we only increase the BVT_count of $(d, 000111)$ and $(e, 000011)$ by one. Table 5 and Table 6 show the result.

Table 5. New HT after appending

Flag	Count
a	3
c	5
d	5
e	4
f	7

Table 6. New BVT after appending

Data	a c d e a c d c
Count	1 1 2 4 2 2 3 2
a	1 0 0 0 1 0 0 0
b	0 0 0 0 0 0 0 0
c	1 1 0 0 1 1 0 1
d	1 1 1 0 1 1 1 0
e	1 1 1 1 0 0 0 1
f	1 1 1 1 1 1 1 1

3.5 Delete the Oldest Transaction from MFI_BVT

In transaction sensitive sliding window model, the oldest transaction T will be deleted from the window when the new transaction comes. First, the count of item of the oldest transaction will be subtract by one in HT, if the count of item in HT is less than one, delete it from HT. Then, the oldest transaction and its suffix subsets will be deleted from MFI_BVT. The algorithm for deleting the oldest transaction from MFI_BVT is described as follows.

```
Algorithm4 DeleteOldest
Input: the oldest transaction T_i,MFI_BVT
Output: the new MFI_BVT
(1) Decrease the count of each item
in T_i by one in HT, and if the count of item in HT is less
than one, delete it from HT.
(2) For the oldest transaction T_i, map T_i into suffix subsets
{(t_1,(x_1, x_2, ...x_m)),(t_2,(x_2, x_3,...x_m),...(t_{m-1},(x_{m-1}, x_m)),
(t_m,x_m)}, and then map those suffix subsets into binary vector.
Next, subtract all of those binary vectors of suffix subsets by
one from MFI_BVT. If the BVT_count of the column is less than one,
delete it from MFI_BVT.
(3) Check out each column in BVT, if the binary vector
|BVT_code_i| equals to 1, delete it from BVT. (4) Perform
pruning and merging strategy to optimize BVT.
```

For example, we delete the oldest transaction $(acdef)$ from example 1. First, Algorithm DeleteOldest decreases the count of items a,c,d,e,f by one in HT. Then, due to the BVT_count of $(a, 1, 101111)$ and $(c, 1, 001111)$ is 1, algorithm DeleteOldest deletes $(a, 1, 101111), (c, 1, 001111)$ and decreases the BVT_count of $(d, 2, 000111)$ and $(e, 3, 000011)$ by one. Next, due to the HT_count of item a is 2 <minsup*$|B_i|$, we set all the row of BVT_code in a to zero, delete a from HT and delete the column which BVT_flage is a. Table 7 and Table 8 show the HT and MFI_BVT after performing Algorithm DeleteOldest.

Table 7. New HT after deleting

Data	Count
c	4
d	4
e	3
f	6

Table 8. New BVT after deleting

Flag	d e c d c
Count	1 3 2 3 2
a	0 0 0 0 0
b	0 0 0 0 0
c	0 0 1 0 1
d	1 0 1 1 0
e	1 1 0 0 1
f	1 1 1 1 1

4 Experiments

The simulation model of our experimental studies is described in Section 4.1. To assess the performance of the algorithms MMFI_DSSW for mining data streams, we conduct two empirical studies based on the synthetic datasets. The scalability of algorithm MMFI_DSSW are examined in Section 4.2. MMFI_DSSW is compared with estDec+[8] algorithm in Section 4.3.

4.1 Simulation Model

We performed the experiments on a 2.47GHz compatible PC with 512MB memory running on Windows XP, and the program is written in Microsoft Visual C++6.0. To evaluate the performance of the algorithm MMFI_DSSW, the test dataset T10.I4.D1000K and T5.I4.D1000K is generated by IBM synthetic data generator[10]. The former synthetic dataset T10.I4 has average transaction size T of 10 items and the average size of frequent itemset I is 4-items, the latter dataset T5.I4 has average transaction size T and average frequent itemset size I are 5 and 4, respectively. Both synthetic datasets have 1,000K transactions, where 1K denotes 1,000. In the experiments, the minimum support threshold is changed from 1% to 0.1%.

4.2 The Scalability of MMFI_DSSW

In this section, memory usage and runtime are examined for mining all maximal frequent itemsets in sliding windows over data streams. The memory usage and runtime in Figure 1 and Figure 2 for both synthetic datasets increase smoothly when minimum support is between 1% and 0.5%, but they increase remarkable when the minimum support is between 0.4% and 0.1%, that is because the memory processes a large number of transformation from itemset to binary vector and logical AND operation for the binary vector. So, all of these indicate the scalability and feasibility of algorithm MMFI_DSSW.

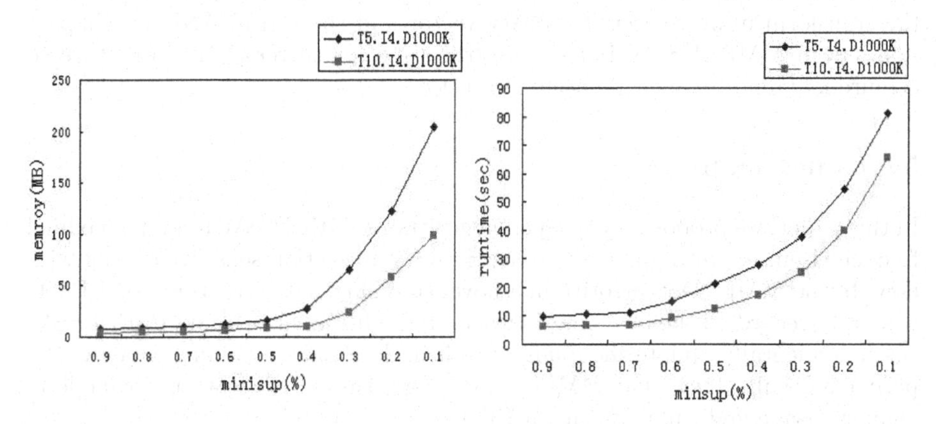

Fig. 1. Memory usages for the compared data

Fig. 2. Runtime for the compared data

4.3 The Comparison for MMFI_DSSW and estDec+

In this section, we compare the algorithm MMFI_DSSW with the estDec+ algorithm [8] with the synthetic data T10.I4.D1000K .The experiments of memory usage are shown in Figures 3, and the processing times are shown in Figures 4. The minimum support threshold is changed form 1% to 0.1%. As shown in these

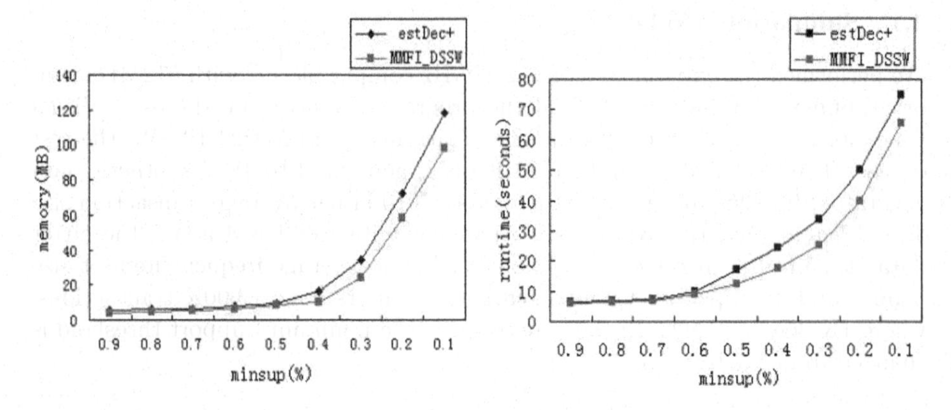

Fig. 3. Memory usages of the compared algorithms

Fig. 4. Runtime of the compared algorithms

experiments, MMFI_DSSW outperforms estDec+ for both memory consumption and CPU cost.

Figure3 shows that, the memory usage of MMFI_DSSW is less than estDec+ obviously with the decrease of the minimum support threshold. Due to we adopt binary vector to store the incoming itemsets, it outperforms estDec+ which applies CP-tree to maintain incoming itemsets. Figure 4 shows that the runtime of MMFI_DSSW also outperforms estDec+. That is because the time for executing the transform from itemsets to binary vector and the logical AND to compare itemsets in MMFI_DSSW is less than the time for traversal tree structure to acquire maximal frequent itemsets in estDec+.

5 Conclusions

In this paper, we propose a single pass algorithm MMFI_DSSW to mine maximal frequent itemsets in data streams based on a transaction sensitive sliding window. In the MMFI_DSS algorithm, a novel summary data structure, MFI_BVT, is used to record all the incoming itemsets. In addition, algorithm MMFI_DSSW can incrementally update the mining result in data streams sliding window. Experiment results show that MMFI_DSSW algorithm is efficient and scalable in memory usage and running time of CPU.

References

1. Li, H, Lee, S, Shan, M: An efficient algorithm for mining frequent itemsets over the entire history of data streams. In: Proceedings of the First International Workshop on Knowledge Discovery in Data Streams, held in conjunction with the 15th European Conference on Machine Learning (ECML 2004) and the 8th European Conference on the Principles and Practice of Knowledge Discovery in Databases (PKDD 2004), Pisa, Italy (2004)

2. Zhi-jun, X., Hong, C., Li, C.: An Efficient Algorithm for Frequent Itemset Mining on Data Streams. In: Perner, P. (ed.) ICDM 2006. LNCS (LNAI), vol. 4065, pp. 474–491. Springer, Heidelberg (2006)
3. Giannella, C., Han, J., Pei, J., Yan, X., Yu, P.S.: Mining Frequent Patterns in Data Streams at Multiple Time Granularities. In: Kargupta, H., Joshi, A., Sivakumar, K. (eds.) Next Generation Data Mining, pp. 191–212. MIT Press, Cambridge, Massachusetts (2003)
4. Lin, C.H., Chiu, D.Y., Wu, Y.H., Chen, A.L.P.: Mining Frequent Itemsets from Data Streams with a Time-Sensitive Sliding Window. In: Proceedings of the Fifth SIAM International on Data Mining, Newport Beach, USA (2005)
5. Teng, W.G., Chen, M.-S., Yu, P.S.: A Regression-Based Temporal Pattern Mining Scheme for Data Streams. In: Proceedings of the 29th VLDB Conference, pp. 93–104. IEEE Press, Berlin, Germany (2003)
6. Chang, J.H., Lee, W.S.: Finding Recent Frequent Itemsets Adaptively over Online Data Streams. In: Proceedings of the Ninth ACM SIGKDD International Conference on Knowledge Discovery and Data Mining, pp. 487–492. ACM Press, Washington, DC, USA (2003)
7. Li, H.-F., Lee, S.-Y., Shan, M.-K.: Online Mining (Recently) Maximal Frequent Itemsets over Data Streams. In: RIDE-SDMA 2005. Proceedings of the 15th International Workshop on Research Issues in Data Engineering: Stream Data Mining and Applications, pp. 11–18. IEEE Press, Tokyo, Japan (2005)
8. Lee, D., Lee, W.: Finding maximal frequent itemsets over online data streams adaptively. In: Proceedings of the fifth IEEE InternationalConference on Data Mining, pp. 266–273. IEEE Press, Houston, USA (2005)
9. Han, J., Pei, J., Yin, Y.: Mining frequent patterns without candidate generation. In: Proceedings of the ACM SIGMOD Conference on Management of Data, pp. 1–12. ACM Press, Dallas, USA (2000)
10. Agrawal, R., Srikant, R.: Fast algorithms for mining association rules. In: Proceedings of the 20th International Conference on VLDB, Santiago, Chile, pp. 487–499 (1994)

Term Consistency Checking of Ontology Model Based on Description Logics*

Changrui Yu[1] and Yan Luo[2]

[1] School of Information Management and Engineering, Shanghai University of Finance and Economics, Shanghai 200433, China
[2] Institute of System Engineering, Shanghai Jiao Tong University, 200052 Shanghai, China
{yucr,yanluo}@sjtu.edu.cn

Abstract. Formalized ontology model and its semantic consistency checking has become one of the highlights in knowledge engineering research. This paper constructs an ontology model based on description logics. The term consistency checking of this model is divided into four types: term satisfiability checking, term subsumption checking, term equivalence checking and term disjointness checking. Then the proof that the four types of checking can be resolved into subsumption checking of term extension is provided in this paper. Moreover, the ontology term consistency algorithm using inference of description logics is proposed and tested by examples.

1 Introduction

Currently, the formal ontology model suiting to consistency checking is still under hot discussion. Some researches of ontology are based on first order logic (FOL), e.g. Ontolingua, CycL, LOOM. Although FOL have a more expressive power, the reasoning process are complex and most of them are even undecidable, which is not suitable for ontology model checking [1]. Description Logics is equipped with a formal and logic-based semantics allowing inferring implicitly represented knowledge from the knowledge that is explicitly contained in the knowledge base [2]. Although DL has a less expressive power than FOL, its inference procedures are more efficient and decidable, which is more suitable for ontology checking: description logic (DL) is used in ConsVISor tool for consistency checking of ontologies [3][4],and temporal description logic (TDL)in [5]. DLs are hence more suitable for ontology checking than first order logic. Especially, it is easy to shift the grammar of DLs to the form of RDF/OWL [6]. Therefore, the ontology models based on DLs are very suitable for concept modeling and knowledge management in Web environment.

The remainder of the paper is organized as follows. In Section 2 we develop an ontology model based on description logics. In Section 3, the term consistency checking of this model is divided into four types: term satisfiability checking, term subsumption checking, term equivalence checking and term disjointness checking. Then the proof that the four types of checking can be resolved into subsumption checking of

* This research work is supported by the Natural Science Fund of China (# **70501022**).

Z. Zhang and J. Siekmann (Eds.): KSEM 2007, LNAI 4798, pp. 496–501, 2007.

term extension is provided. And the ontology term consistency algorithm using inference of description logics is proposed and tested by examples.

2 Ontology Model Based on Description Logics

Definition 1 (Ontology model). Given a terminology description language L, an ontology model is a 5-tuples

$$O=<T, X, TD, XD, TR> \tag{1}$$

where T is a Term Set, X is an Individual Set, TD is a Term Definition Set, XD is an Instantiation Assertion Set, and TR is a Term Restriction Set.

Definition 2 (Term formula). Given term constructor set S, we call the expression, satisfying the syntax rule below, an S-based term formula.

$$D, E \rightarrow C \mid \top \mid \bot \mid \neg C \mid D \sqcap E \mid \forall P.D \mid \exists P.\top \tag{2}$$

Definition 3 (Relationship between term). Given term constructor set S and ontology $O=<T, X, TD, XD, TR>$, D and E are two S-based term formulas. For any interpretation I, there exist the following three relationships between terms:

(i) D is subsumed by E denoted as $D\text{M}E$, if $D^I \subseteq E^I$.
(ii) D and E are equivalent denoted as $D \equiv E$, if $D\text{M}E$ and $E\text{M}D$.
(iii) D and E are disjointness denoted as $D \pitchfork E$, if $D\text{M}\neg E$ and $E\text{M}\neg D$.

Definition 4 (Term Definition Set). Given $O=<T, X, TD, XD, TR>$, Term Definition Set TD is such a set that consists of term definition items subject to the following restrictions, denoted as $TD=\{C_1 \equiv D_1, C_2 \equiv D_2, \ldots, C_n \equiv D_n\}$, where $C_i \in T$, D_i is a term formula, and every term in D_i is from T.

(i) For any i, j $(i \neq j, 1 \leq i \leq n, 1 \leq j \leq n)$, $C_i \neq C_j$ holds.
(ii) If there exist $C_1' \equiv D_1'$, $C_2' \equiv D_2', \ldots, C_m' \equiv D_m'$ in TD, and C_i' occurs in D_{i-1}' $(1 < i \leq m, m \leq n)$, then C_1' must not occur in D_m'.

Definition 9 (Expansion of Instantiation Assertion). Given $O=<T, X, TD, XD, TR>$. $C(a)$ is a class instantiation assertion in XD, and $e(C)$ is an expansion of C with respect to TD. $e(C)(a)$ is said to be the expansion of $C(a)$ with respect to TD. Through transforming each class instantiation assertion into the form of expansion, we can get a new Instantiation Assertion Set XD'. The new set XD' is called the expansion of XD with respect to D, denoted as $e(XD)$.

Definition 10 (Expansion of Term Restriction Set). Given $O=<T, X, TD, XD, TR>$. For convenience, we assume $TR=\{D_1\text{M}E_1, D_2 \equiv E_2, D_3 \pitchfork E_3\}$. If each term relation in TR has been transformed into the expansion form, a new Term Restriction Set $TR'=\{e(D_1)\text{M}e(E_1), e(D_2) \equiv e(E_2), e(D_3) \pitchfork e(E_3)\}$ is obtained. TR' is called the expansion of TR, written as $e(TR)$.

Proposition 1. Given $O=<T, X, TD, XD, TR>$, if TD and TR have a model I in common, then I is also a model of $e(TR)$.

Proposition 2. Given O=<T, X, TD, XD, TR>, where $TD=\{C_1 \equiv D_1, C_2 \equiv D_2, \ldots, C_n \equiv D_n\}$. $T_p=\{B_1, B_2, \ldots, B_m\}$ is the set of primitive terms in TD, and E is a term formula. If I is a model of $e(TR)$, then there must exist a common model I' of both TD and TR, such that $E^{I'}=e(E)^I$.

3 Term Checking of Ontology Model

3.1 Types and Properties of Term Checking

Given term constructor set S and ontology O=<T, X, TD, XD, TR>, D and E are two S-based term formulas, the term checking of ontology models involves the following four types.

(1) Term satisfiability checking

Given O=<T, X, TD, XD, TR> and a term formula D, if there exists an ontology interpretation I, such that $D^I \neq \emptyset$, then D is said to be satisfiable and unsatisfiable otherwise. If there exists a model of TR, such that $D^I \neq \emptyset$, then D is satisfiable with respect to TR. If there exists a common model of both TR and TD, such that $D^I \neq \emptyset$, then D is said to be satisfiable with respect to TR and TD, or else unsatisfiable with respect to TR and TD.

(2) Term subsumption checking

Given O=<T, X, TD, XD, TR> and two term formulas D and E, if $D^I \subseteq E^I$ holds for all the ontology interpretation I, then D is said to be subsumed by E, or E subsumes D, denoted as $\models D$ME. If for all the models I of TR, $D^I \subseteq E^I$ holds, then we say TR entails that D is subsumed by E, denoted as $TR \models D$ME. If $D^I \subseteq E^I$ holds for all the common models I of both TR and TD, then we say TR and TD jointly entails that D is subsumed by E, denoted as $(TR+TD) \models D$ME.

(3) Term equivalence checking

Given O=<T, X, TD, XD, TR> and two term formulas D and E, if $D^I = E^I$ holds for all the ontology interpretation I, then we say D and E are equivalent, written as $\models D \equiv E$. If $D^I = E^I$ holds for all the models I of TR, then we say TR entails that D and E are equivalent, written as $TR \models D \equiv E$. If for all the common models I for both TR and TD, $D^I = E^I$ holds, then we say TR and TD jointly entails that D and E are equivalent, written as $(TR+TD) \models D \equiv E$.

(4) Term disjointness checking

Given O=<T, X, TD, XD, TR> and two term formulas D and E, if $D^I \cap E^I = \emptyset$ holds for all the ontology interpretation I, then we say D and E are disjoint, written as $\models D \pitchfork E$. If in all models I of TR, $D^I \cap E^I = \emptyset$ holds, then we say TR entails that D and E are disjoint, written as $TR \models D \pitchfork E$. If $D^I \cap E^I = \emptyset$ holds in all the common models I for both TR and TD, then we say TR and TD jointly entails that D and E are disjoint, written as $(TR+TD) \models D \pitchfork E$.

Proposition 3 (Reduction to Subsumption). Given O=<T, X, TD, XD, TR>, for any two term formulas D and E:

(i) D is unsatisfiable with respect to TR and TD, if $(TR+TD) \models D$M$\bot$;

(ii) $(TR+TD) \models D \equiv E$, iff $(TR+TD) \models D$ME and $(TR+TD) \models E$MD;

(iii) $(TR+TD) \models D ⩀ E$, iff $(TR+TD) \models (D \sqcap E) \text{M} \bot$.

The fact stated in the propositions implies that the four kinds of term checking can all be reduced to the subsumption.

Proposition 4. Given O=<T, X, TD, XD, TR>. D and E are two term formulas. $e(D)$ is the expansion of D with respect to TD, and $e(E)$ is the expansion of E with respect to TD. We have:

(i) D is satisfiable with respect to TR and TD, iff $e(D)$ is satisfiable with respect to $e(TR)$;

(ii) $(TR+TD) \models D \text{M} E$, iff $e(TR) \models e(D) \text{M} e(E)$;

(iii) $(TR+TD) \models D \equiv E$, iff $e(TR) \models e(D) \equiv e(E)$;

(iv) $(TR+TD) \models D ⩀ E$, iff $e(TR) \models e(D) ⩀ e(E)$.

3.2 Term Consistency Checking Algorithm

Proposition 3 indicates that the four types of term consistency checking can be resolved into subsumption checking. Proposition 4 further indicates that term checking can be realized by checking term expansions. Hence, term checking only requires checking of the subsumption of two term extensions. Accordingly, the process of term consistency checking includes the following four steps:

Step 1. Expansion of term formula, i.e., substituting all defined term involved in term formula D and E with the corresponding primitive terms, and form $e(D)$ and $e(E)$.

Step 2. Normalization of term formula, i.e., shifting the extended term formula $e(D)$ and $e(E)$ to a normalized form. According to the associative law, commutative law and idempotent law of $\sqcap$, and the axiom $\forall P.(C \sqcap D) \equiv (\forall P.C) \sqcap (\forall P.D)$, any S-term formula can be shifted to an S-normal form.

The normalization approach for term formula is shown as follows.

Given term constructor set S and ontology model O=<T, X, TD, XD, TR>, an S-term formula is denoted in the following form.

$$D, E \rightarrow C \mid \neg C \mid \bot \mid D \sqcap E \mid \forall P.D \mid \geqslant m, P \mid \leqslant n, P \tag{4}$$

where C is the atomic class term, P is the atomic property term, and D and E are S-term formulas. If any S-term formula can be shifted to the following form (5) (or $\bot$), then form (5) (or $\bot$) is called an S-normal form.

$$C_1 \sqcap \ldots \sqcap C_l \sqcap \neg D_1 \sqcap \ldots \sqcap \neg D_m \sqcap \forall R_1.A_1 \sqcap \ldots \sqcap \forall R_n.A_n \sqcap \geqslant \alpha_1, P_1 \sqcap \ldots \sqcap \geqslant \alpha_s, P_s \\ \sqcap \leqslant \beta_1, Q_1 \sqcap \ldots \sqcap \leqslant \beta_t, Q_t \tag{5}$$

where $C_1,\ldots,C_l$ are l different atomic class terms, $D_1,\ldots,D_m$ are m different atomic class terms, $R_1,\ldots,R_n$ are n different atomic property terms, $A_1,\ldots,A_n$ are S-normal forms, $\alpha_1,\ldots,\alpha_s$ are s different positive integers, $P_1,\ldots,P_s$ are s different atomic property terms, $\beta_1,\ldots,\beta_t$ are t different positive integers, and $Q_1,\ldots,Q_t$ are t different atomic property terms.

Step 3. Reduction of normal form of terms according to the following reducing rules.

An S-normal form $C_1 \sqcap \ldots \sqcap C_l \sqcap \neg D_1 \sqcap \ldots \sqcap \neg D_m \sqcap \forall R_1.A_1 \sqcap \ldots \sqcap \forall R_n.A_n \sqcap \geqslant \alpha_1,P_1 \sqcap \ldots \sqcap \geqslant \alpha_s,P_s \sqcap \leqslant \beta_1,Q_1 \sqcap \ldots \sqcap \leqslant \beta_t,Q_t$ can be further reduced by a recurrent process according to the following rules:

(i) If there exists i (i=1,…, l) which satisfies $C_i = \bot$, then S-normal form=$\bot$.

(ii) If there exist i, j (i=1,…, l and j=1,…, m) which satisfy $C_i = D_j$, then S-normal form=$\bot$.

(iii) If there exist i, j (i=1,…, n and j=1,…, s) which satisfy $A_i=\bot$, $P_j=R_i$, and $\alpha_j \geq 1$, then S-normal form=$\bot$.

(iv) If there exist i, j (i=1,…, s and j=1,…, t) which satisfy $P_i=Q_j$ and $\alpha_i \geq \beta_j$, then S-normal form=$\bot$.

The reduced S-normal form is either $\bot$ or $C_1 \sqcap \ldots \sqcap C_l \sqcap \neg D_1 \sqcap \ldots \sqcap \neg D_m \sqcap \forall R_1.A_1 \sqcap \ldots \sqcap \forall R_n.A_n \sqcap \geqslant \alpha_1,P_1 \sqcap \ldots \sqcap \geqslant \alpha_s,P_s \sqcap \leqslant \beta_1,Q_1 \sqcap \ldots \sqcap \leqslant \beta_t,Q_t$, where $C_1,\ldots,C_l$, $D_1,\ldots,D_m$ are $l+m$ different atomic class terms, and A_i is likely to be $\bot$.

Step 4. Comparison between structures of normal forms of terms. This step is carried out with a recursive method in order to judge if there exist inclusive relationships between the structures. The comparison rules are shown as follows.

(1) Let D and E be two S-term formulas, $C_1 \sqcap \ldots \sqcap C_l \sqcap \neg D_1 \sqcap \ldots \sqcap \neg D_m \sqcap \forall R_1.A_1 \sqcap \ldots \sqcap \forall R_n.A_n \sqcap \geqslant \alpha_1,P_1 \sqcap \ldots \sqcap \geqslant \alpha_s,P_s \sqcap \leqslant \beta_1,Q_1 \sqcap \ldots \sqcap \leqslant \beta_t,Q_t$ be the S-normal form of D, and $E_1 \sqcap \ldots \sqcap E_x \sqcap \neg F_1 \sqcap \ldots \sqcap \neg F_y \sqcap \forall G_1.B_1 \sqcap \ldots \sqcap \forall G_z.B_z \sqcap \geqslant \pi_1,V_1 \sqcap \ldots \sqcap \geqslant \pi_k,V_k \sqcap \leqslant \varepsilon_1,W_1 \sqcap \ldots \sqcap \leqslant \varepsilon_u,W_u$ be the S-normal form of E. If both of the reduced normal forms of D and E are $\bot$, then there exists an inclusive relationship.

(2) If the reduced normal forms of D and E are not $\bot$, then we carry out the one by one comparison between the five pairs of components of the S-normal forms of D and E. When the following five conditions are tenable, there exists an inclusive relationship between S-term formulas D and E.

i . For each i ($1 \leq i \leq x$), there exists j ($1 \leq j \leq l$) which satifies $E_i=C_j$.
ii . For each i ($1 \leq i \leq y$), there exists j ($1 \leq j \leq m$) which satifies $F_i=D_j$.
iii . For each i ($1 \leq i \leq z$), there exists j ($1 \leq j \leq n$) which satifies $G_i=R_j$ and A_jMB_i.
iv . For each i ($1 \leq i \leq k$), there exists j ($1 \leq j \leq s$) which satifies $V_i=P_j$ and $\pi_i \leq \alpha_j$.
v . For each i ($1 \leq i \leq u$), there exists j ($1 \leq j \leq t$) which satifies $W_i=Q_j$ and $\varepsilon_i \geq \beta_j$.

3.3 Examples

In the following three examples, we assume that the term formulas have already been expanded. Hence we need only focus on the last three steps.

Example. Given term formulas

$$\forall P.\forall P.B \sqcap A \sqcap \forall P.(A \sqcap \forall P.\bot) \tag{3.1}$$

and

$$A \sqcap \forall P.(A \sqcap \forall P.A) \tag{3.2}$$

We now test if there exists any inclusive relationship between (3.1) and (3.2).

In step 2, term formula (3.1) is normalized and shifted to the normal form

$$A \sqcap \forall P.(A \sqcap \forall P.(B \sqcap \bot)) \tag{3.1'}$$

where $(B \sqcap \bot)$ is also an S-normal form. Because $\exists \bot$ and $(B \sqcap \bot)$ is reduced to $\bot$, (3.1′) is shifted to

$$A \sqcap \forall P.(A \sqcap \forall P.\bot) \tag{3.1''}$$

In step 3, the structures of term formulas (3.1″) and (3.2) are compared. At the end of the recursive comparison, we finally need to compare $\bot$ and A. Thus, term formula (3.1) is included by term formula (3.2).

4 Conclusions

Ensuring that ontologies are consistent is an important part of ontology development and testing. Reasoning with inconsistent ontologies may lead to erroneous conclusions.

In this paper, an ontology model is constructed using Description Logics, which is a 5-tuples including Term Set, Individual Set, Term Definition Set, Instantiation Assertion Set and Term Restriction Set. Based on the ontology model, issues of ontology term checking are studied with the conclusion that: the four kinds of term checking, including term satisfiability checking, term subsumption checking, term equivalence checking and term disjointness checking, can be reduced to subsumption checking of term extension. Further, the ontology term consistency algorithm using inference of description logics is proposed and tested by examples.

References

1. Anderson, C.A.: Alonzo Church's contributions to philosophy and intensional logic, Bull. Symbolic Logic 4(2), 129–171 (1998)
2. Baader, F., Calvanese, D., McGuinness, D., Nardi, D. (eds.): The Description Logic Handbook - Theory, Implementation and Applications. Cambridge University Press, Cambridge (2003)
3. Baclawski, K., Kokar, M., Smith, J.: Consistency Checking of RM-ODP Specifications. International Conference on Enterprise Information Systems, Setúbal, Portugal (2001)
4. Baclawski, K., Kokar, M., et al.: Consistency Checking of Semantic Web Ontologies. In: Horrocks, I., Hendler, J. (eds.) ISWC 2002. LNCS, vol. 2342, pp. 454–459. Springer, Heidelberg (2002)
5. Weitl, F., Freitag, B.: Checking semantic integrity constraints on integrated web documents. In: Wang, S., Tanaka, K., Zhou, S., Ling, T.-W., Guan, J., Yang, D.-q., Grandi, F., Mangina, E.E., Song, I.-Y., Mayr, H.C. (eds.) Conceptual Modeling for Advanced Application Domains. LNCS, vol. 3289, pp. 198–209. Springer, Heidelberg (2004)
6. Staab, S., Studer, R.: Handbooks on Ontologies. Springer, Heidelberg (2004)

Design and Realization of Advertisement Promotion Based on the Content of Webpage

Tao Cheng, Shuicai Shi, and Xueqiang Lv

Chinese Information Processing Research Center,
Beijing Information Science and Technology University,
100101 Beijing, China
{cheng.tao,shi.shuicai,lv.xueqiang}@trs.com.cn

Abstract. Online advertisement based on the content becomes the new model as the development of web advertising, and choosing a suitable advertisement which is related to the content of a webpage turns to be the core problem. This paper presents a semantic approach, with a goal of achieving webpage-advertisement matching accurately. The method is implemented and tested, and the result shows that the proposed algorithm is promising.

Keywords: Tongyici Cilin, Thematic Words, Web Data Extraction, Matching Rate.

1 Introduction

With the development and the popularization of Internet, online advertisement was born stealthily, develops rapidly, and becomes mature gradually, and advertisement based on the content of the webpage is the newest model.

In recent years, advertisement based on the content of the webpage not only brings enterprises huge brand returns, but also brings the advertisement promotion companies economic benefits depend on its unique advantages. Take Google for example, its advertising program Google AdSense for content is just one of such online advertisement. According to statistics, most of Google's advertising revenue is from AdSense advertisers, and its advertising revenue through this program in the third quarter of 2005 reached 675 million U.S. dollars.

The core assumption of advertisement based on the content of webpage is: if a user is interested in the content of a webpage, he is likely to have interest in the advertisement whose content is relevant to the webpage. Achieving accurate matching between webpage and advertisement is the core technology of this form of advertising.

Firstly, computer must understand content of a webpage which can be denoted by thematic words. Two important works have been done: thematic text extraction from webpage and thematic words extraction from text.

Then, match the thematic words of webpage and keywords of advertisement. Firstly compute the matching rate, and then display the advertisement with the highest matching rate on the webpage.

Z. Zhang and J. Siekmann (Eds.): KSEM 2007, LNAI 4798, pp. 502–507, 2007.

2 Extract Thematic Text from Webpage

As the carrier of information, most of the webpage are compiled using Hypertext Markup Language (HTML for short). The most remarkable character of HTML is semi-structure. For example, not every beginning tag has the ending tag, and crossing nest is allowed. However, some elements such as 'table' & '/table', 'div' & '/div', et al., are structured at most time.

General approach is: convert the HTML source file to a structured tree which can reflects the structure of the source file, then using heuristic rules to get the thematic information. Typical systems are W4F [1], RoadRunner [2], et al..

However, because of the character of semi-structure, it is difficult to generate a structured tree that fully reflects the structure of webpage. The extraction approach that this paper proposed has two characters.

(1) Generate the structured tree only through the structured tags;
(2) Expand the structured tree by Extend the interspaces.

Interspaces in this paper are the content between the ending tag of previous tree node and the beginning tag of next tree node. Extending the interspaces is to convert the interspaces to new leaf nodes. The following is a simple example of this kind. The Fig.1 is the source file of HTML; the Fig. 2(left)is the structured tree is gained only by computing the structured tags, and the Fig.2(right) is the structured tree after extending interspaces.Apparently, Fig 2(right) has converts the content between 1 and 37 to a new leaf node.

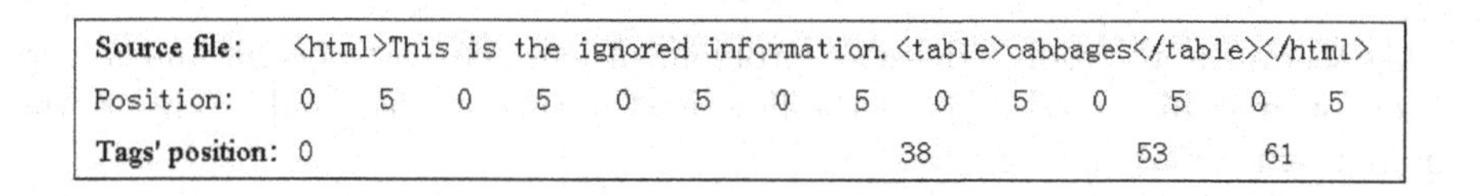

Source file: <html>This is the ignored information.<table>cabbages</table></html>
Position: 0 5 0 5 0 5 0 5 0 5 0 5 0 5
Tags' position: 0 38 53 61

Fig. 1. Examples of Web Source File

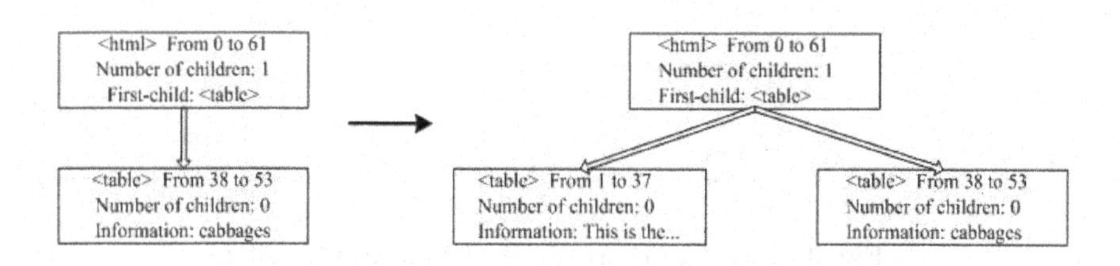

Fig. 2. Example of Expanding the Structured Tree

Finally, identify each area through a set of heuristic rules, such as sum of string with link, sum of string without link, average string with link, variance of string, max string of areas, et al..

3 Extract Thematic Words from Text

In order to extract thematic words from a given article correctly and fully, not only an accurate segmentation are needed as a basis, but also it is important to evaluate the words accurately and fully through the ability to express the purpose of the article[3]. At present, the main methods are: Tang Pei-li's[4] and Cheng Tao's[5]. We use Cheng Tao's method in this paper.

When designing method to calculating weight of words, this paper considered the frequency, position, length, span and related information(similar words, relevant words and lower words).The formula is as fellows.

$$Weight = len \times span \times \Sigma_{i=1}^{k}(f_i \times loc_i) + \Sigma_{j=1}^{j}(fr_j \times kind_j) . \quad (1)$$

Annotation: 'len' is the penalty factor about length; 'span' is related to paragraphs a word spans; 'k' is kind of position, including title, subhead, first or last paragraph, first or last sentence of a paragraph, first or last sentence of first or last paragraph, straight matter and whether behind a clue word (such as: so, therefore, however, in a word, anyway, et al.); 'fi' is the frequency a word appears in the position i; 'loci' is the penalty factor of position i; 'h' is kind of related information; 'frj' is the frequency a word's related information appears in the article; 'kindj' is the penalty factor of whether its related information appears.

4 Choose and Display the Suitable Advertisement

4.1 Advertisement Choosing Approach

This section is to resolve the core problem of this paper: webpage and advertisement matching [6-7]. Fig.4 is the flowchart of this module, and "compute the matching rate" is the core.

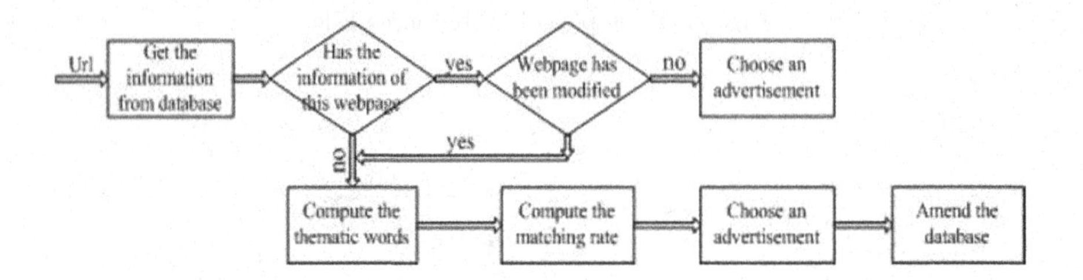

Fig. 3. Main Flowchart of Advertisement Choosing

4.2 Compute the Matching Rate of Advertisement

Many of the Chinese vocabulary are interrelated, such as similar words (China and People's Republic of China), relevant words (chair and table) and upper-lower words (vegetable and cabbage). Therefore, it is not enough to consider situation that a thematic word of webpage equals to a keyword of the advertisement (we call it absolute matching). Similar matching, relevant matching, upper matching and lower matching should be taken into account.

Using expansion version of Tongyici Cilin[8], we can obtain the similar words, the relevant words and the upper&lower words of the thematic words of the webpage. Then compare these words with keywords of advertisement: if there is a keyword of advertisement which is the same as the thematic word itself, it is called absolute matching; if there is a keyword which is the same as the similar word of a thematic word, it is called similar matching. Also we can define relevant matching, upper matching and lower matching in the same way.

The specific formula of computing matching rate is:

$$matchrate = \Sigma_{i=1}^{n} \Sigma_{j=1}^{m} (l_{ij} \times value_j) . \tag{2}$$

Annotation: 'matchrate' is the matching rate between webpage and advertisement; 'n' is the number of keywords of advertisement; 'm' is the number of thematic words of webpage; 'lij' is the penalty factor of matching kind between thematic word j and keyword i; 'value[j]' is the weight of thematic word j.

4.3 The Mechanism of Advertisement Displaying

Advertisement displaying involves three active objects. They are: web user, web server and the end of advertisement promotion. Simply speaking: when a web user request to open a webpage, the corresponding web server sends advertisements which is provided by the end of advertisement promotion to the web user.

5 Experiments

5.1 Thematic Text Extraction from Webpage

The experiment resource is 200 webpages which are artificially chosen from sohu, sina, yahoo, ea al.. The content of these 200 webpages involves sports, economy, education, entertainment, politics and other aspects. Table 1 provides thematic information extraction result, and using accuracy and acceptable rate as evaluation criterion. In this paper, acceptable is: although the result of thematic information extraction has a little noise information, it won't affect the result of thematic words extraction. Definition:

$$acceptableRate = (accurateNum + acceptableNum) \div webpagesNum . \tag{3}$$

The result shows that the acceptable rate is reach up to 95.5%. The accuracy of thematic information extraction is just 83.5%, which may be lower than other methods. However, this arithmetic is simple and the accuracy is acceptable. On the other hand, if only the result of thematic information extraction won't affect the result of thematic words extraction, a small amount of noise is allowed. So we just consider the acceptable rate.

Table 1. Result of Webpages' thematic Information Extraction

Content	number	Right	Accept	Wrong	Accuracy	AcceptableRate
Sports	54	45	8	1	83.3%	98.1%
Economy	17	15	2	0	88.2%	100%
Education	26	21	3	2	80.8%	92.3%
Entertainment	47	41	3	3	87.2%	93.6%
Politics	42	35	5	2	83.3%	95.2%
Others	14	10	3	1	71.4%	92.9%
All	200	167	24	9	83.5%	95.5%

5.2 Thematic Words Extraction from Text

Experimental material is the result of experiment 5.1, a total of 200 articles. Firstly, extract 10 thematic words for every article artificially as correct answer. Then extract 6 words for every article by machine and record the number of correct words extracted by machine. Table.2 is statistics of experiment, using accuracy as the measure of the extraction, and definitions:

$$Accuracy = correctNumber \div MechineExtractNumber. \quad (4)$$

Table 2. Result of Thematic Word Extraction

Content	number	6correct	5correct	4correct	3correct	2&1
Sports	54	9	34	10	1	0
Economy	17	3	13	1	0	0
Education	26	5	18	2	1	0
Entertainment	47	7	32	8	0	0
Politics	42	8	31	2	1	0
Others	14	2	9	2	1	0
All	200	34	137	25	4	0

We can calculate the accuracy through the data of table 3. There are total 1,001 words correctly extracted by machine, and 1,200 words extracted by machine totally. So the accuracy is 1001/1200, that is 83.42%.

5.3 Advertisement Chosen

15,806 advertisements are generated randomly for experiment using 140,333 high-frequency words on the web. Each advertisement has 8.8 keywords averagely. In this experiment, we use the result of experiment 5.2, 10 thematic words for every webpage as experimental material, and the result is as follows.

The detailed column names are: Content and number; Absolute matching; Similar matching; Upper matching; Lower matching; Related matching; Not matching. 'P' is the corresponding percent.

Table 3. Result of Webpages' Segmentation and Identification

Content/Num	A/P	S/P	U/P	L/P	R/P	N/P
Sports/54	45/83.33	6/11.11	2/3.70	0/0	1/1.85	0/0
Economy/17	11/64.71	2/11.76	3/17.65	1/5.88	0/0	0/0
Education/26	18/69.23	1/3.85	0/0	0/0	4/15.38	3/11.54
Entertainment/47	41/87.23	4/8.51	0/0	1/2.13	1/2.13	0/0
Politics/42	36/85.71	3/7.14	1/2.38	2/4.76	0/0	0/0
Others/14	8/57.14	2/14.29	2/14.29	1/7.14	1/7.14	0/0
All/200	159/79.5	18/9.0	8/4.0	5/2.5	7/3.5	3/1.5

6 Conclusions and Future Work

The approaches presented in this paper can achieve a good effect in the webpage-advertisement matching, and have solved the core problem thoroughly. Experiments have confirmed its effectiveness. The next step is to study how to put this advertising model into practice.

References

1. Sahuguet, A., Azavan, F.: Building Intelligent Web Applications Using Lightweight Wappers. Data and Knowledge Engineering 36(3), 283–316 (2001)
2. Crescenzi, V., Mecca, G., Merialdo, P.: RoadRunner: Towards Automatic Data Extraction from Large Web Sites. In: Proceeding of the 26th International Conference on Very Large Database Systems, Rome, Italy, pp. 109–118 (2001)
3. Damle, D., Uren, V.: Extracting significant words from corpora for ontology Extraction. In: Proceedings of the 3rd international conference on Knowledge capture table of contents, pp. 187–188 (2005)
4. Peili, T., Shuming, W., Ming, H.: Algorithm of Thematic Words Extraction from Chinese Texts Based on Semantic. Journal of Jilin University (Information Science Edition) 23(5), 535–540 (2005)
5. Tao, C., Shuicai, S., Xia, W., et al.: Thematic Words Extracting from a Chinese Text Based on Tongyici Cilin. Journal of Guangxi Normal University 25(2), 145–148 (2007)
6. Yih, W.-t., Goodman, J., Carvalho, V.R.: Finding advertising keywords on web pages. In: Proceedings of the 15th international conference on World Wide Web table of contents, New York, NY, USA, pp. 213–222 (2006)
7. Mehta, A., Saberi, A., Vazirani, U., Vazirani, V.: AdWords and Generalized Online Matching. In: FOCS 2005. 46th Annual IEEE Symposium on Foundations of Computer Science, pp. 264–273. IEEE Computer Society Press, Los Alamitos (2005)
8. Jiaju, M., et al.: Tongyici Cilin (2nd edn.), Lexicographical Publishing House, Shanghai, China (1996)

Irregular Behavior Recognition Based on Two Types of Treading Tracks Under Particular Scenes

Ye Zhang, Xiao-Jun Zhang, and Zhi-Jing Liu

Xidian University, Computer Science and Technology,
Xi'an, 710071, China
zhangye@mail.xidian.edu.cn, zxj1004@yahoo.com,
liuprofessor@163.com

Abstract. Visual analysis of human motion from video sequences is one of the most active research topics in the field of computer vision. This research has certain practical value and can be widely applied in some places such as: bank、hotel、garage、government department、large public facility. In this paper we present a novel method for judging irregular behavior based on treading track. Firstly, we use an object detection method in which the background subtraction method and the time difference method are averaged by weights to detect moving body, and then we judge whether someone is suspicious or not on the basis of treading track, it improves the judgment ability of irregular behavior recognition and experiment results have shown that this method gives static performances and good robustness.

Keywords: irregular behavior recognition; time difference method based on background subtraction; closed curve method; spiral line method.

1 Introduction

Body motion analysis [1~2] based on video sequence mainly deals with some study fields such as: computer vision、computer graphics、image processing、pattern recognition and artificial intelligence. It has wide application prospect and potential economy value in the aspects of advanced human-computer interaction、security monitoring、image storage and retrieval based on content [3～5]. Currently the theory study and system exploitation for motion object recognition have becoming more and more mature at home and abroad, however, there is no shaped algorithm to carry through regular and validity judgment for detected object (namely real intelligent processing). Meanwhile the research about judging irregular behavior based on treading track is even scantier. A novel judging irregular behavior method based on treading track is proposed in this paper, first, we use an object detection method in which the background subtraction method and the time difference method are averaged by weights to detect human body, then we judge whether somebody is suspicious or not based on his treading track. Here we mainly study two types of line shape: one is closed curve and the other is spiral line. If somebody's treading track

Z. Zhang and J. Siekmann (Eds.): KSEM 2007, LNAI 4798, pp. 508–513, 2007.

takes on one of these two types of line shape, it demonstrated that he or she wanders around someplace, so we should draw a red rectangle and give an alarm to prompt this person may be suspicious.

2 The Detection of Motion Object

At present, the most popular methods are time difference method [6] and the background subtraction method [7]. The former method has good adaptation to dynamic environment, but it can't extract all the points related to objects inextenso. And the latter can extract object points more inextenso, however, it is excessively alert to the dynamic background changes caused by light and exterior conditions. In order to overcome the aforementioned problems and shortcomings of these two methods, in this paper we propose an object detection method in which the background subtraction method and the time difference method are averaged by weights to realize the precise extraction of moving object contour successfully.

To detect moving object, first, the background model image is computed and the consecutive grayscale images of time k and k +1 is collected. Then, the time difference (T_t) of consecutive grayscale images of time k and k+1 is computed, and so is the time difference of current image and background model image (T_b). T_t, T_b represent difference binary image of the two methods respectively and their expression are shown in formula 1 and formula 2 as follows:

$$T_t(x,y)=\begin{cases}1 & |f_k(x,y)-f_{k+1}(x,y)|>t_t\\ 0 & |f_k(x,y)-f_{k+1}(x,y)|\le t_t\end{cases} \tag{1}$$

$$T_b(x,y)=\begin{cases}1 & |f_k(x,y)-f_{bk}(x,y)|>t_b\\ 0 & |f_k(x,y)-f_{bk}(x,y)|\le t_b\end{cases} \tag{2}$$

Eq.(1) and Eq.(2) are used to process binarination, where t_t and t_b are the thresholds of binary image. According to the above principles the result chart of motion object detection based on time difference method of background subtraction is shown in figure 1 as follows. The contour of moving object extracted by the time difference method based on background subtraction presented in this paper is illustrated in Fig.1 (e). Based on the complexity of background, we choose the coefficients k1=0.8 and k2=0.2 in experiment, where k1 and k2 just denoting a kind of proportional relation and their sum is 1, namely, representing the contribution degree to contour formation of the two methods. We adopt the above values in order to make the best of both the advantages of two methods to segment object from background correctly and the robust experimental results are obtained.

(a) the background image

(b) the current frame image

(c) Extracting contour using background subtraction method

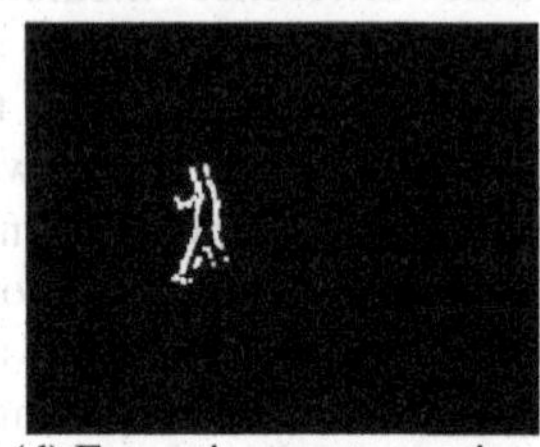

(d) Extracting contour using time difference method

(e) Extracting contour using time difference method based on background subtraction

Fig. 1. The result charts of human body contour extraction

3 Judging Suspicious Person Based on Treading Track

In the following section we mainly discuss two types of treading tracks, one is the closed curve; the other is spiral line, If somebody's treading track takes on one of these two types of line shape, it shows that somebody stays at a specific location (within set scenes) for a long time or wanders around certain small area, so we should draw a red rectangle and give an alarm to arouse people's attention and further watch. Now we discuss these two situations.

3.1 The Treading Track Is Closed Curve

Figure 2 shows the case that the treading track is closed curve, the treading track is formed by choosing a point in each frame from video sequences, and this track passes through points $(x_0, y_0), (x_1, y_1), ..., (x_{n-1}, y_{n-1}), (x_n, y_n)$ respectively, line segment

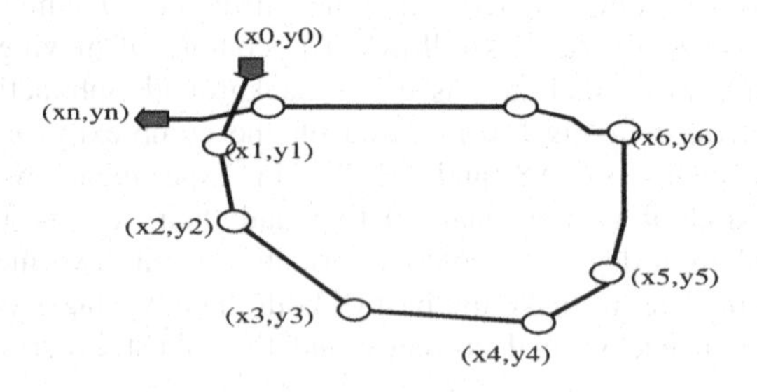

Fig. 2. The treading track is closed curve

in figure is connected based on these recorded track points. While (x_0, y_0) is the starting point of treading track, (x_n, y_n) is the end point of treading track.

Algorithm steps

Step 1: Supposing (x_0, y_0) is the starting point of treading track, (x_n, y_n) is the end point of treading track, and somebody walks to point (x_m, y_m) at present, while m=2,…,n .

Step 2: Judging line segment among point (x_m, y_m), (x_{m-1}, y_{m-1}) and line segments which have passed through before if there exists cross or not (namely :line segment between point (x_0, y_0) and point (x_1, y_1) ,line segment between point (x_1, y_1) and point (x_2, y_2) ,…,line segment between point (x_{m-2}, y_{m-2}) and point (x_{m-1}, y_{m-1})).

Step 3: If line segment among point $(x_m, y_m), (x_{m-1}, y_{m-1})$ is crossed with any a line segment which has passed through before, thus it shows the treading track of this person is closed curve, otherwise m=m+1 and turning to Step 1.

3.2 The Treading Track Is Spiral Line

Figure 3 shows the case that treading track is spiral line; this situation doesn't accord with closed curve, we will adopt other judgment methods.

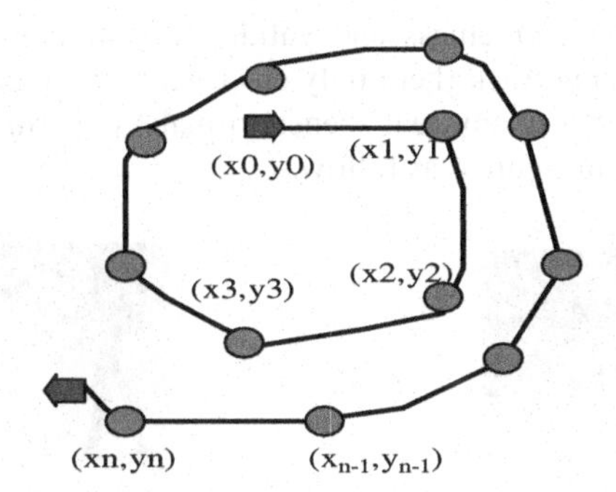

Fig. 3. The treading track is spiral line

Algorithm steps

Step 1: According to formula 3 below we compute the center of somebody's treading track, namely: (X_{zero}, Y_{zero}).

Step 2: Averagely dividing the treading area into eight sub areas with (X_{zero}, Y_{zero}) as its coordinate dot, the angle range of area 0 to area 7 is: $0^0 \sim 45^0$、 $45^0 \sim 90^0$、

$90^0 \sim 135^0$ 、 $135^0 \sim 180^0$ 、 $180^0 \sim 225^0$ 、 $225^0 \sim 270^0$ 、 $270^0 \sim 315^0$ 、 $315^0 \sim 360^0$ respectively.

Step 3: Computing the angle degree of α by the formula 4 as followes, we take turns to judge points $(x_0, y_0), (x_1, y_1), \ldots, (x_n, y_n)$ falling into which sub area according to the angle degree together with the positive and negative of coordinate values.

$$X_{zero} = \frac{x_0 + x_1 + \ldots + x_n}{n+1}, Y_{zero} = \frac{y_0 + y_1 + \ldots + y_n}{n+1} \tag{3}$$

$$\alpha = \arcsin\left(\frac{y_i}{\sqrt{x_i^2 + y_i^2}}\right) \qquad i = 0, 1, \ldots, n \tag{4}$$

Step 4: Based on the calculation and statistical results of step 3, if somebody's track points are more averagely dropped into each sub area, thus we can see this person's treading track is spiral line and we should notify suspicion.

The treading track is recorded by choosing a point in each frame from video sequences and then we carry through judgments as the above algorithms described to each track point, namely, judging while recording track points, thus having achieved the real-time of this algorithm.

4 The Experiment Result and Discussion

In order to validate the correctness and validity of our judging irregular behavior methods based on treading track (here only discussing two types of track line shape), we apply this algorithm in many real scenes to make experiments, the experimental result charts are shown in figure 4 as follow.

(a) The treading track is straight line

(b) The treading track is closed curve

(c) The treading track is spiral line

Fig. 4. The experimental result charts

From figure (a) we can see that when the treading track is straight line, we don't draw a red rectangle and give an alarm, however, when the treading track is closed curve or spiral line as shown in figure (b) and (c) above, it demonstrated that somebody had the suspicion of wandering around someplace, so we should draw a red rectangle and give an alarm to arouse people's attention.

5 Conclusions

In this paper we propose a method using time difference method based on background subtraction to extract motion human body contour accurately from video sequences. And on the basis of it, we discuss the treading track of people, mainly aiming at two types of line shape: closed curve and spiral line. The experimental result has shown this algorithm has the superiority of good judgment to irregular behavior and can be applied in more situations for motion object detection and behavior judgment、 recognition. However, here we only related to two types of track line shape, we can also design other line shape algorithms for judging irregular behavior to different application scenes. The shortages mentioned above need to be further researched.

References

1. Huang, C.-L., Lin, C.-C.: Model-Based Human Body Motion Analysis for MPEG IV Video Encoder. Journal, Proceedings of the IEEE, 435–439 (2001)
2. Molina-Tanco, L., Bandera, J.P, Marfil, R., Sandoval, F.: Real-time human motion analysis for human-robot interaction. Journal, Proceedings of the IEEE, 1402–1407 (2005)
3. Elgammal, A., Duraiswami, R., Harwood, D., et al.: Background and foreground modeling using nonparametric kernel density estimation for visual surveillance. Journal, Proceedings of the IEEE 90(7), 1153–1163 (2002)
4. Collins, R.T., Lipton, A.J., Kanade, T.: Introduction to the special section on video surveillance. Journal, IEEE Transactions on Pattern Analysis and Machine Intelligence 22(8), 745–746 (2000)
5. Road, V.J.: Safety through video detection. In: Journal, Proceedings of 1999 IEEE/IEEJ/JSAI International Conference on Intelligent Transportation Systems, pp. 753–757 (1999)
6. Lipton, A., Fujiyoshi, H., Patil, R.: Moving target classification and tracking from real-time video. In: Proc IEEE Workshop on Applications of Computer Vision, Princeton, NJ, pp. 8–14 (1998)
7. Haritaoglu, I., Harwood, D., Davis, L.S.: W4, Real-time surveillance of people and their activities. Journal, Proc of IEEE Transaction on Pattern Analysis and Machine Intelligence, 809–830 (2000)

Ontology-Based Focused Crawling of Deep Web Sources

Wei Fang, Zhiming Cui, and Pengpeng Zhao

The Institute of Intelligent Information Processing and Application,
Soochow University, Suzhou, 215006, P.R. China
{064027065001,szzmcui}@suda.edu.cn

Abstract. Deep Web sources discovery is one of the critical steps toward the large-scale information integration. In this paper, we present Deep Web sources crawling based on ontology, an enhanced crawling methodology. This focused crawling method based on ontology of Deep Web sources avoids to download a large number of irrelevant pages. Evaluation showed that this new approach has promising results.

Keywords: Ontology, Focused Crawling, Deep Web.

1 Introduction

In the recent years, more and more databases are becoming Web accessible through form-based search interfaces online. However, traditional Web crawler can't realize in this aspect, current searching engines can't get this part of page information. As a result these information are hidden and invisible to users, we call this kind of Web data " Deep Web" or "Invisible Web", "Hidden Web". Information on Deep Web are stored in databases, characterized their abundant information, single theme, high quality, good structure and rapid increasing rate compared to Surface Web [1]. According to a survey of BrightPlant in 2000, the scale of Deep Web is estimated to be around 500 times larger than the surface web. Nearly 450,000 Deep Web sites exist on Internet. The realization of the large-scale information integration of heterogeneous Deep Web sources is an effective way for users to make use of Deep Web information.

In this paper, we focus on building an effective Deep Web Sources Discovery crawler that can autonomously discover and download pages from the Deep Web. An exhaustive full crawl of the Web is a possible approach to this problem, but this would be highly inefficient. As the ratio of query interfaces to Web pages is small, this would lead to unnecessarily crawling a large number of irrelevant pages. Moreover, in order to leave a lot of irrelevance noisy pages out, we propose an ontology-based focused crawling framework for Deep Web. This kind of semantic-based focused crawling, makes use of an ontology to improve decision accuracy. Then, we propose a method of a focused crawling of Deep Web Sources using a page classifier to guide the crawler and focus the search on pages that belong to a specific topic.

Z. Zhang and J. Siekmann (Eds.): KSEM 2007, LNAI 4798, pp. 514–519, 2007.

The main contributions of this paper are: 1. We propose a Deep Web sources discovery method based on ontology using focused crawling techniques. 2. We design a prototype system and validate that our method is feasible and highly effective.

In the remainder of this paper, we start with Section 2 for our Deep Web sources focused crawling strategy, system architecture and relevance computation. In Section 3, we present results of our experiments with the real web data. Section 4 concludes with open issues and future work.

2 Ontology-Based Query Interface Focused Crawling

2.1 System Architecture

Here, we propose an ontology-based Deep Web sources focused crawler framework. Our framework runs a process that consists of two interconnected cycles: ontology cycle and crawling cycle (See Fig.1). The first cycle begins with predefined domain ontology. We define the crawling target interface in the form of an instantiated ontology. This cycle also resets the weights of the concepts which is used to compute the ordering score by learning the information collected in the crawling cycle and proposals for enhancement of the already existing ontology to the user. The crawling cycle comprises executing the Internet crawler, begins with crawler picking the first URL in priority queue, then communicating with the Internet, downloading Deep Web sources interface and storing into database for index. Then it connects to the ontology to determine relevance. The relevance computation is used to choose relevant documents for the user and to focus on links for the further search for relevant documents and Deep Web sources interface available on the Web. The two connected cycles serve as input for the specification of the system architecture.

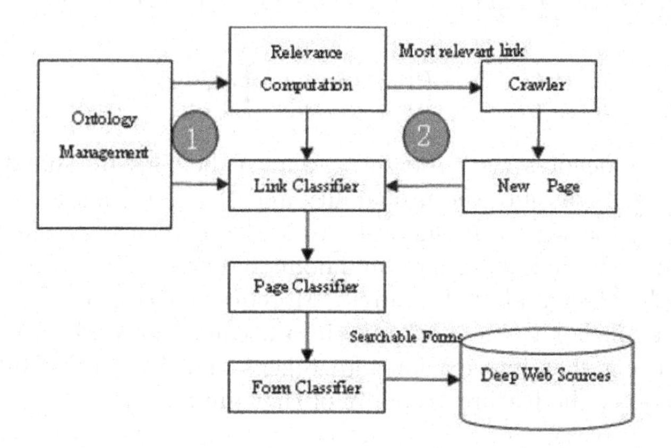

Fig. 1. Ontology Query interface focused crawler Architecture

In addition, to tackle the sparse distribution of forms on the Web, our source focused crawler avoids crawling through irrelevant paths by: limiting the search to a special topic; learning features of links and paths that lead to pages that contain searchable forms; employing appropriate stopping criteria; and computing the similarity of topic. The crawler uses three classifiers to guide its search: the page, the link and the form classifier. The page classifier is trained to classify pages as belonging to topics in a taxonomy. The form classifier is used to filter out useless forms. Once the crawler retrieves a page P, if P is classified as related to a topic, its corresponding forms and links are extracted from it. A form is added to the Deep Web Sources Database if the form classifier decides it is a searchable form, and if it is not already present in the Form Database. The link classifier is trained to identify links that are likely to lead to pages that contain searchable form interfaces in one or more steps. It examines these links extracted from a special topic pages and adds to local link priority queue ordered by their importance if they are in local site, otherwise if links are linking to extra sites then they will be added into site starting link priority queue. If the link is already present in the site queue, it will adjust its priority.

2.2 Link Classifier

The link Classifier aims to identify links that may bring delayed benefit, such as links that eventually lead to pages that contain available forms. Nowadays, there are many automatic text classification methods including: learning distance methods, probability based methods, association-mining methods, support vector machine (SVM), etc. The generic process is to train a classifier using a group of training text and the classifier will be used to classify a new text. In our system we use Naive Bayes algorithm to classify the hyperlink information. Naive Bayes Classifier presumes that the values of the attributes are conditional independent for each other. Suppose that the feature vector of a text is $X=[x_1, x_2 \ldots, x_d]^T$, the probability of X belonging to class C_i can be defined as formula (1):

$$P(C_i|X) = P(C_i)/P(X) * \prod_{j=1}^{d} P(x_j|C_i), \quad (1)$$

where $P(C_i|X)$ denotes the probability that X belongs to class C_i. For each class C_i , we will compute the probability and the result is the class C_i which makes $P(C_i|X)$ the largest. In our system, the link classifier is trained to identify promising links by the link features and domain ontology. Features mainly come from the anchor text and URL address. Moreover, since many links use image instead of the anchor text, so we take into account the path of those images. After segmenting that information into tokens and counting frequency of the words, we can get the feature vector X of that link.

2.3 Page Classifier

Page classifier is also built using Naive Bayes classifier. In the Deep Web Crawler, page classifier is trained with samples obtained in the topic taxonomy of the

Open Directory - similar to other focused crawlers [2]. When the crawler retrieves a page P, the page classifier analyzes the page and assigns to it score which reflects the probability that P belongs to the focus topic. If this probability is greater than a certain threshold θ(e.g.0.5), the crawler regards the page as relevant [3].

2.4 Form Classifier

In order to find Deep Web sources, we need to filter out non-searchable forms, e.g., forms for login, discussion group interfaces, mailing list subscriptions, purchase forms, Web-based email forms. The form classifier is a general (domain-independent) classifier, like in [4], which uses a decision tree to determine whether a form is searchable or not. We select decision trees (the C4.5 classifier) because it has the lowest error rate among the different learning algorithms [3].

2.5 Relevance Computation

This most important component of the overall approach is the relevance computation component. In general the relevance measure is a function which tries to map the content (e.g. form text, hyperlinks, etc.) of Deep Web sources,against the existing ontology to gain an overall relevance-score. The measures and the associated relevance computation strategies are described in detail as following.

Definition 1 (Relevance Function)**.** Relevance score r:=f(i,O_L), with r∈R, the query interface i, and the instantiated ontology O_L.

Definition 2 (Relevance Sets)**.** We distinguish between Single R_s(e), Taxonomic R_t(e), Relation R_r(e), and Total R_0(e) relevance sets. The weight for each entity in the sets is generally 1.0 [5].

The domain ontology (relevance sets) described in Fig 2. From the simple Single relevance measure R_s(vehicle) only lexical entries referring to the core entity vehicle are considered. The Taxonomic relevance measure R_t(vehicle) includes sub-concepts like car and truck. The complex Relational of relevance measure R_r(vehicle) also includes non-taxonomic relations like price, pattern. Their corresponding entities are also included. The widest measure called Total R_0(vehicle) also includes entities being far off the original entity like BWM,F512M. However their weight diminishes with a factor of 0.5 for each distance step. We propose a formula to calculate the relevance contribution of concepts at different layers.

Definition 3 (Ontology Based Topical Relevance Computation)**.** Suppose that S={s_1,s_2...s_n} is the set of Deep Web Sources, c={c_1,c_2...c_t} is the set of topics, with c_i being a topic concept chosen from our domain ontology, and t being the number of chosen topic concepts. Then the topic vector can be described as R=($\mathrm{w}_{1,r}$,$\mathrm{w}_{2,r}$,...,$\mathrm{w}_{t,r}$), where $\mathrm{w}_{i,r}$ is the weight of topic concept c in topic vector. The semantic weight of topic concept formula as following:

$$W_{i,j} = f_{i,j} * w_{i,r}, \tag{2}$$

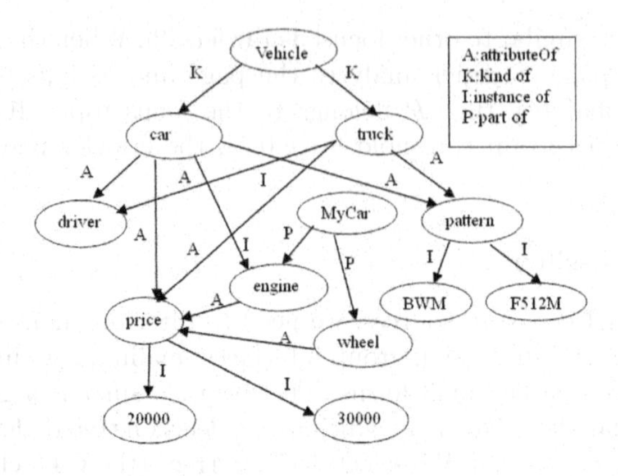

Fig. 2. Relevance Sets(Ontology Layers)

Where $f_{i,j}=tf_{i,j}/max_i tf_{i,j}$, is standard frequency of topic concept c in Deep Web source s_j. It is the quotient of $tf_{i,j}$ which is the frequency of topic concept c in Deep Web source S_j divided by the frequency of topic concept c_i that has the highest occurrence in Deep Web source S_j. Ontology based topical relevance formula can be defined as following [6]:

$$OR(s) = \frac{S_j \bullet R}{|S_j| \times |R|} = \sum_{i=1}^{t} w_{i,j} \times w_{i,r} / (\sqrt{\sum_{i=1}^{t} w_{i,j}^2} \times \sqrt{\sum_{i=1}^{t} w_{i,r}^2}) \quad (3)$$

3 Performance Evaluation

3.1 Training Data Collection

In our experiment, the crawling topic field is "job". We mainly use "Job Thesaurus" and "HowNet" as our ontologies. The training set is collected manually. Getting from some typical job sites (i.e., www.chinahr.com, www.51job.com, etc), we manually extract anchor text, URL address and path of the images. Then we save these information into file job that have Form.txt or no Form.txt according to whether the link is pointing to a Deep Web entrance page or not.

3.2 Experiment Results

In our experiments, we compare the efficiency of our ontology-based crawler against the keywords crawler. In order to integrate search interfaces later according to fields they belong to, we crawl sites in different fields respectively. Table 1 shows the average performance(Precision and Recall) of the two kinds of crawlers. Experimental results show that our strategy is effective and that the efficiency of the Crawler is significantly higher than other crawler.

Table 1. Results For Experiments Performed

	Crawled pages	Keyword crawler	Ontology-based crawler
Recall	5000	83.5%	97.1%
Precision	5000	71.6%	78.7%

4 Conclusions

In this paper we present a new approach for focused crawling: Ontology-Based Deep Web Sources Crawling. This new method combines both semantic and link structure of the Deep Web, can solve the major problem of crawling relevant pages. We introduce system architecture of our crawler that includes link priority queue and relevance computation, which is easy to get exact depth information of pages and avoids downloading a large number of irrelevant pages. Finally, our experimental results show that our strategy ontology based is more effective than crawler keyword based. In the future, discovery of complex Web Service and incorporate more sophisticated features of links may be proposed using our crawling strategies.

Acknowledgments. This research was partially supported by the grants from the Natural Science Foundation of China under grant number 60673092; the Key Project of Ministry of Education of China under grant No.205059; University Doctoral Program Research Program of Ministry of Education of China under Grant No.20040285016; the High-Technology Research Program of Jiangsu Province Under grant No. BG2005019; Higher Education Graduate Research Innovation Program of Jiangsu Province; and the National 211 Project Fund Program.

References

1. Chang, K.C.C., He, B., Li, C., Patel, M., Zhang, Z.: Structured Databases on the Web: Observations and Implications. SIGMOD Record, pp. 61–70 (2004)
2. Chakrabarti, S., Punera, K., Subramanyam, M.: Accelerated focused crawling through online relevance feedback. In: Proceedings of the 11th International World Wide Web Conference, pp. 148–159 (2002)
3. Barbosa, L., Freire, J.: Searching for Hidden-Web Databases. In: WebDb2005. Eight International Workshop On the Web and Databases, Baltimore, Maryland, pp. 1–6 (2005)
4. Cope, J., Craswell, N., Hawking, D.: Automated Discovery of Search Interfaces on the Web. In: Proc. of ADC, pp. 181–189 (2003)
5. Ehrig, M., Maedche, A.: Ontology-Focused Crawling of Web Documents. In: SAC 2003, p. 1"58111"624. ACM, USA (2003)
6. Lin, K., Zheng, J.: Ontology Based Learnable Focused Crawler. Journal of Computational Information System 3(3), 1173–1180 (2007)

Pattern Recognition in Stock Data Based on a New Segmentation Algorithm

Zhe Zhang[1], Jian Jiang[1], Xiaoyan Liu[1], Wing Chiu Lau[1], Huaiqing Wang[1], Shanshan Wang[1], Xinzhu Song[1], and Dongming Xu[2]

[1] Department of Information Systems,
City University of Hong Kong, Hong Kong
[2] University of Queensland, Australia
{50650613,jianjiang5}@student.cityu.edu.hk

Abstract. In trying to find the features and patterns within the stock time series, time series segmentation is often required as one of the fundamental components in stock data mining. In this paper, a new stock time series segmentation algorithm is proposed. This proposed segmentation method contributes to containing both the important data points and the primitive trends like uptrend and downtrend, while most of the current algorithms only contain one aspect of that. The proposed segmentation algorithm is more efficient and effective in reserving the trends and less complexity than those combined split-and-merge segmentation algorithm. The research result shows that patterns found by using the algorithm and prior to the transaction time impact the stock transaction price. Encouraging experiment is reported from the tests that certain patterns appear most frequently before the low transaction price occurrence.

Keywords: Data Mining, Pattern Recognition, and Segmentation.

1 Introduction

Pattern matching is stock data time series is an active area of research in data mining. So the representation of the data is the key to efficient and effective solutions. Shatkay and Zdonik [1] suggest dividing the sequences into meaningful subsequences and representing those subsequences using real-valued functions. Therefore, segmentation is the fundamental component in representing the subsequences and stock data mining.

Many representation of time series have been proposed, including Fourier Transforms [2], minimum message length segmentation [3], and segmentation by piecewise linear representation (PLR) [4]. In [5], a generalized dimension-reduction framework for recent-biased approximations was proposed, aiming at making traditional dimension reduction techniques actionable in recent-biased time series analysis. In [6], the financial time series are segmented based on Perceptually Important Point and interesting and frequently appearing patterns are typically characterized by a few critical points. Among them Perceptually Important Point is a common used algorithm that is customized for financial time series and based on the importance of the data points.

Z. Zhang and J. Siekmann (Eds.): KSEM 2007, LNAI 4798, pp. 520–525, 2007.

However, nearly all of them segment the time series either only consider the turning points or only contain the trend within a period. In order to solve the problem, we propose a new algorithm, which is able to not only contain the important turning points but also contain the principle trend when segmenting the stock time series. More importantly, we find that the patterns prior to the transaction time found by the algorithm impact the transaction price. Certain patterns appear most frequently before the low transaction price occurrence.

The paper is organized as follows. The next section presents the proposed new segmentation algorithm, together with some related algorithms. In Section 3, representations of the common stock trade patterns used in our experiments are given. Then in Section 4, experiments are done on the real stock data to explore the relationship between the patterns and the transaction price. The final section makes the conclusion.

2 Stock Time Series Segmentation

In this section, firstly, we introduce the PIP (Perceptually Important Point) segmentation algorithm that can reserve the critical turning points. Secondly, we introduce the combined split-and-merge segmentation algorithm, which is used to reserve as much trend in the stock time series as possible, based on PIP algorithm. After that, we propose our algorithm, which can reserve most critical turning points as well as the principle trend, in less complexity.

2.1 PIP Segmentation Algorithm

The identification of Perceptually Important Points (PIP) is first introduced in [6]. And the scheme was used to segment the time series following the idea of reordering the sequence P based on the PIP identification process, where the data point identified in an earlier stage is considered as being more important than those points identified afterwards. The distance measurement is depicted in Equation (1). Which means the distance Dis between the test point pi and the line connecting the two adjacent PIPs. Where Pc (xc, yc) is the point with xc=x3 on the line connecting p1(x1,y2) and p2(x2,y2). In the algorithm, the first two PIPs that are found will be the first (x1, y1) and last points (xn, yn) of sequence from 1 to n. The next PIP that is found will be the point in with maximum distance to the first two PIPs.

$$Dis = |y_c - y_i| = (y_1 + (y_n - y_1) \times (x_c - x_1)/(x_n - x_1)) - y_i \quad (1)$$

2.2 Split-and-Merge Segmentation Algorithm Based on PIP

In order to reserve more trends of stock data time series, some researchers try to use two-step segmentation algorithm including split and merge [8]. Regarding that the segmentation in Section 2.1 only contain the critical points and neglect the trend of the time series, we similarly use the two-step segmentation algorithm based on split-and-merge algorithm proposed in [8] and PIP algorithm. In the split phase, we use the PIP principle to segment the stock time series, trying to reserve the critical points as the cutting points. And in the merge phase, we merge the segments by comparing vertical distance measurement with the defined ε, which is computed in Equation (2). In such way, we can merge the segments with similar slopes and reserve a long trend.

$$\varepsilon = \frac{1}{k} \times \sum_{i=0}^{k} |y_i - \hat{y}_i| \tag{2}$$

2.3 New Segmentation Algorithm Based on PIP Approach

In this section, we propose a new segmentation algorithm based on PIP approach, which can reserve more trend than the method in Section 2.1, and in less complexity procedure than the method in Section 2.2. The major principle of the algorithm is that we can select all the points with similar trend and segment them into one part. It can not only find the critical points, but also can find the as much trend as possible that around the critical points. The algorithm is described in detail in Algorithm 1. And the segmentation procedures are shown in Fig. 1.

```
Algorithm 1: Segment time series T by new segmentation
algorithm based on PIP approach

 1: Distance=0, x_k=0, n=0
 2: SP [x_1] = y_1, SP [x_n] =y_n
 3: For     i = x_1: x_n
 4:   SP[i] = y_i,
 5:   Dis =|y_c-y_i|= (y_1+ (y_n-y_1)*(x_c-x_1)/ (xn-x1))-y_i
 6:                 If Dis>Distance
 7:            Then Distance= Dis, x_k =i
 8: End For
 9: For i = x_1: x_n
10: Dis =|y_c-y_i|= (y1+ (y_n-y_1)*(x_c-x_1)/ (x_n-x_1))-y_i
11:            If     |x_i-x_k| <• and |y_i-y_k|< •
12:            Then Point[n] = [x_i, y_i], n=n+1
13:            End If
14: End For
15: For i=0: n
16: Select from Point[n]: x_t1=Max (x_i), x_t2 =Min (x_i)
17: End For
18: Return: S1=T[x_1, x_t1]
20:            S2=T [x_t1, x_t2]
21:            S3=T [x_t2, x_n]
```

In the algorithm, we set a parameter λ as the criteria to select all of the points near the critical points, in order to get the whole segment with similar trend within horizontal area. On the other hand, we set a parameter Δ as the criteria to select all of the points with similar trends within vertical area. It can be calculated in Equation (3). The parameters are defined same as those in section 2.1. And Xk is the average value of Xi. other parameters are defined the same as in section 2.1.

$$\lambda = \frac{1}{k} \times \sum_{i=0}^{k} |x_i - x_k|, \ \Delta = \frac{1}{k} \times \sum_{i=0}^{k} |y_i - y_k| \tag{3}$$

From Fig. 1(c), by using this algorithm, we can segment the stock time series into 3 parts. Specially, this algorithm reserve a long trend in B segment, which represents the whole change trend consists of many critical points instead of just one point.

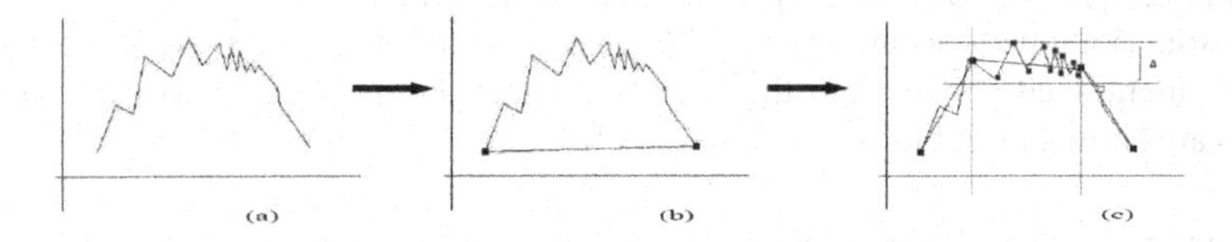

Fig. 1. New PIP-Based Segmentation Procedures

3 Pattern Representation

In order to obtain the objective of finding the relationship between the patterns in bid number sequence and the trading points, the first step is to find the matched patterns, each of which appear prior to the every trading point in the time domain. Developing a time series pattern classification is of fundamental importance for the pattern finding and matching. In this paper, we use several major stock patterns that occurred most frequently defined by Zhang et al.[10], which are based the point number and the combination of slope (positive or negative) between two adjacent points. Suppose that there are four points in sequence from the left to the right: Sp1, Sp2, Sp3, Sp4.

Therefore, the combination of slope (positive or negative) can be illustrated as: $Slope(k_1, k_3) \in \{(+,-),(+,+),(-,+),(-,-)\}$, while k_1 and k_3 denote the slopes of line 1, connecting Sp1 and Sp2 and line 3, connecting Sp2 and Sp3 respectively. The suggested pattern names and the pattern figures are shown in Table 1.

Table 1. Design of 4-Point-Pattern

Pattern name	Up-Trapeziform	Up Flag	Down-Trapeziform	Down Flag
Pattern Description	Sp1<Sp2 AND Sp3>Sp4	Sp1>Sp2 AND Sp3<Sp4	Sp1>Sp2 AND Sp3<Sp4	Sp1>Sp2 AND Sp3>Sp4
Pattern Figure				

4 Experiments and Result Analysis

The used data set is taken form the bid and trade time series of Chicago stock market. 12423 trading points are considered. In the time domain, the patterns can represent both long time series and short time series. Therefore, we change the sequence length, that is, the sequence length applied to our algorithm, to have experiments. The purpose of the experiment is to find the percentage of occurrence of each four-point-pattern that appears prior to the trading points. The first step is to set the length of sequence (e.g.20) and then to find the number and percentage of occurrence of each pattern. The second step is to change the length of sequence, and to find the different numbers and percentages of occurrence for each pattern. The third step is to find the trend in each pattern with the change of sequence length.

The four point patterns are extracted from each segment and classified to one of the patterns mentioned above. And the exactly number and percentage of each matched pattern are shown in Table 2.

Table 2. Percentage of Each Matched 4-point Pattern with Different Sequence Length

Sequence Length	Percentage of Each Matched Pattern			
	Up trapeziform	Bottom trapeziform	Up-flag	Down-flag
20	35.76%	2.62%	21.53%	40.09%
100	28.03%	13.08%	19.63%	39.26%
150	29.41%	13.24%	22.06%	35.29%
250	29.54%	22.72%	13.84%	33.90%

From Table 2 we can find that the Down Flag Patterns occur most frequently before the low trading points occur. The occupancy rate order of Down Flag pattern remains the same in regardless of the change of sequence length (from 20 to 250). In another word, the occurrence of low trading price point results from the occurrence of Down Flag Pattern in a sense.

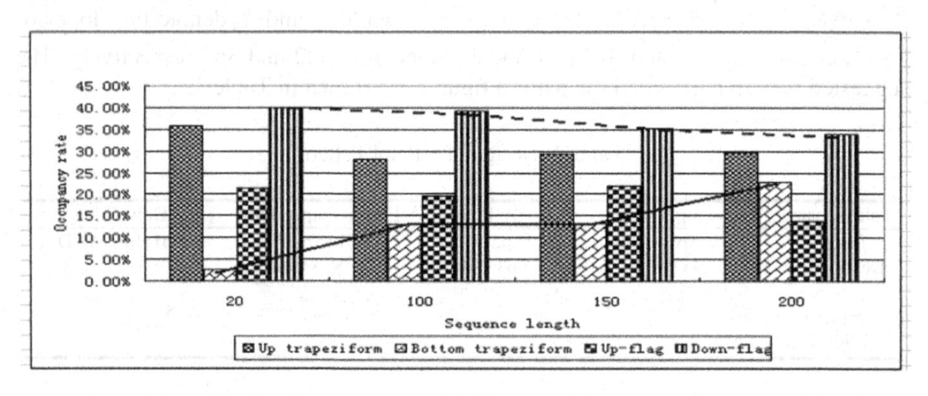

Fig. 2. Trend of Pattern with Different Sequence Length

From Fig.2 we may conclude that with the increase of sequence length, Down Flag pattern's number has a down trend. It means that the trading point is most frequently followed from Down Flag Pattern, particularly in short term.

5 Conclusion

In this paper, we propose a new segmentation algorithm based on PIP algorithm. By using this algorithm, the segmentation is able to not only contain the important turning points but also contain the principle trend. On the other hand, the proposed algorithm is in less complexity than those split-and-merge segmentation algorithms.

We also find that there is really a certain relationship between the patterns prior to the transaction time and the transaction price. We may conclude that the most occurrences of trading point related to the occurrences of Bottom Patterns in a sense; and the trading point is most frequently followed from Bottom Pattern, particularly in short time series.

The significant discovery in the paper makes contribution to the prediction of stock data, especially for the prediction of when a low trading price will appear.

References

1. Shatkay, H., Zdonik, S.: Approximate Queries and Representations for Large Data Sequences. In: 12th ICDE, pp. 536–545 (1996)
2. Agrawal, R., Faloutsos, C., Swami, A.: Efficient similarity search in sequence databases. In: 4th Conference on Foundations of Data Organization and Algorithms, Chicago, USA, pp. 69–84 (1993)
3. Oliver, J.J., Baxter, R.A., Wallace, C.S.: Minimum message length segmentation. In: Wu, X., Kotagiri, R., Korb, K.B. (eds.) PAKDD 1998. LNCS, vol. 1394, pp. 222–233. Springer, Heidelberg (1998)
4. Keogh, E., Smyth, P.: A probabilistic approach to fast pattern matching in time series databases. In: 3rd International Conference of Knowledge Discovery and Data Mining, pp. 24–20 (1997)
5. Zhao, Y.C., Zhang, S.C.: Generalized Dimension-Reduction Framework for Recent-Biased Time Series Analysis. IEEE Trans. Knowl. Data Eng. 18(2), 231–244 (2006)
6. Fu, T.C., Chung, F.L., Ng, C.M.: Financial Time Series Segmentation based on Specialized Binary Tree Representation. In: DMIN 2006. 2006 International Conference on Data Mining, Las Vegas Nevada USA, pp. 26–29 (2006)
7. Chung, F.L., Fu, T.C., Luk, R., Ng, V.: Flexible Time Series Pattern Matching Based on Perceptually Important Points. In: IJCAI Workshop on Learning from Temporal and Spatial Data, pp. 1–7 (2001)
8. Cheong, F.G.P., Xu, Y.J., Lu, H.: The Predicting Power of Textual Information on Financial Markets. IEEE Intelligent Informatics Bulletin 5(1), 1–10 (2005)
9. Ge, X., Smyth, P.: Deformable Markov Model Templates for Time-Series Pattern Matching. In: ACM SIGKDD International Conference on Knowledge Discovery and Data Mining, Boston MA, pp. 81–90. ACM Press, New York (2000)
10. Zhang, Z., Li, J., Wang, S., Wang, H.Q.: Study of Principal Component Analysis on Multi-dimension Stock Data. Chinese Journal of Scientific Instrument 26(8), 2489–2491 (2005)

HMM-Based Korean Named Entity Recognition for Information Extraction

Bo-Hyun Yun

Department of Computer Education, Mokwon University
Doan-dong 800, Seo-ku, Daejon, 302-729, Korea
ybh@mokwon.ac.kr

Abstract. This paper presents the HMM (Hidden Markov Model) based named entity recognition method for information extraction. In Korean language, named entities have the distinct characteristics unlike other languages. Many named entities can be decomposed into more than one word. Moreover, there are contextual relationship between named entities and their surrounding words. There are many internal and external evidences in named entities. To overcome data sparseness problem, we used multi-level back-off methods. The experimental result shows the F-measure of 87.6% in the economic article domain.

1 Introduction

As the amount of free texts in the Web has been increased exponentially, it is very important to extract the appropriate information for the various content service. Such needs give rise to named entity recognition (NER) which extracts words like proper nouns, time and money expressions in the documents. The NER is a classification and identification process of person, location, and organization name (PLO) or numerical expressions. The NER can be divided into the named entity(NE) detection. and the named entity classification. The named entity detection is to catch the named entities in the text. The named entity classification is to label the named entity such as person, organization and location.

In Korean language, NE has the distinct characteristics unlike other languages. Many named entities can be decomposed into more than one word. Moreover, there are contextual relationship among words constructing named entities or between named entities and its surrounding words. There are many internal and external evidences in NEs. Therefore, Korean NER model has to consider these properties.

In this paper, we present the HMM-based Korean NER considering the distinct characteristics in Korean newspaper for information extraction. We investigated the structures of Korean NEs in the sentence and found the combinational regularities of the NE. According to these regularities, we classified words into four classes and recognized Korean NE.

Z. Zhang and J. Siekmann (Eds.): KSEM 2007, LNAI 4798, pp. 526–531, 2007.

2 Related Works

Conventional NER researches have been performed by a stochastic based approach [1,2,6-9] and a rule based approach [4, 10]. In the stochastic criteria, there are Hidden Markov Model[1], Maximum Entropy Model[2, 9], Decision Tree/List Model[6, 7], and Hybrid Model[8].

One of existing Korean NER methods[4,8,10] is the rule based method[4] which uses four information such as word information within Eojeol[1], contextual information, relation between NE and sub-categorization information of a declinable word, and relation information among NEs. The experimental result shows the precision rate of 90.4% and the recall rate of 83.4%. The other is the method[8] using machine learning methods and pattern-selection rules. The method is a hybrid method of NER which combines maximum entropy model, neural network, and pattern-selection rules. The result shows that an F-measure of 84.09% is achieved for the specific domain(Editorials), and an F-measure of 80.27% of general domain(Novels).

3 Characteristics of Korean Named Entity

In Korean, NEs consist of several words; in particular, location and organization names can be decomposed into many words. We analyzed 300 documents that are made up of economic articles, public performance articles, and web pages of the trip guide . In the analysis, we can see that about 66% of NEs can be decomposed into more than one morpheme and 14.4% of NEs more than 4 morphemes.

In addition, we examine the length of NEs in Korean and most of NEs are composed of several words. These examinations indicate that external context of NEs are the very important clue for NER. We know that nouns of 35% are words surrounding of the NE's. About 55% of them are the clue word of the NER. According to the above analysis results, we divided words into four groups by their role in the NER.

- Independent entity (IE): a word that can be a NE by itself. ex1) Paris, Bush, Pentagon, Motorola. ex2) "Kim-dae-jung (president of South Korea)", "Seoul"
- Constituent entity (CE): a word that can not be a NE by itself but can be combined with other nouns. For example, "Co." is one of the CE class. ex1) Co., Soft, Electronics. ex2) "gong-sa (public company)", "gang (river)", "gong-hang (airport)", "geuk-jang (theater)"
- Adjacent entity (AE): a word that can occur in front or back of the NE and can be the clue of NER. "Ms." and "Mr." are included in the AE class. ex1) Mr., CEO, vocalist, governor, district. ex2) "dan-won (member)", "sa-jang (CEO)", "chul-sin (origin)"
- Not entity (NoE): a word that is not an IE, CE or AE. Most of words are included in this class.

[1] The Eojeol is the spacing unit in Korean like a word in English. The Eojeol consists of one or more morphemes. It sometimes corresponds to a word or a phrase in English.

4 Korean Named Entity Recognition

Given a sentence $W = w_1w_2....w_{n-1}w_n$, POS(Part-Of-Speech) tagger finds an optimal POS tag sequence $T = t_1t_2....t_{n-1}t_n$ that maximizes $P(T|W)$. If we use bigram model for tagging, $P(T|W)$ can be represented as the equation (1).

$$P(T|W) \simeq argmax \prod_i P(t_i|t_{i-1})P(w_i|t_i) \qquad (1)$$

Here, t_i is a POS tag of a POS tag sequence and w_i is a word of a sentence. NER can be dealt with a named entity tagging problem that finds an adequate NE tag sequence T. Our NER model is based on an HMM[4] to estimate $P(T|W)$. The HMM-based NER model consists of the following elements.

- N means the state in a set of state $S = s_1, s_2, ..., s_N$. In the POS tagging system, a state corresponds to a POS. In the NER system, a state corresponds to a NE type. N is the number of NE type plus 1(Not a NE) and q_t means a state s in the time t.
- M means the symbol in observation symbol $V = w_1, w_2, ..., w_M$. In the NER, the number of observable symbol is the same as a size of the dictionary.
- A is the transition probability $A = a_{ij}$ defined as follows: $a_{ij} = P(q_{t+1} = S_j|q_t = S_t), 1 \leq i, j \leq N$
- B is the observation probability $B = b_j(k)$ defined as follows: $b_j(k) = P(q_t = w_k|q_t = S_i), 1 \leq j \leq N, 1 \leq k \leq M$
- π means the initial distribution which is the probabilities of i's occurrence of a NE type in the start of the sentence. $\pi = \pi_i$ is defined as follows: $\pi_i = P(q_1 = S_i), 1 \leq i \leq N$

In this paper, we used a trigram in the learning phase. We extracted NEs, prior words, and posterior words in the NE tagged corpus for variable length of NE as shown in (2)[5]. Here, L (Left) and R (Right) means prior and posterior morphemes respectively and n is the length of the NE.

$$...W_{-2}^{L}W_{-1}^{L}W_{1}^{NE}....W_{n}^{NE}W_{1}^{R}W_{2}^{R}.... \qquad (2)$$

Like the learning phase, we use the trigram model in the recognition phase. Thus, the equation (1) can be modified into the equation (3).

$$P(T|W) \simeq argmax \prod_i P(t_i|t_{i-1}t_{i-2})P(w_i|t_i) \qquad (3)$$

Our NER system performs two steps for the NER. In the first phase, we attach all possible sub-entity types to a word and resolve the ambiguity of sub-entity tagging. All words are tagged with the IE, CE, AE or POS. A word can have multiple entity types. For examples, "Jeong-seon" will be a person name or a location name in Korean. "sang-sa" will be an AE_Person (a master sergeant) or a CE_Organization (business affairs).

We used trigrams of the types which are encoded by the sub-entity type function $set(w)$. The function $set(w)$ returns sub-entity types for input word w. If a word w is included in sub-entity types, $set(w)$ will return sub-entity types. Otherwise, $set(w)$ function returns POS. If a morphological analyzer uses 28 POS tags, the number of possible states is 39. With the function $set(w)$, the equation (3) is modified into the equation (4).

$$P(T|W) \approx \prod_{i}^{n} P(set(w_i)|set(w_{i-1})set(w_{-2}))P(w_i|set(w_i)) \quad (4)$$

In the second phase, we find the boundaries of a NE and the type of a NE in the sentence. In this step, we used the NE boundary recognition function $brf(s)$. The boundary recognition function $brf(s)$ returns the type of a sub-entity's boundary for the input word s. For example, 'Person_Start', 'Location_Continue' or 'Organization_Unique' will be returned. With the function $brf(s)$, the equation (4) is modified into the equation (5)

$$P(T|W) \approx argmax \prod_{i}^{n} P(brf(set(w_i))|set(w_{i-1})set(w_{-2}))P(w_i|brf(set(w_i))). \quad (5)$$

5 Experimental Results

As the experimental data, we used 900 NE tagged economic articles for the training and 100 articles for the test. One article has about 7 sentences with 18.7 words. Training corpus includes 1,900 person names, 3,620 location names and 5,230 organization names. We gathered about 68,000 person names, 25,000 location names, and 10,000 organization names in an IE dictionary. The CE dictionary includes 92 location and 121 organization entries; the AE dictionary contains 114 person, 39 location and 33 organization entries.

The basic measures for evaluation of this work are precision, recall, and F-Measure. Precision (P) represents the percentage of the entities that the system recognized which are actually correct. Recall (R) represents the percentage of the correct named entities in the text that the system identified. Both measure are incorporated in the F-measure, $F = 2PR/(P + R)$.

Table 1 shows the result of experiments on variations of the proposed method. In order to overcome data sparseness problem, we use different back-off methods in the classification and the detection step. In the classification step, the bigram model $P(set(w_i)|set(w_{i-1}))$ is adopted. In the detection step, the result of $set(w)$ function performs the mapping of 39 states (E0) to 12 states (E2) by decreasing the number of POS. E0* is the model that performs the back-off by E1 and E2 orderly, when the sparseness problem occurs in E0. Here, SOS means the start of sentence and EOS is the end of sentence. E0* shows the best F-measure of 87.6%.

- E0: 3(IE, CE, AE) * 3(P, L, O) + 28 POS + SOS + EOS : 39 states
- E1: 3(IE, CE, AE) * 3(P, L, O) + 8 POS + SOS + EOS : 19 states
- E2: 3(IE, CE, AE) * 3(P, L, O) + 1(Not_SE) + SOS + EOS : 12 states

Table 1. Results of Named Entity Recognition

Exp. types	Recall	Precision	F-measure
E0	71.2%	90.4%	79.7%
E1	77.3%	87.9%	82.3%
E2	83.3%	88.7%	85.9%
E0*	90.9%	84.5%	87.6%

In our NER system, most of errors are caused from the over-generation problem. The problem decreases the precision rate. For examples, "Sin-yong-ka-deu (Credit card)", "Mi-guk-san (Made in USA)" and "Sang-pum-tu-ja (Product investment)" are recognized as a named entity. "ka-deu (card)", "san (Made in or Mountain)" or "tu-ja (investment)" may be combined with a common noun and form a NE. Most of these words are CE (Constituent Entity). This problem can be alleviated by using mutually exclusive word-level information. The other errors are resulted from data sparseness problem. The problem decreases the recall rate. The organization name, "CJ39-ssyo-ping (CJ39 Shopping)", appears infrequently in the training corpus. For the future works, we try to overcome these problems.

Table 2 shows the F-measure of our method and other approaches. As shown, our method shows the best result against other approaches. Thus, we know that it is important to consider the named entity construction principles in Korean.

Table 2. Comparative Results of Other Approaches

Our Method	[3]'s Method	[4]'s Method	[8]'s Method
87.6%	74%	86.8%	84.09%

6 Conclusion

We proposed the HMM-based NER model considering the named entity construction principles in Korean for information extraction. Many named entities can be decomposed into more than one word. Moreover, there are contextual relationship between the named entity and its surrounding words. Thus, there are many internal and external evidences in NEs. Therefore, we reflected these characteristics of Korean NE. The transition probability is calculated by the encoded trigram from the labeled training data. The observation probability is obtained by the lexical frequency of each entity. Moreover, we presented two kinds of features and back-off model appropriate to Korean.

Experimental result shows that our NER system has the F-measure 87.6% higher than the existing researches in Korean. The classification of four NE subtypes enhances the performance of NER in Korean. Moreover, the merit of our NER system is the speed by using HMM model against the model such as ME, SVM.Our future work is to apply a lot of lexical information to Korean NER model.

References

1. Bikel, D.M., Miller, S., Schwartz, R., Weischedel, R., Nymble: A High-Performance Learning Named-finder. In: Proceedings of the Fifth Conference on Applied Natural Language Processing, pp. 194–201 (1997)
2. Collins, M., Singer, Y.: Unsupervised Models for Named Entity Classification. EMNLP/VLC 1999 (1999)
3. Kim, T.H., Lee, H.S., Ha, Y.S., Lee, M.H., Myaeng, S.H.: Proper Noun Extraction Using Data Sets. In: Proceedings of the 12th Hangul and Korean Information Processing, pp. 11–18 (2000)
4. Lee, K.H., Lee, J.H., Choi, M.S., Kim, G.Ch.: Study on Named Entity Recognition in Korean Text. In: Proceedings of the 12th Hangul and Korean Information Processing, pp. 292–299 (2000)
5. Rabiner, L.R.: A Tutorial on Hidden Markov Models and Selected Applications in Speech Recognition. Proceedings of the IEEE 77(2), 257–286 (1989)
6. Sassano, M., Utsuro, T.: Named Entity Chunking Techniques in Supervised Learning for Japanese Named Entity Recognition. In: Proceedings of the 18th International Conference on Computational Linguistics, pp. 705–711 (2000)
7. Sekine, S., Grishman, R., Shinnou, H.: A Decision Tree Method for Finding And Classifying Names in Japanese Texts. In: Proceedings of the Sixth Workshop on Very Large Corpora (1998)
8. Seon, C.N., Ko, Y., Kim, J.S., Seo, J.Y.: Named Entity Recognition using Machine Learning Methods and Pattern-Selection Rules. In: NLPRS, pp. 229–236 (2001)
9. Uchimoto, K., Ma, Q., Murata, M., Ozakum, H., Isahara, H.: Named Entity Extraction Based on A ME Model and Transformation Rules. In: Processing of the ACL (2000)
10. Fukumoto, J., Shimohata, M., Masui, F., Saski, M.: Description of the Oki System as Used for MET-2. In: Proceedings of 7th Message Understanding Conference (1998)

Activity Recognition Based on Hidden Markov Models

Weiyao Huang, Jun Zhang, and Zhijing Liu

School of Computer Science and Technology
Xidian University, Xi' an, 710071, China
Wyhuang@mail.xidian.edu.cn,
Zhang72jun@163.com, Liuprofessor@163.com

Abstract. Focusing on the application of Intelligent Security Supervisory Control System, this paper proposes a new human activity recognition approach in which the Background Subtraction and the Time-stepping are averaged by weights to implement the precise extraction of moving human contour. In this way, the incompleteness of the extracting objects contour resulting from the color comparability between the human and the background can be resolved. Moreover, an ant colony clustering algorithm is applied to estimate and classify the body posture. Finally, Discrete Hidden Markov Models is used for human posture training, modeling and activity matching to recognize the human motion. Experiment results have shown that this method gives stable performances and good robustness.

Keywords: moving human contour; activity recognition; ant colony clustering algorithm; Hidden Markov Models.

1 Introduction

Human activity understanding and recognition are high-level processes of human activity analysis, and have drawn more and more attention for several decades. Due to the challenge of human activity recognition and its huge application value, more and more research related to human activity recognition have been made in recent years[1,2].

Bobick[3] thought that movement was the basic element of activity, the most fundamental activity, and the basis of high-level activity. For example, dancing is made up of some elementary dance movements. After a dance show, we will find that there are some repetitive movement elements. In the same way, computer may learn the basic elements automatically through searching repetitive movement of continuous human activities, which can be used to segment and label the continuous movements. Due to the fact that continuous human activity is sequences composed of human postures at different times, this paper presents a new moving object detection method, based on the above idea. The Background subtraction and the Time-stepping are on weighted average to extract and detect precise contour of moving body. After that, each sampled contour of moving body is regarded as a basic movement, and ant colony clustering algorithm is applied to perform posture estimation and classification on the information of the extracted human contour. Finally, we use Discrete Hidden

Z. Zhang and J. Siekmann (Eds.): KSEM 2007, LNAI 4798, pp. 532–537, 2007.

Markov Models to generate model for human posture detection, and the recognition of continuous human activity is implemented.

2 Background Modeling

In this paper, an algorithm based on the statistical Gaussian model [4] is applied to estimate the background image which is composed of initialization and update. In an interval time M, the mean and the covariance of the brightness of every pixel are computed as the initially estimated background image, Viz. $B_0 = [\mu_0, \sigma_0^2]$, where

$$\mu_0(x,y) = \frac{1}{M}\sum_{i=0}^{M-1} f_i(x,y) \tag{1}$$

$$\delta_0^2(x,y) = \frac{1}{M}\sum_{i=0}^{M-1}[f_i(x,y) - \mu_0(x,y)]^2 \tag{2}$$

After initializing the estimated background image, with the appearance of new image, the parameters of background image should be updated self-adaptively according to the following formulas. The updated image was $B_t = [\mu_t, \sigma_t^2]$, with

$$\mu_t = (1-\alpha)\mu_{t-1} + \alpha \times f_t \tag{3}$$

$$\delta_t^2 = (1-\alpha)\delta_{t-1}^2 + \alpha(f_t - \mu_t)^2 \tag{4}$$

where α is a given constant and range from 0 to 1.

3 Human Activity Recognition Method

The human activity recognition method presented in this paper consists of three stages. First, the contour of moving body are detected and extracted precisely. Then based on the precise contour information, we perform estimation and classification of human posture. Finally, we do time series analysis and modeling of the detected human activity, and the activity can be recognized accordingly. The main algorithms of these three stages are detailed in the following sections.

3.1 The Precise Recognition of Moving Object

The goal of moving object detection is to extract change regions from background image in sequence images. The most popular methods of object detection are the Time-stepping and the Background Subtraction[5]. The former is of good adaptability to dynamic environments, but it can not extract all the related points of object integrally, while the latter can extract the object points rather integrally, but it is too alert to the dynamic environment change caused by illumination and external conditions. In order to overcome the shortcomings of these two moving object detection algorithms, we propose a weighted average method based on the Time-stepping and the Background Subtraction, and the precise extraction of moving object contour is implanted successfully. Experiments show that this method can also shape an integrated object contour, ensure the connectivity of moving object, and the detection result is quite good, as shown in Fig.1.

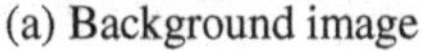
(a) Background image

(b) Current frame image

(c) The extracted contour with the presented method

Fig. 1. The results of human contour extraction

3.2 Human Posture Estimation and Classification

3.2.1 Clustering Algorithm Based on Ant Searching for Food

The process of ant searching for food can be divided into two phases, hunting food and moving food. Each ant can release pheromone on the path it passes, and can apperceive pheromone and pheromone intensity. The more ants pass, the denser the pheromone of the path. Meanwhile, pheromone may volatilize itself as time passes. Ants tend to move towards the direction that the pheromone density is high, hence, the more ants the path pass, the higher probability the subsequent ants choose it will be. And the activity of whole ants colony put up information plus-feedback phenomena. The basic idea of clustering analysis based on ants pheromone trace is detailed below.

Regard data as ants with different attribute, and clustering center as "food source" ants search for, then data clustering process can be consider as the process of ants searching for food source[6]. Assume data object is $X = \{X \mid X_i = (x_{i1}, x_{i2}, \ldots, x_{in}), i = 1, 2, \cdots, N\}$. The steps of the algorithm are as follows.

Step 1: Initialize pheromone of each path, and set them as 0, that is $T_{ij}(0) = 0$.

Step 2: Set the radius of cluster as r , and statistic error as ε , etc..

Step 3: Compute the weighted Euclidean distance d_{ij} between X_i and X_j .

Step 4: Calculate the pheromone of each path by

$$T_{ij}(t) = \begin{cases} 1 & d_{ij} \le r \\ 0 & d_{ij} > r \end{cases} \tag{5}$$

Step 5: The probability of merging object X_i into X_j is computed by

$$p_{ij}(t) = \frac{T_{ij}^{\alpha}(t)\eta_{ij}^{\beta}(t)}{\sum_{s \in S} T_{sj}^{\alpha}(t)\eta_{sj}^{\beta}(t)} \tag{6}$$

Where $S = □X_s □d_{ij} \le r□s = □□□□\cdots$, N}. If $p_{ij}(t)$ is larger than threshold p_0, merge X_i into the area of X_j. η_{ij} is the reciprocal of d_{ij} called visibility. α and β are adjusting factors, which can not only avoid the searching stagnation phenomena resulting from that all ants passing the same path give the same results, but also show again the idea of classical Greedy Algorithm.

3.2.2 Posture Estimation and Classification

In our algorithm, for the extracted binary human contour image, its horizontal and vertical histograms are estimated, which are taken as the input of ant colony clustering algorithm. The near neighbor measure between states Im_1 and Im_2 can be computed by

$$D(\mathrm{Im}_1, \mathrm{Im}_2) = d_1(X_1, X_2) + d_2(Y_1, Y_2) \tag{7}$$

where d_1 and d_2 are Manhattan Distance between horizontal projection and vertical projection, respectively. The horizontal and vertical histograms of an image are compared with the corresponding histograms of another one, and the minimum is taken as contiguity degree. In this way, near neighbor measure can keep invariability for both transform and projection of binary object in scenery. Further, ant colony clustering algorithm cluster the useful training images by near neighbor measure, and based on the established prototypes, we classify new image according to the relative distance between new unknown images.

3.3 Time Series Analysis of Human Posture

HMM [7] defines a limited state set, and each state associates with a probability distribution (multi-dimension in general). The state transition is managed by transfer probability. According to the associated probability distribution, a result or observed value will be generated in a specific state. For observer, only this result is visible, and the state is invisible. Therefore, state is hidden, and the name HMM comes. If the inclusive elements in transfer probability matrix A of HMM are all zero, this HMM is full connected, otherwise is partly connected. Full connected HMM can simulate more complex statistics process, so we use it to recognize human activity. We define each static frame (sample) as a state, take a movement sequence as a combination of every state of different sample frames, and use combination probability as judge rule of activity belongingness to implement matching recognition of human activity.

For specific model, in order to train HMM to recognize its observation sequence $O = O_1, \ldots, O_T$, we can utilize Bayesian theory and Forward Algorithm to estimate the probability of observation sequence $p(O \mid \lambda) = \sum_{i=1}^{N} \alpha_T(i)$, where N is the Markov statuses number in model. Suppose forward variable is the probability of HMM being at state S_i in observation sequence $O_1, \ldots, O_t$ at time t, that is $\alpha_t(\mathrm{i}) = P(O_1, \cdots, O_t, S_t = q_i \mid \lambda)$. We need to adjust parameter λ to let the appearance probability of observation $p(O \mid \lambda)$ be maximal. At this stage, the number of different human postures determines the quantity of HMM coded symbols (the possible state value M). Each activity associates with a HMM, which means HMM number is never equal to the number of different activities. Hence, Maximum Likelihood Estimation Algorithm (Baum-Welch algorithm) is used to estimate λ again and again in order to obtain a better λ'. Then we can figure out the maximum $P_i(i \in 1,2,\cdots,M)$ among $P_1, P_2,\cdots, P_M$, which is used to recognize human gesture of current frame.

4 Experimental Results

This paper validates the effectiveness of presented algorithm through movement analysis of walking (face, side face), running, and squatting. Different human postures included in these four activities are detected from these images through ant colony clustering algorithm. Then we use the sequences composed of basic postures, which are continuous in time, as input of HMM, and utilize them to recognize four activities.

Table 1. Recognition results comparison of HMM and other algorithms

Human activities	*Recognition rate (%)*			*False reject rate (%)*		
	Template match method	DTW	HMM	Template match method	DTW	HMM
Walking(face)	77.4	82.5	89.2	3.6	2.8	2.3
Walking(side face)	72.3	80.7	91.5	4.1	3.2	2.2
Running	69.4	75.1	84.3	4.6	3.7	3.1
Squatting	80.7	85.2	92.6	3.8	3.3	2.5

As shown in Table 1, the template matching method has relatively low recognition rate, and it only fits for matching some simple movements. The recognition decreases when it is used for recognizing walking (side face) and running at some time when they are relatively similar. With the adoption of DTW[8] to perform dynamic warping in time sequence, the total recognition of these four activities has increased, but due to the fact that its spatial robustness is not high, the false accept rate is still high for recognizing walking (side face) and running. The proposed recognition method using HMM improves accuracy over both the template matching method and the improved recognition method using DTW to implement time-normalized process, with average recognition rate increased by 14.45% and 8.525% relatively. Besides, our method can distinguish two different activities with some similar gestures and achieve high recognition rate, through modeling lateral walking and running and adopting Maximum Likelihood (ML) to compute optimum matching value. Experimental results indicate that the proposed method still could achieve high recognition rate even that noise influence exists, the contour is not integral, or gestures are similar. It is also shown that the presented method has strong temporal and spatial robustness, which is an effective and reliable method for activity recognition.

5 Conclusions

The activity recognition method introduced in this paper can be widely used in supervisory and recognition systems, like the surveillance and security systems of hospital, bank, important department offices, etc., also can be utilized in unmanned supervisory and control, unmanned operation fields and so on. Through several experiments comparison, systems adopting the proposed algorithm are practical and feasible, and this algorithm is applicable in moving object tracking and activity recognition in most cases. However, the information of each moving object can not be extracted when they are near or obstructed by each other, and misjudgement may occur. Thus future work focuses on the study of precise multi-object recognition and activity recognition applying the multi-vidicon data fusion technology [9].

References

1. Collins, R.T., Lipton, A.J., Kanade, T.: Introduction to the special section on video supervisory control. IEEE Transactions on Pattern Analysis and Machine Intelligence, 745–746 (2000)
2. Collins, R.T., Lipton, A.J., Kanade, T.: Introduction to the special section on video supervisory control. IEEE Transactions on Pattern Analysis and Machine Intelligence, 745–746 (2000)
3. Bobick, A.F.: Movement, activity, and action: The role of knowledge in the perception of motion. Phil. Trans. Royal Society London B 352(1358), 1257–1265 (1997)
4. Huwer, S., Niemann, H.: Adaptive Change Detection for Real-time Sureillance Applications. In: Proceedings 3rd IEEE International Workshop on Visual Surveillance, pp. 37–46. IEEE Computer Society Press, Los Alamitos (2000)
5. Haritaoglu, I., Harwood, D., Davis, L.: W4: Real-time supervisory control of people and their activities. IEEE Trans. PAMI 22(8), 809–830 (2000)
6. Dorigo, M., Gambardella, L.M.: Ant colony system: a cooperative learning approach to the traveling salesman problem. IEEE Transactions on Evolutionary Computation 1(1), 53–66 (1997)
7. Rabiner, L.: A tutorial on hidden Markov models and selected applications in speech recognition. Proceeding of the IEEE 77(2), 257–286 (1989)
8. Zhao, G.Y., Li, Z.B., Deng, Y.: Human motion recognition and simulation based on retrieval[J] . Journal of Computer Research and Development 43(2), 368–374 (2006) (in Chinese with English abstract)
9. Hernandez, A.I., Garrault, G., Mora, F.: Multisensor fusion for atrial and ventricular activity detection in coronary care monitoring. IEEE Trans. on Biomed. Eng. 46(10), 1186–1190 (1999)

Novel Data Management Algorithms in Peer-to-Peer Content Distribution Networks

Ke Li, Wanlei Zhou, Shui Yu, and Ping Li

School of Engineering and Information Technology
Deakin University, 221 Burwood HWY, VIC 3125, Australia
{ktql,wanlei,syu,pingli}@deakin.edu.au

Abstract. The peer-to-peer content distribution network (PCDN) is a hot topic recently, and it has a huge potential for massive data intensive applications on the Internet. One of the challenges in PCDN is routing for data sources and data deliveries. In this paper, we studied a type of network model which is formed by dynamic autonomy area, structured source servers and proxy servers. Based on this network model, we proposed a number of algorithms to address the routing and data delivery issues. According to the highly dynamics of the autonomy area, we established dynamic tree structure proliferation system routing, proxy routing and resource searching algorithms. The simulations results showed that the performance of the proposed network model and the algorithms are stable.

Keywords: Data Management, Algorithms, PCDN.

1 Introduction

The rapid development of peer-to-peer (P2P) technologies has led to a number of important application systems on the internet, and the system structure of these P2P application systems has changed gradually from Napster-like (centralized inquiry) to Kazaa-like (bias to the free connecting of strong nodes) structures. However, as the arbitrariness of free connecting in a P2P network, making inquiries must rely on flooding, which results a waste of a lot of network resources, therefore, the systematic distensible ability is severely restricted [12].

Content Delivery Network (CDN) is a technology committed to the distribution of contents and resources. It is the virtual network formed by server groups located in different areas, distributing dynamically the content to the edges, handling traffic according to the contents, and the accessing request will be transmitted to the optimal servers, thus enabling users access to the information from the nearest place by the fastest speed [6]. The existing CDN routing technology is mainly based on DNS, which has the problem of high cost of hardware, and has limitations in terms of scalability, reliability and fault-tolerance; therefore, the redirection server is very likely to become a new network bottleneck.

CDN with P2P technologies are highly complementary, and the research on the combination of the two technologies (PCDN) is at the preliminary stage. For example, references [10], [11] reported applications in video streaming using CDN combined

Z. Zhang and J. Siekmann (Eds.): KSEM 2007, LNAI 4798, pp. 538–543, 2007.

with P2P. In this paper, we propose a new PCDN model, and construct the corresponding optimal path selection algorithm based on the new PCDN model. Compared with existing PCDN models, the new model can balance the dynamics of the lower workload network and the service efficiency of the heavy workload network. The new model can guarantee the cost of maintaining routing tables, which are always kept within the affordable range by the dynamic network, and can achieve a higher routing efficiency than that of the Category I and Category II of the structural covered network, therefore, its reliability and hardware cost will be far less than the traditional CDN Network [2], [7].

2 The Optimal Path Selection Algorithm

Based on the dynamics of the lower workload network (through testing Napster and Gnutella networks, the average online time of a node is 60 min [9]), using structural hash algorithms is unwise between the frequently online and offline users. Also each node on and off of the network will destroy the topological structure of the algorithm, hence the cost to calculate frequently the topological structure is staggering. We also need to consider the difference of the network bandwidth of different users -- users with narrow bandwidth are very likely to become data transportation bottlenecks when reconstructing the network topology. Therefore, we propose a new routing algorithm with the concurrent idea of Petri nets [1], [8], at the moment, as the measuring parameters of routing by the techniques of Ant Algorithm [3], [4], [5]. In this paper we present one of the above algorithms. The optimal path selection algorithm is presented as follows:

Definition. *N*=(*S*,*T*;*F*) is a bound prototype network, $\forall t \in T, |{}^{\bullet}t| = 1$. The definition of the *TokenFlag* of *t* is: *TokenFlag*(*s*), and it satisfies the following conditions: $t \in T$, $\exists M : M[t > M'$, $TokenFlag(s') = TokenFlag(s) \circ s$; $s \in {}^{\bullet}t, s' \in t^{\bullet}$, $\circ$ notes a type of connecting relationship. Obviously, if M_0 is the initial ID, then $\forall s \in S$, $TokenFlag(s) = \phi$.

Given a network topology $N^* = (V, E)$, V is the node set, and E is the one-way link set. $s \in V$ is the source nodes, and $d \in \{V - \{s\}\}$ are the destination nodes. Such a network topology can be simulated by a bounded prototype network, the conversion of models are shown in Fig. 1.

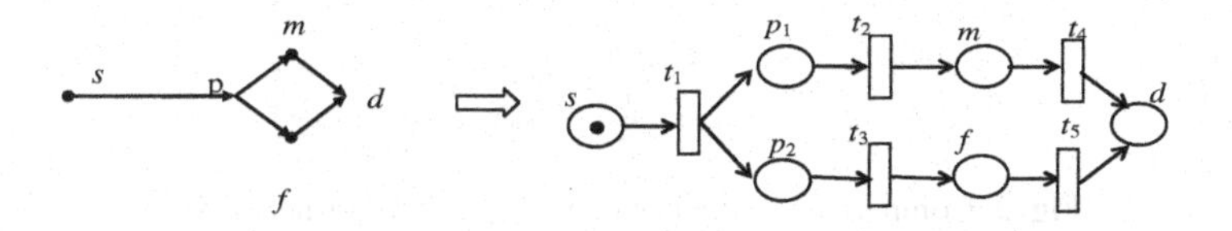

Fig. 1. Convert N^* into N of Petri Network Model

Let $M_0(s)$ =1, exist changed sequences $\sigma_1 = t_1t_2t_4, \sigma_2 = t_1t_3t_5$, there is a different *TokenFlag* within *d* resulted by each changed sequence, finally, $TokenFlag(d) = s \circ p_1 \circ m + s \circ p_2 \circ f$. Each *TokenFlag* corresponds to one route from node *s* to node *d* in N^*. When we consider QoS, we can find an optimal path from these paths. If there are many optimal paths, we choose the path which has the least number of nodes. If there are many second best paths, we choose the path which has the most different nodes which the best path possesses.

3 Performance Evaluations

Simulations were conducted to confirm the proposed model and algorithms. The simulation environment as follows:

The hardware environment:
IBM--CPU: AMD Athlon (tm) 64 processor2800 + 1.81GHz; 512Mb; 120GMb.
Dell--CPU: Pentium(R) 4 + 2.66GHz; 512Mb; 80GMb.

The software environment:
Windows XP professional; OPNET Modeler10.

The simulation results are presented as follows:

- Simulation tests the network performance of CDN network model in the different network workload and different network nodes. In Fig. 2, Application 1 represents the network of 100 light workload nodes; Application 2 represents the network of 100 heavy workload nodes; Application 3 represents the network of 200 heavy workload nodes.

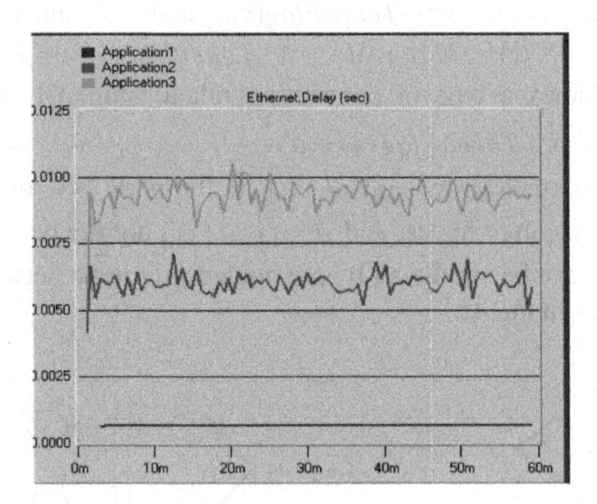

Fig. 2. Comparison of the Delay of Nodes in CDN network

From the experimental results, we can obtain that there is a very serious dependence on the severs performance in CDN network, as network's workload increases, resulting in servers capacity decline, and the entire network delays increased greatly, until the network congestions occur; the servers usually become the network bottlenecks in multi-user instantaneous peak.

- Simulation tests the network performance of traditional P2P network model in different network size (for the paper length limitation, we only present the network performance of heavy workload nodes).

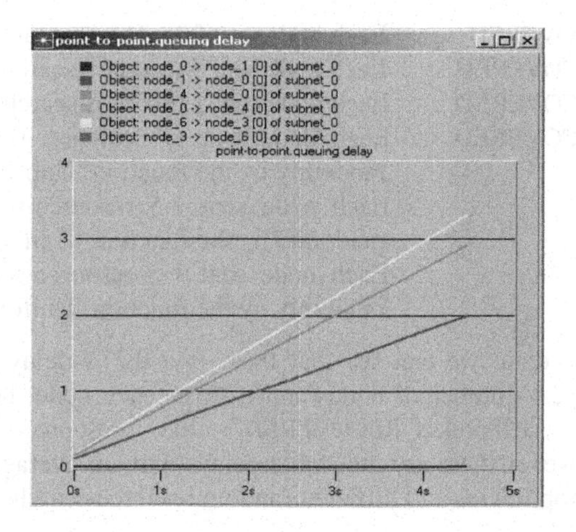

Fig. 3. Delay of Point to Point in P2P Network with Heavy Workload

From the experimental results, we can see that there is a greater reliance on the nodes' performance in the traditional unstructured P2P networks, as the number of nodes and the network workload increase, resulting in increment of the delay of the entire network, or even resulting network congestions.

- Simulation tests the network performance of this new PCDN network model in the different network workloads and different network nodes.

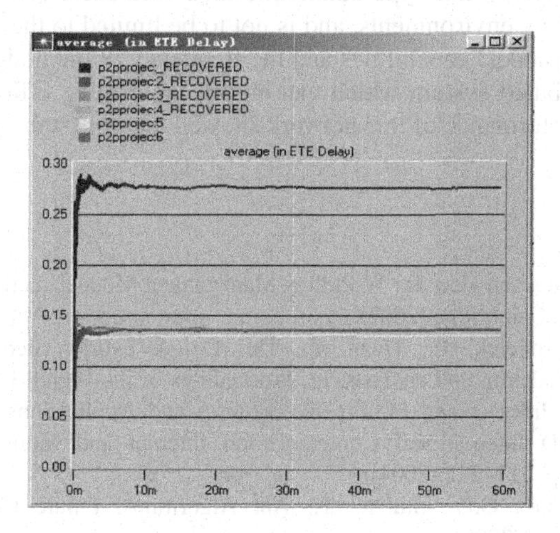

Fig. 4. Delay of Network in Difference Requesting Resource in Various Topologies

Here, we describe the curves in Fig. 4 as follows: P2Pproject_RECOVERED, P2Pproject2_RECOVERED, P2Pproject3_RECOVERED, P2Pproject4_RECOVERED. These curves represent the network topology; all are N_1-10-trees, that is, they represent the case of the largest number of nodes. P2Pproject5 and P2Pproject6 represent the network topology of the covered partly pruning, which represent the case of the middle and less number of nodes, as listed below:

P2Pproject_RECOVERED	Each node sent 20 resource-searches per second;
P2Pproject2_RECOVERED	Each node sent 5 resource-searches per second;
P2Pproject3_RECOVERED	Each node sent 1 resource-search per second;
P2Pproject4_RECOVERED	Each node sent 1-5 resource-searches per second randomly by the function 'Uniform_int';
P2Pproject5	Each node sent 1-5 resource-searches per second randomly by the function 'Uniform_int';
P2Pproject6	Each node sent 1-5 resource-searches per second randomly by the function 'Uniform_int';

From the above data, we can see that there is a 0.27's delay in the N_1 network which has the largest number of nodes and it gives each node the most requests of resource-searches (P2Pproject_RECOVERED, the resource-search is 20/s) compulsively. However there are the stable delays with an average of 0.14s in other different network topologies and different random resource-searches.

4 Conclusions

The proposed PCDN model, together with the algorithms described in this paper, have good adaptability for the dynamic changes of network, whereas the network delay and super-nodes' throughput are not changed significantly along with the change of the number of nodes, and each node can serve the network based on their own abilities. At the same time, this PCDN can also solve the users' dynamic features according to the layered tree-type autonomic area used the PCDN. Apparently, this PCDN can run in any environments, and is not to be limited to the size of the system, the strength of the nodes' capabilities and the frequency of the nodes' on and off; it is a wide area distributed system which can ensure the routing efficiency by dynamic regulation. The performance of this network showed that it is stable.

References

1. van der Aalst, W., van Hee, K.: Workflow Management Models, methods and systems, 1st edn. MIT Press, Cambridge (2002)
2. Dong, Y., Kusmierek, E., Duan, Z., Du, D.: A hybrid client-assisted streaming architecture: Modeling and analysis. In: Proceedings of the Eighth IASTED International Conference on Internet and Multimedia Systems and Applications, Proceedings of the Eighth IASTED International Conference on Internet and Multimedia Systems and Applications, pp. 217–222 (2004)
3. Dorigo, M., Caro, G.D., Stutzle, T.: Ant Algorithms. Future Generation Computer System 16(1), 5–7 (2000)

4. Dorigo, M., Gambardella, L.M.: Ant Colonies for the Traveling Salesman Problem. Biosystem 43(2), 73–81 (1997)
5. Gutjahr, W.J.: A Graph-based Ant System and Its Convergence. Future Generation Computer System 16(8), 873–888 (2000)
6. Tran, M., Tavanapong, W.: On using a CDN's infrastructure to improve file transfer among peers. In: Dalmau Royo, J., Hasegawa, G. (eds.) MMNS 2005. LNCS, vol. 3754, pp. 289–301. Springer, Heidelberg (2005)
7. Hefeeda, M.M., Bhargav, B.K., Yau, D.K.Y.: A hybrid architecture for cost-effective on-demand media streaming. Computer Networks 44(3), 353–382 (2004)
8. Petri, C.A.: Kommunikation mit Automaten. PhD thesis, Institut fur Instrumentelle Mathematik, Bonn, Germany (1962) (in German)
9. Saroiu, S., Gummadi, P.K., Gribble, S.D.: A measurement study of Peer-to-Peer file sharing systems[J]. In: Proceedings of the Multimedia Computing and Networking Conferences, San Jose, California, USA (2002)
10. Khan, S., Schollmeier, R., Steinbach, E.: A performance comparison of multiple description video streaming in peer-to-peer and content delivery networks. In: ICME. 2004 IEEE International Conference on Multimedia and Expo, vol. 1, pp. 503–506. IEEE Computer Society Press, Los Alamitos (2004)
11. Xu, D., Kulkarni, S.S., Rosenberg, C., Chai, H.-K.: Analysis of a CDN-P2P hybrid architecture for cost-effective streaming media distribution. Multimedia Systems 11(4), 383–399 (2006)
12. Xu, Z., Hu, Y.: SBARC: A supernode Based Peer-to-Peer File Sharing System. In: ISCC, Kemer- Antalya, Turky (June 2003)

On-Line Monitoring and Diagnosis of Failures Using Control Charts and Fault Tree Analysis (FTA) Based on Digital Production Model

Hui Peng*, Wenli Shang, Haibo Shi, and Wei Peng

Shenyang Institute of Automation, Chinese Academy of Science,
Shenyang 110014, China
Tel.: +86 024-23970223; +86 024-23970703.
{penghui,shangwl,hbshi,pw}@sia.cn

Abstract. This article presents an efficient Statistic Process Control system architecture for on-line monitoring of manufacturing process. Shenyang Institute of Automation Statistic Process Control system (SIASPC) detects relevant events in Real-time based on digital production model of MES. Failures occur in manufacturing process are diagnosed using control charts and FTA method based on expert knowledge base. The SIASPC has been developed and will be applied to control an Automobile gear-box assembly process.

Keywords: Fault tree analysis (FTA), Fault diagnosis, Knowledge base, Manufacturing execution system (MES), Statistic process control (SPC).

1 Introduction

Over the last decades, the design of on-line statistic process system for manufacturing process has received a great deal of attention. Research on automated on-line monitoring has been done over thirty years. Several approaches to detection, diagnosis and control of failures have been advanced in this period. The majority of these approaches have emerged from developments in the areas of expert systems and model-based diagnosis methods.

A number of models and related algorithms have been proposed over the last years, which include various forms of fault tree models[1,2], decision tree models[3], functional networks[4], system digraphs[5] and qualitative simulation models[6,7]. The application of such models has been demonstrated successfully in a number of domains in industry. Despite the substantial progress in the development of model-based systems, a number of open issues still remain. In this paper, we present an on-line statistic process system architecture based on digital production model.

Fault Tree Analysis (FTA) is an applicable and useful analysis tool. It is used for identifying and classifying faults, and monitoring manufacturing process on-line. The

* Corresponding author.

Z. Zhang and J. Siekmann (Eds.): KSEM 2007, LNAI 4798, pp. 544–549, 2007.

analysis defines a top event, which is a failure or an accident, and then builds the sequence of faults leading to this top event. FTA apply top-down logic in building their models. The analyst views the system from a top-down perspective. Basic events of failure are at the bottom of each failure path.

One important problem of above models is incompletely integration, namely, models is incomplete, inaccurate, non-representative or unreliable [8]. One substantial such difficulty and often the cause of inaccuracies in the models is the lack of sufficient mechanisms for representing the effects that changes in the behaviour or structure of complex dynamic systems have in the fault propagation in those systems. Such changes can in practice confuse a monitoring system and lead to false, irrelevant and misleading alarms. In this paper, an on-line monitoring and diagnosis system is presented based on the digital production models, which represents the fault state dependencies, that is crucial in developing accurate, representative and therefore reliable monitoring models.

This paper will introduce the architecture of on-line monitor and failures diagnosis based on a digital production model of MES. The method to diagnosis of failures using FTA is presented, and a SIASPC-Monitor system is developed.

2 Digital Production Model

Shenyang Institute of Automation Manufacturing Execution system (SIA-MES) provides an extensible, comprehensive solution with powerful functions. SIA-MES presents a clear and full digital product process, which provides a real-time platform for production data, thus actualize integration of MES, ERP and low-level automation system.

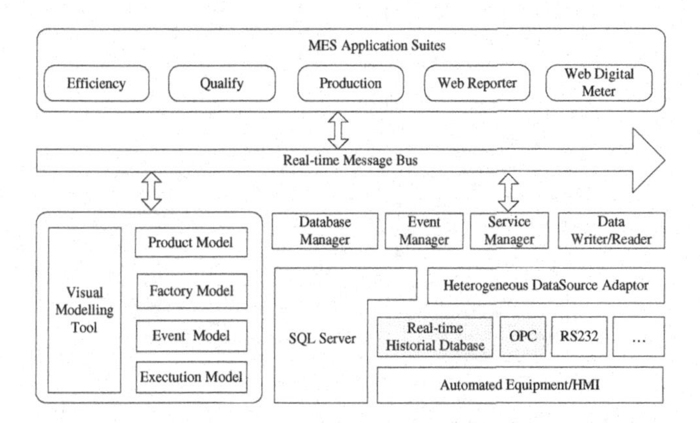

Fig. 1. Structure of SIA-MES

The SIA-MES platform is composed of infrastructure service, production model, visualized modeling tool, real-time message bus, and MES application suites as shown in Figure 1. OOA&D (Object Oriented Analysis and Design) method is used to construct factory model, product model, event model and execution model. Factory resources, enterprise production activities and shop floor business are abstracted and classified, which are described with series special basic semantic meta-object. Related models are expressed with complex objects formed by certain semantic meta-objects. Ultimately, factory resources are described as factory model, production activities are described as event model and enterprise business is described as execution model.

Product model is to define product, material, criterion, formulation and process, and to build assembly BOM (Bill of Material). Assembly BOM includes information of parts, components and processes. For a special product, its assembly BOM provides information of process assembly directory, material racks and feedings.

Factory model is to define factory, equipments, product line and relevant organization mode. On the basic of factory model, production process events are defined. Production event are basic elements to control production activities. Product process can be controlled by established production event.

3 Online Monitoring and Diagnosis of Failures

The general architecture of on-line monitor and failure diagnosis is given in Fig. 2. Product model, factory model, event model and execution model are expressed with complex objects formed by certain semantic meta-objects. Meta-objects are mapped with collected data, so it is possible for automated monitor based on production model to detect, diagnose and control failures in real-time.

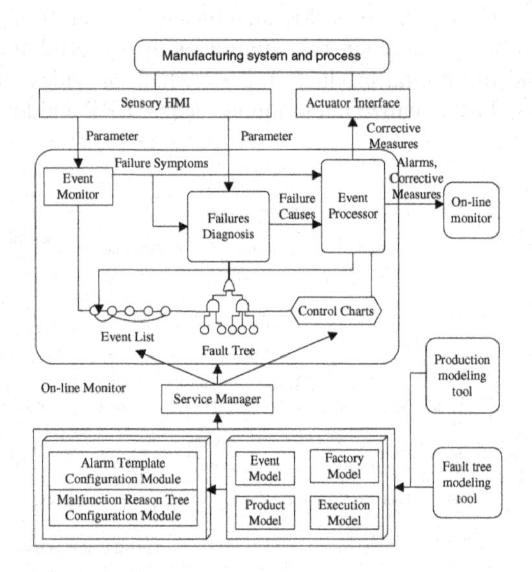

Fig. 2. The architecture of On-line monitor and failures diagnosis

Kernel in the architecture of on-line monitor and failures diagnosis is product/factory/event/execution model and infrastructure services. On the basic of production model, management of process is realized through background event-driven business modules.

SIASPC-Monitor is a real-time quality control system, which implement quality control of manufacturing process based on configured alarm modules in the background, and give an alarm message while failures happen, along with corrective measures. SIASPC-Monitor is a sub-system of SIA-MES, and function modules are configured on the basic of production model. Plant model is established till variables elements in lowest level.

SIASPC-Monitor includes function modules as query variables, variables list, control chart configuration, real-time monitor and chart controls. The step to monitor production variables is shown as follows: (1) Establish production model using visualized modeling tools; (2) Develop SPC modules, including malfunction reason tree configuration and alarm template configuration; (3) Configure malfunction reason tree and alarm template for special field; (4) Improve manufacturing process/procedure with analysis control charts; (5) Monitor manufacturing process on-line using control charts.

4 Diagnosis of Failures Using FTA

Take failures diagnosis of automobile gear-box assembly process for example. FTA method is adopted as Fig. 3. Automobile gearbox can be separated into several modules according to product structure, and relevant fault trees are constructed, which builds expert knowledge base. Fault trees are deducted into some rules by mode of evolutional knowledge-presentation, and these rules build up knowledge base of expert system.

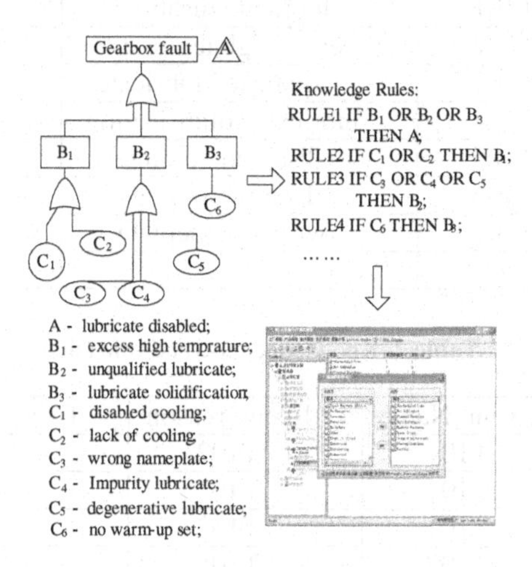

Fig. 3. Fault tree and knowledge base

Take "lubricate disabled" as top node event, which followed by three nodes, and go deep analysis, the whole fault tree is shaped as shown at last. Meanings of each node are explained as Fig. 3. All rules of fault knowledge base are educed by fault tree constructed. To save and manage in database system, rules with "OR" should be decomposed. Rules in Figure 3 are decomposed as follows:

R11 IF B_1 THEN A;
R12 IF B_2 THEN A;

R13 IF B_3 THEN A;
R21 IF C_1 THEN B_1;
R22 IF C_2 THEN B_1;
R31 IF C_3 THEN B_2;
… … …

Each "IF" and "THEN" is followed with a rule element, which is saved as contents of fault knowledge tables. Structure of knowledge tables is shown in Table 1. Notation: Fault_ID as fault code; Fault_name as fault name; NOTE as append content.

Faults of manufacturing process are monitored and controlled on-line in SPC system. Failures are diagnosed as an event happens in integrated production model, and faults detected are handled as an event. Table 2 shows structure of fault knowledge tables. Notation: Rule_ID as rule code; Event_Name as fault name; Event_Reason as fault reason; Event_action as fault corrective measures.

According to above FTA method, diagnosis of all gearbox failures is expressed in knowledge base and rule tables. Visual modeling tool enables well maintenance of knowledge base by operable functions of add new tree, edit tree and delete tree.

Table 1. Knowledge table

Fault_ID	Fault_name	NOTE
T101	lubricate disabled	
T102	excess high temprature	
T103	unqualified lubricate	
T104	lubricate solidification	
T105	disabled cooling	
T106	lack of cooling	
T107	wrong nameplate	
…	…	…

Table 2. Knowledge base and rule table

Rule_ID	Event_Name	Event_reason	Event_action	NOTE
R11	T101	T102		
R12	T101	T102		
R13	T101	T102		
R21	T102	T105		
R22	T102	T106		
R31	T103	T107		
…	…	…	…	…

5 Conclusions

In this paper, we introduced the idea of using production models and knowledge based FTA method for on-line quality monitoring of manufacturing process. We proposed an automated real-time monitor to alarm failures in the manufacturing process. Alarm messages are displayed with the failures causes and corrective measures.

The method for on-line monitoring and diagnosis of failures based on the digital production model offers a practical tool to improve system reliability. The SIASPC system is capable of identifying manufacturing process states by monitoring event signals. Expert knowledge from plant operator is needed in the whole system model design process.

The SPC system has been implemented, which will be applied to failures diagnosis of automobile gearbox assembly. Future work should explore the possibility of automating the fault reasoning by means of Knowledge Discovery and Data Ming. Primary work can be seen in literature [9].

Acknowledgments. The authors acknowledge the financial support of the National Natural Science Foundation of China funding (67646114) and Hi-tech Research and Development Program of China funding (2006AA04Z164). The authors would also like to acknowledge the helpful comments and suggestions of the SIA-MES software group. Their efforts are greatly appreciated.

References

1. Dan, M.S., Joseph, T.: Condition-based fault tree analysis (CBFTA): A new method for improved fault tree analysis (FTA), reliability and safety calculations. Reliability Engineering & System Safety 92(9), 1231–1241 (2007)
2. Montani, S., Portinale, L., Bobbio, A., Codetta-Raiteri, D.: Radyban: A tool for reliability analysis of dynamic fault trees through conversion into dynamic Bayesian networks Reliability Engineering & System Safety (March 2007) (in press)
3. Ungar, L.H., Powell, B.A., Kamens, S.N.: Adaptive networks for fault diagnosis and process control. Computers & Chemical Engineering 14(4-5), 561–572 (1990)
4. Mano, R.M., Raghunathan, R., Venkat, V.: A signed directed graph-based systematic framework for steady-state malfunction diagnosis inside control loops. Chemical Engineering Science 61(6), 1790–1810 (2006)
5. Bouamama, B.O., Medjaher, K., Bayart, M., Samantaray, A.K., Conrard, B.: Fault detection and isolation of smart actuators using bond graphs and external models. Control Engineering Practice 13(2), 159–175 (2005)
6. Gabbar, H.A.: Improved qualitative fault propagation analysis. Journal of Loss Prevention in the Process Industries 20(3), 260–270 (2007)
7. Yiannis, P.: Model-based system monitoring and diagnosis of failures using statecharts and fault trees. Reliability Engineering and System Safety 81, 325–341 (2003)
8. Kim, I.S.: Computerised systems for on-line management of failures. Reliability Engineering System and Safety 44, 279–295 (1994)
9. Zhou, X.M., Peng, w., Shi, H.B.: Improved K-means Algorithm for Manufacturing Process Anomaly Detection and Recognition. Journal of Wuhan University of Technology (Chinese) 28(164), 1036–1041 (2006)

An Improved NN-SVM Based on K Congener Nearest Neighbors Classification Algorithm

Shirong Zhang[1], Kuanjiu Zhou[2], and Yuan Tian[1]

[1] School of Software of Dalian University of Technology
116621 Dalian, P.R. China
zsr.dut@163.com , tianyuan_ca@sina.com

[2] Institute of Systems Engineering, Dalian University of Technology
116024 Dalian, P.R. China
zhoukj@dlut.edu.cn

Abstract. Support vector machine constructs an optimal classification hyperplane by support vectors. While samples near the boundary are overlapped seriously, it not only increases the burden of computation but also decreases the generalization ability. An improved SVM: NN-SVM algorithm was proposed to solve the above problems in literature [1]. NN-SVM just reserves or deletes a sample according to whether its nearest neighbor has same class label with itself or not. However, its generalization ability will be decreased by samples intermixed in another class. Therefore, in this paper, we present an improved NN-SVM algorithm: it prunes a sample according to its nearest neighbor's class label as well as distances between the sample and its k congener nearest neighbors. Experimental results show that the improved NN-SVM is better than NN-SVM in accuracy of classification and the total training and testing time is comparative to that of NN-SVM.

1 Introduction

Support vector machine(SVM) is a relatively new class of machine learning algorithms based on statistical learning theory [2,3,4,5]. It is a hyperplane classifier defined in a kernel induced feature space. It not only has solved certain problems in many learning methods, such as small sample sets, over fitting, high dimension and local minimum, but also has a higher generalization ability than that of artificial neural networks[6]. But SVM also has problems to solve: for example, while samples are overlapped seriously, especially those samples intermixed in another class that may greatly increase the burden of computation and may lead to over learning. Therefore, NN-SVM algorithm is proposed to resolve the above problem: it first prunes the training set, reserves or deletes a sample according to whether its nearest neighbor has same class label with itself or not, then it trains the new set with SVM to obtain a classifier.

The accuracy of classification of NN-SVM could also be increased, as in NN-SVM algorithm it may delete samples which would be support vectors (SVs) because of reserving or deleting a sample just according to its nearest neighbor's

Z. Zhang and J. Siekmann (Eds.): KSEM 2007, LNAI 4798, pp. 550–555, 2007.

class label. In this paper, an improved NN-SVM algorithm is proposed: if a sample's class label is different from that of its nearest neighbor, we don't just delete it. Instead, we calculate the average distance in kernel space between the sample and its k congener nearest neighbors, and the average distance in kernel space between the sample's nearest neighbor and its k congener nearest neighbors (k is a positive integer) and then reserve or delete the sample according to the two average distances.This algorithm is called KCNN-SVM. Through the additional pruning strategy we can delete promiscuous samples more exactly. Experimental results show that the KCNN-SVM has a better performance on accuracy of classification than that of NN-SVM and the total training and classifying time is comparative to that of NN-SVM.

2 An Improved NN-SVM Algorithm Based on K-Congener-Nearest-Neighbors (KCNN-SVM)

As an improved SVM algorithm, NN-SVM first prunes the training set: if a sample's class label is different from that of its nearest neighbor, the sample is deleted. Contrarily, the sample is reserved, and then the algorithm trains the new set with SVM. We find that the pruning strategy of NN-SVM is not suitable for different training sets. For example, there is a two-class classification problem: there is an isolated negative sample in the boundary positive samples set. It's obvious that the negative sample as a noise point in the positive samples set should be deleted. But in the NN-SVM algorithm, those positive samples whose nearest neighbor is the intermixed negative sample would be deleted, although those positive samples may contribute to constructing the classification plane. In this situation, the accuracy of classification of NN-SVM algorithm would decrease because it has deleted samples mistakenly.

To resolve similar problems in NN-SVM algorithm discussed above, we present an improved pruning strategy: Find the nearest neighbor point x_{ne} of every sample x_i in the training set and compare class label of x_{ne} with that of x_i. If they have the same label, do nothing; if they are opposite, calculate the average distance dis_i between x_i and its k congener nearest neighbors and the average distance dis_{ne} between x_{ne} and its k congener nearest neighbors (k is a positive integer). If $dis_i > dis_{ne}$, delete x_i; contrarily, delete x_{ne}.

Distances in NN-SVM are calculated in original space. But SVM always constructs an optimal separating hyperplane in kernel space. Therefore, KCNN-SVM calculates all distances in kernel space. In order to calculate distances in kernel space, the definition of kernel function [7] is introduced below:

Definition 1. *(kernel or positive definite kernel) let χ be a subset of R^n, and $K(x,y)$ on $\chi \times \chi$ is called a kernel function(kernel or positive definite kernel), if there exists a mapping ϕ from χ to Hilbert space H:*

$$\chi \rightarrow H \ , \ x \rightarrow \phi(x)$$

which satisfies $K(x,y) = \phi(x) \cdot \phi(y)$. And $\phi(x) \cdot \phi(y)$ is a inner product between two vectors in Hilbert space H.

Distance $D(x, y)$ in kernel space between x and y could be calculated through Definition 1 (x and y are vectors in original space):

$$D(x,y) = \sqrt{\|\phi(x) - \phi(y)\|^2} = \sqrt{K(x,x) + K(y,y) - 2K(x,y)} \qquad (1)$$

From (1),distances in kernel space only depend on dot products in original space. Now if there is a "kernel function" K such that $K(x, y) = \phi(x) \cdot \phi(y)$, we would only need to use K in the improved algorithm to calculate distances. The "kernel function" K could be that used in SVM, such as polynomial kernel.

In the pruning strategy above, k congener nearest neighbors of a sample are obtained through calculating Euclidean distance in original space, however, data in original space have been mapped to a high dimensional feature space to construct the classification hyperplane in SVM. So Definition 2 is introduced for describing neighbors of a sample in kernel space:

Definition 2. *(KNN(x)) let x be a sample in the original training set, KNN(x) is defined as a set which contains the k congener nearest neighbors of x. Those neighbors of x are obtained through calculating Euclidean distance in the kernel space by (1).*

With the Definition 1 and 2, the particular learning steps of improved NN-SVM algorithm are summarized as follows:

Suppose there is a training set: (x_1, y_1), (x_2, y_2), $\cdots$,(x_m, y_m), $x_i \in R^n$, $y_i \in${1,-1},i=1, 2,$\cdots$,m. Let the new training set after pruning be T, $T = \emptyset$.

a. Calculate distances d_{ij} (i,j=1,2,$\cdots$,m) between every sample x_i in the training set and other samples except itself through (1).

b. Find the nearest neighbor point x_{ne} of every sample x_i through d_{ij}.

c. Compare the class label of x_{ne} with that of x_i, if they are the same, do nothing, if they are opposite, the average distance $dis_i = \frac{1}{k}\sum_{x_p \in KNN(x_i)} D(x_i, x_p)$ and the average distance $dis_{ne} = \frac{1}{k}\sum_{x_q \in KNN(x_{ne})} D(x_{ne}, x_q)$ are calculated through (1). If $d_i > d_{ne}$, x_i is marked to be deleted; contrarily, x_{ne} is marked to be deleted.

d. Append samples in training set not deleted to T.

In the improved algorithm, while class label of x_{ne} is different from that of x_i, KCNN-SVM should find KNN(x_i) and KNN(x_{ne}). So comparing with the time complexity of NN-SVM algorithm which is $o(m^2)$, time complexity of the improved algorithm increases $o(s \cdot m)$ (s is times of class label of x_{ne} is different from that of x_i, m is the size of training set and $s \ll m$). Although time of calculating increases a little, the improved algorithm estimates the intermixed degree of a sample in another class through calculating average distance in kernel space between it and its k congener nearest neighbors. Therefore, KCNN-SVM can delete promiscuous samples more exactly and avoid deleting samples that may contribute to classification, so the accuracy of classification of KCNN-SVM algorithm may increase, and experimental results verify our notion. While the value of k increases, KCNN-SVM can prune samples more exactly. But the accuracy of classification would not increase evidently when the value of k is too large. It just add burden of computation for finding the k congener nearest neighbors of a sample. So we should find a proper k for different training sets and $k = 2$ or $k = 3$ could be a good choice through experimental results.

In addition, we find that while $k = 0$ in step c of KCNN-SVM algorithm, the improved algorithm degenerates to the original NN-SVM algorithm. So NN-SVM algorithm is just a special instance of the KCNN-SVM algorithm.

3 Experimental Results

Experiments were performed on a 1.5Ghz Pentium4 machine with 640MB memory. Libsvm [8] was also used in experiments. Three data sets used as follows (DS1 and DS2 are downloaded from "http://ida.first.fraunhofer.de/projects/bench", and DS3 is downloaded from "http://www.csie.ntu.edu.tw"):

Data Set1(DS1), banana, it is a 2-dimensional, two-class classification problem. Training set has 400 patterns and test set has 4900 patterns.

Data Set2(DS2), ringnorm, it is a 20-dimensional, two-class classification problem. Training set has 400 patterns and test set has 7000 patterns.

Data Set3(DS3),it is a 123-dimensional, two-class classification example. Training set of "a3a" has 3185 patterns and test set has 29376 patterns.

Gaussian kernel is used for three data sets above and we get (2) from (1) to calculate distances:

$$D(x,y) = \sqrt{\|\phi(x) - \phi(y)\|^2} = \sqrt{2 - 2K(x,y)} \tag{2}$$

For DS1($c = 10$), we can find the changing of samples' distribution comparing before pruning with after pruning through Fig.1 and Fig.2.

KCNN-SVM could exactly delete samples that intermixed in another class seriously. The result obtained from DS1 is shown by Table 1. The test accuracy

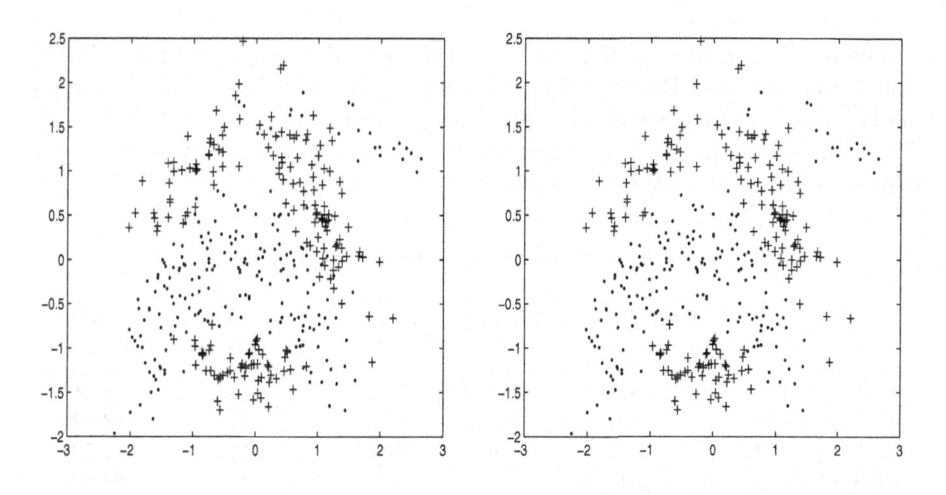

Fig. 1. DS1 before pruned

Fig. 2. DS1 pruned by KCNN-SVM(k=3)

Table 1. Comparing performances on DS1

	Number of samples	Training time(ms)	Testing time(ms)	Number of classifying exactly	Test accuracy
SVM	400	371	891	4340	88.57%
NN-SVM	344	200	591	4324	88.24%
KCNN-SVM(k=1)	353	213	611	4333	88.43%
KCNN-SVM(k=2)	360	220	630	4349	88.76%
KCNN-SVM(k=3)	359	215	620	4351	88.80%

of NN-SVM is below that of SVM. However, while the value of k increases in KCNN-SVM, it could delete samples more exactly and its test accuracies are even higher than that of SVM ($k = 3$ is the best in KCNN-SVM algorithm). The training processes of both NN-SVM and KCNN-SVM are faster than that of SVM because of having deleted samples that are hard to classify. Furthermore, NN-SVM and KCNN-SVM have much faster testing processes.

For DS2($c = 10$), Result obtained is shown by Table 2.

Table 2. Comparing performances on DS2

	Number of samples	Training time(ms)	Testing time(ms)	Number of classifying exactly	Test accuracy
SVM	400	597	2483	6856	97.94%
NN-SVM	251	370	1362	6672	95.31%
KCNN-SVM(k=1)	266	391	1513	6832	97.60%
KCNN-SVM(k=2)	260	382	1453	6860	98.00%

The result on DS2 is similar to that of DS1. In order to test on higher dimension, DS3 is used. For DS3($c = 100$), the result is shown by Table 3. The accuracy of classification of KCNN-SVM is just a little higher than that of SVM($k = 3$). The main reason may be that distances of high dimension are calculated by Euclidean distance. It may cause errors that would influence test accuracies. But

Table 3. Comparing performances on DS3

	Number of samples	Training time(ms)	Testing time(ms)	Number of classifying exactly	Test accuracy
SVM	3815	13797	53541	24621	83.81%
NN-SVM	2441	3315	17707	24564	83.62%
KCNN-SVM(k=1)	2640	4998	23855	24587	83.70%
KCNN-SVM(k=2)	2615	4507	21393	24619	83.80%
KCNN-SVM(k=3)	2611	4457	19787	24635	83.86%

accuracies of classification of KCNN-SVM algorithm are still higher than that of NN-SVM algorithm. In addition, the training and testing time of KCNN-SVM and NN-SVM is much faster than that of SVM.

4 Conclusions

An improved NN-SVM(KCNN-SVM) based on k congener nearest neighbors is proposed in this paper. Experimental results show that accuracies of classification of KCNN-SVM are higher than that of NN-SVM and SVM on both low and high dimensional data sets. Training and testing time of KCNN-SVM is faster than that of SVM and is comparative to that of NN-SVM. To further improve the generalization ability of KCNN-SVM algorithm on some high dimensional data sets, we need to find some proper dimension reduction methods and kernel functions while calculating distance in kernel space.

Acknowledgement. The research is supported by the National Natural Science Funds of China (70431001).

References

1. Li, H., Wang, C.: An improved support vector machines: NN-SVM. Chinese Journal of Computers 26(8), 1015–1020 (2003)
2. Vapnik: Statistical Learning Theory. John Wiley and Sons, West Sussex (1998)
3. Vapnik: An overview of statistical learning theory. IEEE Transactions on Neural Networks 10(5), 988–999 (1999)
4. Burges, C.J.C.: A tutorial on support vector machines for pattern recognition. Data Mining and Knowledge Discovery 2(2), 121–167 (1998)
5. Bennett, K.: Combining support vector and mathematical programming methods for classification. In: Advances in Kernel Methods: Support Vector Learning, pp. 307–326. MIT Press, Cambridge (1999)
6. Joachims, T.: Text categorization with support vector machines: learning with many relevant features. In: Nédellec, C., Rouveirol, C. (eds.) ECML 1998. LNCS, vol. 1398, pp. 137–142. Springer, Heidelberg (1998)
7. Boser, B.E., Guyon, I.M., Vapnik: A training algorithm for optimal margin classifiers. In: Proceedings of the 5th Annual ACM Workshop on Computational Learning Theory, pp. 144–152. ACM Press, Pittsburgh (1992)
8. Chang, C.-C., Lin., C.-J.: LIBSVM: a library for support vector machines (2001), `http://www.csie.ntu.edu.tw/~cjlin/libsvm`

Handling Contradictions in Default Theories*

Zhangang Lin and Zuoquan Lin

School of Mathematical Sciences
Peking University, Beijing 100871, China
{zglin,lz}@is.pku.edu.cn

Abstract. Reiter's default logic can not tolerate contradictions in default theories. In the paper we modify the definition of default extension to permit conflicts between justifications of used defaults and default conclusions, which are called incoherences, to ensure the existence of extension. Then desired extensions are selected as preferred ones according to the criterion that incoherences should be minimal in a preferred extension. Besides, the underlying reasoning system is shifted to four-valued logic to deal with inconsistencies in axioms of default theories.

1 Introduction

To model commonsense reasoning, Reiter's default logic [1] augments classical logic by *defaults* of form $\frac{\alpha:\beta_1,\cdots,\beta_n}{\gamma}$, where α is the *prerequisite* and β_i a *justification* for $1 \leq i \leq n$. In the paper each default is assumed to have only one justification. As to default $\delta = \frac{\alpha:\beta}{\gamma}$, define $pre(\delta) = \alpha$, $just(\delta) = \beta$, and $con(\delta) = \gamma$. For a set of defaults S, define $Pre(S) = \{pre(\delta)|\delta \in S\}$, $Just(S) = \{just(\delta)|\delta \in S\}$, and $Con(S) = \{con(\delta)|\delta \in S\}$. Knowledge in default logic is represented by a *default theory* (D, W), where D is a set of defaults and W a set of formulas. Here we consider only *finite default theory* in which D and W are both finite. A *default extension* of a given default theory which can be viewed as a belief set is defined as follows.

Definition 1. *Let $T = (D, W)$ be a default theory. For any set of formulas S, $\Gamma(S)$ is the least set of formulas such that*

1. $\Gamma(S) = \{\alpha | \Gamma(S) \models \alpha\}$.
2. $W \subseteq \Gamma(S)$.
3. *If* $\frac{\alpha:\beta}{\gamma} \in D$, $\Gamma(S) \models \alpha$ *and* $\neg\beta \notin S$, *then* $\gamma \in \Gamma(S)$.

Γ is called the belief revision operator w.r.t. T. A set of formulas E is a default extension of T iff $E = \Gamma(E)$.

Despite its simple syntax and powerful expressivity, Reiter's default logic is not satisfactory enough in two aspects. On one hand, as to a default theory (D, W),

* This work is partially supported by NSFC (grant number 60373002 and 60496322) and NKBRPC (grant number 2004CB318000).

Z. Zhang and J. Siekmann (Eds.): KSEM 2007, LNAI 4798, pp. 556–561, 2007.

if there are contradictions in W, then it would have only one *trivial default extension* containing all formulas. Such contradictions are called *inconsistencies*. To solve the problem, some researchers [2,3] shift the underlying reasoning system of default logic to paraconsistent logic [4,5], multi-valued logic in particular [6,7,8]. On the other hand, if there are contradictions between justifications of used defaults and the derived default conclusions for each *closed process*, which are called *incoherences*, then the default theory would have no default extensions [9]. To solve the problem, the definition of extension has to be modified [10,11,12]. Instead of trying to guarantee the existence of extension, some others try to find the condition of the existence of default extension [1,10,13].

These works are good attempts to overcome contradictions in default theories, but they all fail to handle inconsistencies and incoherences simultaneously.

Hinted by the trick—first guarantee the existence, then select desired ones—in preferential four-valued logic [6], in the paper we modify the definition of extension such that incoherences are permitted in extension to guarantee the existence of extension and take preferential four-valued logic as the underlying reasoning system to overcome inconsistencies. Then desired extensions are selected as preferred ones according to the criterion that incoherences in a preferred extension should be minimal. With this approach, we manage to overcome all contradictions in default theories.

The rest of the paper is structured as follows. In Section 2, we briefly review preferential four-valued logic. In Section 3 we represent our system. Some properties of it are also investigated here. We conclude the paper in Section 4.

2 Preferential Four-Valued Logic

To reason paraconsistently, Belnap's four-valued logic [7] consists of four truth values: $FOUR = \{t, f, \top, \bot\}$(also written as $(1,0), (0,1), (1,1), (0,0)$ respectively). Intuitively, they represent true, false, inconsistent and in lack of information respectively. Connectives in classical logic are defined in $FOUR$ as follows: $\neg(x,y) \stackrel{\text{def}}{=} (y,x)$, $(x_1,y_1) \wedge (x_2,y_2) \stackrel{\text{def}}{=} (x_1 \wedge x_1, y_1 \vee y_2)$, $(x_1,y_1) \vee (x_2,y_2) \stackrel{\text{def}}{=} (x_1 \vee y_1, y_1 \wedge y_2)$, $(x_1,y_1) \rightarrow (x_2,y_2) \stackrel{\text{def}}{=} \neg(x_1,y_1) \vee (x_2,y_2)$.

A *four-valued valuation* is a function that assigns a truth value from $FOUR$ to each atomic formula. Any valuation can be extended to complex formulas in the canonical manner. A valuation is a *model* of S if, for each formula $\psi \in S$, $v(\psi) \in \{t, \top\}$. Let S be a set of formulas and ϕ a formula. We say S entails ϕ in $FOUR$, namely $S \models^4 \phi$, if for each model M of S, $M(\phi) \in \{t, \top\}$.

Although $\models^4$ has some nice properties, it is too cautious to reason with it. To enhance its reasoning capability, it is natural to reduce the number of considered models when drawing conclusions according to some criteria, one of which is the concerned models should be most classical [6].

Definition 2. *Let M and N be both four-valued models of the set of formulas S. M is* more classical *than N, denoted $M \leq_c N$, if for any formula ϕ, $M(\phi) = \top$ or $M(\phi) = \bot$ implies $N(\phi) = \top$ or $M(\phi) = \bot$. M is a* most classical model *of S if S has no model more classical than M.*

Let S be a set of formulas. S most classically entails α, denoted $S \models^4_{mc} \alpha$, if for each most classical model M of S, $M(\alpha) \geq_t t$. Just as expected, each finite set of formulas has at least one most consistent four-valued model, thus we can always get useful information from finite set of formulas. Moreover, as to a consistent set of formulas, $\models$ and $\models^4_{mc}$ are identical [6].

Because of the nice properties of $\models^4_{mc}$, it is chosen as the underlying logic of our system to overcome inconsistencies. For simplicity, define $Cn^4_{mc}(S) = \{\alpha | S \models^4_{mc} \alpha\}$. Note that Cn^4_{mc} is nonmonotinic due to the preferences in four-valued models.

3 Extensions Based on Four-Valued Logic

In Reiter's default logic, default extension only records what can be believed, whereas in the new system, to identify incoherences, the formulas used as justifications have to be recorded as well. Thus newly defined extension is of the form (E, J), in which E represents what should be believed and J is the set of the formulas used as justifications. Unlike Reiter's default logic, there may exist incoherences between E and J, i.e. $\neg E \cap J \neq \emptyset$.

The fact that preferential four-valued logic is nonmonotonic makes our system different with Reiter's default logic in the following two aspects. First, the application of an applicable default may invalidate the prerequisite of a used default and, just like incoherence, cause the nonexistence of extension. To avoid this, in the new system the prerequisites of used defaults has to be recorded as well. Second, in Reiter's default logic, the revised belief set is minimal subject to the three conditions in Definition 1. But in the new system the cardinalities of belief sets should not be compared directly. Instead we make sure the number of used defaults to be minimal subject to that applicable defaults should be used.

For the above two reasons, we choose the set of used defaults as the operand and result of belief revision operator. For convenience, write $Cn^4_{mc}(W \cup Pre(S) \cup Con(S))$ as $In(S)$ in a default theory $T = (D, W)$, where $S \subseteq D$.

Definition 3. *Let $T = (D, W)$ be a default theory. A default $\frac{\alpha:\beta}{\gamma} \in D$ is applicable to $S' \subseteq D$ w.r.t. $S \subseteq D$ in T if $\alpha \in In(SS)$ for some $SS \subseteq S'$, $\neg\beta \notin In(S)$ or $Just(S)$ contains a formula equivalent to β.*

3.1 Alternative Extension

Definition 4. *Let $T = (D, W)$ be a default theory and $B \subseteq D$ a set of defaults. $\Gamma(B)$ is the least set of defaults B' such that*

- *(Cond 1) $B' \subseteq D$.*
- *(Cond 2) If $\frac{\alpha:\beta}{\gamma} \in D$ is applicable to B' w.r.t. B in T, then $\frac{\alpha:\beta}{\gamma} \in B'$.*

If $B = \Gamma(B)$, then $(In(B), Just(B))$ is an alternative extension *of T generated by B.*

We may prove Γ is well defined just as in [1].

Theorem 1. *If (E, J) is an alternative extension of default theory T, then $GD(E, J, T) = \{\frac{\alpha:\beta}{\gamma} \in D | \alpha \in E, \beta \in J\}$ is the fixpoint of Γ generating (E, J).*

For the above reason, we call $GD(E, J, T)$ the *the set of generating default* for (E, J) w.r.t. T. Just as we stated, the sizes of alternative extensions is defined by the cardinalities of the sets of used defaults. Thus we have

Definition 5. *Let (E_1, J_1) and (E_2, J_2) be alternative extensions of default theory T. If $GD(E_1, J_1, T) \subseteq GD(E_2, J_2, T)$, then (E_1, J_1) is smaller than (E_2, J_2), denoted $(E_1, J_1) \leq (E_2, J_2)$. If $(E_1, J_1) \leq (E_2, J_2)$ and $(E_1, J_1) \neq (E_2, J_2)$, then (E_1, J_1) is strictly smaller than (E_2, J_2), denoted $(E_1, J_1) < (E_2, J_2)$.*

The following theorem makes alternative extension more intuitive.

Theorem 2. *Let $T = (D, W)$ be a default theory. For the sets of formulas E and J, define $G_0 = \emptyset$, and for $i \geq 0$*

$$G_{i+1} = \{\delta | \delta = \frac{\alpha : \beta}{\gamma} \in D \text{ is applicable to } \bigcup_{k=0}^{i} G_k \text{ w.r.t. } GD(E, J, T)\}$$

Then (E, J) is an alternative extension of T iff $E = In(\bigcup_i G_i)$ and $J = Just(\bigcup_i G_i)$.

Minimality does not hold as to alternative extension.

Theorem 3. *If (E_k, J_k) are alternative extensions of default theory $T = (D, W)$ for $k = 1, 2, \cdots$, then T has the least alternative extension (E, J) larger than (E_k, J_k) for each k such that $E = In(\bigcup_i G_i)$ and $J = Just(\bigcup_i G_i)$, where $G_0 = \bigcup_k GD(E_k, J_k, T)$, and for $i \geq 0$*

$$G_{i+1} = \{\delta | \delta = \frac{\alpha : \beta}{\gamma} \in D \text{ is applicable to } \bigcup_{k=0}^{i} G_k \text{ w.r.t. } \bigcup_{k=0}^{i} G_k\}$$

The above theorem indicates that all the alternative extensions of a default theory comprise a *complete upper semilattice* w.r.t. $\leq$.

3.2 The Existence of Alternative Extension

Since the applicable subsequent defaults can not invalidate justifications of the prior defaults. Thus we immediately have Lemma 1 and Theorem 4.

Lemma 1. *For default theory (D, W), if $D_1 \subseteq D_2 \subseteq D$, then $\Gamma(D_1) \subseteq \Gamma(D_2)$.*

Theorem 4 (Semimonotonicity). *If $T = (D, W)$ and $T' = (D', W)$ are two default theories, where $D \subseteq D'$, and (E, J) is an alternative extension of T, then T' has an alternative extension (E', J') such that $(E, J) \leq (E', J')$.*

From the semimonotonicity of alternative extension(or directly from Lemma 1), we have the existence of alternative extension.

Theorem 5. *Each default theory has at least one alternative extension.*

Thus all the inconsistencies and incoherences can be resolved with alternative extension, as shown in the following two examples.

Example 1. Consider default theory $T = (D, W)$, where $D = \{\frac{C:D}{D}\}$ and $W = \{A, \neg A, B, B \to C\}$. Since W is inconsistent, T has only a trivial default extension Contrarily, The new Γ has $\{\frac{C:D}{D}\}$ as its fixpoint. Thus T has a nontrivial alternative extension $(Cn^4_{mc}(\{A, \neg A, B, B \to C, C, D\}), \{D\})$.

Example 2. Consider default theory $(\{\frac{:B}{\neg A}\}, \{A\})$, $(\{\frac{:B}{C}, \frac{:D}{\neg C}\}, \emptyset)$, and $(\{\frac{:\neg B}{A}\}, \{A \to B\})$. It can be verified that they have no default extension. But they do have $(Cn^4_{mc}(\{A, \neg A\}), \{B\})$, $(Cn^4_{mc}(\{C, \neg C\}), \{B, D\})$, and $(Cn^4_{mc}(\{A, A \to B\}), \{\neg B\})$ as their alternative extensions, in which the third has an incoherence $\neg B$.

3.3 Preferred Extension

It is of course too credulous to permit incoherences in an extension. As the following example shows, since justification conflicts are permitted, some counterintuitive alternative extensions arise.

Example 3. Consider default theory $(\{\frac{:\neg A}{B}\}, \{A\})$. According to the definition of alternative extension, it has two alternative extensions $(Cn^4_{mc}(\{A\}), \emptyset)$ and $(Cn^4_{mc}(\{A, B\}), \{\neg A\})$, in which the latter is counterintuitive because it has an incoherence $\neg A$ which is unnecessary.

Inspired by the trick in preferential four-valued logic that conflicts in a belief set should be minimal, we try to exclude those counterintuitive alternative extensions according to the criterion that incoherences in a "good" extension should be minimal, i.e. an extension has incoherences when it has to.

Definition 6. *Let T be a default theory. An alternative extension (E^*, J^*) of T is a* preferred extension *of T if $\neg E^* \cap J^*$ is minimal in $\{\neg E \cap J | (E, J)$ is an alternative extension of $T\}$.*

Let us continue Example 3. Since $\neg Cn^4_{mc}(\{A\}) \cap \emptyset = \emptyset$, $\neg Cn^4_{mc}(\{A, B\}) \cap \{\neg A\} = \{\neg A\}$, the former is a preferred extension of the given default theory, which is identical with Reiter's default logic.

Theorem 6 (Existence of Preferred Extension). *Each finite default theory has at least one preferred extension.*

By Example 3, we note preferred extensions are identical to default extensions as to default theories having consistent default extensions.

Theorem 7. *E is a consistent default extension of default theory T iff T has a preferred extension (E, J) such that $E \cap \neg J = \emptyset$.*

Note that semimonotonicity does not hold as to preferred extension from Theorem 7. Also note that minimality does not hold as to preferred extensions either. See the following example.

Example 4. Consider default theory $T = (\{\frac{:\neg B}{\neg A}\}, \{A, A \rightarrow B\})$. It can be verified that $(E_1, J_1) = (Cn^4_{mc}(\{A, A \rightarrow B\}), \emptyset)$ and $(E_2, J_2) = (Cn^4_{mc}(\{A, A \rightarrow B, \neg A\}), \{\neg B\})$ are both preferred extensions of T and $(E_1, J_1) \leq (E_2, J_2)$.

4 Conclusion

Our main contribution in the paper is to provide a default logic that generalizes Reiter's default logic and can overcome all contradictions in default theories. Thus the new system has potential applications in commonsense reasoning in presence of inconsistencies and incoherences.

Although the new system has some nice properties, it seems more impractical to reason with it than Reiter's default logic. In the research, we find that it seems possible to get the alternative and preferred extensions of a default theory by computing the default extensions of transformations of the given default theory. The future work will focus on the issue.

References

1. Reiter, R.: A logic for default reasoning. Artificial Intelligence 13, 81–132 (1980)
2. Han, Q., Lin, Z.: Paraconsistent default reasoning. In: 10th International Workshop on Non-Monotonic Reasoning, pp. 197–203 (2004)
3. Yue, A., Lin, Z.: Default logic based on four valued semantics. Chinese Journal of Computer Science 28(9), 1447–1458 (2005)
4. daCosta, N.: Theory of inconsistent formal systems. Notre Dame Journal of Formal Logic 15, 497–510 (1974)
5. Besnard, P., Torsten, S.: Signed system for paraconsistent reasoning. Journal of Automated Reasoning 20(1-2), 191–213 (1998)
6. Arieli, O., Avron, A.: The value of the four values. Artificial Intelligence 102(1), 97–141 (1998)
7. Belnap, N.: A useful four-valued logic. Modern uses of multiple-valued logic, pp. 30–56 (1977)
8. Ginsberg, M.L.: Multivalued logics: a uniform approach to reasoning in artificial intelligence. Computational Intelligence 4, 265–316 (1988)
9. Antoniou, G., Sperschneider, V.: Operational concepts of nonmonotonic logics, part 1: Default logic. Artificial Intelligence 8(1), 3–16 (1994)
10. Lukaszewicz, W.: Considerations on default logic: an alternative approach. Computational Intelligence 4(1), 1–16 (1988)
11. Schaub, T.: On constrained default theories. In: ECAI, pp. 304–308 (1992)
12. Brewka, G.: Cumulative default logic: in defense of nonmonotonic inference rules. Artif. Intell. 50(2), 183–205 (1991)
13. Zhang, M.: A new research into default logic. Inf. Comput. 129(2), 73–85 (1996)

Algorithm for Public Transit Trip with Minimal Transfer Times and Shortest Travel Time

Gang Hou[1], Kuanjiu Zhou[2], and Yang Tian[1]

[1] School of Software of Dalian University of Technology
116621 Dalian, P.R. China
hg.dut@163.com , tianyuan_ca@sina.com
[2] Institute of Systems Engineering, Dalian University of Technology
116024 Dalian, P.R. China
zhoukj@dlut.edu.cn

Abstract. Considering passengers' travel psychoanalysis, the public traffic model of optimum route is proposed, which primary goal is minimal transfer times and the second goal is shortest travel time. On the integrated analysis of the space relationship between stations, the public traffic network's searching model is proposed. Basing on the searching model, the minimal transfer times problem is solved by translating to the problem of the shortest route between two stations. The whole shortest route algorithm of the minimal transfer times is issued, which can efficiently solve the travel project of all the least minimal times between two stations. Finally, basing on the analysis of public traffic travel time chain, the time evaluating function is provided to calculate the public traffic travel time.

1 Introduction

The research on the theory of public traffic travel optimum transfer problem includes the mathematic description of public traffic network and the design of optimum route algorithm. In the aspect of public traffic network description, Anez uses the allelomorph chart to describe the traffic network which can cover public traffic routes[1]; Choi discusses the method that how to use the GIS techniques to produce public traffic routes and stop from the geography datum[2]. Huang Zhengdong researches the relationship between public traffic entity and basic road network in GIS, which provides the foundation of the mathematic description of public traffic network[3]. In the aspect of optimum route algorithm, some researchers bring forward the algorithm from different point of view, because the index of 'optimum' differs in thousands ways. Considering the characteristics of public traffic network, Zhang Guowu proposes a kind of several shortest route algorithm on the basis of popularization of Floyd algorithm[4]; Koncz proposes the static public traffic network several routes selection algorithm with the least transfer times as the primary goal and the shortest travel distance as the second goal[5]; Yang Xinmiao designs the route selection model with the least transfer times and the shortest travel distance as goals[6].

The passengers' travel psychoanalysis investigative statistic of correlative literatures shows that people will consider transfer times, travel time and travel

Z. Zhang and J. Siekmann (Eds.): KSEM 2007, LNAI 4798, pp. 562–567, 2007.

distance orderly when they take buses [6]. Therefore, basing on the statistic result, it is proposed that the public traffic model of optimum route with the least transfer times as primary goal and the shortest travel time as second goal. On the integrated analysis of the space relationship between stations, the searching model is proposed. Basing on the searching model, the least transfer times problem is solved by translating to the problem of the shortest route between two stations. It also improves classical Dijkstra algorithm when the edge value is 1, which reduces the algorithm's time complexity. It is designed the prior node algorithm on the shortest route and the whole shortest route algorithm which can solve the whole shortest routes between two spots efficiently in $O(n^2)$ time complexity. Finally, basing on the analysis of public traffic travel time chain, the time evaluating function is provided to calculate the public traffic travel time and find the best travel project.

2 Public Traffic Network Modeling

Public traffic network includes three entities: bus line, bus stop and bus station (the station that different buses stop at the same place). We will find less transfer routes if we only transfer at the bus station when considering transfer problem, because we neglect one situation that we can walk a short distance during transfer process. So we should put several stops or stations in proper distance which we may transfer at into one transfer node. According to the analysis above, we describe the public traffic network searching model as followings:

Definition 1. *$G = (V, E, w^z_{i,j})$ is a simple directional graph endowed with value, thereinto, $V = \{v_i | 1 \leq i \leq n\}$ is transfer node's gather, n is the transfer node's number; $E = \{e_j | 1 \leq j \leq m\}$ is the edge's gather, m is the edge's number. $w^z_{i,j}$ is the value of edge,$w^z_{i,j}$ represents that they are nonstop or not between two nodes, $w^z_{i,j} = 1$ when two nodes are nonstop, otherwise the value is 0. The shortest route in graph G is the route which $w^z_{i,j}$'s value is the smallest between two nodes.*

3 Searching Algorithm

3.1 The Improved Dijsktra Algorithm with Edge Value Is One

The step of the algorithm is as follow:

(1) set $d(v_0) = 0$, $d(v_i) = \infty$ $(i = 1, 2n)$, $K = 0$, $S = \{v_0\}$, $S_0 = \{v_0\}$;

(2) The Kth step, $S_{K+1} = Next_{S_K} \bigcap \bar{S}$, $d(v_i) = K + 1 (i \in S_{K+1})$, $S = S \bigcup S_{K+1}$;

(3) If $|S| = |V(G)|$ or $v_d \in S$, the algorithm end; otherwise, $K = K + 1$, turn to step (2).

When the algorithm has been executed K times, we have that $S_K = \{v_i | d(i) = K\}$ and $S = \{v_i | d(i) \leq K\}$. v_0 represents the beginning node; v_d represents the ending node; v_i represents one node in graph G; $Next_{S_K}$ represents all the subsequence nodes of nodes in node gather S_K.

Algorithm analysis: The basal step of the algorithm is step (2). It can gather the node into S by level sequence. Every time the new joint nodes are the subsequence of the nodes which joint S last time. It equally use the BFS to search in graph G until the terminal node has jointed gather S. If graph G is stored by neighboring table, according to the knowledge of data structure, the time complexity of the algorithm is $O(n + e)$. In graph G, n is the number of nodes and e is the number of edges.

3.2 The Prior-Node Algorithm

Theorem 1. *In graph G which value of edges are all positive number, the shortest route between two points must be shorter or call even than the shortest route between them in graph G' which is the sub-graph of graph G. According to the knowledge of graph theory, it is apparently right.*

Thought of the algorithm: According to Theorem 1, the shortest route between v_o and v_i we find in G' must be shorter or equal than the shortest route between them in graph G. If the length of the two route is equal, then the route we find in G' is also the shortest route between v_o and v_i in graph G and we can use this route to find another prior node of v_i.We repeat the operation above until there is no new prior node of v_i. The step of the prior-node algorithm is as follow:

$Dijkstra(G, v_o, v_i, P); K = dist(P);$
$v^1_{Prev(v_i)} = PrevNode(P, v_i);$
$PrevList(v_i) = v^1_{Prev(v_i)};$
$for(j = 1; j < m; j++)$
$\{$
$G = G - e(v^j_{Prev(v_i)}, v_i);\ \ Dijkstra(G, v_o, v_i, P);\ \ if(dist(P) > K)break;$
$else\ \{v^{j+1}_{Prev(v_i)} = PrevNode(P, v_i); PrevList(v_i) = PrevList(v_i) \bigcup v^{j+1}_{Prev(v_i)};\}$
$\}$

Algorithm analysis: The time complexity of the circulation in the algorithm is $O(n + e)$ and it need to be executed m times. m is the in-degree of node (Generally speaking m is much smaller than e. In the worst condition m is equal to e). Therefore the time complexity of the algorithm is $O(e(n+e))$ in the worst condition.

3.3 All-Shortest-Route Algorithm

Theorem 2. *In graph G which edge value is positive number, if $P(i)$ is the shortest route between v_1 and v_i, and $v_j \in P(i)$, then $P(j)$is the shortest route between v_1 and v_j.*

Proof. Reduction to absurdity. If $P(j)$ isn't the shortest v_1 between and v_j, then there is a shortest route between them, suppose it's $P'(j)$. Obviously $d'(j) < d(j)$. So $d'(i) = d'(j) + d(j, i) < d(i) = d(j) + d(j, i)$. It's contradict that $P(i)$ is the shortest route.

Thought of the algorithm: According to theorem 2, we find that each node we find in the shortest route, the route between it and the starting node is also the shortest route between them. All this routes are in the shortest route between the starting node and the ending node. We use the linear-table PrevList(v_i) to store the prior node of v_i. There are two portions in PrevList(v_i), named Nodename and PrevNodePoint. They represent the name of node and the pointer point its own linear-table. The algorithm steps are as follow:

(1) Input the starting node v_0 and the ending node v_d, initialize the root of the shortest routing tree, PrevList(v_d)=NULL; NodeName=v_d; PrevList(v_d)= PrevNodePoint.

(2) We use the prior node algorithm to find all the prior nodes of v_i in the shortest route and then put them into the linear-table PrevList(v_i) in turn (We start the step with v_d). So we can find all the child-node of v_i in the shortest route tree.

(3) We use the nodes which is stored in PrevList(v_i) as the father nodes and find their prior nodes by prior-node algorithm. Then we store them into their own linear-table.

(4) Repeat step (2)and step (3),until PrevList(v_i)=v_0. According to the PrevNodePoint of each node, we can build a shortest route tree. It can help us to find a shortest route by tracing the root tree from leaf node to the root node.

Algorithm analysis: The all-shortest-route algorithm is basing on the prior node algorithm. Each time we find a prior node, we need execute Dijkstra algorithm one time. There are e number edges in graph G. So, in the worst condition, all the edges in graph G are in the shortest route. Therefore the algorithm time complexity is $O(e(n+e))$, equaling to $O(n^2)$.

4 The Time Evaluating Function

In the whole passengers travel process, the total travel time T_K include the time walking to station, the time leaving station, the time waiting in station, the time on bus and the transfer time between different lines. We define the whole time make up of all the time above as a passengers travel time-line. Every passenger travel on a bus contains a whole travel time-line. The different part of the travel time-line has it own different proportion value. Therefore T_K isn't that we plus different time part easily. Using the proportion value of different time part, we can actually analysis the different choosing action that the passenger will have when they meet different condition in their travel process. We describe the total travel time T_K as follow:

$$T_K = W_w \cdot T_K^w + W_a \cdot T_K^a + W_i \cdot TK^i + n \cdot W_t \cdot T_K^s \tag{1}$$

In the equation (1), T_K^w is the walking time which contains walking to station time and leaving station time; T_K^a is the time waiting on the starting station; T_K^i is the time spending on the bus; T_K^s is the transfer time which contains transfer waiting time and transfer walking time; n is the transfer times; W_w, W_a, W_i, W_t are the proportion value of different time part.

The walking time is related with walking distance D_K^w and walking speed V_w; the waiting on the station time is related with the frequency of the bus sending; the time spending on bus is deciding by the distance of the bus line D_K^i the speed of the bus V_i and the time waiting on the journey T_K^s. The transfer time is related with the distance of different transfer station D_K^t and the time T_K^a we wait on the station that we have transferred to. Therefore, we can describe the travel time as follow:

$$T_K = \frac{W_w \cdot f_w \cdot D_K^w}{V_w} + \frac{W_a \cdot f_a}{F} + W_i \cdot (\frac{D_K^i}{V_i} + T_K^s) + n \cdot W_t \cdot (\frac{f_w \cdot D_K^t}{V_w} + \frac{f_a}{F}) \quad (2)$$

In the equation (2), f_w is the no-beeline-coefficient of the route which we walked; f_a is the modify-parameter of the waiting time. The literature[7] gives the advice parameter value. We give a Time Evaluating Function to calculate the travel time. But when we calculate the travel time, we need specifically public traffic data of different city.

5 Experiment

Taking Dalian public traffic data as an example: there are 89 bus lines and 376 stations with different names. Through producing the transfer nodes by congregating the bus stations, there are 314 bus stations left. We calculate the whole bus taking projects among 314 stations above by the algorithm which takes 5 minutes on P4 computer. It need to be executed $N = 314 * (314-1)/2 = 49141$ times. The results are represented at table 1:

Table 1. Result of the algorithm executes with Dalian public traffic data

Transfer Time	Nonstop	One Time Transfer	Twice Transfer	More than Twice
Result	3636	26418	18473	614
Proportion	7.4%	53.76%	37.59%	1.25%

Through the result above, we can calculate the average transfer times of Dalian public traffic, and the result is $n = 1 * 53.76\% + 2 * 37.59\% = 1.2894$ times. The calculate result 1.2894 is closed to 1.2 which is the data Dalian public traffic average transfer times according to a related reference[8]. In the same time, in all the bus taking projects that the algorithm finds the one time transfer and the two times transfer are in a high proportion. The projects that we need to transfer more than two times is only taking 1.25%. It is also same with the reality condition. We can give conclusions through the experimentation as follow:

(1) The public traffic network modeling method that we bring forward in this paper can reflect the reality transfer condition of the public traffic network.

(2) The searching model and the searching algorithm can calculate efficiency bus taking project in a certain time and it can also find several transfer projects between two stations.

6 Conclusions

In the aspect of solving the minimal transfer times, the problem is solved by translating it to the shortest route problem between two stations basing on the searching model. It also improves classical Dijkstra algorithm when the edge value is 1, which reduces the algorithm's time complexity. It is designed the prior node algorithm on the shortest route and the whole shortest route algorithm which can solve the whole shortest routes between two spots efficiently in $O(n^2)$ time complexity. In the aspect of transfer time calculating, basing on the analysis of public traffic travel time chain, the time evaluating function is provided to calculate the public traffic travel time. Finally, through the experiment, it is proved that the modeling project and the searching algorithm are feasible.

Acknowledgement. The research is supported by the National Natural Science Funds of China (70431001).

References

1. Anez, J., Barrat, T., Perez, B.: Dual graph representation of transport networks. Transportation Research Part B 30(3), 209–216 (1996)
2. Choil, K, Jan, G.W.: Development of a transit network from a street map database with spatial analysis and dynamic segmentation. Transportation Research Part C (8), 129–146 (2000)
3. Huang, Z.: A transit trip guidance system based on detailed data representation. Engineering Journal of Wuhan University 36(3), 69–75 (2003)
4. Zhang, G., Qian, D.: The Arithmetic for Solving Multiple Shortest Paths in Public Transportation Newwork. Systems Engineering-Theory and Practice 12(4), 22–26 (1992)
5. Koncz, N., Greenfeld, J., Mouskos, K.: A Strategy for Solving Static Multiple Optimal Path Transit Network Problems. Journal of Transportation Engineering 122(3), 218–225 (1996)
6. Yang, X., Wang, W., Ma, W.: GIS-Based Public Transit Passenger Route Choice Model. Journal of Southeast University(Natural Science Edition) 30(6), 87–91 (2000)
7. Heinz, S., Michael, F.: Optimal strategies: a new assignment model for transit networks. Transportation Research 23B(2), 83–102 (1989)
8. Mao, Y., Wu, D.: The Transfer Issue of Public Transportation in Dalian. Public Traffic (4), 15–16 (2003)

Typed Category Theory-Based Micro-view Emergency Knowledge Representation

Qingquan Wang and Lili Rong

Systems Engineering Institute, Dalian University of Technology, 116024 Dalian, China
dlwqq@hotmail.com, llrong@dlut.edu.cn

Abstract. The emergency knowledge transferred is a special product which consists of knowledge pieces that can be reorganized according to requirements of various decision-making scenarios. Using typed-category theory we propose a new knowledge representation method for combining indispensable semantic information into a new categorical knowledge structure, knowledge piece category. The knowledge pieces are micro-ontologies enabling enhanced relationships among concepts in micro-view. Finally, we use knowledge matching to illuminate the application of knowledge pieces.

Keywords: Emergency Decision-Making Support; Typed-Category Theory; Knowledge Piece; Knowledge Matching.

1 Introduction

Quick and effective decision-making is crucial in emergency responses [1]. The speed and quality of acquiring knowledge is one of the main factors involved in effective decision-making. Knowledge management among emergency departments composes an agile knowledge supply chain (AKSC) [2]. In AKSC, distributed knowledge must be reorganized according to various pending problems.

Category theory is a relatively young branch of mathematics designed to describe various structural concepts from different mathematical fields in a uniform way [3], as pointed out by Hoare [4], "Category theory is quite the most general and abstract branch of pure mathematics". The application of category theory has been developed with the improvement of computer science since 1980. Category theory application reached a new milepost with the publication of "Categories and Computer Science" written by Walters in 1991. Category theory is the foundation of computer science typically in the fields of computer programming [5], software engineering [6], semantics [7], and artificial intelligence [8]. Recently, some researchers have been associating knowledge management with category theory. For example, category theory has been introduced into object-oriented domain models [9], semantic network [10] and ontology [11]. In the last five years, category theory has also been introduced in the field of knowledge engineering, especially in ontology engineering, in ontology merging with pushout [12], the IFF category theory (meta) ontology [13], the geometry of knowledge with category [14], meta-ontology [15], the representation of complex data structures using ontologies in the semantic web [16], ontological

Z. Zhang and J. Siekmann (Eds.): KSEM 2007, LNAI 4798, pp. 568–574, 2007.

methodology of domain analysis in [17], ACO (structure of meta ontology) [18], mathematic base for ontology structure [19] and knowledge representation [20].

The main contributions of a seminal paper written by Ruqian Lu [21] are the combination of category theory and knowledge science. Category theory is introduced in knowledge science as a mathematical foundation. This improved category theory provides abstract knowledge representation and processing. Compared to traditional category theory, in typed pseudo-category all morphisms have types, and the composition of morphisms is not necessarily a morphism.

Knowledge pieces are special outputs of knowledge representation that have structure, character and granularity. Generally knowledge pieces are a kind of structural knowledge representation from a micro- view. At first we have no knowledge, then gradually acquire pieces of knowledge by observing data, and eventually acquire complete knowledge for solving problems. Knowledge pieces can be acquired, delivered, created and produced.

The goal of this paper is to research the expanded definitions of types and composition rules of typed category theory in a micro-view and also the approach of the application of structural knowledge representation and knowledge matching.

2 Emergency Knowledge Piece – Categorical View

This section gives detailed definitions and characteristics of a knowledge piece category (KP category) and some relevant explanations in knowledge processing.

Definition 1. A KP category is a Typed Category $\boldsymbol{K} = (O, M, T, R)$ which consists of a class of objects O; a class of morphisms M; a class of morphism types T; and a class of composition rules R, defined as follows.

1. $M = \{M(a,b,t) \mid a,b \in O, t \in T\}$ is the set of morphisms on the objects, each morphism f can be written as

$$f : a \xrightarrow{t} b \text{ or } f_t(a,b),\ (a,b \in O, t \in T), \tag{1}$$

which a is the domain of f, b the codomain of f, t the type of f, respectively written as *dom* (f), cod (f) and *typ* (f). Morphisms between two objects form a morphism set.

2. For each object a, an identity morphism has domain a and codomain a, Written as ID_a.

3. For each type $t \in T$, t = (*sou*, *tar*, *val*) is a triple, which *sou* is the restriction of this type of morphism's domain; *tar* is the restriction of this type of morphism's codomain; *val* is the value of type.

The triple can determine a unique type of morphism.

4. For each pair types t and s, there may have a rule in R while $t.tar = s.sou$ or $s.sou$ is the subobject (subclass) of $t.tar$. denoted as

$$r = t \times s,\ (r.sou = t.sou,\ r.tar = s.tar). \tag{2}$$

5. For each pair of morphisms $f_t(a, b)$ and $g_s(b, c)$, $dom(g) = cod(f)$), there exists a composition morphism

$$g_s \circ f_t : a \xrightarrow{r} c,\ (r = t \times s), \tag{3}$$

when there is a composition rule $r = comb\ (t, s)$ in R.

6. And for each morphism $f_t(a, b)$, there are the identity compositions.

$$f \circ ID_a = f \text{ , } ID_b \circ f = f \text{ .} \tag{4}$$

Type of each identity morphism is unit type, written as u^T.

7. For each set of morphisms $f_t(a, b)$, $g_s(b, c)$ and $h_s(c, d)$, associativity of them is

$$h_r \circ (g_s \circ f_t) = (h_r \circ g_s) \circ f_t \text{,} \tag{5}$$

The following explanation gives definitions used in structural knowledge representation. Object class O represents concepts in a certain domain, morphism class M the relationships among O; morphism type class T the restrictions on M; composition rule class R the relationship compositions of M.

In brief, the concept set O with defined typed relationships M constitute the intension of knowledge piece, its structure; and the subset of types T and subset of composition rules R form the extension of knowledge piece. In categorical view, knowledge piece relate to another special category which take types of morphisms as objects.

Definition 2. A KP category $\boldsymbol{K}$ is a discrete category if all the morphisms are identities.

Actually, the knowledge matching based on keywords belongs to a kind of matching between discrete categories.

Definition 3. A KP category $\boldsymbol{K}$ is a connected category if there is one morphism at least for each pair of objects. In fact, if we get a knowledge piece with the character of connected category, we can obtain sufficient semantic information from the composition rules R.

Definition 4. A KP category $\boldsymbol{K} = (O, M, T, R)$ is a subcategory of $\boldsymbol{K'} = (O', M', T', R')$: For each object o in O, $o \subseteq o'$ (o is the subclass of o' or they are equal); $M \subseteq M'$ (M is the subset of M' on O); $R \subseteq R'$ (R is the subset of R' on T).

Definition 5. A KP category $\boldsymbol{K}$ is a full subcategory of $\boldsymbol{K'}$ if a selection of objects on $\boldsymbol{K'}$ together with all the morphisms between them.

In knowledge processing, being subcategory means extracting partial structure and semantic information of another category without distortion.

Definition 6. For each morphism $f_t(a, b)$, there is an inverse $g_s(b, a)$ if

$$g_s(b,a) \circ f_t(a,b) = ID_a \text{ and } f_t(a,b) \circ g_s(b,a) = ID_b$$
$$\text{where } (t \times s = u^T) \subseteq R \text{ .} \tag{6}$$

Inverse of $f_t(a, b)$ is denoted as

$$f^{-1}{}_{t^{-1}}(b,a) \text{ .} \tag{7}$$

A morphism can have at most one inverse.

Definition 7. For each morphism $f_t(a, b)$, if it has an inverse, the morphism is said to be isomorphism, and then a and b are said to be isomorphic.

Definition 8. $\boldsymbol{K} = (O, M, T, R)$ is a KP category. Its opposite category $\boldsymbol{K}^{OP} = (O, M^{OP}, T^{OP}, R^{OP})$ is defined as

1. The object class O of $\boldsymbol{K}$ is same as that of $\boldsymbol{K}^{OP}$.
2. For each morphism $f_t(a, b)$ in M, there is only one inverse in M^{OP}.
3. For each type of $t = (sou, tar, val)$ in T, there is only one antonymous type

$$t^{-1} = (tar, sou, val^{-1}) \text{ in } T^{1}. \tag{8}$$

4. For each rule $r = comb\ (t, s)$ in R, there is also a converse rule a type $r^{-1} = comb\ (t^{-1}, s^{-1})\ t$ in R^{OP}.

The inverses of morphisms usually mean passivity or antonym.

Definition 9. Given two categories, $\boldsymbol{K_1} = (O_1, M_1, T_1, R_1)$ and $\boldsymbol{K_2}$, (O_2, M_2, T_2, R_2), a functor F: $\boldsymbol{K_1} \longrightarrow \boldsymbol{K_2}$ maps each morphism of $\boldsymbol{K_1}$ onto a morphism of $\boldsymbol{K_2}$. Functor F is defined as follows.

1. F is a homomorphism from M_1 to M_2.
2. F associates each type t in T_1 with a type Ft in T_2.
3. R associates each rule r in R_1 with a rule Fr in R_2.

F preserves identities. i.e. if ID_x is a identity morphism in $\boldsymbol{K_1}$, then $F(ID_x)$ is a identity morphism in $\boldsymbol{K_2}$.

F also preserves compositions. i.e.

$$F(f \circ g) = F(f) \circ F(g). \tag{9}$$

The matching in knowledge science is implemented using the matching between two knowledge pieces. Knowledge matching is the basal work of intelligent information retrieval, ontology construction and knowledge acquisition.

A matching between two KP can be achieved not only by keywords, but also by their semantic information and categorical structure. Because semantic information is correspondingly sufficiency in a local KP category, a knowledge matching includes meanings of concepts, types of relationships, rules of compositions, and structures of category. Following this approach, we can abstract a structure away from the contents of information.

Category theory creates relationships among categories using functors. However it is impossible to map each object and morphism between two KP categories. In the processing of knowledge matching, we are interested in the overlap of two KP categories, the degree of their overlapping and the methods of this overlapping. To achieve the matching purpose, we introduce an unusual concept Greatest Common Subcategory (GCS). If we got the GCS of two categories, we would have obtained the genuine quantitative value of KP category similarity. Therefore we have the definition of GCS in typed category theory as follows:

Definition 10. Given two KP categories, $\boldsymbol{K_1} = (O_1, M_1, T_1, R_1)$ and $\boldsymbol{K_2} = (O_2, M_2, T_2, R_2)$, the common subcategory $\boldsymbol{K_{12}} = (O_{12}, M_{12}, T_{12}, R_{12})$ is defined as follows:

1. O_{12} is the sub-object class, denoted as

$$O_{1,2} = \{o \mid o \in (O_1 \cap O_2) \cup O', O' \subset \overline{O_1 \cap O_2}\} \tag{10}$$

where O' is special object class; for each object a, $a \in O_1 \cap O'$, there exists at least one morphism $f_t(a, b)$, $b \in O_2 \cap O'$, which is a inherited relationship.

The size of O_{12} depends on the precision required through their inherited relationships in concept hierarchy. More specific concepts in concept hierarchy used means more precise, but lower knowledge matching value. Therefore the precision also needs to adjust the degree of concept abstraction.

2. M_{12} is the subset of intersection between M_1 and M_2 with the limitation of its types and objects, denoted as

$$\{f \mid typ(f) \subseteq typ(M_1 \cap M_2) \wedge dom(f) \subseteq O_1 \cap O_2 \wedge cod(f) \subseteq O_1 \cap O_2\}. \tag{11}$$

M_{12} induces the identity morphisms for each object in O_{12}, i.e. it may have a discrete subcategory in $\boldsymbol{K}_{12}$. If all morphisms in $\boldsymbol{K}_{12}$ are identity morphisms, the KP matching should be keywords matching.

3. T_{12} is the subset of the $T_1 \cup T_2$.

4. R_{12} is the subset of the $R_1 \cup R_2$.

Satisfying the largest sizes of classes and morphisms, the common subcategory is a greatest common subcategory (GCS).

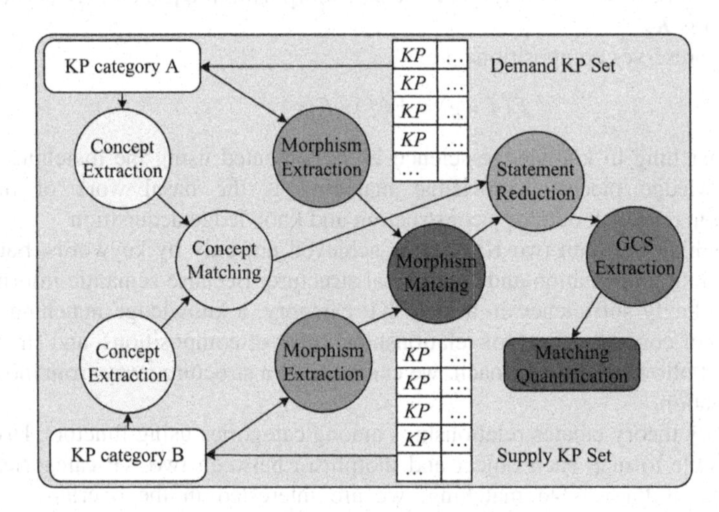

Fig. 1. Processing of Knowledge Matching

Figure 1 shows the processing of matching quantification. Firstly the concepts in these two KP are compared not only by their equivalences but also inherited relationships i.e. including concept combination; secondly morphisms are compared for acquiring same types of relationships, thirdly, statements are sorted and pared through the composition rules, and finally GCS is extracted which can be a criterion to quantize knowledge matching. Such matching can be achieved between emergency knowledge pieces of demand and supply, and also among the homologous sources.

3 Conclusion and Future Work

Based on typed category theory proposed by Ruqian Lu in 2005, we presented a new approach of knowledge representation through a micro-view of categorical knowledge structure, knowledge pieces. Categorical knowledge pieces are proved to facilitate emergency knowledge reorganization according to requirements of various decision-making scenarios. In the narrow sense, a knowledge piece is a special micro-ontology with relevant sufficient relationships among concepts. First we gave knowledge piece a formal definition in a categorical view in this paper. And then we applied this approach in knowledge similarity matching through the greatest common subcategory. Knowledge matching is a basic work in the fields of emergency ontology construction and emergency knowledge reorganization.

The research in this paper is basic for emergency knowledge management based on category theory. Future study will interest in knowledge reasoning on the categorical structure using categorical constructs, like pushout and colimit to reorganize emergency knowledge for support-making.

Acknowledgement

This research is supported by the Natural Science Foundation of China (Grant No, 70571011, 70431001).

References

1. Rong, L.L.: Reorganizing the Knowledge in Government Documents for Quick Response. In: KSS'2005, Laxenburg, Austria (2005)
2. Wang, Q.Q., Rong, L.L.: Agile Knowledge Supply Chain for Emergency Decision-Making Support. In: Shi, Y., van Albada, G.D., Dongarra, J., Sloot, P.M.A. (eds.) ICCS 2007. LNCS, vol. 4490, pp. 178–185. Springer, Heidelberg (2007)
3. ter Hofstede, A.H.M., Lippe, E., van der Weide, T.P.: Applications of a Categorical Framework for Conceptual Data Modeling. Acta Informatica 34, 927–963 (1997)
4. Hoare, C.A.R.: Notes on an Approach to Category Theory for Computer Scientists. Constructive Methods in Computing Science 55, 245–305 (1989)
5. Buneman, P., Naqvi, S., Tannen, V., Wong, L.: Principles of Programming with Complex Objects and Collection Types. Theoretical Computer Science 149(1), 3–48 (1995)
6. Wermelinger, M., Fiadeiro, J.L.: Connectors for mobile programs. IEEE Transactions on Software Engineering 24(5), 331–341 (1998)
7. Fiadeiro, J., Maibaum, T.: Categorical Semantics of Parallel Program Design, Science of Computer Programming. Science of Computer Programming 28(2), 111–138 (1997)
8. Zhozhikashvili, A.V., Stefanyuk, V.L.: Category-theoretical patterns for problems of artificial intelligence. Journal of Computer and Systems Sciences International 38(5), 677–687 (1999)

9. DeLoach, S.D., Hartrum, T.C.: A Theory-Based Representation for Object-Oriented Domain Models. IEEE Transactions on Software Engineering 26, 500–517 (2000)
10. Lenisa, M., Power, J., Watanabe, H.: Category Theory for Operational Semantics. Theoretical Computer Science 327, 135–154 (2004)
11. Krotzsch, M., Hitzler, P., Ehrig, M., Sure, Y.: Category Theory in Ontology Research: Concrete Gain from an Abstract Approach, http://www.aifb.uni-karlsruhe.de/WBS/phi/pub/
12. Hitzler, P., Krtzsch, M., Ehrig, M., Sure, Y.: What Is Ontology Merging -A Category-Theoretical Perspective using Pushouts[J]. American Association for Artificial Intelligence (2005)
13. Kent, R.E.: The IFF Category Theory Ontology (2002), http://www.ontologos.org/Papers/IJCAI2001
14. Brazhnik, O.: Databases and the geometry of knowledge. Data & Knowledge Engineering 61(2), 207–227 (2007)
15. Johnson, M., Dampney, C.N.G.: On Category Theory as a (meta) Ontology for Information Systems Research. ACM 377(4), 1–10 (2001)
16. Colomb, R.M., Dampney, C.N.G.: An approach to ontology for institutional facts in the semantic web. Information and Software Technology 47(12), 775–783 (2005)
17. Poli, R.: Ontological Methodology, Int. J. Human-Computer Studies 56, 639–664 (2002)
18. Herre, H., Heller, B.: General Formal Ontology (GFO), http://www.onto-med.de/en/theories
19. Healy, M.J., Caudill, T.P.: Ontologies and Worlds in Category Theory: Implications For Neural Systems. Axiomathes 16, 165–214 (2006)
20. Oles, F.J.: An application of lattice theory to knowledge representation. Theoretical Computer Science 249, 163–196 (2000)
21. Lu, R Q.: Towards a mathematical theory of knowledge. Journal of Computer Science and Technology 20, 751–757 (2005)

A Chinese Time Ontology*

Chunxia Zhang[1], Cungen Cao[2], Yuefei Sui[2], and Zhendong Niu[1]

[1] School of Computer Science and Technology, Beijing Institute of Technology, Beijing 100081, China
[2] Institute of Computing Technology, Chinese Academy of Sciences, Beijing 100080, China
cxzhang@bit.edu.cn, cgcao@ict.ac.cn, yfsui@ict.ac.cn, zniu@bit.edu.cn

Abstract. Temporal knowledge representation and identification are crucial in many application areas, such as Semantic Web, knowledge sharing and reusing, and natural language processing. Building time ontologies has been viewed as a promising means to solve this task. However, practical time ontology is closely related with specific nations and cultures, though they may share a common part. In this paper, we present a Chinese time ontology for web services and knowledge systems which involve Chinese temporal entities and temporal properties. First we define a core component called the base time ontology. Upon this base time ontology, we build the other part of the Chinese time ontology, which includes temporal measurement and representation in Chinese specific ways, and transformation between temporal entities in the Chinese time ontology and the DAML time ontology. We argue that the base time ontology is not only a basic and integral part of the Chinese time ontology, but also a base for constructing other time ontologies as well.

Keywords: Time ontology, semantic web, Chinese time ontology, temporal representation, temporal transformation.

1 Introduction

Representation and reasoning about time plays an important role in many application areas such as the Semantic Web, knowledge sharing and reusing, and natural language processing including information retrieval, information extraction, question answering, and machine translation. A time ontology, whether formal or informal, or whether explicit or implicit, is demanded to realize the Semantic Web, which aims to provide automated web services based on descriptions of the content and capabilities of Web resources. Currently, much effort has been paid on building explicit time ontologies, such as the DAML Time Ontology [1-4], KSL-time [5], time of DAML-S

* The first and fourth authors are supported by the Natural Science Foundation (grant no. 60705022), the Program for New Century Excellent Talents in Universities of China, and Beijing Institute of Technology Basic Research Foundation (grant no.411002). The second and third authors are supported by the Natural Science Foundation (grant no. 60273019, 60496326, 60573063, and 60573064) and the National 973 Program (grant no. 2003CB317008 and G1999032701).

Z. Zhang and J. Siekmann (Eds.): KSEM 2007, LNAI 4798, pp. 575–580, 2007.

[6], the TimeML Specification [7], the temporal portion of Teknowledge's IEEE Standard Upper Ontology [8], Kestrel [9], and Time and Dates of Cycorp's knowledge base [10].

In the history, Chinese people had their own ways of representing temporal entities. The current Chinese calendar uses both the Gregorian calendar and the traditional Chinese calendar. The former is a kind of solar calendar, and is used almost everywhere in the world, while the latter is a type of lunisolar calendar, and is mainly used in Chinese societies.

In building time ontologies, our observation indicates that time ontologies are closely related with specific nations or cultures, though they may share a common part. This is true if nations, e.g. the Chinese nation, have a long history. For example, in our work on agricultural knowledge acquisition and analysis [11], we found that ancient Chinese farmers used a timing system for farming (e.g. what crops and animals were planted and raised during what time?). Although no one exactly knows when the timing system was created, it is still one of the dominant timing systems in Chinese societies.

Present time ontologies such as the DAML time ontology have two shortcomings: (1) they do not consider calendars' affect on the time ontology. Since calendars are the basis of computing durations of time units and their relationships, and people build different methods of time counting and temporal representation on the foundation of different calendars. (2) They do not analyze special temporal representations, which are concerned with a country's culture or history.

In this paper, we present the Chinese time ontology (CTO) based on the current Chinese calendars: the Gregorian calendar and the traditional Chinese calendar. The ontology is intended for web services and knowledge systems, which involve Chinese temporal entities and temporal properties. We build our CTO in two steps. First, we build a core component, called *the base time ontology*, based on Gregorian calendar. Second, upon this base ontology, we develop the other part of CTO, which includes temporal measurement and representation in Chinese idiosyncratic ways, and transformation between temporal entities in CTO and the DAML time ontology. We will also argue that the base time ontology is not only a basic and integral part of the Chinese time ontology, but also a base for constructing other time ontologies.

The rest of this paper is organized as follows. Section 2 presents the base time ontology. Section 3 introduces temporal measurement and representation of CTO, and the conversion between temporal entities in the CTO and the DAML time ontology. Section 4 summaries the paper.

2 A Base Time Ontology

In this section we introduce the base time ontology. The ontology consists of topological temporal relations, clocks and calendars, which refers to the DAML time ontology [1-3], and Vila's work [12]. Here, we choose the Gregorian calendar as the foundation for the base time ontology, since we aim to develop the base time ontology as a common ontology for different time ontologies. Temporal primitives are instants and intervals, because the instant- and period-based time theory is more natural and intuitive, and can describe temporal knowledge more conveniently than a pure instant- or period-based time theory [12,13]. Axioms below give temporal relations, clocks and calendars.

1. Predicates instant(t) and interval(T) means that t and T are an instant and interval, respectively. Predicate temporal-entity(x) is to represent that x is a temporal entity.
   a) $\forall t$ (instant(t)$\rightarrow$temporal-entity(t))
   b) $\forall T$ (interval(T)$\rightarrow$temporal-entity(T))
   c) $\forall x$ (temporal-entity(x)$\rightarrow$ instant(x)$\vee$interval(x))
2. There is a before relation on instants, which is irreflexive, asymmetric, transitive and linear. Predicate before(t_1,t_2) indicates that t_1 is before t_2.
   a) $\forall t_1$ ($\neg$before(t_1,t_1))
   b) $\forall t_1$ $\forall t_2$ (before(t_1,t_2)$\rightarrow\neg$before(t_2,t_1))
   c) $\forall t_1$ $\forall t_2$ $\forall t_3$ (before(t_1,t_2)$\wedge$before(t_2,t_3)$\rightarrow$before(t_1,t_3))
   d) $\forall t_1$ $\forall t_2$ (before(t_1,t_2)$\vee$before(t_2,t_1)$\vee$(t_1=t_2))
3. startFn(T) and endFn(T) are two functions, representing the starting and ending instants of T, respectively. Axioms c) and d) say that there exist exactly a starting instant and an ending instant for each interval.
   a) $\forall T$ $\forall t$ ((startFn(T)=t)$\rightarrow$interval(T)$\wedge$instant(t))
   b) $\forall T$ $\forall t$ ((endFn(T)=t)$\rightarrow$ interval(T)$\wedge$instant(t))
   c) $\forall T$ $\forall t_1$ $\forall t_2$ ((startFn(T)=t_1)$\wedge$(startFn(T)=t_2)$\rightarrow$$t_1$=$t_2$)
   d) $\forall T$ $\forall t_1$ $\forall t_2$ ((endFn(T)=t_1)$\wedge$(endFn(T)=t_2)$\rightarrow$$t_1$=$t_2$)
   e) $\forall T$ $\forall t_1$ $\forall t_2$ ((startFn(T)=t_1)$\wedge$(endFn(T)=t_2)$\rightarrow$before(t_1,t_2))
4. Predicate inside(t,T) denotes that t is an instant within T.

   inside(t,T) $\leftrightarrow$ (t=startFn(T))$\vee$(t=endFn(T))$\vee$
   (before(startFn(T), t)$\wedge$before(t, endFn(T)))
5. Temporal relations between intervals can be classified into 7 types of relations.
   a) precedes(T_1,T_2)$\leftrightarrow$before(endFn(T_1), startFn(T_2))
   b) meets(T_1,T_2)$\leftrightarrow$endFn(T_1)=startFn(T_2)
   c) overlaps(T_1,T_2)$\leftrightarrow$before(startFn(T_2),endFn(T_1))$\wedge$before(startFn(T_1),startFn(T_2)) $\wedge$before(endFn(T_1), endFn (T_2))
   d) starts(T_1,T_2)$\leftrightarrow$(startFn(T_1)=startFn(T_2))$\wedge$before(endFn(T_1), endFn(T_2))
   e) during(T_1,T_2)$\leftrightarrow$before(startFn(T_2), startFn(T_1))$\wedge$before(endFn(T_1),endFn(T_2))
   f) finishes(T_1,T_2) $\leftrightarrow$(endFn(T_1)=endFn(T_2))$\wedge$before(startFn(T_1), startFn(T_2))
   g) equal(T_1,T_2)$\leftrightarrow$(startFn(T_1)=startFn(T_2))$\wedge$(endFn(T_1)=endFn(T_2))
6. durationFn(T, u) is the amount of time within T as measured in unit u, and TemporalUnit(U) shows that U is a temporal unit as a duration.
   a) durationFn(T, u)$\rightarrow$Interval(T)$\wedge$TemporalUnit(u)
7. The Gregorian calendar uses a tropical year as the basic cycle. Astronomically, a year, month and day in the calendar is a solar year, solar month and solar day. For a solar year y, a solar month m and a solar day d, we have
   a) startFn(y)=0:00am on the solar day when T happens, where T means the event that the sun returns to the vernal equinox
   b) endFn(a solar year)=12:00pm on the solar day before the solar day on which the sun returns to the vernal equinox next time
   c) durationFn(y, d)=365.2422
   d) durationFn(T, y)$\times$12=durationFn(T, m))

3 Chinese Time Ontology

Our main task in developing the Chinese time ontology is to define instants, intervals and their interpretations in the traditional Chinese calendar. Whether a time is an instant or interval is generally a pragmatic question. For example, we could consider the date January 1, 2007 as an instant, and say that an event occurs on that day; or we could consider the date as an interval, and say that an event occurs during that day.

In the traditional Chinese timing system, there are nine principal time units: huajia(花甲), ji(纪), lunar year, quarter, lunar month, xun(旬), lunar day, shichen(时辰), and geng(更). A timing unit u as a calendar interval has its starting and ending instants, while as a duration it can be any interval whose length is durationFn(u). It is pointed out that a day in the traditional Chinese calendar is also a solar day, but the day differs from a day in the Gregorian calendar in starting and ending instants. So we use *a lunar day* for a day in the traditional Chinese calendar.

a) startFn(a solar day)=0:00am
b) startFn(a lunar day)=11:00pm on a solar day before the solar day.

3.1 Representation and Transformation of Lunar Years

This subsection will introduce three representing systems for lunar years: *the Stem-Branch System, the Twelve Animals Zodiac and Terrestrial Branch System*, and *the Emperor's Title and Reign Title system.* The first system combines ten sequential celestial stems with twelve sequential terrestrial branches as shown in Table 1. (1) The sixty stem-branch combinations of celestial stems and terrestrial branches constitute a cycle of 60 lunar years, starting from *Jiazi* (a combination of Jia(甲) and Zi(子)) and ending on *Guihai*. Each combination actually denotes a class of lunar years. For example, A.D.2007 and 2067 are instances of *Dinghai*. (2) The Animals Zodiac and Terrestrial Branch System forms a cycle of 12 lunar years. Each animal represents a class of lunar years with an identical terrestrial branch. For example, A.D. 2007 and A.D.2019 are instances of the *Zodiac Class Pig*, because terrestrial branches of these years are exactly the same, i.e., *Hai*. (3) In Chinese history, most emperors used their titles or reign titles to represent lunar years. Each emperor in a dynasty started counting lunar years by his or her first year of reign as the first lunar year using of the emperor's title and reign titles.

Table 1. The 10 Celestial Stems and 12 Terrestrial Branches in the Stem-Branch System

Celestial Stems	Jia	Yi	Bing	Ding	Wu	Ji	Geng	Xin	Ren	Gui		
	1	2	3	4	5	6	7	8	9	10	11	12
Terrestrial Branches	Zi	Chou	Yin	Mao	Chen	Si	Wu	Wei	Shen	You	Xu	Hai

In order to use our Chinese time ontology practically, an instant *t* is required to be identifiable. For example, if t refers to an hour/shichen in a solar/lunar day, then the solar/lunar year, month and day to which the hour/shichen belongs is given in t. Based on this, we introduce get-year(t, solar/lunar), get-month(t, solar/lunar), get-day(t, solar/lunar), get-hour(t, solar), and get-shichen(t, lunar). For a given instant t in the

Gregorian calendar, we can calculate its corresponding instant in the traditional Chinese calendar, based on TCT_1, TCT_2, and TCT_3.

TCT_1. ¬before(y, A.D.1) ∧ y=get-year(t, solar)
→stem-branch-no(y, 60–mod(3-mod(y, 60), 60))

TCT_2. before(y, B.C.1) ∧ y=get-year(t, solar)
→stem-branch-no(y, 60–mod(mod(y, 60)+2, 60))

TCT_3. solar-year(y) ∧ stem-branch-no(y, n)→celestial-stem-no(y, 10–mod(10-n, 10))∧terrestrial-branch-no(y, 12–mod(12–n, 12))

Here, the predicate solar-year(y) indicates y is a solar year, the function mod(x, n) calculates the remainder of an integer x divided by n. The predicate stem-branch-no(y, n) means the stem-branch combination of y, as a lunar year, or month, or day, is the nth index of the stem-branch counting cycle. Accordingly, we define celestial-stem-no(y, n) and terrestrial-branch-no(y, n).

3.2 Representation and Transformation of Lunar Months, Days and Hours

There are two methods of naming months for a given lunar year: *the ordinal numbers* and *the Stem-Branch System*. From the solar year y, we can compute the Stem-Branch representation of lunar months in y based on TCT_4, TCT_5, and TCT_6.

TCT_4. m=get-month(t, lunar) ∧ y=get-year(t, lunar) ∧ celestial-stem-no(y, n)
→celestial-stem-no(m, mod(2×mod(n, 5)+m, 10))

TCT_5. m=get-month(t, lunar) →terrestrial-branch-no(m, mod(m+2, 12))

TCT_6. m=get-month(t, lunar) ∧ (mod(y, 5)=(z+5)/2)
→ terrestrial-branch-no(m, mod(z+m-1, 10))

Three main ways for representing days are: *the ordinal number approach*, *the Stem-Branch system* and *the phases of the moon approach*. The day representing system of phases of the moon uses several special phases during each lunar month to represent days. These phases include *Shuo*(朔, new moon), *Wang*(望, full moon), *Jiwang*(既望, the day after new moon) and *Hui*(晦, the last day of a lunar month). Meanings of these phrases are given below, where the function get-moon-phase(t) represents the moon phase used in each instant t.

y=get-year(t, lunar) ∧ m=get-month(t, lunar) ∧ mp=get-moon-phase(t)
→(mp=Shuo → get-day(t, lunar)=1) ∨(mp=Wang → get-day(t, lunar)=15)∨
(mp=Jiwang→ get-day(t, lunar)=16) ∨
(mp=Hui → get-day(t, lunar)=get-last-lunar-day(y, m)))

There exist three common ways of expressing hours: *the color of the sky approach*, *the Stem-Branch system* and *the ordinal number approach*. A day is averagely divided into twelve segments, and each segment is called a *Shichen*. The twelve Shichen are named *Yeban*(夜半), *Jiming*(鸡鸣), *Pingdan*(平旦), *Richu*(日出), *Shishi*(食时), *Yuzhong*(隅中), *Rizhong*(日中), *Ridie*(日昳), *Bushi*(晡时), *Riru*(日入), *Huanghun*(黄昏), and *Rending*(人定). For example, let t be an instant in which Jiming occurs, we have

(sc=get-shichen(t)) ∧ (sc=Jiming)
↔startFn(sc)=1:00am∧endFn(sc)=3:00am ∧ duration(sc, hour)=2

The Stem-Branch system is also used to denote hours. Each Shichen corresponds to a fixed terrestrial branch. The celestial stem of a Shichen can be determined by the celestial stem of the day in which the Shichen is, as shown in TCT_7.

TCT_7. (sc =get-shichen(t)) $\wedge$(d=get-day(t, lunar)) $\wedge$celestial-stem-no(d, n) $\rightarrow$celestial-stem-no(sc, mod(2$\times$mod(n, 5)-1, 10))

The representing system of the Gregorian calendar can be treated as the "*time communicator*" for temporal entities represented by various calendars. That is, the temporal entity in the traditional Chinese calendar can be converted into one in the Gregorian calendar, and it can further be transformed into one in other calendars.

4 Conclusion

In a nation with a long history, the time ontology is generally not purely based on the Gregorian calendar; therefore a special treatment is necessary in developing a time ontology for its relevant web services and knowledge systems. In this paper, we introduced, in two steps, a Chinese time ontology for knowledge systems involving Chinese temporal entities or objects. First, we developed a base time ontology based on the Gregorian timing system. Then we introduced a Chinese time ontology, which consists of temporal measurement, representation, and conversion between temporal entities in the Gregorian and the traditional Chinese timing systems. Our development method of time ontology could be used in other nations, and especially the base time ontology can be reused for other time ontologies.

References

1. Hobbs, J.R., Pan, J.: An Ontology of Time for the Semantic Web. ACM Transactions on Asian Language Information Processing 1, 66–85 (2004)
2. DAML-Time Homepage, http://www.cs.rochester.edu/~ferguson/daml
3. Hobbs, J.R.: A DAML Ontology of Time (2002), http://www.cs.rochester.edu/~ferguson/daml/daml-time-nov2002.txt
4. Bohlen, M.H., Gamper, J.: How Would You Like to Aggregate Your Temporal Data? In: 13th International Symposium on Temporal Representation and Reasoning, pp. 1–136 (2006)
5. KSL-Time (2000), http://www.ksl.stanford.edu/ontologies/time
6. DAML-S 0.9 Draft Release (2003), http://www.daml.org/services/daml-s/0.9/
7. Pustejovsky, J.: Time and the Semantic Web. In: 12th International Symposium on Temporal Representation and Reasoning, pp. 5–8 (2005)
8. Time Ontology Needed (2002), http://ext4-www.ics.forth.gr/mail/ontoweb-sig1/0012.html
9. Kestrel: Time Ontology (2000), http://www.kestrel.edu/DAML/2000/12/TIME.daml
10. Time and Dates (2002), http://www.cyc.com/cycdoc/vocab/time-vocab.html
11. Cao, C.G., et al.: Progress in the Development of National Knowledge Infrastructure. Journal of Computer Science and Technology 5, 523–534 (2002)
12. Vila, L.: A Theory of Time and Temporal Incidence based on Instants and Periods. In: 3rd International Workshop on Temporal Representation and Reasoning, pp. 21–28 (1996)
13. Allen, J.F.: Time and Time Again: The Many ways to Represent Time. The International Journal of Intelligent Systems 6, 341–356 (1991)

Toward Patterns for Collaborative Knowledge Creation

Haoxiang Xia[1], Zhaoguo Xuan[1], Taketoshi Yoshida[2], and Zhongtuo Wang[1]

[1] Institute of Systems Engineering, Dalian University of Technology, Dalian, 116024 China
{hxxia, zgxuan, ztwang}@dlut.edu.cn
[2] Graduate School of Knowledge Science, Japan Advanced Institute of Science and Technology, Ishikawa, 923-1292 Japan
yoshida@jaist.ac.jp

Abstract. The current organizational knowledge creation models are largely limited by the insufficiency for practically guiding the real-world knowledge creation processes. To overcome this, the authors suggest investigating workable knowledge creation patterns in groups, organizations, and inter-organizational communities, inspired by the ideas of problem-solving methods in knowledge engineering and design patterns in software engineering. The ideas of patterns are further connected to the concept of schemas in cognitive psychology, indicating that the patterns actually reflect a fundamental means for human intelligence to handle the complexities of real-world problem- solving. To promote further research on knowledge creation patterns, some basic considerations are given in the present paper. First, the collaborative knowledge creation patterns are classified into two categories, i.e. the pattern of problem solutions and the patterns about how the collaborative groups being organized. Second, a rough dynamic model of pattern development is given.

Keywords: Collaborative knowledge creation; patterns; schema.

1 Introduction

Following Nonaka and Takeuchi's [1] famous work on the "knowledge-creating company", organizational knowledge creation has become a hotspot in the management community. Notable progresses have been made to understand how a company obtains its core capabilities and sustainable competitiveness through knowledge conversion and creation and organizational learning (e.g. [2][3]). All these efforts are doubtlessly valuable. However, as contended by Zhu [4], they emphasize too much on theorizing concepts and models, whilst it is to some extent neglected whether the proposed models are applicable in real-world contexts.

To overcome the practical limitations of the existing organizational knowledge creation theories, in the present paper we suggest exploiting situation-specific knowledge creation patterns. The basic observation motivating our suggestion is that there are different knowledge creation modes specific to different problem situations; thus the "universal" knowledge creation theories may not be directly applicable in the

Z. Zhang and J. Siekmann (Eds.): KSEM 2007, LNAI 4798, pp. 581–586, 2007.

real world until they can be specialized to fit the actual problem situations. Therefore, it is of practical importance to identify and specify the recurring problem situations and the corresponding knowledge creation modes or patterns to ease the actual knowledge creation processes. This notion is fundamentally consistent with the ideas of the "problem-solving methods" or PSM [5] in knowledge engineering (KE) and the "design patterns" [6] in software engineering (SE). Especially, in the history of artificial intelligence, there was a shift of the research focus from searching "general problem solvers" in 1950s and 60s to collecting situation-specific "problem-solving methods" in 1980s and 90s. We suspect that a similar shift may happen in knowledge management, from pursuing generic knowledge creation models to exploring the situation-specific patterns of organization knowledge creation.

Hence, in this paper we try to give a glimpse at the concepts of "problem-solving methods" in knowledge engineering and "design patterns" in software engineering. This observation leads to a further argument that patterns reveal a very fundamental means for human intelligence to handle the complexity of real-world problem-solving, regarding patterns as externalization of schemas in cognitive psychology. With respect to this viewpoint, we pose the perspective that discovering and utilizing the knowledge creation patterns would possibly form a worthwhile direction for studying collaborative knowledge creation, and give some primitive considerations on this direction. Here by collaborative knowledge creation we refer to knowledge creation in teams, organizations and inter-organizational communities, which covers more situations than the term "organizational knowledge creation" does.

2 About Patterns in Knowledge and Software Engineering

As stated in the introduction section, our suggestion to explore situation-specific knowledge creation patterns is inspired by the developments of "problem-solving methods" in knowledge engineering and "design patterns" in software engineering. To promote further discussion, in this section we give a short description for the basic ideas behind these two concepts. Furthermore, we connect these two concepts to the concept of schema in cognitive psychology, indicating that the pattern idea reflects a basic means for human intelligence to handle the complexities of problem solving.

The idea of problem-solving methods was firstly conceived as "the abstract patterns that constitute a common, reusable problem solving procedure for a task such as diagnosis in different domains" [7]. Early efforts on exploiting PSM may include "heuristic classification" [8], "Propose and Revise" [9] and the "generic tasks" by Chandrasekaran [10], etc. Since the late 1980s, more work on the development and utilization of PSM has emerged (e.g. [11][12][5]) in the KE community. The essential ideas behind PSM are on one hand to reuse the patterns of situation-specific reasoning procedures to different application domains so as to ease the development and maintenance of expert systems, on the other hand to facilitate knowledge acquisition by modeling expert knowledge according to the task and PSM structures, instead of modeling how the expert thinks.

"Design patterns" in software engineering was stemmed from the architect Christopher Alexander's [13] concepts of patterns and pattern languages in building and city planning. Facing the complexities of software development, software engineers learned from Alexander and tried to summarize the past successful experiences in some uniform formats in order to reuse them in future applications. A landmark contribution of "design pattern" was Gamma et. al.'s book [6], which specified 23 concrete design patterns recurring in very different programming problems; their patterns, amongst the others', have been widely adopted in the software community. Design patterns and other types of software patterns have become a major focus in the software engineering community since the 1990s; and the idea of patterns in programming has been further extended to the full lifecycle of software development (e.g. [14]).

The conceptual roots behind PSM and design patterns are essentially identical, i.e. to explicitly represent the successful experiences in the past designs in order to reuse the experiences in new applications, and to ease the transfer of the expertise among human groups. Such ideas are actually simple and intuitionistic; however, they reflect the very basic means of problem-solving by human intelligence and may have far-reaching implications beyond architecture and knowledge and software engineering. In our understanding, the concept of patterns can be related to the concept of schemas in cognitive psychology, which has had profound impacts on this academic discipline since Piaget's cognitive development theory [15]. As summarized by Marshall [16], schemas are major knowledge structures influencing the way people acquire and store information; and they provide fundamental mechanisms for long-term memory and problem-solving. The evolution of schemas finally causes an individual's cognitive development. Analogously, viewing patterns as "schemas" of a group cognitive system, we suspect that they are the substrate of group cognition:

1) Patterns form and evolve during repetitive collaborative problem-solving processes, as schemas are developed during repetitive personal problem-solving.
2) Being generated during collaborative problem-solving, patterns can be regarded as the basis of long-term "memory" of a group cognitive system, analogous to schemas being the basis of personal long-term memory.
3) Patterns reflect a basic way for handling the complexity of real-world problems. patterns store the successful experiences of a group or social cognitive system. When a new problem that fits a particular pattern is encountered, the pattern can be activated to solve that problem.
4) Patterns provide a basic means for structured communications an knowledge transfer within a group or community.

From knowledge creation point of view, accumulation of patterns is thus important for the success of knowledge creation in groups, organizations and other sorts of communities.

3 Considerations on Knowledge Creation Patterns

With patterns being viewed as the foundation of group cognition, it can be inspired that the actual knowledge creation may be smoothened by summarizing, transferring and reusing the fundamental knowledge creation patterns that fit specific problem situations. Thus, we conjecture that the studies on knowledge creation patterns may possibly grow to form a new research direction of knowledge management. However, to form such a research direction, much work needs to be done. In this section, we just attempt to give our very primitive thinking on developing this research direction.

First, one key question is from which aspects the patterns for collaborative knowledge creation can be discovered. To answer this, a categorization of the patterns is necessary. In the context of collaborative knowledge creation, we believe that two categories of patterns should be of concern, namely the patterns of solutions of specific problem types and the patterns about how people to be organized to solve a particular type of problems. The previously-noted software patterns in SE and PSM in KE can be regarded as examples of the first category of patterns. To a more general level, some basic strategies and methods of general problem solving may be regarded as candidates of knowledge creation patterns. For example, one candidate pattern is "depth-first searching", which is about locating some particular items in a tree-like structure by using the depth-first traversing process. Another candidate pattern is "decomposition and assembly". To solve a problem with this pattern, people firstly divide a complex problem into several simpler sub-problems and solve them separately, and then assemble the sub-solutions to form the overall solution. In contrast to the previous general-level "problem solution patterns", more concrete patterns are to be identified to fit more specific problem situations.

The second category of patterns (i.e. organizational patterns) are largely neglected in the pattern-centered movements in SE and AI. However, the exploitation of effective organizational modes is worthwhile. A few organizational models of software development teams may potentially be the patterns of this category. One example is the team development mode adopted by the Google company. Such a "self-organized team", as we call it, is formed around a new idea proposed by an employee. The idea may attract a few other interested employees to join and thus to form an agile team to realize the original idea by constructing a software product. This team organization mode may potentially be adopted by the R&D projects in other industries than software and thus it can be regarded as a candidate of organizational patterns. Thorough investigations of the organizational and managerial modes in different projects of different business sectors may lead to more applicable organizational patterns for collaborative knowledge creation.

Our second consideration is about the dynamics of the development and use of knowledge creation patterns. In our understanding, knowledge creation patterns are on one hand developed through the accumulation of experiences in the solving of the real-world problems; on the other hand, the patterns are ultimately used to assist real-world problem-solving and knowledge creation activities. A cyclic process of pattern development may exist. Furthermore, as patterns are abstract solutions to fit generic problem situations, in many cases some mid-level constructs may be needed to bridge the abstract patterns and the concrete knowledge creation processes. We call such mid-level constructs as "frameworks". The major difference between "patterns" and "frameworks" lies in that patterns are domain-independent, while frameworks are commonly bounded to the specific problem domains. The distinction of abstraction levels of patterns, frameworks, and the actual knowledge creation processes can be compared to the distinction of the abstract problem-solving methods, the expert system shell, and the actual expert systems. With the mediation of the "frameworks", a dual learning cycle exists in the development and use of the collaborative knowledge creation patterns, as illustrated by the following figure.

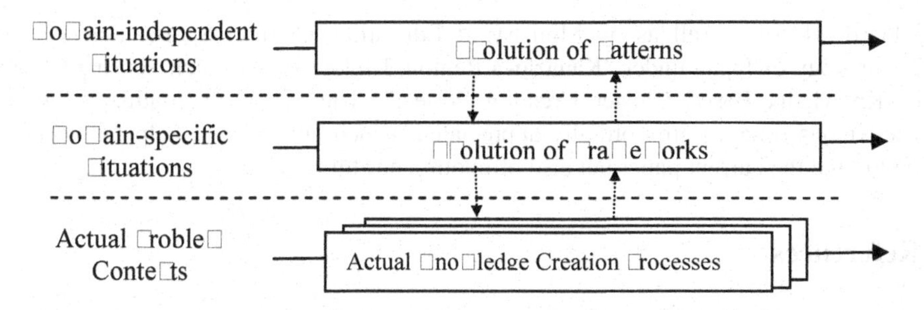

Fig. 1. Illustration of Pattern-based Knowledge Creation

Figure 1 depicts a rough dynamic model of pattern development and use. In this model two hidden processes are associated with an actual knowledge creation process, namely the evolution process of frameworks and the evolution process of patterns. Consequently, there are two learning cycles that links these three processes, i.e. the learning cycle between the actual knowledge creation process and the evolution process of frameworks, and the learning cycle between the evolution process of frameworks and the evolution process of patterns. When a pattern forms, it can direct the generation of a set of frameworks, and furthermore facilitate the actual knowledge creation processes through the functions of the generated frameworks. On the other hand, the actual knowledge creation processes may often encounter new problem situations that are not fully covered by the existing frameworks and patterns; thus the accumulating experiences in the solving of real-world problems would in turn cause the gradual changes of frameworks and patterns. When an encountered new problem type cannot be fitted by the self-adjustment of the existing patterns, the dynamic process may eventually lead to the creation of a new pattern.

4 Conclusions

In this paper the authors advocate to explore the patterns for facilitating collaborative knowledge creation, inspired by the developments of the "design patterns" in software engineering and the problem-solving methods in knowledge engineering. We believe this is a worthwhile direction for knowledge management. However, the aim of the present paper is more on postulating the problem than on offering a solution. What is discussed is just a rough vision and a very primitive step toward "pattern-based collaborative knowledge creation"; and much work needs to be done to develop this research field. For the further work in the near future, the authors would first, as a proof-of-concept research, carry out some case studies in the R&D contexts, especially in some R&D projects of software development and automobile design, in order to extract applicable collaborative knowledge creation patterns. Furthermore, we would explore a systematic set of knowledge creation patterns or a pattern language for facilitating knowledge creation in specific problem situations.

Acknowledgments. This work is in part supported by National Natural Science Foundation of China under Grants 70431001 (as Key Project), 70301009, and

70620140115; as well as by Ministry of Education, Culture, Sports, Science and Technology of Japan under "Kanazawa Region, Ishikawa High-Tech Sensing Cluster of Knowledge-Based Cluster Creation Project". The insightful comments by the anonymous referees are cordially appreciated, which are of great help for not only improving the current paper but also promoting our future work.

References

1. Nonaka, I., Takeuchi, H.: The Knowledge-Creating Company. Oxford University Press, New York (1995)
2. Cook, S.D., Brown, J.S.: Bridging Epistemologies: the Generative Dance between Organizational Knowledge and Knowing. Organization Science 10, 381–400 (1999)
3. Wierzbicki, A.P., Nakamori, Y.: Creative Space: Models of Creative Processes for the Knowledge Civilization Age. Springer, Berlin (2005)
4. Zhu, Z.: Needed: Pragmatism in KM. In: KSS 2006. 7th International Symposium on Knowledge and Systems Sciences, pp. 271–272. Global-Link Publisher, Hong Kong (2006)
5. Fensel, D.: Problem-Solving Methods: Understanding, Description, Development, and Reuse. Springer, Berlin (2000)
6. Gamma, E., Helm, R., Johnson, R., Vlissides, J.: Design Patterns: Elements of Reusable Object-Oriented Software. Addison-Wesley, Reading, MA (1995)
7. Clancey, W.: Model Construction Operators. Artificial Intelligence 53, 1–115 (1992)
8. Clancey, W.: Heuristic Classification. Artificial Intelligence 27, 289–350 (1985)
9. Marcus, S., Stout, J., McDermott, J.: VT: An Expert Elevator Designer That Uses Knowledge-based Backtracking. AI Magazine 9, 95–111 (1988)
10. Chandrasekaran, B.: Generic Tasks in Knowledge-Based Reasoning: High-level Building Blocks for Expert Systems Design. IEEE Expert 1(3), 23–30 (1986)
11. Wielinga, B.J., Schreiber, A.T., Breuker, J.A.: KADS: A Modelling Approach to Knowledge Engineering. Knowledge Acquisition 4, 5–53 (1992)
12. Eriksson, H., Shahar, Y., Tu, S., Puerta, A., Musen, M.: Task Modeling with Reusable Problem-Solving Methods. Artificial Intelligence 79, 293–326 (1995)
13. Alexander, C.: The Timeless Way of Building. Oxford University Press, New York (1979)
14. Yacoub, S., Ammar, R.H.: Pattern-Oriented Analysis and Design: Composing Patterns to Design Software Systems. Addison-Wesley, Reading, MA (2003)
15. Piaget, J.: The Children's Conception of the World. Routledge, London (1929)
16. Marshall, S.P.: Schemas in Problem Solving. Cambridge Univ. Press, Cambridge, UK (1995)

Service-Mining Based on Knowledge and Customer Databases

Yan Li[1], Peng Wen[2], Hu Wang[3], and Chunqiang Gong[3]

[1] The Department of Mathematics and Computing, [2] Faculty of Engineering and Surveying,
The University of Southern Queensland,
QLD 4350, Australia
{liyan, pengwen}@usq.edu.au
[3] School of Information System, Wuhan University of Technology,
Wuhan, P.R. China
{wangh, gcq}@263.net

Abstract. This paper addresses a service-mining technique and applies this technique to improve the services of vehicle service centers. We propose a service-mining system and its data structure to discover the most important services required through analyzing service records, feedback records and the available products. The system can improve the quality of mining automatically by updating mining strategies regularly.

Keywords: Service-mining, Behavior prediction, Vehicle service, Customer Relationship Management.

1 Introduction

Currently, many application systems, such as enterprises resource planning (ERP), customer relationship management (CRM), run in large companies are lack of the capabilities to mine from their database with some special service guidelines to find out the best service opportunity and the best item of service for their customers. Although technologies of data mining provide us ways to find the relevant knowledge from massive data, it is not suitable to those companies who must provide their services regularly and the required services totally depend on the conditions of their products. What we want to achieve in this research is to find a method to predict the best time and the best item of service for customers according to their driving behaviors and the service provided previously [1].

This paper proposes a system framework that discovers the most important services required by analyzing and combining specific products and an expert system with service records and feedback records. Moreover, the system can improve its mining quality by updating strategies regularly.

The rest of the paper is organized as follows. Section 2 describes the structure of the service-mining system (SMS) and Section 3 analyses the method to mining the services based on knowledge and customer databases. Finally, the paper is concluded in Section 4.

Z. Zhang and J. Siekmann (Eds.): KSEM 2007, LNAI 4798, pp. 587–592, 2007.

2 System Structure and Database

An SMS is a system that predicts a suitable time and an item of service through the driving behaviour of a customer. The predicting comes from the data processed based on customer database. This can be adjusted by the system itself as the new service records are added into customer database. Figure 1 shows the system structure.

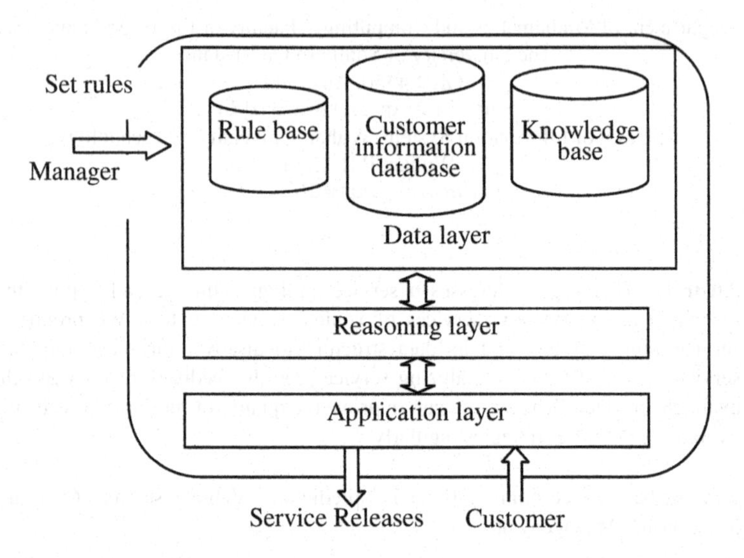

Fig. 1. The structure of SMS

An SMS consists of data layer, reasoning layer and application layer. Data layer is mainly composed of three databases: rule base, customer information and knowledge databases. The function of reasoning layer is to analyze and process the information in the rule base and customer information database using the service-mining model to discover potential service patterns [2], for example, when a part of a customer's vehicle needs to be replaced or checked. The system can inform a customer about a upcoming service within three days to one week in advance. In the meantime, the reasoning layer needs to analyze the service feedback from customers, to update the predicting model and the knowledge in the database with a particular structure [3]. An accurate customer's classification will help the system to predict a customer's driving behavior. Application layer provides user interfaces for different functions, collects customer's feedback, and releases services by means of some digital communications [4], such as e-mail, SMS, telephone, fax, etc.

3 The Service Mining Method

The service mining method involves several components, such as the evaluation of services and prediction of conditions for a service. In this paper, we describe how to predict customer's vehicle mileage and control the quality of services.

3.1 Prediction Model

The prediction model is used to find out what is the most possible time and the most suitable service a service provider can provide to their customers. The time and the service vary depending upon the information of the customers and their vehicles. If two customers own different vehicles, or if they purchase their vehicles at different time, or they have different driving behaviors and road conditions, the service and the time are also totally different. As a service provider has normally more than tens of thousand customers, it is no doubt that they must work on having automatic solutions for their customers if they want to provide them a suitable service in time. Their solutions should be able to update over time. Therefore, a prediction model is essential to determine the two important pieces of information: the service time and the service type.

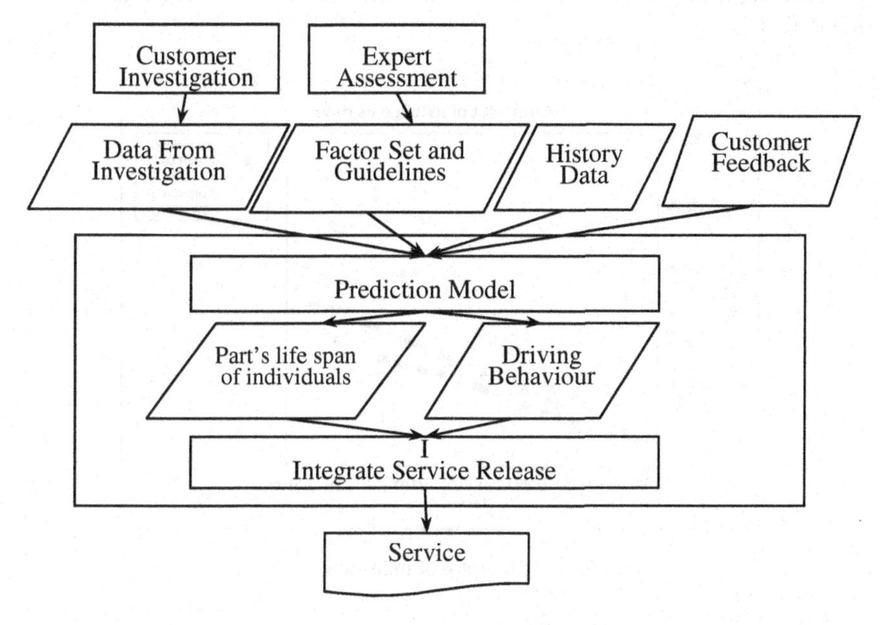

Fig. 2. The process for mining service

The process for mining service as shown in Fig. 2 has the following steps:

1. To obtain customer's driving behavior by analyzing the customer feedback information;
2. To obtain the most possible life-spans of the vehicle parts for different customers by referring the standard life-spans of the parts and the driving behavior of individual customers.
3. To obtain the most possible time a particular part of a vehicle requests to be maintained or replaced according to the result in Step 2 and the vehicle's maintenance history.
4. To issue the service suggestion to the service instructor and the customer.

3.2 The Customer Driving Mileage Function

There are many ways to estimate the values of the mileage per day for customers. The simplest way is to use a one-variable linear regression [5]. In the customer service database, a series of service history records for a particular customer can be easily filtered for regression. Every time when a customer sends his vehicle for service, the mileage and the time are recorded. As the relationship of mileage and time represents as a linear function, it is easily to predict that what time a particular part of the customer's vehicle reaches its maintaining or replacing time. Because the function is different from customer to customer, the function for a particular customer needs to be saved in the database for late prediction.

Fig. 3 shows the relationship between mileage and time for four customers whose history records are randomly picked up in our database. The one-variable linear regression algorithm is then applied to the data. The simulation is carried out in Minitab R14.

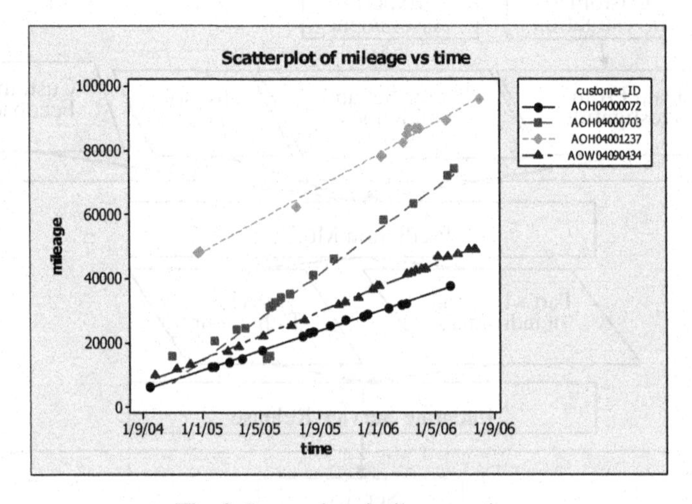

Fig. 3. Scatterplot of mileage vs time

As shown in Fig. 3 the regression lines are different depending on the service history records of individual customers. When some new services are added in, the regression lines are updated automatically.

3.3 To Discover the Comprehensive Factor for All Customers

As mentioned before, all drivers have their special driving experiences. Some of them always drive on the rough surfaces of roads, some like sudden braking. Those driving experiences have the great influences on the part's life span of a vehicle. This is why some drivers have to replace their tyres earlier than others.

It is assumed that there are n factors affecting m parts of a vehicle to different degrees (the standard values are given by experts), however, a customer's driving behavour makes the levels of the influence change. At the very beginning, we set all factors for a customer as value 1. It means that we think the standard values of all factors are suitable for those new customers because there are no maintenance records in our database available to mining for the purpose of the factor updates. As the new maintenance records are added into database, we can analyze them using our model to adjust the values of the factors for a particular customer.

If *i, j, k* represent the number of parts, factors, and customers. The comprehensive factor CF_{ki} can be determined by the following equation:

$$CF_{ki} = \frac{\sum_{j=1}^{n} c_{kj} w_{ij}}{\sum_{j=1}^{n} p_{ij} w_{ij}}. \tag{1}$$

Where, c_{kj} is a parameter based on customer's driving behavior through customer's feedback. w_{ij} is the initial weight value of part *i* under factor *j*. p_{ij} is an index related to the standard life-span of part *i*. *n* is the total parts a service provider is able to serve.

3.4 Integrated Service Release

Integrated service releases use production specifications, sometimes called *if-then* rules. The rules can take various forms, for example, *if* <condition>, *then* <action>. The *'if clause'* consists of a set of conditions. These conditions are much more complex than a symbol or a simple statement. There may be many conditions, all of which must be established as "*true*" in the current state of the machine. And, the *'then clause'* is a sequence of action schema whose variables are bound by the values that are passed from the bindings that have been established in the process of matching the variables that occurred in the *'if clause.'* A typical rule in SMS is as follows:

If $T_i = \frac{L_i}{r} + t_i$, then the service for a customer's vehicle: part *i* needs to be checked or replaced at the time T_i according to our prediction model. Here, T_i is the date of the next time part *i* will be checked or replaced. L_i means the most possible life-span of part *i* which is predicted by the prediction model. r stands for the value of average mileage per day for a particular customer. So, $\frac{L_i}{r}$ represents part's life-span measured by time. t_i is the date of the last time that part *i* is checked or replaced. By using this approach, it is not difficult to predict which part needs to be checked or replaced and at what time depending on the detailed history service records about a particular vehicle. The service provider has a capability to inform their customers in advance in order to reduce the driving risks of a driver who drives the vehicle but is unaware of his vehicle is under a bad condition.

The typical example is that a customer replaced a tyre at 10 May 2006. According to the questionnaire he fills in, SMS estimates that the tyre he replaces should have an extra 20% mileage longer than its standard life-span according to the SMS analyzing results of the driver's driving behavior. And SMS also predicts that the average mileage of the driver is 74.6 km per day. SMS will then estimate that the date to check the tyre again is about 15 February 2007, and the date to replace the tyre is about 1st April 2007. The system will issue the reminders to the customer in advance.

During this period of time, if there are some new service records being inputted into the database, and if those new records will influence the driving behavior or average mileage/day estimation, SMS will adjust the knowledge database over time, update the service schedule again.

4 Conclusion

In this paper, we proposed and developed an SMS system. By analyzing customer driving history records, we can get the customer's driving behavior first, then, to analyse the life-spans of the parts which are highly affected by customer's driving behaviors. By using the rule-matching, we discover the relationship of the mileage and time for a customer. We discover a more accurate view of service mining for an individual customer. The SMS, therefore, can make decision about: whose vehicle is at what time to have what kind of service. As the model being modified again and again, a more accurate model can be established.

In this system, we identify customer's driving behaviors by analyzing customer feedback records and service records. These behaviors help us to find out the life-spans of vehicle components.

Based on the above method, mining services and self-learning can be achieved. The service mining system, which is based on knowledge and customer databases, can be also applied to CRM. From the service mining to self-improving of service quality, the system implements intellectual services through the whole process automatically. The modeling and prediction of customer behaviors are key factors in the system. It requires continuing efforts to improve the accuracy of models and algorithms.

References

1. Song, H., Wang, H.: Digitalization Strategy of Automobile Maintenance Service Enterprises. Journal of Wuhan Technologies (information & management engineering) 26(6), 9–11 (2004) (Chinese)
2. Yang, H., Lin, Y., Zeng, T.: Data Analysis of corporate After-Service in Business Selling. Mathematics in Practice and Theory 35(7), 74–83 (2005)
3. Michon, J.A.: A critical review of driver behavior models: What do we know, what should we do? In: Schwingr, Evans, L.A. (eds.) Human behavior and Traffic Safety, pp. 487–525. Plenum Press, New York (1985)
4. Lian, Y., Fassino, M., Baldasare, P.: Predicting Customer Behavior via Calling Links. In: Proceedings of International Joint Conference on Neural Networks, Motreal, Canada, vol. 4, pp. 2555–2560 (2005)
5. Fang, P.: Comment On "Application of After-service Data". Mathematics in Practice and Theory 35(7), 98–105 (2005)
6. Mor, E., Minguillon, J.: An empirical evaluation of classifier combination schemes for predicting user navigational behavior. In: ITCC'2003. Information Technology: Coding and Computing [Computers and Communications], Proceedings International Conference, pp. 467–471 (2003)

A Multi-criteria Decision Support System of Water Resource Allocation Scenarios

Jing He[1,2], Yanchun Zhang[1], and Yong Shi[2]

[1] School of Computer Science and Mathematics, Victoria University, PO Box 14428, Melbourne, VIC 8001, Australia
hejing@gucas.ac.cn, yzhang@csm.vu.edu.au
[2] Research Center on Fictitious Economy and Data Science, Chinese Academy of Sciences, Beijing 100080 P.R. China
yshi@gucas.ac.cn

Abstract. With water resource allocation criteria, the three main components are economic, ecological and sociological aspects, which are often conflicting with each other. Water resource allocation problem is analyzed via the development of a multi-criteria decision support system. The main techniques used are (1) multi criteria decision making (MCDM),(2) geographic information system (GIS) (3) multi-regional computable general equilibrium (CGE) (4) Expert System (ES). The Decision Support System (DSS) proposed in this paper is an effective combination of GIS, CGE, ES, where the three modules have the joint interface and are connected by middlewares. The above three modules are united as a whole Multi-criteria Decision Support System of Water Resource (WRMCDSS).

1 Introduction

Water is one of the most resource for human development. Although combination of them is rare to be used, earlier Multi-criteria decision making (MCDM), geographic information system (GIS), multi-regional computable general equilibrium (CGE), Expert System (ES) have been applied for water resource management respectively in the past fifty years. The multi-criteria decision making (MCDM)functionality implemented allows the decision support system (DSS) to model user's preference and to aggregate the performances of DSS with regard to the decision criteria. It is considered for the first time in this paper to unite them under a whole architecture.

This paper will put forward a new multi-criteria decision support system of water resource to find the tradeoff solution from conflicting objectives. The Decision Support System (DSS) proposed in this paper is an effective combination of geographic information system (GIS), multi-regional computable general equilibrium (CGE), Expert System (ES), where the three modules have the joint interface and are connected by middlewares. There are some previous work to deal with the three main challenging issue in the development of decision support system (DSS) for water resource [1] such as Modeling, data integration,

Z. Zhang and J. Siekmann (Eds.): KSEM 2007, LNAI 4798, pp. 593–598, 2007.

data extraction and schema transformation. Multi-criteria decision support system (MCDSS) is participatory, technically and scientifically informed. It has been established by scholars that water resources problems are going to be more complex in the future world wide [2]. Multi-criteria (MC) analysis has been established as one of the tools for solving complex water resource problems [2], [3], [4]. WRMCDSS proposed in this paper can be carried out at international level. This is especially the case when a water course crosses several national boundaries. WRMCDSS can also be carried out at national, regional, district and at river village level.

This paper is organized as follows. In section 1 we give relevant literature review relating to the topics contained in proposed approach. In section 2 we present underlying assumptions and justifications for proposed methodology. Basic model for selected decision support system in Yellow River is described in Section 3. Brief discussion is provided with regard to the results obtained by some other multi-criteria methods. The conclusion in section 4 closes the paper.

2 Description of Components in WRMCDSS

The main idea of WRMCDSS based on CGE, GIS, ES is as follows: constructing the three structure model based on the database system. The first is to record the economical data in CGE model, the second is to record the spatial data in GIS model, the third is to record the knowledge data in ES model. Through the measure of MCDSS for time and space, the system can fit the change of structure and content. This paper proposes a figure to explain the mapping relationship in time and space among different database shown in Fig. 1.

The best contribution from GIS to DSS is the establishment of spatial data base, spatial data analysis and visual information query, statistics, and search. The best contribution from CGE to DSS is the establishment of economical data base, economical data analysis and economical information query, statistics, and search.

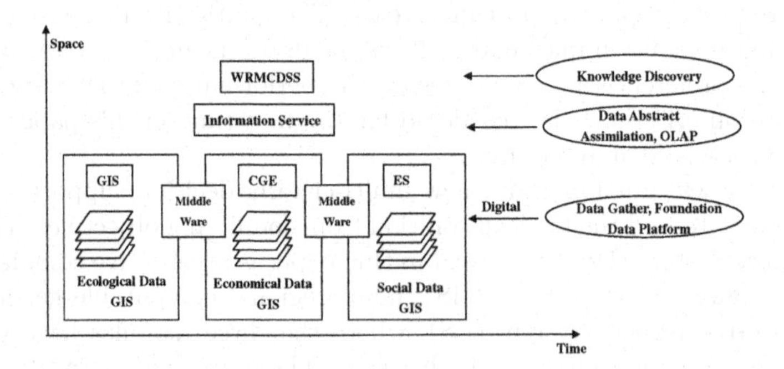

Fig. 1. Mapping Relationship in Time and Space among CGE, GIS, ES

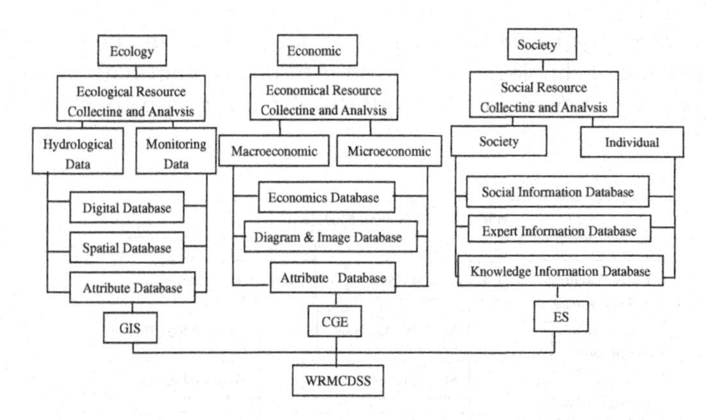

Fig. 2. System General Architecture

The professional software environment to realize the WRMCDSS system needs a set of GIS, CGE, DSS development tools. At the same time hard ware must match the software including some working station and advance PC. At present time, some DSS system supply some self-defined denotation way of knowledge and strategy for reasoning, explaining, controlling. So we adopt advanced language C^{++} based on MapInfo [5] to develop WRMCDSS. Fig. 2 shows System General Architecture.

The stored data in different data warehouses is all kind of temporal, spatial, multimedia, economic data. Linking up between different module and reading data are the important work to improve the efficiency of software. Many economic index data of CGE can be gained directly for GIS, such as drainage area, gross extent of water conservancy establishment, and other topographical parameters.

The design of structure is the key point which can determine whether the DSS function can be realized. There are two main approaches to combine GIS, CGE, ES in WRMCDSS, which are tight-coupling and loose-coupling. Tight coupling is to use the expression and inference mechanism to transform the inner data model and data structure of GIS, CGE, ES, then a real intelligent water resource management system come into being. Loose-coupling can be called as the parallel combination. It is the outside combination of GIS, CGE, ES, DSS. The four system has the joint interface and are connected by middle ware. The loose-coupling approach is used in this paper.

GIS system server uses Mapinfo develop platform of company Mapinfo. It arranges storied spatial and diagrammatic data, collect information, construct the system according to river valley region and district & county. Sybasel 1.0 is used for database, client end is based on Visual C^{++} as basic developing platform. Diagram part use Mapobject 2.0. Attribute data use SQL Anywhere 6.5. Systematic GIS platform use Mapinfo which can guarantee the advanced open, extending character of the system. It permits system extend to a real distributed, multiple client visiting, safe WRMCDSS which can meet the sharing-information requirements in different water resource management.

The MCDSS constructed based on GIS and CGE can solve integration of these information technologies in the water resources information, provide data and application system share. The WRMCDSS function module can be shown in Fig. 3.

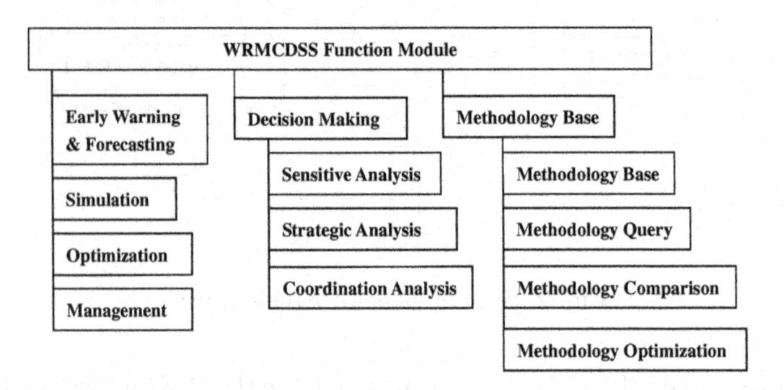

Fig. 3. WRMCDSS Function Module

MCDM used in this decision support system is to regulate withdrawals and other treatments in a planned manner to meet a balance among economic, ecological, sociological objects. From economic aspects, our WRMCDSS is based on social, economical, environmental account for regional water resource. By means of CGE theory, it accomplishes the prediction of economic-demanding quantity of water based on building water-demanding index of each macro-economic department. The key character of GIS in WRMCDSS is to connect the related data with the geographical location. Then the related data can be expressed and analyzed from spatial point of view. GIS in this WRMCDSS supplies a better approach to comprehend the dynamic interfacing, inter-dependent, inter-influent effect among different regional water resource. This expert system in WRMCDSS can collect the opinions of experts from different research field. Obviously, there is not unique computer framework in which all physical information on water resource management topics could fit. Experts knowledge which belongs or is closely related to administration, law, business and other non-technical disciplines should be handled in this components as expert system.

3 Basic Model

3.1 Basic Database

The WRMDSS is developed based on MapInfo develop platform. There are 8 geographical attributes of Yellow River in one year such as average precipitation, assuming amount, consuming water amount, waste water discharge, length of water quality evaluation, II and III water quality, IV water quality, V and above V quality. A data example from 1998-2001 year Geographical Attributes Data of Yellow River is shown such as natural regional area, administration regional area, precipitation of natural regional comparison with average data of the past ten years, run-off measure of major station comparison with average data of the past ten years, retain water variable of large-scale, middle-scale reservoir, waste and foul water discharging quantity histogram, water pollution, importing sand quantity of major Hydrometric Station trunk stream tributary. The economics attributes database is shown such as income absolute value income index, life payout food, pure income, pure income index, life consumption, food consumption living at city, living at country, wage absolute value, wage index, deposit amount, deposit per capita.

Only two basic database is shown in this paper beyond other economic, ecological, social database. Then we can get the regional water resource Input Output table from our WRMCDSS. Most National data come from statistical bureau of China [6], some regional Shanxi and Gansu Province come from statistical bureau of Shanxi [7] and Gansu [8] Province. Then we can get the static equilibrium model from our DSS system.

3.2 Static Equilibrium Model

A basic framework of the multi-object static equilibrium model can be presented as follows:

$$Min(f^{(1)}(x), f^{(2)}(x), ..., f^{(k)}(x))^T \quad (1)$$
$$s.t. \begin{cases} g_i(x) \leq 0 & (i = 1, 2, ..., m) \\ h_j(x) = 0 & (j = 1, 2, ..., l) \end{cases}$$

$f^{(1)}(X), f^{(2)}(X), f^{(3)}(X), ..., f^{(k)}(x)$ represent the economic, ecological, sociological objectives. $g_i(x)$, $h_j(x)$ is the basic constraints to establish the general equilibrium model of realtime water resource system. Object and basic relationship among objects is extracted and purified from economics, environment, society. These objects unites as an organic whole according to their inner relationships, thus they construct as objective conditions in the multiple objective analysis model. These relationships not only construct the constraints to each single-object, but also reflect the restrict relationship between each object. Linear and non-linear equation are used to describe the above relationship in our model.

3.3 Grave Model Based on GIS

We can get the distance from regional A to regional B easily from GIS. The intermediate product from regional p to regional q can be calculated as follows [9]:

$$T_{ij} = \alpha \frac{D_j S_i}{TDS} \frac{(GDP_i)^{\gamma_1} (GDP_j)^{\gamma_2}}{(d)^{\beta}} \quad (2)$$

T_{ij} is the intermediate product from regional p to regional q, D_j is the gross demand of regional j, S_i is the gross supply of regional i, TDS is the sum of regional gross supply (or demand), GDP is the variable reflects the regional economic scale, d is the distance from i to j, α, β, γ are parameters.

Then we can get the multi-regional water resource IO table.

3.4 Dynamic General Equilibrium Model

The dynamic CGE model in this paper is the regional CGE model. The main objective is to inquire the sustainable developing path of water resource. During those days the planning and management of the water resources were for single uses. As time passed on it was recognized that resources were integrated and therefore, the need for longer range planning that would include basins development, or evaluate social-economic consequences and environmental impacts of management strategies.

4 Conclusion

The proposed Decision Support System (DSS) is an effective combination of GIS, CGE, ES, where the three modules have the joint interface and are connected by middlewares. This is especially the case when this WRMCDSS is serviced for analysis and development of management polices for sustainable management of water resources in both Australia and China.

Acknowledgments

The work is partially supported by ARC Discovery Project DP0345710, VU New Research Directions Grant, and NSFC No 70602034, 70531040, 70472074, 70621001.

References

1. `http://e-research.csm.vu.edu.au/`
2. Matondo, J.I.: A Comparison Between Conventional and Integrated Water Resources Planning and Management. Physics and Chemistry of the Earth 27, 831–838 (2002)
3. Srdjevic, B., Medeiros, Y.D.P., Faria, A.S.: An Objective Multi-criteria Evaluation of Water Management Scenarios. Water Resources Management 18, 35–54 (2004)
4. Ozelkan, E.C., Duckstein, L.: Analysing Water Resources Alternatives and Handling Criteria by Multi Criterion Decision Techniques. Journal of environmental management 48(11), 69–96 (1996)
5. `http://www.mapinfo.com.cn/`
6. `http://www.stats.gov.cn/`
7. `http://www.stats-sx.gov.cn/tjgb/ndgb/200606110088.htm`
8. `http://www.gs.stats.gov.cn/`
9. Shangtong, L., Jianwu, H.: A Three-regional Computable General Equilibrium (CGE) Model for China. In: Presented at 15th International Input-Output Conference, Beijing (2005)

Gödel, Escher, Bach and Super-expertise

Pamela N. Gray[1], Xenogene Gray[2], and Deborah Richards[2]

[1] Centre for Research in Complex Systems, Charles Sturt University, Bathurst, Australia
[2] Macquarie University, North Ryde, Australia
pgray@csu.edu.au,
{xengray, richards}@ics.mq.edu.au

Abstract. A major problem in knowledge acquisition is expert combinatorics. A human expert, such as a lawyer, deals with combinatorics restricted to the client's case in hand; only part of the full combinatorics is worked out. An artificial expert, with computational intelligence, processes all possible user cases consistently; it has Super-expertise that can process any case much quicker and more expediently than a human expert. This paper considers the nature and limits of Super-expertise, with some reference to the early visions of artificial intelligence of Hofstadter, in order to develop programming epistemology as methodology that may solve many of the knowledge acquisition problems that have produced the Feigenbaum bottleneck. An application of a fifth generation language, a Superexpert system shell called eGanges, which was designed according to a computational epistemology of a legal expert, is used to illustrate this development of programming epistemology.

Keywords: Combinatorics, eGanges, Epistemology, Expert Systems, Super-expertise.

1 Hofstadter Visions of AI

Shortly after the demonstration of the earliest expert systems, Hofstadter [1] addressed some of the major difficulties in human reasoning that would have to be accommodated in artificial intelligence, by reference to the artistic works of the music of Bach and the art of Escher. The music of Bach consisted of repetition with slight variation on each repetition; the patterns in the music were combinatoric and exploited, systematically, the possible combinatorial variations of a limited number of notes. Escher's famous upstairs/downstairs drawing which conforms to the constraints of artistic perspective at each step from upstairs to downstairs, and back again, portrays the potentially infinite recursion that a closed system might have. Both the works of Bach and the works of Escher, in their chosen art form, actualize difficult reasoning. The question is: can programmers actualize the difficult aspects of human intelligence in their computational manifestation?

Then Hofstadter considers the new difficulty of Gödel's Incompleteness theorem which asserts that in an evolving world, it is not possible to reason reliably with the

Z. Zhang and J. Siekmann (Eds.): KSEM 2007, LNAI 4798, pp. 599–604, 2007.

potential of new factors always immanent. Completeness and accuracy can not co-exist in an infinite domain. Both Bach and Escher conformed to Gödel's theorem because they used closed systems; Bach had limited notes and Escher's stairs did not end in infinity, although the recursion they suggested was potentially infinite within their closed system.

2 Gödel and Combinatorics

This paper is concerned with a knowledge engineering visualisation of expert combinatorics and the constraints which operate it, particularly in the large and complex legal domain. Applications of the fifth generation shell, eGanges, are used to illustrate this visualization and some of the constraints which automate its combinatorics through the user interface. The user interface provides a communication system which is an integral part of the combinatoric system of eGanges permitting the expert to provide the minutiae of the required knowledge acquisition with their respective logic status. The minutiae, or elements, required is minimized to the conditions which are required for, or are consistent with, the attainment of one outcome, called the Positive Final Result. For example, we may wish to know if a certain court order would be made.

There may be both conjunctions and disjunctions of these Positive elements, so that there is an initial finite set of combinatorics, the Positive combinatorics; in the legal domain there is often some legal choice of the means to a Positive Final Result. A second set of combinatorics arises when all the elements are contradictories of the Positive elements; these are the Negative elements that have Negative combinatorics. De Morgan's rules determine the visualisation of Positive conjunctions as Negative disjunctions, and Negative conjunctions as Positive disjunctions.

As eGanges was designed on the basis of a computational legal epistemology [2], it accommodates a third set of elements, namely, the Uncertain elements, where user input indicates uncertainty about all the necessary and sufficient conditions for the Positive Final Result; this is a third set of combinatorics, the Uncertain combinatorics. A fourth set of combinatorics is also provided to cover the explosion of possible combinations of Positive, Negative and Uncertain elements, within the closed finite system of the given Positive elements, their possible contradictories and uncertainties.

A fifth set of combinatorics arises from the environmental potential of new elements that are not yet within the defined system of elements. This fifth set is potentially infinite.

In the legal domain, a compliance system deals with the first four sets of combinatorics; if a system provides for the fifth set of combinatorics, then it is a judicial expert system. Gödel's theorem invalidates the combinatorics of an automated judge as the fifth set is potentially infinite.

3 Superexpertise

A human expert, such as a lawyer, deals with combinatorics restricted to the client's case in hand; only part of the full combinatorics is worked out. An artificial expert, with computational intelligence, processes all possible user cases consistently; it can be said to have *Super-expertise* that can process any case much quicker and more expediently than a human expert. It was difficult to knowledge engineer a Super-expert system because it was difficult to acquire the expert's knowledge of all four sets of combinatorics as well as distinguish the fifth set of combinatorics which was not automatable; in the legal domain, jurisprudence had not enhanced legal knowledge with its combinatoric framework. This was realised in the assessment of the Feigenbaum bottleneck by Quinlan [3], p.168:

> Part of the bottleneck is perhaps due to the fact that the expert is called upon to perform tasks that he does not ordinarily do, such as setting down a comprehensive roadmap of some subject.

Super-expertise requires management of the expert sets of combinatorics and should dispel much of the problem of the Feigenbaum bottleneck. In the knowledge engineering of Super-expertise, this indicates the need for a sound programming epistemology. The programmer must translate the expert epistemology into programming epistemology. As indicated by Hofstadter, the programmer must actualise the expert epistemology. Combinatorics is the essential epistemology of Super-expertise, but further development of programming epistemology is also required, as dictated by the computational expert epistemology [2].

eGanges is a 5GL shell that automates the first four sets of expert combinatorics so that compliance applications can be constructed by an expert. Further expert epistemology is reflected in the visualization, interface, and processing constraints of the shell. The visualization is a two dimensional River map as shown in the Rivers window of the eGanges interface shown in Figure 1. Each tributary of the River is a rule with the inference arrow before the consequent indicating flow. In this application, elements of the Australian Spam Act are mapped to show how compliance with the Act is achieved. The application was constructed by Philip Argy, who is President of the Australian Computer Society and also Senior Partner for IT in the leading Australian law firm, Mallesons Stephen Jaques, Pamela N. Gray and Xenogene Gray, in 2004.

Programming epistemology is a concept that cannot be developed fully in this short paper. However, some introduction is given, limited to the epistemology of the fifth generation language, Superexpert shell, eGanges, which was based on a computational legal expert epistemology [2]. What can be learned from this single exercise may provide some generic programming epistemology that remains useful for other domains of expertise. Programming epistemology ensures the validity of expert combinatorics; Superexpert systems are a category of expert systems which have this epistemological validity. If a Superexpert system is based on a computational domain epistemology, confirmed at the outset by the domain expert, this facilitates the expert verification of a completed expert system; it circumvents the problem for the expert of understanding the programming epistemology as a necessary part of the verification process. With a growing number of multi-discipline or hybrid experts, this is now feasible.

For further research papers on eGanges see http://www.grayske.com

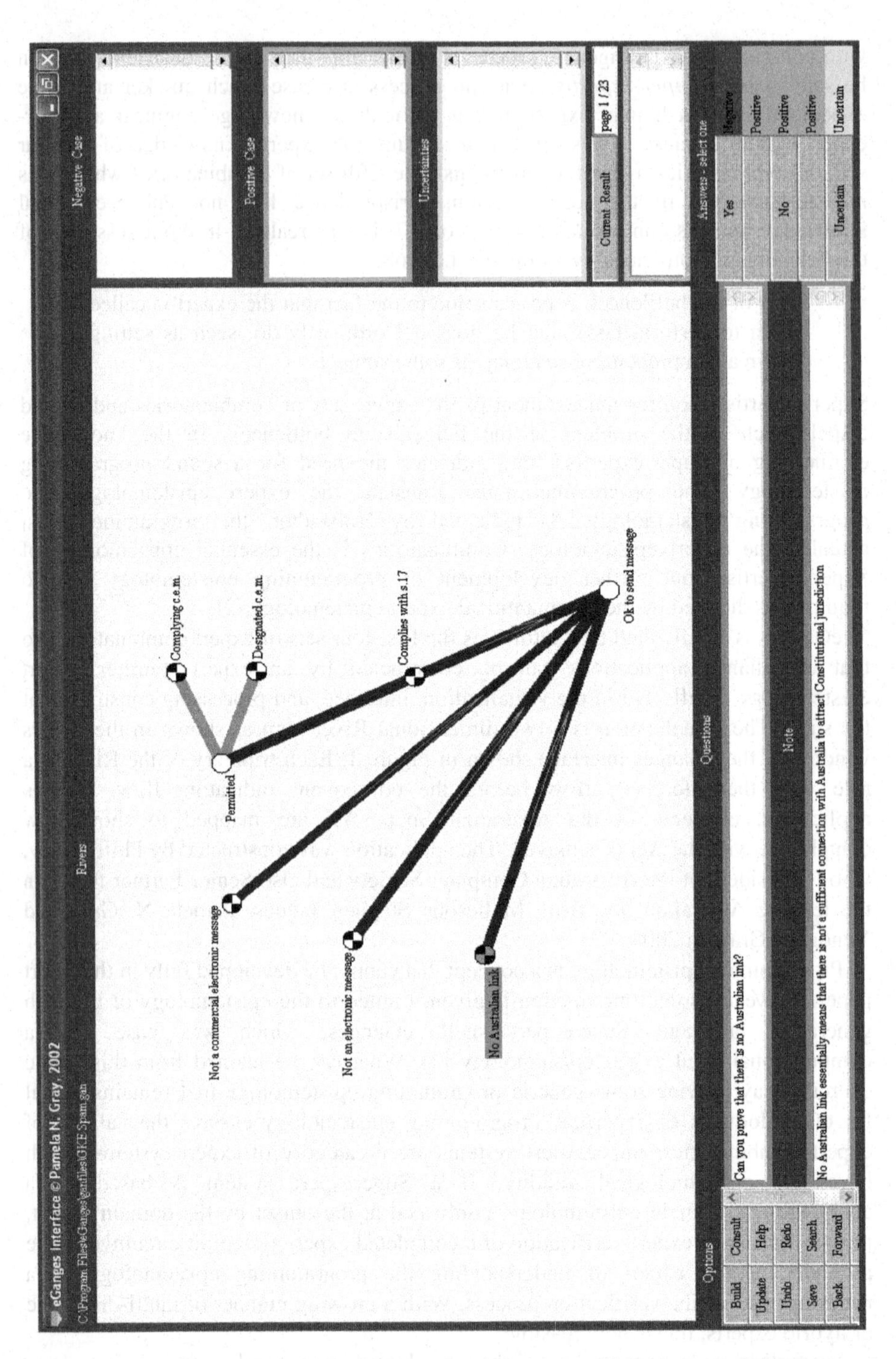

Fig. 1. In the eGanges Interface - Initial map of Spam Act © P.N. Gray, P.N. Argy and X. Gray, 2004

4 Combinatorics – An eGanges Example

eGanges programming epistemology uses McCarthy's [4] concept of circumscription to partition elements of the fifth set of combinatorics from elements of the first four sets of combinatorics. In addition it uses circumscription to partition deductive premises from inductive and abductive premises of the expert. This permits the management of non-monotonic logic in the domain. Massively extended and complex deduction is thus isolated for automation. Since deduction is necessary reasoning, it is a form of logic suitable for automation.

eGanges is a 5GL because its visualization (using the industry definition) is processed by a set of constraints (AI definition), namely the constraints of the four sets of expert combinatorics. Figure 2 illustrates the nesting of the visualization that may be as deep as the knowledge requires. "No Australian link" is both an antecedent in Figure 1 and also a consequent in Figure 2. Maps represent major deductive premises that can be processed as user input provides minor deductive premises; user input takes the form of answers to the questions that establish each element or node. Potential answers are labeled to indicate that the answer establishes a negative, positive or uncertain element. There are three possible positive answers where any of the three available answers is consistent with the Positive Final Result.

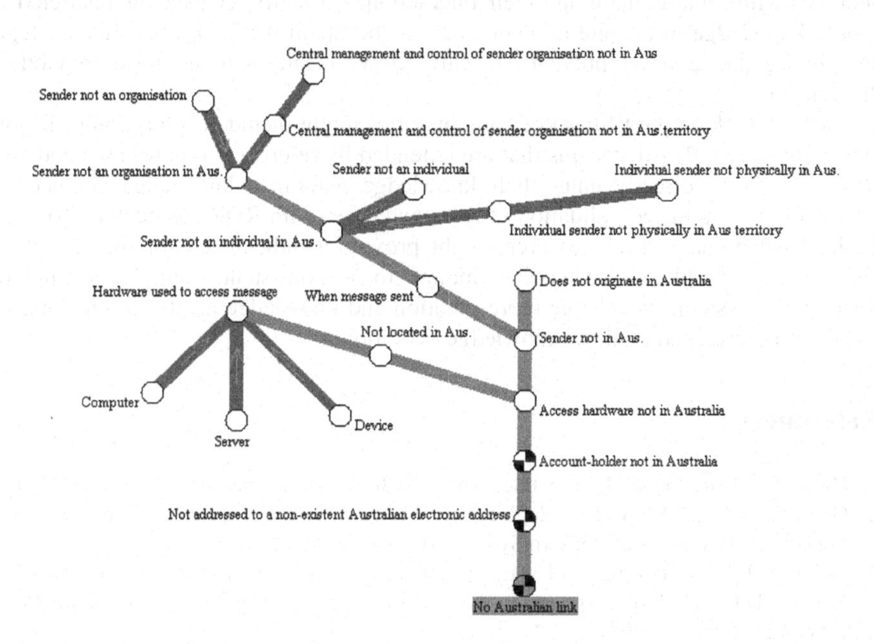

Fig. 2. Submap of "No Australian link" from Spam Act © P.N. Gray, P.N. Argy and X. Gray, 2004

As user input is given the blank node changes to the colour of the selected answer showing any pattern of combination of elements. The colours correspond to the feedback windows (red for negative elements, blue for positive elements and yellow for uncertain elements). As answers are given, labels of the nodes appear in the appropriate feedback windows, subject to the heuristic constraints of the combinatoric processing. For instance, a negative element that is one of a set of disjunctions will initially appear in the positive feedback window with "(Neg)" before it. When all the elements in the disjunction are established as negative, then all the "(Neg)" points will be moved from the positive feedback window to the Negative feedback window and automatically establish failure of the Positive Final Result.

The computational epistemology of 3D legal logic poses a visualisation of a spherical logic structure ([2], p.218, [5], p.278) which is comprehensive and determines the eGanges heuristics. A River map from the sphere is processed according to the requirements of the sphere. This sphere of legal logic provides the legal domain representation of the combinatoric constraints (c.f. Escher's stairs).

5 Conclusion and Further Work

The eGanges Superexpert system shell addresses the combinatorics issue by guiding the domain expert to articulate and organize their knowledge via a fifth generation communication system which visualizes and constrains the knowledge entered. The shell facilitates the capture of the positive, negative and uncertain knowledge elements within the domain and their interaction. Currently, eGanges is restricted to handle knowledge in a finite domain, such as the Spam Act and other fieldsof legal compliance and conflict prevention. This allows eGanges to conform to Gödel's theorem.

Further work on the fifth set of combinatorics might include exploration of Ripple Down Rules (RDR) [6] systems that are extended by reference to new cases and also allow experts to directly enter their knowledge without a knowledge engineering intermediary; automated validation of rules and cases with RDR has been unexplored in the legal domain. Such research might provide a judicial aid for the remaining combinatoric problem that could amount to a consistent, though incomplete, Superexpert system; knowledge representation and knowledge acquisition techniques could be incorporated to allow knowledge evolution.

References

1. Hofstadter, D.R.: Gödel, Escher, Bach. Penguin Books, Harmondsworth, England (1979)
2. Gray, P.N.: Legal Knowledge Engineering Methodology for Large Scale Expert Systems, PhD thesis, University of Western Sydney, Sydney, Australia (2007)
3. Quinlan, J.R.: Discovering rules by induction from large collections of examples. In: Mitchie, D.E. (ed.) Expert systems in the micro-electronic age, Edinburgh University Press, Edinburgh, Scotland (1979)
4. McCarthy, J.: Circumscription: a form of non-monotonic reasoning. Artificial Intelligence 13, 27–39 (1980)
5. Gray, P.N.: Artificial legal intelligence. Dartmouth Publishing Co., Aldershot (1997)
6. Compton, P., Jansen, R.: Knowledge in context: a strategy for expert system maintenance. In: AI 1988: Proceedings of the second Australian Conference in Artificial Intelligence, Adelaide, pp. 283–297 (1988)

Extracting Features for Verifying WordNet

Altangerel Chagnaa[1], Cheol-Young Ock[1], and Ho-Seop Choe[2]

[1] School of Computer Engineering and Information Technology,
University of Ulsan, South Korea
{golden1, okcy}@mail.ulsan.ac.kr
[2] Information System Development Team, Korean Institute of Science and Technology
Information, Daejeon, South Korea
hschoe@kisti.re.kr

Abstract. WordNet is a semantic lexicon for the English language and many countries have been developing their own WordNet. Almost, all of the WordNets are manually built and unfortunately these WordNets are not verified and are being used in many knowledge-based applications. In this paper we aimed at the clustering based verification of a manually built lexical taxonomy WordNet, namely the Korean WordNet, U-WIN. For this purpose two kinds of clustering methods are used: K-Means approach and ICA based approach. As a result the ICA based approach gives better result, and it shows very effective characteristic for extracting features.

Keywords: WordNet, evaluation, taxonomy, and clustering.

1 Introduction

WordNet [1] is a semantic lexicon for the English language that has became an important and useful resource for various knowledge-based applications. Since then, many countries have been developing their own WordNets and almost, all of the WordNets are manually built. And they have evaluation problems, which means until now there is no framework developed for verifying such a lexical taxonomy. Thus they are applied to many knowledge-based applications without verification. The verification is very important to not only the manually built WordNet, also such other manually built lexical resources.

In this paper we aimed at the clustering based verification of the manually built lexical taxonomy U-WIN. Various association measures (mutual information (MI), t-score and log-likelihood ratio) and two types of clustering algorithms (the traditional K-Means clustering algorithm and the ICA based clustering) are used in the experiment. In the feature based case, the ICA based clustering gives better result than traditional K-Means algorithm and it has its natural property of selecting appropriate features.

The remainder of this paper is organized as follows. In section 2, there is a brief introduction of the Korean WordNet U-WIN and section 3 shows the clustering based verification of U-WIN. Section 4 explains about the extraction of features from the data. Finally we end the paper with the conclusion and future works.

Z. Zhang and J. Siekmann (Eds.): KSEM 2007, LNAI 4798, pp. 605–610, 2007.

2 Korean WordNet, U-WIN

Korean WordNet, namely User-Word Intelligent Network (U-WIN) [2] is being developed at the University of Ulsan since 2002. It is aimed to be a large scale lexical knowledge base which is useful for various fields as linguistics, Korean information processing, information retrieval, machine translation and semantic web etc.

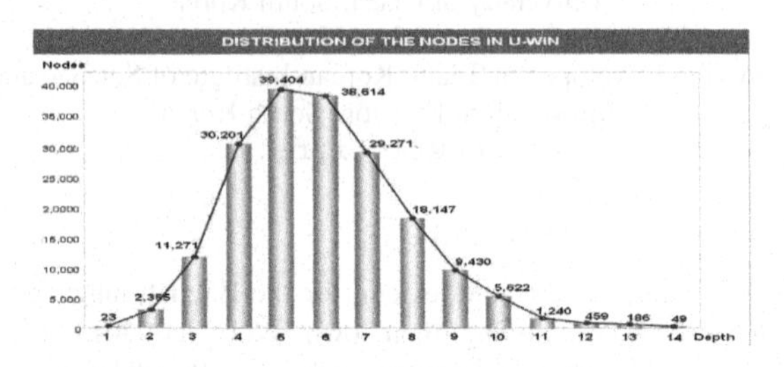

Fig. 1. Distribution of nodes in U-WIN

The base knowledge used for constructing U-WIN is the Korean Standard Dictionary that contains very detailed information about Korean words and their senses. U-WIN has many kinds of semantic and conceptual relations. Here we have used only Subclass_of relations (IS_A and Kind_of relations) in the taxonomy. There are 23 top-level nodes. Taxonomy goes down to the depth of 14 levels and the distribution of the nodes in each depth is shown in the figure above.

3 Verification of U-WIN Based on Clustering

3.1 Feature Selection

We ran experiments for the word senses in each level in the taxonomy. From the corpus we have extracted each word's related verbs and selected them as its features. In this work we only used object-verb relation as noun and verb syntactic relation.

Because of the small size of sense tagged corpus, we have sparseness problem [3]. To overcome this problem, we adjusted the features of the parent nodes as following. If the parent doesn't have the feature that child nodes have, then add the feature to the parent features and don't alter the parent's own features. Another reason for doing this is we assume that parent node is a concept, not only a word.

3.2 Answer Set and the Evaluation Measurement

The answer set (key groups of the clustering) is created by just cutting from the position above the selected words (parent node of the words) including the parent node in a taxonomy; i.e. words which have the same parent are in the same group of the key set (see figure 2 for more detail). Table 1 shows the statistics after creation of the answer set and the column Clusters shows the number of groups in the answer set.

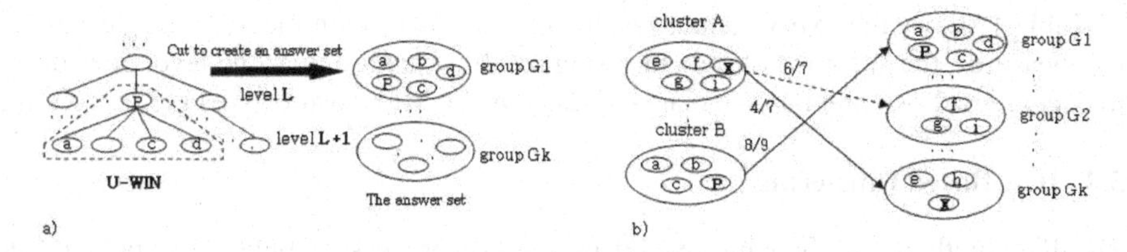

Fig. 2. Figures for: a) Creating the clustering key for the words at level L in U-WIN; b) An example of selecting a corresponding group for the cluster using F-Measure. *Cluster A* has the highest score to *Group G2* among the other groups (6/7=0.857), but *Group G2* is not selected as the winning group for *Cluster A*. Instead *Group Gk* is selected as the corresponding group, because it contains the parent node of the cluster; so the score for this cluster is 4/7=0.571.

The words in 4 top-level nodes (THING, ACTION, LIVING THING and SHAPE) of the U-WIN are selected in the experiment. Most of the word senses in the syntactic relations found from the dictionary explanation are in these top-level nodes. For comparing the result set of the clustering to the answer set, F-measure [6] is used.

For each cluster in the answer set, we selected one group in the result (key) set which contains the corresponding parent node. For a more detailed example, see the above figure 2b. And the overall clustering result is the average of the F-Measures of the clusters.

3.3 K-Means Clustering

With the traditional K-Means clustering 3 kinds of measures are selected as the measure of the relationship between a noun and a verb (lexical association measure): mutual information (MI) [5], t-score [6] and log-likelihood ratio; thus each noun is expressed as a vector.

Table 1. The statistics of the data and the experimental result

SET	Depth	Nouns	NV pairs	Features	Clusters	K-Means clustering			ICA
						MI	T-score	Log-like	MI
THING	4	301	2791	731	71	0.478	0.473	0.491	0.589
	5	495	3181	813	137	0.538	0.548	0.51	0.605
	6	485	2657	711	161	0.629	0.608	0.6	0.676
	7	350	1647	492	124	0.644	0.632	0.651	0.688
	8	205	1004	362	77	0.677	0.671	0.674	0.738
ACTION	4	249	1673	642	71	0.584	0.588	0.561	0.652
	5	308	1332	376	113	0.627	0.622	0.615	0.671
	6	287	1218	370	99	0.654	0.636	0.631	0.706
	7	154	687	257	64	0.721	0.703	0.712	0.743
	8	67	281	133	31	0.83	0.767	0.804	0.833
LIVING THING	4	79	1031	507	11	0.404	0.377	0.351	0.457
	5	190	1598	585	58	0.576	0.591	0.57	0.631
	6	182	698	237	73	0.752	0.747	0.745	0.786
	7	117	543	215	44	0.718	0.715	0.726	0.802
	8	78	309	146	32	0.771	0.771	0.767	0.781
SHAPE	4	182	1260	414	54	0.586	0.543	0.586	0.624
	5	169	1063	393	56	0.61	0.632	0.58	0.767
	6	97	617	238	35	0.657	0.66	0.618	0.673
	7	60	351	161	24	0.731	0.729	0.725	0.868
	8	35	118	71	16	0.85	0.716	0.783	0.879

Table 1 shows the experimental result for each association measure. Regardless of the measures, the accuracy of the clustering result is increasing as the level goes down to deeper level. Among the association measures the MI measure gives the best result.

3.4 ICA Based Clustering

The ICA method was first introduced into signal processing field, especially in the blind source separation tasks. Later on it has been used in variety of application areas (face recognition, document analysis, linguistic feature extraction [4] etc.).

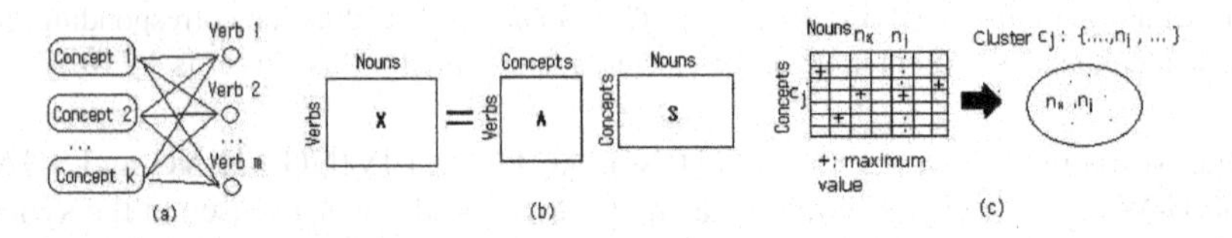

Fig. 3. The illustration of the ICA model from concept views: a) verbs act like microphones, indicating the mixture of concepts with their noun usage. b) The model in matrix notation. c) Cluster creation from the ICA result.

Here, the noun and verb contextual information is supposed as a mixture of independent components and the ICA algorithm decomposes it into separate statistically independent sources. Assume that these separated sources are separate concepts. As a result every noun is expressed by a weight vector of the concepts which shows the membership to the concepts.

In this experiment, each noun is assigned to only one concept (or cluster) which has the maximum weight i.e. each noun is assigned to its corresponding one cluster (fig. 3c). Afterward, resulting clusters are compared to the answer set created early in section 3.2. Because of the superior performance in the previous experiment, we selected MI as the association measure in ICA based clustering.

3.5 Comparison

The last column in the table 1 shows the result of the experiment in each level (or depth) of the U-WIN and for each association measure.

Roughly, the number of features is decreasing as the level goes down to deeper ones. Naturally the number of features will decrease as the number of words decreases, but we can also see that the average number of feature per word tends to decrease. It proves that in the taxonomy, deeper nodes are more specific senses of the words and upper level nodes are more general senses of the words.

The ICA based clustering outperforms the traditional K-Means algorithm in all cases, thus it shows that the ICA based clustering is more suitable for the feature based framework. Because the ICA models a cluster by a combination of features based on statistical properties of the nouns, but K-Means clustering uses only the Euclidean distance between objects as their similarity. In the following section we will show with the characteristic of ICA in feature extraction with examples.

4 Extracting Features

Here we show the feature extraction of ICA in more detail with the examples from the experiment. Note that, the sign in the ICA result is arbitrary.

4.1 Decomposing Data

The following figure shows the decomposition of the experiment data in THING set at level 4, with 301 nouns and 731 verbs. The input matrix of 731x301 size is decomposed into 71 independent components (i.e. clusters).

To assign a noun to a cluster, the last matrix is used (fig.3b). For each noun, select only one cluster that the noun has the maximum absolute value. For the extracted features, the middle matrix shows the feature weights to the clusters.

4.2 Feature Comparison

In this subsection the comparison of the features extracted from data to the key feature (the feature of the parent node in corresponding key group) is shown.

Below figure shows some examples from the ICA clustering result. The features of the concept extracted from the data are matching well to the key features as shown in the figure 8a, and the F-measure between them is 1.0.

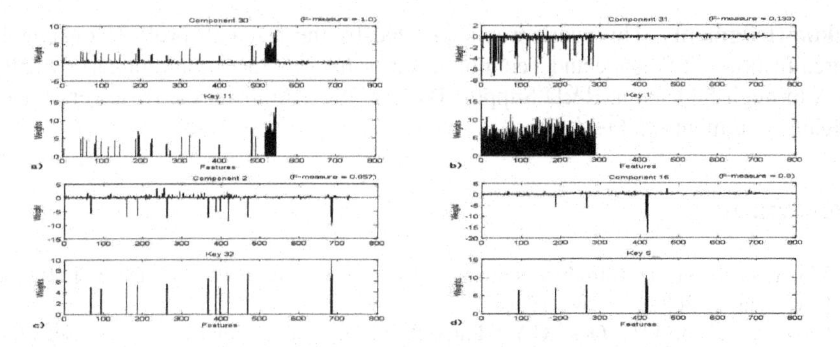

Fig. 4. Examples from the experiment. In each graph, upper plot shows the bar graph of the feature weights in the extracted cluster (component) and lower plot shows the bar graph of the feature values of the parent node which corresponds to the cluster above.

In some cases as shown in the figure 8b, which have very low accuracy (F-measure is 0.133) the key itself has very broad range of feature set. In this case ICA finds the appropriate feature set, however the nouns in the key are likely to be assigned to other cluster because of the broad range of feature set.

From the comparison of figure 8c and 8d, the both key has the feature number 94 (the nearest bar to 100 on the plot) but the ICA method didn't extract it as the feature of the cluster in both. This verb '만들다' is very frequent verb in Korean and it means 'to make' in English. At this time this verb is recognized as insignificant feature.

In most cases, as shown above, ICA method extracts the significant features successfully from the data.

5 Conclusion and Future Works

This paper presented the clustering based verification of the manually built lexical taxonomy, U-WIN. Words in each level of the taxonomy and their parent (parent nodes) are clustered and compared with a corresponding answer set. Words with a same parent node and their parent node form a group of the answer set in the current level. Various association measures and two clustering algorithms are used in the experiment. This approach is not specific for U-WIN but it is also applicable for other taxonomies as well (without modification).

The experimental result shows that among the association measures mutual information (MI), t-score and log-likelihood ratio, the mutual information gives the best result. Regardless of association measures and clustering algorithm, the accuracy of clustering result is increasing as the level goes down to deeper level. Beside the decrease of the feature as the taxonomy level goes down do deeper ones, also the average number of feature per word tends to decrease. It shows that in the taxonomy, deeper nodes are more specific senses of the words and upper level nodes are more general senses of the words.

The ICA based clustering outperforms the traditional K-Means algorithm and it shows that the ICA based clustering is more suitable for the feature based framework. This is because ICA method estimates the appropriate combination of features based on the statistical property of independence (not the classic Euclidean distance).

Acknowledgments. This work was supported by the research project consigned by Korea Institute of Science and Technology Information (KISTI) and in part by MIC & IITA through IT leading R&D Support Project. We also thank the reviewers for their valuable comments and reviews.

References

1. Miller, G.A., et al.: Introduction to WordNet: an on-line lexical database. J. Int. Jour. Lexicography 3(4), 235–244 (1990)
2. Choi, H.S., Im, J.H., Ock, C.Y.: Large Scale Korean Intelligent Lexical Network. J. Hangeul 273, 125–151 (2006)
3. Chagnaa, A., et al.: On the Evaluation of Korean WordNet. In: Matousek, V., Mautner, P. (eds.) TSD 2007. LNCS (LNAI), vol. 4629, pp. 123–130. Springer, Heidelberg (2007)
4. Honkela, T., Hyvarinen, A., Vayrynen, J.: Emergence of Linguistic Features: Independent Analysis of Contexts. In: Neural Computation and Psychology Workshop 9, Plymouth, UK (2005)
5. Church, K., Hanks, P.: Word Association Norms, mutual Information and Lexicography. J. Comp. Ling. 16, 22–29 (1990)
6. Manning, C., Schutze, H.: Foundations of Statistical Natural Language Processing. MIT Press, Cambridge, MA (1999)

A Hybrid Approach for Learning Markov Equivalence Classes of Bayesian Network*

Haiyang Jia, Dayou Liu, Juan Chen, and Xin Liu

College of Computer Science and Technology, Key Laboratory for Symbolic Computation and Knowledge Engineering of Ministry of Education, Jilin University, Changchun, Jilin China 130012
{jiahy,dyliu,chenjuan,liuxin}@jlu.edu.cn

Abstract. Bayesian Networks is a popular tool for representing uncertainty knowledge in artificial intelligence fields. Learning BNs from data is helpful to understand the casual relation between the variable. But Learning BNs is a NP hard problem. This paper presents a novel hybrid algorithm for learning Markov Equivalence Classes, which combining dependency analysis and search-scoring approach together. The algorithm uses the constraint to perform a mapping from skeleton to MEC. Experiments show that the search space was constrained efficiently and the computational performance was improved.

Keywords: Bayesian network, Structural learning, Markov equivalence class, condition independence test.

1 Introduction

Bayesian Networks (BNs, also called Belief Networks) is a popular tool for representing uncertainty knowledge in artificial intelligence fields; Learning BNs from data is helpful to understand the casual relationship between variables. But it has been proved that learning BNs is a NP hard problem[1]. So improving the efficiency of learning is an important problem.

This paper presents a novel algorithm for learning Markov Equivalence Classes of BNs, which combining dependency analysis and search-score approach together. Experiment and quantitative analysis prove that the search space was constrained and the efficiency was improved. The remainder of this paper is organized as following: Section 2 introduces some background knowledge. Section 3 describes the algorithm in detail, section 4 describes the experiment and section 5 draw the conclusion.

* Supported by NSFC Major Research Program 60496321, National Natural Science Foundation of China under Grant Nos. 60373098, 60573073, 60603030, 60503016 the National High-Tech Research and Development Plan of China under Grant No. 20060110Z2037, the Major Program of Science and Technology Development Plan of Jilin Province under Grant No. 20020303, the Science and Technology Development Plan of Jilin Province under Grant No. 20030523, European Commission under Grant No. TH/Asia Link/010 (111084).

Z. Zhang and J. Siekmann (Eds.): KSEM 2007, LNAI 4798, pp. 611–616, 2007.

2 Background

Notation: Capital letter X,Y,Z notate a random variable; lowercase x,y,z notate the value of a random variable; bold capital letter **X,Y,Z** notate a set of random variable; when discussing about sets, the capital letter V,E notate a set, lowercase v,e notate the element in a set. (v_1,v_2) notate a unordered pair, $<v_1,v_2>$ notate ordered pair.

There are two approaches for BNs structural learning: One is score-search based approach, which defines the task as an optimization problem; the other is constraint based approach, which defines it as a constraint satisfaction problem.

Two BNs are said to be equivalent if the joint distributions that can be represented by one can also be represented with the other.

Definition 1. v_i, v_j is a *V structure* at v_k in G= (V, E) iff.: $v_i, v_j, v_k, \in V$, $<v_i, v_k> \in E$, $<v_j, v_k> \in E$, $<v_i, v_j>, <v_j, v_i> \notin E$,short for $V(v_i, v_j \mid v_k)$.

Definition 2. The *skeleton* of a DAG is the undirected graph resulting from ignoring the directionality of every edge.

Theorem 1. Two BNs B_1 and B_2 are *Markov Equivalence* iff. B_1 and B_2 have the same skeleton and same V structure [3, 4].

All of the DAG that equivalent to each other is a *Markov equivalence class*, short for MEC.

3 Skeleton Based Learning Algorithm

This paper presents a hybrid algorithm for Learning MEC base on the space of skeleton (short for LEBS). There are two phases of LEBS: (I) perform a complete 0 order CI test (short for CI_0) and part of 1 order CI test (short for CI_1) to get a constrain on the skeleton and three sets for identify MEC; (II) search MEC in the space of skeleton. (Short for S-space)

3.1 Conditional Independence Test

BNs structure can be viewed as a encoding for a group of conditional independence relationships among the nodes. The main idea of constraint-based algorithms is using conditional independence test (CI test) to identify the conditional independence among the nodes. We use the value of mutual information and the conditional mutual information to perform CI test:

$$I(A,B) = \sum_{a,b} P(a,b) \log \frac{P(a,b)}{P(a)P(b)}, \quad I(A,B \mid C) = \sum_{a,b,c} P(a,b,c) \log \frac{P(a,b \mid c)}{P(a \mid c)P(b \mid c)}$$

We suppose A is (conditional) independent of B when $I<\varepsilon$, short for Ind (A, B) and Ind (A, B|C). There is an edge between A and B in the skeleton iff. $\neg Ind(A,B \mid C)$, for all possible C [5]. So undirected edge between A and B in skeleton SK means A, B can influence each other, and such influence exists under any condition set. CI test in this phase described in the following two algorithms:

```
Algorithm CV0(Data.CV,UN_0,SKC_0)
/*performing CI_0 test with sample set Data, output the order 0 skeleton
  contraint SKC_0 and two sets CV and UN_0*/
INIT[initional SKC_0 with a completed graph,CV,UN_0 with an empty set]
    SKC_0 = {(v_i,v_j) | ∀v_i,v_j ∈ V ∧ v_i ≠ v_j}.
    CV = Φ.  UN_0 = Φ.
SKC[perform CI_0 test, if Ind(v_i,v_j) delete (v_i,v_j) from SKC_0]
    FOR ∀v_i,v_j ∈ V ∧ v_i ≠ v_j
      IF Ind(v_i,v_j)
        SKC_0=SKC_0-(v_i,v_j).
GCV[for each pair of nodes v_j,v_k, which is the neighbour of v_i in SKC_0,
    if (v_j,v_k) ∉ SKC_0 add the triple into CV, else add it into UN_0]
    FOR ∀v_i ∈ V
      FOR ∀(v_i,v_j),(v_i,v_k) ∈ SKC_0 ∧ v_j ≠ v_k
        IF (v_j,v_k) ∉ SKC_0
          CV = CV + (v_i,v_j,v_k).
        ELSE
          UN_0 = UN_0 + (v_i,v_j,v_k) ■
```

Fig. 1. Algorithm to perform CI_0, the output CV will be used in phase II, the others will be used by algorithm NV1

```
Algorithm NV1(Data,SKC_0,UN_0.NV,UN,SKC_1)
/*Perform CI_1 test, the input parameters SKC_0 and UN_0 are the output of CV0*/
CI1[perform CI_1 on the triple in UN_0 if Ind(v_j,v_k | v_i) delete(v_j,v_k) from SKC_1,
    and add the triple into NV, otherwise add it into UN]
    SKC_1 = SKC_0.
    FOR ∀(v_i,v_j,v_k) ∈ UN_0 ∧ (v_i,v_k),(v_i,v_j) ∈ SKC_1
      IF Ind(v_j,v_k | v_i)
        SKC_1 = SKC_1 - (v_j,v_k).
        NV = NV + (v_i,v_j,v_k).
      ELSE
        UN = UN + (v_i,v_j,v_k) ■
```

Fig. 2. Algorithm to perform CI_1, the output NV UN SKC_1 will be used in phase II

3.2 Search in S-Space

The main task of the phase II is searching the best MEC in a constrained S-space. From the theorem 1 we can get: given a skeleton, if all V structure can be identified, then we can identify a unique MEC. The propositions below perform a mapping from the space of skeleton to the space of MEC (short for E-space):

Definition 3. Call a graph *V structured partially directed acyclic graph* (VPDAG) corresponding a MEC, iff:

1) VPDAG and MEC have the same skeleton;
2) all direct edge in VPDAG participate at least one V structure in MEC;
3) V structures in MEC are same as the V structure in VPDAG.

Definition 4. In a skeleton SK, call V_k V_j is a *possible V structure* at V_i, iff. $(V_i,V_j)\in SK$, $(V_i,V_k)\in SK$, $(V_j,V_k)\notin SK$, short for $PV(V_j,V_k \mid V_i)$.

Proposition 1. Given a skeleton, if $(V_i,V_j,V_k)\in CV \wedge PV(V_j,V_k \mid V_i)$, then $V(V_j,V_k \mid V_i)$.

Proof: Since $PV(V_j,V_k \mid V_i)$ it has that V_i,V_j,V_k has four possible local structures: $V_j \rightarrow V_i \leftarrow V_k$; $V_j \rightarrow V_i \rightarrow V_k$; $V_j \leftarrow V_i \leftarrow V_k$; $V_j \leftarrow V_i \rightarrow V_k$. The last three belong to same MEC, in such MEC if V_i is unknown, then V_j,V_k influence each other through V_i, which means $\neg Ind(V_j,V_k)$; Since $(V_i,V_j,V_k)\in CV$ it has $Ind(V_j,V_k)$; such MEC (last three DAG) can be excluded, so $V(V_j,V_k \mid V_i)$. ∎

Proposition 2. Given a skeleton, if $(V_i,V_j,V_k)\in NV \wedge PV(V_j,V_k \mid V_i)$, then $\neg V(V_j,V_k \mid V_i)$.

Proof: If $V(V_j,V_k \mid V_i)$ then V_j,V_k influence each other through V_i when V_i is known, which means $\neg Ind(V_j,V_k \mid V_i)$; Since $(V_i,V_j,V_k)\in NV$ it has $Ind(V_j,V_k \mid V_i)$; so the local structure is impossible to be a V structure, that is $\neg V(V_j,V_k \mid V_i)$. ∎

Proposition 3. Given a skeleton, if $(V_i,V_j,V_k)\in UN$ and $PV(V_j,V_k \mid V_i)$, then there must be a path between V_j,V_k in the skeleton besides $(V_i,V_j),(V_i,V_k)$; $V(V_j,V_k \mid V_i)$ or $\neg V(V_j,V_k \mid V_i)$ can not be identified by CI_0 or CI_1.

Proof: Since $(V_i,V_j,V_k)\in UN$ it has $\neg Ind(V_j,V_k)$ and $\neg Ind(V_j,V_k \mid V_i)$:

1. If $V(V_j,V_k \mid V_i)$, then V_j,V_k can not influence each other through V_i when V_i is unknown, since $\neg Ind(V_j,V_k)$, then there must be another path between V_j,V_k in the skeleton besides $(V_i,V_j),(V_i,V_k)$, through which V_j,V_k influence each other;
2. If $\neg V(V_j,V_k \mid V_i)$, then V_j,V_k can not influence each other through V_i when V_i is known, since $\neg Ind(V_j,V_k \mid V_i)$, then there must be another path between V_j,V_k in the skeleton besides $(V_i,V_j),(V_i,V_k)$, through which V_j,V_k influence each other;

Under such condition $V(V_j,V_k \mid V_i)$ or $\neg V(V_j,V_k \mid V_i)$ can not be identified by CI_0 test or CI_1 test (which need higher order CI test). ∎

Based on the Proposition above, given a skeleton SK, following rules can be used to identify the V structure at V_i:

N (V_i) is the neighbors of V_i in SK, for $\forall V_j,V_k \in N(V_i) \wedge V_j \neq V_k$.

(1) If $PV(V_j,V_k \mid V_i)$ and $(V_i,V_j,V_k) \in CV$, then $V(V_j,V_k \mid V_i)$;
(2) If $PV(V_j,V_k \mid V_i)$ and $(V_i,V_j,V_k) \in NV$, then $\neg V(V_j,V_k \mid V_i)$;
(3) If $PV(V_j,V_k \mid V_i)$ and $(V_i,V_j,V_k) \in UN$ and there is a path between V_j,V_k in the skeleton besides $(V_i,V_j),(V_i,V_k)$, then generate two sub graph, one is $V(V_j,V_k \mid V_i)$ the other is $\neg V(V_j,V_k \mid V_i)$;
(4) If $\neg PV(V_j,V_k \mid V_i)$, it has that V_i,V_j,V_k is a complete graph, then V_j,V_k can not make a V structure at V_i;

The algorithm for generating DAG with a given VPDAG was given in [6]. The score of MEC can be calculated locally, the paper [7] presents a set of operator on MEC, the change in score by applying each operator can be calculated locally.

4 Experiment

In this section we describe the experiments carried out to evaluate LEBS. The experiment was divided into two groups. The purpose of first group is to evaluate the decrease of search space; the purpose of second group is to evaluate the computational efficiency of LEBS. The data set used during the experiment is Asia network, which presents the probability of a patient having tuberculosis, lung cancer or bronchitis respectively based on different factors.

After phase I, there are 9 edges left in SKC_1. Through the algorithm in [8], all DAG and MEC which satisfy SKC_1 can be generated. By applying CV UV and UN constraints, the E-space was decreased to 28.5%. Fig. 3. presents the decrease in D-space and E-space.

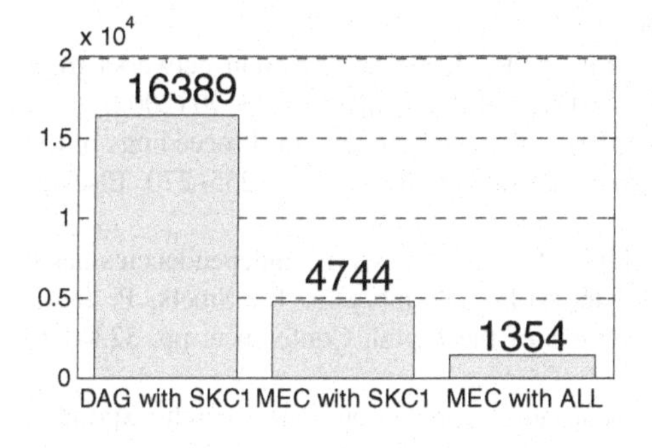

Fig. 3. Experiment result for evaluate the reducing of search space

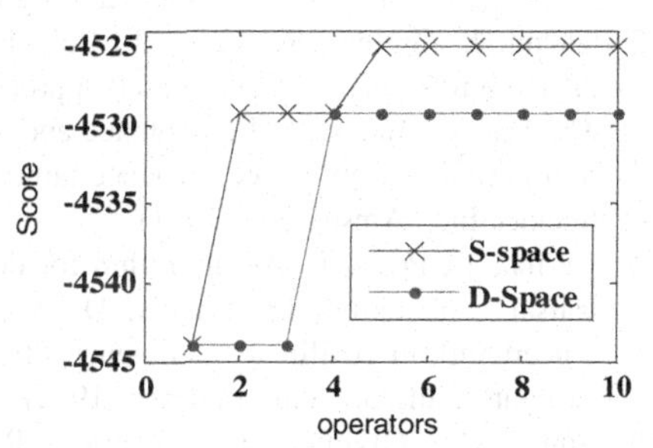

Fig. 4. Score changed by state transfer operator

In the second group of experiment, we evaluate the number of CI test and the score changed by search operator.

In phase I, CI_0 test is $O(n^2)$. The times of CI_1 in NV1 not only depend on the output of CV0, but also depend on the nodes order in which CI_1 performed. For example: $(V_i,V_j,V_k)\in UN_0$, if either (V_i,V_j) or (V_i,V_k) was eliminated in SKC_1 before calling $I(V_j,V_k \mid V_i)$, then such CI_1 is unneeded, because any skeleton in phase II satisfy $\neg PV(V_j,V_k \mid V_i)$. We test all possible node sequence for CI_1. The max, min and mean times for CI_1 are 44, 28 and 32.7. We use the number of V_i appear in UN_0, only when V_i is the first element of the triple, to sort the nodes. For descend (ascent) order the CI_1 is 25(40) times, so we use this heuristic information to perform CI_1.

In the phase II, we use greedy search strategy. In D-space, the state transfer operator is deleting edge and reversing edge. In S-space, the state transfer operator is deleting edge. For each edge the operator was applied (if more than one operator can be applied, the one that increase the score of the state most was applied). Fig. 4. represents the max score changed with each operator applied in both S-space and D-space.

5 Conclusion

The algorithm of this paper combining score-search based and constraint based approach for learning MEC. By using the constraint information in a more sufficient method, the search space was mapped from skeleton space to a constrained space of MEC. The size of such search space is reduced not only by eliminating edge, but also use the result to identify V structure. V structure is the main identity of MEC, so LEBS can search in E-space more efficiently.

References

1. Chickering, D.M., Heckerman, D., Meek, C.: Large-sample learning of Bayesian networks is NP-hard. Journal of Machine Learning Research 5, 1287–1330 (2004)
2. Cheng, J., Greiner, R., Kelly, J., Bell, D., Liu, W.R.: Learning Bayesian networks from data: An information-theory based approach. Artificial Intelligence 137, 43–90 (2002)
3. Thomas, V., Judea, P.: Equivalence and synthesis of causal models. In: Proceedings of the Sixth Annual Conference on Uncertainty in Artificial Intelligence, pp. 255–270. Elsevier Science Inc., Amsterdam (1991)
4. Verma, T., Pearl, J.: An algorithm for deciding if a set of observed independencies has a causal explanation. In: Dubois, D., Wellman, M.P., D'Ambrosio, B., Smets, P. (eds.) Uncertainty in Artificial Intelligence Proceedings of the Eighth Conference, pp. 323–330. Morgan Kaufman, San Francisco (1992)
5. Spirtes, P., Glymour, C., Scheines, R.: Causation, prediction, and search. Springer, Heidelberg (1993)
6. Dor, D., Tarsi, M.: A simple algorithm to construct a consistent extension of a partially oriented graph. Cognitive Systems Laboratory, UCLA, Computer Science Department (1992)
7. Chickering, D.M.: Learning Equivalence Classes of Bayesian-Network Structure. Journal of Machine Learning Research 2, 445–498 (2002)
8. Barbosa, V.C., Szwarcfiter, J.L.: Generating all the acyclic orientations of an undirected graph. Information Processing Letters 72, 71–74 (1999)

A WSMO-Based Semantic Web Services Discovery Framework in Heterogeneous Ontologies Environment

Haihua Li[1], Xiaoyong Du[1,2], and Xuan Tian[1]

[1] School of Information, Renmin University of China, China
[2] Key Laboratory of Data Engineering and Knowledge Engineering, MOE
{lihhjs, duyong, tianxuan}@ruc.edu.cn

Abstract. Nowadays, WSMO (Web Service Modeling Ontology)[1] has received great attention of academic and business communities, since its potential to achieve dynamic and scalable infrastructure for web services is extracted. Therefore, we design an ontology-based Semantic Web Services (SWSs) discovery framework based on WSMO so as to searching dynamically web services located at different nodes. Also, we provide several SWSs matching techniques based on this framework.

1 Introduction

Semantic heterogeneity naturally arises in the application of SWS, since different nodes may have a variety of differences in the levels of explicitness and formalization of concepts in their ontology specifications. In this scenario, the performance of SWSs discovery may be impaired. Take for example, how to search service resources from both IBM's UDDI[2] registry and Microsoft's UDDI registry. Currently, the main services selection approaches [2][4] improve their performance to some extent, but they are performed mainly on single UDDI registry and they do not essentially resolve service resources share on different nodes (e.g. different UDDI registries). Also, compared with other distributed system like Peer-to-Peer or Grids [7][8][9], web services have their own characters, e.g., their own structural and behavioral characters. Thus, we focus on web services discovery.

Nowadays, the major initiatives in the area of SWS are WSMO, OWL-S[3], METEOR-S [5] and IRS-II [3]. However, only WSMO among them explicitly consider semantic interoperability problems for web services. Currently, WSMO is still a conceptual model for describing web services semantically and only a small amount of work has been done on web services selection, discussed mainly in [11]. Therefore, we design a WSMO-based semantic web services discovery framework to resolve services discovery on different nodes and provide some

[1] http://www.wmso.org/
[2] http://www.uddi.org/
[3] http://www.daml.org/services/owl-s/1.0/

Z. Zhang and J. Siekmann (Eds.): KSEM 2007, LNAI 4798, pp. 617–622, 2007.

services selection techniques based on it. Noticeablely, our approach also provides a way for other dynamic resources selection.

2 A WSMO-Based Services Discovery Model

As shown in Fig. 1, web service providers or consumers have their own knowledge backgrounds, which are represented by their own Web Service Ontology Repositories respectively. In terms of personal preferences, providers publish their web services on one or several selected UDDI registries. Then, these UDDI registries re-annotate the above web services using different ontological descriptions respectively. With respect to consumers, they select their services by submitting service requirements into one or several UDDI registries. Each UDDI registry compares the coming request against its services by applying its own services matching approach. Note that compared with traditional UDDI registry, we define the so-called semantic enhanced UDDI registry whose services are all well annotated with lists of concepts.

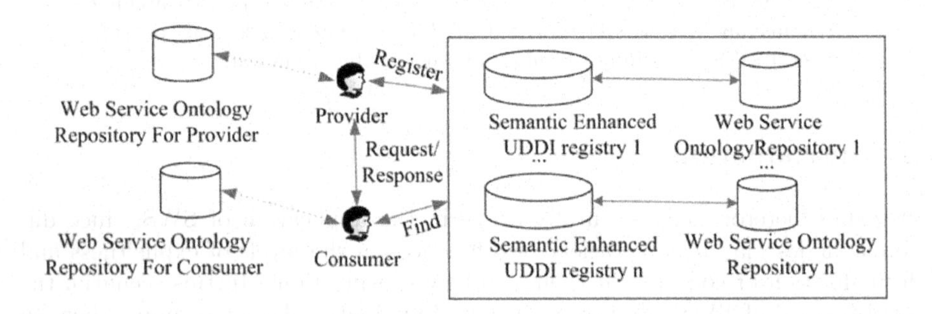

Fig. 1. A WSMO-based Services Discovery Model

3 Services Matching

Based on the graph theory to depict a dictionary (e.g. WordNet), a term can be represented as a node and a semantic relationship (e.g. is-A relationship) between terms (i.e. nodes) can be represented as an edge. Also, we assume that if there exist synonym and equivalence relationships between terms then these terms will be put in a same node. Accordingly, a dictionary is described as: $DGraph = (T, E)$ where T is the set of terms, E is the set of edges. Then, the semantic similarity between two terms is defined as:

$$Sim_{if}\left(t_1, t_2\right) = log\frac{hypos\left(t_1, t_2\right)}{Max}, \qquad (1)$$

where $t_1, t_2 \in T$, $hypos(t_1, t_2)$ represents the layer number of node p, which is the most immediate common parent of t_1 and t_2, and Max is the maximum layer number in $DGraph$. Note that we currently consider the semantic relationships mainly including equivalence relationship, synonymy relationship and is-A relationship, since other semantic relationships (e.g. meronyms) will provide limited effect on semantic similarity [10], and the information content of a concept includes both its linguistic and structural information (we adopt information content-based approach). In the following section, gc and ws are user goal and matched web service respectively, which both are represented by concepts.

3.1 Goal Concept Representation

Given a goal $gk = (k_1, k_2, \ldots, k_m)$, the matrix GCR (Goal-Concept-Relevance) represents the semantic similarity degrees between keywords in gk and concepts in Ontology O [12], and defined as: $GCR_{m*N} = (r_{ij})_{m*N}$ where r_{ij} represents the semantic similarity degree between keyword k_i and concept c_j, and $N = \|O\|$. Then, we obtain vector GV (Goal Vector), that is, $GV = (cw_1, cw_2, \ldots, cw_N)$ where $cw_j = \sum_{i=1}^{m} r_{ij}$. Finally, we define vector $gc = \{c_i | i = 1 \ldots m'\}$, where $cw_i \geq \delta \, (0 \leq \delta \leq 1)$ and δ is a threshold assigned by our system. Take for example, if a user wants to search *paper retrieval* services and submits a query like $gk = \{paper\ retrieval\}$, then $gc = \{paper, essay, report, retrieval, search, discovery, selection\}$. Also, we will pre-store the concepts representations of commonly used keywords in order to optimize its performance.

3.2 Semantic-Based Web Service Matching Approach

Function-based Services Matching. Firstly, we introduce an approach for matching of two concept sets, which is depicted as:

$$Semsim\,(CS_1, CS_2) = \frac{2\sum_{i=1}^{p} Sim_{if}\,(c_{1i}, c_{2a})}{p+q} \tag{2}$$

Where CS_1 and CS_2 are two concept sets respectively, $CS_1 = \{c_{1i} | i = 1 \ldots p\}$, $CS_2 = \{c_{2j} | j = 1 \ldots q\}$, for a given i, a satisfies that:

$Sim_{if}\,(c_{1i}, c_{2a}) = \max\{Sim_{if}\,(c_{1i}, c_{2j})\,| j = 1 \ldots q\}$. Note that compared with traditional similarity measures between two vectors or words (e.g. cosine), we adopt ontology or lexicon based approaches while they generally use corpora based statistical approaches. Thus, function-based service matching is defined as:

$$SWS\_F\,(gc.F, ws.F) = Semsim\,(gc.F, ws.F) \tag{3}$$

where $gc.F$ and $ws.F$ are their functional descriptions by concepts respectively, e.g. $ws.F = \{paper\ search\}$.

Structure-based Services Matching. Considering services structures, a service is composed by several operations which are associated with each other, and its structure can be defined as $ws.S = \{op_i | 1 \ldots u\}$. Operation $op = \{D, I, O\}$, where D is the text description of its function represented with concepts, I is its input information and O is its output information (I and O are its behavioral information). Also, we assume that operations of a web service is independent each other. Thus, the structure-based services matching is defined in the following part.

$$SWS\_S\,(gc.S, ws.S) = \frac{2\sum_{i=1}^{s} Semsim\,(op_{1i}.D, op_{2a}.D)}{s + s'} \quad (4)$$

where $gc.S = \{op_{1i} | i = 1 \ldots s\}$, $ws.S = \{op_{2j} | j = 1 \ldots s'\}$, for a given i, a satisfies that:

$Semsim\,(op_{1i}.D, op_{2a}.D) = \max\{Semsim\,(op_{1i}.D, op_{2j}.D)\,|j = 1 \ldots s'\}$, $gc.S$ and $ws.S$ are their structural descriptions respectively, e.g., the text description of operation op_0: $op_0.D = \{paper\ search\}$.

Behavior-based Services Matching. Each operation op has its own input and output information which are composed by several parameters. And for each parameter $p = \{N, T\}$, where N is its parameter name, T is its data type. Then, the matching of parameters p, p' is defined as:

$$PSim\,(p, p') = \omega Semsim\,(p.N, p'.N) + (1 - \omega) Semsim\,(p.T, p'.T) \quad (5)$$

where $p.N$, $p'.N$, $p.T$, $p'.T$ are all represented by concepts, ω is a threshold assigned by our system. Due to space constraints, the introduction of behavior-based services matching is omitted, since it is very similar to the structure-based service matching.

As described above, in semantic heterogeneity environments, web services matching can be defined as:

$$\begin{aligned} SimWS\,(gc, ws) = {} & \mu_1 SWS\_F\,(gc.F, ws.F) + \\ & \mu_2 SWS\_S\,(gc.S, ws.S) + \mu_3 SWS\_B\,(gc.B, ws.B) \end{aligned} \quad (6)$$

where SWS_B represents the behavior-based service matching, $gc.B$ and $ws.B$ are their behavioral descriptions by concepts respectively. μ_1, μ_2, μ_3 ($\mu_1 + \mu_2 + \mu_3 = 1$) are the thresholds given by our system.

4 Experimental Evaluations

We have created domain ontology A and B as our simulation ontologies using our good knowledge and rich experiences in areas of domain ontology design and evolvement [13].

4.1 Cross-Ontology Service Matching Evaluations

We chose 10 students of our computer science department and 16 pairs of goals and web services. Firstly, each student read the descriptions of both goals and web services. Then, different computational similarities were given using different approaches respectively. Finally, the students ranked the above similarity results with their satisfactions from 1 to 5, who were not informed of which approach is used for a specific similarity score. Also, the correlations were measured using the Pearson Product Moment Correlation. Table 1 shows that the correlations of human evaluations and different approaches. Observingly, these correlations are more than 58 % when $\mu_1 \geq 0.5$, while the effects in other scenarios are not very satisfactory. But, behavioral features play very limited role. Thus, we should consider functional and structural features as a whole in services search.

Table 1. Correlations for Different Web Services Matching Approaches and Human Evaluations

Approach	μ_1	μ_2	μ_3	Correlation
Function-based Approach/Simple Model	1	0	0	0.588
Structure-based Approach	0	1	0	0.345
A hybrid Approach-1	0.5	0.5	0	0.691
A hybrid Approach-2	0.67	0.33	0	0.702
A hybrid Approach-3	0.33	0.67	0	0.527
A hybrid Approach-4	0.34	0.33	0.33	0.403
General Model	0.67	0.33	0	0.762
Advanced Model	0.67	0.22	0.11	0.807

4.2 Different Web Services Matching Model Evaluations

Since such computation is very time-consuming, three different matching models are proposed in terms of user preferences:

1. Simple Model: This model is adopted to consider only the functional properties of web services.
2. General Model: The matching process in this model is performed using a hybrid approach based on the functional and structural features of web services.
3. Advanced Model: This model is implemented by considering all aspects including the functional, structural and behavioral features of web services.

Similar to section 4.1, we also chose 10 students and 16 pairs of goals and web services. Also, we set μ_1, μ_2, μ_3 in General Model and Advanced Model in terms of our testing statistics. Shown as Table 1, the correlation of human judgments and our Advanced Model is more than 80% although its matching implementation is very complex compared to other models, while the other model measurements have less correlation (0.588 and 0.762 respectively).

5 Conclusions

In this paper, we design a WSMO-based web services discovery framework and provide SWSs matching techniques on this framework. In the future, we will continue our researches in the area of SWS, e.g., researches on semantic association relationships. I believe that there are many efforts still to be made on it.

Acknowledgements

This work is partly supported by the National Natural Science Foundation of China under Grant NO. 60603020, 60496325 and 60573092.

References

1. Berners-Lee, T., Hendler, J.: The semantic web. Scientific American (2001)
2. Akkiraju, R., Goodwin, R., Doshi, P., Roeder, S.: A Method for Semantically Enhancing the Service Discovery Capabilities of UDDI. In: IJCAI-03 Workshop on Information Integration on the Web (August 2003)
3. Motta, E., Domingue, J., Cabral, L., Gaspari, M.: IRS-II: A Framework and Infrastructure for Semantic Web Services. In: the 2nd International Semantic Web Conference (2003)
4. Dong, X., Halvey, A., Madhavan, J., Nemes, E., Zhang, J.: Similarity Search for Web Services. In: the 30th VLDB Conference, Toronto, Canada (2004)
5. Patil, A., Oundhakar, S., Sheth, A., Verma, K.: METEOR-S Web service annotation framework. In: the 13th Int'l World Wide Web Conference, pp. 553–562. ACM Press, New York (2004)
6. Medjahed, B., Bouguettaya, A.: A Dynamic Foundational Architecture for Semantic Web Services. Distributed and Parallel Databases 17(2), 179–206 (2005)
7. Tangmunarunkit, H., Decker, S., Kesselman, C.: Ontology-based resource matching in the grid - the grid meets the semantic web. In: SemPGRID 2003, Budapest (May 2003)
8. Castano, S., Ferrara, A., Montanelli, S., Pagani, E., Rossi, G.: Ontology-addressable contents in P2P networks. In: WWW'03 1st SemPGRID Workshop, Budapest, Hungary (May 2003)
9. Castano, S., Ferrara, A., Montanelli, S.: Matching techniques for resource discovery in distributed systems using heterogeneous ontology descriptions. In: ITCC (2004)
10. Rodriguez, M., Egenhofer, M.: Determining Semantic Similarity among Entity Classes from Differenct Ontologies. IEEE Transactions on knowledge and data engineering 15(2), 442–456 (2003)
11. Kerrigan, M.: Web Service Selection Mechanisms in the Web Service Execution Environment (WSMX). In: SAC 2006. The 21st Annual ACM Symposium on Applied Computing, Dijon, France (April 2006)
12. Gruber, T.R.: A translation approach to portable ontology specifications. Technical Report, Knowledge System Laboratory, 92–71 (1993)
13. Hu, H., Zhao, Y.Y., et al.: Cooperative ontology development environment CODE and a demo semantic web on economics. In: APWeb 2005. The Seventh Asia Pacific Web Conference (2005)

Cardinal Direction Relations in 3D Space*

Juan Chen, Dayou Liu, Haiyang Jia, and Changhai Zhang**

College of Computer Science and Technology, Jilin University, Changchun 130012, China
Key Laboratory of Symbolic Computation and Knowledge Engineering of Ministry of Education, Jilin University, Changchun 130012, China
changhai@jlu.edu.cn

Abstract. The existing 3D direction models approximate spatial objects either as a point or as a minimal bounding block, which decrease the descriptive capability and precision. Considering the influence of object's shape, this paper extends the planar cardinal direction (CD) into 3D space and obtains a new model called TCD (three-dimensional cardinal direction). Base on the smallest cubic TCD relations and original relations, explain the correlations between basic TCD relations and block algebra. Then according to the results in block algebra, two novel ways to compute the inverse and composition of basic TCD relations are proposed. And an $O(n^4)$ algorithm to check the consistency of a set of basic TCD constraints over simple blocks is given.

1 Introduction

Modeling spatial relations is an indispensable part of Qualitative Spatial Reasoning (QSR) [1]. Direction as one of most important relations has received lots of attentions. According to different spatial objects, existing models can be divided into two classes: point based, such as cone or projection based methods [2] and double-cross method [3]; region based, such as rectangle algebra [4], direction relation matrix [5] and cardinal direction (CD) [6]. It should be noted most current models concentrate on planar objects except 3D-double cross model [7] and block algebra (BA) [8]. But these models approximate spatial objects either as a point or as a minimal bounding block (MBB), which decrease the description capability and precision. Based on this, we introduce new qualitative symbols to represent 3D direction then obtain a new 3D direction model, called TCD (three-dimensional cardinal direction).

The rest of paper is organized as follows. Section 2 introduces basic definitions of TCD. Section 3 elaborates the correlations between TCD and BA. Section 4 and 5 give the novel ways of inversing and composing basic TCD relations, Section 6 discusses the constraint satisfaction problems (CSP). Section 7 is the conclusions and future works.

* Supported by NSFC Major Research Program 60496321, Basic Theory and Core Techniques of Non Canonical Knowledge; National Natural Science Foundation of China under Grant Nos. 60373098, 60573073, 60603030 the National High-Tech Research and Development Plan of China under Grant No. 2003AA118020, the Major Program of Science and Technology Development Plan of Jilin Province under Grant No. 20020303, the Science and Technology Development Plan of Jilin Province under Grant No. 20030523.
** Corresponding author.

Z. Zhang and J. Siekmann (Eds.): KSEM 2007, LNAI 4798, pp. 623–629, 2007.

2 Basic Definitions

Consider the three-dimensional Euclidean space. Blocks are defined as non-empty and bounded sets of points in it. a is a simple block *iff* it is homeomorphic to the closed unite orb $\{(x, y, z)| \ x^3+y^3+z^3 \leq 1\}$. All simple blocks form a set, denoted by *BOD*. If a block is connected, then its projection on each axis forms single intervals on each axis. $inf_x(a)$ and $sup_x(a)$ are the greatest lower bound and smallest upper bound of the projection of a on x-axis. Similarly $inf_y(a)$, $sup_y(a)$, $inf_z(a)$ and $sup_z(a)$. They form the minimal bounding box of a -- $MBB(a)$, and divide the space into 27 direction tiles, as shown in Fig. 1, where superscript U, B, M denote the top, middle and bottom of a, respectively. Following the definitions of atomic CD relations in [6], The 27 atomic TCD relations can be defined by the projections on x, y and z axis. A basic TCD relation is a non-empty subset of $\{NW^U, \dots, SE^U, NW^M, \dots, SE^M, NW^B, \dots, SE^B\}$; as shown in Fig. 2 $dir(a, b)=N^U{:}O^U{:}O^M$ All basic TCD relations form a set of *JEPD* (jointly exhaustive and pairwise disjoint) relations, denoted by $\mathcal{D}$.

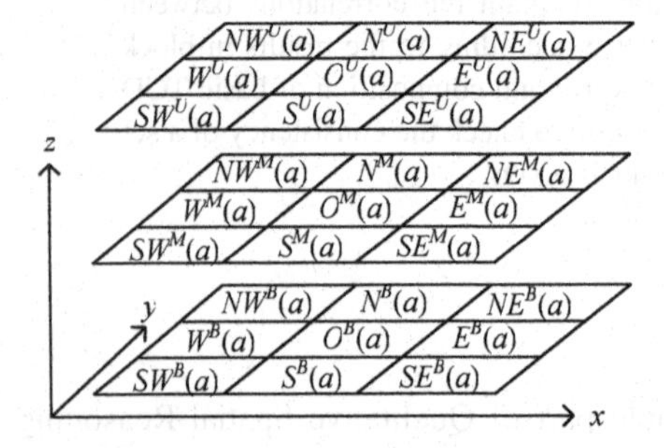

Fig. 1. The 27 atomic TCD relations

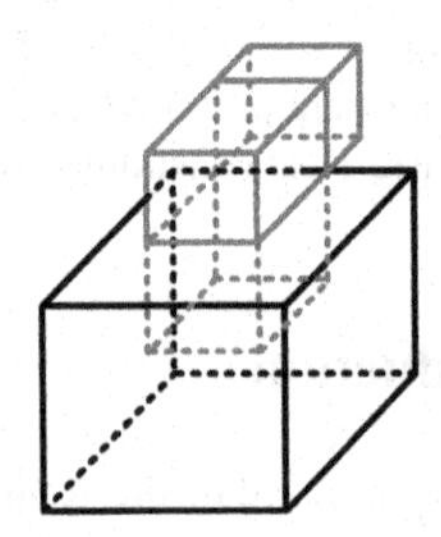

Fig. 2. $dir(a, b)=N^U{:}O^U{:}\ O^M$

3 Correlations Between TCD and BA

Balbiani [8] introduced BA as the n-dimensional extension of temporal interval algebra (IA) [9]. The considered objects are the blocks whose sides are parallel to the axes of some orthogonal basis in n dimension Euclidean space. Suppose n=3, the basic relations between these objects are defined from the basic interval relations, $\mathcal{A}_{int}=\{p, pi, m, mi, e, d, di, s, si, f, fi, o, oi\}$, in the following way: $\mathcal{A}_{cub}=\{(r_x, r_y, r_z)| \ r_x, r_y, r_z \in \mathcal{A}_{int}\}$. Each relation of $2^{\mathcal{A}cub}$ can be seen as the union of its basic relations. For arbitrary R, $T \in 2^{\mathcal{A}cub}$, it has: (1) the inverse relation $R^{\smile}=\cup\{(r_x^{\smile}, r_y^{\smile}, r_z^{\smile})| \ r\in R\}$; (2) the composition $R\circ T=\cup\{(r_x\circ t_x) \times(r_y\circ t_y)\times(r_z\circ t_z) \mid r\in R, t\in T\}$.

Definition 1. Basic TCD relation r is *cubic iff* there exist two rational cubes a and b such that $dir(a, b)=r$ is satisfied; otherwise, it is *non-cubic*.

According to the definitions of atomic TCD relations, the projections of arbitrary blocks on each axis must satisfy a certain interval relation, as shown in Table 1, where $\pi=\{x, y, z\}$ represents corresponding axis.

Table 1. The six groups of basic interval relation corresponding to cardinal direction relation

Interval	Projection relation	interval	Projection relation
$\{p, m\}$	$sup_\pi(a) \leq inf_\pi(b)$	$\{pi, mi\}$	$sup_\pi(b) \leq inf_\pi(a)$
$\{o, fi\}$	$inf_\pi(a) < inf_\pi(b) < sup_\pi(a) \leq sup_\pi(b)$	$\{oi, si\}$	$inf_\pi(b) < inf_\pi(a) < sup_\pi(b) \leq sup_\pi$
$\{e, d, s, f\}$	$inf_\pi(b) \leq inf_\pi(a) \wedge sup_\pi(a) \leq sup_\pi(b)$	$\{di\}$	$inf_\pi(a) < inf_\pi(b) \wedge sup_\pi(b) < sup_\pi(a)$

Let $BR(r)=BR_x(r)\times BR_y(r)\times BR_z(r)$ be the block relation of basic cubic TCD relation r and $CD(T)=\cup\{CD(t)|t\in T\}$ be the set of cubic TCD relations for block relation T. Such as $BR(N^U{:}N^M)=\{e, d, s, f\}\times\{pi, mi\}\times\{oi, si\}$ and $CD(\{s, o\}\times\{f\}\times\{o\})= CD((s, f, o))\cup CD((o, f, o))=\{O^B{:}O^M, W^B{:}O^B{:}W^M{:}O^M\}$.

Definition 2. Basic TCD relation $r=r_1{:}\ldots{:}r_m$ includes $t=t_1{:}\ldots{:}\ t_n$, *iff* $\{t_1,\ldots, t_n\}\subseteq\{r_1, \ldots, r_m\}$, denoted by $t\subseteq r$.

Definition 3. Basic TCD relation r is the smallest cubic direction (SCD) of t, denoted by $SCD(t)$, *iff* it is cubic and the smallest relation including t.

Definition 4. Basic relations whose smallest cubic direction is r are called the original relations of r, denoted by $ORG(r)$.

Base on above definitions, $\mathcal{D}$ can be divided into 6^3=216 equivalence classes. Each element in each class equals in its SCD, while each basic BA relation corresponds to a basic cubic TCD relation which is the SCD of a set of basic TCD relations. Thus TCD is correlated with BA.

4 Inversing

Definition 5. The inverse of $r\in\mathcal{D}$ denoted by $inv(r)$ is a TCD relation belongs to $2^{\mathcal{D}}$ which satisfies $\forall\, a, b\in BOD$, $dir(a, b)\in inv(r)$ holds *iff* dir$(b, a)=r$ holds.

Theorem 1. Let $r=r_1{:}\ldots{:}r_n$ be a basic relation, $q=SCD(r)$, then the following holds:

$$inv(r)=\{\cup_t ORG(t){:}\ t\in CD(BR\ (q)^\smile)\}$$

Proof: $\forall\, u\in inv(r)$, there must exist a, $b\in BOD$ such that $dir(a, b)=r\wedge dir(b, a)=u$ holds, then $dir(MBB(a), MBB(b))=q\wedge dir(MBB(b), MBB(a))=SCD(u)$ is satisfied. Let $t=SCD(u)$, it has $t\in CD(BR(q)^\smile)$. According to definitions of SCD relation and original relation, $u\in ORG(t)$. Conversely, $\forall\, t\in CD(BR(q)^\smile)$, there must exist cubes a, $b\in BOD$ such that $dir(a, b)=q\wedge dir(b, a)=t$ is satisfied. Assume $\forall\ u=u_1{:}\ldots{:}u_m\in ORG(t)$. Form block $b_0=b\cap(u_1(a)\cup\ldots\cup u_m(a))$ and $a_0=a\cap(r_1(b)\cup\ldots\cup r_n(b))$ then $dir(a_0, b_0)= r\wedge dir(b_0, a_0)=u$ holds, i.e., $u\in inv(r)$. So $inv(r)=\{\cup_t ORG(t){:}\ t\in CD(BR(q)^\smile)\}$holds. ■

5 Composing

Definition 6. Let r_1 and r_2 be two basic relations, their composition $r_1 \circ r_2$ is a set of basic relations satisfying: for each $t \in r_1 \circ r_2$, there exist blocks a, b, $c \in BOD$ such that $dir(a, b)=r_1 \wedge dir(b, c)=r_2 \wedge dir(a, c)=t$ is satisfied.

Theorem 2. For arbitrary basic relations r_1 and r_2, $r_1 \circ r_2 = r_1 \circ SCD(r_2)$ holds.

Lemma 1. Let r_1 be a atomic relation, r_2 be a basic cubic relation and $t \in r_1 \circ r_2$, then the following implication holds:

$$(\forall b, c \in BOD)(dir(b, c)=r_2 \rightarrow \exists a\, (dir(a, b)=r_1 \wedge dir(a,c)=t).$$

Proof: If $\forall b, c \in BOD$ satisfying $dir(b, c)=r_2$, assume $\beta_1(c),..,\beta_{27}(c)$ are the 27 tiles formed by $MBB(c)$, there must exist a maximal subset $\{\delta_1(c),\dots,\delta_m(c)\} \subseteq \{\beta_1(c),\dots, \beta_{27}(c)\}$, such that $\forall i \in \{1\dots m\}$, $\delta_i(c) \cap r_1(b) \neq \varnothing \wedge \forall \delta_j(c) \in \{\beta_1(c),\dots,\beta_{27}(c)\} \backslash \{\delta_1(c),\dots,\delta_m(c)\}$, $\delta_j(c) \cap r_1(b) = \varnothing$ holds. Let $t=t_1:\dots:t_k$ and $\delta=\delta_1:\dots:\delta_m$, if $t \in r_1 \circ r_2$ then $t \subseteq \delta$ holds, i.e., $\exists a \in BOD$ such that $a \in r_1(b) \wedge a \cap t_i(c) \neq \varnothing$, $1 \leq i \leq k$. Otherwise, there exists t_p, $1 \leq p \leq k$ such that $t_p(c) \in \{\beta_1(c),\dots, \beta_{27}(c)\} \backslash \{\delta_1(c),\dots,\delta_m(c)\}$, $a \cap t_p(c) \neq \varnothing$ holds. Since $a \in r_1(b)$ then $r_1(b) \cap t_p(c) \neq \varnothing$ contradicts $r_1(b) \cap t_p(c) = \varnothing$. ■

Theorem 3. Let $r_1=r_{11}:\dots:r_{1k}$ be basic relation and r_2 be a basic cubic relation, then $r_1 \circ r_2 = \sigma(r_{11} \circ r_2, \dots, r_{1k} \circ r_2)$ holds.

To facilitate illustration, let $\sigma(P_1,..,P_m)$ represents all rational basic cardinal directions formed by the cross joining the relation sets, such as $\sigma(\{O^M\},\{W^M{:}W^B,\ W^B\}, \{SW^M, O^B\})=\{\ W^M{:}O^M{:}\ SW^M{:}W^B,\ O^M{:}W^M{:}O^B{:}W^B,\ O^M{:}O^B{:}W^B\}$.

Theorem 4. Let r_1 be an atomic relation and r_2 be a basic cubic relation, then $r_1 \circ r_2 = \{\ \cup_t ORG(t){:}\ t \in CD(BR(r_1) \circ BR(r_2))\}$ holds.

Proof: $\forall u \in r_1 \circ r_2$, there must exist a, b, $c \in BOD$, such that $dir(MBR(a), MBR(b))= r_1 \wedge dir(MBR(b), MBR(c))=r_2$.Then $MBB(a)\ BR(r_1)\ MBB(b) \wedge MBB(b)\ BR(r_2)\ MBB(c)$ holds. So $dir(MBB(a), MBB(c)) \in BR(r_1) \circ BR(r_2)$, i.e., $SCD\ (u) \in CD(BR(r_1) \circ BR(r_2))$. Let $t=SCD(u)$ then $u \in ORG(t)$. Conversely, $\forall t \in CD\ (BR(r_1) \circ BR(r_2))$, there must exist cubes a, b, $c \in BOD$ such that $dir(a, b)=r_1 \wedge dir(b, c)=r_2 \wedge dir(a, c)=t$. Assume $\forall u=u_1:\dots:u_k \in ORG(t)$, form a simple body $a_0=a \cap u_1(c) \cap \dots \cap u_k(c)$, it have $dir(a_0, b)= r_1 \wedge dir(b, c)=r_2 \wedge dir(a_0, c)=u$. So $u \in r_1 \circ r_2$ holds. ■

To sum up, for arbitrary basic relation $r_1=r_{11}:\dots:r_{1k}$ and r_2, following equations must hold:

- $r_1 \circ r_2 = r_1 \circ SCD(r_2)$
- $r_1 \circ r_2 = \sigma(r_{11} \circ SCD(r_2), \dots, r_{1k} \circ SCD(r_2))$
- $r_1 \circ r_2 = \sigma(ORG(t_1), \dots, ORG(t_k))$, where $t_i = CD(BR(r_{1i}) \circ BR(SCD(r_2)))$, $1 \leq i \leq k$.

6 Reasoning with Basic TCD Constraints

The most common CSP in QSR is checking the consistency of a set of given spatial constraints with the form of *xRy*, where *x*, *y* denote spatial objects and *R* is the relation between them. If the consistency of this set of constraints can be decided in polynomial time, it is called tractable. This section will give an algorithm to check the consistency of a set of the basic TCD constraints over simple blocks.

Helly's Topological Theorem: Let $\mathcal{F}$ be a finite family of closed sets in *d* dimension such that the intersection of every *k* members of $\mathcal{F}$ is a cell, for $k \leq d$, and it is nonempty for $k = d + 1$. Then $\cap\mathcal{F}$ is a cell.

When d=3, $\cap\mathcal{F}$ is a cell means that the intersection belongs to *BOD*. To check the basic TCD constraint network, first is to extend the two-dimensional predicate NTB (non-trivially block) in [6] to 3D space, i.e., $inf_x(a)<sup_x(a) \wedge inf_y(a)<sup_y(a) \wedge inf_z(a)< supz_y(a)$; then follow the proof of theorem 1 and notations in [10], we get:

Theorem 5. Let $\mathcal{F}_a=\{\Sigma_a^{b1}, \ldots, \Sigma_a^{bk}\}$ be the family of sets of the maximal solutions for component variables of *a* w.r.t any of its reference variables b_i in an path-consistent basic TCD network $\mathcal{N}$ over *BOD*. If for every subfamily $\mathcal{B}_3=\{\Sigma_a^{b1}, \Sigma_a^{b2}, \Sigma_a^{b3}\}\subseteq\mathcal{F}_a$, NTB holds for $\mathcal{B}_3$ and $\cap\mathcal{B}_3$ is a simple block, then NTB holds for $\mathcal{F}_a$ and $\cap\mathcal{F}_a \in BOD$.

Proof: Since every subfamily $\mathcal{B}_3 =\{\Sigma_a^{b1}, \Sigma_a^{b2}, \Sigma_a^{b3}\}$the intersection $I_3 =\Sigma_a^{b1} \cap \Sigma_a^{b2} \cap \Sigma_a^{b3}$ is a simple block and NTB holds for $\mathcal{B}_3$, I_3 is the maximal partial solution for variable *a* w.r.t variables b_1, b_2 , b_3, so that all basic TCD constraints C_a^{bi} from the sub-network $\mathcal{N}_3$ with variables *a*, b_1, b_2 , b_3, are satisfied. Hence, if add a new set Σ_a^{b4} to $\mathcal{B}_3$, forming a subfamily $\mathcal{B}_4$, then $\mathcal{B}_4$ also satisfies predicate NTB. Otherwise Σ_a^{b4} contains a convex subset η which corresponds to a maximal solution-block for some component variable a_i^{b4} of *a* w.r.t b_4, so that η has not a non-trivial block intersection with some solution-block of other component variable a_i^{bp} of *a* w.r.t variable b_p of the sub-network $\mathcal{N}_3$ restricted to variables *a*, b_1, b_2, b_3. This means that the constraints between 3 variables, namely *a*, b_p and b_4, are not satisfied. But this contradicts the assumption that $\mathcal{N}$ is path-consistent. Hence NTB holds for $\mathcal{B}_4$.

Since $\mathcal{N}$ is path-consistent, then for every subfamily $\mathcal{B}_2\subseteq\mathcal{F}_a$, $\cap\mathcal{B}_2 \in BOD$ must hold. Otherwise $(dir(a, b_i) \circ dir(b_i, b_j))\cap dir(a, b_i)= \varnothing$, or $(dir(a, b_j) \circ dir(b_j, b_i))\cap dir(a, b_j)=\varnothing$ which contradict the assumption $\mathcal{N}$ is path-consistent. From the hypothesis it has$\cap\mathcal{B}_3 \in BOD$, and we prove $\cap\mathcal{B}_4\neq\varnothing$. Therefore,$\cap\mathcal{F}_a \in BOD$ must hold by Helly's topological theorem. Follow the proving procedure of first part, it is easy to prove NTB also holds for $\mathcal{F}_a$. ■

So we get the algorithm BOD-CON to check the consistency of basic TCD constraints over connected blocks, whose complexity is $O(n^4)$.

Algorithm: BOD-CON
Input: the basic TCD constraint network $\mathcal{N}$,
Output: 'consistent' if $\mathcal{N}$ can be satisfied; 'inconsistent', otherwise.

- Step 1: enforcing path-consistence: apply path-consistent algorithm to $\mathcal{N}$, if $\mathcal{N}$ is not path-consistent return 'Inconsistent'
- Step 2: Translating TCD constraints into order constraints(similar to step 1 in SK-CON[6])
- Step 3: Finding the maximal solution of the order constraints (similar to step 2 in SK-CON[6])
- Step 4: enforcing predicate NTB and simple block to the solution
  - For each variable $a \in V$ Do
    - For every tuple (b_1, b_2 , b_3,) of variables in V Do
      - If$\{\Sigma_a^{b1}, \Sigma_a^{b2}, \Sigma_a^{b3}\}$ doesn't satisfy NTB or $\Sigma_a^{b1} \cap \Sigma_a^{b2} \cap \Sigma_a^{b3} \notin BOD$
        Then return 'Inconistent'
- Step 5: Return 'Consistent'.

7 Conclusions

In this paper we extend planar CD relations to 3D space, consequently, propose TCD direction model. Base on SCD and original relations, correlations between TCD and BA are explained. Then two novel ways to compute the inverse and composition of basic TCD relations are given. And an $O(n^4)$ consistency checking algorithm is presented.

For future work, since temporal IA is the base of BA, and BA is correlated with TCD; introducing a temporal dimension into TCD to form a spatio-temporal direction model is a working direction. Due to the well-formed reasoning property of BA can be used to compute the inverse and composition of TCD relations. It is worth to observe that the same consistency problem we have considered may be solved with a non-constructive method, i.e., an algorithm that uses the operations of the algebra.

References

1. Cohn, A.G., Hazarika, S.M.: Qualitative spatial representation and reasoning: An overview. Fundamenta Informaticae. 43, 2–32 (2001)
2. Frank, A.U.: Qualitative spatial reasoning about cardinal directions. In: Mark, D., White, D. (eds.) Proc. of the 7th Austrian Conf. on Artificial Intelligence, pp. 157–167. Morgan Kaufmann, Baltimore (1991)
3. Freksa, C.: Using orientation information for qualitative spatial reasoning. In: Frank, A.U., Formentini, U., Campari, I. (eds.) Theories and Methods of Spatio-Temporal Reasoning in Geographic Space. LNCS, vol. 639, pp. 162–178. Springer, Heidelberg (1992)
4. Balbiani, P., Condotta, J.-F., Cerro, L.F.d.: A Model for Reasoning about Bidmensional Temporal Relations. In: Cohn, A.G., Schubert, L.K., Shapiro, S.C. (eds.) Proc. of Principles of Knowledge Representation and Reasoning (KR) Principles of Knowledge Representation and Reasoning, pp. 124–130. Morgan Kaufmann, Trento (1998)

5. Goyal, R., Egenhofer, M.J.: Similarity of Cardinal Directions. In: Jensen, C.S., Schneider, M., Seeger, B., Tsotras, V.J. (eds.) SSTD 2001. LNCS, vol. 2121, pp. 36–55. Springer, Heidelberg (2001)
6. Skiadopoulos, S., Koubarakis, M.: On the consistency of cardinal direction constraints. Artificial Intelligence. 163, 91–135 (2005)
7. Pacheco, J.M., Escrig, T., Toledo, F.: Integrating 3D Orientation Models. In: Escrig, M.T., Toledo, F.J., Golobardes, E. (eds.) CCIA 2002. LNCS (LNAI), vol. 2504, pp. 88–100. Springer, Heidelberg (2002)
8. Balbiani, P., Condotta, J.-F., Cerro, L.F.d.: A tractable subclass of block algebra: constraint propagation and preconvex relations. In: Barahona, P., Alferes, J.J. (eds.) EPIA 1999. LNCS (LNAI), vol. 1695, pp. 75–89. Springer, Heidelberg (1999)
9. Allen, J.: Maintaining Knowledge about Temporal Intervals. Communications of the ACM 26, 832–843 (1983)
10. Navarrete, I., Morales, A., Sciavicco, G.: Consistency Checking of Basic Cardinal Constraints over Connected Regions. In: Veloso, M.M. (ed.) 20th International Joint Conference on Artificial Intelligence, Hyderabad, India, pp. 495–500 (2007)

Knowledge Integration on Fresh Food Management

Yoshiteru Nakamori[1], Yukihiro Yamashita[1], and Mina Ryoke[2]

[1] School of Knowldge Science, Japan Advanced Institute of Science and Technology
1-1 Asagidai, Nomi, Ishikawa 923-1292, Japan
[2] School of Business Sciences, University of Tsukuba
3-29-1 Otsuka, Tokyo 112-0012, Japan
{nakamori, yukiyama}@jaist.ac.jp, ryoke@gssm.otsuka.tsukuba.ac.jp

Abstract. This study examines the issue of demand forecasting in fresh food management based on a methodology for knowledge integration and creation called the *i*-System. More specifically, we divide the domain of the study into "managerial knowledge," "purchase behavior," and "demand forecasting model/system," and endeavor to integrate the findings of each area of the study. Finally, we propose a fresh food management system based on the findings.

Keywords: Demand forecasting, fresh food management, *i*-System, managerial knowledge, KJ method, purchasing factors.

1 Introduction

Demand forecasting is an important management skill at grocery supermarkets which deal with perishable foods. Demand forecasting closely relates to sales rate, and demand forecasting errors directly influence the management performance of the company. Because perishable foods generally have short shelf life and are difficult to preserve, errors in demand forecasting result in excess product orders or production, that lead to added preservation/freshness management and personnel costs and losses due to waste and forced discounting. In addition, demand in excess of stock results in lost opportunity and customer trust. Therefore, demand forecasting is an important management skill at supermarkets, and minimizing the loss due to waste, forced discounting, and lost opportunity through effective demand forecasting plays an important role in the management activities.

Studies on demand forecasting are thought to have developed in various fields such as system engineering and information science, based mainly on the results of research utilizing mathematic and statistical models (e.g. moving average model, simple linear regression analysis, and multiple linear regression analysis). In such research, researchers narrow down the objects to forecast more specifically and approach demand forecasting problems in various fields such as retail business [1][2], production business [3][4], mail-order business [5][6],

Z. Zhang and J. Siekmann (Eds.): KSEM 2007, LNAI 4798, pp. 630–635, 2007.

textile/apparel business [7][8], electricity demand [9][10][11], and water demand [12][13]. In addition, the spread of information systems such as POS (Point Of Sales) and FSP (Frequent Shoppers Program) has increased the data, in addition to sales data, that can be manipulated and research on demand forecasting in the retail business has been progressing.

Because current approaches fail to provide adequate solutions to these problems, new approaches are required. Research on demand forecasting has developed mainly in the scientific and technological fields. Recently, however, attempts have been made to reconsider demand forecasting from the management perspective [14], an approach that may yield potential solutions to the problems that have plagued the traditional approach. With this in mind, this study assumes the necessity of constructing a demand forecasting system that combines a traditional system engineering approach, one that forecasts demand from past sales data, with a knowledge management approach in order to develop solutions to demand forecasting problems.

This study examines the issue of demand forecasting in fresh food management based on a methodology for knowledge integration and creation called the *i*-System. More specifically, we divide the domain of the study into "managerial knowledge," "purchase behavior," and "demand forecasting model/system," and endeavor to integrate the findings of each area of the study. Finally, we propose a fresh food management system based on the findings.

2 Methodology

This study employs the "Methodology for Knowledge Integration and Creation (*i*-System) [15]" to combine a system engineering type of approach with a knowledge management type of approach to address demand forecasting problems. The *i*-System is a "system theory that integrates and creates knowledge" combining Western structure - ability paradigm and Eastern dialectic thought. With the assumption that this structure is comprised of the scientific-actual front, the social-relational front, and the cognitive-mental front, and that the abilities of the actors under each structure are the intelligence ability, the involvement ability, and the imagination ability, this methodology develops the union of theory and practice, in which the integration power of knowledge and the intervention ability of the leaders and analysts in relation to the structure and abilities listed above become inseparable.

First, the problem and the result desired as a goal are set.

- Problem Setting: For demand forecasting problems, a new approach combining a system engineering approach with a knowledge management approach is needed.
- Knowledge: Construct a fresh food management system.

The problem is divided into the below-listed three fronts and approaches are tested. That is, knowledge regarding scientific truth, relations with society, and individual awareness is collected for each front and integrated:

- Cognitive-mental front: Survey and analyze issues that are considered when the amount and kinds of products to be produced are decided by the manager, who has the authority to decide such, and the management method related to daily sales.
- Social-relational front: Collect and analyze issues that consumers consider when purchasing products and consumer opinion that is not reflected in POS data.
- Scientific-actual front: Just as in past research approaches, construct a demand forecasting model that forecasts the sales amount under certain conditions based on past sales data and a fresh food management system. Note that for this front, a system is proposed based on the outcome of the cognitive-mental front and the social-relational front.

Using outcomes from the three above-listed fronts, we attempt to construct a demand forecasting management system that possesses the expertise of a system engineering approach as well as the expertise of a knowledge management approach.

3 Managerial Knowledge

In order to acquire managerial knowledge regarding prepared food department operation, such as decision making regarding the amount and types of prepared food to be produced and daily sales methods, we conducted interviews twice with managers responsible for decisions regarding the types and amount of prepared food to be produced.

The first survey was administered to ten managers working in Tokyo area supermarkets who have the power of decision regarding the types and amount of prepared food to be produced, and included a wide variety of items designed to elicit information concerning basic matters in the prepared food department, including issues to be considered when deciding on the types and amount of prepared food to be produced and measures taken to reduce lost opportunity and loss from waste. The period of research was from March 2nd, 2005 to March 23rd, 2005. The interview time was approximately one hour per person. The second survey was a follow-up interview of 3 managers who had participated in the initial survey. Follow-up questions concerned detail regarding prepared food department operation and were based on the results of the first survey. The survey period was from November 22nd, 2005 to November 24th, 2005, and the interview time was approximately 1 hour and 30 minutes per person.

We used the Semi-Structured Interview Method [16] from the Interview Guide. The Semi-Structured Interview Method is a surveying method wherein open ended question topics regarding the content that the researcher wishes to research are prepared in advance and the researcher actively asks questions in concordance with those question topics. However, it is also a survey method wherein the order of the questions and topics may be changed based on the conditions and atmosphere of the interview, allowing specific follow-up questions to be added on the spot in the course of conversation.

After making a transcript of the interview, information from the interview, such as considerations in deciding of the amount of prepared food products to produce and the types of prepared food products to produce and the daily sales method, were extracted and coded. The coded data was then organized, analyzed, and structured via the KJ Method [17], and the management methods of the prepared food managers were converted into diagram form. The diagramed management method of the prepared food managers was evaluated at the time of the second interview and amendments were added.

The results of analysis utilizing the KJ method show that, based in the "category of information which forms the foundation for prepared food product production and sales," the management of the prepared food departments is executed with reference to four categories: "decisions regarding prepared food product type and amount," "prepared food sales," "the time period from close to closing to closing time," and "tasks after closing."

4 Consumer Purchasing Behavior

In order to ascertain matters which consumers consider when deciding on prepared food purchases and consumer opinion which is not reflected in POS data, we implemented a web questionnaire related to prepared food purchasing. We conducted a web survey of 1,000 males and females from 16 to 69 years of age living in the 23 Special Wards of Tokyo who make prepared food purchases 2 or more times per week at a supermarket. The survey elicited responses to questions concerning matters considered when making prepared food purchases and preferences concerning prepared food purchasing. The survey period was from November 11th, 2005 to November 15th, 2005. In order to more accurately grasp actual consumer tendencies, we corrected the distribution of respondents for each question by age group in line with the distribution of actual Tokyo population by age group based on the FY 2000 national census issued by the Ministry of Internal Affairs and Communications.

We created survey items based on the results of interviews with prepared food managers. The survey was structured to begin with basic questions such as survey participant profile (age, gender, region, and family structure), frequency of prepared food purchase per one-week period, and names of prepared foods that the participant frequently makes purchases, and divided matters which the consumer considers in making prepared food purchases into items evaluated on a 5-point Likert scale from "very important consideration" to "not a consideration at all" and items which the respondent could comment freely about, such as times when prepared food purchases are made impulsively and preference concerning prepared food purchases.

Qualitatively comparing the survey results of managerial knowledge and consumer purchasing behavior, we evaluate the validity of managerial knowledge. Regarding "weather and climate," "in-store events," "yearly and regional events," and "television," though the individuals involved in sales consider these factors to affect the production amount and produced type, consumers do not seem to give much consideration to these factors when making prepared food purchases according to the purchasing factor results. Because of differences in opinion between managers and consumers, confirmation via objective data such as POS is required in the future. Regarding the "day of the week," though it was not a major consideration for prepared food purchases, we found a trend for more prepared food purchases during the weekend period as compared to weekdays. Thus, the opinion on this matter is the same for both managers and consumers.

5 Fresh Food Management System

Based on the results of the study on the managerial knowledge of prepared food departments and the study on consumer purchasing behavior pertaining to prepared food purchasing, we have developed a fresh food management system. As for a model which forecasts demand from past data, we used K-representatives [18], which is one form of clustering in which the concept of fuzziness is introduced into the traditional clustering technique. Specifically, we grouped data possessing the same conditions, estimated the distribution from the past sales amount data under those conditions, and forecasted the sales amount.

Via the results of the study on the managerial knowledge of prepared food departments and the study on consumer purchasing behavior pertaining to prepared food purchasing, it is thought that several important explanatory variables that increase the accuracy of demand forecasting results have been identified. However, it is not always possible to obtain data on these variables, a reason cited above for the low performance of demand forecasting in the real world. In order to respond to this problem, we designed the management system for this study, based on the experience and knowledge of fresh food product managers, in which the distribution calculated by a demand forecasting model could be altered freely, and constructed it so that the risk relationship of waste loss vs. lost opportunity would be indicated in line with the alteration.

6 Summary and Future Outlook

In order to develop a new fresh food management system, this study employed the *i*-System. The study fronts were divided into managerial knowledge study, consumer purchasing behavior study, and demand forecasting model and system study. In the managerial knowledge study, we administered an interview survey to prepared food department managers and gathered managerial knowledge for the prepared food department. In the consumer purchasing behavior study, we administered a questionnaire survey to consumers, surveyed the purchasing factors for prepared foods, and then qualitatively compared managerial knowledge and consumer purchasing behavior. In the demand forecasting model and system study, we constructed a system which is able to actively reflect the knowledge of human beings, unlike the way of thinking in traditional demand forecasting models and systems.

However, because the system constructed in this study exhibits a high degree of dependence on user experience and knowledge, it is necessary to allow the study results of managerial knowledge and consumer purchasing behavior to be reflected to a greater degree in the system. Moreover, regarding demand forecasting, the validation of forecasting accuracy and consideration of using other demand forecasting models are also necessary.

References

1. Hashimoto, I.: Estimated sale in large scale retailer (5). Aichi Institute of Technology research report B, 31, 61–66 (1996)
2. Nanjo, M., et al.: Research on sales amount forecast system in retail trade (1). Hiroshima Prefectural University bulletin 16, 49–61 (2004)
3. Namihira, H.: Research of tire demand forecast. Japanese management engineering association magazine 45, 539–545 (1995)
4. Noju, M.: Demand forecast and seasonal variation of beer. Operations Research 43, 426–430 (1998)
5. Matsuda, Y., Ebihara, J.: Forecasting Model in the Mail - Order Industry. Unisys technology review 71, 52–69 (2001)
6. Itsuki, R., et al.: Development and evaluation of expert system for demand forecast support in mail. The Inst. of Electrical Engineers thesis magazine C 118, 242–248 (1998)
7. Thomassey, S., Fiordaliso, A.: A hybrid sales forecasting system based on clustering and decision trees. Decision Support Systems 42, 408–421 (2006)
8. Thomassey, S., Happiette, M.: A neural clustering and classification system for sales forecasting of new apparel items. Applied Soft Computing 7, 1177–1187 (2007)
9. Tanaka, H., Hasegawa, J., Ito, M.: Power demand estimate the next day that depends on heavy regression analysis and hierarchical neural net. Operations Research 41, 499–503 (1996)
10. Kawai, K., Ono, M.: Development of electric power demand forecast system that applies neural net work. Operations Research 41, 481–486 (1996)
11. Haida, T., Muto, S.: Development of the maximum demand forecast support system based on heavy recurrence technique. Operations Research 41, 476–480 (1996)
12. Pulido-Calvo, I., Montesinos, P., Roldan, J., Ruiz-Navarro, F.: Linear regressions and neural approaches to water demand forecasting in irrigation districts with telemetry systems. Biosystems Engineering 97, 283–293 (2007)
13. Gato, S., Jayasuriya, N., Roberts, P.: Temperature and rainfall thresholds for base use urban water demand modeling. Journal of Hydrology 337, 364–376 (2007)
14. Nakano, M.: Consideration from demand forecast-management aspect as systematic knowledge creation. Osaka International University bulletin 17, 147–162 (2003)
15. Nakamori, Y., Zhu, Z.: Exploring a Sociologist Understanding for the *i*-System 1, 1–8 (2004)
16. Ito, T., et al. (eds.): Knowing while moving, thinking while involving -practice of qualitative research in psychology-. Nacanishiya publication Ltd (2005)
17. Kawakita, J.: Conception method - for the creativity development. Chuokoron-sha (2003)
18. Ohn, M.S., Huynh, V.N., Nakamori, Y.: A Clustering Algorithm for Mixed Numeric and Categorical Data. Journal of Systems Science and Complexity 16, 562–571 (2003)

An Approach to Knowledge Transferring in Science-Policy Process

Mitsumi Miyashita and Yoshiteru Nakamori

School of Knowledge Science, Japan Advanced Institute of Science and Technology,
1-1 Asahidai, Nomi, Ishikawa 923-1292, Japan
{mitsumi, nakamori}@jaist.ac.jp

Abstract. In this paper, we propose the roles of "intermediary actors" between scientists and policy makers for effective knowledge transferring in science-policy process. We adapted the Quinn's competing value framework to classify their roles into eight types. That is, intermediary actors in science-policy process play roles as monitors, coordinators, directors, producers, innovators, brokers, facilitators and mentors. We developed a list of 40 functions with each role consisting of five functions. Based on several case examples setting intermediary actors playing one of the eight roles between scientists and policy makers, we found that the intermediary actors play not only the designated roles, but also other roles for knowledge transferring. In all cases, intermediary actors play most of facilitator and mentor functions. These functions make effective knowledge transferring, that is knowledge sharing among scientists and policy makers.

Keywords: Science-policy process, knowledge transferring, knowledge sharing.

1 Introduction

From agenda setting to implementation, environmental policies in areas as diverse as air quality, climate change, water quality and land use; all depend on environmental monitoring and research to set emission limits, establish safe levels of exposure, evaluate the fate of pollutants in the ecosystem, and many other decisions at the local, national and international level [1]. The data that support this process are often complex, ambiguous, dispersed across multiple monitoring networks maintained by different organizations, provided one by one in many narrow technical papers, developed with competing theories, and presented with jargon that is not clearly understood by policy makers. The culture of science that generates and analyzes the data is very different from the culture of politics that uses the resulting knowledge for decision making. Environmental problems like climate change or water quality are not scientific problems or political problems alone, but interdisciplinary problems that require a unified science-policy solution.

Figure 1 shows the science-policy process which defines the conditions that facilitate the use of scientific data for policy [2]. The scientists convert raw data from monitoring networks into information. They interpret this information into scientific knowledge.

Z. Zhang and J. Siekmann (Eds.): KSEM 2007, LNAI 4798, pp. 636–641, 2007.

Then, scientific knowledge is transformed (or translated) into policy-relevant scientific knowledge by the collaborative works with scientists and policy makers. As the knowledge is provided to policy makers, they use it as one factor among many others in their decision making. It is important to note that scientific data is only one source of information that is a strong science-based component. The science-policy process represents a path from scientific data to the policy knowledge in a form that increases the likelihood that it will be used appropriately. The solid arrows indicate the path that goes by only scientists or policy makers and the dotted arrow indicates the path that goes by two different kinds of actors, scientists and policy makers. The difficulty in the dotted arrow path comes not only from the limitation of their knowledge but also from the different characteristics between scientists and policy makers. Scientists and policy makers are generally marked by very distinct behaviors and attributes. These differences contribute to some of the difficulties associated with transforming scientific knowledge into policy-relevant scientific knowledge.

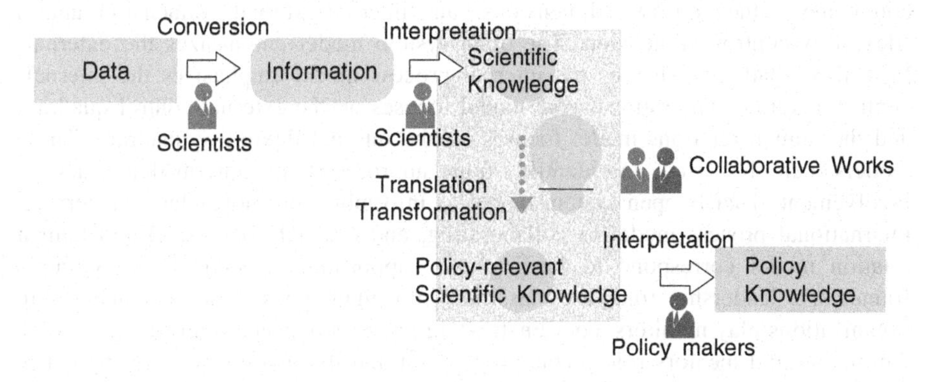

Fig. 1. Process from data to policy makers' knowledge [2]

In view of this, we propose the roles of "intermediary actors" between scientists and policy makers for effective knowledge transferring (Fig. 2) in science-policy process.

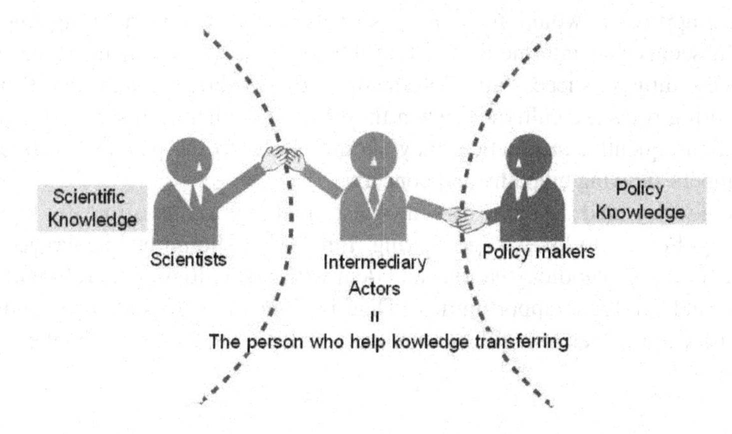

Fig. 2. Intermediary actor setting between scientists and policy makers

2 Roles of Intermediary Actors

There are mainly four different approaches of intermediary actors with involvement in knowledge transferring between scientists and policy makers, which are leading, informing, supporting and collaborating. Leading means that the intermediary actor gives scientists and policy makers chances to work together. Informing means that the intermediary actor brings scientific knowledge to policy makers. There are no direct collaboration works among scientists and policy makers, but contacted by intermediary actors individually. Supporting means that the intermediary actor has no power to control the project but help the collaboration of scientists and policy makers for knowledge sharing. Collaborating means that the intermediary actor involves in a project, and shares knowledge with scientists and policy makers.

The competing value framework can be extensively used in the discipline of leadership [3]. The framework comprises four models or quadrants created by competing values on two dimensions: an 'internal-external' dimension and a 'flexibility-control' dimension. The open system model emphasizes the external-flexibility quadrant, whereas the internal process model emphasizes the internal-control quadrant. The rational goal model focuses on the external-control quadrant and the human relations model focuses on the internal-flexibility quadrant. These models can be used to identify four approaches of intermediary actors' involvement. That is open system model as informing for knowledge transferring, international process model as collaborating, and rational goal model and human relation model correspond to leading and supporting, respectively. In Quinn's framework, leadership roles are classified into eight types. That is, managers in organizations play monitors, coordinators, directors, producers, innovators, brokers, facilitators and mentors roles. This proposition and its instrument were applied in this study.

A list of 40 functions shown in left side of Table 1 was developed under the eight types of actors. Each type consists of five functions. The right side indicates how intermediary actors involve to the relations between scientists and policy makers. The functions and relations show that mentor and facilitator roles, categorized in supporting approach, would be most positively associated with group knowledge sharing. It seems that intermediary actors are required to play mentor role to share knowledge (cultures) successfully. Intermediary actors should ensure that facilitating and mentoring roles are cultivated when they work together with scientist and policy makers. Consequently, in practice, they should attend to the personal difficulties of their members showing empathy and concern.

The innovator and broker roles are also making a big effort to knowledge transferring, but not to knowledge sharing. Innovation should be encouraged so that intermediary actors handle situations and deal with difficulties creatively with an eye to future and strategic opportunities. That is, innovator should also help policy makers (decision makers) to continuously adapt to changes in the external environment.

Table 1. List of functions in each type of intermediary actors and relations with scientists and policy makers

Type	Functions	Relations
Producer	1 Working productively 2 Utilizing available resources 3 Managing time 4 Managing stress 5 Fostering a productive work environment	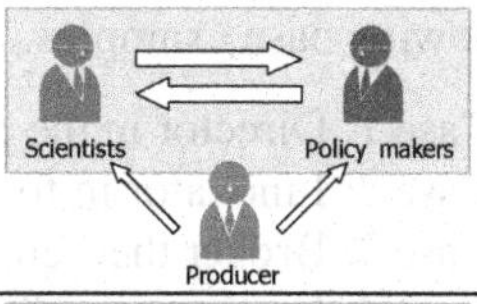 Scientists, Policy makers, Producer
Director	1 Setting goals and objectives 2 Preparing rules & policies 3 empowering members to implement a vision 4 Making decisions 5 Maintaining and sharing a vision	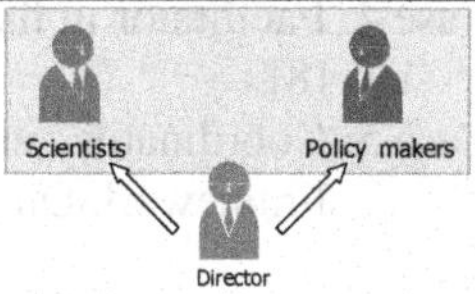 Scientists, Policy makers, Director
Innovator	1 Maintaining flexibility 2 Thinking creatively 3 Shaping ideas into solutions 4 Managing change 5 Translating solutions into practical terms	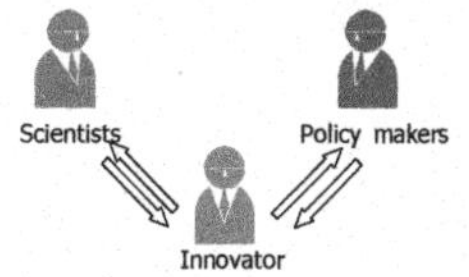 Scientists, Policy makers, Innovator
Broker	1 Building a power-base 2 Negotiating agreement and commitment 3 Maintaining a positive image and reputation 4 Maintaining a network of external contacts 5 Presenting ideas	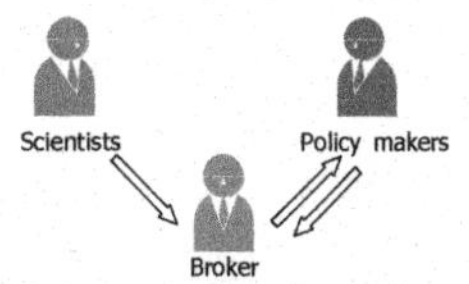 Scientists, Policy makers, Broker
Mentor	1 Awaring of individual needs 2 Communicating effectively 3 Training & educating to make wider understandable area 4 Motivating others to participate in decision making 5 Listening & caring actively	 Scientists, Policy makers, Mentor
Facilitator	1 Building teamwork 2 seeking consensus 3 Managing conflict 4 Defining roles and expectations 5 Encouraging expression of ideas	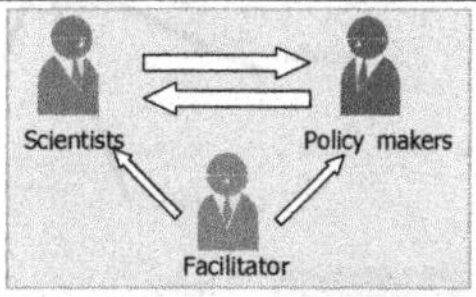 Scientists, Policy makers, Facilitator
Coordinator	1 Managing projects 2 Coordinating staff effectively 3 Solving problems 4 Recognizing members for their contributions 5 Sharing and exchanging relevant information	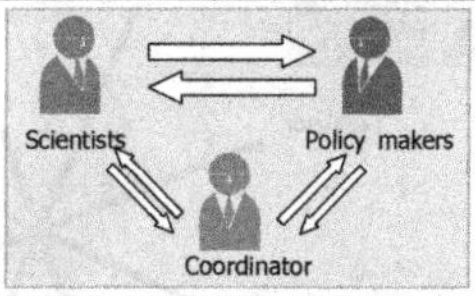 Scientists, Policy makers, Coordinator
Monitor	1 Providing a sense of stability and continuity 2 Auditing and analyzing the effort 3 Monitoring individual performance 4 Managing collective performance 5 Collecting and distributing information	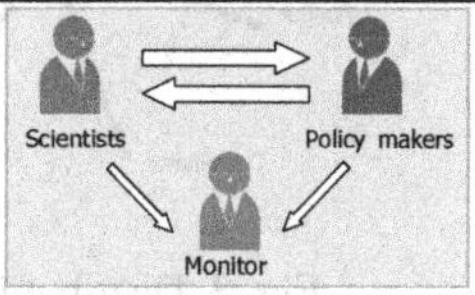 Scientists, Policy makers, Monitor

3 Case Examples

We used the function lists as a checklist to clarify intermediary actor roles in the following case examples.

Case 1: Director in the project of Coastal Water Program [4]
Case 2: Innovator in forest policy in California [5].
Case 3: Broker theoretically defined in literatures [6, 7].
Case 4: Facilitator in the Quantifying and Understanding the Earth System program [8].
Case 5: Coordinator in Science-Policy Interfacing in Support of the Water Framework Directive Implementation [9].

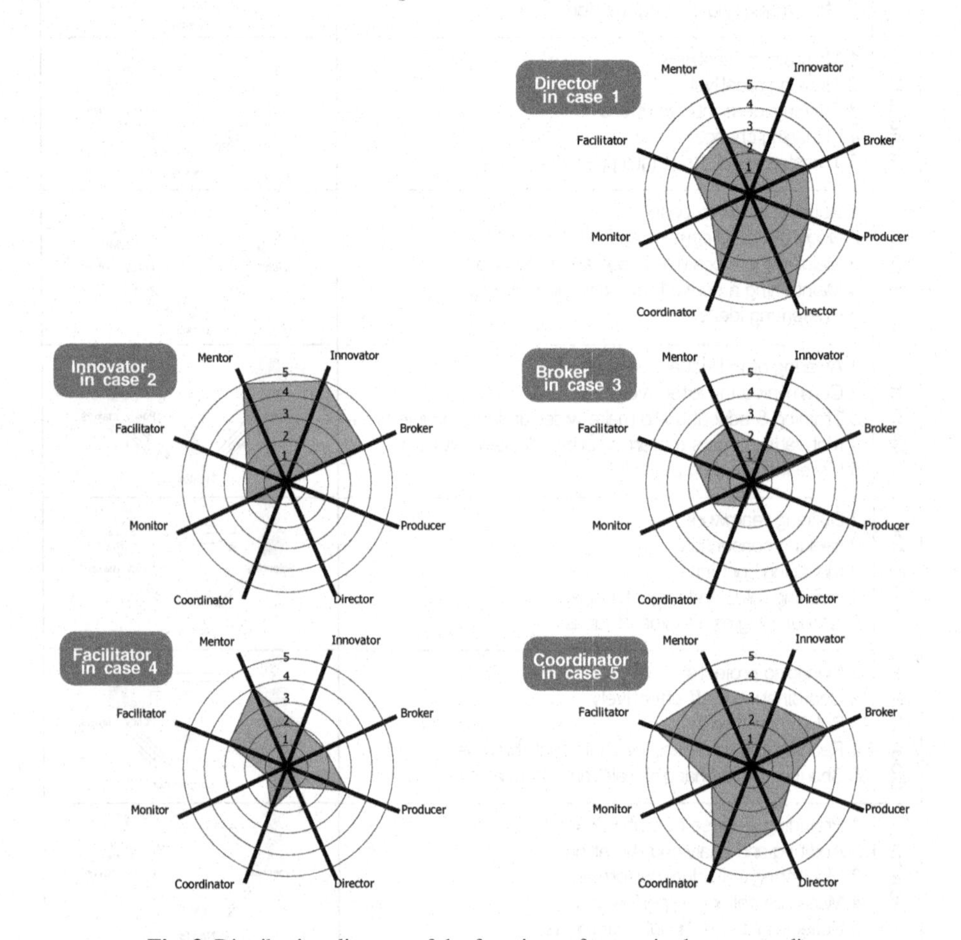

Fig. 3. Distribution diagram of the functions of actors in the case studies

Figure 3 shows the distribution diagram of the functions of actors in the case studies. The numbers represented by 1 to 5 in each type indicates the number of functions which intermediary actors play. When the number is small, the intermediary actor plays few functions as its type; meanwhile, the intermediary actor plays most functions when the number is large. Intermediary actors play most of functions categorized in the same role as they called. We also found that the intermediary actors play not only the designated roles, but also other roles. Most of facilitator and mentors roles are played by intermediary actors in all cases. It means that their roles make big effort for successful knowledge transferring from scientific knowledge to policy-relevant knowledge.

4 Conclusion

We proposed to set persons called "intermediary actors" for effective knowledge transferring in science-policy process. In view of this, this paper addressed to suggest the roles of intermediary actors. Since there are four different approaches for knowledge transferring which are leading, informing, supporting and collaborating, we adapted the Quinn's competing value framework to classify into eight types. That is, intermediary actors in science-policy process play roles as monitors, coordinators, directors, producers, innovators, brokers, facilitators and mentors. We developed a list of 40 functions, with each role consisting of five ones. Based on several case studies setting intermediary actors playing one of the eight roles between scientists and policy makers, we found that the intermediary actors play not only the designated roles, but also other roles for knowledge transferring. There are no position called monitor, mentor, broker and producer, but those roles are covered by other actors called in different name. In all cases, intermediary actors play most of facilitator and mentor functions. These functions make effective knowledge transferring, that is knowledge sharing among scientists and policy makers. On the other hand, the roles of innovator and broker are effective for knowledge transferring but not for knowledge sharing.

References

1. Engel-Cox, J.A., Hoff, R.M.: Science-policy data compact: use of environmental monitoring data for air quality policy. Environmental Science & Policy 8(2), 115–131 (2005)
2. Miyashita, M., Nakamori, Y.: Knowledge Management in Science-Policy Process. In: Proc. of the 7th International Symposium on Knowledge and Systems Sciences, pp. 265–270 (2006)
3. Quinn, R.E., Fearman, S.E., Thompson, M.P., McGrath, M.R.: Becoming a Master Manager: A Competency Framework. John Willey & Sons, Inc., New York (1996)
4. United States Commission on Ocean Policy, Commission Meetings Documents, http://www.oceancommission.gov/meetings/welcome.html
5. Abstracts: Blodgett Forest Research Symposium (2001), http://www.cnr.berkeley.edu/forestry/research/blodsym2001.html
6. United Nations System-Wide Earthwatch: Information for decision-making and Earthwatch, http://earthwatch.unep.net/about/docs/csd95str.htm
7. Canadian Health Services Research Foundation: The Third Community: Knowledge brokers, Research and Policy, Canadian Health Services Research Foundation, Ottawa (2004)
8. Natural Environmental Research Council: Science into policy: Taking part in the process, NERC (2000)
9. WFD Implementation Report, http://www.spi-water.eu/

Trust Analysis of Web Services Based on a Trust Ontology

Manling Zhu[1,2] and Zhi Jin[1]

[1] Academy of Mathematics and System Science, CAS, Beijing, China
[2] Graduate University of Chinese Academy of Sciences, Beijing, China
{zhumanling,zhijin}@amss.ac.cn

Abstract. This paper proposes a formalism for the Trust requirements modeling framework, which can be used as a means of studying the trustworthiness of service-oriented environments. We argue that a modeling framework, representing the underlying concepts and formal reasoning rules, can describe the Trust domain as well as support the dynamic analysis of trustworthiness. This model offers better understanding to the Trust relationships and will assist individuals in making rational decisions by computing trust level of potential interactors.

Keywords: Trust Ontology, Web Services, Formal Reasoning.

1 Introduction

As the open service-oriented information systems are widely used, trust is becoming one of the main issues in these increasingly networked environments. Techniques for system analysis and design have been focused primarily on addressing functional requirements, assuming that all parties are trusted [3]. However, in today's environments, we consider that web services are software agents which have their own requirements to fulfill, and certain core of competence, as well as common and special know-how knowledge on how to fulfill their own requirements [2]. Since the parties communicating frequently in multiple ways often do not have enough knowledge on each other, it is inevitable for software agents to consider the trustworthiness of the potential partners.

This paper develops a kind of trust ontology which is a conceptualization of trust related beliefs, and the underlying relations among them. There is also another effort on trust ontology for service-oriented environment [5] which puts much focus on the different types of trusted objects but did not anatomize the trust beliefs in agents' mind. We also have worked on proposing a preliminary trust model for service [6], but in this paper we improve on the trust dimensions and present a method for software agents to compare the beliefs respectively by a computational measurement based on the trust preference they own.

2 Concept Model of Trust

Trust is intuitively considered as a term in sociology. Quite similar with the social attitude in human society, the interaction between software agents indicates

Z. Zhang and J. Siekmann (Eds.): KSEM 2007, LNAI 4798, pp. 642–648, 2007.

their social nature, including cognitive attributes and human belief. Our trust model adopts social trust view [1], and describe the specific domain by proposing a generic formal trust ontology through eliciting trust related concepts, the associations among the concepts and the constraints.

2.1 Trust Concepts

Trust concepts are divided into concepts on belief and on Trust reasoning.

Concepts on Belief
Castelfranchi [1] extended the trust beliefs in D. Gambetta's social trust view [4] which is a reference for the trust dimensions in our concept model of trust.

Considering the most elementary case of trust: it is useful to identify what ingredients are really basic in one's mind of trust. A trustee must be capable of the service, and will act for trustor. Then the trustor has an important category of core beliefs on trustee:

1. *Competence Belief:* trustor believes trustee has the ability or power to do what trustor needs done.
2. *Intention Belief:* trustor believes trustee is motivated and intends to do the action. This makes trustee predictable.

In trustor's mind, it is also necessary to consider whether a trustee is honest, whether he is persistent about his intention, whether his behavior is predictable, and whether he is of good repute held by public. Then the trustor has another category of confidence beliefs on trustee:

3. *Integrity Belief:* trustor believes trustee makes good faith agreements, tells the truth, and fulfills promises. This is the trust in trustee's personality.
4. *Persistence Belief:* trustor believes trustee is stable enough in his intention, that he has no serious conflicts with it, or that he is predictable by character.
5. *Predictability Belief:* trustor believes trustee's actions are consistent enough that trustor can forecast him in a given situation.
6. *Reputation Belief:* trustor believes trustee gets credits for his past acts from public, that trustee is of good repute.

Concepts on Trust Reasoning
We extract the concepts on trust reasoning process from web service trust domain. At first we give what we mean of the agent service world. As shown in Figure 1, the agent service world consists of a number of agents. Every agent is capable of providing services, and may request for other services. Then the requestor considers the potential providers' trustworthiness through reasoning.

The basic concepts related to trust are represented as follows:

- An **agent** is an active entity. $Agent(a)$ holds iff a is an agent. All the agents form an agent world, which is called a set of agents. In a trust relationship, we can differentiate agents into two roles:
  - **Trustor** is an agent who initiates the trust relationship.
  - **Trustee** is an agent who receives the trust estimation.
- A **goal** is a condition or state of affairs in the world. $Goal(g)$ holds iff g is a goal. All the goals form a set of goals.

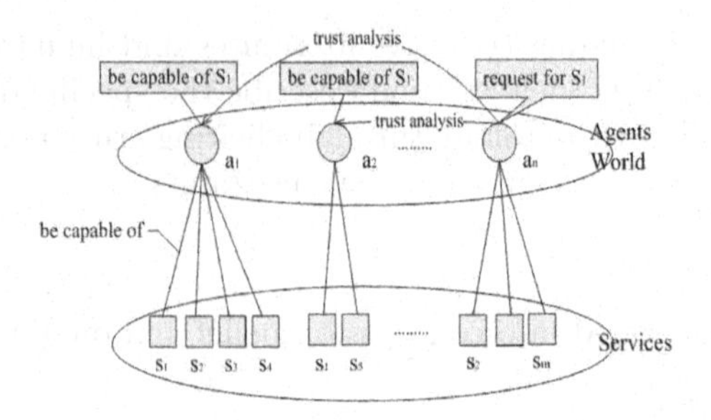

Fig. 1. The General Structure of The Agent Service World

- A **service** is a set of functions. $Service(s)$ holds iff s is a service.
- An **action** is an acting behavior which is the composing element of a service. $Action(act)$ holds iff act is an action.
- A **consequence** is a non-zero utility. Consequences of an action are those action effects, and all the consequences of an action are certain. The consequence set of a service is the union of the consequences of its sequence of actions. $Consequence(c)$ holds iff c is a consequence.
- A **resource** is a physical or informational entity. $Resource(e)$ holds iff e is a resource.
- A **know-how** is the knowledge and skill required to guide one's behaviors. $Know-how(k)$ holds iff k is a know-how knowledge and skill.

2.2 Associations

Besides classifying concepts, an ontology provides the associations between the ontological categories. Each association relates a source concept to a target concept. We group the associations into five parts depending on the different source concepts. Some of these kind of associations are listed in Table 1. In the formation of each association, the left side is called the source category and the right side is called the sink category.

2.3 Concept Constraints

We give trust reasoning rules which we group into six parts depending on the six different beliefs. Each group consists of the rules for reasoning relevant belief.

a and b are different agents. act is an action and c is the consequence of doing act. $\{act_i^s\}(i \in [1,n])$ are a set of actions composing service s. $s_j(j \in [1,m])$ are services. $g_j(j \in [1,m])$ are the goals to be achieved by doing service $s_j(j \in [1,m])$.

About Competence Belief. If an agent has the resource of an action and knows how to use the resource to act, then he is capable of doing the action. If an agent is able to do all the actions composing a service and has the composing knowledge of the service, then he can do this service.

Rule 1. $\boldsymbol{has\_resource}(a, e, act) \wedge \boldsymbol{has\_know\text{-}how}(a, k, act) \Rightarrow \boldsymbol{can\_action}(a, act)$

Table 1. The Main Associations Categories of Trust Ontology

Associations	formation	description
has_goal	Agent→Goal	Each Agent has some Goals to be achieved.
can_action	Agent→Action	An Agent has the capability of do-
intend_action	Agent→Action	An Agent intends to do an Action.
commit_service	Agent→Service	An Agent promises to provide a Service.
achieve_goal	Agent→Goal	An Agent achieves a Goal.
achieved_by_service	Goal → Service	A Goal is achieved by a Service.
consist_of	Service→Action	A Service consists of a sequence of Actions.
bring_about	Action→Consequence	An Action brings about a Consequence.
has_competence_part	Trust→Competence	Trust has Competence Belief and Intention
has_intention_part	Trust→Intention	Belief as its core belief part.

Rule 2. $\forall i \in [1,n]$, $\boldsymbol{can\_action}(a, act_i^s) \wedge \boldsymbol{has\_know\text{-}how}(a, k, act_i^s) \wedge \boldsymbol{consist\_of}(s, \{act_i^s\}) \Rightarrow \boldsymbol{can\_service}(a, s)$

About Intention Belief. If an agent wants to have a consequence, and he knows that an action brings about the consequence, then he intends to do the action. If an agent knows that an action brings about the consequence and tells another agent about that, then the second one knows that information. If an agent commits to provide a service, that is he commits to do all the actions composing this service, then he intends to do these actions.

Rule 3. $\boldsymbol{know\_by\_action}(a, c, act) \wedge \boldsymbol{want\_consequence}(a, c) \Rightarrow \boldsymbol{intend\_action}(a, act)$

Rule 4. $\boldsymbol{know\_by\_action}(b, c, act) \wedge \boldsymbol{tell\_about\_consequence}(b, a, \boldsymbol{bring\_about}(act, c)) \Rightarrow \boldsymbol{know\_by\_action}(a, c, act)$

Rule 5. $\forall i \in [1,n]$, $\boldsymbol{commit\_service}(a, s) \wedge \boldsymbol{consist\_of}(s, \{act_i^s\}) \Rightarrow \boldsymbol{commit\_action}(a, act_i^s) \Rightarrow \boldsymbol{intend\_action}(a, act_i^s)$

About Integrity Belief. If an agent commits to do actions he can not perform, that is he tells lie, then the agent is not honest.

Rule 6. $\exists i \in [1,n]$, $\boldsymbol{commit\_service}(a, s) \wedge \neg \boldsymbol{can\_action}(a, act_i^s) \wedge \boldsymbol{consist\_of}(s, \{act_1^s, ..., act_n^s\}) \Rightarrow \neg \boldsymbol{Integrity}(a)$

Rule 7. $\exists i \in [1,n]$, $\boldsymbol{tell\_about\_competence}(a, b, \boldsymbol{can\_action}(a, act_i^s)) \wedge \neg \boldsymbol{can\_action}(a, act_i^s) \wedge \boldsymbol{consist\_of}(s, \{act_1^s, ..., act_n^s\}) \Rightarrow \neg \boldsymbol{Integrity}(a)$

About Persistence Belief. If an agent can and commits to do a service, but actually he does not perform it, then this agent is not persistent.

Rule 8. $\exists i \in [1,n]$, $\boldsymbol{can\_service}(a, s) \wedge \boldsymbol{commit\_service}(a, s) \wedge \boldsymbol{consist\_of}(s, \{act_1^s, ..., act_n^s\}) \wedge \neg \boldsymbol{perform\_action}(a, act_i) \Rightarrow \neg \boldsymbol{Persistence}(a)$

About Predictability Belief. If an agent achieves a goal by doing a service, but does not achieve another goal by the relevant service, then he is not predictable.

Rule 9. $\exists j,k \in [1,m]$, $\boldsymbol{achieve\_goal}(a,g_j) \wedge \boldsymbol{achieved\_by\_service}(g_j,s_j) \wedge \neg$ $\boldsymbol{achieve\_goal}(a,g_k) \wedge \boldsymbol{achieved\_by\_service}(g_k,s_k) \Rightarrow \neg \boldsymbol{Predictability}(a)$

About Reputation Belief If an agent achieves a goal by doing expected service, then his reputation will increase and vice versa.

Rule 10. $\boldsymbol{commit\_goal}(a,g) \wedge \boldsymbol{commit\_service}(a,s)$
$\wedge \boldsymbol{achieved\_by\_service}(g,s) \wedge \boldsymbol{achieve\_goal}(a,g) \Rightarrow\uparrow \boldsymbol{Reputation}(a)$
$\boldsymbol{commit\_goal}(a,g) \wedge \boldsymbol{commit\_service}(a,s)$
$\wedge \boldsymbol{achieved\_by\_service}(g,s) \wedge \neg \boldsymbol{achieve\_goal}(a,g) \Rightarrow\downarrow \boldsymbol{Reputation}(a)$

3 The Process of Trust Analysis

We classified six trust beliefs into core beliefs and confidence beliefs according to the roles they play in trust. In particular, *Reputation Belief* is represented as an accumulated value because of the trustee's past actions.

3.1 Reputation Score

The *Reputation Belief* acquired from other agents' rating makes it is a vector using discrete values. We may adopt a credit scoring scheme to quantify the reputation belief and define a function of *Reputation Belief* to each agent, whose input is whether he achieves a goal or not when he acts as a trustee, and output is the credit value of an agent. And the computing strategy should encourage agents to act successfully and avoid failures. That is the strategy makes the credit value increase slowly and decrease fast. Also, if an agent fails to achieve an expected goal, the trustee's credit will decrease sharply according to his current credit value. That is if his credit value is high then the decreasing value is large and vice versa. So we give a computing procedure, in which $Credit\_Value$ is the function of credit score of an agent, $\alpha \in (0,1), \beta \in [0,1)$.

Procedure Credit Scoring
Input successOrFailure
Output Credit_Value
    if successOrFailure==ture
        Credit_Value=Credit_Value+1
    else Credit_Value=Credit_Value-(1+$\lceil$Credit_ Value $*\alpha+\beta\rceil$)

The whole circle of reputation scoring is three steps:

-When an agent enters a system, his credit value is 0.
-Whenever an agent does a service, run **Procedure** Credit Scoring.
-If the output is no more than -3, then this agent will be distrusted.

3.2 Trust Dimension Rules

In our trust dimension rules, $\{act_i^s\}(i \in [1,n])$ are a set of actions composing service s, and the concepts and associations are those we described above.

Trust Dimension Rule 1. $\forall i \in [1, n]$
$(\neg \textit{\textbf{can\_service}}(a, s) \vee \neg \textit{\textbf{intend\_action}}(a, act_i^s)) \wedge \textit{\textbf{consist\_of}}(s, \{act_i^s\})$
$\Rightarrow \neg \textit{\textbf{trust\_by\_service}}(*, a, s)$
Trust Dimension Rule 2. $\forall i \in [1, n]$
$\textit{\textbf{can\_service}}(a, s) \wedge \textit{\textbf{intend\_action}}(a, act_i^s) \wedge \textit{\textbf{consist\_of}}(s, \{act_i^s\})$
$\wedge (\neg \textit{\textbf{Integrity}}(a) \vee \neg \textit{\textbf{Persistence}}(a) \vee \neg \textit{\textbf{Predictability}}(a)$
$\vee \downarrow \textit{\textbf{Reputation}}(a)) \Rightarrow \downarrow \textit{\textbf{trust\_by\_service}}(*, a, s)$

As agents in the open environment may hold different preference to appraise trust, it is necessary to consider the trustors' preference among four confidence beliefs respectively. For example, an agent's whole preference may be:

Integrity Belief $\succeq$ *Persistence Belief* $\succeq$ *Predictability Belief* $\succeq$ *Reputation Belief*

That is agents sort trust dimensions by their significance from their perspective. After sorting, agents can compare different potential trustees' trust beliefs in turn, and select the most trusted one.

After one performance, the trustor adds the credit evaluation of the interactor to his credit value so that the reputation is updated. Then the agent does the next service selection according to the trust ontology with that updated reputation.

4 Conclusions

Our model, adopting social trust view, presents a formalism of service trust framework supporting analysis, reasoning and computing. Trust Ontology we proposed offers better understanding to the trust relations. As the model of trust is computable by a process of trust measurement, it will assist agents in making rational operation decisions in service-oriented environment.

Acknowledgments. The authors wish to thank the anonymous reviewers for their constructive comments and suggestions. This work was supported by the National Natural Science Fund for Distinguished Young Scholars of China under Grant No. 60625204, the Key Project of National Natural Science Foundation of China under Grant No. 60496324, the National 863 High-tech Project of China under Grant No. 2006AA01Z155.

References

1. Castelfranchi, C., Falcone, R.: Principles of trust for MAS: cognitive anatomy, social importance, and quantification. In: ICMAS 1998. Proceedings of the International Conference of Multi-Agent Systems, pp. 72–79 (1998)
2. Liu, L., Chi, C., Jin, Z., Yu, E.: Towards A Service Requirements Ontology Based on Knowledge and Intention. Technical Report. Knowware Group (2006)

3. Yu, E., Liu, L.: Modeling Trust for System Design Using the i* Strategic Actors Framework. In: Falcone, R., Singh, M., Tan, Y.-H. (eds.) Trust in Cyber-societies. LNCS (LNAI), vol. 2246, pp. 175–194. Springer, Heidelberg (2001)
4. Gambetta, D.: Can We Trust Trust? In: Gambetta, D. (ed.) Trust: Making and Breaking Cooperative Relations. Blackwell, Oxford Press, pp. 213–237 (1990)
5. Hussain, F.K., Chang, E., Dillon, T.S.: Trust Ontology for Service-Oriented Environment. In: AICCSA 2006. ACS/IEEE International Conference on Computer Systems and Applications, pp. 320–325. IEEE Computer Society Press, Los Alamitos (2006)
6. Zhu, M., Liu, L., Jin, Z.: A Social Trust Model for Services. In: 11th Australian Workshop on Requirements Engineering (2006)

Building Bilingual Ontology from WordNet and Chinese Classified Thesaurus

He Hu[1,2] and Xiaoyong Du[1,2]

[1] School of Information, Renmin University of China, 100872, Beijing, China
[2] Key Lab of Data and Knowledge Engineering, MOE, 100872, Beijing, China
{hehu,duyong}@ruc.edu.cn

Abstract. Both WordNet and Chinese Classified Thesaurus(CCT) are widely used in information retrieval and management systems. In this paper we propose a novel approach for building bilingual ontologies based on these existing knowledge bases, WordNet and CCT. The bilingual ontology has the merit that contains both domain related and general purpose semantic information coverage from these two complementary knowledge sources. A lattice based similarity measure and assessment algorithm is used for aligning and merging these two knowledge bases.

1 Introduction

Traditionally, Knowledge is considered as structured information, and the most important aspect of knowledge is to eliminate the "unstructureness" of information. Ontology is a suitable representation for knowledge. Domain ontology represents a domain of discourse, and contains relationships such as the definition of classes, relations, and axiom functions. With the increasing interest on semantic web, ontology and ontology-based systems have attracted more and more attention in computer science community[1].

Bilingual ontologies are extremely useful for applications such as cross-lingual information retrieval and machine translation; Reusing existing resources is becoming more and more attractive to ontology users and developers. Both WordNet [2] and Chinese Classified Thesaurus(CCT)[3] are widely used in information retrieval and management systems. Theses resources are carefully hand-crafted knowledge bases; they are very valuable as they can aid the process of building general purpose or domain oriented concepts hierarchies. In this paper we propose an approach for building bilingual ontologies based on these existing knowledge bases, WordNet and CCT. The bilingual ontology has the merit that contains both domain related and general purpose semantic information coverage from these two complementary knowledge sources.

2 Related Work

Carpuat et al. [4] propose a language independent, corpus-based method for automatically creating a bilingual ontology from two existing ontologies. It's

Z. Zhang and J. Siekmann (Eds.): KSEM 2007, LNAI 4798, pp. 649–654, 2007.

mainly a statistical approach, which is different with our metric-based alignment approach in this paper.

Wong S.H.S.[5] studies concept organisation in WordNet 1.5 and EuroWordNet 2. Based on the arbitrariness in concept classification observed in these wordnets, She argues that concept formation in natural languages is a plausible means to improve concept relatedness in lexical databases. Wong's work is in a systematic style; she analysis some shortcomings in WordNet and suggests the adoption of more natural semantic relatedness.

In building the bilingual ontology, a key issue is how to compute semantic similarity. Several distance-based methods for computing semantic similarity were proposed in[6,7]. We employ a lattice based metric and use Vector Space Model (VSM) [8] to compute the semantic similarity between WordNet synsets hierarchies and different levels of CCT terms by exploiting the hierarchical structure.

3 Lattice Based Ontology Comparison and Merging

3.1 Ontology Lattice

We mainly exploit the hierarchical relationships(hypernym in WordNet, subClassOf,BorderTerm and NarrowerTerm in CCT); these relationships organize synsets and terms of the two knowledge bases into a lattice structure. An ontology can be formally represented as a hierarchy of a concept space.

In WordNet, English words are organized into synonym sets groups, each representing one underlying lexicalized concept. The synsets are often organized in a hierarchy structure. Both the WordNet and CCT system follows a hierarchy structure, which can be formally depicted with a lattice structrue.

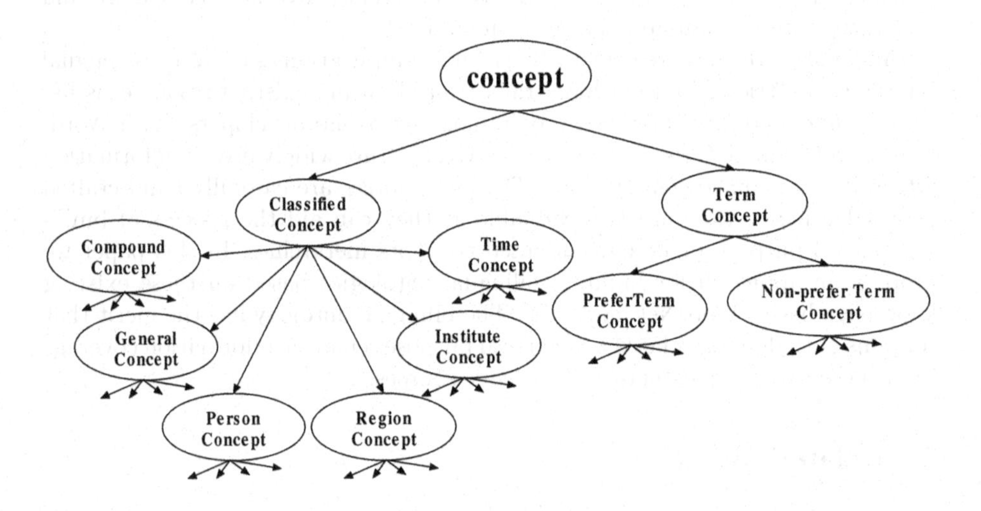

Fig. 1. Concept hierarchy from CCT

Definition. Ontology Lattice: For any particular domain $\mathcal{D}$, and a hierarchy relation $\mathcal{H}$, we use $\prec$ to represent the $\mathcal{H}$ relation: for any tow concepts C_1, C_2 satisfies $\mathcal{H}$ (C_1, C_2). We have $C_1 \prec C_2$. $(\mathcal{D}, \prec)$ forms an ontology lattice.

Popular ontology relations such as part-of, instance-of, kind-of(subclass-of) and attribte-of are all kind of hierarchy, relation $\mathcal{H}$; and they form a ontology lattice. The ontology lattice, cut this way, can be treated as a direct acyclic graph (dag), and be represented by matrix form.

Algorithm 1. Find Identical Concepts (FIC)

Input: Two lattices concepts A_i, B_j,ε_{sim}
Output: True or False
Begin
1: **if** $(Lex - Sim(A_i, B_j) < \varepsilon_{sim})$ **then**
2: RETURN True;
3: **end if**
4: **if** $SubConcept(A_i) \subseteq SuperConcept(B_j)$ **then**
5: RETURN True;
6: **end if**
7: **if** $SubConcept(B_j) \subseteq SuperConcept(A_i)$ **then**
8: RETURN True;
9: **end if**
10: **if** $Related(A_i, B_j)$ **then**
11: RETURN True;
12: **end if**
13: RETURN False;
End.

Algorithm 2. Lattice Alignment Algorithm (LAA)

Input: Two lattices **A**, **B**; m, n is the number of **A** and **B**'s nodes.
Output: a new lattices **A**, **B**
Begin
1: Loop in all A and B's nodes using ***Find-Identical-Concepts***.
2: Let k = number of **A** and **B**'s identical nodes.
3: For nodes in A which are identical with B's nodes,change it's number to corresponding B node number,
4: For all the other nodes, give them new numbers from n+1 to n+m-k.
5: For all nodes in B, the nodes numbers remain unchanged, new(empty) columns and rows are added from m+1 to n+m-k.
6: return lattices **A**, **B**;
End.

3.2 Alignment Algorithm

In most cases, the matrixes of Chinese and English ontologies (WordNet and CCT) are not in the same dimensions; to use some matrix comparison methods such as VSM, the matrixes must be transformed into the comparison space of the same dimension. Fig.2 demonstrates the transformation process. The numbers in bold indicate they are original columns or rows in lattice matrix before alignment; the other numbers (not in bold) indicate extended(empty) columns or rows filling by zeros.

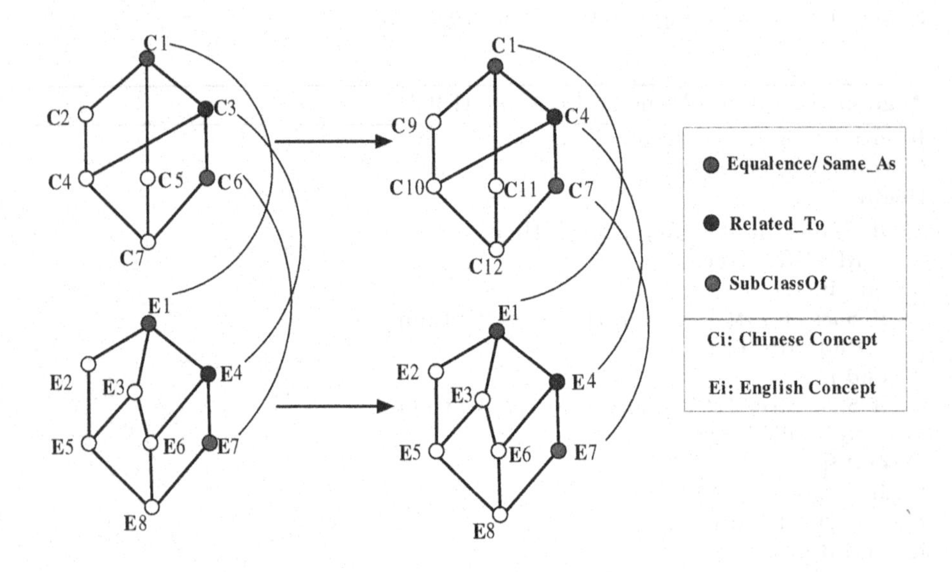

Fig. 2. Lattice Alignment of Chinese and English Hierarchies

After the matrixes being transformed into the same dimension space(in Fig. 3), the traditional model of VSM can be used. The measure of two Ontology lattices A and B is formulated as: Diff(A,B)=(Vect(A) · Vect(B)) / |Vect(A) · Vect(B)|.

Lex-Sim follows the edit distance formulated by Levenshtein[9] which is a well-established method for weighting the difference between two strings. It measures the minimum number of token insertions, deletions, and substitutions required to transform one string into another using a dynamic programming algorithm. Lex-Sim is defined by the following formula:

$$Lex - Sim(A_x, B_y) = max(0, \frac{|min(A_x|,|B_y|) - ed(A_x,B_y)}{min(|A_x|,|B_y|)}),$$

where $ed(A_x,B_y)$ is the edit distance between A_x and B_y.

The matrices generated from the alignment have some unique properties: the equivalent concepts (1,4 and 7 in the figure) are aligned at the same positions in the two matrices; all the other concepts are placed at different positions in matrices; the hierarchy relations are represented by matric values between concepts of corresponding rows and columns; with the value one indicating existence of the relationship, zero indicating non-existence of the relationship between two concepts;

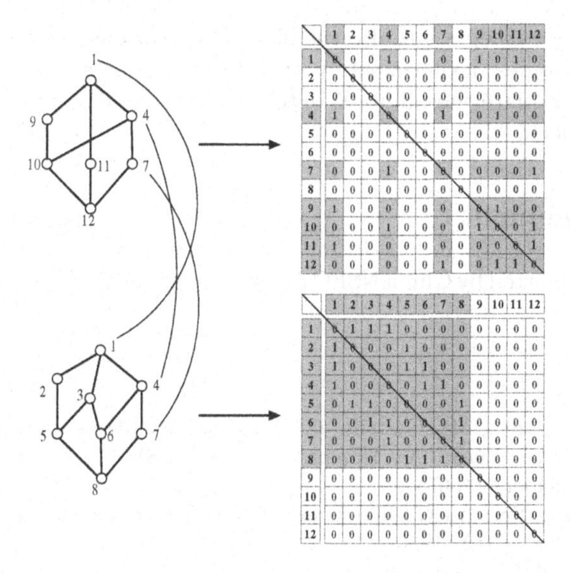

Fig. 3. Lattice Alignment and their VSM Representations

3.3 Experiments

Let n be the number of **A**'s nodes, and m be the number of **B**'s nodes; the matrix alignment algorithm need n*m loops of **Find-Identical-Concepts**. Let k be the average number of children nodes for **A** and **B**, the lattice comparison complexity will be n $\times$ m $\times$ 2k. We can see the algorithm is polynomial in time complexity, which indicates its' efficiency in real world applications.

By utilizing the lattice-based metric approach, we can combine the two concepts hierarchies of WordNet and CCT. We carry out experiments with some general purpose ontologies and domain ontologies respectively, in particular the spatiotemporal ontology and economic ontology. The spatiotemporal ontology and economic ontology includes the basic concepts and relations used in each field. The concepts hierarchy of these ontologies is based on "Chinese Library Classification" (CLC), a classification method proposed by CLC editors committee, which classifies the controlled vocabulary in some domains and is adopted widely for organizing literature resources. Spatiotemporal and economic ontologies have been created; the classes and properties are organized in hierarchy. We aligned and incorporate these CLC data with WordNet synset conocepts in our ontology building system.

4 Conclusion

We have presented a novel lattice based approach for building bilingual ontologies based on two existing knowledge bases, WordNet and CCT. The lattice based approach has unique advantages as it provides a unified method which takes into account not only concepts but also the hierarchy relationships between concepts in the two ontologies. Experiments with ConAnnotator annotation system are conducted for evaluate the bilingual ontologies, which show the efficiency of the proposed approach.

Acknowledgement

This work is supported by China NSFC programs (60496325, 60573092, 60603020).

References

1. Vallet, D., Fernndez, M., Castells, P.: An Ontology-Based Information Retrieval Model. In: Gómez-Pérez, A., Euzenat, J. (eds.) ESWC 2005. LNCS, vol. 3532, Springer, Heidelberg (2005)
2. Miller, G.: WordNet: An On-line Lexical Database. International Journal of Lexicography 3(4), 235–312 (1990)
3. Jun, W.: Digital Library Knowledge Organization based on Classification and Subject Thesauri. Journal of Library Science in China 30(151), 41–45 (2004)
4. Carpuat, M., Ngai, G., Fung, P., Church, K.W.: Creating a Bilingual Ontology: A Corpus-Based Approach for Aligning WordNet and HowNet. In: The Proceedings of GWC 2002, the 1st Global WordNet conference, Mysore, India (2002)
5. Wong, S.H.S.: Fighting Arbitrariness in WordNet-like Lexical Databases C A Natural Language Motivated Remedy. In: Sojka, Pala, Smrz, Fellbaum, Vossen (eds.) The proceedings of GWC 2004, the 2nd Global WordNet conference (2004)
6. Ganesan, P., Garcia-Molina, H., Widom, J.: Exploiting Hierarchical Domain Structure to Compute Similarity. ACM Trans. Inf. Syst. 21(1), 64–93 (2003)
7. Andreasen, T., Bulskov, H., Knappe, R.: On Ontology-based Querying. In: IJCAI. 18th International Joint Conference on Artificial Intelligence, Ontologies and Distributed Systems, pp. 53–59 (2003)
8. Salton, G., Lesk, M.: Computer evaluation of indexing and text prcessing. Journal of the ACM 12(1), 8–36 (1968)
9. Levenshtein, I.V.: Binary codes capable of correcting deletions, insertions, and reversals. Cybernetics and Control Theory 10(8), 707–710 (1966)

An Ontology-Based Framework for Building Adaptable Knowledge Management Systems

Yinglin Wang[1], Jianmei Guo[1], Tao Hu[2], and Jie Wang[3]

[1] Computer Science and Engineering Department, Shanghai Jiao Tong University, Shanghai 200030, China
ylwang@sjtu.edu.cn
[2] College of Tourism, Hainan Univ., Hainan 570228, China
[3] Dept. of Civil and Environmental Engineering, Stanford University, CA 94305, USA

Abstract. A framework of an adaptive knowledge management system is put forward. While the invariable infrastructure of the framework is coded in the system, the other user-specific features are supposed to be customized or defined later in the deployment phase by the end users. Moreover, a model for integrating the knowledge management process and the business process is presented. In this model, three spaces, i.e., the task space, the knowledge space and the process space, are proposed. The relationships of these spaces and the functions of the integrated system are discussed. Based on the framework, a system named ReKM has been developed and deployed in enterprises.

Keywords: knowledge management system, knowledge reuse, ontology, adaptable system.

1 Introduction

Knowledge and the ability of learning are the headspring of innovation and development in enterprises [1]. But in practice, trying to build a knowledge management (KM) system for an enterprise is a challenge. This is because the requirement of an enterprise often changes. Users often change their minds after they finish a prototype. So the knowledge management systems that can adapt to on-demand requirements are urgently needed [2].

From the viewpoint of the users, an ideal KM system should at least have the following capabilities:

- Structural modeling capability: support semantic knowledge modeling and reuse.
- Cooperative modeling capability: to integrate the KM process with the business processes in a teamwork environment.
- Adaptable modeling capability: adaptable to the changes of the requirement.

However, the KM systems that have all the above three capabilities are still beyond our reach. As to the commercial KM systems, the group wares, email groups and

Z. Zhang and J. Siekmann (Eds.): KSEM 2007, LNAI 4798, pp. 655–660, 2007.

forums do not have the structural modeling property. Besides, some of the contents contained in these systems may be incorrect because they did not be evaluated through professional evaluation processes [3]. On the other hand, the structured knowledge modeling approaches, such as ontology editing tools [4] [5], do not support process modeling for the teamwork environment yet.

In this paper we propose a framework for creating adaptable KM systems. We demonstrate ways of using ontology structure to model the knowledge. And then we discuss the ways of combing the KM processes and the business processes together.

In the following sections the framework is introduced first. The ontology model and the mechanism of adaptability are discussed. Then the integration model is depicted.

2 The Framework of the Methodology

Although the concepts and objects may be different in different enterprises, the underlying structure is common and almost invariable. So the common structure are extracted to compose the framework.

The key idea embodied in the framework is to separate the features that are general and almost unchangeable from those features that are apt to change later. In this frame-work objects are grouped into classes. Classes are the basic elements in the domain ontology. The workflow facility is used as the logic framework to model collaborative activities in the KM processes. In the framework illustrated in Fig 1, domain ontology plays a key role to set up a semantic bridge between different modules of the system.

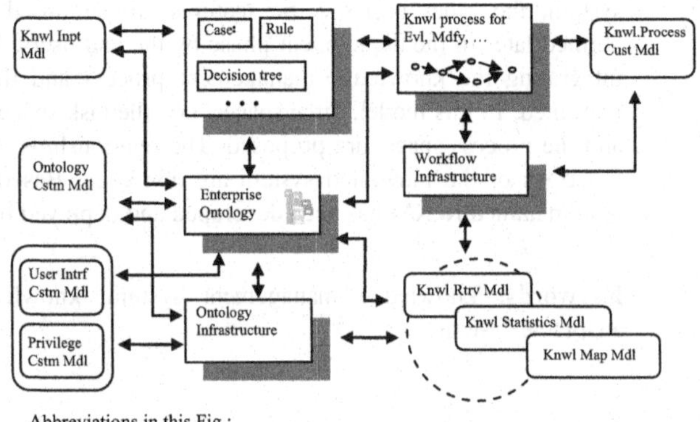

Fig. 1. The adaptable framework

3 The Adaptability Through Ontology Customization

3.1 The Ontology Model Used in ReKM

Ontology is a conceptualization which is an abstract, simplified view of the world that we wish to represent for some purpose [6]. The ontology model we adopted is defined as follows.

Definition: An ontology model includes five different parts: set of classes, set of enumeration types, set of relationships between classes, set of relationships between objects, set of axioms. It is as follows.

OntologyModel = (S_class, S_Enum, S_RelC, S_RelO, S_Axiom)

S_class indicates the set of all classes in the domain.

S_Enum indicates the set which contains all the enumeration data types. Enumeration data is a way to classify other objects in the world from a specific angle of view. For the richness of representation, here we suggest that sometimes the set of enumeration values of an enumeration type could be categorized in a hierarchy based on different granularities. E.g., all the enumeration values of *color* may be organized in a hierarchy. In the first level the values are listed as *White, Black, Red, Blue* etc., and in the second level *Red* may be subdivided into *Pink, Carmine,* etc, which could be divided further. The detailed values in the lower levels are called sub name of the abstract value, e.g., *Pink* is a sub name of *Red*, written as *Sub-Name('Pink','Red').*

S_RelC is the set of relationships between classes, such as generalization-specialization relationships, aggregation relationships etc.

S_RelO is the set of relationships among objects except properties of the classes. Relationships show the loose or random connections among the objects. E.g., we may define a relationship, named *occurr_before*, between two events, such as follows.

occur_before(Event e1,Event e2) iff e1.time < e2.time;

Moreover, other relationships may be defined via the relationships which have defined, e.g.,

1. *occur_after(Event e1,Event e2) iff occur_before(Event e1,Event e2);*
2. *occur_in_middle_of(Event e1, Event e2, Event e3) iff occur_before(Event e2,Event e1)* $\wedge$ *occurr_before(Event e1,Event e3).*

Sometimes relationships between two objects may not be deduced by other data. For instance, in the above example, if the exact occurrence times of the events didn't be defined, the relationship *occur_before* may need to be inputted to the system.

S_Axiom indicates the set of all the axioms. An axiom is a self-evident principle or universally recognized truth in a given domain. If the relation *occur_before(Event e1,Event e2)* has been defined as before, then the following axiom is true.

occur_before(e1,e3) $\leftarrow$ *occur_before(e1,e2)* $\wedge$ *occur_before(e2,e3).*

3.2 The Adaptation Mechanism

If the domain ontology could be customized by the end users, then adaptability would be possible. The customization process is as follows:

1. Firstly, edit the domain ontology via the ontology customization module;
2. Then, customize the interfaces for each of the classes which have been defined in the first step.
3. Store the customized interface styles in profiles for each kind of the interfaces of each class.

The framework is illustrated in Fig. 2. Through the customization process, the specific class structure of knowledge in the domain is supplied by the users. Then the modules which have been encoded will play their roles in the knowledge capture and reuse processes later.

For instance, when a user wants to retrieve some knowledge items that satisfy a condition, the user firstly identify a class in the class hierarchy (tree), and then the system will generate a retrieval interface according to the class definition and the interface style profile. Then the retrieval condition is defined by the user through the retrieval interface. And then the result will return to the user.

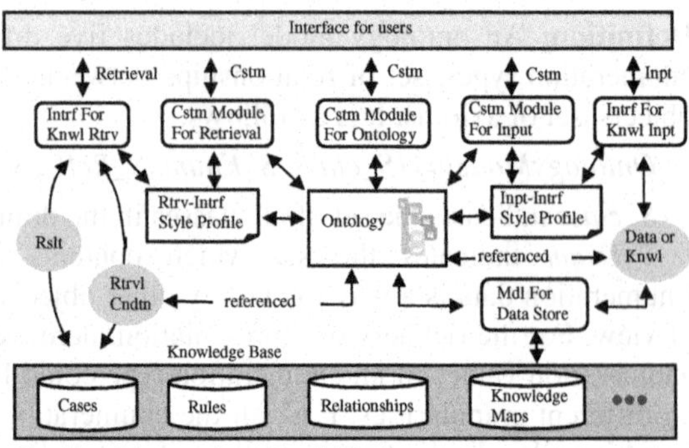

Fig. 2. The adaptation mechanism illustration

The above process is a universal process for retrieval, which can be used for any kind of domain ontology. During the deployment phase at different enterprises, the ontology should be customized by end users to deal with changes.

Basically, we use the case based method to represent knowledge items. Each case is an object of a class in the ontology. As class definition can be defined and even can be changed later by the users, the system becomes adaptable.

4 The Integration Model of Knowledge and Business Processes

Besides the modeling of the knowledge structure, the knowledge processes should also be modeled to support the teamwork environment [7]. Thus the integration issue between the knowledge management process and the business process emerges.

We put forward an abstract integration model as follows.

Integration Model:

IM={KS,PS,TS, RE}.

This model contains three spaces: the knowledge space *KS*, the process space *PS* and task space *TS*. They respectively indicates the set of all the knowledge items, the set of all the processes (including activities), and the set of all the tasks in a given domain. In *TS* and *PS* both the decomposition and generalization relationships may exits among their elements. In fact the tasks in *TS* play roles of the connection between the knowledge items and the processes (or activities). *RE* contains all the relationships between the elements of *TS*, *KS* and *PS*. They can be customized by the end users. As the space is limited, here we omit the related definitions.

Here we highly recommend that the users classify their tasks in a hierarchy based on the ontology infrastructure. In fact, we allow the users to define the task dimensions, e.g., a task may be described by its professional field (e.g., mechanical field, electrical field, system integration field, etc.), modality classification (design, drawing, analysis, etc.) and the related object (e.g., parts or product). The values in a dimension can be organized in hierarchy. For instance, *design* may be divided into *novel design* and *design change*, and drawing may be divided into drawing via

AutoCAD, drawing via UG, etc. The dimensions and hierarchical structure in *ST* space provide the basis for the integration.

Through carefully study of the requirements, we found at least four types of functions which should be integrated. They are as follows.

1. Process reuse: The process which solves a kind of problems can be reused later.
2. Active knowledge pushing service: The right knowledge items need to be pushed automatically to the right persons in the right time.
3. Knowledge evaluation: Knowledge should be checked through professional evaluation processes.
4. Knowledge capture: The raw knowledge items could be captured via routine business processes.

The above four functions can be implemented based on the abstract integration model. It is shown in Fig 3. Firstly, when a task is to be completed, the user is supposed to enquiry the system for the processes that can fulfill the task. Secondly, during the execution phase of business processes the knowledge items that are essential for fulfilling the task will be sent to the participant according to the relationships previously defined. Thirdly, every time a new knowledge item is found and submitted, its relationship with the tasks in *TS* should be provided by the submitter. Then a knowledge evaluation process is followed. Fourthly, sometimes what is acquired from a problem solving process can be regarded as raw knowledge or as hypo-knowledge. This kind of business process can be regarded as a knowledge capture process in the meantime.

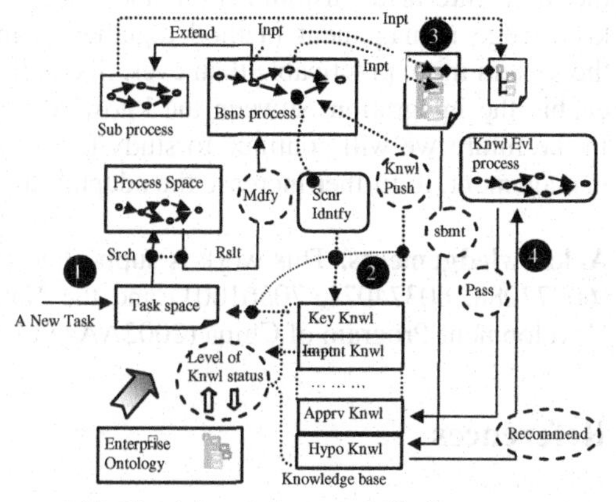

The abbreviations are the same as in Fig. 1 except: Scnr: scenario; Idntfy: Identify; sbmt: submit

Fig. 3. Function illustration of the integration model

The three spaces can be customized in the beginning of the deployment phase by the users or they can evolve later during the business execution processes. Thus adaptability is obtained.

5 Case Study and Conclusions

Fig. 4 is a simple example of the lessons learned cases in a bus company. The figure shows a scenario that a desiner submit an enquiry for the learsons learned cases in the case base. The designer chose the "lessons learned" class at first, and then the interface was automaticlly generated based on the customization profile and the domain ontology. Then the designer inputed the enquiry condition of the lessons learned cases on the "air compressor" or on the "engine". Then the system searched the case base and returned the list of the results. Finally the designer chose a case from the list to see its detail. The following is another enquiry example in a pest knowledge managemnet system of agriculture.

"select an instance x from the class pest_event where symptomsOf(symptom y, pest_event x) ∧ symptoms(symptom z, pest_event x) ∧ y. symptomName = 'BrownSpeckle' ∧ y. OnPosition = 'Leaf' ∧ z. symptomName ='DryShrink' ∧ y .OnPosition = 'Leaf'∧ Before (y, z)".

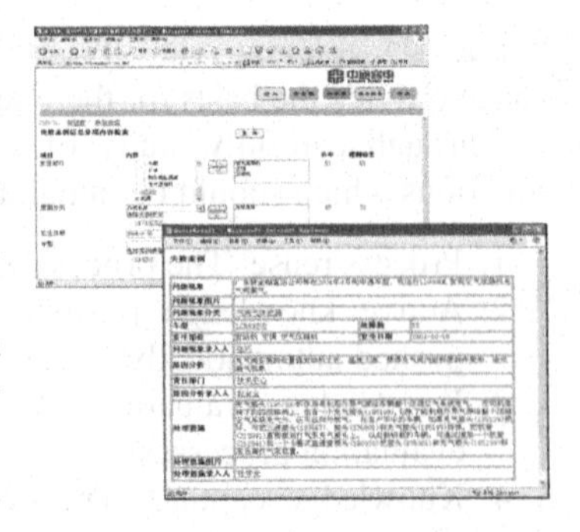

Fig. 4. A retrieved result

We have two main contributions in this paper. Firstly, we combined the ontology customization, the user interface customization and case based knowledge management method together to make the system more practicable than ever. Secondly, we proposed an integration model to enable the integration between the knowledge processes and the business processes. In the future we will continue to study the issue about ontology evolution or change management, to further increase the adaptability of the KM system.

Acknowledgements. This work is supportted by the Natural Science Funds of China (60773088, 60374071, 70561001) and the National High Technology Research and Development Program of China (2002AA411420) .

References

1. Leiponen, A.: Managing Knowledge for Innovation: The Case of Business-to-Business Services. Journal of Product Innovation Management 23(3), 238–258 (2006)
2. Sutcliffe, A., Sutcliffe, A.G.: The Domain Theory: Patterns for Knowledge and Software Reuse. CRC Press, Boca Raton (2002)
3. Borghoff, U., Pareschi, R.: Information Technology for Knowledge Management. Springer, New York (1998)
4. Chang, E., Sidhu, A., Dillon, T., Sidhu, B.: Ontology-based Knowledge Representation for Protein Data. In: Dillon, T., Yu, X., Chang, E. (eds.) INDIN 2005: 3rd International Conference on Industrial Informatics, Frontier Technologies for the Future of Industry and Business, Perth, pp. 535–539. IEEE, Perth, WA (2005)
5. Hogeboom, M., Lin, F., Esmahi, L., Yang, C.S.: Constructing Knowledge Bases for e-Learning Using Protege 2000 and Web Services. In: AINA 2005. 19th International Conference on Advanced Information Networking and Applications, pp. 215–220. IEEE Computer Society, Los Alamitos (2005)
6. Genesereth, M.R., Nilsson, N.J.: Logical Foundations of Artificial Intelligence. Morgan Kaufmann, San Mateo (1987)
7. Wang, Y.L., Wang, J., Zhang, S.S.: Collaborative Knowledge Management by Integrating Knowledge Modeling and Workflow Modeling. In: IRI 2005. The 2005 IEEE International Conference on Information Reuse and Integration, pp. 13–18. Las Vegas, Piscataway, (2005)

Knowledge Engineering Technique for Cluster Development

Pradorn Sureephong[1], Nopasit Chakpitak[1], Yacine Ouzroute[2], Gilles Neubert[2], and Abdelaziz Bouras[2]

[1] College of Arts, Media and Technology, Chiang Mai University. 239, Huawkaer Rd. T.Suthep A.Mueang 50200 Chiang Mai, Thailand
dorn@camt.info, nopasit@camt.info

[2] LIESP Laboratory, University Lumiere Lyon 2. 160 Boulevard de l'Université 69676 Bron Cedex, Lyon, France
firstname.lastname@univ-lyon2.fr

Abstract. After the concept of industry cluster was tangibly applied in many countries, SMEs trended to link to each other to maintain their competitiveness in the market. The major key success factors of the cluster are knowledge sharing and collaboration between partners. This knowledge is collected in form of tacit and explicit knowledge from experts and institutions within the cluster. The objective of this study is about enhancing the industry cluster with knowledge management by using knowledge engineering which is one of the most important method for managing knowledge. This work analyzed three well known knowledge engineering methods, i.e. MOKA, SPEDE and CommonKADS, and compare the capability to be implemented in the cluster context. Then, we selected one method and proposed the adapted methodology. At the end of this paper, we validated and demonstrated the proposed methodology with some primary result by using case study of handicraft cluster in Thailand.

Keywords: Knowledge Engineering, Industry Cluster, CommonKADS, Knowledge Management System.

1 Introduction

The knowledge-based economy is affected by the increasing use of information technologies. Most of industries try to use available information to gain competitive advantages[1]. From the study of ECOTEC in 2005[2] about the critical success factors in cluster development, first two critical success factors are *collaboration* in networking partnership and *knowledge creation* for innovative technology in the cluster which are about 78% and 74% of articles mentioned as success criteria accordingly. This knowledge is created through various forms of local inter-organizational collaborative interaction [3].

Study of Yoong and Molina [4] assumed that one way of surviving in today's turbulent business environment for business organizations is to form strategic

Z. Zhang and J. Siekmann (Eds.): KSEM 2007, LNAI 4798, pp. 661–666, 2007.

alliances or mergers with other similar or complementary business companies. Thus, grouping as a cluster seems to be the best solution to increase the competitiveness for companies[5][6]. Although, many literatures claimed that knowledge is very important for cluster development but there is no empirical method to initiate or improve knowledge sharing for cluster.

Developing knowledge-based application creates difficulties to knowledge engineers[7]. Knowledge-based project cannot be handle by general software engineering methodology. The lifecycle of knowledge based application and software application is different in many aspects. In order to achieve the objective of knowledge engineering, Knowledge-Based Engineering (KBE) application lifecycle [8] focuses on these six critical phases as shown in figure 1.

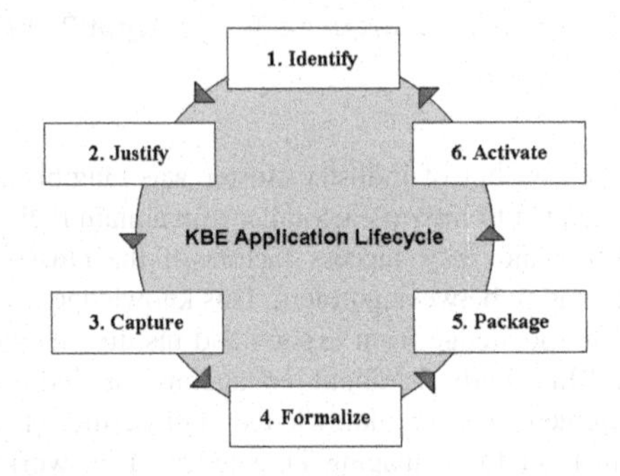

Fig. 1. Knowledge based engineering application lifecycle

1.1 Knowledge Engineering Techniques

Actually, knowledge is not a new idea [9] [10], philosophers and scholars had been studying it for centuries. There are many knowledge engineering techniques which are used for solving problem. However, we choose well known techniques that widely used in many projects for this study, i.e. MOKA, SPEDE and CommonKADS. These techniques were applied in different projects in various domains. All these methods are based on this KBE application lifecycle [8] which focuses on six critical phases as shown in fig. 1. We will analyze their capacity to be implemented in the cluster context.

MOKA (*M*ethodology and tools *O*riented to *K*nowledge based engineering *A*pplications) aims to help the structure side of the capture process (Fig. 1) in KBE application lifecycle. It focuses on two levels of representation - an informal and a formal model. These models provide the means of recording the structure behind the knowledge - including not only things about the product and design process but about the design rationale as well. Informal model is assembled from five categories of knowledge types; described on forms, known as ICARE forms (*I*llustrations, *C*onstraints, *A*ctivities, *R*ules and *E*ntities)[11].

SPEDE (*S*tructured *P*rocess *E*licitation *D*emonstrations *E*nvironment) provides an effective means to capture, validate and communicate vital knowledge to provide business benefit [12]. The concept of this methodology is to develop a guide that suitable for using with a variety of BPR/BPI projects, while at the same creating a means of enhancing process. This was made possible by breaking the activities into generally acknowledged high level BPR/BPI stages and providing a sequence at that level. The concept of this methodology is called *Swim Lane*. SPEDE provides Knowledge Acquisition (KA) tools to facilitate and assist the process in knowledge context.

CommonKADS (*Common Knowledge Acquisition and Design System*) is a methodology to support structured knowledge engineering. It provided *CommonKADS model suite* for creating requirements specification for knowledge system. The organization, task, and agent models analyze the organizational environment and the corresponding critical success factors for a knowledge system. The knowledge and communication models yield the conceptual description of problem-solving functions and data that were handled and delivered by a knowledge system. The design model converts it into a technical specification that is the basics for software system implementation [13].

1.2 Knowledge Engineering Technique Selection

In the study, we used knowledge-based engineering lifecycle and their provided tools as our criteria to select knowledge engineering technique for this study. The result of the comparison was shown in Table. 1.

Table 1. Three methods compared with Knowledge Based Engineering Lifecycle

KBE Lifecycle	MOKA	SPEDE	CommonKADS
1. Identify	-	Understand the project	Context Level
2. Justify	-	Understand the project	OTA Model
3. Capture	Informal Model	Design the process	Concept Level
4. Formalize	Formal Model	Evaluate the new process	Concept Level
5. Package	-	Communicate Process	Artifact Level
6. Activate	-	-	-

MOKA focuses on capturing and formalizing knowledge [11] to solve the specific engineering problems. On the other hand, we found that both SPEDE and CommonKADS techniques support KBE lifecycle from knowledge identifying to packaging phase. Then, we consider in the detail of each models of SPEDE and CommonKADS. SPEDE provided swim lane flowcharts as tools for each processes. SPEDE technique mainly focused on business process improvement in knowledge context. CommonKADS provided models and templates for each level. These templates support knowledge engineer for knowledge elicitation in different knowledge task, i.e. *analytic tasks* and *synthetic tasks*. These help knowledge engineer to be able to apply this technique in different type of knowledge problems. From two criteria, support KBE lifecycle and provided tools, CommonKADS technique is suitable for applying with industry cluster problems.

2 The Proposed Methodology

The concept of proposed methodology is adopted from CommonKADS methodology which was divided into three levels called *CommonKADS model suite*, i.e. Context Level, Concept Level and Artifact Level. However, managing structured knowledge in the industry cluster is different from single organization in many aspects because of characteristic of the organization. For example, there is no single policy maker in the cluster. So, KE could not utilize Context level's worksheets to assess single company for developing KMS for all companies within the cluster.

2.1 Context Level

This level contain organization, task and agent model. The main objectives of this level are giving the scope and clear view of the organization, knowledge intensive tasks and actors who involved in the task. It also provides the impact assessment, changes and consensus for knowledge engineering project. Knowledge Engineer (KE) should start at the most influence association in the cluster. Due to, association always be a group of potential companies in the industry and able to set policy/direction for the industry. KE could utilize Organization Mode (OM) worksheets for interviewing with associations. Then, the outputs from OM are knowledge intensive tasks from broken down process and agents who are related to each task. Then, KE could interview with experts in each task by using TM and AM worksheet. Finally, KE validate the result of each module with association again to assess impact and changes with OTA worksheet.

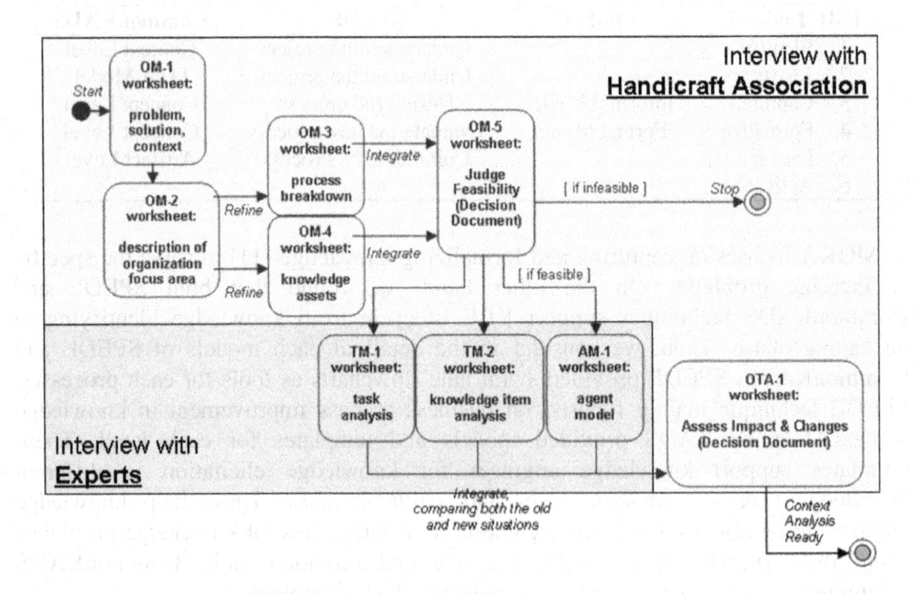

Fig. 2. A road map for carrying out knowledge oriented organization and task analysis

2.2 Concept Level

The second level contain knowledge and communication model. This level is related with capturing and formalizing phases in KBE lifecycle. The main objectives of this to the level are to explicate in detail the types and structures used in performing task and to model the communicative transaction between the agents involved [13]. We acquired Task knowledge from context level which composed of task goal, decomposition and control. Inference and Domain knowledge could be obtained from knowledge elicitation process.

2.3 Artifact Level

The last level is the artifact level, contain design model. This level is related to packaging phase in KBE lifecycle or application development. The main objectives of this level are to give the technical system specification, such as architecture, implement platform, software modules, representation constructs, and computational mechanisms needed to implement the functions laid down in the knowledge and communication models [13].

3 Validation and Results

The initial investigations have been done with 10 firms within the two biggest handicraft associations in Thailand and Northern Thailand. *NO*rthern *H*andicraft *M*anufacturer and *EX*porter (NOHMEX) association is the biggest handicraft association in Thailand which includes 161 manufacturers and exporters. Another association which is the biggest handicraft association in Chiang Mai is named Chiang Mai Brand. It is a group of qualified manufacturers who have capability to export their products and passed requirements of Thailand's ministry of commerce. Until end of 2006, there are 99 authorized enterprises to use Chiang Mai brand on their products[14].

At the beginning of this study, CommonKADS was used as a knowledge engineering methodology in the context level (organization model, task model, and agent model) in order to understand organization environment and corresponding critical success factors for knowledge system.

As shown in Fig. 2, Organization Model (OM-1 to OM-5), we found that handicraft cluster has its own vision as *"Knowledge sharing hub for handicraft exporter"*. And, companies defined their problems, such as intellectual property problem, lack of collaboration, CDA development, product innovation, and product exporting. However, this cluster has many opportunities and solutions as well. We used "product exporting" and "product innovation" as our mock-up problems due to these problems is knowledge intensive and feasible in business and technical aspect.

From the Task Model (TM-1 to TM-2), we analyzed the feasibility of each tasks that related to product exporting and product innovation processes. This model makes it possible to rank and prioritize the different knowledge-improvement scenarios.

Agent Model (AM-1) proposed organizational recommendations, improvements, and actions. From the experts' point of view, they proposed actions for solving product exporting and product innovation problems as follow,

1) Develop information system that provides knowledge from experts about product selection, marketing information, or economic data from government organization.
2) Archive past lesson learn or experiences with in electronic forum
3) Create best practice of each task and store in knowledge-based system
4) Increasing the collaboration and information sharing within the cluster.
5) Create tools to support the capability of cluster development agency (CDA) to facilitate the cluster.

References

1. Malmberg, A., Power, D.: (How) do (firms in) cluster create knowledge. In: DRUID Summer Conference 2003 on creating, sharing and transferring knowledge, Copenhagen, pp. 12–14 (June 12-14, 2004)
2. dti: A Practical Guide to Cluster Development.: Report to Department of Trade and Industry and the English RDAs by Ecotec Research & Consulting (2005)
3. Malmberg, A., Power, D.: On the roal of global demand in local innovation processes: Rethinking Regional Innovation and Change. In: Shapiro, P., Fushs, G. (eds.) Kluwer Academic Publishers, Dordrecht (2003)
4. Young, P., Molina, M.: Knowledge Sharing and Business Clusters. In: 7th Pacific Asia Conference on Information Systems, pp. 1224–1233 (2003)
5. Porter, M.E.: Competitive Advantage of Nations. Free Press, New York (1990)
6. Porter, M.E.: On Competition. Harvard Business School Press, Boston (1998)
7. Lodbecke, C., Van Fenema, P., Powell, P.: Co-opetition and Knowledge Transfer. The DATA BASE for Advances in Information System 30(2), 14–25 (1999)
8. Preston, S., Chapman, C.: Knowledge Acquisition for Knowledge-based Engineering Systems. Int. J. Information Technology and Management 4(1) (2005)
9. Davenport, T.H., Prusak, L.: Working Knowledge: How Organizations Manage What They Know. Haward Business School Press, Boston, MA (1998)
10. Klung, J., Stein, W., Licht, T.: Knowledge Unplugged. Palgrave, Great Britain (2001)
11. Stroke, M.: MOKA: methodology for knowledge-based engineering applications. Bury St Edmunds, Professional Engineering Publishing, UK (2001)
12. Shadbolt, N., Milton, N.: From knowledge engineering to knowledge management. British Journal of Management 10(4), 309–322 (1999)
13. Schreiber, A., Th., A.H., Anjewerden, A., de Hoog, R., Shadbolt, N., van de Velde, W., Wielinga, B.: Knowledge Engineering and Management: The CommonKADS Methodology. The MIT Press, Cambridge (1999)
14. Sureephong, P., Chakpitak, N., Ouzrout, Y., Neubert, G., Bouras, A.: Economic based Knowledge Management System for SMEs Cluster: case study of handicraft cluster in Thailand. In: SKIMA Int. Conference, pp. 15–10 (2006)

Author Index

Lecture Notes in Artificial Intelligence (LNAI)

Vol. 4827: A. Gelbukh, Á.F. Kuri Morales (Eds.), MICAI 2007: Advances in Artificial Intelligence. XXIV, 1234 pages. 2007.

Vol. 4798: Z. Zhang, J.H. Siekmann (Eds.), Knowledge Science, Engineering and Management. XVI, 669 pages. 2007.

Vol. 4795: F. Schilder, G. Katz, J. Pustejovsky (Eds.), Annotating, Extracting and Reasoning about Time and Events. VII, 141 pages. 2007.

Vol. 4790: N. Dershowitz, A. Voronkov (Eds.), Logic for Programming, Artificial Intelligence, and Reasoning. XIII, 562 pages. 2007.

Vol. 4788: D. Borrajo, L. Castillo, J.M. Corchado (Eds.), Current Topics in Artificial Intelligence. XI, 280 pages. 2007.

Vol. 4775: A. Esposito, M. Faundez-Zanuy, E. Keller, M. Marinaro (Eds.), Verbal and Nonverbal Communication Behaviours. XII, 325 pages. 2007.

Vol. 4772: H. Prade, V.S. Subrahmanian (Eds.), Scalable Uncertainty Management. X, 277 pages. 2007.

Vol. 4766: N. Maudet, S. Parsons, I. Rahwan (Eds.), Argumentation in Multi-Agent Systems. XII, 211 pages. 2007.

Vol. 4755: V. Corruble, M. Takeda, E. Suzuki (Eds.), Discovery Science. XI, 298 pages. 2007.

Vol. 4754: M. Hutter, R.A. Servedio, E. Takimoto (Eds.), Algorithmic Learning Theory. XI, 403 pages. 2007.

Vol. 4737: B. Berendt, A. Hotho, D. Mladenic, G. Semeraro (Eds.), From Web to Social Web: Discovering and Deploying User and Content Profiles. XI, 161 pages. 2007.

Vol. 4733: R. Basili, M.T. Pazienza (Eds.), AI*IA 2007: Artificial Intelligence and Human-Oriented Computing. XVII, 858 pages. 2007.

Vol. 4724: K. Mellouli (Ed.), Symbolic and Quantitative Approaches to Reasoning with Uncertainty. XV, 914 pages. 2007.

Vol. 4722: C. Pelachaud, J.-C. Martin, E. André, G. Chollet, K. Karpouzis, D. Pelé (Eds.), Intelligent Virtual Agents. XV, 425 pages. 2007.

Vol. 4720: B. Konev, F. Wolter (Eds.), Frontiers of Combining Systems. X, 283 pages. 2007.

Vol. 4702: J.N. Kok, J. Koronacki, R. Lopez de Mantaras, S. Matwin, D. Mladenič, A. Skowron (Eds.), Knowledge Discovery in Databases: PKDD 2007. XXIV, 640 pages. 2007.

Vol. 4701: J.N. Kok, J. Koronacki, R. Lopez de Mantaras, S. Matwin, D. Mladenič, A. Skowron (Eds.), Machine Learning: ECML 2007. XXII, 809 pages. 2007.

Vol. 4696: H.-D. Burkhard, G. Lindemann, R. Verbrugge, L.Z. Varga (Eds.), Multi-Agent Systems and Applications V. XIII, 350 pages. 2007.

Vol. 4694: B. Apolloni, R.J. Howlett, L. Jain (Eds.), Knowledge-Based Intelligent Information and Engineering Systems, Part III. XXIX, 1126 pages. 2007.

Vol. 4693: B. Apolloni, R.J. Howlett, L. Jain (Eds.), Knowledge-Based Intelligent Information and Engineering Systems, Part II. XXXII, 1380 pages. 2007.

Vol. 4692: B. Apolloni, R.J. Howlett, L. Jain (Eds.), Knowledge-Based Intelligent Information and Engineering Systems, Part I. LV, 882 pages. 2007.

Vol. 4687: P. Petta, J.P. Müller, M. Klusch, M. Georgeff (Eds.), Multiagent System Technologies. X, 207 pages. 2007.

Vol. 4682: D.-S. Huang, L. Heutte, M. Loog (Eds.), Advanced Intelligent Computing Theories and Applications. XXVII, 1373 pages. 2007.

Vol. 4676: M. Klusch, K.V. Hindriks, M.P. Papazoglou, L. Sterling (Eds.), Cooperative Information Agents XI. XI, 361 pages. 2007.

Vol. 4667: J. Hertzberg, M. Beetz, R. Englert (Eds.), KI 2007: Advances in Artificial Intelligence. IX, 516 pages. 2007.

Vol. 4660: S. Džeroski, L. Todorovski (Eds.), Computational Discovery of Scientific Knowledge. X, 327 pages. 2007.

Vol. 4659: V. Mařík, V. Vyatkin, A.W. Colombo (Eds.), Holonic and Multi-Agent Systems for Manufacturing. VIII, 456 pages. 2007.

Vol. 4651: F. Azevedo, P. Barahona, F. Fages, F. Rossi (Eds.), Recent Advances in Constraints. VIII, 185 pages. 2007.

Vol. 4648: F. Almeida e Costa, L.M. Rocha, E. Costa, I. Harvey, A. Coutinho (Eds.), Advances in Artificial Life. XVIII, 1215 pages. 2007.

Vol. 4635: B. Kokinov, D.C. Richardson, T.R. Roth-Berghofer, L. Vieu (Eds.), Modeling and Using Context. XIV, 574 pages. 2007.

Vol. 4632: R. Alhajj, H. Gao, X. Li, J. Li, O.R. Zaïane (Eds.), Advanced Data Mining and Applications. XV, 634 pages. 2007.

Vol. 4629: V. Matoušek, P. Mautner (Eds.), Text, Speech and Dialogue. XVII, 663 pages. 2007.

Vol. 4626: R.O. Weber, M.M. Richter (Eds.), Case-Based Reasoning Research and Development. XIII, 534 pages. 2007.

Vol. 4617: V. Torra, Y. Narukawa, Y. Yoshida (Eds.), Modeling Decisions for Artificial Intelligence. XII, 502 pages. 2007.

Vol. 4612: I. Miguel, W. Ruml (Eds.), Abstraction, Reformulation, and Approximation. XI, 418 pages. 2007.

Vol. 4604: U. Priss, S. Polovina, R. Hill (Eds.), Conceptual Structures: Knowledge Architectures for Smart Applications. XII, 514 pages. 2007.

Vol. 4603: F. Pfenning (Ed.), Automated Deduction – CADE-21. XII, 522 pages. 2007.

Vol. 4597: P. Perner (Ed.), Advances in Data Mining. XI, 353 pages. 2007.

Vol. 4594: R. Bellazzi, A. Abu-Hanna, J. Hunter (Eds.), Artificial Intelligence in Medicine. XVI, 509 pages. 2007.

Vol. 4585: M. Kryszkiewicz, J.F. Peters, H. Rybinski, A. Skowron (Eds.), Rough Sets and Intelligent Systems Paradigms. XIX, 836 pages. 2007.

Vol. 4578: F. Masulli, S. Mitra, G. Pasi (Eds.), Applications of Fuzzy Sets Theory. XVIII, 693 pages. 2007.

Vol. 4573: M. Kauers, M. Kerber, R. Miner, W. Windsteiger (Eds.), Towards Mechanized Mathematical Assistants. XIII, 407 pages. 2007.

Vol. 4571: P. Perner (Ed.), Machine Learning and Data Mining in Pattern Recognition. XIV, 913 pages. 2007.

Vol. 4570: H.G. Okuno, M. Ali (Eds.), New Trends in Applied Artificial Intelligence. XXI, 1194 pages. 2007.

Vol. 4565: D.D. Schmorrow, L.M. Reeves (Eds.), Foundations of Augmented Cognition. XIX, 450 pages. 2007.

Vol. 4562: D. Harris (Ed.), Engineering Psychology and Cognitive Ergonomics. XXIII, 879 pages. 2007.

Vol. 4548: N. Olivetti (Ed.), Automated Reasoning with Analytic Tableaux and Related Methods. X, 245 pages. 2007.

Vol. 4539: N.H. Bshouty, C. Gentile (Eds.), Learning Theory. XII, 634 pages. 2007.

Vol. 4529: P. Melin, O. Castillo, L.T. Aguilar, J. Kacprzyk, W. Pedrycz (Eds.), Foundations of Fuzzy Logic and Soft Computing. XIX, 830 pages. 2007.

Vol. 4520: M.V. Butz, O. Sigaud, G. Pezzulo, G. Baldassarre (Eds.), Anticipatory Behavior in Adaptive Learning Systems. X, 379 pages. 2007.

Vol. 4511: C. Conati, K. McCoy, G. Paliouras (Eds.), User Modeling 2007. XVI, 487 pages. 2007.

Vol. 4509: Z. Kobti, D. Wu (Eds.), Advances in Artificial Intelligence. XII, 552 pages. 2007.

Vol. 4496: N.T. Nguyen, A. Grzech, R.J. Howlett, L.C. Jain (Eds.), Agent and Multi-Agent Systems: Technologies and Applications. XXI, 1046 pages. 2007.

Vol. 4483: C. Baral, G. Brewka, J. Schlipf (Eds.), Logic Programming and Nonmonotonic Reasoning. IX, 327 pages. 2007.

Vol. 4482: A. An, J. Stefanowski, S. Ramanna, C.J. Butz, W. Pedrycz, G. Wang (Eds.), Rough Sets, Fuzzy Sets, Data Mining and Granular Computing. XIV, 585 pages. 2007.

Vol. 4481: J. Yao, P. Lingras, W.-Z. Wu, M. Szczuka, N.J. Cercone, D. Ślęzak (Eds.), Rough Sets and Knowledge Technology. XIV, 576 pages. 2007.

Vol. 4476: V. Gorodetsky, C. Zhang, V.A. Skormin, L. Cao (Eds.), Autonomous Intelligent Systems: Multi-Agents and Data Mining. XIII, 323 pages. 2007.

Vol. 4460: S. Aguzzoli, A. Ciabattoni, B. Gerla, C. Manara, V. Marra (Eds.), Algebraic and Proof-theoretic Aspects of Non-classical Logics. VIII, 309 pages. 2007.

Vol. 4457: G.M.P. O'Hare, A. Ricci, M.J. O'Grady, O. Dikenelli (Eds.), Engineering Societies in the Agents World VII. XI, 401 pages. 2007.

Vol. 4456: Y. Wang, Y.-m. Cheung, H. Liu (Eds.), Computational Intelligence and Security. XXIII, 1118 pages. 2007.

Vol. 4455: S. Muggleton, R. Otero, A. Tamaddoni-Nezhad (Eds.), Inductive Logic Programming. XII, 456 pages. 2007.

Vol. 4452: M. Fasli, O. Shehory (Eds.), Agent-Mediated Electronic Commerce. VIII, 249 pages. 2007.

Vol. 4451: T.S. Huang, A. Nijholt, M. Pantic, A. Pentland (Eds.), Artifical Intelligence for Human Computing. XVI, 359 pages. 2007.

Vol. 4442: L. Antunes, K. Takadama (Eds.), Multi-Agent-Based Simulation VII. X, 189 pages. 2007.

Vol. 4441: C. Müller (Ed.), Speaker Classification II. X, 309 pages. 2007.

Vol. 4438: L. Maicher, A. Sigel, L.M. Garshol (Eds.), Leveraging the Semantics of Topic Maps. X, 257 pages. 2007.

Vol. 4434: G. Lakemeyer, E. Sklar, D.G. Sorrenti, T. Takahashi (Eds.), RoboCup 2006: Robot Soccer World Cup X. XIII, 566 pages. 2007.

Vol. 4429: R. Lu, J.H. Siekmann, C. Ullrich (Eds.), Cognitive Systems. X, 161 pages. 2007.

Vol. 4428: S. Edelkamp, A. Lomuscio (Eds.), Model Checking and Artificial Intelligence. IX, 185 pages. 2007.

Vol. 4426: Z.-H. Zhou, H. Li, Q. Yang (Eds.), Advances in Knowledge Discovery and Data Mining. XXV, 1161 pages. 2007.

Vol. 4411: R.H. Bordini, M. Dastani, J. Dix, A.E.F. Seghrouchni (Eds.), Programming Multi-Agent Systems. XIV, 249 pages. 2007.

Vol. 4410: A. Branco (Ed.), Anaphora: Analysis, Algorithms and Applications. X, 191 pages. 2007.

Vol. 4399: T. Kovacs, X. Llorà, K. Takadama, P.L. Lanzi, W. Stolzmann, S.W. Wilson (Eds.), Learning Classifier Systems. XII, 345 pages. 2007.

Vol. 4390: S.O. Kuznetsov, S. Schmidt (Eds.), Formal Concept Analysis. X, 329 pages. 2007.

Vol. 4389: D. Weyns, H. Van Dyke Parunak, F. Michel (Eds.), Environments for Multi-Agent Systems III. X, 273 pages. 2007.

Vol. 4386: P. Noriega, J. Vázquez-Salceda, G. Boella, O. Boissier, V. Dignum, N. Fornara, E. Matson (Eds.), Coordination, Organizations, Institutions, and Norms in Agent Systems II. XI, 373 pages. 2007.

Vol. 4384: T. Washio, K. Satoh, H. Takeda, A. Inokuchi (Eds.), New Frontiers in Artificial Intelligence. IX, 401 pages. 2007.